Sulfur Metabolism in Phototrophic Organisms

Advances in Photosynthesis and Respiration

VOLUME 27

Series Editor:

GOVINDJEE
University of Illinois, Urbana, Illinois, U.S.A.

Consulting Editors:
Julian EATON-RYE, *Dunedin, New Zealand*
Christine H. FOYER, *Harpenden, U.K.*
David B. KNAFF, *Lubbock, Texas, U.S.A.*
Anthony L. MOORE, *Brighton, U.K.*
Sabeeha MERCHANT, *Los Angeles, California, U.S.A.*
Krishna NIYOGI, *Berkeley, California, U.S.A.*
William PARSON, *Seattle, Washington, U.S.A.*
Agepati RAGHAVENDRA, *Hyderabad, India*
Gernot RENGER, *Berlin, Germany*

The scope of our series, beginning with volume 11, reflects the concept that photosynthesis and respiration are intertwined with respect to both the protein complexes involved and to the entire bioenergetic machinery of all life. *Advances in Photosynthesis and Respiration* is a book series that provides a comprehensive and state-of-the-art account of research in photosynthesis and respiration. Photosynthesis is the process by which higher plants, algae, and certain species of bacteria transform and store solar energy in the form of energy-rich organic molecules. These compounds are in turn used as the energy source for all growth and reproduction in these and almost all other organisms. As such, virtually all life on the planet ultimately depends on photosynthetic energy conversion. Respiration, which occurs in mitochondrial and bacterial membranes, utilizes energy present in organic molecules to fuel a wide range of metabolic reactions critical for cell growth and development. In addition, many photosynthetic organisms engage in energetically wasteful photorespiration that begins in the chloroplast with an oxygenation reaction catalyzed by the same enzyme responsible for capturing carbon dioxide in photosynthesis. This series of books spans topics from physics to agronomy and medicine, from femtosecond processes to season long production, from the photophysics of reaction centers, through the electrochemistry of intermediate electron transfer, to the physiology of whole organisms, and from X-ray crystallography of proteins to the morphology or organelles and intact organisms. The goal of the series is to offer beginning researchers, advanced undergraduate students, graduate students, and even research specialists, a comprehensive, up-to-date picture of the remarkable advances across the full scope of research on photosynthesis, respiration and related processes.

The titles published in the Series are listed at the end of this volume.

Sulfur Metabolism in Phototrophic Organisms

Edited by

Rüdiger Hell
University of Heidelberg,
Germany

Christiane Dahl
University of Bonn, Germany

David Knaff
University of Texas, Lubbock,
USA

and

Thomas Leustek
Rutgers University, New Brunswick,
USA

Library of Congress Control Number: 2007939823

ISBN 978-1-4020-6862-1 (HB)
ISBN 978-1-4020-6863-8 (e-book)

Published by Springer,
P.O. Box 17, 3300 AA Dordrecht, The Netherlands.

www.springer.com

Cover Image: Glutathione is the most important low molecular weight thiol in nearly all living cells with numerous functions including stress tolerance and redox signal transduction. Shown is the in situ labelling of glutathione in the cytosol of plant cells (green colour) using a fluorescent dye.

Left panels: heterotrophic cells from suspension culture of *Arabidopsis thaliana*. Red indicates staining of cell walls and integral plasmamembranes with propidium iodide. Right panels: stoma cells of *Nicotiana tabacum*. Red colour indicates autoflorescence from chlorophyll. Dark areas within cells are caused by vacuoles. For detailed description see chapter 24 by A.J. Meyer and M. Fricker.

The photograph was generously provided by Dr. Andreas J. Meyer, University of Heidelberg, Germany, and Dr. Mark Fricker, University of Oxford, UK.

Printed on acid-free paper

From the Series Editor

Advances in Photosynthesis and Respiration
Volume 27: Sulfur Metabolism in Phototrophic Organisms

I am delighted to announce the publication, in *Advances in Photosynthesis and Respiration* (AIPH) Series, of a book *Sulfur Metabolism in Phototrophic Organisms*. Four distinguished authorities (Rüdiger Hell, Christiane Dahl, David Knaff, and Thomas Leustek) have edited this Volume: Hell is an authority from the University of Heidelberg (Germany), Dahl from the University of Bonn (Germany), Leustek from Rutgers University (USA), and Knaff from Texas Tech University(USA). Taken together, these editors have a unique combination of expertise in Biochemistry, Microbiology, Plant Physiology, and Molecular Biology. Their joint efforts have provided a comprehensive overview of *Sulfur Metabolism in Phototrophic Organisms*. The editors have presented a authoritative decisive volume on sulfur metabolism in both phototrophic eukaryotes and prokaryotes. This book is unique in my Series since it is the first time, we have covered the metabolism of this important element in photosynthetic systems.

Published Volumes (2007–1994) in the Series are:

- *Volume 25* (2006): ***Chlorophylls and Bacteriochlorophylls: Biochemistry, Biophysics, Functions and Applications***, edited by Bernhard Grimm, Robert J. Porra, Wolfhart Rüdiger, and Hugo Scheer, from Germany and Australia. 37 Chapters, 603 pp, Hardcover. ISBN: 978-1-4020-4515-8
- *Volume 24* (2006): ***Photosystem I: The Light-Driven Plastocyanin: Ferredoxin Oxidoreductase***, edited by John H. Golbeck, from USA. 40 Chapters, 716 pp, Hardcover. ISBN: 978-1-4020-4255-3
- *Volume 23* (2006): ***The Structure and Function of Plastids***, edited by Robert R. Wise and J. Kenneth Wise, from USA. 27 Chapters, 575 pp, Hardcover. ISBN: 978-1-4020-4060-3; Softcover (2007) ISBN: 978-1-4020-6570-5
- *Volume 22* (2005): ***Photosystem II: The Light-Driven Water:Plastoquinone Oxidoreductase***, edited by Thomas J. Wydrzynski and Kimiyuki Satoh, from Australia and Japan. 34 Chapters, 786 pp, Hardcover. ISBN: 978-1-4020-4249-2
- *Volume 21* (2005): ***Photoprotection, Photoinhibition, Gene Regulation, and Environment***, edited by Barbara Demmig-Adams, William W. III Adams and Autar K. Mattoo, from USA. 21 Chapters, 380 pp, Hardcover. ISBN: 978-1-4020-3564-7
- *Volume 20* (2006): ***Discoveries in Photosynthesis***, edited by Govindjee, J. Thomas Beatty, Howard Gest and John F. Allen, from USA, Canada and UK. 111 Chapters, 1304 pp, Hardcover. ISBN: 978-1-4020-3323-0
- *Volume 19* (2004): ***Chlorophyll a Fluorescence: A Signature of Photosynthesis***, edited by George C. Papageorgiou and Govindjee, from Greece and USA. 31 Chapters, 820 pp, Hardcover. ISBN: 978-1-4020-3217-2
- *Volume 18* (2005): ***Plant Respiration: From Cell to Ecosystem***, edited by Hans Lambers and Miquel Ribas-Carbo, from Australia and Spain. 13 Chapters, 250 pp, Hardcover. ISBN: 978-1-4020-3588-3
- *Volume 17* (2004): ***Plant Mitochondria: From Genome to Function***, edited by David Day, A. Harvey Millar and James Whelan, from Australia. 14 Chapters, 325 pp, Hardcover. ISBN: 978-1-4020-2399-6
- *Volume 16* (2004): ***Respiration in Archaea and Bacteria:Diversity of Prokaryotic Respiratory Systems***, edited by Davide Zannoni, from Italy. 13 Chapters, 310 pp, Hardcover. ISBN: 978-1-4020-2002-5
- *Volume 15* (2004): ***Respiration in Archaea and Bacteria: Diversity of Prokaryotic Electron Transport Carriers***, edited by Davide Zannoni, from Italy. 13 Chapters, 350 pp, Hardcover. ISBN: 978-1-4020-2001-8

- *Volume 14* (2004): ***Photosynthesis in Algae***, edited by Anthony W. Larkum, Susan Douglas and John A. Raven, from Australia, Canada and UK. 19 Chapters, 500 pp, Hardcover. ISBN: 978-0-7923-6333-0
- *Volume 13* (2003): ***Light-Harvesting Antennas in Photosynthesis***, edited by Beverley R. Green and William W. Parson, from Canada and USA. 17 Chapters, 544 pp, Hardcover. ISBN: 978-0-7923-6335-4
- *Volume 12* (2003): ***Photosynthetic Nitrogen Assimilation and Associated Carbon and Respiratory Metabolism***, edited by Christine H. Foyer and Graham Noctor, from UK and France. 16 Chapters, 304 pp, Hardcover. ISBN: 978-0-7923-6336-1
- *Volume 11* (2001): ***Regulation of Photosynthesis***, edited by Eva-Mari Aro and Bertil Andersson, from Finland and Sweden. 32 Chapters, 640 pp, Hardcover. ISBN: 978-0-7923-6332-3
- *Volume 10* (2001): ***Photosynthesis Photobiochemistry and Photobiophysics***, by Bacon Ke, from USA. 36 Chapters, 792 pp, Softcover: ISBN: 978-0-7923-6791-8. Hardcover: ISBN: 978-0-7923-6334-7
- *Volume 9* (2000): ***Photosynthesis: Physiology and Metabolism***, edited by Richard C. Leegood, Thomas D. Sharkey and Susanne von Caemmerer, from UK, USA and Australia. 24 Chapters, 644 pp, Hardcover. ISBN: 978-0-7923-6143-5
- *Volume 8* (1999): ***The Photochemistry of Carotenoids***, edited by Harry A. Frank, Andrew J. Young, George Britton and Richard J. Cogdell, from UK and USA. 20 Chapters, 420 pp, Hardcover. ISBN: 978-0-7923-5942-5
- *Volume 7* (1998): ***The Molecular Biology of Chloroplasts and Mitochondria in Chlamydomonas***, edited by Jean David Rochaix, Michel Goldschmidt-Clermont and Sabeeha Merchant, from Switzerland and USA. 36 Chapters, 760 pp, Hardcover. ISBN: 978-0-7923-5174-0
- *Volume 6* (1998): ***Lipids in Photosynthesis: Structure, Function and Genetics***, edited by Paul-André Siegenthaler and Norio Murata, from Switzerland and Japan. 15 Chapters, 332 pp, Hardcover. ISBN: 978-0-7923-5173-3
- *Volume 5* (1997): ***Photosynthesis and the Environment***, edited by Neil R. Baker, from UK. 20 Chapters, 508 pp, Hardcover. ISBN: 978-0-7923-4316-5

- *Volume 4* (1996): ***Oxygenic Photosynthesis: The Light Reactions***, edited by Donald R. Ort, and Charles F. Yocum, from USA. 34 Chapters, 696 pp, Softcover: ISBN: 978-0-7923-3684-6. Hardcover: ISBN: 978-0-7923-3683-9
- *Volume 3* (1996): ***Biophysical Techniques in Photosynthesis***, edited by Jan Amesz and Arnold J. Hoff, from The Netherlands. 24 Chapters, 426 pp, Hardcover. ISBN: 978-0-7923-3642-6
- *Volume 2* (1995): ***Anoxygenic Photosynthetic Bacteria***, edited by Robert E. Blankenship, Michael T. Madigan and Carl E. Bauer, from USA. 62 Chapters, 1331 pp, Hardcover. ISBN: 978-0-7923-3682-8
- *Volume 1* (1994): ***The Molecular Biology of Cyanobacteria***, edited by Donald R. Bryant, from USA. 28 Chapters, 916 pp, Hardcover. ISBN: 978-0-7923-3222-0

Further information on these books and ordering instructions can be found at <http://www.springeronline.com> under the Book Series 'Advances in Photosynthesis and Respiration.' Table of Contents of Volumes 1–25 can be found at <http://www.life.uiuc.edu/govindjee/photosynSeries/ttocs.html>. Special discounts are available to members of the International Society of Photosynthesis Research, ISPR (<http://www.photosynthesisresearch.org/>).

Other Volumes, at the typesetters are:

- ***Biophysical Techniques II*** (Editors: Thijs Aartsma and Jörg Matisyk) (expected to be Volume 26) ; and
- ***The Purple Phototrophic Bacteria*** (Editors: C. Neil Hunter, Fevzi Daldal, Marion Thurnauer and J. Thomas Beatty).

About Volume 27: Sulfur Metabolism in Phototrophic Organisms

Sulfur Metabolism in Phototrophic Organisms has 24 authoritative Chapters, and is authored by 55 international authorities from 10 countries (Australia, Canada, Denmark, Germany, Israel, Italy, Japan, Netherlands, United Kingdom, United States of America). It is a truly an international book and the editors deserve our thanks and our congratulations this gift for our future.

Sulfur is one of the most versatile elements in life due to its reactivity in different oxidation and reduction states. In phototrophic organisms, the redox properties of sulfur in proteins, and of sulfur-containing metabolites, are particularly important in the interaction between the reductive assimilation processes of photosynthesis and reactive oxygen species that arise as by-products of electron transport chains. Thiol groups in proteins and metabolites are targets of reactive oxygen species, resulting in potential damage and at the same time giving rise to redox signal cascades that trigger repair reactions and adaptation to environmental stress. Further, reduced sulfur compounds play a prominent role as electron donors for photosynthetic carbon dioxide fixation in anoxygenic phototrophic sulfur bacteria. Interest in the investigation of the multiple functions of sulfur-related processes has exponentially increased in recent years, especially in molecular and cellular biology, biochemistry, agrobiotechnology and ecology. This book provides, for the first time, in-depth and integrated coverage of the functions of sulfur in phototrophic organisms including bacteria, plants and algae; it bridges gaps between biochemistry and cellular biology of sulfur in these organisms, and of biology and environments dominated by them. This book is designed to be a comprehensive resource on sulfur in phototrophic organisms for advanced undergraduate and graduate students, beginning researchers and teachers in the area of photosynthesis, bacterial energy metabolism, biotechnology, plant nutrition, plant production and plant molecular physiology.

The readers can easily find the titles and the authors of the individual chapters in the Table of Contents of this book. Instead of repeating this information here, I prefer to thank each and every author by name (listed in alphabetical order) that reads like a "Who's Who in Sulfur Metabolism in Phototrophic Organisms":

Christoph Benning; Don Bryant; Meike Burow; Leong-Keat Chan; Christiane Dahl; Luit De Kok; Mark D. Fricker; Niels-Ulrik Frigaard; R. Michael Garavito; Jonathan Gershenzon; Mario Giordano; Arthur Grossman; Robert Hänsch; Thomas E. Hanson; Günther Hauska; Malcolm J. Hawkesford; Rüdiger Hell; Cinta Hernández-Sebastià; Holger Hesse;

Rainer Hoefgen; Josef Hormes; Timothy J. Hurse; Johannes Imhoff; Ulrike Kappler; Patrick Keeling; Jürg Keller; David Knaff; Stanislav Kopriva; Thomas Leustek; Frédéric Marsolais; Ralf R. Mendel; Andreas Meyer; Hartwig Modrow; Rachael Morgan-Kiss; Alessandra Norici; Jörg Overmann; Lolla Padmavathi; Nicola Patron; Marinus Pilon; Elizabeth A. H. Pilon-Smits; Alexander Prange; Simona Ratti; Thomas Rausch; John A. Raven; Kazuki Saito; Yosepha Shahak; Nakako Shibagaki; Mie Shimojia; Hideki Takahashi; Michael Tausz; Luc Varin; Markus Wirtz; Ute Wittstock; Hong Ye; Fangjie Zhao.

Timeline for Sulfur in Phototrophic Organisms

A timeline of discoveries is one way to assess the progress and the status of a field. See for example:

- Howard Gest H and Robert E. Blankenship (2005) Time line of discoveries: anoxygenic photosynthesis. In: Govindjee, Beatty JT, Gest H and Allen JF (eds) Discoveries in Photosynthesis. Advances in Photosynthesis and Respiration, vol 20. Springer, Dordrecht, pp 51–62.
- Govindjee and Krogmann DW (2005) Discoveries in oxygenic photosynthesis (1727–2003): a perspective. In: Govindjee, Beatty JT, Gest H and Allen JF (eds) Discoveries in Photosynthesis. Advances in Photosynthesis and Respiration, vol 20. Springer, Dordrecht, pp 63–105.

Below, we provide a timeline for sulfur in phototrophic organisms (courtesy of Rüdiger Hell):

1836: Christian Gottfried Ehrenberg describes the discovery of the first purple sulfur bacteria "*Monas okenii*" and "*Ophidomonas jenensis*"

1861 and **1882**: Sulfate is found to be an essential plant mineral nutrient by German plant physiologists J.A.L.W. Knoop and J. v. Sachs using hydroponic cultures

1872 and **1875**: The German botanist founder of bacterial systematics, Ferdinand Cohn, unequivocally proved that the cellular inclusions of the above purple bacteria consisted of elemental sulfur

1883: Theodor W. Engelmann discovered and described action spectrum of photosynthesis in purple bacteria

1919: Johannes Buder confirmed Engelmann's theories that the purple bacteria perform a photo synthetic metabolism. They assimilate carbon dioxide or organic compounds anaerobically in the light

1931: The definite proof for the phototrophic nature of the purple and green sulfur bacteria, by the Dutch American microbiologist Cornelis B. van Niel, working in Pacific Grove, CA: For the phototrophic sulfur bacteria the reducing agent is hydrogen sulfide (or sulfur or thiosulfate) and for the photosynthesis of plants, algae and cyanobacteria, it is water, with the oxidation products being molecular sulfur (and/or sulfate) and molecular oxygen, respectively

1966: Research on enzymology of oxidative sulfur metabolism started with reports of adenosine phosphosulfate reductase in purple sulfur bacteria (Harry D. Peck, O.W. Thiele and Hans G. Trüper)

1973: Jerome Schiff begins to identify the eukaryotic sulfate assimilation pathway in Euglena

1988 and **1995**: First genetics in green (John Ormerod) and purple sulfur bacteria (Christiane Dahl)

1996: Tom Leustek and John Wray isolate genes for adenosine phosphosulfate reductase, the decisive enzyme in plant sulfate reduction

2002: The first genome sequence of a green sulfur bacterium by Jonathan A. Eisen and coworkers

Books, that are relevant to Sulfur metabolism, arranged chronologically since 1970 (courtesy of Rüdiger Hell).

Researchers and lecturers who wish to examine all published material in the area of sulfur metabolism in plants, algae and bacteria would benefit by consulting all of the following books listed below.

1970: A.B. Roy and P.A. Trudinger (eds.) (1970) The Biochemistry of Inorganic Compounds of Sulphur. Combridge University Press, London.

1984: A. Müller and B. Krebs (eds.) (1984) Sulfur, its significance for chemistry, for the geo-, bio- and cosmosphere and technology. Elsevier, Amsterdam.

1998: E. Schnug (ed.) (1998) Sulphur in Agroecosystems. Series: Nutrients in Ecosystems, Vol. 2. Springer, New York.

2000: P. Lens and L.H. Pol (2000) Environmental technologies to treat sulfur pollution. IWA Publishing, London.

2003: R. Steudel (ed.) (2003) Elemental sulfur and sulfur-rich compounds I & II. Series: Topics in Current Chemistry, Vols. 230 & 231. Springer, New York.

Y. P. Abrol and A.C. Ahmad (eds.) (2003) Sulphur in plants. Springer, New York.

2005: K. Saito, L.J. DeKok, M.J. Hawkesford, E. Schnug, A. Sirko and H. Rennenberg (eds.) (2005) Sulfur Transport and Assimilation in Plants in the Post-Genomic Era. Backhuys Publ., Leiden.

2007: C. Dahl and C.G. Friedrich (eds.) (2007) Microbial sulfur metabolism. Springer, New York.

Hawkesford M and De Kok LJ (2007) Sulfur in Plants. An Ecological Perspective. Springer, London.

Future AIPH and Other Related Books

The readers of the current series are encouraged to watch for the publication of the forthcoming books (not necessarily arranged in the order of future appearance):

- *Abiotic Stress Adaptation in Plants: Physiological, Molecular and Genomic Foundation* (Editors: Ashwani Pareek, Sudhir K. Sopory, Hans J. Bohnert and Govindjee);
- *C-4 Photosynthesis and Related CO_2 Concentrating Mechanisms* (Editors: Agepati S. Raghavendra and Rowan Sage);
- *The Chloroplast Biochemistry, Molecular Biology and Bioengineering, Part 1 and Part 2* (Editors: Constantin Rebeiz, Hans Bohnert, Christoph Benning, Henry Daniell, Beverley R. Green, J. Kenneth Hoober, Hartmut Lichtenthaler, Archie R. Portis and Baishnab C. Tripathy);
- *Photosynthesis: Biochemistry, Biophysics, Physiology and Molecular Biology* (Editors: Julian Eaton-Rye and Baishnab Tripathy);
- *Photosynthesis In Silico: Understanding Complexity from Molecules to Ecosystems* (Editors: Agu Laisk, Ladislav Nedbal and Govindjee).

In addition to these contracted books, we are in touch with prospective Editors for the following topics, among others:

- Ecophysiology
- Molecular Biology of Cyanobacteria II
- Interaction of Photosynthesis and Respiration in Cyanobacteria
- ATP Synthase and Proton Translocation
- Interactions between Photosynthesis and other Metabolic Processes in Higher Plants
- Genomics, Proteomics and Evolution
- Biohydrogen Production
- Lipids II
- Carotenoids II
- Green Bacteria and Heliobacteria
- Global Aspects of Photosynthesis
- Artificial Photosynthesis

Readers are encouraged to send their suggestions for these and future Volumes (topics, names of future editors, and of future authors) to me by E-mail (gov@uiuc.edu) or fax (1-217-244-7246).

In view of the interdisciplinary character of research in photosynthesis and respiration, it is my earnest hope that this series of books will be used in educating students and researchers not only in Plant Sciences, Molecular and Cell Biology, Integrative Biology, Biotechnology, Agricultural Sciences, Microbiology, Biochemistry, and Biophysics, but also in Bioengineering, Chemistry, and Physics.

I take this opportunity to thank and congratulate all the four editors (Rüdiger Hell, Christiane Dahl, David Knaff, and Thomas Leustek) for their outstanding and painstaking editorial work. I thank all the 55 authors (see the list above) of this book in our AIPH Series: without their authoritative chapters, there would be no such Volume.

I thank Rüdiger Hell for his contribution to the Timeline of Sulfur in plant, algal and bacterial metabolism and for providing the list of books on this topic, cited above. We owe Jacco Flipsen and Noeline Gibson (both of Springer) thanks for their friendly working relation with us that led to the production of this book. Thanks are also due to Jeff Haas (Director of Information Technology, Life Sciences, University of Illinois at Urbana-Champaign, UIUC), Evan DeLucia (Head, Department of Plant Biology, UIUC) and my dear wife Rajni Govindjee for constant support.

October 24, 2007
Govindjee
Series Editor, *Advances in Photosynthesis and Respiration*
University of Illinois at Urbana-Champaign
Department of Plant Biology
Urbana, IL 61801-3707, USA
E-mail: gov@uiuc.edu
URL: http://www.life.uiuc.edu/govindjee

Left to right: Govindjee (Series Editor), Julian Eaton-Rye (Editor of a future volume in the Series), and Thomas Wydrzynski (Editor of Volume 22 in the Series). Photo taken at the 14th International Congress on Photosynthesis, July, 2007.

Govindjee, born in 1932, obtained his B.Sc. (Chemistry, Biology) and M.Sc. (Botany, Plant Physiology) in 1952 and 1954, from the University of Allahabad, India. He learned his Plant Physiology from Prof. Shri Ranjan, who was a former student of Felix Frost Blackman. For his Ph.D. (in Biophysics, 1960), he studied under the pioneers of photosynthesis: Profs. Robert Emerson and Eugene Rabinowitch, both at the University of Illinois at Urbana-Champaign (UIUC), IL, U.S.A. He considers Eugene Rabinowitch as his main role model and his mentor.

Govindjee is best known for his research on the mechanisms of excitation energy transfer, light emission, the primary photochemistry and the electron transfer in Photosystem II (PS II). His research, with many collaborators, has included the discovery of a short-wavelength form of chlorophyll (Chl) a functioning in the Chl b-containing system, now called PS II; of the two-light effects in Chl a fluorescence and in NADP (nicotinamide dinucleotide phosphate) reduction in chloroplasts (Emerson Enhancement). Further, he has worked on the existence of different spectral fluorescing forms of Chl a and the temperature dependence of excitation energy transfer down to 4 K; basic relationships between Chl a fluorescence and photosynthetic reactions; unique role of bicarbonate on the acceptor side of PS II, particularly in the protonation events involving the Q_B binding region; the theory of thermoluminescence in plants; picosecond measurement on the primary photochemistry of PS II; and the use of Fluorescence Lifetime Imaging Microscopy (FLIM) of Chl a fluorescence in understanding photoprotection against excess light.

Govindjee's current focus is on the 'History of Photosynthesis Research,' in 'Photosynthesis Education,' and in the 'Possible Existence of Extraterrestrial Life.' He is the founding Historical Corner Editor of 'Photosynthesis Research', and the founding Series Editor of 'Advances in Photosynthesis and Respiration'. He has served on the faculty of the UIUC for ~40 years. Since 1999, he has been Professor Emeritus of Biochemistry, Biophysics and Plant Biology at the same institution. His honors include: Fellow of the American Association of Advancement of Science (AAAS); Distinguished Lecturer of the School of Life Sciences, UIUC; Fellow and Life member of the National Academy of Sciences (India); President of the American Society for Photobiology (1980–1981); Fulbright Scholar and Fulbright Senior Lecturer; Honorary President of the 2004 International Photosynthesis Congress (Montréal, Canada); the 2006 recipient of the Lifetime Achievement Award from the Rebeiz Foundation for Basic Biology, and most recently, the 2007 recipient of the 'Communications Award' of the International Society of Photosynthesis Research (ISPR).

Further information on Govindjee and his research can be found at his web site: http://www.life.uiuc.edu/govindjee. He can always be reached by e-mail at gov@uiuc.edu.

Contents

Part I: General Chapters: Sulfate Activation and Reduction, Biosynthesis of Sulfur Containing Amino Acids

Part III: Sulfur in Phototrophic Prokaryotes

Part V: Specific Methods

Preface

Sulfur is one of the most versatile elements in life due to its reactivity in different oxidation and reduction states. In phototrophic organisms, the redox properties of sulfur in proteins, and of sulfur-containing metabolites, are particularly important for the mediation between the reductive assimilation processes of photosynthesis and reactive oxygen species that arise as byproducts of electron transport chains in chloroplasts and mitochondria. Further, reduced sulfur compounds play a prominent role as electron donors for photosynthetic carbon dioxide fixation in anoxygenic phototrophic sulfur bacteria. Indeed, sulfur together with iron may be assumed to have contributed to early electron transport processes in protolife. Reductive atmospheric conditions before the invention of oxygenic photosynthesis probably promoted the functional evolvement of processes based on the redox properties of the different oxidation states of sulfur. When the atmospheric environment became oxidative as a consequence of the effective development of photosynthesis in bacteria and later in algae and plants, the cellular milieu remained reductive. Reduced sulfur moieties such as cysteine and methionine changed into targets of abundant reactive oxygen species that were derived from respiration or photosynthesis. It is still a matter of speculation whether defense reactions to prevent oxidation led to the development of redox signal transduction. Interest in the investigation of such dualistic functions of sulfur-related processes has exponentially increased in recent years, especially in molecular and cellular biology, biochemistry, agrobiotechnology and ecology.

Sulfur metabolism is thus of truly elemental importance for life, but complex in its biochemical, biophysical and metabolic functions. The editors of this book come from different backgrounds (biochemistry, microbiology, plant physiology, molecular biology) and joined their efforts to provide a comprehensive overview of *Sulfur Metabolism in Phototrophic Organisms*. The aim was to present a definitive volume on sulfur metabolism in phototrophic eukaryotes (plants, algae) and prokaryotes (anoxygenic and oxygenic phototrophic bacteria). The most recent informations available in these fields have been compiled. Contributors had the freedom to develop their topics in depth, and we feel that nothing significant in this field has been neglected.

The book is organized along four major themes: (1) sulfate activation and reduction, biosynthesis of sulfur containing amino acids, (2) sulfur in plants and algae, (3) sulfur in phototrophic prokaryotes, and (4) ecology and biotechnology. For each theme several chapters contributed by leading experts in the field are combined to yield a detailed picture of the current state of art.

A comprehensive volume combining the aspects of sulfur metabolism not only in eukaryotic phototrophs but also in phototrophic bacteria has never been published. Here, we try to emphasize the common characteristics of the processes of assimilatory sulfate reduction in eukaryotic and prokaryotic phototrophs. Ample parallels are also found in the uptake of sulfate into the cell, sulfation pathways, in the biosynthetic pathways of sulfur-containing amino acids, or sulfur donation reactions in the biosynthesis of other sulfur-containing cell constituents.

This book provides for the first time a comprehensive and integrated knowledge of the functions of sulfur in phototrophic organisms including bacteria, plants and algae; it bridges gaps between biochemistry and cellular biology of sulfur in these organisms, and of biology and environments dominated by them. Specialized functions of sulfur in animals in reductive environments such as sediments have purposely been omitted. As the contents of this book have been compiled for plant scientists, agronomists and microbiologists alike, we see an ideal opportunity to unite the common interest in sulfur meta-bolism. Many plant scientists might not be aware of the important role sulfur compounds play in anoxygenic sulfur bacteria where they serve as photosynthetic electron donors. In order to provide the reader specializing in plant issues with a comprehensive introduction into and overview about various groups of anoxygenic phototrophic bacteria, we included chapters about systematics and ecology of anoxygenic phototrophic bacteria. In other chapters the role of sulfur compounds for these organisms and the biochemistry of dissimilatory sulfur transformations in this group are described in depth.

We hope this book will be of use for advanced students as well as teachers and researchers in the fields of photosynthesis, bacterial energy metabolism, plant nutrition, plant production and plant molecular physiology and stimulates further research to close the gaps of our knowledge on sulfur in plants.

We have many people to thank. First of all, we would like to take this opportunity to acknowledge all authors for providing uniformly excellent chapters. Each author is a leading authority in his/her field and has generously offered the time and effort to make this book a success. All authors were willing to look beyond their research specialties to give the reader a more encompassing view. In addition, they put up with the inevitable delays in publication. We also acknowledge the help received from Noeline Gibson and Jacco Flipson (of Springer). And, of course, we wish to thank the series editor Govindjee. His patience, good humor and professional competence helped us immensely.

Rüdiger Hell
University of Heidelberg
Heidelberg Institute of Plant Sciences
Im Neuenheimer Feld 360
69120 Heidelberg
Germany
rhell@hip.uni-heidelberg.de

Christiane Dahl
University of Bonn
Institute of Microbiology & Biotechnology
Meckenheimer Allee 168
53115 Bonn
Germany
ChDahl@uni-bonn.de

Thomas Leustek
Rutgers University
Biotechnology Center for Agriculture and the Environment
Department of Plant Biology and Pathology
59 Dudley Road, Room 328
New Brunswick, New Jersey 08901-8520
USA
leustek@AESOP.Rutgers.edu

David Knaff
Department of Chemistry and Biochemistry and Center for Biotechnology and Genomics
Texas Tech University
Lubbock, Texas 79409-1061
USA
david.knaff@ttu.edu

Rüdiger Hell

Rüdiger Hell was born in a small village in the west of Germany and enjoyed rural life and excellent early education on nearly everything except biology. He developed a strong interest in the special life of plants and, after civil service, studied biology at the technical University of Darmstadt, Germany. He specialized in the combination of biochemistry and ecology in the hope this would help him to understand how plants function. He graduated with a diploma thesis on Crassulacean Acid Metabolism (CAM) in cell cultures in 1985. During his PhD research, he studied under the supervision of his excellent teacher Prof. Ludwig Bergmann at the University of Cologne, Germany. He worked out the biosynthesis of glutathione in plants and demonstrated the still valid concept of redox control and feedback sensitivity of the first enzyme of the glutathione pathway. After graduation in 1989, he moved into the regulatory interactions between chloroplasts and the cytosol. The German Science Foundation (DFG, Deutsche Forschungsgemeinschaft) provided him a fellowship from 1990 to 1993 to work with Prof. Wilhelm Gruissem at the University of California, Berkeley. There he was introduced into the powerful techniques of molecular biology and plant genetic transformation to study the control of chloroplast RNA stability by nucleus-encoded RNA-binding proteins. The inspiring atmosphere by wonderful colleagues in the laboratory and the scientific and cultural spirit at Berkeley enabled him to extend experiments to the regulation of metabolic reactions that are distributed between chloroplasts and the cytosol. This became his major research area and led him to choose primary sulfur metabolism for investigation which at that time was hardly understood.

In 1993 he returned to Germany to take a position as research group leader at the Department of Plant Physiology at the Ruhr-University Bochum, where he was allowed to independently follow his ideas on cloning genes of plant sulfur metabolism. The aim was to establish the regulatory mechanism that control this pathway and relate it to the network of primary assimilatory pathways of carbon and nitrogen. Again funded by DFG, he set up a research program and quickly provided evidence for the significance of protein–protein interaction in the cysteine synthase complex. Within five years of intense teaching and research, the license of university teaching (Habilitation) was achieved in 1998. In order to broaden his research interest Rüdiger Hell accepted a position as head of the Molecular Mineral Assimilation group at the Leibniz-Institute of Plant Genetics and Crop Plant Research (IPK) at Gatersleben in what used to be Eastern Germany. Until 2003 he taught plant physiology at the University of Halle and used the excellent research facilities and highquality plant research community at IPK to investigate sulfate assimilation, iron allocation and nitrogen use efficiency in *Arabidopsis thaliana* as well as in crop plants.

In 2002, he received invitations for professorships in Heidelberg and in Düsseldorf, Germany. In 2003, he accepted the chair of Molecular Biology of Plants at the University of Heidelberg, Germany's oldest University and particularly strong in biosciences. In recent years, his laboratory has demonstrated that the cysteine synthase protein complex is a sensor for sulphide in plant cells, that glutathione is a part of signal transduction of cellular redox state, and that sulfur metabolism has an important and special role in mounting the plant's defence against pathogens. Rüdiger Hell has served as Dean of the Faculty of Biosciences at Heidelberg from 2005 to 2007, and continues to regularly review for national and international funding agencies and the most important journals in plant sciences.

Rüdiger Hell's work is funded by DFG and the German Academic Exchange Service (DAAD, Deutscher Akademischer Austauschdienst). He is indebted to the lively and excellent fellows in his research group and the Heidelberg Institute of Plant Sciences. He owes incredibly much to his wife, Dr. Helke Hillebrand, who always supported him and advises him best when life becomes tough.

Further information can be found at the Heidelberg Institute of Plant Sciences web site: http:// www.bot.uni-heidelberg.de. He can be reached by e-mail at rhell@hip.uni-heidelberg.de.

Christiane Dahl

Christiane Dahl is a lecturer of Microbiology in the Institute of Microbiology & Biotechnology at the Rheinische Friedrich-Wilhelms Universität in Bonn, Germany. She studied Biology at University of Bonn from 1982 to 1987 and received her diploma for a thesis on "Comparative enzymology of sulfur metabolism in the purple sulfur bacterium *Thiocapsa roseopersicina* under chemo- and photolithoautotrophic conditions". She received her PhD in 1992 for a thesis on a much "hotter" topic, the biochemistry of dissimilatory sulfate reduction in the hyperthermophilic archaeon *Archaeoglobus fulgidus*. Both theses were supervised by her much esteemed academic teacher Hans G. Trüper. During the work for the PhD, a research visit to the laboratory of Nicholas M. Kredich at the Department of Biochemistry, Duke University, Durham, North Carolina, proved most rewarding. In 1993 she accepted a position as a research group leader, at the Institute of Microbiology & Biotechnology in Bonn. She then switched back to anoxygenic phototrophic sulfur bacteria, to study their oxidative sulfur metabolism in more depth. Up to that point, biochemical techniques had been used for analysis of the various enzymatic steps involved. Now, she introduced targeted genetics to explore the mechanisms of reduced sulfur compound oxidation. *Allochromatium vinosum* was the first purple sulfur bacterium for which methods like conjugational transfer of plasmids, insertional mutagenesis and in frame deletion of genes, epitope tagging and complementation of

genes were employed. In the meantime, this purple sulfur bacterium had been developed into a model for studying oxidative sulfur metabolism. In 1999, she completed her Habilitation in Microbiology at the University of Bonn. The expertise of her group includes methods in molecular genetics and biology, protein biochemistry and analysis of inorganic sulfur compounds. Current research focuses on the uptake of insoluble elemental sulfur by bacterial cells, the oxidation of thiosulfate, and the oxidation of intracellularly stored sulfur by the so-called Dsr system with sulfite reductase as a component of major importance. She also initiated the complete genome sequencing of *Allochromatium vinosum* within the Community Sequencing Program 2008 of the United States DOE (Department of Energy) Joint Genome Institute. Together with Cornelius G. Friedrich (Dortmund, Germany) she organized the first "International Symposium on Microbial Sulfur Metabolism" held in June 2006 in Münster, Germany. Since 1998, she is a member of the Editorial Board of FEMS Microbiology Letters and is also a member of the Scientific Committee for the International Symposium on Phototrophic Prokaryotes.

Dahl's research is supported by the generous funding by the Deutsche Forschungsgemeinschaft and by the Deutscher Akademischer Austauschdienst. Her studies have been facilitated by the enthusiastic collaboration of students and technical assistants in her group and past and ongoing cooperation with many research groups, both in

Germany and overseas. The latter include productive studies on the Sox system for thiosulfate oxidation with Ulrike Kappler (Brisbane, Australia) and Cornelius G. Friedrich (Dortmund, Germany), speciation of stored sulfur via XANES (X-ray Absorption Near Edge Spectrum) spectroscopy with Alexander Prange (Mönchengladbach, Germany, and Baton Rouge, LA), detection of siroamide in *Allochromatium vinosum* with Russ Timkovich (Tuscaloosa, AL), the solution structure of DsrC via Nuclear Magnetic Resonance (NMR) with John Cort (Richland, WA), the crystal structure of DsrEFH (Dong-Hae Shin, Seoul, Korea) and on sulfur globule proteins with Dan C. Brune (Tempe, AZ). Christiane Dahl is deeply indebted to her mentor Hans G. Trüper. He incited her interest in sulfur metabolism and his never ceasing support and continuous encouragement is invaluable to her. She appreciates very much the support of her husband, Achim Heidrich, who always provides encouragement in all her endeavours.

Further information can be found at her web site: http://www.ifmb-a.uni-bonn.de/ag-dahl.htm. She can be reached by e-mail at ChDahl@uni-bonn.de.

David Knaff

David Knaff is a native New Yorker and received his early education in New York City. He was lucky enough to attend the Bronx High School of Science, where he not only received a superb education, but also had the good fortune to work with inspiring teachers who introduced him to the rewards of scientific inquiry. After receiving a B.S. in chemistry from the Massachusetts Institute of Technology (MIT) in 1962, he obtained M.S. (in 1963) and Ph.D. (in 1966) degrees, both in chemistry, from Yale University with the support of a National Science Foundation (NSF) pre-doctoral fellowship. His Ph.D. research, under the direction of Prof. Jui Wang, focused on developing model systems for photosynthetic energy conservation using illuminated solutions of metal-containing porphyrins in non-aqueous polar solvents. This work stimulated his interest in all aspects of photosynthesis and, despite the fact that he had never had any formal training in biochemistry, he was able to obtain, in 1966, a three-year National Institutes of Health (NIH) post-doctoral fellow in the Department of Cell Physiology at the University of California at Berkeley, where he received his first direct training in the biochemistry of photosynthetic systems under the mentorship of Prof. Daniel I. Arnon. He remained at Berkeley until 1976 as a Cell Physiology research scientist and was fortunate enough to share laboratory space with Richard (Dick) Malkin, Peter Schürmann and Bob Buchanan, who were wonderful colleagues and who introduced him to areas of biophysics and of plant and bacterial photosynthesis beyond those encountered in his work with Prof. Arnon.

In 1976, he moved to Lubbock, Texas to assume a faculty position in the Department of Chemistry and Biochemistry at Texas Tech University, where he is currently the Paul W. Horn Professor of Chemistry and serves as Co-Director of the Texas Tech Center for Biotechnology and Genomics. He served six years, from 1990 through 1996, as chair of his academic department. His early work in photosynthesis focused on biophysical and biochemical studies of membrane-bound electron transfer chains in plant chloroplasts, cyanobacteria and in purple and green anoxygenic photosynthetic bacteria. More recently, the emphasis in his research group has shifted to the study of soluble enzymes that catalyze electron transfer reactions involved in the early stages of sulfate and nitrate assimilation in oxygenic photosynthetic organisms, with a particular emphasis on enzymes that use reduced ferredoxin as an electron donor. His interest in ferredoxin-dependent enzymes led, with considerable assistance from Peter Schürmann and Bob Buchanan, to opening a new area of investigation in his lab at Texas Tech in redox-regulation of enzyme activity and gene expression in systems related to the chloroplast ferredoxin/thioredoxin system. The thioredoxin work produced extremely rewarding collaborations with the research groups of

Jean-Pierre Jacquot, Myroslawa Miginiac-Maslow and Jean-Marc Lancelin (supported in its early days by a joint NSF/CNRS (Centre National de la Recherche Scientifique) grant) in France, and with Carl Bauer's group at Indiana University in the U.S.

David Knaff, who is currently half-way through his second five-year term as editor-in-chief of the journal *Photosynthesis Research*, has published more than 200 refereed scientific articles and is co-author, with William (Bill) Cramer, of "Energy Transduction in Biological Membranes". His work continues to be supported generously by U.S federal agencies (the Department of Energy and the Department of Agriculture) and by the Robert A. Welch Foundation.

Further information can be found at the Department of Chemistry and Biochemistry, University of Texas, Lubbock, web site: http://www.depts. ttu.edu/chemistry. David Knaff can be reached by e-mail at david.knaff@ttu.edu.

Thomas Leustek

Thomas Leustek is a professor at the Biotechnology Center for Agriculture and the Environment of Rutgers University, New Brunswick, NJ. He had the privilege to spend and enjoy most of his academic career in the New Jersey/New York metropolitan area. In 1981, he received his B.A. from Newark College of Arts and Sciences, Newark, New Jersey. Early on he was interested in plant sciences. In 1987, he obtained his Ph.D. working on nutritional aspects of plants. During his period as a postdoctoral fellow (1986–1988) at the Roche Institute of Molecular Biology in Nutley, New Jersey, he worked on protein folding and heat stability in *E. coli* and *Rhodospirillum* under the supervision of Prof. Herbert Weissbach, who became an important teacher and mentor for him; Weissbach introduced him into the world of Biochemistry. He stayed on at the Roche Institute as a research fellow (1988–1991) and extended the work on heat-shock proteins to plant cells.

In 1991 he accepted an offer to join the Plant Science Department at Rutgers University, New Brunswick as an Assistant Professor. There, he developed a comprehensive research program on the plant assimilatory sulfate reduction pathway. He was among the first to clone genes and cDNAs and developed functional complementation of *E. coli* to perfection. The reward was the identification of adenosine-phosphosulfate reductase, the enigmatic central enzyme of the sulfate reduction pathway. His discoveries closed a long-standing biochemical controversial discussion on the nature of electron transfer to free or bound reaction intermediates and helped to establish the now accepted mechanisms in plants and different groups of bacteria. As an Associate Professor and since 2001 as a Professor at the newly founded Biotechnology Center for Agriculture, he also worked intensively on the improvement of crop plants with respect to their nutritional value. More recently, he has embarked on a new challenge to the biochemists: the biosynthesis of essential amino acids in plants. At Rutgers, he is Director of the Plant Biology Graduate Program; he has received several awards for his outstanding teaching and sustained research.

Tom Leustek has received numerous honors and serves on a number of panels, such as the editorial boards of *The Plant Journal* and *Photosynthesis Research*. He advises scientific panels including USDA (United States Department of Agriculture) and NRI (National Resources Institute) but also biotech companies and is a regular reviewer for the most important journals in plant biology.

Further information can be found at a web site: http://www.cook.rutgers.edu/~biotech/faculty/index.html. Thomas Leustek can be reached by e-mail at leustek@AESOP.Rutgers.edu.

Author Index

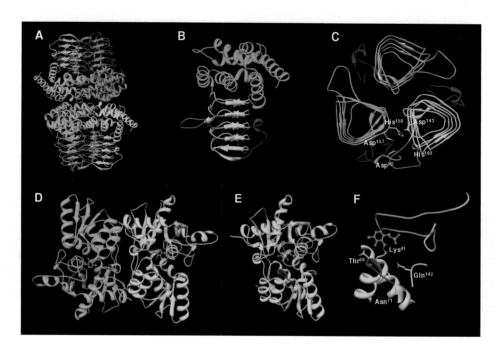

Fig. 2. Three-dimensional structures of prokaryotic SAT and OAS-TL. The figure depicts the overall homomeric quaternary structure of SAT (A) and OAS-TL (D). In both cases the monomers are shown in different colours in order to illustrate the sterical position of the monomers. Each SAT monomer (B) consists of an α-helical domain (green), a β-sheet domain (golden) that included a random coil loop (red) and the C-terminal tail (dark blue), which is responsible for CSC formation. Please note that the last 11 amino acid residues of the C-terminal tail are missing in the structural presentation, since they were not solved by X-ray crystallography. (C) Top view on an SAT trimer, where α-helical domains are deleted for better overview. The active site of SAT lies in between the β-sheet domain of two monomers, thus a SAT 'dimer of trimers' (A) contains six independent active sites. The most important residues for binding of substrates and the enzymatic activity are named and shown in sticks. Each OAS-TL monomer (E) of the native OAS-TL dimer (B) contains one PLP (red) that is bound via an internal Schiff base by Lys[41]. The asparagine loop, which is important for substrate binding and induction of the global conformational change of OAS-TL during catalysis is shown in dark blue. The green colour indicates the β8a–β9a loop that seems to be important for interaction with SAT. Both structural elements are in close proximity to the active site of OAS-TL (F). Important residues for binding of substrates and the enzymatic activity are highlighted in F. The software UCSF Chimera V1.2422 (University of California, San Francisco) was used to visualize SAT (pdb: 1T3d) and OAS-TL (pdb: 1OAS). See Chapter 4, p. 67.

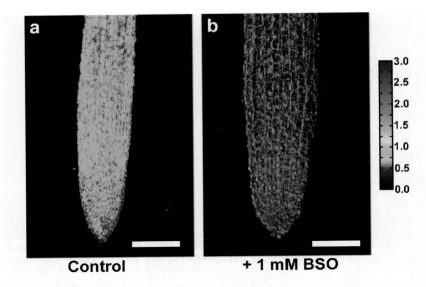

Control   **+ 1 mM BSO**

Fig. 6. Inhibition of glutathione biosynthesis causes oxidation of roGFP2. Arabidopsis seeds transformed with roGFP2 were germinated on Agar plates with or without 1 mM BSO for 7 days and imaged for the redox status of roGFP2 and GSH levels. All images are maximum projections from stacks of serial optical sections. The ratio images show analysis of control roots (a) and roots grown on 1 mM BSO (b). The color scale shows the pseudocolor coding for reduced (blue) and oxidized roGFP2 (red). Bars = 100 μm. See Chapter 24, p. 493.

Chapter 1

Introduction to Sulfur Metabolism
in Phototrophic Organisms

Christiane Dahl[*]

Institut für Mikrobiologie & Biotechnologie, Rheinische Friedrich-Wilhelms-Universität Bonn, Meckenheimer Allee 168, D-53115 Bonn, Germany

Rüdiger Hell

Department of Molecular Biology of Plants, Heidelberg Institute of Plant Sciences, University of Heidelberg, Im Neuenheimer Feld 360, D-69120 Heidelberg, Germany

Thomas Leustek

Rutgers University, Biotechnology Center for Agriculture and the Environment, Department of Plant Biology and Pathology, 59 Dudley Road, Room 328, New Brunswick, New Jersey 08901-8520, USA

David Knaff

Department of Chemistry and Biochemistry and Center for Biotechnology and Genomics, Texas Tech University, Lubbock, Texas 79409-1061, USA

[*]Corresponding author, Phone: +49 228 732119, Fax: +49 228 737576, E-mail: ChDahl@uni-bonn.de

Rüdiger Hell et al. (eds.), Sulfur Metabolism in Phototrophic Organisms, 1–14.
© 2008 *Springer.*

the purple nonsulfur bacteria *Rhodovulum sulfidophilum* and *Rhodopseudomonas palustris*. It also occurs as intermediate in assimilatory sulfate reduction, although at low concentrations due to its toxicity. Linked to organic molecules this oxidation state is found in substantial amounts in sulfolipids that occur in most phototrophic organisms and a few nonphotosynthetic organisms (see Benning et al., this book). Cellular sulfite and, indirectly, SO_2, are oxidized to sulfate in animals and plants (see Hänsch and Mendel, chapter 12).

D. Elemental Sulfur

Sulfur is the element with the largest number of allotropes (about 30), but only a few are found in nature and occur in biological systems. The homocyclic, orthorhombic crystalline α-sulfur (α-S_8) (*cyclo*-octasulfur) is the most thermodynamically stable form at standard conditions (Roy and Trudinger, 1970; Steudel, 1996a; Steudel, 1996b). Other sulfur rings from S_6 to S_{20} have been synthesized as pure substances, of these, S_6, S_7, and S_{12} have also been detected in samples of biological origin (Steudel, 1987; Steudel, 2000). Commercially available sulfur consists mainly of S_8 rings, polymeric sulfur and traces of S_7 rings. The bright yellow color originates from the small amount of S_7 rings (Steudel and Holz, 1988). Elemental sulfur occurs in huge deposits (e.g., in USA, Mexico, Poland) mostly together with calcite, clay, anhydrite or with gypsum and can be found in volcanic areas (Falbe and Regitz, 1995; Dahl et al., 2002). Polymeric sulfur consists of very long chains of almost all sizes (Steudel, 2000). Regardless of the molecular size, all sulfur allotropes are hydrophobic, are not wetted by water and have very low solubilities in water (Steudel and Eckert, 2003). Biologically produced sulfur deposited in bacterial sulfur globules occurs in different speciations depending on metabolic properties of the organisms and the environmental conditions: long sulfur chains very probably terminated by organic residues (mono- or bis-organyl polysulfanes) in purple and green sulfur bacteria, *cyclo*-octasulfur in chemotrophic sulfur oxidizers like *Beggiatoa alba* and *Thiomargararita namibiensis* and long chain polythionates in the aerobically grown acidophilic sulfur oxidizer *Acidithiobacillus ferrooxidans* (Prange et al., 2002; Dahl and Prange, 2006). However,

the discovery of elemental sulfur in plants was a surprise at its time (Cooper et al., 1996; Cooper and Williams, 2004). Today, it is evident that plants are able to produce either preformed or induced elemental sulfur as a defense compound against attack by fungal pathogens (see Burow et al., chapter 11). In contrast to phototrophic bacteria, the pathway leading to the deposition of sulfur globules in defined plant tissues is still unknown. However, the S^0 oxidation state occurs during the biosynthesis of iron–sulfur clusters in bacteria as well as in plants (see Padmavathi et al., chapter 7).

E. Polysulfides and Polysulfanes

Polysulfides (S_n^{2-}) can be produced either by reaction of elemental sulfur with sulfide or by a partial oxidation of sulfide, e.g., by phototrophic or chemotrophic sulfur oxidizing bacteria (Bryantseva et al., 2000). Polysulfide anions can exist only in neutral or alkaline solutions (pH > 6). The higher the pH of polysulfide mixtures, the shorter are the chains lengths (the chain length varies from two to eight sulfur atoms, depending on the pH). Aqueous polysulfide solutions are yellow/orange and are subject for rapid autoxidation when exposed to air, forming thiosulfate (Steudel et al., 1986). Polysulfane (H–S_n–H) mixtures are almost insoluble in water and homogeneous samples decompose slowly, forming H_2S and S_8 rings.

F. Sulfide

Hydrogen sulfide, the most reduced form of inorganic sulfur, occurs naturally in large amounts in underground deposits of "sour" natural gas. Other sources are minerals in soils and rocks or sulfide-enriched springs. It is also released during decay of organic matter. Furthermore, it is the main product of sulfate respiration. In water, H_2S is partly ionized. Except at the highest pH values, the concentration of sulfide ions S^{2-} is negligible (Philips and Philips, 2000). Sulfide is the preliminary end product of assimilatory sulfate reduction. At cellular pH the equilibrium between H_2S and HS^- requires efficient and immediate fixation into organic compounds to avoid losses to the gas phase and reduce the risk of deleterious redox processes with protein disulfide bridges. The β-replacement reaction with activated serine

(O-acetylserine) is therefore comparable to the fixation of ammonia in nitrate assimilation, both with respect to biosynthetic metabolism and importance for cell life (see chapter 4 Hell and Wirtz, this book).

G.   Organic Polysulfanes

Mixtures of (bis-)organic polysulfanes ($R–S_n–R$) with sulfur chains of different chain lengths (up to 13 atoms have been reported in pure substances [Steudel, 2000]) are light yellow, strong refractive oily liquids, especially in the presence of higher homologous compounds and can exist in many different conformational isomers. The solubility of organic polysulfanes (chain-like or cyclic) in water or organic solvents depends on the nature of the organic portions of the molecules.

The chemistry of organic polysulfanes with $n > 2$ has been reviewed by Steudel and Kustos (1994). A great variety of polysulfanes appear to be generated by living organisms and are widely distributed (Steudel and Kustos, 1994). A well-known example is lentionine $C_2H_4S_5$ occurring in the shiitake mushroom (*Lenthinus edodes*), which forms a heterocycle with a disulfane and a trisulfane bridge between two methylene groups. Numerous symmetric and nonsymmetric disulfanes and trisulfanes occur in plants, such as 2-butylpropenyldisulfane and other disulfanes in asfetida oil, allylmethyl and propylmethyl di- and trisulfanes in garlic and onions and various di-, tri- and tetrasulfanes in seeds of the neem tree (*Azadirachta indica*) (Kelly and Smith, 1990 and references therein).

H.   Relevant Organic Sulfur Compounds

Sulfur contents in phototrophic organisms can be highly variable depending on lifestyle. These variations are smaller between plant species. Leaf material from plants rich in secondary sulfur compounds such as the *Brassicaea* contain, under regular mineral supply, about 300 mmol/kg total S and normal plants such as sugar beet about 50 mmol/kg. In both cases sulfate contributes 30%–50% of total S (Houba and Uittenbogaard, 1994). Due to the special chemistry of sulfur an immense number of sulfur-containing organic compounds are found in nature with highly variable functions (Mitchell, 1996;

Beinert, 2000). Apart from primary organic sulfur compounds that are common to most living organisms (sulfur amino acids, several vitamins, iron–sulfur clusters, etc.), plants are able to synthesize so-called secondary sulfur compounds of huge chemical heterogeneity and taxonomic restriction. They may function in plant defence against pathogens, insects and other herbivores, in allelopathy, positive plant–insect interactions (scent), signalling and others (Harborne, 1988). Prominent examples of defence compounds are the glucosinolates (*Brassicaceae*), alliins (*Allium*) and thiophenes (*Asteraceae*) (see Burow et al., this book). Glucosinolates and alliins are degraded during tissue disruption and give rise to breakdown products with antimicrobial, but also in several cases human health promoting properties (Hell and Kruse, 2007). The herbivore deterrent pungent taste of the breakdown products are well known also for their culinary properties (e.g., mustard and garlic). The lachrymatory factor (propanthial S-oxide) from onion that makes the eyes water is a by-product of the enzyme-mediated breakdown of alliins (Imai et al., 2002). Highly specialized functions are also found in species-specific compounds such as turgorin (4-O-(β-D-glucopyranosyl-6′-sulfate)), the signalling substance in leaf movement of the Mimosa plant (Varin et al., 1997). However, sulfation of metabolites appears to be a way of targeted activation/inactivation of signals, such as the plant hormones jasmonate, phytosulfokines (see Hernandez-Sebastiá et al., chapter 6) and lipo-polysaccharides in plant/*Rhizobia* symbiosis (Roche et al., 1991). The demonstrated role of targeted redox processes on disulfide bridges in proteins is a still unresolved area of research. Originally discovered in the context of photosynthesis, the roles of thioredoxins, glutaredoxins and peroxiredoxins becomes increasingly clear for sensing and transduction of environmentally-induced redox changes (see Meyer and Rausch, chapter 9; Meyer and Fricker, chapter 24). The *E. coli* transcription factor OxyR is activated by reversible disulfide bond formation and post-translationally modulated in its activity by S-nitrosylation, S-hydroxylation and S-glutathionylation. OxyR may thus be just one first example for the transduction of external oxidative processes to cellular responses in other organisms (Kim et al., 2002).

IV. The Natural Sulfur Cycle

The sulfur cycle in nature is extremely complex, not only because sulfur exists in many different oxidation states (see above) but also because both chemical and enzyme-catalyzed reactions are involved in many sulfur transformations. Organisms participate in the sulfur cycle in two fundamentally different ways: the assimilation of sulfur compounds serves for the biosynthesis of sulfur-containing cell constituents, while during dissimilation sulfur compounds serve as electron donors or acceptors for energy-yielding processes (Fig.1). The dissimilatory part of the sulfur cycle is restricted to prokaryotes. Recent reviews on chemical, geochemical, biochemical and biotechnological aspects of the sulfur cycle have been provided (Brüser et al., 2000; Middelburg, 2000; Steudel, 2000; Lens and Kuenen, 2001). Cyanobacteria, algae and plants use electrons from photosynthesis to contribute the vast part of sulfate assimilation and subsequent formation of reduced sulfur-containing cell constituents. On a global scale the oceans and land vegetation probably contribute approximately equal parts to sulfate assimilation (see chapter by Giordano, Norici, Ratti and Raven, chapter 20). Substantial input of atmospheric sulfur (SO_2) from burning of fossil energy resources was observed until the early 1990s but due to installation of sulfur filters this input has declined in the past decade (Schnug, 1998). A small, but steadily increasing input of sulfate is observed by fertilization of agroecosystems to compensate for decreasing atmospheric intake and to maintain high productivity of crops (see chapter by Zhao, Tausz and De Kok, chapter 21).

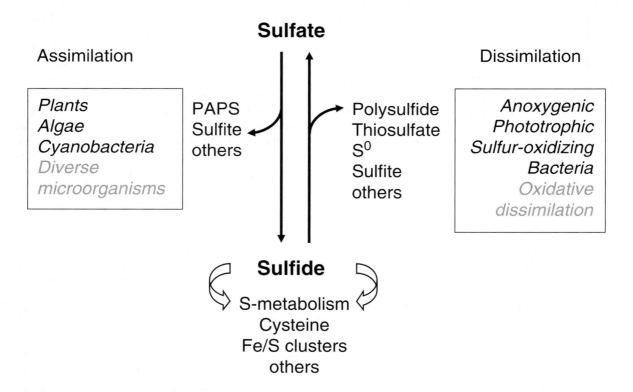

Fig. 1. Schematic overview of the sulfur cycle with emphasis on phototrophic organisms. General assimilatory and dissimilatory processes are indicated in grey letters. Aerobic and anaerobic chemolithotrophic bacteria are also able to dissimilate sulfide for energy production. Animals carry out metabolic dissimilation and are able to activate sulfate to 3′-phosphoadenosine 5′-phosphosulfate (PAPS) but cannot reduce and assimilate sulfate. Geochemical and atmospheric processes are omitted.

A. Reductive Steps

1. Sulfate Assimilation and the Part Taken by Phototrophs in This Process

The absence of external reduced sulfur compounds that can be incorporated into organic matter makes the assimilatory reduction of sulfate to sulfide an essential requirement. Some phototrophic bacteria are very much specialized for living in habitats with reduced sulfur compounds and such bacteria may lack a sulfate reduction pathway (e.g., members of the *Chlorobiaceae*). Usually assimilatory sulfate reduction is a highly regulated process. While many studies have been carried out on eukaryotes, sulfate reduction pathways are clearly also induced under sulfur-limited conditions in bacteria (Kredich, 1987) including anoxygenic phototrophic bacteria (Haverkamp and Schwenn, 1999; Neumann et al., 2000).

The importance of assimilatory reduction for primary production of biomass including human nutrition sparked intensive investigation of this pathway in algae and plants. Today it is generally accepted that sulfate uptake, reduction and fixation of sulfide are the most regulated steps in this process (see chapters by Hawkesford, (2) Kopriva et al., (3) and Hell and Wirtz (4), Höfgen and Hesse, (5) this book). The mechanism of activation and reduction of the chemically quite inert sulfate molecule received particular attention. In plants and algae sulfate in the plastids is first activated to form a sulfonucleotide from ATP (adenosine-phosphosulfate; APS), catalyzed by ATP sulfurylase. Due to instability of APS and the unfavourable reaction equilibrium ($\Delta G' = +45\,kJ/mol$) the backward reaction was believed to be prevented by cleavage of pyrophosphate by abundant pyrophosphatase activity and further activation by APS kinase to PAPS (Leustek et al., 2000). Indeed PAPS is very important for sulfate transfer reactions to metabolites, peptides and proteins (see chapter by Hernández-Sebastià, Marsolais and Varin, chapter 6). But it turned out to be only the primary substrate for reduction in diverse eubacteria (including enterobacteria), archaea and fungi, but not in plants and algae. Despite the before mentioned considerations, APS appears to be the major substrate for assimilatory reduction according to comprehensive biochemical taxonomic analyses of the

responsible enzyme (see Kopriva et al., chapter 3). The signature of the plant, algae and cyanobacteria type sulfonucleotide (APS) reductase was seen in a two domain structure consisting of a reductase domain with an $[4Fe–4S]^{2+}$ cluster and a C-terminal thioredoxin/glutaredoxin domain, using glutathione as electron donor. In contrast, the bacterial sulfonucleotide (PAPS) reductase was believed to have only the reductase domain without iron–sulfur cluster and to use thioredoxin as electron donor. Today the situation indicates several variants of assimilatory sulfonucleotide reductases: (1) the *E. coli* CysH-like enzymes reduces PAPS using thioredoxin or glutaredoxin as electron donor but no $[4Fe–4S]^{2+}$ cluster and is found in many eubacteria, archaea and fungi (Berendt et al., 1995); (2) another CysH-like group reduces APS rather than PAPS, uses thioredoxin as electron donor, has an $[4Fe–4S]^{2+}$ cluster and was found in *Pseudomonas aeruginosa* (Bick et al., 2000); (3) yet another protein with strong homology to CysH is from *Bacillus subtilis* and is able to reduce APS and PAPS (Berndt et al., 2004); (4) the plant-type sulfonucleotide reductase (e.g., *Arabidopsis thaliana, Enteromorpha intestinalis*) uses APS, carries an $[4Fe–4S]^{2+}$ cluster and contains a C-terminal glutaredoxin-like domain (Bick et al., 1998; Gao et al., 2000); (5) a reductase from the moss *Physcomitrella patens* that shares high similarity with CysH-like proteins, has no iron–sulfur cluster but prefers APS rather than PAPS (Kopriva et al., 2007). This functional variability seems independent from oxygenic or phototrophic lifestyles and reflects the complex evolutionary development of assimilatory sulfate reduction (see chapter 03 by Kopriva et al.).

2. Dissimilatory Sulfate Reduction

Under anoxic conditions sulfate, and elemental sulfur are used as electron acceptors of anaerobic respiratory processes. The sulfate- and sulfur-reducing bacteria are a metabolically versatile group of microorganisms, belonging to many different families and genera. The ability to reduce sulfate as a terminal electron acceptor is found within the Gram-negative and the Gram-positive bacteria, the genus *Thermodesulfobacterium*, and also in some *Archaea* (genus *Archaeoglobus*). Dissimilatory sulfur reduction is found for many sulfate reducers and the specialists of the genus *Desulfuromonas* and is widespread within the hyperthermophilic archaea. Some sulfur respires can in addition use sulfite, thiosulfate, organic sulfoxides, inorganic polysulfides and/or organic disulfanes as terminal electron acceptors. All these processes lead to the release of large amounts of sulfide. At the interface of oxic and anoxic zones, chemical oxidation of sulfide can lead to the formation of thiosulfate and polythionates.

Just as during assimilatory sulfate reduction, sulfate has first to be activated at the expense of two ATP equivalents per sulfate molecule. This activation is catalyzed by ATP sulfurylase and yields APS which is reduced directly to sulfite by the iron–sulfur flavoprotein APS reductase (Brüser et al., 2000). The latter bares no resemblance to the APS/PAPS reductases of the assimilatory pathway. The last step, reduction of sulfite to sulfide is catalyzed by dissimilatory sulfite reductase, an enzyme containing iron–sulfur clusters and sirohemeamide as prosthetic groups (Matias et al., 2005). While the direct electron-donating compounds are neither known for APS reductase nor for sulfite reductase, electrons are very probably delivered to these enzymes through multi-subunit transmembrane complexes (Heidelberg et al., 2004; Matias et al., 2005).

B. Oxidative Steps

Reduced inorganic sulfur compounds such as sulfide, polysulfides, sulfur, sulfite, thiosulfate, and various polythionates allow energy generation by aerobic and anaerobic chemotrophic and phototrophic sulfur-oxidizing bacteria (Brune, 1995; Friedrich, 1998; Dahl and Prange, 2006), some archaea (Stetter, 1996), and even of some eukarya (Grieshaber and Völkel, 1998). Most sulfur oxidation observed in members of the *Eukarya* is mediated by lithoautotrophic bacterial symbionts (Nelson and Fisher, 1995). Although in general sulfate is the major oxidation product of reduced sulfur compound oxidation, other end products may be formed depending on the organism. The ability to use the various sulfur compounds also varies. Sulfur oxidation pathways are therefore found to be widely variable and may involve different nonpolymeric and polymeric intermediates.

Anoxygenic phototrophic sulfur bacteria require light as an energy source and use reduced sulfur compounds as electron-donating substrates for photosynthetic CO_2 reduction. They represent an assemblage of predominantly aquatic bacteria that are able to grow under anoxic conditions by photosynthesis, without oxygen production. The most striking and common property of these bacteria is the ability to carry out light-dependent, bacteriochlorophyll-mediated processes, a property shared with cyanobacteria, prochlorophytes, algae and green plants. The major pigments are bacteriochlorophylls a, and b in the purple bacteria, c, d, and e in the green bacteria and g in the Gram-positive heliobacteria. Photosynthesis in anoxygenic phototrophic bacteria depends on anoxic or oxygen-deficient conditions, because the synthesis of the photosynthetic pigments and the formation of the photosynthetic apparatus are repressed by oxygen.

Although the various anxyogenic phototrophic bacteria have many common characteristics with regard to photosynthesis they form an extremely heterogeneous group of bacteria. Based on phenotypic characteristics like the occurrence of different bacteriochlorophylls and carotenoids and based on phylogenetic analyses, one can distinguish between the green sulfur bacteria (phylum *Chlorobi*), the anoxygenic filamentous phototrophs (also termed green gliding bacteria or green nonsulfur bacteria) of the family *Chloroflexaceae*, the purple sulfur bacteria (*Chromatiaceae* and *Ectothiorhodospiraceae*), the purple "nonsulfur" bacteria (affiliated with the *Alphaproteobacteria* and the *Betaproteobacteria*) and the heliobacteria. On the basis of their 16S rRNA sequences, the purple bacteria belong to the *Proteobacteria*, while *Chloroflexus* and relatives and the green sulfur bacteria, respectively, form separate lines of descent (Woese et al., 1985). The heliobacteria (Madigan and Ormerod, 1995) are related with certain Gram-positive bacteria (Beer-Romero and Gest, 1987). The aerobic anoxygenic phototrophic bacteria are a relatively recently discovered bacterial group. These Gram-negative, marine bacteria contain bacteriochlorophyll a, form intracytoplasmic membrane systems, and contain reaction centers which resemble those of the purple bacteria. In contrast to the purple bacteria, oxygen stimulates the synthesis of bacteriochlorophyll and carotenoids in these organisms.

The sulfur oxidation properties and pathways are described in detail elsewhere (chapters by Shahak and Hauska, Dahl, Frigaard and Bryant, Chan, Morgan-Kiss and Hanson, this book).

Cyanobacteria are equipped with two photosystems and normally perform a plant-like oxygenic photosynthesis. However, under anoxic conditions many cyanobacteria are able to use sulfide as an alternative electron donor. *Oscillatoria limnetica* oxidizes sulfide to elemental sulfur (Cohen et al., 1975) whereas *Microcoleus chtonoplastes* (de Wit and van Gemerden, 1987) forms thiosulfate as the end product of sulfide oxidation. More recent studies showed that thiosulfate is the most common product of sulfide oxidation by cyanobacteria and that sulfite or polysulfide can be formed as additional products depending on the organism (Rabenstein et al., 1995).

References

Abrams WR and Schiff JA (1973) Studies on sulfate utilization by algae. 11. An enzyme-bound intermediate in the reduction of adenosine-5′-phosphosulfate (APS) by cell-free extracts of wild-type *Chlorella* and mutants blocked for sulfate reduction. Arch Microbiol 94: 1–10

Abrol YP and Ahmad A (2005) Sulphur in Plants. Kluwer Academic Publishers, London

Barbosa-Jefferson VL, Zhao FJ, Mcgrath SP and Magan N (1998) Thiosulphate and tetrathionate oxidation in arable soils. Soil Biol Biochem 30: 553–559

Barrett EL and Clark MA (1987) Tetrathionate reduction and production of hydrogen sulfide from thiosulfate. Microbiol Rev 51: 192–205

Bavendamm W (1924) Die farblosen und roten Schwefelbakterien des Süß- und Salzwassers. In: Kolkwitz R (ed) Pflanzenforschung (2), pp 1–156. Verlag G. Fischer, Jena

Beer-Romero P and Gest H (1987) *Heliobacillus mobilis*, a peritrichously flagellated anoxyphototroph containing bacteriochlorophyll g. FEMS Microbiol Lett 41: 109–114

Beinert H (2000) A tribute to sulfur. Eur J Biochem 267: 5657–5664

Berendt U, Haverkamp T and Schwenn JD (1995) Reaction mechanism of thioredoxin: 3′-phospho-adenylylsulfate reductase investigated by site-directed mutagenesis. Eur J Biochem 233: 347–356

Berndt C, Lillig CH, Wollenberg M, Bill E, Mansilla MC, de Mendoza D, Seidler A and Schwenn JD (2004) Characterization and reconstitution of a 4Fe–4S adenylyl sulfate/phosphoadenylyl sulfate reductase from *Bacillus subtilis*. J Biol Chem 279: 7850–7855

Bick JA, Aslund F, Chen YC and Leustek T (1998) Glutaredoxin function for the carboxyl-terminal domain of the plant-type 5′-adenylylsulfate reductase. Proc Natl Acad Sci USA 95: 8404–8409

Bick JA, Dennis JJ, Zylstra GJ, Nowack J and Leustek T (2000) Identification of a new class of 5′-adenylylsulfate (APS) reductases from sulfate-assimilating bacteria. J Bacteriol 182: 135–142

Brown KA (1982) Sulfur in the environment: a review. Environ Pollut 3B: 47–80

Brune DC (1989) Sulfur oxidation by phototrophic bacteria. Biochim Biophys Acta 975: 189–221

Brune DC (1995) Sulfur compounds as photosynthetic electron donors. In: Blankenship RE, Madigan MT and Bauer CE (eds) Anoxygenic Photosynthetic Bacteria, (Advances in Photosynthesis, Vol.2) pp 847–870. Kluwer Academic Publishers, Dordrecht

Brunold C and Schiff JA (1976) Studies of sulfate utilization by algae. 15. Enzymes of assimilatory sulfate reduction in *Euglena* and their cellular localization. Plant Physiol 57: 430–436

Brüser T, Lens P and Trüper HG (2000) The biological sulfur cycle. In: Lens P and Pol LH (eds) Environmental Technologies to Treat Sulfur Pollution, pp 47–86. IWA Publishing, London

Bryantseva IA, Gorlenko VM, Kompantseva EI, Tourova TP, Kuznetsov B and Osipov GA (2000) Alkaliphilic heliobacterium *Heliorestis baculata* sp. nov. and emended description of the genus *Heliorestis*. Arch Microbiol 174: 283–291

Buder J (1915) Zur Kenntnis des *Thiospirillum jenense* und seine Reaktionen auf Lichtreize. Jahrb f wiss Bot 56: 529–584

Buder J (1919) Zur Biologie des Bakteriopurpurins und der Purpurbakterien. Jahrb f wiss Bot 58: 525–628

Bunker HJ (1936) A Review of the Physiology and Biochemistry of the Sulfur Bacteria. HM Stationery Office, London

Cohen Y, Padan E and Shilo M (1975) Facultative anoxygenic photosynthesis in the cyanobacterium *Oscillatoria limnetica*. J Bacteriol 123: 855–861

Cohn F (1875) Untersuchungen über Bakterien II. Beitr z Biol d Pflanzen 1: 141–207

Cooper RM, Resende MLV, Flood J, Rowan MG, Beale MH and Potter U (1996) Detection and cellular localization of elemental sulfur in disease-resistant genotypes of *Theobroma cacao*. Nature 379: 159–162

Cooper RM and Williams JS (2004) Elemental sulphur as an induced antifungal substance in plant defence. J Exp Bot 55: 1947–1953

Dahl C and Friedrich CG (2007) Microbial Sulfur Metabolism. Springer, Berlin, Heidelberg, New York

Dahl C and Prange A (2006) Bacterial sulfur globules: occurrence, structure and metabolism. In: Shively JM (ed) Inclusions in Prokaryotes, pp 21–51. Springer, Heidelberg

Dahl C, Prange A and Steudel R (2002) Natural polymeric sulfur compounds. In: Steinbüchel A (ed) Miscellaneous Biopolymers and Biodegradation of Synthetic Polymers, pp 35–62. Wiley-VCH, Weinheim

De Kok LJ and Schnug E (2005) Proceedings of the 1st Sino-German workshop on aspects of sulfur nutrition in plants. Landbauforschung Völkenrode – FAL Agricult Res special issue No 283

de Wit R and van Gemerden H (1987) Oxidation of sulfide to thiosulfate by *Microcoleus chtonoplastes*. FEMS Microbiol Ecol 45: 7–13

Denger K, Laue H and Cook AM (1997) Thiosulfate as a metabolic product: the bacterial fermentation of taurine. Arch Microbiol 168: 297–301

Dick WA (1992) Sulfur cycle. In: Lederberg J (ed) Encyclopedia of Microbiology, pp 123–133. Academic, San Diego

Ehrenberg CG (1838) Die Infusionsthierchen als vollkommene Organismen. Ein Blick in das tiefere organische Leben der Natur. Leopold Voss-Verlag, Leipzig

Engelmann TW (1882) Ueber Licht- und Farbenperception niederster Organismen. E Pflüger Arch f Physiol 29: 387–400

Falbe J and Regitz M (1995) Römpp Chemie Lexikon, 9th edn. Thieme, Stuttgart

Falkowski PG (2006) Evolution: tracing oxygen's imprint on Earth's metabolic evolution. Science 311: 1724–1725

Friedrich CG (1998) Physiology and genetics of sulfuroxidizing bacteria. Adv Microb Physiol 39: 235–289

Friedrich CG, Bardischewsky F, Rother D, Quentmeier A and Fischer J (2005) Prokaryotic sulfur oxidation. Curr Opin Microbiol 8: 253–259

Friedrich CG, Rother D, Bardischewsky F, Quentmeier A and Fischer J (2001) Oxidation of reduced inorganic sulfur compounds by bacteria: emergence of a common mechanism? Appl Environ Microbiol 67: 2873–2882

Gao Y, Schofield OME and Leustek T (2000) Characterization of sulfate assimilation in marine algae focusing on the enzyme 5'-adenylylsulfate reductase. Plant Physiol 123: 1087–1096

Giordano M, Norici A and Hell R (2005) Sulfur and phytoplankton: acquisition, metabolism and impact on the environment. New Phytologist 166: 371–382

Grieshaber MK and Völkel S (1998) Animal adaptations for tolerance and exploitation of poisonous sulfide. Ann Rev Physiol 60: 33–53

Gutierrez-Marcos JF, Roberts MA, Campbell EI and Wray JL (1996) Three members of a novel small gene-family from *Arabidopsis thaliana* able to complement functionally an *Escherichia coli* mutant defective in PAPS reductase activity encode proteins with a thioredoxin-like domain and "APS reductase" activity. Proc Natl Acad Sci USA 93: 13377–13382

Harborne JB (1988) Introduction to Ecological Biochemistry. Academic, London

Haverkamp T and Schwenn JD (1999) Structure and function of a *cysBJIH* gene cluster in the purple sulfur bacterium *Thiocapsa roseopersicina*. Microbiology 145: 115–125

Hawkesford M (2004) Sulphur metabolism in plants. J Exp Bot 55: (404) special issue

Hawkesford M and De Kok LJ (2007) Sulfur in Plants. An Ecological Perspective. Springer, London

Heidelberg JF, Seshadri R, Haveman SA, Hemme CL, Paulsen IT, Kolonay JF, Eisen JA, Ward N, Methe B, Brinkac LM, Daugherty SC, Deboy RT, Dodson RJ, Durkin AS, Madupu R, Nelson WC, Sullivan SA, Fouts D, Haft DH, Selengut J, Peterson JD, Davidsen TM, Zafar N, Zhou L, Radune D, Dimitrov G, Hance M, Tran K, Khouri H, Gill J, Utterback TR, Feldblyum TV, Wall JD, Voordouw G and Fraser CM (2004) The genome sequence of the anaerobic, sulfate-reducing bacterium *Desulfovibrio vulgaris* Hildenborough. Nat Biotechnol 22: 554–559

Hell R (1997) Molecular physiology of plant sulfur metabolism. Planta 202: 138–148

Hell R and Kruse C (2007) Sulfur in biotic interactions of plants. In: Hawkesford M and De Kok LJ (eds) Sulfur in Plants. An Ecological Perspective, pp 197–224. Springer, London

Hell R and Leustek T (2005) Sulfur metabolism in plants and algae – a case study for an integrative scientific approach. Photosyn Res 86: 297–298

Hesse H and Hoefgen R (2003) Molecular aspects of methionine biosynthesis. Trends Plant Sci 8: 259–262

Houba VJG and Uittenbogaard J (1994) Chemical Composition of Various Plant Species. Wageningen University, The Netherlands

Imai S, Tsuge N, Tomotake M, Nagatome Y, Sawada H, Nagata T and Kumagai H (2002) An onion enzyme that makes the eyes water – A flavoursome, user-friendly bulb would give no cause for tears when chopped up. Nature 419: 685

Kanno N, Nagahisa E, Sato M and Sato Y (1996) Adenosine 5'-phosphosulfate sulfotransferase from the marine macroalga *Porphyra yezoensis* Ueda (Rhodophyta): Stabilization, purification, and properties. Planta 198: 440–446

Kelly DP and Smith NA (1990) Organic sulfur compounds in the environment. Adv Microb Ecol 11: 345–385

Kim SO, Merchant K, Nudelman R, Beyer WF, Keng T, DeAngelo J, Hausladen A and Stamler JS (2002) OxyR: a molecular code for redox-related signaling. Cell 109: 383–396

Kopriva S, Fritzemeier K, Wiedemann G and Reski R (2007) The putative moss 3'-phosphoadenosine phosphosulfate reductase is a novel form of adenosine 5'-phosphosulfate reductase without iron sulfur cluster. J Biol Chem 282: 22930–22938

Kredich NM (1987) Biosynthesis of cysteine. In: Neidhardt FC, Ingraham TL, Low KB, Magasanik B, Schaechter M and Umbarger HE (eds) *Escherichia coli* and *Salmonella typhimurium*: Cellular and Molecular Biology, pp 419–428. American Society of Microbiology, Washington, DC

Lens PNL and Kuenen JG (2001) The biological sulfur cycle: novel opportunities for environmental biotechnology. Water Sci Technol 44: 57–66

Leustek T, Martin MN, Bick JA and Davies JP (2000) Pathways and regulation of sulfur metabolism revealed through molecular and genetic studies. Annu Rev Plant Physiol Plant Mol Biol 51: 141–165

Li J and Schiff JA (1991) Purification and properties of adenosine 5′-phophosulfate sulfotransferase from *Euglena*. Biochem J 274: 355–360

Luther GW, Church TM, Scudlark JR and Cosman M (1986) Inorganic and organic sulfur cycling in salt-marsh pore waters. Science 232: 746–749

Madigan MT and Ormerod JG (1995) Taxonomy, physiology and ecology of heliobacteria. In: Blankenship RE, Madigan MT and Bauer CE (eds) Anoxygenic Photosynthetic Bacteria, (In Advances in Photosynthesis, Vol. 2) pp 17–30. Kluwer Academic Publishers, Dordrecht

Matias PM, Pereira IAC, Soares CM and Carrondo MA (2005) Sulphate respiration from hydrogen in *Desulfovibrio* bacteria: a structural biology overview. Prog Biophys Molec Biol 89: 292–329

Middelburg J (2000) The geochemical sulfur cycle. In: Lens P and Hulshoff Pol W (eds) Environmental Technologies to Treat Sulfur Pollution, pp 33–46. IWA Publishing, London

Mitchell SC (1996) Biological Interactions of Sulfur Compounds. Taylor and Francis, London

Mothes K and Specht W (1934) Über den Schwefeltoffwechsel der Pflanzen. Planta 22: 800–803

Müller A and Krebs B (1984) Sulfur – Its Significance for Chemistry, for the Geo-, Bio- and Cosmosphere and Technology. Elsevier, Amsterdam

Müller C (1870) Chemisch-Physikalische Beschreibung der Thermen von Baden in der Schweiz (Canton Aargau). Zehnder, Baden

Müller OF (1786) Animalcula Infusoria Fluviatilia Et Marina, Quae Detexit, Systematice Descripsit Et Ad Vivum Delineari Curavit. Mölleri, Hauniae

Nadson GA (1906) The morphology of the lower algae. III *Chlorobium limicola* Nads., the green chlorophyll bearing microbe (in Russian). Bull Jard Bot St Petersb 6: 190–194

Nelson DC and Fisher CR (1995) Chemoautotrophic and methanoautotrophic endosymbiontic bacteria at deep-sea vents and seeps. In: Karl DM (ed) Deep-Sea Hydrothermal Vents, pp 125–167. CRC Press, Boca Raton, FL

Neumann S, Wynen A, Trüper HG and Dahl C (2000) Characterization of the *cys* gene locus from *Allochromatium vinosum* indicates an unusual sulfate assimilation pathway. Mol Biol Reports 27: 27–33

Norici A, Hell R and Giordano M (2005) Sulfur and primary production in aquatic environments: an ecological perspective. Photosyn Res 86: 409–417

Perty M (1852) Zur Kenntnis kleinster Lebensformen nach Bau, Funktionen, Systematik, mit Spezialverzeichnis der in der Schweiz Beobachteten. Jent und Reinert, Bern

Philips D and Philips SL (2000) High temperature dissociation constants of HS⁻ and the standard thermodynamic values for S^{2-}. J Chem Eng Data 45: 981–987

Postgate JR (1968) The sulphur cycle. In: Nickless G (ed) Inorganic Sulphur Chemistry, pp 259–279. Elsevier, Amsterdam

Prange A, Chauvistré R, Modrow H, Hormes J, Trüper HG and Dahl C (2002) Quantitative speciation of sulfur in bacterial sulfur globules: X-ray absorption spectroscopy reveals at least three different speciations of sulfur. Microbiology 148: 267–276

Rabenstein A, Rethmeier J and Fischer U (1995) Sulphite as intermediate sulphur compound in anaerobic sulphide oxidation to thiosulphate by marine cyanobacteria. Z Naturforsch 50c: 769–774

Rausch T and Wachter A (2005) Sulfur metabolism: a versatile platform for launching defence operations. Trends Plant Sci 10: 503–509

Roche P, Debelle F, Maillet F, Lerouge P, Faucher C, Truchet G, Denarie J and Prome JC (1991) Molecular basis of symbiotic host specificity in *Rhizobium meliloti: nodH* and *nodPQ* genes encode the sulfation of lipo-oligosaccharide signals. Cell 67: 1131–1143

Roy AB and Trudinger PA (1970) The Biochemistry of Inorganic Compounds of Sulfur. Cambridge University Press, London

Saito K, De Kok LJ, Stulen I, Hawkesford MJ, Schnug E and Sirko A (2005) Sulfur Transport and Assimilation in the Postgenomic Era. Backhuys Publishers, Leiden

Schippers A and Sand W (1999) Bacterial leaching of metal sulfides proceeds by two indirect mechanisms via thiosulfate or via polysulfides and sulfur. Appl Environ Microbiol 65: 319–321

Schmidt A (1973) Sulfate reduction in a cell-free system of *Chlorella*. The ferredoxin-dependent reduction of a protein-bound intermediate by a thiosulfonate reductase. Arch Microbiol 93: 29–52

Schmidt A and Jäger K (1992) Open questions about sulfur metabolism in plants. Annu Rev Plant Physiol Plant Mol Biol 43: 325–349

Schnug E (1998) Sulfur in Agroecosystems. Series: Nutrients in Ecosystems, Vol. 2. Springer, New York

Schwenn JD (1989) Sulphate assimilation in higher plants – a thioredoxin-dependent PAPS-reductase from spinach leaves. Z Naturforsch C 42: 93–102

Setya A, Murillo M and Leustek T (1996) Sulfate reduction in higher plants: Molecular evidence for a novel 5′-adenylylsulfate reductase. Proc Natl Acad Sci USA 93: 13383–13388

Stetter KO (1996) Hyperthermophilic prokaryotes. FEMS Microbiol Rev 18: 149–158

Steudel R (1982) Homocyclic sulfur molecules. Topics Curr Chem 102: 149–176

Steudel R (1985) Neue Entwicklungen in der Chemie des Schwefels und des Selens. Nova Acta Leopoldina 264: 231–246

Steudel R (1987) Sulfur homocycles. In: Haiduc I and Sowerby DB (eds) The Chemistry of Inorganic Homo- and Heterocycles, pp 737–768. Academic Press, London

Steudel, R (1996a) Das gelbe Element und seine erstaunliche Vielseitigkeit. Chemie in unserer Zeit 30: 226–234

Steudel R (1996b) Mechanism for the formation of elemental sulfur from aqueous sulfide in chemical and microbiological desulfurization processes. Ind Eng Chem Res 35: 1417–1423

Steudel R (1998) Chemie der Nichtmetalle, 2nd edn. W. de Gruyter, Berlin

Steudel R (2000) The chemical sulfur cycle. In: Lens P and Hulshoff Pol W (eds) Environmental Technologies to Treat Sulfur Pollution, pp 1–31. IWA Publishing, London

Steudel R (2003a) Elemental Sulfur and Sulfur-Rich Compounds I. Springer, Berlin

Steudel R (2003b) Elemental Sulfur and Sulfur-Rich Compounds II. Springer, Berlin

Steudel R and Albertsen A (1999) The chemistry of aqueous sulfur sols – models for bacterial sulfur globules? In: Steinbüchel A (ed) Biochemical Principles and Mechanisms of Biosynthesis and Biodegradation of Polymers, pp 17–26. Wiley-VCH, Weinheim

Steudel R and Eckert B (2003) Solid sulfur allotropes. In: Steudel R (ed) Elemental Sulfur and Sulfur-Rich Compounds, pp 1–79. Springer, Berlin

Steudel R and Holz B (1988) Detection of reactive sulfur molecules (S_6, S_7, S_9, S_μ) in commercial sulfur, in sulfur minerals, and in sulfur metals slowly cooled to $20\,°C$. Z Naturforsch B 43: 581–589

Steudel R and Kustos M (1994) Organic polysulfanes. In: King RB (ed) Encyclopedia of Inorganic Chemistry, pp 4009–4038. Wiley, Sussex

Steudel R, Holdt G and Nagorka R (1986) On the autoxidation of aqueous sodium polysulfide. Z Naturforsch 41b: 1519–1522

Suter M, von Ballmoos P, Kopriva S, den Camp RO, Schaller J, Kuhlemeier C, Schürmann P and Brunold C (2000) Adenosine 5′-phosphosulfate sulfotransferase and adenosine 5′-phosphosulfate reductase are identical enzymes. J Biol Chem 275: 930–936

Takano B and Watanuki K (1988) Quenching and liquid chromatographic determination of polythionates in natural water. Talanta 35: 847–854

Trüper HG (1984a) Microorganisms and the sulfur cycle. In: Müller A and Krebs B (eds) Sulfur, Its Significance for Chemistry, for the Geo-, Bio-, and Cosmosphere and Technology, pp 351–365. Elsevier Science Publishers B.V., Amsterdam

Trüper HG (1984b) Phototrophic bacteria and their sulfur metabolism. In: Müller A and Krebs B (eds) Sulfur, Its Significance for Chemistry, for the Geo-, Bio-, and Cosmosphere and Technology, pp 367–382. Elsevier Science Publishers B.V., Amsterdam

Trüper HG (1989) Physiology and biochemistry of phototrophic bacteria. In: Schlegel HG and Bowien B (eds) Autotrophic Bacteria, pp 267–281. Science Tech Publishers, Madison

Trüper HG and Fischer U (1982) Anaerobic oxidation of sulphur compounds as electron donors for bacterial photosynthesis. Phil Trans R Soc Lond B 298: 529–542

van Niel BC (1936) On the metabolism of the *Thiorhodaceae*. Arch Microbiol 7: 323–358

van Niel CB (1931) On the morphology and physiology of the purple and green sulfur bacteria. Arch Microbiol 3: 1–112

Varin L, Chamberland H, Lafontaine JG and Richard M (1997) The enzyme involved in sulfation of the turgorin, gallic acid 4-O-(β-D-glucopyranosyl-6′-sulfate) is pulvini-localized in *Mimosa pudica*. Plant J 12: 831–837

Warming E (1875) Om nogle ved Danmarks kyster levede bakterier. Vidensk Medd Dan Naturhist Foren Khobenhavn 20–28: 3–116

Winogradsky SN (1887) Über Schwefelbakterien. Bot Ztg 45: 489–508

Woese CR, Stackebrandt E, Macke TJ and Fox GE (1985) The phylogenetic definition of the major eubacterial taxa. Syst Appl Microbiol 5: 327–336

Chapter 2

Uptake, Distribution and Subcellular Transport of Sulfate

Malcolm J. Hawkesford
Rothamsted Research, Harpenden, Hertfordshire, AL5 2JQ, UK

Summary

Sulfate transport in and out of the cell in all phototrophic organisms, and in addition in eukaryotes, influx and efflux from the vacuole, which acts as a cellular sulfate store, and into the plastid, the site of sulfate reduction, is facilitated by multiple transporters. In addition, in vascular plants these transport processes are coordinated to facilitate management of sulfur fluxes between organs, from roots to shoots and to generative tissues. In prokaryotes, uptake into the cell is driven predominantly by an ABC-transporter, which is the product of at least four genes and energized directly by ATP. In eukaryotes, a family of H$^+$-sulfate co-transporters (SulP) has been characterized which fulfills transport roles at least for uptake into the cell and efflux from the vacuole. In addition, differential expression of this gene family in vascular plants enables selective movement between tissues dependent on developmental cues and sulfate availability. A vital transport step is the transport into the plastid, for which a transporter has only been identified in algae and for which no vascular plant homologue is known. The expression of many sulfate transporters responds to availability and demand for sulfur, and transduction mechanisms which control this are beginning to be elucidated. The significance of a C-terminal STAS domain in SulP transporters is still unclear.

Correspondance, details, Fax: 00 44 1582 763010, E-mail: malcolm.hawkesford@bbsrc.ac.uk

I. Introduction

The sulfur demands of phototrophic organisms including plants, algae and cyanobacteria are met primarily by the acquisition of sulfate from the environment. Specific transport systems have evolved to both take up and to fulfill internal requirements for the transport of sulfate. In addition, most phototrophic organisms including plants, have a wide range of transporters for organic molecules including sulfur-containing amino acids and in some instances these may contribute to the sulfur economy of the organism. In multi-cellular organisms such transporters are important for the distribution of reduced sulfur compounds such as glutathione (Rennenberg et al. 1979; Zhang et al. 2004) or S-methylmethionine (Bourgis et al. 1999), for example from sink to source tissues. Plants are also able to utilize atmospheric H_2S absorbed via the leaves and assimilate this directly into cysteine, but this is seldom a major physiological consideration (De Kok et al. 2002). The importance of the transport steps is underlined by pathway analysis which indicates that sulfate uptake into the cell is the major regulated step (Vauclare et al. 2002).

Sulfate uptake into plants was determined to be an enzymatic process catalyzed by transporters, and was a saturable process with a high affinity for sulfate (19 μM), was pH dependent and was competitively inhibited by selenate (Leggett and Epstein 1956). The cloning of sulfate transporter genes and the subsequent detailed analyses of transport phenomena performed in the last decade, more than 40 years later, have verified the accuracy of this pioneering study.

Two major transport systems for sulfate have been described: the prokaryotic ABC-type transporter which includes a periplasmic sulfate binding protein and which occurs predominantly in prokaryotes including cyanobacteria, and the SulP family of H^+-cotransporters which predominate in vascular plants. Genomic information has indicated that members of the SulP family also exist in prokaryotes transporting a diverse set of substrates. Although many ABC-type transporters

---

Abbreviations: ABC–ATP-binding cassette; OAS–*O*-acetylserine; SLIM1–sulfur limitation 1 (transcription factor); STAS–sulfate transporter and anti-sigma factor antagonist; SURE–sulfur-responsive element

exist in all eukaryotes, evidence for a role in sulfate transport in eukaryotes is restricted to one specific example in *Chlamydomonas reinhardtii*. The different transporters and their occurrence will be dealt with in turn.

II. Sulfate Transport in Cyanobacteria

The well-characterized sulfate permease of cyanobacteria, in common with other prokaryotic sulfate transporters (Kertesz 2001), belongs to the ABC-type (ATP-binding cassette) consisting of a multi-subunit transporter including a periplasmic substrate binding protein (probably present in substantial molar excess in the periplasmic space), two channel-forming membrane intrinsic proteins and a cytoplasm-located ATP-binding protein. The genes (see Table 1) that encode the proteins which are components of the complex are up-regulated under sulfur-deficient conditions, to aid in scavenging for available sulfur. Additionally, in an adaptation to low sulfur availability, the sbpA protein which is stoichiometrically the most abundant subunit, has no S-containing amino acid residues, as is the case for enteric bacteria. Although considerably enhancing sulfate uptake, sbpA is not absolutely essential for transport. In addition there is a regulatory gene, *cysR*, required for expression (Green et al. 1989; Kohn and Schumann 1993; Laudenbach and Grossman 1991). In eukaryotes, including *Arabidopsis*, a large family of ABC-transporter homologues with a wide substrate specificity (but as far as is known, not sulfate) exist as domains of single proteins rather than as separate proteins (Sanchez-Fernandez et al. 2001).

In addition, the availability of complete genomic sequences for a number of cyanobacteria indicates the existence of multiple transporters belonging to the SulP group (for detailed description of members of this group belonging to the vascular plants, see section IV.A) which could theoretically transport sulfate. Phylogenetic analysis indicates a distinct clustering of the cyanobacterial sequences and a divergence from the plant and other eukaryotic groups (Saier et al. 1999) which would be expected in these evolutionary distant organisms. Analysis of the prokaryotic sequences themselves, shows distinct clades which may be related to functionality.

Table 1. Gene designation of ABC-sulfate transporter subunits.

Subunit	Cyanobacterial designation	Bacterial gene	*Chlamydomonas* chloroplast transporter	*Marchantia* chloroplast gene
Membrane pore	*cysT*	*cysT*	*SulP*	*mbpY*
Membrane pore	*cysW*	*cysW*	*SulP2*	
ATP binding protein	*cysA*	*cysA*	*Sabc*	*mbpX* (similar to *Arabidopsis* NAP3 (Sanchez-Fernandez et al. 2001)
Periplasmic sulfate binding component	*sbpA*	*cysP*	*Sbp* (not periplasmic but associated with complex)	

The cyanobacterial SulP sequences themselves form a diverse group, and for example one clade within the group seems to be involved in bicarbonate transport (Price et al. 2004) and another clade has one example, LtnT, found only in a few cyanobacteria, which is unable to transport sulfate but has a low affinity nitrate transporter activity (Maeda et al. 2006). Some bacterial homologues of the bicarbonate transporter group possess a C-terminal carbonic anhydrase domain (Felce and Saier 2004), which increases the inter-conversion of carbon dioxide and bicarbonate. No such fusion proteins appear in the cyanobacterial genomes studied to date.

III. Sulfate Transport in Algae and Aquatic Plants

Many of the characteristics of sulfate transport by higher plants were first described in studies on algae, partially due to their amenability for experimentation: *Lemna minor* (Neuenschwander et al. 1991; Thoiron et al. 1981), *Lemna paucicostata* (Datko and Mudd 1984a, b), *Lemna gibba* (Lass and Ullrich-Eberius 1984), *Hydrodictyon reticulatum* (Rybová et al. 1988), *Chlorella pyrenoidosa* (Vallee and Jeanjean 1968a, b), *Chlorella vulgaris* (Passera and Ferrari 1975), *Chlamydomonas reinhardtii* (Yildiz et al. 1994). Evidence for a 3H⁺-sulfate cotransport mechanism was obtained in *Lemna gibba* (Lass and Ullrich-Eberius 1984). The high affinity transport system was shown to require high proton concentrations suggesting a mechanism of a proton-sulfate co-transport system. This H⁺-cotransport mechanism for sulfate transport was subsequently supported by studies on vascular plants which demonstrated a pH dependency of sulfate transport in *Brassica napus* vesicles (Hawkesford et al. 1993) and by the low pH stimulation of transport seen in yeast expressing the high affinity sulfate transporter from the legume, *Stylosanthes hamata* (Smith et al. 1995a).

Compartmental analysis in *Lemna minor* indicated the importance of the vacuole as an internal store of sulfate (Thoiron et al. 1981), although a store in which sulfate only turned over slowly (Datko and Mudd 1984b). An increased capacity for sulfate uptake under sulfur limiting conditions was reported in all instances, in common with the situation for higher plants (see below). It was suggested that both saturating high affinity and low affinity transport systems operated, but only the high affinity system was induced by sulfur limiting conditions (Datko and Mudd 1984b). An indication of the importance of the cysteine precursor, *O*-acetylserine (OAS) in determining sulfate transport expression (see section VII for a more detailed discussion and subsequent studies with vascular plants), in common with earlier bacterial studies (Kredich 1993), was provided in feeding experiments in *Lemna minor* (Neuenschwander et al. 1991).

In spite of these early studies and the clear ecological importance of algae in the aquatic environment (Norici et al. 2005) little further molecular characterization of the transporters has occurred with the exception of *Chlamydomonas rheinhardtii*. In this species a gene family of seven sulfate transporters of the SulP type has been identified (Pollock et al. 2005) but not further characterized, and there is a detailed characterization of a chloroplast transporter of the ABC-type (see section V). In addition, *Chlamydomonas rheinhardtii* has proved exceptionally useful for elucidation of genes involved in sensing and signaling sulfur status (see section VIII).

IV. A Family of Sulfate Transporters in Higher Plants

A. The H⁺-Sulfate Co-Transporter Family (SulP)

The first putative sulfate transporter sequence to be identified from plants was by homology to the *Neurospora crassa* sulfate transporter (Ketter and Marzluf 1988; Sandal and Marcker 1994). This was the soybean nodule-specific protein, encoded by the *GMAK170* gene, a protein of 486 amino acids, which was considerably shorter than the *Neurospora* sequence (788 amino acids). In this same report the human DRA gene (Hästbacka et al. 1994) was shown also to have homology to these sequences and this was subsequently characterized as a human H^+-sulfate transporter and part of a family of mammalian sulfate transporters, which were distinct from mammalian Na-coupled sulfate transporters (Bissig et al. 1994). No evidence has been forthcoming to demonstrate that the *GMAK170* gene product is able to function as a sulfate transporter.

The cloning of confirmed plant sulfate transporter cDNAs was achieved first by functional complementation of a sulfate transporter-deficient yeast mutant (Smith et al. 1995b) with plant cDNA libraries (Smith et al. 1995a; Smith et al. 1997; Takahashi et al. 1996). A homologous yeast sulfate transporter was also isolated using this mutant (Smith et al. 1995b). The first yeast sulfate transporter-deficient mutants were isolated by selection using selenate/chromate toxic analogues (Breton and Surdin-Kerjan 1977; Cherest et al. 1997; Smith et al. 1995b). Subsequently yeast deletion mutants have been essential tools for verification and functional characterization of cloned putative sulfate transporters. A similar selenate-resistance strategy was used to isolate transporter mutants of *Arabidopsis* (Shibagaki et al. 2002).

Initially three sulfate transporters were isolated from the tropical legume *Stylosanthes hamata*, showing either high or low affinity for sulfate in the yeast expression system. Subsequently a high affinity type was isolated in barley using an identical approach (Smith et al. 1997). In each case the cDNA libraries were made from mRNA isolated from root tissues of sulfur-starved plants, a condition known to induce maximal activity (see below). The availability of cDNA sequences facilitated the isolation and subsequent characterization of a number of cDNAs for sulfate transporters from *Arabidopsis* (Kataoka et al. 2004b; Shibagaki et al. 2002; Takahashi et al. 1996; Takahashi et al. 2000; Takahashi et al. 1997; Vidmar et al. 2000). Subsequently sulfate transporters have been cloned and analyzed from a range of plant species including maize (Bolchi et al. 1999; Hopkins et al. 2004), potato (Hopkins et al. 2005), tomato (Howarth et al. 2003), *Brassica* (Buchner et al. 2004b; Heiss et al. 1999), wheat (Buchner et al. 2004a), rice (Godwin et al. 2003) and *Sporobolus stapfianus* (Ng et al. 1996).

The availability of the fully sequenced genomes of *Arabidopsis* and rice and indicated the existence around 14 genes sequences showing homology to the SulP sulfate transporters in each genome. Phylogenetic analysis of all known sequences clearly separates the plant, yeast, fungi and mammalian kingdoms (Saier et al. 1999), however analyses of sequence homology of the plant species alone, suggests at least five distinct clades within the plant family. A typical phylogenetic analysis of amino acid sequences of *Arabidopsis* and rice putative sulfate transporters is shown (Fig. 1). In most cases all species have a similar distribution of isoforms between the clades, although exactly corresponding homologues are only identifiable for closely related species (for example for *Arabidopsis* and *Brassica*, or for rice and wheat). In some species, for example wheat, not only does polyploidy increase the complexity of expressed isoforms, but also there have been recent duplication events resulting in additional very similar isoforms (Buchner et al. 2004a). There remains debate as to the significance of the multiple members of the family with regard to redundancy or individual specialization of isoforms (Hawkesford 2003).

The clade which forms Group 1 includes many well studied transporters and often comprises three genes, for example as found in *Arabidopsis* (*AtSULTR1;1–3*). Within Group 1, the monocotyledonous species are distinct from the *Arabidopsis* clade preventing the alignment to the direct corresponding homologues (Fig. 1, and wheat and maize data, not shown). Many Group 1 sulfate transporters have been expressed in yeast and most have high substrate affinities (K_m) for sulfate in this heterologous expression system, for

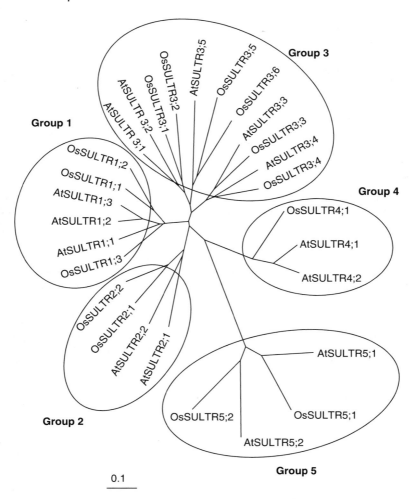

Fig. 1. Unrooted phylogenetic tree of the rice (*Oryza sativa*) and *Arabidopsis* members of the SulP sulfate transporter family. Phylogenetic representation of the plant sulfate transporter amino acid sequences showing subdivision into 5 putative Groups. Accession numbers: *Arabidopsis*: AtSULTR1;1, AB018695; AtSULTR1;2, AB042322; AtSULTR1;3, AB049624; AtSULTR2;1, AB003591; AtSULTR2;2, D85416; AtSULTR3;1, D89631; AtSULTR3;2, AB004060; AtSULTR3;3, AB023423; AtSULTR3;4, B054645; AtSULTR3;5, AB061739; AtSULTR4;1, AB008782; AtSULTR4;2, AB052775; AtSULTR5;1, NP_178147; AtSULTR5;2, NP_180139; rice: OsSULTR1;1, AF493790; OsSULTR1;2, AAN59764.1; OsSULTR1;3, BAC98594; OsSULTR2;1, AAN59769; OsSULTR2;2, AAN59770; OsSULTR3;1, NP_921514; OsSULTR3;2, AAN06871; OsSULTR3;3, AK104831; OsSULTR3;4, AK067270; OsSULTR3;5, NM_192602; OsSULTR3;6, NM_191791; OsSULTR4;1, AF493791; OsSULTR5;1, BAC05530; OsSULTR5;2, BAB03554. Alignments were performed using ClustalX program (Thompson et al. 1997) version 1.81 and the tree was drawn using the Treeview32 program (Page 1996).

example AtSULTR1;1 has a K_m of 1.5–3.6 µM (Takahashi et al. 2000; Vidmar et al. 2000) and AtSULTR 1;2 a K_m of 6.9 µM (Yoshimoto et al. 2002). Many of the transporters in this Group are highly expressed in the root tissues, and furthermore are highly regulated by sulfur supply, with massively increased transcript abundance when plants are sulfur-deficient, for example (Buchner et al. 2004b). Studies of cellular expression patterns indicate expression in root tips, root hairs, exodermal, cortical and endodermal layers with less expression in the central vascular region (Rae and Smith 2002; Shibagaki et al. 2002; Takahashi et al. 2000; Yoshimoto et al. 2002). A notable exception is AtSULTR1;3 whose expression appears to be specific to the phloem in both root and cotyledons (Yoshimoto et al. 2003). In tomato contrasting expression occurs for the two isoforms reported, with one isoform showing general expression, highest in exodermis and endodermis and the other with primarily exodermal expression (Howarth et al. 2003). Sulfate uptake, but not transfer to the shoot was impaired in selenate resistant mutants with a lesions in *AtSULTR1;2* (Shibagaki et al. 2002); only mutations of *AtSULTR1;2* confer selenate resistance in *Arabidopsis* (El Kassis et al. 2007). Taken together, it is clear that Group 1 transporters are responsible for primary sulfate acquisition in plant roots.

Typically two sulfate transporter isoforms for any individual species are found in Group 2 (Fig. 1). One of the original *Stylosanthes hamata* isoforms belongs to this group (Smith et al. 1995a). Affinities (K_m) for sulfate are generally lower than for Group 1: ShST3, 99 μM (Smith et al. 1995a); AtSULTR2;1, 0.41 mM (Takahashi et al. 2000) although one report indicates 5 μM (Vidmar et al. 2000); AtSULTR2;2, > 1.2 mM (Takahashi et al. 2000). Group 2 sulfate transporters are regulated by sulfur nutrition, although not generally so dramatically as for the Group 1 sulfate transporters. One or the other isoforms is generally more strongly regulated, but not in a consistent pattern: in root tissues AtSULTR2;2 is more regulated than 2;1 (Takahashi et al. 2000), however the pattern for the closely homologous *Brassica* isoforms is reversed (Buchner et al. 2004b). Both isoforms are usually expressed throughout the plant (Buchner et al. 2004b) but tend to be localized to vascular tissues (Takahashi et al. 2000). It is likely that this Group of transporters contribute to translocation of sulfate within the plant vascular systems. Variation between species, for example as quoted above, may reflect different developmental stages or a need for more precision in localization of analysis.

The Group 3 clade is relatively large and diverse with five *Arabidopsis* and six rice sulfate transporters (Fig. 1). At least three and possibly four sub-clades are apparent, each containing both rice and *Arabidopsis* examples, indicating relatively ancient gene duplications. Only one isoform, AtSULTR3;5 has been successfully expressed and characterized in yeast, however transport was only observed when co-expressed with AtSULTR2;1. The observed K_m for sulfate was 503 μM in the co-expression system compared to 545 μM for AtSULTR2;1 alone, along with an approximate threefold increase in V_{max}, (Kataoka et al. 2004a). It is proposed that a hetero-dimer is required for activity of AtSULTR3;5 and for maximal activity of AtSULTR2;1. *In planta* AtSULTR2;1 and 3;5 are co-expressed in the root xylem parenchyma and pericycle cells, and although AtSULTR3;5 is constitutively expressed with no regulation by sulfur nutrition, a role for the dimer in enhancing uptake during deficiency via interaction with the inducible SULTR2;1 is suggested. In *Lotus japonicus* root nodules, the 3;5 homologue (SST1), which complements a yeast sulfate transporter deficient mutant for growth on sulfate media, functions as a sulfate transporter across the symbiosome membrane and mutant analysis indicates that this sulfate import is crucial for nitrogen fixation (Krusell et al. 2005). None of the five *Brassica* Group 3 transporters are sulfur-regulated although tissue specificity of isoform expression varies greatly (Buchner et al. 2004b).

In *Arabidopsis* and *Brassica* there are two Group 4 isoforms (Buchner et al. 2004b; Kataoka et al. 2004b), however in rice and wheat there only appears to be one (Buchner and Hawkesford, unpublished). Early localization data entailing identification of putative transit sequence and utilizing a partial sulfate transporter-green fluorescent protein fusion protein, indicated a chloroplast membrane localization (Takahashi et al. 1999). There is an absolute requirement to transport sulfate into the chloroplast, the site of reduction and the mechanism for this transport has never been confirmed. Subsequent additional analysis with full length sulfate transporter:green fluorescent protein constructs showed a more conclusive tonoplast membrane localisation for the *Arabidopsis* Group 4 sulfate transporters (Kataoka et al. 2004b). Expression was highest in roots tissues in both *Arabidopsis* and *Brassica* (Buchner et al. 2004b; Kataoka et al. 2004b), was induced by sulfur-deficiency (particularly SULTR4;2) and at least in the case of *Arabidopsis* was localized to pericycle and xylem parenchyma cells. Analysis of *Arabidopsis* double knockout plants, and critically of vacuoles isolated from these plants indicated a role in sulfate efflux from the vacuole tissue. For example, vacuoles isolated from the double knockout (4;1 and 4;2) line contained more sulfate than the wild-type, and this was decreased in lines over-expressing a 4;1

construct. However, no direct demonstration of transport, for example in yeast mutants, has been achieved. Irrespective of specific sulfate transport function, *in planta* studies indicated that sulfate efflux from the root vacuoles was dependent on the presence of functional Group 4 proteins. It is hypothesized that this would optimize channeling of sulfate toward the xylem vessels and thus expression of the Group 4 transporters in roots would have a potential role in regulating root to shoot transport (Kataoka et al. 2004b).

Group 5 is quite distinct from the other isoforms (sequences are quite dissimilar to other sulfate transporters) and typically contains two isoforms for any given species, and the two isoforms are also quite dissimilar to one another. The most striking observation is that Group 5 sulfate transporters are truncated sequences and possess little N or C-terminal regions beyond the transmembrane domain. Some secondary structure predictions suggest fewer membrane spanning helices although there is no reason to suspect such a divergence from the rest of the family. Green fluorescent protein fusion constructs localize Group 5 members to internal membranes and it would be tempting to speculate a role in vacuolar loading, although studies with knock out mutants have failed to give a clear phenotype (Buchner, Takahashi and Hawkesford, unpublished). There are no reports indicating sulfate transport, either *in planta* or in expression systems such as yeast. The possibility remains that these transporters have a substrate other than sulfate.

B. Structure

Predicted protein sizes for eukaryotic members of this family are in the range 500–700 amino acids and although predictions of secondary structure vary widely depending upon prediction method used and sequences analyzed, a consensus of 10–12 transmembrane spanning helices is predicted. One possible model is shown in Fig. 2.

Structure–function relationships have been examined by site-directed mutagenesis of the *Stylosanthes* ShST1 transporter and subsequent analysis of localization and function in yeast mutants (Howitt 2005; Khurana et al. 2000; Loughlin et al. 2002; Shelden et al. 2001, 2003). Mutations in the human DTDST transporter, known to be responsible for diastrophic dysplasia disease (Hästbacka

et al. 1994), and involving conserved residues in transmembrane helices 9 and 11 when introduced into ShST1 affect either transport activity or trafficking to the plasma membrane (Khurana et al. 2000). Similarly three semi-conserved proline residues, unusually predicted to be in transmembrane helices 1–3 (notable for short extra membrane sequence linking loops) were also critical for transporter function (Shelden et al. 2001). Charged residues, which may influence topology, or be involved in ion binding or in ion channel function, have been mutated systematically: evidence was obtained for pairing of residues which would indicate tertiary structure arrangements of the transmembrane regions (Shelden et al. 2003). ShST1 contains five cysteine residues (non-conserved) and a cysteine-less variant was shown to have transport characteristics indistinguishable from the wild type (Howitt 2005); this variant will be useful for future topology analysis using combinations of mutagenesis introducing cysteine residues and probing with sulfhydryl reagents.

Homology of the carboxyl terminal region of most eukaryotic SulP transporters to bacterial anti-sigma factor antagonists, for example the *Bacillus subtilis* SPOIIAA, has defined this region as the STAS (sulfate transporter and anti-sigma factor antagonist) domain (Aravind and Koonin 2000). SpoIIAA protein is involved in nutrient regulation of sporulation; dephosphorylation of a serine activates the protein, enabling interaction with a second protein with the net result of the release of the sigma factor which induces sporulation. In the DRA transporter, the STAS domain is involved in a protein-protein interaction with the cystic fibrosis transmembrane regulator resulting in a mutual activation (Ko et al. 2004). All plant sulfate transporters examined with the notable exception of the Group 5 transporters possess this domain. To examine and test the possibility that the STAS domain may have a regulatory role in plants or may be involved in trafficking to or stability in the plasma membrane, both chimaeric and deletion constructs (Shibagaki and Grossman 2004) together with site-directed mutagenesis (Rouached et al. 2005) have been performed. Deletion of the STAS domain prevented trafficking to the plasma membrane, and heterologous chimaeras had a deleterious effect on transport kinetics (Shibagaki and Grossman 2005). Mutations

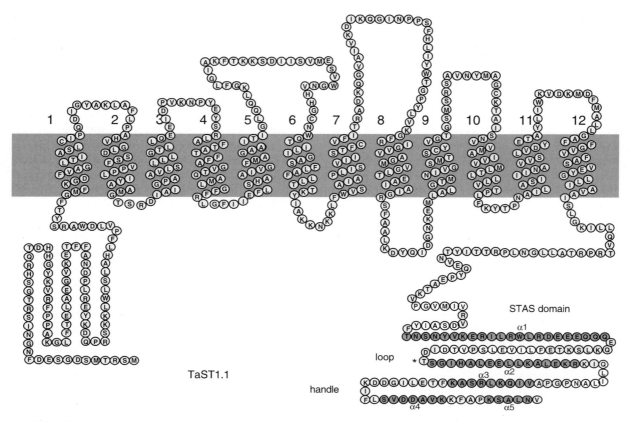

Fig. 2. Possible topology of a wheat sulfate transporter. The suggested topology of a sulfate transporter (TaST-B1.1b, AJ512820 (Buchner et al. 2004a)) from *Triticum aestivum*. Consensus positions of 12 trans-membrane helices were determined from an alignment of six plant sulfate transporters (wheat (Q8H0K4, Q9XGB6, Q8H2D7), maize (Q9AT12), barley (Q40008) and *Stylosanthes* (P53391)) using the Aramemmon programme (http://aramemnon.botanik.uni-koeln.de/). Arrangements of extra-membrane loops are arbitrary although the STAS domain is depicted schematically. Within the STAS domain, alpha helices (α1–5) are shaded dark grey and beta sheet structures have no shading. A conserved phosphorylatable threonine in the loop region is indicated by an asterisk.

of the equivalent phosphorylatable residue (compared to the SpoIIAA protein) in the plant sulfate transporter (a threonine, marked with an asterisk in Fig. 2) resulted in a complete loss of activity of sulfate transport (Rouached et al. 2005). It is hypothesized that there are protein:protein interactions, mediated via the STAS domain, which are required for sulfate transporter function in plants.

V. Transport into the Chloroplast

There is an absolute requirement for transport of sulfate across the chloroplast (or plastid in the root) inner membrane into the stroma, the site for reduction and assimilation of sulfate into cysteine.

Sulfate uptake into isolated chloroplasts has saturable kinetics with a K_m of around 2.5–3 mM and with a V_{max} of 0.7–13 µmol per mg Chl per hour (Gross et al. 1990; Mourioux and Douce 1979). Furthermore transport was competitively inhibited by phosphate, and therefore it was suggested that the triose-phosphate/phosphate translocator was responsible for sulfate uptake in addition to its primary phosphate transport function. The rates were much lower than for phosphate reflecting the greater requirement for phosphate in CO_2 fixation. The triose-phosphate/phosphate translocator has been cloned (Flügge et al. 1989) but no conclusive evidence has been presented that this transporter is responsible for sulfate uptake into the chloroplast *in vivo*.

A multi-subunit ABC-type transporter has been shown to be present in the chloroplast membrane of *Chlamydomonas reinhardtii* (Chen and Melis 2004; Chen et al. 2005; Chen et al. 2003; Melis and Chen 2005). It is proposed that the transporter comprises a heterodimer of transmembrane proteins constituting the 'pore', two associated cytosolic sulfate binding proteins and two stroma located ATP binding subunits attached to each of the membrane subunits (Melis and Chen 2005). Homologues to the nuclear gene for the membrane pore proteins (termed confusingly *SulP* but with no relation to the SulP family) have been identified in many prokaryotes, including cyanobacteria (homologous to the *cysT* gene) and in some eukaryotes, for example the liverwort, *Marchantia polmorpha*, in which it is a chloroplast gene. Corresponding nuclear genes have been identified for the other subunits in *Chlamydomonas*, designated *SulP2, Sbp* and *Sabc*, respectively (Table 1). Available evidence supports a role for the *SulP* gene product in chloroplast sulfate transport in *Chlamydomonas* as, for example, it is induced by sulfur-deficiency and antisense transformants are impaired in sulfate transport (Chen and Melis 2004; Chen et al. 2005). No such genes have been found in vascular plants (Chen et al. 2003).

It has been suggested that one of the subtypes of the SulP family (Group 4, see above for discussion on this topic) may be responsible for plastid sulfate influx (Takahashi et al. 1999), but this transporter has subsequently been shown to be a tonoplast membrane protein (Kataoka et al. 2004b).

Proteome analysis of the plastid membrane of *Arabidopsis* indicates the occurrence of 19 ABC-type transporters, any of which theoretically may be responsible for sulfate uptake into higher plant plastids (Weber et al. 2005).

VI.   Fluxes of Sulfate around the Plant

Fluxes of sulfate into and around a stylized vascular plant are shown in Fig. 3. Initial sulfate uptake into the symplast may be at several sites, potentially occurring through root hairs, at the root periphery or having passed through the cell walls of the cortex (apoplastic route), at a site near to the endodermis which acts as an apoplastic barrier. Within the root symplasm there may be cell to cell transfer via plasmodesmata (symplastic

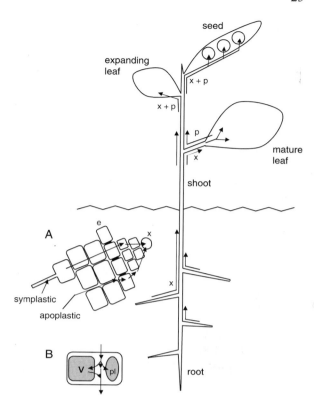

Fig. 3. Routes of sulfate movements *in planta*. Schematic representation of major fluxes of sulfate from the soil solution to the root xylem (x) and pathways to various sink tissues. Fluxes around the plant in xylem or phloem (p) are indicated by arrows. The inset (A) shows possible symplastic and apoplastic fluxes from cell to cell within the root cross section. The endodermal layer (e) is shown with the Casparian barrier. Inset B indicates theoretical fluxes of sulfate across individual membranes in a typical cell: initially into the cell, both into and subsequently out of the vacuole (v) as a temporary store, into the plastid (pl) for reduction, or exported from the cell.

route), however at least one efflux step is required prior to xylem loading. Vascular plants distribute sulfur around the plant, regulated in response to changing demands; this is at least partially achieved by the flexibility provided by the gene family of transporters (SulP family) which have different kinetic properties and different patterns of expression in response to tissue, developmental and environmental cues. Distribution and redistribution of sulfur pools during development in both vegetative and generative tissues have been described in soybean and wheat. Initial distribution occurs via the xylem, however sulfate is preferentially distributed to young expanding leaves. Sulfate is redistributed from mature leaves to roots (Bell et al. 1995; Rennenberg et al. 1979), younger leaves (Sunarpi and Anderson 1996) or generative sinks (seeds). As redistribution occurs during development, patterns of expression are modified, in some cases only in specific cells, for example to facilitate xylem to phloem transfer enabling preferential flux to young leaves rather than mature leaves (Anderson 2005). Redistribution is an important process during grain filling in cereals and coincides with redistribution of resources (N, C and S) occurring during leaf senescence. Sulfate in the vacuoles of mature leaves is an important store of sulfur and redistribution is particularly important under sulfur-limiting conditions, although the efficiency for this may vary between species. For example in *Macroptilium atropurpureum* sulfate pools were only slowly redistributed from mature leaves (Clarkson et al. 1983) and in *Brassica napus* there was a time delay before sulfate was redistributed from mature to young leaves (Blake-Kalff et al. 1998). It has been suggested that remobilization from the vacuole (the major internal store of sulfate) may be limiting (Bell et al. 1994). In response to sulfur-limitation, transporters involved in primary uptake in the root (BSULTR1;1 and 1;2), those involved in low affinity internal transport (2;1) and in vacuolar efflux (4;1) show increased transcript abundance. Similarly in stem tissues BSULTR2;1 shows increased expression in response to sulfur-limitation, as does BSULTR1;1 and 1;2, and in the leaves BSULTR1;1, 1;2, 4;1 and 4;2 increase in expression. Increased expression of the vacuolar efflux transporters correlates with unloading tissue sulfate during sulfur-stress, that is utilizing stored reserves, whereas the increased expression of BSULTR1;1 and 1;2 observed in leaf tissue may reflect a localized requirement for enhanced uptake into a cell, for example in young tissues.

Most research has focused on uptake into cells but as outlined above, there is also a requirement for efflux. There is little data to indicate how this is achieved at the molecular level, whether involving discrete transporters, specific or non-specific ion channels (Frachisse et al. 1999; Roberts 2006) or whether the SulP family catalyzes these effluxes. This is an area requiring further examination.

VII. Regulation by Availability and Demand

Increased capacity for sulfate uptake has already being described for many eukaryotic algae (see above). A marked induction of sulfate transport capacity under sulfur insufficient conditions is also seen in cell cultures (Smith 1975), in plant roots including barley (Lee 1982) and *Macroptilium atropurpureum* (Clarkson et al. 1983) and in isolated *Brassica napus* vesicles (Hawkesford et al. 1993).

Uptake into barley increases at least 10–15-fold, and this appears to be an increase in V_{max} rather than an effect on K_m. This increase is interpreted as increased transporter abundance rather than modification of transporter kinetics. Increased transporter capacity in isolated vesicles also indicated an increase specifically in a membrane component as being responsible for the increased transport capacity (Hawkesford et al. 1993). Transcript abundance has been determined with the availability of gene probes and in most cases a massive increase in transcript abundance upon sulfur-limitation has been observed (Smith et al. 1995a; Smith et al. 1997). Upon sulfur re-supply the abundance of transcripts reduced rapidly (usually within a few hours) indicating a rapid sensing and transduction pathway for the repression (Buchner et al. 2004b; Smith et al. 1997). Using an antibody to a synthesized polypeptide, plasma membrane-located sulfate transporter protein was detected and seen to change in abundance in parallel with changes in whole plant sulfate uptake capacity (Hawkesford and Wray 2000; Hopkins et al. 2005). A notable exception to this simple regulatory model was noted in potato (Hopkins et al. 2005), where only a small transient increase in uptake capacity was seen in spite of a large change in specific mRNA abundance. In all cases the change in transcript abundance appears to be very great in contrast to the measured changes in transporter activity. This may indicate additional post transcriptional levels of control.

A model for repression of transporter activity by sulfate or reduced sulfur compounds and de-repression under sulfur limiting conditions when these compounds are depleted has been suggested. This is based on evidence of external supply of sulfate or cysteine inhibiting sulfate uptake in cultured tobacco cells (Smith 1976) or of methionine inhibiting sulfate uptake in *Chlorella* (Passera and Ferrari 1975), or glutathione inhibiting uptake in tobacco cells or roots (Herschbach and Rennenberg 1994; Rennenberg et al. 1988; Rennenberg et al. 1989). Analysis of mRNA abundance indicates that substantial regulation is at the level of transcription (Smith et al. 1997) and efforts have been made to determine the identity of the regulatory molecule more precisely. For example, feeding experiments on sulfur-depleted maize seedlings showed that cysteine but not glutathione repressed sulfate transporter expression in the presence of the glutathione synthesis inhibitor butathionine sulphoximine (Bolchi et al. 1999). Similar experiments in *Arabidopsis*, again using butathionine sulphoximine, indicated, in contrast, that glutathione synthesis was required for repression (Lappartient et al. 1999) and furthermore glutathione in the phloem acted as a shoot to root signal indicating shoot demand for sulfur (Lappartient and Touraine 1996). A refinement of the model includes OAS, the precursor for cysteine and the direct link to C/N metabolism, as a positive effector of sulfate transporter gene expression (Neuenschwander et al. 1991; Smith 1977; Smith et al. 1997). This dual model is based on the proposed regulatory model for the cysteine regulon in *Escherichia coli* (Kredich 1993) and has been adopted as a basic working model in plants (Hawkesford et al. 2003), although no homologues of the substrate-interacting *trans*-acting factors found in *E.coli* are known in plants. External supply of OAS to plant roots resulted in increased sulfate transporter transcript abundance in parallel with increased tissue cysteine and glutathione, suggestive of a dominant influence of OAS (Smith et al. 1997). Whilst over-expression of enzymes leading to OAS accumulation in leaf tissues resulted in sulfate transporter transcript induction in the roots, measurements in the same study of bulk tissue OAS accumulation after prolonged sulfur-limitation did not correlate with transcript abundance or uptake activity (Hopkins et al. 2005). The role of OAS in regulatory mechanisms remains to be confirmed.

VIII.   Signal Transduction Pathway

Expression of many components of the sulfate transport system in cyanobacteria, algae and plants respond to availability of or demand for sulfur.

The sensing may be via metabolic intermediates which accumulate or are depleted (section VII) and these levels must then be transduced to changes in gene expression via *trans*-acting factors in a manner similar to that found in bacteria (Kredich 1993) or yeast (Thomas and Surdin-Kerjan 1997). Three mutants (*sac1, 2, 3*) affecting responses to sulfur supply were identified in *Chlamydomonas* (Davies et al. 1994). Sac1 is a membrane protein and may be involved in sensing and is critical in the down-regulation of photosynthesis during sulfur limitation (Davies et al. 1996) and up-regulation of ATP sulfurylase (Yildiz et al. 1996); no close homologue has been found in vascular plants. *Sac2* mutants also have weakened induction of sulfur-deficiency induced genes, but the nature of the *sac2* gene is unknown. The sac3 protein is a Snf1-like protein kinase and the mutant lacks any control of the arylsulfatase gene and fails to up-regulate high affinity sulfate transport in response to sulfur-limitation (Davies et al. 1999). The closest plant homologue is SNRK2.3 and mutations in this gene in *Arabidopsis* reduced induction of the ATSULTR2;2 gene in sulfur-limited conditions (Kimura et al. 2006). Cytokinins may be independently involved in the regulation of the transporters as cytokinin treatment down-regulated AtSULTR1;1 and 1;2, and was dependent on the CRE1/WOL/AHK4 receptor (Maruyama-Nakashita et al. 2004b). A seven base-pair *cis*-acting element, termed SURE (<u>su</u>lfur-<u>re</u>sponsive element), has been identified which occurs in the promoter region of many sulfur responsive genes, including the transporters (Maruyama-Nakashita et al. 2005) but also in non-responsive gene promoters; therefore additional, as yet unidentified *cis*-elements must be required to confer specificity. A transcription factor, SLIM1, identified in a screen for sulfur deficiency-non-responsive mutants, belonging to the EIL family (ethylene-insensitive-like), has been shown to be required for induction of *SULTR1;2* and other sulfate transporter gene expression under low sulfur conditions (Maruyama-Nakashita et al. 2006). Interestingly, although also involved in activating expression of glucosinolate degradation pathway genes, sulfur-regulation of APS reductase gene expression appears to be independent of SLIM1. A specific requirement for a protein phosphatase as an upstream regulatory factor for *SULTR;1* induction of expression by sulfur deficiency has also been proposed (Maruyama-Nakashita et al. 2004a). All of these observations have yet to be integrated into a coherent model of a signal transduction pathway.

IX. Perspective

Since the early descriptions of sulfate transport in cereals and in algae, substantial progress has been made in elucidating the complexity of sulfate transport systems. A major task remains to understand regulation of expression of isoforms in relation to changing availability and demand, particularly in complex vascular plants. In vascular plants the roles for many of the SulP isoforms are elusive and a focus for continuing research. Elucidation of transporter structure may provide insights into regulation and selectivity, for example between sulfate and selenate. An intriguing area concerns the significance of the STAS domain and possible interactions with other proteins in the cell. With the predominant emphasis having been on uptake, and to some extent subcellular transport, the molecular basis of efflux mechanisms, which may be catalyzed by channels or members of the SulP family, and which are responsible for cell to cell transfer and xylem loading are virtually unknown.

Acknowledgements

Rothamsted Research is grant-aided by the United Kingdom Biotechnology and Biological Sciences Research Council.

References

Anderson JW (2005) Regulation of sulfur distribution and redistribution in grain plants. In: Saito K, De Kok LJ, Stulen I, Hawkesford MJ, Schnug E, Sirko A, Rennenberg H (eds) Sulfur Transport and Assimilation in Plants in the Post Genomic Era, pp 23–31. Backhuys Publishers, Leiden

Aravind L, Koonin EV (2000) The STAS domain – a link between anion transporters and antisigma-factor antagonists. Curr Biol 10: R53–R55

Bell CI, Cram WJ, Clarkson DT (1994) Compartmental analysis of $^{35}SO_4^{2-}$ exchange kinetics in roots and leaves

of a tropical legume *Macroptilium atropurpureum* cv. Siratro. J Exp Bot 45: 879–886

Bell CI, Clarkson DT, Cram WJ (1995) Partitioning and redistribution of sulfur during S-stress in *Macroptilium atropurpureum* cv. Siratro. J Exp Bot 46: 73–81

Bissig M, Hagenbuch B, Stieger B, Koller T, Meier PJ (1994) Functional expression cloning of the canalicular sulfate transport system of rat hepatocytes. J Biol Chem 269: 3017–3021

Blake-Kalff MMA, Harrison KR, Hawkesford MJ, Zhao FJ, McGrath SP (1998) Distribution of sulfur within oilseed rape leaves in response to sulfur deficiency during vegetative growth. Plant Physiol 118: 1337–1344

Bolchi A, Petrucco S, Tenca PL, Foroni C, Ottonello S (1999) Coordinate modulation of maize sulfate permease and ATP sulfurylase mRNAs in response to variations in sulfur nutritional status: stereospecific down-regulation by L- cysteine. Plant Mol Biol 39: 527–537

Bourgis F, Roje S, Nuccio ML, Fisher DB, Tarczynski MC, Li CJ, Herschbach C, Rennenberg H, Pimenta MJ, Shen TL, Gage DA, Hanson AD (1999) S-methylmethionine plays a major role in phloem sulfur transport and is synthesized by a novel type of methyltransferase. Plant Cell 11: 1485–1497

Breton A, Surdin-Kerjan Y (1977) Sulfate uptake in *Saccharomyces cerevisiae*: biochemical and genetic study. J Bacteriol 132: 224–232

Buchner P, Prosser IM, Hawkesford MJ (2004a) Phylogeny and expression of paralogous and orthologous sulphate transporter genes in diploid and hexaploid wheats. Genome 47: 526–534

Buchner P, Stuiver CEE, Westerman S, Wirtz M, Hell R, Hawkesford MJ, De Kok LJ (2004b) Regulation of sulfate uptake and expression of sulfate transporter genes in *Brassica oleracea* as affected by atmospheric H_2S and pedospheric sulfate nutrition. Plant Physiol 136: 3396–3408

Chen HC, Melis A (2004) Localization and function of SulP, a nuclear-encoded chloroplast sulfate permease in *Chlamydomonas reinhardtii*. Planta 220: 198–210

Chen HC, Yokthongwattana K, Newton AJ, Melis A (2003) SulP, a nuclear gene encoding a putative chloroplast-targeted sulfate permease in *Chlamydomonas reinhardtii*. Planta 218: 98–106

Chen HC, Newton AJ, Melis A (2005) Role of SulP, a nuclear-encoded chloroplast sulfate permease, in sulfate transport and H_2 evolution in *Chlamydomonas reinhardtii*. Photosynth Res 84: 289–296

Cherest H, Davidian J-C, Thomas D, Benes V, Ansorge W, Surdin-Kerjan Y (1997) Molecular characterization of two high affinity sulfate transporters in *Saccharomyces cerevisiae*. Genetics 145: 627–635

Clarkson DT, Smith FW, Vandenberg PJ (1983) Regulation of sulfate transport in a tropical legume, *Macroptilium atropurpureum*, cv Siratro. J Exp Bot 34: 1463–1483

Datko AH, Mudd SH (1984a) Responses of sulfur-containing-compounds in *Lemna paucicostata* Hegelm. 6746 to changes in availability of sulfur sources. Plant Physiol 75: 474–479

Datko AH, Mudd SH (1984b) Sulfate uptake and its regulation in *Lemna paucicostata* Hegelm. 6746. Plant Physiol 75: 466–473

Davies JP, Yildiz F, Grossman AR (1994) Mutants of Chlamydomonas with aberrant responses to sulfur deprivation. Plant Cell 6: 53–63

Davies JP, Yildiz FH, Grossman A (1996) Sac1, a putative regulator that is critical for survival of *Chlamydomonas reinhardtii* during sulfur deprivation. Embo J 15: 2150–2159

Davies JP, Yildiz FH, Grossman AR (1999) Sac3, an Snf1-like serine threonine kinase that positively and negatively regulates the responses of Chlamydomonas to sulfur limitation. Plant Cell 11: 1179–1190

De Kok LJ, Stuiver CEE, Westerman S, Stulen I (2002) Elevated levels of hydrogen sulfide in the plant environment: nutrient or toxin. In: Omasa K, Saji H, Youssefian S, Kondo N (eds) Air Pollution and Biotechnology in Plants, pp 201–203. Springer-Verlag, Tokyo

El Kassis E, Cathala N, Rouached H, Fourcroy P, Berthomieu P, Terry N, Davidian J-C (2007) Characterization of a selenate-resistant Arabidopsis mutant. Root growth as a potential target for selenate toxicity. Plant Physiol 143: 1231–1241

Felce J, Saier MH (2004) Carbonic anhydrases fused to anion transporters of the SuIP family: Evidence for a novel type of bicarbonate transporter. J Molec Microbiol Biotech 8: 169–176

Flügge UI, Fischer K, Gross A, Sebald W, Lottspeich F, Eckerskorn C (1989) The triose phosphate-3-phosphoglycerate phosphate translocator from spinach chloroplasts: nucleotide sequence of a full-length cDNA clone and import of the in vitro synthesized precursor protein into chloroplasts. Embo J 8: 39–46

Frachisse JM, Thomine S, Colcombet J, Guern J, Barbier-Brygoo H (1999) Sulfate is both a substrate and an activator of the voltage-dependent anion channel of Arabidopsis hypocotyl cells. Plant Physiol 121: 253–261

Godwin RM, Rae AL, Carroll BJ, Smith FW (2003) Cloning and characterization of two genes encoding sulfate transporters from rice (*Oryza sativa* L.). Plant Soil 257: 113–123

Green LS, Laudenbach DE, Grossman AR (1989) A region of a cyanobacterial genome required for sulfate transport. Proc Natl Acad Sci U S A 86: 1949–1953

Gross A, Bruckner G, Heldt HW, Flugge UI (1990) Comparison of the kinetic properties, inhibition and labeling of the phosphate translocators from maize and spinach mesophyll chloroplasts. Planta 180: 262–271

Hästbacka J, de la Chapelle A, Mahtani MM, Clines G, Reeve-Daly MP, Daly M, Hamilton BA, Kusumi K, Trivedi B, Weaver A, Coloma A, Lovett M, Buckler A, Kaitila I, Lander ES (1994) The diastrophic dysplasia gene encodes a novel sulfate transporter: positional cloning by fine-structure linkage disequilibrium mapping. Cell 78: 1073–1087

Hawkesford MJ (2003) Transporter gene families in plants: the sulphate transporter gene family – redundancy or specialization? Physiol Plant 117: 155–163

Hawkesford MJ, Wray JL (2000) Molecular genetics of sulphate assimilation. Adv Bot Res Incorporating Adv Plant Pathol 33: 159–223

Hawkesford MJ, Davidian J-C, Grignon C (1993) Sulfate proton cotransport in plasma-membrane vesicles isolated from roots of *Brassica napus* L – increased transport in membranes isolated from sulfur-starved plants. Planta 190: 297–304

Hawkesford MJ, Buchner P, Hopkins L, Howarth JR (2003) The plant sulfate transporter family: specialized functions and intergration with whole plant nutrition. In: Davidian J-C, Grill D, De Kok LJ, Stulen I, Hawkesford MJ, Schnug E, Rennenberg H (eds) Sulfur Transport and Assimilation in Plants: Regulation, Interaction and Signaling, pp 1–10. Backhuys Publishers, Leiden

Heiss S, Schäfer HJ, Haag-Kerwer A, Rausch T (1999) Cloning sulfur assimilation genes of *Brassica juncea* L.: cadmium differentially affects the expression of a putative low-affinity sulfate transporter and isoforms of ATP sulfurylase and APS reductase. Plant Mol Biol 39: 847–857

Herschbach C, Rennenberg H (1994) Influence of glutathione (GSH) on net uptake of sulfate and sulfate transport in tobacco plants. J Exp Bot 45: 1069–1076

Hopkins L, Parmar S, Bouranis DL, Howarth JR, Hawkesford MJ (2004) Coordinated expression of sulfate uptake and components of the sulfate assimilatory pathway in maize. Plant Biol 6: 408–414

Hopkins L, Parmar S, Bļaszczyk A, Hesse H, Hoefgen R, Hawkesford MJ (2005) O-acetylserine and the regulation of expression of genes encoding components for sulfate uptake and assimilation in potato. Plant Physiol 138: 433–440

Howarth JR, Fourcroy P, Davidian J-C, Smith FW, Hawkesford MJ (2003) Cloning of two contrasting high-affinity sulfate transporters from tomato induced by low sulfate and infection by the vascular pathogen *Verticillium dahliae*. Planta 218: 58–64

Howitt SM (2005) The role of cysteine residues in the sulphate transporter, SHST1: Construction of a functional cysteine-less transporter. Biochim Biophys Acta-Biomembr 1669: 95–100

Kataoka T, Hayashi N, Yamaya T, Takahashi H (2004a) Root-to-shoot transport of sulfate in Arabidopsis. Evidence for the role of SULTR3;5 as a component of low-affinity sulfate transport system in the root vasculature. Plant Physiol 136: 4198–4204

Kataoka T, Watanabe-Takahashi A, Hayashi N, Ohnishi M, Mimura T, Buchner P, Hawkesford MJ, Yamaya T, Takahashi H (2004b) Vacuolar sulfate transporters are essential determinants controlling internal distribution of sulfate in Arabidopsis. Plant Cell 16: 2693–2704

Kertesz MA (2001) Bacterial transporters for sulfate and organosulfur compounds. Res Microbiol 152: 279–290

Ketter JS, Marzluf GA (1988) Molecular cloning and analysis of the regulation of cys-14 +, a structural gene of the sulfur regulatory circuit of Neurospora crassa. Molec Cell Biol 8: 1504–1508

Khurana OK, Coupland LA, Shelden MC, Howitt SM (2000) Homologous mutations in two diverse sulphate transporters have similar effects. FEBS Lett 477: 118–122

Kimura T, Shibagaki N, Ohkama-Ohtsu N, Hayashi H, Yoneyama T, Davies JP, Fujiwara T (2006) Arabidopsis SNRK2.3 protein kinase is involved in the regulation of sulfur-responsive gene expression and O-acetyl-L-serine accumulation under limited sulfur supply. Soil Sci Plant Nutr 52: 211–220

Ko SBH, Zeng WZ, Dorwart MR, Luo X, Kim KH, Millen L, Goto H, Naruse S, Soyombo A, Thomas PJ, Muallem S (2004) Gating of CFTR by the STAS domain of SLC26 transporters. Nature Cell Biology 6: 343–350

Kohn C, Schumann J (1993) Nucleotide sequence and homology comparison of two genes of the sulfate transport operon from the cyanobacterium Synechocystis sp PCC 6803. Plant Mol Biol 21: 409–412

Kredich NM (1993) Gene-regulation of sulfur assimilation. In: De Kok LJ, Stulen I, Rennenberg H, Brunold C, Rauser WE (eds) Sulfur Nutrition and Assimilation in Higher Plants – Regulatory Agricultural and Environmental Aspects, pp 37–47. SPB Academic, The Hague

Krusell L, Krause K, Ott T, Desbrosses G, Krämer U, Sato S, Nakamura Y, Tabata S, James EK, Sandal N, Stougaard J, Kawaguchi M, Miyamoto A, Suganuma N, Udvardi MK (2005) The sulfate transporter SST1 is crucial for symbiotic nitrogen fixation in Lotus japonicus root nodules. Plant Cell 17: 1625–1636

Lappartient AG, Touraine B (1996) Demand-driven control of root ATP sulfurylase activity and SO_4^{2-} uptake in intact canola. The role of phloem-translocated glutathione. Plant Physiol 111: 147–157

Lappartient AG, Vidmar JJ, Leustek T, Glass ADM, Touraine B (1999) Inter-organ signaling in plants: regulation of ATP sulfurylase and sulfate transporter genes expression in roots mediated by phloem-translocated compound. Plant J 18: 89–95

Lass B, Ullrich-Eberius CI (1984) Evidence for proton/sulfate cotransport and its kinetics in Lemna gibba G1. Planta 161: 53–60

Laudenbach DE, Grossman AR (1991) Characterization and mutagenesis of sulfur-regulated genes in a cyanobacterium – evidence for function in sulfate transport. J Bacteriol 173: 2739–2750

Lee RB (1982) Selectivity and kinetics of ion uptake by barley plants following nutrient deficiency. Ann Bot 50: 429–449

Leggett JE, Epstein E (1956) Kinetics of sulfate absorption by barley roots. Plant Physiol 31: 222–226

Loughlin P, Shelden MC, Tierney ML, Howitt SM (2002) Structure and function of a model member of the SulP transporter family. Cell Biochem Biophys 36: 183–190

Maeda S, Sugita C, Sugita M, Omata T (2006) Latent nitrate transport activity of a novel sulfate permease-like protein of the cyanobacterium Synechococcus elongatus. J Biol Chem 281: 5869–5876

Maruyama-Nakashita A, Nakamura Y, Watanabe-Takahashi A, Yamaya T, Takahashi H (2004a) Induction of SULTR1;1 sulfate transporter in Arabidopsis roots involves protein phosphorylation/dephosphorylation circuit for transcriptional regulation. Plant Cell Physiol 45: 340–345

Maruyama-Nakashita A, Nakamura Y, Yamaya T, Takahashi H (2004b) A novel regulatory pathway of sulfate uptake in Arabidopsis roots: implication of CRE1/WOL/AHK4-mediated cytokinin-dependent regulation. Plant J 38: 779–789

Maruyama-Nakashita A, Nakamura Y, Watanabe-Takahashi A, Inoue E, Yamaya T, Takahashi H (2005) Identification of a novel cis-acting element conferring sulfur deficiency response in Arabidopsis roots. Plant J 42: 305–314

Maruyama-Nakashita A, Nakamura Y, Tohge T, Saito K, Takahashi H (2006) Arabidopsis SLIM1 is a central transcriptional regulator of plant sulfur response and metabolism. Plant Cell 18: 3235–3251

Melis A, Chen HC (2005) Chloroplast sulfate transport in green algae – genes, proteins and effects. Photosynth Res 86: 299–307

Mourioux D, Douce R (1979) Transport du sulfate à travers la double membrane limitante, ou enveloppe, des chloroplastes d'épinard. Biochimie 61: 1283–1292

Neuenschwander U, Suter M, Brunold C (1991) Regulation of sulfate assimilation by light and O-acetyl-L-serine in Lemna minor-L. Plant Physiol 97: 253–258

Ng AY-N, Blomstedt CK, Gianello R, Hamill JD, Neale AD, Gaff DF (1996) Isolation and characterisation of a lowly expressed cDNA from the resurrection grass Sporobolus stapfianus with homology to eukaryotic sulfate transporter proteins (Accession No. X96761) (PGR96-032). Plant Physiol 111: 651

Norici A, Hell R, Giordano M (2005) Sulfur and primary production in aquatic environments: an ecological perspective. Photosynth Res 86: 409–417

Page RDM (1996) Treeview: An application to display phylogenetic trees on personal computers. Comp Applic Biosci 12: 357–358

Passera C, Ferrari G (1975) Sulfate uptake in two mutants of Chlorella vulgaris with high and low sulfur amino acid content. Physiol Plant 35: 318–321

Pollock SV, Pootakham W, Shibagaki N, Moseley JL, Grossman AR (2005) Insights into the acclimation of Chlamydomonas reinhardtii to sulfur deprivation. Photosynth Res 86: 475–489

Price GD, Woodger FJ, Badger MR, Howitt SM, Tucker L (2004) Identification of a SulP-type bicarbonate transporter in marine cyanobacteria. Proc Natl Acad Sci U S A 101: 18228–18233

Rae AL, Smith FW (2002) Localisation of expression of a high-affinity sulfate transporter in barley roots. Planta 215: 565–568

Rennenberg H, Schmitz K, Bergmann L (1979) Long-distance transport of sulfur in *Nicotiana tabacum*. Planta 147: 57–62

Rennenberg H, Polle A, Martini N, Thoene B (1988) Interaction of sulfate and glutathione transport in cultured tobacco cells. Planta 176: 68–74

Rennenberg H, Kemper O, Thoene B (1989) Recovery of sulfate transport into heterotrophic tobacco cells from inhibition by reduced glutathione. Physiol Plant 76: 271–276

Roberts SK (2006) Plasma membrane anion channels in higher plants and their putative functions in roots. New Phytol 169: 647–666

Rouached H, Berthomieu P, El Kassis E, Cathala N, Catherinot V, Labesse G, Davidian J-C, Fourcroy P (2005) Structural and functional analysis of the C-terminal STAS (sulfate transporter and anti-sigma antagonist) domain of the *Arabidopsis thaliana* sulfate transporter SULTR1.2. J Biol Chem 280: 15976–15983

Rybová R, Nešpůrkova L, Janáček K (1988) Sulfate ion influx and efflux in *Hydrodictyon reticulatum*. Biol Plant 30: 440–450

Saier MH, Eng BH, Fard S, Garg J, Haggerty DA, Hutchinson WJ, Jack DL, Lai EC, Liu HJ, Nusinew DP, Omar AM, Pao SS, Paulsen IT, Quan JA, Sliwinski M, Tseng TT, Wachi S, Young GB (1999) Phylogenetic characterization of novel transport protein families revealed by genome analyses. Biochim Biophys Acta 1422: 1–56

Sanchez-Fernandez R, Davies TGE, Coleman JOD, Rea PA (2001) The *Arabidopsis thaliana* ABC protein superfamily, a complete inventory. J Biol Chem 276: 30231–30244

Sandal NN, Marcker KA (1994) Similarities between a soybean nodulin, *Neurospora crassa* sulphate permease II and a putative human tumour suppressor. Trends Biochem Sci 19: 19

Shelden MC, Loughlin P, Tierney ML, Howitt SM (2001) Proline residues in two tightly coupled helices of the sulphate transporter, SHST1, are important for sulphate transport. Biochem J 356: 589–594

Shelden MC, Loughlin P, Tierney ML, Howitt SM (2003) Interactions between charged amino acid residues within transmembrane helices in the sulfate transporter SHST1. Biochemistry 42: 12941–12949

Shibagaki N, Grossman AR (2004) Probing the function of STAS domains of the Arabidopsis sulfate transporters. J Biol Chem 279: 30791–30799

Shibagaki N, Grossman AR (2005) Approaches using yeast cells to probe the function of STAS domain in SULTR1;2. Sulfur Transport and Assimilation in Plants in the Post Genomic Era, pp 33–35. Backhuys Publishers, Leiden

Shibagaki N, Rose A, McDermott JP, Fujiwara T, Hayashi H, Yoneyama T, Davies JP (2002) Selenate-resistant mutants of *Arabidopsis thaliana* identify Sultr1;2, a sulfate transporter required for efficient transport of sulfate into roots. Plant J 29: 475–486

Smith FW, Ealing PM, Hawkesford MJ, Clarkson DT (1995a) Plant members of a family of sulfate transporters reveal functional subtypes. Proc Natl Acad Sci U S A 92: 9373–9377

Smith FW, Hawkesford MJ, Prosser IM, Clarkson DT (1995b) Isolation of a cDNA from *Saccharomyces cerevisiae* that encodes a high-affinity sulfate transporter at the plasma-membrane. Mol Gen Genet 247: 709–715

Smith FW, Hawkesford MJ, Ealing PM, Clarkson DT, Vanden Berg PJ, Belcher AR, Warrilow AGS (1997) Regulation of expression of a cDNA from barley roots encoding a high affinity sulphate transporter. Plant J 12: 875–884

Smith IK (1975) Sulfate transport in cultured tobacco cells. Plant Physiol 55: 303–307

Smith IK (1976) Sulfate transport in tobacco cells. Plant Physiol 57: 95–95

Smith IK (1977) Evidence for O-acetylserine in *Nicotiana tabacum*. Phytochemistry 16: 1293–1294

Sunarpi, Anderson JW (1996) Distribution and redistribution of sulfur supplied as [^{35}S]sulfate to roots during vegetative growth of soybean. Plant Physiol 110: 1151–1157

Takahashi H, Sasakura N, Noji M, Saito K (1996) Isolation and characterization of a cDNA encoding a sulfate transporter from *Arabidopsis thaliana*. FEBS Lett 392: 95–99

Takahashi H, Yamazaki M, Sasakura N, Watanabe A, Leustek T, De Almeida Engler J, Engler G, Van Montagu M, Saito K (1997) Regulation of sulfur assimilation in higher plants: A sulfate transporter induced in sulfate-starved roots plays a central role in *Arabidopsis thaliana*. Proc Natl Acad Sci U S A 94: 11102–11107

Takahashi H, Asanuma W, Saito K (1999) Cloning of an Arabidopsis cDNA encoding a chloroplast localizing sulphate transporter isoform. J Exp Bot 50: 1713–1714

Takahashi H, Watanabe-Takahashi A, Smith FW, Blake-Kalff M, Hawkesford MJ, Saito K (2000) The roles of three functional sulphate transporters involved in uptake and translocation of sulphate in *Arabidopsis thaliana*. Plant J 23: 171–182

Thoiron A, Thoiron B, Demarty M, Thellier M (1981) Compartmental analysis of sulfate transport in *Lemna minor* L taking plant-growth and sulfate metabolization into consideration. Biochim Biophys Acta 644: 24–35

Thomas D, Surdin-Kerjan Y (1997) Metabolism of sulfur amino acids in *Saccharomyces cerevisiae*. Microbiol Mol Biol Rev 61: 503–532

Thompson JD, Gibson TJ, Plewniak F, Jeanmougin F, Higgins DG (1997) The CLUSTAL_X windows interface: flexible strategies for multiple sequence alignment aided by quality analysis tools. Nucleic Acids Res 25: 4876–4882

Vallée M, Jeanjean R (1968a) Sulphate transport system of *Chlorella pyrenoidosa* and its regulation. II. Regulation of entry. Biochim Biophys Acta 150: 607–617

Vallée M, Jeanjean R (1968b) Sulphate transport system of *Chlorella pyrenoidosa* and its regulation. I. Kinetics of permeation. Biochim Biophys Acta 150: 599–606

Vauclare P, Kopriva S, Fell D, Suter M, Sticher L, von Ballmoos P, Krähenbühl U, den Camp RO, Brunold C (2002) Flux control of sulphate assimilation in *Arabidopsis thaliana*: adenosine 5'-phosphosulphate reductase is more susceptible than ATP sulphurylase to negative control by thiols. Plant J 31: 729–740

Vidmar JJ, Tagmount A, Cathala N, Touraine B, Davidian J-C (2000) Cloning and characterization of a root specific high-affinity sulfate transporter from *Arabidopsis thaliana*. FEBS Lett 475: 65–69

Weber APM, Schwacke R, Flugge UI (2005) Solute transporters of the plastid envelope membrane. Ann. Rev. Plant Biol. 56: 133–164

Yildiz FH, Davies JP, Grossman AR (1994) Characterization of sulfate transport in *Chlamydomonas reinhardtii* during

sulfur-limited and sulfur-sufficient growth. Plant Physiol 104: 981–987

Yildiz FH, Davies JP, Grossman A (1996) Sulfur availability and the *SAC1* gene control adenosine triphosphate sulfurylase gene expression in *Chlamydomonas reinhardtii*. Plant Physiol 112: 669–675

Yoshimoto N, Takahashi H, Smith FW, Yamaya T, Saito K (2002) Two distinct high-affinity sulfate transporters with different inducibilities mediate uptake of sulfate in Arabidopsis roots. Plant J 29: 465–473

Yoshimoto N, Inoue E, Saito K, Yamaya T, Takahashi H (2003) Phloem-localizing sulfate transporter, Sultr1;3, mediates re-distribution of sulfur from source to sink organs in Arabidopsis. Plant Physiol 131: 1511–1517

Zhang MY, Bourbouloux A, Cagnac O, Srikanth CV, Rentsch D, Bachhawat AK, Delrot S (2004) A novel family of transporters mediating the transport of glutathione derivatives in plants. Plant Physiol 134: 482–491

Chapter 3

Phylogenetic Analysis of Sulfate Assimilation and Cysteine Biosynthesis in Phototrophic Organisms

Stanislav Kopriva*
John Innes Centre, Norwich, NR4 7UH, UK

Nicola J. Patron and Patrick Keeling
*Canadian Institute for Advanced Research,
Department of Botany, University of British Columbia, 3529-6270 University Boulevard,
Vancouver, BC, V6T 1Z4, Canada*

Thomas Leustek
*Rutgers University, Biotechnology Center for Agriculture
and the Environment, Department of Plant Biology and Pathology,
59 Dudley Road Room 328, New Brunswick, New Jersey 08901-8520, USA*

*Corresponding author, Fax: +44 1603 450014, E-mail: stanislav.kopriva@bbsrc.ac.uk

Rüdiger Hell et al. (eds.), Sulfur Metabolism in Phototrophic Organisms, 31–58.
© 2008 *Springer.*

Summary

Sulfur is an essential nutrient for all organisms. The majority of sulfur in nature is found in inorganic form of sulfate, which has to be reduced and incorporated into bioorganic compounds. Assimilatory sulfate reduction occurs in various chemotrophic bacteria and fungi and in photosynthetic organisms, but is missing in animals and most prokaryotic and eukaryotic obligate parasites. Despite its central position in plant primary metabolism, the question of evolution of the pathway and origin of plant genes involved in sulfate assimilation has never been addressed. We have therefore made use of the vast amount of available sequence data to perform a phylogenetic analysis of sulfate assimilation genes from a range of lineages of photosynthetic organisms including photosynthetic bacteria, primary symbionts such as plants, green and red algae and various secondary and tertiary symbionts. The analysis revealed very complicated relations between the different lineages and different evolutionary histories of the individual genes of the pathway. Whereas, for example, plant sulfite reductase is clearly of a cyanobacterial origin, the other genes in the pathway, although being plastidial are, unusually, not of cyanobacterial origin. The clear separation between adenosine phosphosulfate- and phosphoadenosine phosphosulfate-reducing organisms seen in previous analyses has been lost with the inclusion of genes from diatom and cryptomonad secondary symbiont algae. In fact, a new variant of the key enzyme of sulfate assimilation, adenosine 5′-phosphosulfate reductase, lacking an iron sulfur cofactor, has been discovered. In addition, many interesting fusion proteins between various components of the pathway were uncovered in the newly sequenced algal genomes which open new exciting opportunities to improve the efficiency of the pathway or some of its reactions. In the chapter, protein phylogenies of seven enzymes of the pathway will be discussed in detail with relation to distribution of enzyme variants among prokaryotic and eukaryotic lineages, origin of plant genes, and the origin of genes in algae with secondary and tertiary plastids.

I. Introduction

A. Sulfate Assimilation

Sulfur occurs in nature in various oxidation states in inorganic, organic, and bioorganic compounds. Sulfur can readily change its oxidation state, there-

fore oxidized sulfur compounds serve as terminal electron acceptors in respiration of sulfate reducing bacteria and reduced sulfur compounds support chemotrophic or phototrophic growth of many bacteria and Achaea as electron donors. Sulfur is an essential nutrient for all organisms as it is part of cysteine and methionine, sulfur containing proteogenic amino acids with frequent catalytic and structural functions, coenzymes and prosthetic groups such as iron–sulfur clusters, thiamine, lipoic acid, coenzyme A, etc. In these metabolites sulfur is in a reduced form, however, the major form of sulfur available in nature is in the oxidized form of sulfate. Many microorganisms, algae, plants and fungi (but not animals) are able to take the sulfate up, reduce it and incorporate in amino acids in the pathway of

---

Abbreviations: APR – adenosine 5′-phosphosulfate reductase; APS – adenosine 5′-phosphosulfate; ATPS – ATP sulfurylase; CBS – cystathionine β-synthase; EST – expressed sequence tag; GSH – glutathione; LGT – lateral gene transfer; OAS – O-acetylserine; OASTL – O-acetylserine (thiol)lyase; PAPS – 3′-phosphoadenosine 5′-phosphosulfate; PPase – inorganic pyrophosphatase; SAT – serine acetyltransferase; SiR – sulfite reductase

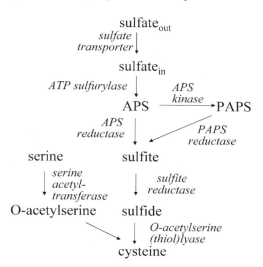

sulfate$_{out}$

sulfate transporter

sulfate$_{in}$

ATP sulfurylase   *APS kinase*

APS $\xrightarrow{}$ PAPS

APS reductase   *PAPS reductase*

serine   sulfite

serine acetyl-transferase   *sulfite reductase*

O-acetylserine   sulfide

O-acetylserine (thiol)lyase

cysteine

Fig. 1. Assimilatory sulfate reduction in plants.

assimilatory sulfate reduction (Fig. 1). In this pathway the entry of sulfate into cells is facilitated by sulfate transporters. The chemically stable sulfate has to be activated before reduction by adenylation with ATP to adenosine 5′-phosphosulfate (APS) in a reaction catalyzed by ATP sulfurylase (ATPS). APS can be directly reduced to sulfite by APS reductase (APR) or phosphorylated by APS kinase to form 3′-phosphoadenosine 5′-phosphosulfate (PAPS). PAPS is reduced to sulfite or serves as a source of activated sulfur for various sulfonation reactions modifying proteins, saccharides or secondary compounds. PAPS reductase uses thioredoxin as an electron donor, whereas APS reductase can be either thioredoxin-dependent or react with glutathione (see section III.D). APS dependent pathway is more frequent, being found in plants, algae, and most bacteria, while the PAPS reductase seems to be restricted to fungi, some enteric γ-proteobacteria and many (but not all) cyanobacteria. Sulfite produced by APR or PAPS reductase is then reduced to sulfide by sulfite reductase (SiR). The reaction requires the transfer of six electrons provided by NADPH in chemotrophic organisms or by ferredoxin in phototrophs, thus linking sulfate assimilation with photosynthesis. Sulfide is then incorporated into the amino acid skeleton of O-acetylserine or O-acetylhomoserine (in fungi) to form cysteine or homocysteine respectively in a reaction catalyzed by O-acetyl(homo)serine-(thiol)lyase (OASTL). The activated precursor is provided from serine and acetyl-coenzyme A by serine acetyltransferase (SAT). SAT and OASTL form a multienzyme complex often called cysteine synthase. The pathway of plants was subjected to many comprehensive reviews (Leustek et al., 2000; Hawkesford and Wray, 2000; Saito, 2004; Kopriva and Rennenberg, 2004; Rausch and Wachter, 2005; Kopriva, 2006) and is described in great detail elsewhere in this book.

It should be noted that sulfate is reduced in many bacteria and Archaea in a dissimilatory manner as an electron acceptor for respiration. The dissimilatory sulfate reduction often utilizes the same enzyme activities as sulfate assimilation, i.e., ATP sulfurylase, APS reductase, and sulfite reductase, but the structure of these enzymes is very different (Hansen, 1994; Stahl et al., 2002) and will therefore not be discussed here in detail. The biochemistry and molecular biology of dissimilatory sulfur metabolism will be described in section IV.

B.   Photosynthetic Organisms

This chapter focuses on the evolution of sulfate assimilation in plants and other photosynthetic organisms. The rationale for this limitation is firstly the clear physiological link of sulfate assimilation to photosynthesis and, secondly, the vast amount of new sequence data available for photosynthetic organisms, including various algae that have not yet been thoroughly exploited. Photosynthesis uses energy from light to produce chemical energy and is therefore essential for life on Earth. Photosynthesis occurs in plants, algae, and several groups of photosynthetic bacteria. Whereas in plants and algae photosynthesis is confined to chloroplasts, in photosynthetic bacteria it occurs directly in the cytoplasm. The photosynthesizing bacteria can be separated in two major groups. Cyanobacteria possess chlorophyll and are capable of oxygen generation whereas other photoautotrophic bacteria absorb light with the help of bacteriochlorophyll and exist in an anoxygenic environment.

The ability for phototrophic life in bacteria is not limited to a phylogenetically distinct group. Bacteriochlorophyll dependent photosynthesis is found in five classes of bacteria: (1) green sulfur bacteria (*Chlorobium, Chlorobaculum,* and *Pelodictyon*), (2) *Chloroflexus* division called also green non-sulfur bacteria (*Chloroflexus, Roseiflexus*), (3) *Heliobacteriaceae*, within the low GC Gram positive bacteria, (4) purple sulfur bacteria of γ-proteobacteria (*Chromatium, Thiocapsa, Ectothiorhodospira*) and (5) purple non-sulfur bacteria spread between α- (e.g., *Rhodobacter, Rhodopseudomonas, Roseovarius, Rhodospirillum*) and β-proteobacteria (*Rho-*

doferax, Rubrivivax) (Imhoff, 1992). These bacteria use hydrogen, sulfide, thiosulfate, or ferrous iron as electron donors to support their anoxic, phototrophic growth. A special physiological group is capable of aerobic photosynthesis, it includes species such as *Roseobacter, Roseococcus,* or *Porphyrobacter,* which use light energy only as a supplement and are not capable of photoautotrophic growth (Yurkov and Beatty, 1998). More details on taxonomy of anoxygenic photosynthetic bacteria are available in Chapter 14.

Cyanobacteria are a morphologically diverse group of bacteria. Because of their ability to photosynthesize they have been described in the past as 'blue-green algae' (Stafleu et al., 1972) and their classification is still based more on phenotypical characters than molecular data. Cyanobacteria are divided into five major orders (*Chroococcales, Nostocales, Oscillatoriales, Pleurocapsales,* and *Stigonematales*) (Waterbury, 2001). Cyanobacteria are the accepted ancestors of all plastids (Nelissen et al., 1995; Cavalier-Smith, 2002). The primary endosymbiosis which gave rise to primary plastids is thought to have been a single event after which much of the cyanobacterial genome was transferred to the nucleus and the glaucophytes diverged, followed by red and green algae, the latter of which also gave rise to plants (McFadden, 2001; Martin et al., 2002; Rodriguez-Ezpeleta et al., 2005). This theory is strongly corroborated by phylogenetic analyses that show strong support for cyanobacteria and plants and algae as sister taxa in phylogenies based on the majority of plastid proteins (Martin et al., 2002; Chu et al., 2004). The great diversity of today's algal world is, however, due to further endosymbiosis events in which red or green algae were engulfed by eukaryotic hosts and gave rise to secondary and tertiary plastids (Archibald and Keeling, 2002; Bhattacharya et al., 2004; Patron et al., 2006). These complex plastids are surrounded by three or more membranes (Schwartzbach et al., 1998) and require new mechanisms of protein transport (Sulli et al., 1999; Waller et al., 2000; van Dooren et al., 2001; Patron et al., 2005). Euglenoids and chlorarachniophytes have secondary plastids of a green algal origin, though neither plastid nor host cell are directly related. Heterokonts, haptophytes, cryptophytes, apicomplexans, and dinoflagellates (collectively, chromalveolates) all possess secondary endosymbionts derived from red algae and although the bulk of molecular evidence supports a single red-algal endosymbiotic event (Fast et al., 2001; Archibald and Keeling, 2002; Yoon et al., 2002a; Cavalier-Smith, 2003; Harper and Keeling, 2003; Bhattacharya et al., 2004; Patron et al., 2004; Waller et al., 2006), the relatedness of the plastid is still a matter of contention (Bodyl, 2004; Sánchez Puerta et al., 2005). To add to the complexity, in some dinoflagellates, such as *Kryptoperidinium* or *Karlodinium,* the secondary plastid was replaced by a plastid from another secondary alga, resulting in tertiary plastid (Chesnick et al., 1997; Yoon et al., 2002b; Patron et al., 2006). A consequence of these complex evolutionary histories is that the genomes of such secondary and tertiary symbionts are a mosaic drawn from many sources, which makes analyses of evolutionary origins of single genes a very difficult task.

C. Genomics of Photosynthetic Organisms

The technological progress in high throughput DNA sequencing created vast amount of data, most of which is publicly available for analysis. This is especially true for bacterial genomes, where since the completion of the first genome of *Haemophillus influenzae* (Fleischmann et al., 1995) 599 bacterial genomes have been fully sequenced (NCBI Entrez Genome Project, November 2007). To date there are complete sequences for 29 cyanobacterial genomes, as well as many photosynthetic bacteria: 4 green sulfur bacteria (*Chlorobium tepidum, C. chlorochromatii, C. phaeobacteroides, Pelodictyon luteolum*), a green non-sulfur bacterium *Chloroflexus aurantiacus,* 12 species and strains of purple non-sulfur bacteria (such as *Rhodobacter sphaeroides, Rhodopseudomonas palustris, Rhodospirillum rubrum, Rhodoferax ferrireducens,* or *Rubrivivax gelatinosus*), and 6 aerobic phototrophic bacteria (e.g., *Erythrobacter litoralis, Roseobacter* sp., and *Roseovarius nubinhibens*). Many plant and algal genomes have also been completely sequenced, e.g., *Arabidopsis thaliana* (The Arabidopsis Genome Initiative, 2000), *Oryza sativa* (Goff et al., 2002; Yu et al., 2002), *Populus trichocarpa* (http://genome.jgi-psf.org/Poptr1/Poptr1. home.html), *Physcomitrella patens* (http://moss. nibb.ac.jp/), *Selaginella moellendorffii* (http://www. jgi.doe.gov/ sequencing/why/CSP2005/selaginella. html), *Chlamydomonas reinhardtii* (http://genome. jgi-psf.org/Chlre3/Chlre3.home.html; Grossman, 2005), *Cyanidioschyzon merolae* (http://merolae. biol.s.u-tokyo.ac.jp/; Matsuzaki et al., 2004), and *Thalassiosira pseudonana* (Armbrust et al., 2004). In addition, a large number of EST projects have been initiated for several diverse algae such as *Bigelowiella natans*, *Cyanophora paradoxa*, *Glaucocystis nostochinearum*, *Heterocapsa triquetra*, *Karlodinium micrum*, and *Pavlova lutheri*, within the Protist EST Program (http://amoebidia.bcm.umontreal.ca/public/pepdb/welcome.php). Clearly, this wealth of sequence information enables now to study the evolution of metabolic pathways and the plasticity of photosynthetic organisms.

II. Occurrence of Sulfate Assimilation in Different Taxa

Although sulfate assimilation is essential for autotrophic growth and synthesis of cysteine and methionine, the pathway seems to be readily dispensable when the lifestyle of organism allows. It is absent in all metazoans, which satisfy their need for reduced sulfur by ingestion of sulfur containing amino acids cysteine or methionine. Interestingly, whereas plants can synthesize methionine from cysteine and not vice versa, animals possess the enzymes for the reverse transsulfuration pathway to synthesize cysteine from methionine and not the methionine biosynthetic genes (Leustek et al., 2000; Stipanuk, 2004). Therefore, methionine is an essential amino acid for animal nutrition. Another group of organisms lacking sulfate assimilation pathway are numerous bacteria and protists adapted to parasitism. In bacterial species that have undergone a significant genome reduction, the sulfate assimilation operon is almost invariably lost (Sakharkar et al., 2004; Nozaki et al., 2005). This is enabled by the adaptation of the nutrition of these parasites for metabolites provided by the host. Eukaryotic parasitic protists such as certain apicomplexans and the ciliates, interestingly the sister taxa to dinoflagellates, as well as parasitic fungi such as microsporidia, presumably scavenge reduced sulfur compounds from their host (Fulton and Grant, 1956; Payne and Loomis, 2006). A third group of organisms usually lacking sulfate assimilation are Archaea and bacteria using dissimilatory sulfide (or thiosulfate) oxidation or sulfate reduction for respiration and energy conversion. The habitats of such organisms always contain sulfide (Perez-Jimenez and Kerkhof, 2005), therefore, there is no need for sulfate assimilation to sustain cysteine biosynthesis. In total, from the 551 completely sequenced bacterial genomes 300

revealed a capacity for sulfate assimilation judged by the presence of the *cysH* gene encoding APS/PAPS reductase and the same was true for 5 out of 48 sequenced Archaea.

On the other hand, sulfate assimilation is present in plants, algae, and fungi. In plants and primary algae (green, red, and glaucophytes) the reductive part of the pathway, i.e., APR and SiR, is confined to plastids, whereas ATPS and APS kinase are present in plastids and in the cytosol and cysteine synthesis takes place in the plastids, cytosol, and mitochondria (Leustek et al., 2000; Rotte and Leustek, 2000; Koprivova et al., 2001; Wirtz et al., 2004). There is not enough information on localization of the pathway in secondary and tertiary symbionts, but in many cases, e.g., the stramenopile (diatom) *Thalassiosira pseudonana*, the genes seem to encode proteins with pre-sequences that suggest a chloroplast location. The only remarkable exception is *Euglena gracilis*, which possesses a sulfate-reducing pathway in the mitochondria (Brunold and Schiff, 1976; Saidha et al., 1988). Yeasts and other fungi seem to reduce sulfate in the cytosol.

III. Phylogenetic Analysis of Sulfate Assimilation Genes

The recent increase in availability of sequence information from various algae and plants enables a wide sampling of taxa for thorough phylogenetic analysis of photosynthetic organisms. The questions of the origin of plant genes and metabolic diversity of plants and algae can be now addressed with much better resolution. In addition, the possibility of using the genomic information for phylogenomics allows for the reconstruction of the evolutionary history of organisms (Delsuc et al., 2005), which is especially important for the complex algal lineages. Since sulfate assimilation is essential for plants and algae and is directly linked with photosynthesis we have undertaken a detailed phylogenetic analysis of the genes involved in the pathway.

A. Sulfate Transporters

1. Types of Sulfate Transporters

Before sulfate can be reduced it has to be taken up into the cells through the plasma membrane and, in organisms that reduce sulfate in the plastid, further transport across the membranes of cellular organelles is also required. In addition, in plants a long distance sulfate transport from roots to the shoots requires additional transport steps between cells of different tissues. The transport of sulfate is facilitated by transporters via a proton coupled co-transport (Hawkesford et al., 1993), or anion exchange and sodium co-transport in animals (Markovich, 2001). Because of the various transport steps, a large number of specific transporters with different affinities for sulfate exist in plants (reviewed in Buchner et al., 2004, see also Chapter 2). On the other hand, most microorganisms possess only a simple sulfate uptake system (Sirko et al., 1990; Laudenbach and Grossman, 1991; Kertesz, 2001). The ATP dependent uptake is accomplished by a permease complex composed from three cytoplasmic membrane components CysA, CysT, and CysW and a sulfate binding protein SbpA in the periplasmic space (SulT family transporter). In addition to this ABC type of transporter, some bacteria possess a sulfate transporter from the major facilitator superfamily (MFS; Kertesz, 2001). Yeast and filamentous fungi possess two isoforms of high affinity sulfate transporters (Cherest et al., 1997; Van de Kamp et al., 2000). These are single membrane proteins of 80–95 kDa with 10 or 11 predicted transmembrane domains. *P. chrysogenum* with a mutation in the SutB transporter are unable to grow on sulfate as sole sulfur source, whereas *S. cerevisiae* with both transporters disrupted are able to grow at very high sulfate concentrations (30 mM) indicating a presence of a low affinity sulfate uptake system, at least in yeast. Animals also contain multiple sulfate uptake systems: three sodium dependent transporters NaSi-1, SUT-1, and NaDC-1 and at least five anion exchangers (Markovich, 2001).

In higher plants sulfate transporters are encoded by a large family of 14 genes in Arabidopsis and 15 genes in rice. The genes can be divided into five groups according to sequence similarity (see Chapter 2). The individual genes have distinct affinities for sulfate and tissue-specific expression. Surprisingly in the completely sequenced genomes of the lower plants *P. patens* and *S. moellendorffii* only genes from groups 1 and 4, i.e., high affinity and vacuolar transporters, respectively, were identified. The green alga *Chlamydomonas* contains three 'plant-type' sulfate transporter genes and another nuclear gene encoding a SulP transporter responsible for sulfate transport into the chloroplast. SulP is closely related to the ABC transport system from cyanobacteria and similar genes are found in chloroplast genomes of several other green algae and the lower plants *Mesostigma viride, Marchantia polymorpha,* and *Anthoceros formosae.* The transporter responsible for sulfate uptake into plastids of higher plants still remains to be identified. In the complete genomes of the red alga *C. merolae* and the stramenopile, *T. pseudonana,* which has a secondary plastid of red-algal origin, only genes encoding a bacterial type MFS sulfate transporter are found. The prokaryotic MSF transporters are found in several other algae with plastid of red-algal origin, such as the haptophytes *Pavlova lutheri* and *Isochrysis galbana* and the dinoflagellate *Karlodinium micrum,* additionally a 'plant-type' transporter was found in *P. lutheri.*

2. Phylogenetic Analysis

An analysis of the full-length sequences reveals most of the streptophytes form a well-supported clade and, although *Arabidopsis* and *O. sativa* isoforms form five moderate- to well-supported sub-clades indicating that these duplicated in the ancestors of tracheophytes, the multiple isoforms from the lower plants *P. patens* and S. *moellendorffii* seem to have duplicated independently within those lineages (Fig. 2). The transporters of the green-lineage group together with strong support but are not related to cyanobacteria, as is often the case for transporters. In contrast, the red-lineage plastid transporters from *C. merolae* and *T. pseudonana* fall within a strongly supported clade of bacteria including cyanobacteria. The transporters from *P. patens* and S. *moellendorffii,* which cluster with group 1 transporters when only plant sequences are analyzed are positioned basal to the node separating low affinity (group 2) and high affinity (group 1) transporters of higher plants. Thus, their function as high or low affinity transporters has to be verified experimentally.

B. ATP Sulfurylase

1. Biochemical Properties

ATP sulfurylase (EC 2.7.7.4) catalyzes activation of sulfate by adenylation to APS. The formation of

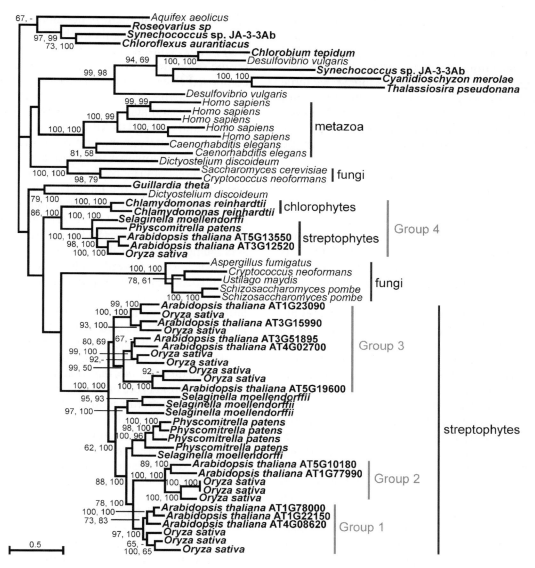

Fig. 2. Protein maximum likelihood phylogeny for plant type sulfate transporters. The tree was inferred using PhyML with nine categories of rates. Numbers at nodes indicate bootstrap support over 60 from (top to bottom or left to right) maximum likelihood, as determined by PhyML, and weighted neighbor joining. Analysis used 285 characters, the gamma-shaped parameter 1.797 and proportion of invariable sites 0.000. Photosynthetic organisms are typed in bold.

APS is an energetically very unfavorable process, which has to be driven forwards by consumption of the reaction products by subsequent reactions. APS is thus reduced to sulfite by APR or phosphorylated to PAPS by APS kinase. However, the hydrolysis of the second reaction product of ATPS, pyrophosphate, by inorganic pyrophosphatase (PPase) also contributes to shifting of reaction equilibrium towards APS synthesis.

ATPS is found also in Metazoa, in which, although unable to reduce sulfate, it plays an essential role in the synthesis of PAPS as a donor of activated sulfate for various sulfotransferases (Varin et al., 1997; Honke and Taniguchi, 2002). In these species ATPS and APS kinase are fused, creating a single bifunctional protein, PAPS synthetase (Rosenthal and Leustek, 1995; Venkatachalam, 2003). Similarly, in *Mycobacteria* and *Rhizobia*

ATPS and APS kinase are essential for the synthesis of PAPS to enable the sulfation of virulence and Nod factors, respectively (Folch-Mallol et al., 1996; Mougous et al., 2006). ATPS is also required in anaerobic sulfate-reducing bacteria, where APS serves as terminal electron acceptor for respiration, and in several classes of chemolitotrophic bacteria, which use ATPS to synthesize ATP from APS and pyrophosphate (Kappler and Dahl, 2001).

2. Types of ATPS in Different Organisms

ATPS from different organisms have very different molecular structures in terms of subunits, however the basic subunit in some of them is conserved and readily alignable. Plant ATPS is a homotetramer of 52–54 kDa polypeptides (Murillo and Leustek, 1995). Activity has been detected in both chloroplasts and the cytosol (Lunn et al., 1990). In *Arabidopsis*, the overall foliar ATPS activity continually declined during plant growth. During this time the more abundant chloroplast ATPS activity decreased, while cytosolic activity grew (Rotte and Leustek, 2000). ATPS thus seems to have different functions in the two compartments: sulfate assimilation in the plastids and activation of sulfate for synthesis of sulfated compounds in the cytosol (Rotte and Leustek, 2000). ATPS is encoded by small multigene families in all plant species analyzed to date. cDNAs encoding chloroplast and cytosolic isoforms of ATPS were isolated from potato (Klonus et al., 1994). On the other hand, four isoforms of ATPS were isolated from *Arabidopsis*, all of them containing a chloroplast transit peptide (Murillo and Leustek, 1995; Hatzfeld et al., 2000a) and so activity in the cytosol is unaccounted for. No information is available for the biochemical properties of ATPS from algae. However, sequence analysis of the two ATPS isoforms found in the genome of the diatom *T. pseudonana* revealed one fused to APS kinase and PPase forming a polyprotein, which indicates an increased efficiency of catalysis compared with ATPS alone. As already discussed, metazoan ATPS is part of PAPS synthetase and functions as a single protein of 56 kDa (Venkatachalam, 2003). Most animals possess two copies of the gene. ATPS from yeast and fungi is also fused to APS kinase; however, the protein is a homohexamer of 59 to 64 kDa subunits. The

APS kinase-like domain is not functional and is the site of allosteric regulation by PAPS (MacRae et al., 2001; Ullrich et al., 2001). Bacterial ATPS on the other hand, consists of four heterodimers composed from 35 kDa CysD and 53 kDa CysN subunits (Leyh et al., 1988). CysD belongs to the ATP pyrophosphatase family of proteins and is thus distantly related to APR and PAPS reductase. CysD is the catalytic subunit of bacterial ATPS. Its activity is, however, energetically dependent on hydrolysis of GTP by the CysN subunit. CysD and CysN are therefore almost invariably linked in a single operon, which often contains the APS kinase encoding gene, CysC. *Mycobacteria* and *Rhizobia*, which require PAPS for sulfation, possess multiple copies of the CysD/N operon, the latter species as NodP and NodQ on the symbiotic plasmid. Interestingly, the bacterial CysN gene originated from an archaeal or eukaryotic translation elongation factor 1α (EF-1α) by lateral gene transfer (LGT) and acquired new function (Inagaki et al., 2002). No gene homologues of CysD or CysN (apart from EF-1α) are found among eukaryotes, so that this form of ATPS is limited to prokaryotes and some Archaea, which in turn most probably acquired the sulfate assimilation operon by LGT from proteobacteria. A further ATPS can be found in the sulfate reducing bacteria where, similar to plants, it is a homo-oligomer of single-function subunits (Sperling et al., 1998).

3. Phylogenetic Analysis

ATPS sequences were obtained from sequenced genomes and several EST projects. Multiple isoforms were found in most plant and algal species. Maximum likelihood analysis resulted in a complex tree with many unexpected relationships (Fig. 3). Whereas the different isoforms of plant ATPS (plastidic and cytosolic) cluster together and are therefore results of relatively recent gene duplications these were, very surprisingly, unrelated to the chlorophyte sequences. Instead they are very strongly related to the ATPS part of the PAPS protein found in Metazoa.

The chlorophytes fall deep in a well-supported clade that also contains many alpha-proteobacteria, algae and protozoans with secondary-plastids and fungi. It is by no means certain, but certainly likely that the origin of these isoforms in eukaryotes is

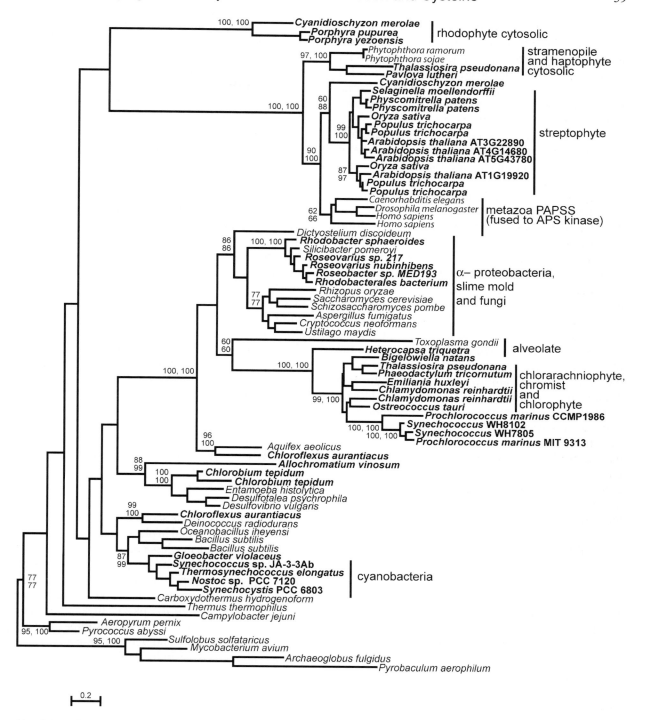

Fig. 3. Protein maximum likelihood phylogeny for ATP sulfurylase. The tree was inferred using PhyML with nine categories of rates. Numbers at nodes indicate bootstrap support over 60 from (top to bottom or left to right) maximum likelihood, as determined by PhyML, and weighted neighbor joining. Analysis used 433 characters, the gamma-shaped parameter 1.514 and proportion of invariable sites 0.059. Photosynthetic organisms are typed in bold.

the mitochondria. Also in this clade are several strains of *Prochlorococcus/Synechoccus*, known to exchange genetic data by LGT via phage with high frequency (Lindell et al., 2004; Sullivan et al., 2005; Coleman et al., 2006). All other cyanobacteria are elsewhere suggesting that these cyanobacteria obtained this isoform from a eukaryote.

The spread of the Viridiplantae suggests that there were ancestrally two isoforms in the ancestor of green algae and plants, one from an endosymbiont and one from the host. The plant lineage kept the host copy while the green algal lineage kept the plastid copy. Since there is little sequence data from the basal chlorophyte lineages, from before the divergence of plants, this cannot be proven at present.

Algae in the red plastid lineage also encoded multiple isoforms of ATPS of differing evolutionary origins. The dissimilatory ATPS of anaerobic sulfate reducing and sulfur oxidizing bacteria cluster with genes from some Gram positive bacteria, e.g., *Bacillus subtilis*, which does not possess the CysD/N system. Another surprising finding is that the ATPS from *Entamoeba histolytica* clustered among the bacterial taxa. The role for ATPS in *Entamoeba* is not obvious, since this eukaryote does not possess APS or PAPS reductase and is thus not able to reduce sulfate. However, *Entamoeba* produces monoethyl sulfate and 3-cholesteryl sulfate (Bakker-Grunwald and Geilhorn, 1992), therefore, the ATPS, which is fused to APS kinase, is probably responsible for production of the activated sulfate for these sulfations.

C. APS Kinase

1. Biochemical Properties

APS kinase (EC 2.7.1.25) catalyzes the transfer of phosphate from ATP to APS to form PAPS. PAPS is an important metabolite, not only as a form of activated sulfate for reduction in fungi and some heterotrophic bacteria but also for the sulfation of various metabolites. Although APS kinase does not participate in the pathway of sulfate assimilation in plants and algae, it interacts with the pathway by competing for APS with APR. The enzyme from *Penicillium chrysogenum* was crystallized as homodimer of 24 kDa subunits (MacRae et al., 2000). In plants, APS kinase is localized in the chloroplast and the cytosol.

Four genes encoding APS kinase are found in the *Arabidopsis* genome, all located on different chromosomes. Three of these genes code for proteins with chloroplast transit peptides, the forth, located on chromosome 3, likely encodes the isoform responsible for cytosolic activity. Very little is known about the biochemical properties and functions of the individual plant APS kinases.

APS kinase belongs to the group of P-loop-fold proteins that hydrolyze or bind nucleoside triphosphates, such as kinases and sulfotransferases (Leipe et al., 2003). Within this large group of proteins it belongs more specifically to the DxD group of kinases together with gluconate kinase, shikimate kinase, and dephosphocoenzyme A kinase. APS kinases from different organisms are well conserved. APS kinase is often found fused to ATP sulfurylase, which catalyzes the synthesis of its substrate. These fusion proteins contain the APS kinase domain at N-terminus (Metazoa) or at C-terminus (some bacteria and filamentous fungi) and can even be fused with yet another protein, as in *T. pseudonana*. Whereas in Metazoa the APS kinase is active in the synthesis of PAPS, in filamentous fungi the APS kinase domain is inactive and functions in regulation of ATPS activity (MacRae et al., 1998). The PAPS in fungi such as *P. chrysogenum*, is synthesized from APS by a second APS kinase paralogue.

2. Phylogenetic Analysis

Analysis of APS kinase sequences revealed that the multiple isoforms found in plants and green algae are specifically related, the duplications having occurred at the base of the tracheophytes (Fig. 4). The position of the clade, however, is entirely unsupported, as is much of the backbone of the tree and so it is not clear if the ancestry is from the plastid (cyanobacterial) or from the host. Likewise the relationships between the fungal, metazoan and various proteobacterial lineage, though being individually supported, cannot be resolved at present.

D. APS and PAPS Reductase

1. APS and PAPS Dependent Sulfate Assimilation – an Historical Overview

Since the sulfate assimilation pathway was first resolved in the enteric bacteria, *Escherichia coli*

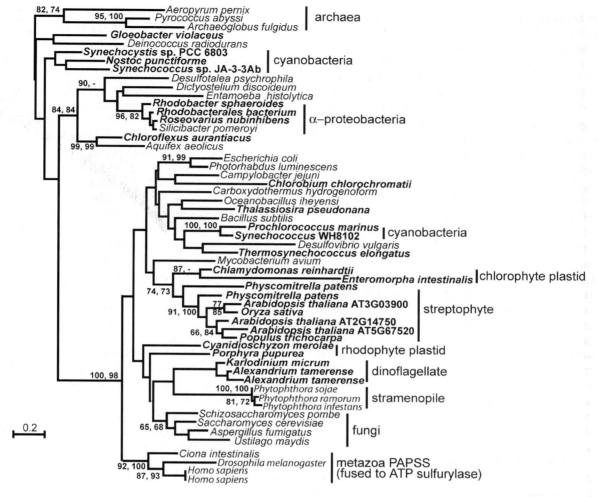

Fig. 4. Protein maximum likelihood phylogeny for APS kinases. The tree was inferred using PhyML with nine categories of rates. Numbers at nodes indicate bootstrap support over 60 from (top to bottom or left to right) maximum likelihood, as determined by PhyML, and weighted neighbor joining. Analysis used 151 characters, the gamma-shaped parameter 1.211 and proportion of invariable sites 0.086. Photosynthetic organisms are typed in bold.

and *Salmonella typhimurium* (Kredich, 1971), it was thought that both activation steps, adenylation of sulfate to APS with subsequent phosphorylation to PAPS, are generally required for sulfate reduction. However, studies with the green alga *Chlorella* revealed that APS can be directly utilized to form sulfite bound to a thiol carrier and the corresponding enzyme was named APS sulfotransferase (Tsang et al., 1971; Schmidt, 1972). The APS sulfotransferase activity was detected in a variety of plants and photosynthetic bacteria (Schmidt, 1975; Schmidt and Trüper, 1977); therefore it was considered to represent the major

sulfate reducing enzyme in photosynthetic organisms in contrast to enteric bacteria and yeast, which seemed to utilize PAPS for reduction (Brunold, 1990, Schmidt and Jäger, 1992). Since APS and PAPS are very similar substrates and the reaction conditions for their reduction are also very similar, complementation of *E. coli* mutants deficient in *cysH* was chosen as a method to clone plant APS and/or PAPS reducing enzyme(s). Three homologous cDNA clones were obtained from *A. thaliana* encoding a protein similar to *E. coli* PAPS reductase with a C-terminal thioredoxin-like extension (Gutierrez-Marcos et al., 1996;

Setya et al., 1996). Since the enzyme produced sulfite from APS but not from PAPS, it was named APS reductase (Setya et al., 1996). The cDNA was shown to correspond to the previously characterized APS sulfotransferase but biochemical analysis of its reaction products revealed that it produces free sulfite and is therefore a reductase (Suter et al., 2000). In attempts to show that APR is the only plant enzyme responsible for sulfate reduction to sulfite the single copy gene encoding APR was disrupted in the moss *Physcomitrella patens* by homologous recombination (Koprivova et al., 2002). This resulted in loss of APR activity but not the ability of such plants to grow on sulfate as the sole sulfur source and thus led to cloning of putative PAPS reductase from this moss species (Koprivova et al., 2002). On the other hand, phylogenetic analysis found that many *cysH* genes, from various bacterial lineages, are more related to plant APR than to PAPS reductase from *E. coli*. These observations led to the discovery of bacterial assimilatory APS reductase, which lacks the thioredoxin-like domain and requires thioredoxin for activity (Abola et al., 1999; Bick et al., 2000; Kopriva et al., 2002; Williams et al., 2002).

2. Biochemical Properties of APS and PAPS Reductases

APR catalyzes a thiol-dependent two-electron reduction of APS to sulfite (Suter et al., 2000). In higher plants it is localized exclusively in the plastids (Koprivova et al., 2001). The enzyme was first purified from the green alga *Chlorella* as a protein of molecular weight greater than 300 kDa (Schmidt, 1972). Li and Schiff (1991) purified APR from *Euglena gracilis* as a tetramer of 25 kDa subunits, whereas APR from *Porphyra yezoensis* consisted of 43 kDa subunits (Kanno et al., 1996). The molecular mass of native *Lemna minor* APR (Suter et al., 2000) and recombinant APR2 from *A. thaliana* (Kopriva and Koprivova, 2004) was estimated to be 91 kDa, revealing that plant APR is a dimer of ca. 45 kDa subunits. Assimilatory APR from several bacteria is either a dimer or monomer of approximately 30 kDa subunits (Kim et al., 2004; Carroll et al., 2005a). Both plant and bacterial assimilatory APR possess an atypical diamagnetic and asymmetric $[Fe_4S_4]$ cluster (Kopriva et al., 2001, 2002; Carroll et al., 2005a). APR reactions can be divided into

three steps. In the first the N-terminal part of the protein reacts with APS resulting in a stable reaction intermediate with sulfite bound to the only conserved Cys residue between APS and PAPS reductases (Weber et al., 2000). In the second step the intermediate is released by the C-terminal thioredoxin-like domain in plant APR or by free thioredoxin in case of bacterial APR. Finally, the thioredoxin or thioredoxin-like domain are reduced by thioredoxin reductase or GSH (Weber et al., 2000; Kopriva and Koprivova, 2004; Carroll et al., 2005b).

PAPS reductase is comprised of two 28 kDa subunits devoid of any prosthetic groups. It contains a single conserved cysteine residue which is responsible both for the dimerisation and for enzyme activity. In the first step of the reaction mechanism a reduced PAPS reductase binds PAPS, reduces it to sulfite and is oxidized in the process. The return to reduced state is achieved by reaction with reduced thioredoxin or glutaredoxin (Lillig et al., 1999).

APS reductase is found also in dissimilatory sulfate reducers and sulfur oxidizing bacteria and Archaea. The enzyme is a heterodimer of a 75 kDa FAD containing α-subunit and a 20 kDa β-subunit binding two $[Fe_4S_4]$ clusters (Fritz et al., 2002). Again, there is no sequence similarity to the assimilatory APR.

3. Distribution of APS and PAPS Reductases in Different Organisms

Biochemical data from various APS and PAPS reductases pointed to a clear association of the ability to reduce APS in the presence of an FeS cluster (for details see Kopriva and Koprivova, 2004). Most importantly, a bifunctional APS/PAPS reductase from *Bacillus subtillis* possesses the FeS cluster. After chemical removal of the cofactor APS, but not PAPS, reduction is abolished (Berndt et al., 2004), which strongly corroborates the functional link between the cluster and APS reduction. The cofactor is bound to the protein by three or four cysteine residues, which are found as two invariant Cys pairs in almost all APR proteins (Kopriva et al., 2002; Kim et al., 2004; Kopriva and Koprivova, 2004; Carroll et al., 2005a). These cysteine pairs thus serve as a sequence marker to distinguish APS from PAPS reductases (Kopriva et al., 2002; Kopriva and

Koprivova, 2004). Analysis of 599 sequenced prokaryotic genomes revealed that, based on the presence of the two Cys pairs, 202 species (or strains) possess APS reductase whereas 98 contain PAPS reductase, the bifunctional APS/PAPS reductase from *B. subtillis*, and possibly other Firmicutes, being counted among the APRs. The few Archaea capable of assimilatory sulfate reduction contain an APR. On the other hand, fungi encode PAPS reductase exclusively. PAPS reductase is confined to only two bacterial lineages: γ-proteobacteria and cyanobacteria, both lineages, however, also contain species with APR. Species of cyanobacteria, such as *Synechocystis* and *Synechococcus*, have long been known to reduce PAPS, but certain species, e.g., *Plectonema*, are known to be APS reducing (Schmidt, 1977). Most of the sequenced cyanobacteria possess the PAPS reductase gene, however, our recent analysis of several cyanobacteria revealed that the reduction of APS is much more common in these species than would seem from the genomic sequences (Kopriva S, Wiedemann G, unpublished). This is very important when considering the origin of the APS reductase in the plastids of algae and plants.

The putative PAPS reductase from the moss *P. patens*, which also contains an APS reductase, is more similar to fungal and bacterial PAPS reductases than to plant APS reductases and does not possess the two Cys pairs. It also lacks the C-terminal thioredoxin-like domain. Surprisingly, a detailed biochemical analysis revealed that although the enzyme reduces PAPS, it is far more active with APS. Since this protein does not bind an FeS cluster, it represents a novel form of APS reductase (Kopriva et al., 2007). This novel 'APR-B', although able to complement the loss of 'normal' APR in *P. patens* APR knock-outs (Koprivova et al., 2002), is substantially less catalytically efficient *in vitro* than the FeS cluster possessing APR. Orthologs of this novel APS reductase were found in a further lower plant lineage, in the spike-mosses *Selaginella lepidophylla* and *S. moellendorffii*, suggesting that its presence in *Physcomitrella* is not a result of a recent horizontal gene transfer to that species specifically. In *P. patens* there is evidence of expression of both APR and APR-B, but in *S. lepidophylla* transcripts have only been found for APR-B. The genomic copy of APR in the related

S. moellendorffii contains several base changes in the active site suggesting that the protein may be inactive (Kopriva S, unpublished). Other lower plants, such as *Equisetum* or ferns, possess an APS reductase (Kopriva S, unpublished) and all chlorophyte sequences to date support the presence of an APS reductase only.

The diatom *T. pseudonana* reduces APS at a very high rate, up to 100-fold higher than plants (Gao et al., 2000; Kopriva S, unpublished). Nevertheless, the only two genes homologous to APS reductase identified in the complete genome are more similar to PAPS reductase (or the novel APR-B from lower plants) and lack the FeS binding Cys pairs. However, like the APS reductase sequences of plants, they also contain C-terminal thioredoxin-like domains (Kopriva S, unpublished). Similar genes were found in several other algae, the haptophyte *Emiliania huxlei*, the chlorarachniophyte *Bigelowiella natans*, the diatom *Fragilariopsis cylindrus*, and the dinoflagellate *Heterocapsa triquetra*. Although the enzymatic activity of these proteins has not yet been confirmed *in vitro*, like *T. pseudonana*, many of these species have been shown to reduce APS (Gao et al., 2000; Kopriva S, unpublished), and it seems that they also belong to the new class B of APS reductase. Without the two cysteine pairs an FeS cluster cannot be bound to these proteins and so this novel class of APR probably does not require the cofactor.

4. Phylogenetic Analysis

In Fig. 5 the combined APR and PAPR phylogenetic tree is presented without species names and only the major lineages are indicated to show the relationship and distribution of these isoforms. However, a greater number of characters can be included with separate analyses of the two enzymes and so the two classes were also analyzed independently (Figs. 6 and 7). Because the new APR-B enzyme found in some algae and lower plants is more related to the PAPR enzymes, these are included in the analysis of this enzyme. Although previous analyses resolved a split between APS and PAPS reductases (Kopriva et al., 2002; Kopriva and Koprivova, 2004), this is not the case in the maximum likelihood analysis. Both the combined tree and the individual APR and PAPR trees are rooted with three Archaeal genes of

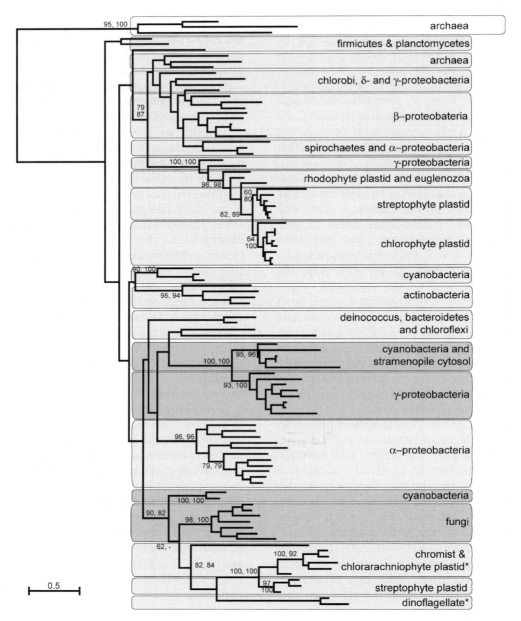

Fig. 5. Protein maximum likelihood phylogeny for APS and PAPS reductases. The tree was inferred using PhyML with nine categories of rates. Numbers at nodes indicate bootstrap support over 60 from (top to bottom or left to right) maximum likelihood, as determined by PhyML, and weighted neighbor joining. Analysis used 184 characters, the gamma-shaped parameter 1.27 and proportion of invariable sites 0.014. Photosynthetic organisms are typed in bold. Light and dark shading marks APS and PAPS reductases, respectively.

unconfirmed function that show weak similarity to both APS reductase and also to the bacterial CysD subunit of ATPS.

In the APR tree, all plants and algae with primary plastids are specifically related (Fig. 6). Related to these is the APR found in the

mitochondria of *E. gracilis*. It is possible that this species, which contains a secondary plastid of green algal origin, relocated the sulfate-reducing pathway, which it likely obtained from this endosymbiont, to the mitochondria.

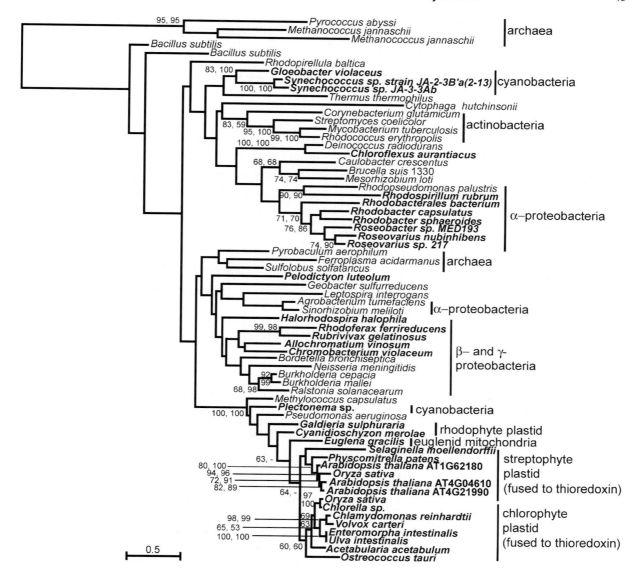

Fig. 6. Protein maximum likelihood phylogeny for APS reductases. The tree was inferred using PhyML with nine categories of rates. Numbers at nodes indicate bootstrap support over 60 from (top to bottom or left to right) maximum likelihood, as determined by PhyML, and weighted neighbor joining. Analysis used 149 characters, the gamma-shaped parameter 1.159 and proportion of invariable sites 0.018. Photosynthetic organisms are typed in bold.

The sequence from the cyanobacteria *Plectonema*, although related with strong support to plants and primary algae, does unambiguously fall at the base of the plastid clade and, though it is likely, cannot be confirmed as the source of the plastidial APR. The other lineages of bacteria form monophyletic clades, although these are, for the most part, unsupported. The position of the Archaea with APR, which is suggested to have arisen by lateral transfer, although being nested within the proteobacteria is unsupported and not related to any particular group. The origin of the enzyme in this lineage is also unresolved at present.

Within the PAPR tree the eukaryote sequences from fungi and algae with secondary plastids, as well as those from the mosses and spike mosses cluster together with the exception of a couple of divergent sequences from *G. theta* and *T. pseudo-*

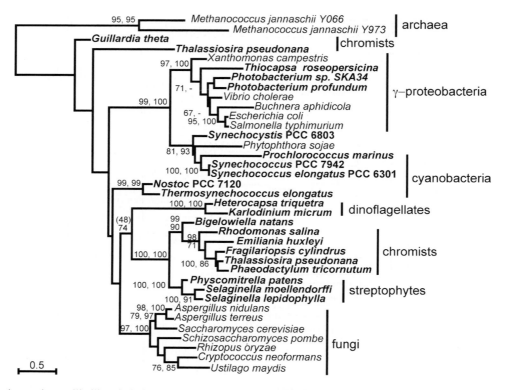

Fig. 7. Protein maximum likelihood phylogeny for PAPS reductases. The tree was inferred using PhyML with nine categories of rates. Numbers at nodes indicate bootstrap support over 60 from (top to bottom or left to right) maximum likelihood, as determined by PhyML, and weighted neighbor joining. Analysis used 149 characters, the gamma-shaped parameter 1.559 and proportion of invariable sites 0.030. Photosynthetic organisms are typed in bold.

nana (Fig. 7). The relationship with the fungi is not supported but the chromist and alveolate algae relationship is moderately supported. Within this larger 'algal' clade, the chromists are strongly related to the chlorarachniophyte *B. natans*, and also to the streptophyte isoforms. While *B. natans* is known to contain multiple enzymes more related to the red- than green-algal lineage (Archibald et al., 2003), this is extremely interesting. The possibility that the green lineage, after the divergence of streptophytes, obtained an APR-B from a chromist alga cannot be excluded.

The cyanobacteria are related to the γ-proteobacteria, to the exclusion of all eukaryotes and so the origin of the algal APR-B is not plastidial (unsurprising since primary plastids contain APR) and they are likely of host (eukaryotic) origin. It is evident however, that the new APS reductase isoform evolved from a PAPS reductase and changed the substrate specificity to APS while retaining the FeS independent reaction mechanism. This seems to be the most efficient way of

reducing activated sulfate, as the energy costs for building the cofactor are eliminated as well as the necessity of a second activation step by ATP, which is the price organisms possessing PAPS reductase pay for saving the cluster. It seems that during the evolution of plants, at least for some time, two genes were present but the APR-B was lost soon after branching of the lineage leading to spike-mosses. Since all chlorophyte sequences to date support the presence of an APS reductase only, it cannot be ruled out that the PAPS reductase, which gave rise to APR-B was introduced to the plant lineage by LGT before the divergence of bryophytes and then lost again after the divergence of the lycopodiophytes.

E. Sulfite Reductase

1. Biochemical Properties

Sulfite reductase catalyzes the six electron reduction of sulfite to sulfide. The electron donor for the reduction is ferredoxin (Fd) in plants (EC 1.8.7.1) or NADPH in bacteria (EC 1.8.1.2). Plant SiR contains one siroheme and one [4Fe–4S] cluster per monomer as cofactors (Krueger and Siegel, 1982). The SiR protein structure and sequence is very similar to Fd-nitrite reductase, which catalyzes an equivalent step, a six electron reduction of nitrite to ammonia, in nitrate assimilation. The similarity of the enzymes is corroborated by the fact that both enzymes can reduce sulfite and nitrite, albeit with different efficiency (Krueger and Siegel, 1982). Indeed, in *Arabidopsis* SiR contains 19% identical amino acids with nitrite reductase indicating that these genes may have the same evolutionary origin. SiR activity was localized exclusively to plastids both in photosynthetic and non-photosynthetic organs (Brunold and Suter, 1989). Interestingly, SiR is abundant in the nucleoids of pea chloroplasts and is able to compact chloroplast DNA (Sato et al., 2001), which might represent a new function apart from sulfate assimilation.

2. Types of SiR in Different Organisms

The reaction mechanism of sulfite reduction requires siroheme, therefore this cofactor, as well as the iron sulfur cluster, are found in SiRs from all sources. The overall structure of the SiR enzyme, however, differs substantially in various species. Plant SiR is a monomeric protein of 65 kDa containing no prosthetic groups other than the siroheme and FeS cluster (Nakayama et al., 2000). SiR is encoded by a single copy gene in *Arabidopsis*, in contrast to other enzymes of the pathway (Bork et al., 1998). However, two or more SiR isoforms are present in other plant species (Yonekura-Sakakibara et al., 1998). In contrast, the bacterial NADPH-SiR is an oligomer of eight 66 kDa flavoprotein subunits (CysJ) and four 64 kDa siroheme and [4Fe–4S] cluster binding hemoproteins (CysI) (Crane et al., 1995). SiR in yeasts and other fungi are again different, composed from α and β subunit of 116 and 167 kDa respectively. Similar to bacterial SiR, the fungal enzyme requires siroheme, FAD and FMN (Kobayashi and Yoshimoto, 1982).

In addition to the assimilatory sulfite reductase, a dissimilatory sulfite reductase (Dsr) (EC 1.8.99.1) is present in organisms reducing sulfite to sulfide during anaerobic respiration, such as *Desulfovibrio* species and *Archaeoglobus fulgidus*

(Dahl et al., 1993). Dsr is also present in the sulfur oxidizing bacteria *Thiobacillus denitrificans* and *Allochromatium vinosum*, where it catalyzes the reverse reaction, oxidation of sulfide to sulfite (Stahl et al., 2002). The enzyme is a tetramer of two α and two β subunits with a molecular mass between 40 and 45 kDa, again, containing a siroheme linked to a [4Fe–4S] cluster. Dsr from *Desulfovibrio* and some other species contains a third polypeptide of ca. 12 kDa but its function is not known (Pierik et al., 1992). There is no sequence similarity between the Dsr subunits and any of the assimilatory SiR forms.

3. Phylogenetic Analysis

Since they all are readily alignable, we compared sulfite reductase from plants and algae together with the siroheme binding SiR subunits from bacteria and fungi with several nitrite reductase (NiR) sequences from different sources (Fig. 8). The analysis shows that the two enzymes (SiR and NiR) arose from an ancient gene-duplication in Eubacteria, before the primary endosymbiosis that gave rise to plastids. A relationship to cyanobacteria in both the SiR and NiR branches supports an endosymbiont origin for both SiR and NiR in all plastids; both primary and secondary, red and green. The non-photosynthetic stramenopile, *Phytophthora*, however, is more related to the fungi SiR and likely encodes a protein of host origin. Like the fungi *Phytophthora* possesses only homologues of SiR, NiR in this lineage being of an independent, unrelated origin. The SiR is closely related to proteobacteria but shows no supported relationship to α-proteobacteria and so a mitochondrial origin cannot be assumed. The photosynthetic bacteria encode a SiR protein that is only marginally related to other photosynthetic organisms and may represent an ancient form of assimilatory SiR.

F. Serine Acetyltransferase

1. Biochemical Properties

Serine acetyltransferase (SAT, EC 2.3.1.30) catalyzes the acetylation of L-serine with acetyl-CoA to form *O*-acetylserine in the first step of L-cysteine synthesis. SAT forms a complex called cysteine synthase with the subsequent enzyme, OASTL, which is essential for its enzymatic activity

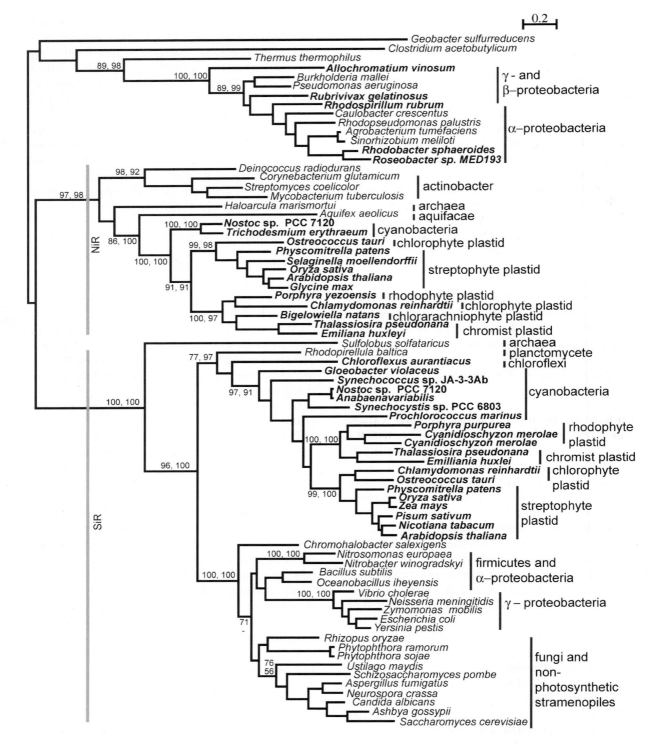

Fig. 8. Protein maximum likelihood phylogeny for sulfite and nitrite reductases. The tree was inferred using PhyML with nine categories of rates. Numbers at nodes indicate bootstrap support over 60 from (top to bottom or left to right) maximum likelihood, as determined by PhyML, and weighted neighbor joining. Analysis used 379 characters, the gamma-shaped parameter 1.58 and proportion of invariable sites 0.006. Photosynthetic organisms are typed in bold.

(Kredich, 1996; Bogdanova and Hell, 1997; Droux et al., 1998). The complex, however, does not facilitate the reaction by substrate channeling, because the OASTL in complex is not functional and the synthesis of cysteine is accomplished by a free OASTL (Droux et al., 1998; Wirtz et al., 2001). SAT is a hexamer of 29 kDa subunits. It is a member of the hexapeptide acyltransferase family of enzymes, folding in a characteristic left-handed parallel β-helix domain (Olsen et al., 2004). SAT is an important regulatory step in sulfate assimilation as it undergoes a feedback regulation by cysteine. Its product, OAS, directly, or after spontaneous isomerization to N-acetylserine, is a potent transcriptional regulator of sulfate assimilation genes (Kredich, 1996; Leustek et al., 2000). Since in plants cysteine synthesis takes place in all three compartments capable of protein synthesis isoforms of SAT and OASTL are found in the plastids, cytosol, and mitochondria. Consequently, SAT is encoded in multigene families of 3–5 members (Kawashima et al., 2005).

However, not all organisms assimilating sulfate possess a functional SAT. The amino acid acceptor of sulfide in budding yeast is O-acetylhomoserine, which is synthesized by homoserine transacetylase (Met2; Born et al., 2000). Although catalyzing a very similar reaction, there is no sequence similarity between Met2 and SAT. Surprisingly however, despite there being no SAT homologue encoded in the *S. cerevisiae* genome, SAT activity was clearly detected *in vitro* (Takagi et al., 2003). Since expression of *E. coli* SAT, but not OASTL, was capable of reverting cysteine auxotrophs it seems that the physiological role of this activity (if it is not an artifact) is other than in cysteine synthesis (Takagi et al., 2003). SAT is, however, found in *Schizosaccharomyces pombe*, which thus seems to possess both pathways of cysteine synthesis, and in the parasites *Entamoeba histolytica* and *E. dispar*. The biochemical and molecular properties of SAT and OASTL are described in detail in Chapter 4.

2. Phylogenetic Analysis

All plastid containing organisms are contained in a well-supported clade, distinct from the cyanobacteria (Fig. 9). SAT in these organisms can be confirmed to be of host, rather than cyanobacterial endosymbiont origin. This clade, however, also contains proteobacteria. The α-proteobacteria do not branch at the base of the eukaryotes, most of the internal branches of this clade are not well supported and the monophyly of streptophytes is not recovered. A mitochondrial origin, and subsequent re-targeting to plastids in lineages that obtained plastids, cannot be, therefore, ruled out for eukaryotes.

Outside of the main eukaryotic clade, the non-photosynthetic protists *Entamoeba* and *Trypanosoma* show a specific relationship to δ-proteobacteria. Neither group is well sampled, however, and this relationship may not hold with more taxa.

G.   O-Acetylserine (thiol)lyase

1. Biochemical Properties

O-acetylserine (thiol)lyase (EC 4.2.99.8) catalyzes the formation of cysteine from OAS and sulfide. The enzyme belongs to the family of β-replacement enzymes utilizing pyridoxal-5′-phosphate as cofactor. OASTL is a homodimer of 35 kDa subunits found inactive in the cysteine synthase complex modulating the activity of SAT or, in the active form, as a free enzyme (Droux et al., 1998; Wirtz et al., 2001). In bacteria, OASTL is encoded by two genes *cysK* and *cysM*. Whereas CysK is specific for *O*-acetylserine and is predominantly expressed under aerobic growth, CysM has a broader substrate specificity comprising many different β-substituted amino acids (Maier, 2003) and is highly expressed at anaerobic conditions and in presence of thiosulfate (Kredich, 1996). In plants and algae isoforms of OASTL, similar to SAT, are localized in the cytosol, mitochondria, and plastids in all organs (see Chapter 4). Accordingly, at least three genes are present in genomes of higher plants (Jost et al., 2000). Since sulfate assimilation is specific to plastids, the mitochondrial and cytosolic SAT and OASTL may have alternative functions. Indeed, the primary function for the mitochondrial OASTL was proposed to be detoxification of cyanide, acting as a β-cyanoalanine synthase (Hatzfeld et al., 2000b; Maruyama et al., 2000; Warrilow and Hawkesford, 2000).

In yeast, Metazoa, and protozoan parasites, cysteine is synthesized by transsulfuration from homocysteine in two steps catalyzed by cystathionine β-synthase (CBS) and cystathionine γ-lyase

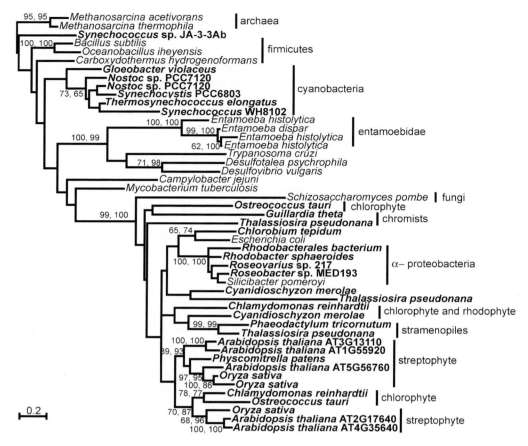

Fig. 9. Protein maximum likelihood phylogeny for serine acetyltransferases. The tree was inferred using PhyML with nine categories of rates. Numbers at nodes indicate bootstrap support over 60 from (top to bottom or left to right) maximum likelihood, as determined by PhyML, and weighted neighbor joining. Analysis used 157 characters, the gamma-shaped parameter 1.125 and proportion of invariable sites 0.055. Photosynthetic organisms are typed in bold.

(Ono et al., 1999; Nozaki et al., 2005). Homocysteine can be produced from methionine, but in yeast it is the primary product of sulfate assimilation formed by O-acetylhomoserine thiollyase (OAHSTL; Ono et al., 1999). OAHSTL is very similar to OASTL and is also able to catalyze cysteine synthesis from OAS and sulfide. But since the SAT homologues in yeast are unable to synthesize OAS, the OAHSTL does not play any direct role in Cys synthesis (Takagi et al., 2003). CBS is therefore found in most eukaryotes, including all fungi, and all metazoa except nematodes. It is however, lost in the Viridaeplantae lineage. Surprisingly, rather than CBS, OASTL were found in the nematodes *Caenorhabditis elegans* and *C. briggsae* and also in *Dictyostelium discoideum*. No other genes of sulfate assimila-

tion are present in these genomes. On the other hand, genes encoding ATPS, SAT and OASTL were found in the parasite *Entamoeba histolytica*, which is not capable of sulfate assimilation since it is lacking the reducing enzymes (P)APR and SiR. Since it is impossible to distinguish between CBS and OASTL at the primary sequence level, both were included in the phylogenetic analysis. A more detailed description of cysteine synthesis and its regulation is given in Chapter 4.

2. Phylogenetic Analysis

OASTL proteins of Viridaeplantae and rhodophytes are all related (Fig. 10). However they are part of a large and unsupported clade that also contains many prokaryotes. A cyanobacterial

Fig. 10. Protein maximum likelihood phylogeny for O-acetylserine (thiol)lyases and cystathionine β-synthases. The tree was inferred using PhyML with nine categories of rates. Numbers at nodes indicate bootstrap support over 60 from (top to bottom or left to right) maximum likelihood, as determined by PhyML, and weighted neighbor joining. Analysis used 359 characters, the gamma-shaped parameter 1.088 and proportion of invariable sites 0.006. Photosynthetic organisms are typed in bold.

origin seems most likely since the bulk of non-photosynthetic eukaryotes are in other, better supported, clades. Within this clade, quite separate to other Metazoa, are the OASTL homologues encoded by nematodes. These are within the plastid-lineage and, although their position in unsupported, an LGT origin cannot be ruled out. The only other non-photosynthetic eukaryotes are Phytophtora, these proteins do not show any supported relationship to their photosynthetic stramenopile relatives and so the origin of this isoform is also unclear.

There is an entirely separate and highly supported clade consisting of both auto- and heterotrophic chromalveolates as well as the tiny green alga *Ostreococcus* and the chlorarchniophyte *B. natans*. There is no experimental information for any of these isoforms and so they may have a novel function. Interestingly two, entirely unrelated members of this clade, the tertiary plastid-containing dinoflagellate *K. micrum* and also B. natans have a glutaredoxin moiety fused to the C terminus of the molecule.

The eukaryotic clade of CBS is well supported although its position within the tree is not, and so the root of the divergence cannot be known. It is however present in very basal eukaryotes such as the Lobosea and Mycetazoa.

The final well-supported clade consists primarily of stramenopiles and fungi, but also α-proteobacteria and a few cyanobacteria, the latter of which are quite separate from the larger cluster of their relatives. Again there is no specific relationship between the eukaryotes and the α-proteobacteria but, given that there is some evidence that this isoform is mitochondrial in fungi, an endosymbiont origin remains a possibility.

IV. Protein Complexes and Fusions

Sulfate assimilation seems to be especially prone to tinkering with domain and protein fusions. Its central role in core metabolism across the kingdoms of life is also demonstrated by ability of structurally unrelated proteins to catalyze the same reaction, indicating parallel evolution in different lineages. This was clearly demonstrated during the discussion of individual sulfate assimilation proteins (see section III.B–D). Altogether at least three unrelated proteins are capable of the adenylation of

sulfate, three major APS/PAPS reducing enzymes exist, and five completely different enzymes reduce sulfite to sulfide. On top of this, many subtle variations in domain structure and fusions result in a great variety of sulfur assimilation genes and enzymes. The greatest part of this subtle variation was unraveled only very recently due to the progress in algal genomics. It was long known that ATPS is often fused with APS kinase, as in animal PAPS synthetase or in the rhizobial NodP/Q system (see section III.B). The finding of ATPS fusion with inorganic pyrophosphatase in several algae is, however, a very interesting variation in attempt to increase efficiency of this enzyme. A very special case is the fusion between ATPS and APR in the dinoflagellate *Heterocapsa triquetra*. This protein likely catalyzes the reduction of sulfate to sulfite, since no other paralogues of ATPS or APR have been detected to date. This protein fusion supports the idea that in other organisms, such as plants, the sulfate assimilation enzymes form a multienzyme complex. Clearly, cysteine synthase is such a complex (see Chapter 4), but it is possible that also ATPS, APR, and SiR form a complex, preferably associated with thylakoid membranes. Association of sulfate reducing enzymes, perhaps including also cysteine synthase, would increase the efficiency of the individual reactions by substrate channeling. This is particularly important for ATPS which is very inefficient in the forward reaction. In addition the channeling would prevent the escape of the toxic intermediates sulfite and/or sulfide as well as facilitating the transfer of reducing equivalents from photosynthesis. Indeed, immunogold localization of APR in several species of *Flaveria* plants showed a frequent association of the label with the thylakoids (Koprivova et al., 2001). On the other hand, by Western blot analysis APR was shown to be present in stroma fraction of pea chloroplasts (Prior et al., 1999). The sulfate assimilation metabolon, however, remains to be demonstrated.

The APR found in the green and red algae (except *C. merolae*) and plants is clearly a fusion between an APS/PAPS reductase and a thioredoxin. The APR from *Euglena*, which is otherwise similar, and specifically related to the APR from chlorophytes, does not contain the thioredoxin extension. Although the sequence of the C-terminal domain is more related to thioredoxin, the domain clearly functions as a glutaredoxin which is compatible with GSH being the most probable *in vivo* electron donor (Bick et al., 1998; Kopriva and Koprivova, 2004). The fusion between ancestral APR and thioredoxin leading to the 'plant-type' APR thus brought the enzyme and its cofactor together on one polypeptide. The event most likely occurred before the divergence of red and green algae, with subsequent loss of the domain in the lineage leading to *C. merolae* (although another Cyanidale, *G. sulfuraria*, retained the fusion), and independently after the secondary endosymbiosis event gave rise to *Euglena*. Another piece into the fusion mosaic is the novel form of OASTL from *K. micrum* and *B. natans*, which are fused at their C-terminus to glutaredoxin (Patron et al., 2006). Since cysteine synthesis does not require electron transport, the function of this fusion is unclear. It might be however be involved in redox regulation of the enzyme (Patron et al., 2006).

V. Conclusions

A. Origin of Plant Sulfate Assimilation

With a single exception in *Euglena*, sulfate assimilation is localized in plastids. As with other plastidic pathways, e.g., the Calvin cycle (Martin and Schnarrenberger, 1997), or heme biosynthesis (Obornik and Green, 2005), cyanobacterial origin of the genes involved can be anticipated. However, surprisingly, most of the genes of sulfate assimilation cannot be related to cyanobacteria. The only genes for which symbiotic origin can be proposed with confidence are sulfite reductase and OASTL. Serine acetyltransferase, on the other hand, seems to be derived from the host. For other genes the evolutionary history cannot be reliably reconstructed. Great difficulty by interpretation of the data is caused by a frequent lack of monophyly of well defined taxa, e.g., cyanobacteria in ATPS and PAPS reductase trees. Plant APS reductase is most closely related to several γ-proteobacteria but also to a cyanobacterium *Plectonema*. Existence of two distinct isoforms of APR in several lower plants shows that at some point in their evolution plants possessed both genes, but it is not possible to assign one of them to the host and the other to symbionts with confidence. Interestingly, ATPS is probably the first plant gene identified which is more related to

Metazoa than to Chlorophytes. The origin of plant sulfate assimilation thus probably will remain an unsolved question.

B. Sulfate Assimilation in Secondary/Tertiary Symbionts

Inclusion of the secondary and tertiary symbionts in the analysis revealed a great number of unique genes and gene variants. Among them the fusions of APS kinase, ATPS, and PPase and of OASTL with glutaredoxin, as well as the new form of APS reductase are of great interest and potential use for improving sulfur use efficiency of crop plants. As expected, the algae possessed genes from different origins, reflecting their evolutionary history. Often both gene copies from the host and the symbionts are retained and possibly the encoded proteins are differentially localized. Our understanding of sulfate assimilation in these species, although improved by this phylogenetic analysis, still requires biochemical studies of the organisms and the respective enzymes.

Acknowledgements

Research in SK's laboratory at JIC is supported by the Biotechnology and Biological Sciences Research Council (BBSRC). Funding from German Research Council DFG, Genome Canada/Genome Atlantic, and Swiss National Science Foundation is also acknowledged.

References

Abola AP, Willits MG, Wang RC and Long SR (1999) Reduction of adenosine-5'-phosphosulfate instead of 3'-phosphoadenosine-5'-phosphosulfate in cysteine biosynthesis by *Rhizobium meliloti* and other members of the family *Rhizobiaceae*. J Bacteriol 181: 5280–5287

Archibald JM and Keeling PJ (2002) Recycled plastids: a green movement in eukaryotic evolution. Trends Genet 18: 577–584

Archibald JM, Rogers MB, Toop M, Ishida K and Keeling PJ (2003) Lateral gene transfer and the evolution of plastid-targeted proteins in the secondary plastid-containing alga *Bigelowiella natans*. Proc Natl Acad Sci USA 100: 7678–7683

Armbrust EV et al. (2004) The genome of the diatom *Thalassiosira pseudonana*: ecology, evolution, and metabolism. Science 306: 79–86

54   Stanislav Kopriva et al.

Bakker-Grunwald T and Geilhorn B (1992) Sulfate metabolism in *Entamoeba histolytica*. Mol Biochem Parasitol 53: 71–78

Berndt C, Lillig CH, Wollenberg M, Bill E, Mansilla MC, de Mendoza D, Seidler A and Schwenn JD (2004) Characterization and reconstitution of a 4Fe–4S adenylyl sulfate/phosphoadenylyl sulfate reductase from *Bacillus subtilis*. J Biol Chem 279: 7850–7855

Bhattacharya D, Yoon HS and Hackett JD (2004) Photosynthetic eukaryotes unite: endosymbiosis connects the dots. Bioessays 26: 50–60

Bick JA, Aslund F, Chen Y and Leustek T (1998) Glutaredoxin function for the carboxyl-terminal domain of the 'plant-type' 5'-adenylylsulfate reductase. Proc Natl Acad Sci USA 95: 8404–8409

Bick JA, Dennis JJ, Zylstra GJ, Nowack J and Leustek T (2000) Identification of a new class of 5'-adenylylsulfate (APS) reductases from sulfate-assimilating bacteria. J Bacteriol 182: 135–142

Bodyl A (2004) Evolutionary origin of a preprotein translocase in the periplastid membrane of complex plastids: a hypothesis. Plant Biol 6: 513–518

Bogdanova N and Hell R (1997) Cysteine synthesis in plants: protein–protein interactions of serine acetyltransferase from *Arabidopsis thaliana*. Plant J 11: 251–262

Bork C, Schwenn JD and Hell R (1998) Isolation and characterization of a gene for assimilatory sulfite reductase from *Arabidopsis thaliana*. Gene 212: 147–153

Born TL, Franklin M and Blanchard JS (2000) Enzyme-catalyzed acylation of homoserine: mechanistic characterization of the *Haemophilus influenzae met2*-encoded homoserine transacetylase. Biochemistry 39: 8556–8564

Brunold C (1990) Reduction of sulfate to sulfide. In: Rennenberg H, Brunold C, De Kok LJ and Stulen I (eds) Sulphur nutrition and sulphur assimilation in higher plants, pp 13–31. SPB Academic Publishing, The Hague, The Netherlands

Brunold C and Schiff JA (1976) Studies of sulfate utilization by algae.15. Enzymes of assimilatory sulfate reduction in *Euglena* and their cellular localization. Plant Physiol 57: 430–436

Brunold C and Suter M (1989) Localization of enzymes of assimilatory sulfate reduction in pea roots. Planta 179: 228–234

Buchner P, Takahashi H and Hawkesford MJ (2004) Plant sulphate transporters: co-ordination of uptake, intracellular and long-distance transport. J Exp Bot 55: 1765–1773

Carroll KS, Gao H, Chen H, Leary JA and Bertozzi CR (2005a) Investigation of the iron–sulfur cluster in *Mycobacterium tuberculosis* APS reductase: implications for substrate binding and catalysis. Biochemistry 44: 14647–14657

Carroll KS, Gao H, Chen H, Stout CD, Leary JA and Bertozzi CR (2005b) A conserved mechanism for sulfonucleotide reduction. PLoS Biol 3: e250

Cavalier-Smith T (2002) Chloroplast evolution: secondary symbiogenesis and multiple losses. Curr Biol 12: R62–R64

Cavalier-Smith T (2003) Genomic reduction and evolution of novel genetic membranes and protein-targeting machinery in eukaryote–eukaryote chimeras (meta-algae). Philos Trans R Soc Lond B Biol Sci 358: 109–134

Cherest H, Davidian JC, Thomas D, Benes V, Ansorge W and Surdin-Kerjan Y (1997) Molecular characterization of two high affinity sulfate transporters in *Saccharomyces cerevisiae*. Genetics 145: 627–635

Chesnick JM, Kooistra W, Wellbrock U and Medlin LK (1997) Ribosomal RNA analysis indicates a benthic pennate diatom ancestry for the endosymbionts of the dinoflagellates *Peridinium foliaceum* and *Peridinium balticum* (Pyrrhophyta). J Euk Microbiol 44: 314–320

Chu KH, Qi J, Yu ZG and Anh V (2004) Origin and phylogeny of chloroplasts revealed by a simple correlation analysis of complete genomes. Mol Biol Evol 21: 200–206

Coleman ML, Sullivan MB, Martiny AC, Steglich C, Barry K, DeLong EF and Chisholm SW (2006) Genomic islands and the ecology and evolution of *Prochlorococcus*. Science 311: 1768–1770

Crane BR, Siegel LM and Getzoff ED (1995) Sulfite reductase structure at 1.6 A: evolution and catalysis for reduction of inorganic anions. Science 270: 59–67

Dahl C, Kredich NM, Deutzmann R and Trüper HG (1993) Dissimilatory sulphite reductase from *Archaeoglobus fulgidus*: physico-chemical properties of the enzyme and cloning, sequencing and analysis of the reductase genes. J Gen Microbiol 139: 1817–1828

Delsuc F, Brinkmann H and Philippe H (2005) Phylogenomics and the reconstruction of the tree of life. Nat Rev Genet 6: 361–375

Droux M, Ruffet ML, Douce R and Job D (1998) Interactions between serine acetyltransferase and O-acetylserine (thiol) lyase in higher plants – structural and kinetic properties of the free and bound enzymes. Eur J Biochem 255: 235–245

Fast NM, Kissinger JC, Roos DS and Keeling PJ (2001) Nuclear-encoded, plastid-targeted genes suggest a single common origin for apicomplexan and dinoflagellate plastids. Mol Biol Evol 18: 418–426

Fleischmann RD et al. (1995) Whole-genome random sequencing and assembly of *Haemophilus influenzae Rd*. Science 269: 496–512

Folch-Mallol JL, Marroqui S, Sousa C, Manyani H, Lopez-Lara IM, van der Drift KM, Haverkamp J, Quinto C, Gil-Serrano A, Thomas-Oates J, Spaink HP and Megias M (1996) Characterization of *Rhizobium tropici* CIAT899 nodulation factors: the role of nodH and nodPQ genes in their sulfation. Mol Plant Microbe Interact 9: 151–163

Fritz G, Roth A, Schiffer A, Buchert T, Bourenkov G, Bartunik HD, Huber H, Stetter KO, Kroneck PM and Ermler U (2002) Structure of adenylylsulfate reductase from the hyperthermophilic *Archaeoglobus fulgidus* at 1.6-A resolution. Proc Natl Acad Sci USA 99: 1836–1841

Fulton JD and Grant PT (1956) The sulphur requirements of the erythrocytic form of *Plasmodium knowlesi*. Biochem J 63: 274–282

Gao Y, Schofield OM and Leustek T (2000) Characterization of sulfate assimilation in marine algae focusing on the enzyme 5′-adenylylsulfate reductase. Plant Physiol 123: 1087–1096

Goff SA et al. (2002) A draft sequence of the rice genome (*Oryza sativa* L. ssp. *japonica*). Science 296: 92–100

Grossman AR (2005) Paths toward algal genomics. Plant Physiol 137: 410–427

Gutierrez-Marcos JF, Roberts MA, Campbell EI and Wray JL (1996) Three members of a novel small gene-family from *Arabidopsis thaliana* able to complement functionally an *Escherichia coli* mutant defective in PAPS reductase activity encode proteins with a thioredoxin-like domain and"APS reductase" activity. Proc Natl Acad Sci USA 93: 13377–13382

Hansen TA (1994) Metabolism of sulfate-reducing prokaryotes. Antonie Van Leeuwenhoek 66: 165–185

Harper JT and Keeling PJ (2003) Nucleus-encoded, plastid-targeted glyceraldehyde-3-phosphate dehydrogenase (GAPDH) indicates a single origin for chromalveolate plastids. Mol Biol Evol 20: 1730–1735

Hatzfeld Y, Lee S, Lee M, Leustek T and Saito K (2000a) Functional characterization of a gene encoding a fourth ATP sulfurylase isoform from *Arabidopsis thaliana*. Gene 248: 51–58

Hatzfeld Y, Maruyama A, Schmidt A, Noji M, Ishizawa K and Saito K (2000b) beta-cyanoalanine synthase is a mitochondrial cysteine synthase-like protein in spinach and Arabidopsis. Plant Physiol 123: 1163–1171

Hawkesford MJ and Wray JL (2000) Molecular genetics of sulphate assimilation. Adv Bot Res 33: 159–223

Hawkesford MJ, Davidian J-C and Grignon C (1993) Sulfate proton cotransport in plasma vesicles isolated from roots of *Brassica napus* L. Increased transport in membranes isolated from sulphur starved plants. Planta 190: 297–304

Honke K and Taniguchi N (2002) Sulfotransferases and sulfated oligosaccharides. Med Res Rev 22: 637–654

Imhoff JF (1992) Taxonomy, phylogeny and general ecology of anoxygenic phototrophic bacteria. In: Carr NG and Mann NH (eds) Biotechnology handbook: photosynthetic prokaryotes, pp 53–92. Plenum Press, London, New York

Inagaki Y, Doolittle WF, Baldauf SL and Roger AJ (2002) Lateral transfer of an EF-1 alpha gene: Origin and evolution of the large subunit of ATP sulfurylase in eubacteria. Curr Biol 12: 772–776

Jost R, Berkowitz O, Wirtz M, Hopkins L, Hawkesford MJ and Hell R (2000) Genomic and functional characterization of the oas gene family encoding O-acetylserine (thiol) lyases, enzymes catalysing the final step in cysteine biosynthesis in *Arabidopsis thaliana*. Gene 253: 237–247

Kanno N, Nagahaisa E, Sato M and Sato Y (1996) Adenosine 5′-phosphosulfate sulfotransferase from the marine macroalga *Porphyra yezoensis* Ueda (Rhodophyta): stabilization, purification, and properties. Planta 198: 440–446

Kappler U and Dahl C (2001) Enzymology and molecular biology of prokaryotic sulfite oxidation. FEMS Microbiol Lett 203: 1–9

Kawashima CG, Berkowitz O, Hell R, Noji M and Saito K (2005) Characterization and expression analysis of a serine acetyltransferase gene family involved in a key step of the sulfur assimilation pathway in Arabidopsis. Plant Physiol 137: 220–230

Kertesz MA (2001) Bacterial transporters for sulfate and organosulfur compounds. Res Microbiol 152: 279–290

Kim SK, Rahman A, Bick JA, Conover RC, Johnson MK, Mason JT, Hirasawa M, Leustek T and Knaff DB (2004) Properties of the cysteine residues and iron–sulfur cluster of the assimilatory 5′-adenylyl sulfate reductase from *Pseudomonas aeruginosa*. Biochemistry 43: 13478–13486

Klonus D, Hofgen R, Willmitzer L and Riesmeier JW (1994) Isolation and characterization of two cDNA clones encoding ATP-sulfurylases from potato by complementation of a yeast mutant. Plant J 6: 105–112

Kobayashi K and Yoshimoto A (1982) Studies on yeast sulfite reductase. IV. Structure and steady-state kinetics. Biochim Biophys Acta 705: 348–356

Kopriva S (2006) Regulation of sulfate assimilation in Arabidopsis and beyond. Ann Bot 97: 479–495

Kopriva S and Koprivova A (2004) Plant adenosine 5′-phosphosulphate reductase: the past, the present, and the future. J Exp Bot 55: 1775–1783

Kopriva S and Rennenberg H (2004) Control of sulphate assimilation and glutathione synthesis: interaction with N and C metabolism. J Exp Bot 55: 1831–1842

Kopriva S, Büchert T, Fritz G, Weber M, Suter M, Benda R, Schaller J, Feller U, Schürmann P, Schünemann V, Trautwein AX, Kroneck PMH and Brunold C (2001) Plant adenosine 5′-phosphosulfate reductase is a novel iron–sulfur protein. J Biol Chem 276: 42881–42886

Kopriva S, Büchert T, Fritz G, Suter M, Benda R, Schünemann V, Koprivova A, Schürmann P, Trautwein AX, Kroneck PMH and Brunold C (2002). The presence of an iron–sulfur cluster in adenosine 5′-phosphosulfate reductase separates organisms utilising adenosine 5′-phosphosulfate and phosphoadenosine 5′-phosphosulfate for sulfate assimilation. J Biol Chem 277: 21786–21791

Kopriva S, Fritzemeier K, Wiedemann G and Reski R (2007) The putative moss 3′phosphoadenosine 5′phosphosulfate reductase is a novel form of adenosine 5′phosphosulfate reductase without iron sulfur cluster. J Biol Chem 282: 22930–22938

Koprivova A, Melzer M, von Ballmoos P, Mandel T, Brunold C and Kopriva S (2001) Assimilatory sulfate reduction in C_3, C_3–C_4, and C_4 species of *Flaveria*. Plant Physiol 127: 543–550

Koprivova A, Meyer A, Schween G, Herschbach C, Reski R and Kopriva S (2002) Functional knockout of the adenosine 5'phosphosulfate reductase gene in *Physcomitrella patens* revives an old route of sulfate assimilation. J Biol Chem 277: 32195–32201

Kredich NM (1971) Regulation of L-cysteine biosynthesis in *Salmonella typhimurium*. J Biol Chem 246: 3474–3484

Kredich NM (1996) Biosynthesis of cysteine. In: Neidhart FC et al. (eds) *Escherichia coli* and *Salmonella typhimurium*. Cellular and molecular biology, pp 514–527. ASM Press, Washington, DC

Krueger RJ and Siegel LM (1982) Spinach siroheme enzymes: isolation and characterization of ferredoxin-sulfite reductase and comparison of properties with ferredoxin-nitrite reductase. Biochemistry 21: 2892–2904

Laudenbach DE and Grossman AR (1991) Characterization and mutagenesis of sulfur-regulated genes in a cyanobacterium: evidence for function in sulfate transport. J Bacteriol 173: 2739–2750

Leipe DD, Koonin EV and Aravind L (2003) Evolution and classification of P-loop kinases and related proteins. J Mol Biol 333: 781–815

Leustek T, Martin MN, Bick JA and Davies JP (2000) Pathways and regulation of sulfur metabolism revealed through molecular and genetic studies. Annu Rev Plant Physiol Plant Mol Biol 51: 141–165

Leyh TS, Taylor JC and Markham GD (1988) The sulfate activation locus of *Escherichia coli* K12: cloning, genetic, and enzymatic characterization. J Biol Chem 263: 2409–2416

Li JY and Schiff JA (1991) Purification and properties of adenosine 5'-phosphosulphate sulphotransferase from *Euglena*. Biochem J 274: 355–360

Lillig CH, Prior A, Schwenn JD, Åslund F, Ritz D, Vlamis-Gardikas A and Holmgren A (1999) New thioredoxins and glutaredoxins as electron donors of 3'-phosphoadenylylsulfate reductase. J Biol Chem 274: 7695–7698

Lindell D, Sullivan MB, Johnson ZI, Tolonen AC, Rohwer F and Chisholm SW (2004) Transfer of photosynthesis genes to and from *Prochlorococcus* viruses. Proc Natl Acad Sci USA 101: 11013–11018

Lunn J, Droux M, Martin J and Douce R (1990) Localization of ATP sulfurylase and O-acetylserine (thiol)lyase in spinach leaves. Plant Physiol 94: 1345–1352

MacRae IJ, Rose AB and Segel IH (1998) Adenosine 5'-phosphosulfate kinase from *Penicillium chrysogenum* site-directed mutagenesis at putative phosphoryl-accepting and ATP P-loop residues. J Biol Chem 273: 28583–28589

MacRae IJ, Segel IH and Fisher AJ (2000) Crystal structure of adenosine 5'-phosphosulfate kinase from *Penicillium chrysogenum*. Biochemistry 39: 1613–1621

MacRae IJ, Segel IH and Fisher AJ (2001) Crystal structure of ATP sulfurylase from *Penicillium chrysogenum*: insights into the allosteric regulation of sulfate assimilation. Biochemistry 40: 6795–6804

Maier TH (2003) Semisynthetic production of unnatural L-alpha-amino acids by metabolic engineering of the cysteine-biosynthetic pathway. Nat Biotechnol 21: 422–427

Markovich D (2001) Physiological roles and regulation of mammalian sulfate transporters. Physiol Rev 81: 1499–1533

Martin W and Schnarrenberger C (1997) The evolution of the Calvin cycle from prokaryotic to eukaryotic chromosomes: a case study of functional redundancy in ancient pathways through endosymbiosis. Curr Genet 32: 1–18

Martin W, Rujan T, Richly E, Hansen A, Cornelsen S, Lins T, Leister D, Stoebe B, Hasegawa M and Penny D (2002) Evolutionary analysis Arabidopsis, cyanobacterial, and chloroplast genomes reveals plastid phylogeny and thousands of cyanobacterial genes in the nucleus. Proc Natl Acad Sci USA 99: 12246–12251

Maruyama A, Ishizawa K and Takagi T (2000) Purification and characterization of beta-cyanoalanine synthase and cysteine synthases from potato tubers: are beta-cyanoalanine synthase and mitochondrial cysteine synthase same enzyme? Plant Cell Physiol 41: 200–208

Matsuzaki M, et al. (2004) Genome sequence of the ultrasmall unicellular red alga *Cyanidioschyzon merolae* 10D. Nature 428: 653–657

McFadden GI (2001) Primary and secondary endosymbiosis and the origin of plastids. J Phycol 37: 951–959

Mougous JD, Senaratne RH, Petzold CJ, Jain M, Lee DH, Schelle MW, Leavell MD, Cox JS, Leary JA, Riley LW and Bertozzi CR (2006) A sulfated metabolite produced by stf3 negatively regulates the virulence of *Mycobacterium tuberculosis*. Proc Natl Acad Sci USA 103: 4258–4263

Murillo M and Leustek T (1995) Adenosine-5'-triphosphate-sulfurylase from *Arabidopsis thaliana* and *Escherichia coli* are functionally equivalent but structurally and kinetically divergent: nucleotide sequence of two adenosine-5'-triphosphate-sulfurylase cDNAs from *Arabidopsis thaliana* and analysis of a recombinant enzyme. Arch Biochem Biophys 323: 195–204

Nakayama M, Akashi T and Hase T (2000) Plant sulfite reductase: molecular structure, catalytic function and interaction with ferredoxin. J Inorg Biochem 82: 27–32

Nelissen B, Van de Peer Y, Wilmotte A, and De Wachter R (1995) An early origin of plastids within the cyanobacterial divergence is suggested by evolutionary trees based on complete 16S rRNA sequences. Mol Biol Evol 12: 1166–1173

Nozaki T, Ali V and Tokoro M (2005) Sulfur-containing amino acid metabolism in parasitic protozoa. Adv Parasitol 60: 1–99

Obornik M and Green BR (2005) Mosaic origin of the heme biosynthesis pathway in photosynthetic eukaryotes. Mol Biol Evol 22: 2343–2353

Olsen LR, Huang B, Vetting MW and Roderick SL (2004) Structure of serine acetyltransferase in complexes with

CoA and its cysteine feedback inhibitor. Biochemistry 43: 6013–6019

Ono BI, Hazu T, Yoshida S, Kawato T, Shinoda S, Brzvwczy J and Paszewski A (1999) Cysteine biosynthesis in *Saccharomyces cerevisiae*: a new outlook on pathway and regulation. Yeast 15: 1365–1375

Patron NJ, Rogers MB and Keeling PJ (2004) Gene replacement of fructose-1,6-bisphosphate aldolase supports the hypothesis of a single photosynthetic ancestor of chromalveolates. Eukaryot Cell 3: 1169–1175

Patron NJ, Waller RF, Archibald JM and Keeling PJ (2005) Complex protein targeting to dinoflagellate plastids. J Mol Biol 348: 1015–1024

Patron NJ, Waller RF and Keeling PJ (2006) A tertiary plastid uses genes from two endosymbionts. J Mol Biol 357: 1373–1382

Payne SH and Loomis WF (2006) Retention and loss of amino acid biosynthetic pathways based on analysis of whole-genome sequences. Eukaryot Cell 5: 272–276

Perez-Jimenez JR and Kerkhof LJ (2005) Phylogeography of sulfate-reducing bacteria among disturbed sediments, disclosed by analysis of the dissimilatory sulfite reductase genes (dsrAB). Appl Environ Microbiol 71: 1004–1011

Pierik AJ, Duyvis MG, van Helvoort JMLM, Wolbert RBG and Hagen WR (1992) The third subunit of desulfoviridin-type dissimilatory sulfite reductases. Eur J Biochem 205: 111–115

Prior A, Uhrig JF, Heins L, Wiesmann A, Lillig CH, Stoltze C, Soll J and Schwenn JD (1999) Structural and kinetic properties of adenylyl sulfate reductase from *Catharanthus roseus* cell cultures. Biochim Biophys Acta 1430: 25–38

Rausch T and Wachter A (2005) Sulfur metabolism: a versatile platform for launching defence operations. Trends Plant Sci 10: 503–509

Rodriguez-Ezpeleta N, Brinkmann H, Burey SC, Roure B, Burger G, Loffelhardt W, Bohnert HJ, Philippe H and Lang BF (2005) Monophyly of primary photosynthetic eukaryotes: green plants, red algae, and glaucophytes. Curr Biol 15: 1325–1330

Rosenthal E and Leustek T (1995) A multifunctional *Urechis caupo* protein, PAPS synthetase, has both ATP sulfurylase and APS kinase activities. Gene 165: 243–248

Rotte C and Leustek T (2000) Differential subcellular localization and expression of ATP sulfurylase and 5′-adenylylsulfate reductase during ontogenesis of Arabidopsis leaves indicates that cytosolic and plastid forms of ATP sulfurylase may have specialised functions. Plant Physiol 124: 715–724

Saidha T, Na SQ, Li JY and Schiff JA (1988) A sulphate metabolising centre in *Euglena* mitochondria. Biochem J 253: 533–539

Saito K (2004) Sulfur assimilatory metabolism. The long and smelling road. Plant Physiol 136: 2443–2450

Sakharkar KR, Dhar PK and Chow VT (2004) Genome reduction in prokaryotic obligatory intracellular parasites of humans: a comparative analysis. Int J Syst Evol Microbiol 54: 1937–1941

Sánchez Puerta MW, Bachvaroff TV and Delwiche CF (2005) The complete plastid genome sequence of the haptophyte *Emiliania huxleyi*: a comparison to other plastid genomes DNA Res 12: 151–156

Sato N, Nakayama M and Hase T (2001) The 70-kDa major DNA-compacting protein of the chloroplast nucleoid is sulfite reductase. FEBS Lett 487: 347–350

Schmidt A (1972) On the mechanism of photosynthetic sulfate reduction. An APS-sulfotransferase from *Chlorella*. Arch Microbiol 84: 77–86

Schmidt A (1975) Distribution of the APS-sulfotransferase activity among higher plants. Plant Sci Lett 5: 407–415

Schmidt A (1977) Assimilatory sulfate reduction via 3′-phosphoadenosine-5′-phosphosulfate (PAPS) and adenosine-5′-phosphosulfate (APS) in blue-green algae. FEMS Microbiol Lett 1: 137–140

Schmidt A and Jäger K (1992) Open questions about sulfur metabolism in plants. Annu Rev Plant Physiol Plant Mol Biol 43: 325–349

Schmidt A and Trüper HG (1977) Reduction of adenylylsulfate and 3′-phosphoadenylylsulfate in phototrophic bacteria. Experientia 33: 1008–1009

Schwartzbach SD, Osafune T and Loffelhardt W (1998) Protein import into cyanelles and complex chloroplasts. Plant Mol Biol 38: 247–263

Setya A, Murillo M and Leustek T (1996) Sulfate reduction in higher plants: molecular evidence for a novel 5′-adenylsulfate reductase. Proc Natl Acad Sci USA 93: 13383–13388

Sirko A, Hryniewicz M, Hulanicka D and Bock A (1990) Sulfate and thiosulfate transport in *Escherichia coli* K-12: nucleotide sequence and expression of the cysTWAM gene cluster. J Bacteriol 172: 3351–3357

Sperling D, Kappler U, Wynen A, Dahl C and Truper HG (1998) Dissimilatory ATP sulfurylase from the hyperthermophilic sulfate reducer *Archaeoglobus fulgidus* belongs to the group of homo-oligomeric ATP sulfurylases. FEMS Microbiol Lett 162: 257–264

Sulli C, Fang Z, Muchhal U and Schwartzbach SD (1999) Topology of *Euglena* chloroplast protein precursors within endoplasmic reticulum to Golgi to chloroplast transport vesicles. J Biol Chem 274: 457–463

Sullivan MB, Coleman ML, Weigele P, Rohwer F and Chisholm SW (2005) Three prochlorococcus cyanophage genomes: signature features and ecological interpretations. PLoS Biology 3: 790–806

Suter M, von Ballmoos P, Kopriva S, Op den Camp R, Schaller J, Kuhlemeier C, Schürmann P and Brunold C (2000) Adenosine 5′-phosphosulfate sulfotransferase and adenosine 5′-phosphosulfate reductase are identical enzymes. J Biol Chem 275: 930–936

Stafleu FA, Bonner CEB, McVaugh R, Meikle RD, Rollins RC, Ross R and Voss EG (eds) (1972) International Code of Botanical Nomenclature. A. Oosthoek, Utrecht, The Netherlands

Stahl DA, Fishbain S, Klein M, Baker BJ and Wagner M (2002) Origins and diversification of sulfate-respiring microorganisms. Antonie Van Leeuwenhoek 81: 189–195

Stipanuk MH (2004) Sulfur amino acid metabolism: pathways for production and removal of homocysteine and cysteine. Annu Rev Nutr 24: 539–577

Takagi H, Yoshioka K, Awano N, Nakamori S and Ono B (2003) Role of *Saccharomyces cerevisiae* serine O-acetyltransferase in cysteine biosynthesis. FEMS Microbiol Lett 218: 291–297

The Arabidopsis Genome Initiative (2000) Analysis of the genome sequence of the flowering plant *Arabidopsis thaliana*. Nature 408: 796–815

Tsang ML, Goldschmidt EE and Schiff JA (1971) Adenosine-5′-phosphosulfate (APS[35]) as an intermediate in the conversion of adenosine-3′-phosphate-5′-phosphosulfate (PAPS[35]) to acid volatile radioactivity. Plant Physiol 47 Suppl 20

Ullrich TC, Blaesse M and Huber R (2001) Crystal structure of ATP sulfurylase from *Saccharomyces cerevisiae*, a key enzyme in sulfate activation. EMBO J 20: 316–329

Van de Kamp M, Schuurs TA, Vos A, Van der Lende TR, Konings WN and Driessen AJM (2000) Sulfur regulation of the sulfate transporter genes sutA and sutB in *Penicillium chrysogenum*. Appl Environ Microbiol 66: 4536–4538

van Dooren GG, Schwartzbach SD, Osafune T and McFadden GI (2001) Translocation of proteins across the multiple membranes of complex plastids. Biochim Biophys Acta 1541: 34–53

Varin L, Marsolais F, Richard M and Rouleau M (1997) Sulfation and sulfotransferases 6: biochemistry and molecular biology of plant sulfotransferases. FASEB J 11: 517–525

Venkatachalam KV (2003) Human 3′-phosphoadenosine 5′-phosphosulfate (PAPS) synthase: biochemistry, molecular biology and genetic deficiency. IUBMB Life 55: 1–11

Waller RF, Patron NJ and Keeling PJ (2006) Phylogenetic history of plastid-targeted proteins in the peridinin-containing dinoflagellate *Heterocapsa triquetra*. Int J Syst Evol Microbiol 56: 1439–1447

Waller RF, Reed MB, Cowman AF and McFadden GI (2000) Protein trafficking to the plastid of *Plasmodium falciparum* is via the secretory pathway. EMBO J 19: 1794–1802

Warrilow AGS and Hawkesford MJ (2000) Cysteine synthase (O-acetylserine (thiol) lyase) substrate specificities classify the mitochondrial isoform as a cyanoalanine synthase. J Exp Bot 51: 985–993

Waterbury JB (2001) The cyanobacteria—isolation, purification, and identification. In: Dworkin M et al. (eds) The prokaryotes: an evolving electronic resource for the microbiological community, 3rd edition, release 3.7, November 2, 2001, Springer, New York, http://link.springer-ny.com/link/service/books/10125/

Weber M, Suter M, Brunold C and Kopriva S (2000) Sulfate assimilation in higher plants: characterization of a stable intermediate in the adenosine 5′-phosphosulfate reductase reaction. Eur J Biochem 267: 3647–3653

Williams SJ, Senaratne RH, Mougous JD, Riley LW and Bertozzi CR (2002) 5′-Adenosinephosphosulfate lies at a metabolic branch point in mycobacteria. J Biol Chem 277: 32606–32615

Wirtz M, Berkowitz O, Droux M and Hell R (2001) The cysteine synthase complex from plants. Mitochondrial serine acetyltransferase from *Arabidopsis thaliana* carries a bifunctional domain for catalysis and protein–protein interaction. Eur J Biochem 268: 686–693

Wirtz M, Droux M and Hell R (2004) O-acetylserine (thiol) lyase: an enigmatic enzyme of plant cysteine biosynthesis revisited in *Arabidopsis thaliana*. J Exp Bot 55: 1785–1798

Yonekura-Sakakibara K, Ashikari T, Tanaka Y, Kusumi T and Hase T. (1998) Molecular characterization of tobacco sulfite reductase: enzyme purification, gene cloning, and gene expression analysis. J Biochem 124: 615–621

Yoon HS, Hackett JD, Pinto G and Bhattacharya D (2002a) The single, ancient origin of chromist plastids. Proc Natl Acad Sci USA 99: 15507–15512

Yoon HS, Hackett JD and Bhattacharya D (2002b) A single origin of the peridinin- and fucoxanthin-containing plastids in dinoflagellates through tertiary endosymbiosis. Proc Natl Acad Sci USA 99: 11724–11729

Yu J et al. (2002) A draft sequence of the rice genome (*Oryza sativa* L. ssp. *indica*). Science 296: 79–92

Yurkov VV and Beatty JT (1998). Aerobic anoxygenic phototrophic bacteria. Microbiol Molec Biol Rev 62: 695–724

Chapter 4

Metabolism of Cysteine in Plants and Phototrophic Bacteria

Rüdiger Hell* and Markus Wirtz
Department of Molecular Biology of Plants, Heidelberg Institute of Plant Sciences, University of Heidelberg, Im Neuenheimer Feld 360, D-69120 Heidelberg, Germany

Summary

The sulfur amino acids cysteine and methionine function in many basic and essential processes of life. For cysteine this includes structural, catalytic, regulatory and metabolic functions. The special redox chemistry of sulfur and the thiol group in particular proved to be a versatile tool during evolution, not the least in electron transport processes in association with iron. Plants are primary producers and carry out assimilatory sulfate reduction to first synthesize cysteine that subsequently forms the backbone for methionine formation. This reaction sequence seems to be conserved in all phototrophic organisms. The position of cysteine biosynthesis between assimilation of inorganic sulfate and metabolization of organic

---

*Corresponding author, Phone: +49 6221 54 6284, Fax: +49 6221 54 5859, E-mail:rhell@hip.uni-hd.de

Rüdiger Hell et al. (eds.), Sulfur Metabolism in Phototrophic Organisms, 59–91.
© 2008 *Springer.*

sulfide makes it a prime target for coordination of both complex processes. It is thus a mediator between supply and demand in sulfur metabolism of a cell.

Much attention has been paid to cysteine biosynthesis in plants, while less is known about the pathway in algae, cyanobacteria and purple bacteria. Recent evidence indicates that the two enzymes of cysteine synthesis, serine acetyltransferase and O-acetylserine-(thiol)-lyase are highly conserved between these groups and, at least in plants, form a reversible protein complex. This so-called cysteine synthase complex has been suggested to act as sensor for sulfide in cells and to be part of a regulatory loop that maintains cysteine homeostasis between sulfate reduction and cysteine consumption. Kinetic studies of the properties of the enzymes together with structural modelling of the proteins in the cysteine synthase complex as well studies using transgenic plants strongly support this unique regulatory system in plants.

The degradation of cysteine is still an under-investigated subject. Possible alternative routes of thiol transfer and sulfide release are compiled here and discussed with respect to their putative functions in S-transfer reactions, cysteine degradation, detoxification reactions and iron-sulfur cluster biosynthesis.

I. Introduction

Cysteine biosynthesis in phototrophic organisms represents the process of integration of reduced sulfur produced by assimilatory sulfate reduction into the organic form of an amino acid. In contrast, some organisms including mammals and some fungi, first assimilate reduced sulfur into methionine and then synthesize cysteine via trans-sulfurylation (Hell, 1997; Wirtz and Droux, 2005; see chapter by Höfgen and Hesse, this book). Thus, primary producers in general first add reduced sulfur to activated serine, i.e. O-acetylser-ine, as acceptor. In a second step sulfide replaces acetate and cysteine is released. From cysteine all downstream products including methionine, cofactors (coenzyme A, biotin, thiamine, lipoic acid), iron-sulfur clusters and many other compounds are produced (Hell, 1997; Beinert, 2000). The importance of the fixation of reduced sulfur in phototrophic organisms is therefore comparable to the assimilation of reduced nitrogen by the glutamine synthetase/glutamate-oxo-glutar-ate-aminotransferase system. In plants, algae and most bacteria, as far as the latter have been investigated in this respect, the synthesis of cysteine occurs in a two-step reaction sequence catalyzed by two enzymes: Serine acetyltransferase (SAT; also termed serine transacetylase; EC 2.3.1.30) and O-acetylserine-(thiol)-lyase (OAS-TL; EC 2.5.1.47; common synonyms are O-acetylserine

sulfhydrylase and cysteine synthase). The biochemical pathway of cysteine synthesis was first described in an enzymological sense by the landmark work of N. M. Kredich and coworkers using *Salmonella typhimurium* (Kredich and Tomkins, 1966). They designated the SAT encoding gene as *cysE* and the OAS-TL encoding genes *cysK* and *cysM* and discovered the protein association between SAT and OAS-TL, the cysteine synthase complex (CSC; Kredich et al., 1969).

II. Cysteine Synthesis in Prokaryotes

Most prokaryotes use the two-step pathway encoded by *cysE*-like and *cysK/M*-like genes for cysteine biosynthesis according to genome databases, but exceptions exist, for instance in the genus Bacillus and probably also other Taxa. Even archaebacteria from habitats that are rich in reduced sulfur rely on these genes and have been shown to encode functional gene products similar to prokaryotes (Kitabatake et al., 2000).

A. The Enterobacteria Regulatory System

Cysteine synthesis is best investigated in *S. typhimurium* and *E. coli* with respect to structural components and regulation (Kredich, 1996) and forms the basis for investigation of the process in phototrophic organisms. Cysteine synthesis in enterobacteria is part of the cys-regulon that comprises structural genes for sulfate uptake, reduction and cysteine synthesis. Uptake of sulfate is mediated by a sulfate permease, which is also the case in cyanobacteria. It is a multisubunit ABC-type transport system that differs structurally

---

Abbreviations: CAS – β-cyanoalanine synthase; CDes – cysteine desulfhydrase; IC50 – 50% inhibition constant; NAS – N-acetylserine, OAS – O-acetylserine; OAS-TL – O-acetylserine (thiol) lyase; PLP – pyridoxal 5′-phosphate; SAT - serine acetyl-transferase

completely from the sulfate/proton cotransporters found in plants (see chapter by Hawkesford in this book). Interestingly, sequence similarity suggests a relationship to the sulfate transport systems (*SulP*) in the chloroplast envelope of the green alga *Chlamydomonas* and the moss *Marchantia* (Chen et al., 2003; Melis and Chen, 2005). Such sequences are, however, missing in the genomes of vascular plants according to databases and seem to have been lost during evolution of higher photoautotrophs. The assimilatory reduction pathway proceeds via 5′-adenosine-phosphosulfate (APS), 3′-phosphoadenosine-phosphosulfate (PAPS), sulfite and sulfide, most likely without any bound intermediate despite the instability and potential toxicity of these intermediates. The biochemical functions and evolution of the central enzyme in this pathway, assimilatory (P)APS reductase are reviewed in this book (see chapter by Kopriva et al.). SAT activity requires acetyl-coenzyme A (acetyl-CoA) and serine to form *O*-acetylserine (OAS) as energetically activated form of serine. OAS-TL activity is typically encoded by two genes: *cysK* and *cysM*. The *cysK* gene product uses OAS and sulfide to form cysteine (type A OAS-TL), while the *cysM* product also accepts thiosulfate as substrate (type B OAS-TL). The reaction between *O*-acetylserine and thiosulfate produces *S*-sulfocysteine, which is converted into cysteine by a so far uncharacterized mechanism. It has been proposed that the *cysM* gene is only effectively operating under anaerobic growth conditions (Kredich, 1996).

Since enterobacteria are usually well supplied with reduced sulfur or even sulfur amino acids from their environment, it is just consequent that the sensing of free cysteine concentrations is at the center of the cys-regulon, taking care that the energy consuming uptake and reduction of inorganic sulfate is only activated in situations where the cellular demand for cysteine exceeds external availability. The regulation is based on *cysE*, the only gene with constitutive expression in the regulon. The encoded SAT protein has exclusively been found in association with OAS-TL and is highly sensitive to feedback inhibition by cysteine (K_I = 1.1 μM; Kredich et al., 1969). Thus, in the presence of sufficient free cysteine in the cell SAT activity is shut down and hardly any OAS is produced. If cysteine or sulfide levels drop SAT begins to form OAS, the substrate

for OAS-TL. However, lack of sulfide results in accumulation and chemical disproportionation of OAS to *N*-acetylserine (NAS). NAS, in turn, is an activator for CysB, a transcriptional activator of promoters of the cys-regulon genes. Activation of sulfate uptake and reduction brings back sulfide into the cell that is consumed by OAS-TL to form cysteine. As soon as cysteine levels reach sufficient concentrations, SAT is feedback inhibited and lack of OAS results in shut-down of the regulon. The cys-regulon represents a typical repressor/operator mechanism of prokaryotic gene regulation (reviewed by Kredich, 1996). It seems evident that phototrophic bacteria developed at least partly different regulatory mechanisms due to their lifestyle.

B. Phototrophic Bacteria

Cysteine synthesis in bacteria with oxygenic photosynthesis is much less investigated compared to enterobacteria. The cyanobacterium *Synechococcus* sp. PCC 7942 (formerly *Anacystis nidulans*) contains three genes involved in cysteine synthesis. Two are encoded on an endogenous plasmid and are transcriptionally regulated by sulfate availability. The gene *srpG* encodes a type A OAS-TL whereas *srpH* encodes SAT. Both genes show significant sequence homology to the corresponding *E. coli* genes and are able to functionally complement *E. coli cysK cysM* and *cysE* mutant strains that are auxotrophic for cysteine. DNA hybridization indicates a second copy of srpG in *Synechococcus* sp. PCC 7942, possibly encoding for a type B OAS-TL enzyme (Nicholson et al., 1995; Nicholson and Laudenbach, 1995). Sulfate activation and reduction proceed by the same enzymes in cyanobacteria and enterobacteria (Niehaus et al., 1992). However, the only known regulatory protein in cyanobacterial sulfur metabolism, CysR, appears not be functionally equivalent to cysB in enterobacteria and thus its role in cysteine synthesis is much less defined. CysR expression itself is regulated by sulfur availability and the encoded protein is assumed to bind DNA due to a helix–loop–helix motif and similarity to other prokaryotic regulatory proteins. Inactivation of CysR results in loss of induction of numerous genes under sulfur limiting conditions. This includes the sulfate permease system (see chapter by Hawkesford, this book), a periplasmic protein of unknown function and the *srpG* and *srpH* genes. Since CysR is not required for growth when sulfate or thiosulfate are the only sulfur sources, it may be involved in regulating growth on other sulfur-containing compounds such thiocyanate (Laudenbach et al., 1991).

An important aspect of sulfur availability for microorganisms in general from a genetic as well as an ecological point of view is the reduction of growth yield under conditions of sulfur limitation. If methionine and cysteine could be replaced in the organism's major proteins for sulfur-free amino acids, the saving in sulfur would potentially increase nutrient use efficiency and competitiveness. Sulfate permease, which is abundantly produced by sulfur-starved enterobacteria and cyanobacteria lacks Met and Cys residues. Moreover, the cyanobacterium *Calothrix* sp. PCC 7601 harbours sulfur-depleted versions of the photosynthesis-related protein phycocyanin and its auxiliary polypeptides that are specifically expressed under conditions of sulfur limitation. The elevated synthesis of these proteins positively affects the sulfur budget of these cells (Mazel and Marliere, 1989).

III. Cysteine Synthesis in Plants and Algae

A. Subcellular Compartmentation and Functional Genomic Organisation

Similar to prokaryotes, the synthesis of cysteine in plants and algae can be divided into three steps: (1) assimilatory sulfate reduction to reduced sulfide, (2) provision of carbon and nitrogen containing acceptor for sulfur, (3) fixation of reduced sulfur into the organic backbone to produce cysteine, the first stable form of bound reduced sulfur. In contrast to sulfide production and the synthesis of most amino acids in plants and algae, cysteine synthesis takes place not only in plastids but also in the cytosol and the mitochondria. Consequently, small nuclear gene families encode SAT and OAS-TL proteins that are imported into these compartments (Fig. 1). Plants and at least green algae appear to be similarly organized in this respect. *Euglena* differs fundamentally in its subcellular organization with sulfate reduction being confined to mitochondria, while little is

known about the locations of cysteine synthesis. Because *Euglena* is a taxonomically distant, facultative photosynthetic organism, the reader is conferred to specialized literature (Saidha et al., 1988; Schiff et al., 1993).

The reasons for compartmentation of cysteine synthesis have been subject of speculation, but clear evidence as to its function is still missing. Lunn et al. (1990) suggested that endomembranes may be impermeable for cysteine transport, pointing out that each compartment with the capacity for protein biosynthesis would therefore require its own cysteine biosynthesis. While this hypothesis is intriguing, it seems awkward to assume that all amino acid acids can be transported into and inside the cell by broad-spectrum amino acid transporters, but not cysteine. Specialized functions in sulfur metabolism, such as assimilation in plastids, regulation in the cytosol and degradation in the mitochondria, have been suggested (Hell, 1997). However, mitochondria are the site of sulfur amino acid degradation in animals via sulfite oxidase, but this molybdoenzyme enzyme is located in the peroxisomes in plants (see chapter by Hänsch and Mendel, this book). A different possibility is provided by the chemical properties of sulfide, the endproduct of assimilation. Sulfide always exists in equilibrium between HS^- and H_2S at around neutral pH values. H_2S is believed to be membrane permeable and thus freely available to all compartments as substrate for OAS-TL (Fig. 1). The volatility of H_2S could require cysteine synthesis in the cytosol to capture this gas and avoid costly losses of reduced sulfur. In turn, mitochondria might need to be protected from H_2S in order to avoid interference with the electron transport chain or even discharge of the proton gradient by the HS^- equilibrium, similar to the toxicity of NH_3/NH_4^+. On the other hand, it should not be overlooked that mitochondria as well as chloroplasts require substantial amounts of elemental S in iron-sulfur clusters of their electron transport chains. This partially reduced oxidation state of sulfur is believed to be derived from cysteine via carrier-mediated steps (see Padmavathi et al., this book).

In contrast to such special functions the three subcellular locations could of course be the consequence of sheer redundancy. However, the targeted overexpression of an enzymatically inactive SAT in the cytosol of tobacco cells

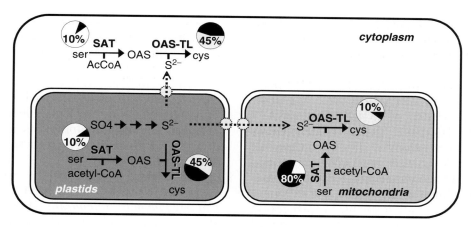

Fig. 1. Subcellular distribution of serine acetyltransferase and *O*-acetylserine(thiol)lyase activity in higher plants. The figure represents a schematic view on a higher plant cell. De novo cysteine synthesis takes place in the cytoplasm (white), the mitochondria (grey) and the plastids (dark grey). Important metabolites are depicted in the reaction pathways, reaction steps are shown as arrows. Enzymes are named above the reaction in bold and shadowed letters. In a subcellular compartment, the percentages of total SAT and OAS-TL activities are shown in white and grey pie chart diagrams, respectively. For better understanding the percentage of total activity is also indicated in the pie chart. The localisation of OAS-TL activity was revealed in spinach (Lunn et al., 1990) and confirmed in Datura innoxia (Kuske et al., 1996). Subcellular distribution of SAT was analysed in Pisum sativum (Ruffet et al., 1995). Transport of metabolites across membranes is indicated as dashed arrows.

resulted in strong changes in cysteine and glutathione contents, suggesting an important and specific role of cysteine synthesis for each compartment (Wirtz and Hell, 2007). In addition, the nearly ubiquitous and constitutive expression of mRNAs encoding SAT and OAS-TL as revealed by *Arabidopsis thaliana* gene expression profile databases (see https://www.genevestigator.ethz.ch/) supports a general requirement in all cell types under all conditions, rather than redundancy.

Contradictory to these observations, the absence of functional OAS-TL activity has been reported for mitochondria of spinach (*Spinacea oleracea* L.). Warrilow and Hawkesford (2000) showed in enzymological experiments that OAS-TL activity from spinach mitochondria in fact is the result of a side activity of β-cyanoalanine synthase. This enzyme has been found in cytosolic and/or mitochondrial compartments in several plant species; it catalyzes a partial backward reaction of OAS-TL (section V.E) and is, as OAS-TL, a structural member of the superfamily of β-substituting alanine synthases (bsas). *Arabidopsis thaliana* carries independent and true OAS-TL and β-cyanoalanine synthase enzymes in its mitochondria (Hatzfeld et al., 2000; Jost et al., 2000) and so far no other plant species has been reported with similar properties as spinach. Further genome

analyses will allow to fully evaluate mitochondrial cysteine synthesis.

The role of the different compartments is further obscured by the relative distribution of enzymatic activities. These have been determined from cellular fractions for all three compartments of spinach and pea (Droux, 2003; Droux, 2004) and are complemented by a number of localization studies on either SAT or OAS-TL with other species (Lunn et al., 1990; Rolland et al., 1992; Ruffet et al., 1995; Kuske et al., 1996). In general, it appears that about 45% of OAS-TL activity is associated with plastids, 45% with the cytosol, and the remainder with mitochondria. In contrast, 80% of total SAT activity is found in mitochondria, 10% in plastids and 10% in the cytosol (Fig. 1). Thus, measurable maximal activities of the two consecutive steps of cysteine synthesis differ substantially with respect to the three subcellular locations. Unless substrate availability or unknown factors strongly affect these activities, the exchange of OAS and probably also cysteine has to be assumed between cytosol and the organelles.

B.   Gene Families and Features of Gene Products

The model plant *Arabidopsis thaliana* is by far the best investigated higher organism with respect to cysteine synthesis (Hell et al., 2002). The Arabidopsis SAT gene family consists of five members which have been named with respect to their location at the chromosomes as AtSAT1 (At1g55920, L34076), AtSAT2 (At2g17640, L78444,), AtSAT3 (At3g13110, U22964), AtSAT4 (At4g35640, AF331847) and AtSAT5 (At5g56760, U30298). Other nomenclatures from various descriptions of cDNAs confuse this organization. A unified nomenclature termed Serat for serine acetyltransferase has been suggested (Kawashima et al., 2005), but currently does not comprise all available sequences in the public databases. AtSAT2 and 4 were added very recently to the SAT gene family based upon their sequence homology and their availability to complement cysteine-auxotrophy of *E. coli* mutants lacking endogenous SAT activity (Kawashima et al., 2005). However, AtSAT2 and 4 are much less expressed than AtSAT1, 3 and 5 with respect to mRNA levels. K_M values of purified recombinant or native plant SATs range from 1 to 3 mM for serine and 0.01 to 0.3 mM for acetyl-CoA, corresponding to the assumed cellular concentrations of these molecules (Ruffet et al., 1994; Noji et al., 1998; Wirtz et al., 2000; Wirtz et al., 2001). To a very limited extent, also propionyl-CoA, but not buthionyl-CoA, may be accepted as a substrate (R. Hell, unpublished). In contrast, the affinities of AtSAT2 and AtSAT4 are 10 to 100-fold lower for both substrates. These properties cast some doubts on a significant contribution of these isoforms to cysteine synthesis (Kawashima et al., 2005).

The localization of all Arabidopsis SAT isoforms has been verified using SAT-GFP-fusion proteins. AtSAT2, 4 and 5 are located in the cytosol and AtSAT1 and AtSAT3 are targeted to plastids and mitochondria (Sun et al., 2004; Kawashima et al., 2005). The localization of AtSAT1 appears not entirely clear due to a report with variable development-dependent plastid or cytosolic localization (Noji et al., 1998). It should be mentioned that AtSAT1 protein was found in association with the nucleus, in nuclear proteome analysis (Bae et al., 2003). However, SAT proteins are of low abundance, as shown by biochemical purification of SAT from spinach chloroplasts (Ruffet et al., 1994). Therefore, immuno-localisation with isoform-specific antibodies might be required to confirm the localization of AtSATs.

Activities and isoforms of SATs have been described for subcellular compartments of plants of different taxonomy. This refers to a plastid SAT isoform from spinach (SAT56; (Noji et al., 2001b) as well as cDNAs encoding cytosolic (SAT7) and organelle-localized (SAT1) SATs from tobacco (Wirtz and Hell, 2003a). In pea and in spinach, cytosolic, chloroplastic and mitochondrial SAT activities were characterized (Droux, 2003; Droux, 2004).

With respect to algae, only a plastid-localized SAT and OAS-TL have been reported for *Chlamydomonas reinhardtii*. Both genes are strongly upregulated in response to sulfur deficiency (Ravina et al., 2002). The availability of complete genome sequences of algae of distant taxonomic origin including *C. reinhardtii* (greenalgae), *Thalassiosira pseudonana* (a diatom) *Cyanidioschyzon merolae* (unicellular red algae) will allow detailed comparisons of the evolutionary origin and compartmental organization of genes of cysteine synthesis. *C. reinhardtii* probably carries three SAT and four OAS-TL encoding genes. The effect of upregulation depends on the presence of an active *Sac1* gene (Ravina et al., 2002). The *sac* mutants of *C. reinhardtii* are described in detail in this book (see chapter Shibagaki and Grossman). Sac1 shows similarity to a Na^+/SO_4^{2-} cotransporters and may function as a sulfate sensor. Many, but not all transcripts related to sulfur assimilation depend on Sac1 for their expression (Zhang et al., 2004).

OAS-TL proteins are, in contrast to SATs, more variable in their enzymatic functions. OAS-TL has a considerable variability for the nucleophilic reactant (section IV.B) and may be involved in the synthesis of secondary products in some plant species (Murakoshi et al., 1986; Warrilow and Hawkesford, 2000; Maier, 2003). OAS-TLs belong to the pyridoxal-phosphate-dependent superfamily of β-substituting alanine synthases (bsas). This family also includes β-cyanoalanine synthase (CAS), based on high amino acid homology. CAS is believed to function in the detoxification of cyanide using cysteine as substrate and releasing β-cyano alanine and sulfide (section V.E). The OAS-TL gene family of Arabidopsis consists of 9 genes that encode for 8 functionally transcribed OAS-TL-like proteins (Jost et al., 2000). OAS-TL A (At4g14880), B (At2g43750) and C (At3g59760) are the most expressed authentic OAS-TLs in this family and the proteins have been localized in the cytosol, the plastid and the mitochondria, respectively (Hell et al., 1994; Hesse and Hoefgen, 1998; Hesse et al., 1999; Wirtz et al., 2004). The combination of subcellular localisation and biochemical studies of recombinant AtcysD1 (At3g04940), AtcysD2 (At5g28020) suggest that these gene products are authentic OAS-TL isoenzymes of low abundance and cytosolic localization. In contrast, AtcysC1 (At3g61440) is predicted to be localized in mitochondria and to act as a CAS *in vivo* (Hatzfeld et al., 2000; Yamaguchi et al., 2000). The functions of the remaining OAS-TL like proteins (At3g03630 and At5g28030) in plant sulfur metabolism are currently unknown.

With respect to function the careful, enzyomological analysis of the three major expressed OAS-TLs in Arabidopsis revealed that their affinities for sulfide are much higher than previously described (Wirtz and Hell, 2003b; Wirtz et al., 2004). The K_M values range from 3 to $6 \mu M$ and correspond to the assumed sulfide levels in plant and form the basis for the observed rates of cysteine synthesis *in planta* (Schmidt and Jäger, 1992). The true enzyme substrate is likely to be HS^-, which constitutes about 50% of total sulfide at pH around pH 7 (Wirtz et al., 2004). Although sulfide concentrations have apparently never been quantified exactly in plant tissues, they are generally believed to be in the low μM range (Hell et al., 2002). If this assumption holds true, sulfide availability would not limit rates of cysteine synthesis under sufficient sulfur supply. In contrast, the affinity of OAS-TL for the second substrate, OAS, was confirmed to be rather low. The K_M values for OAS are approximately 10 to 100 times higher than the concentrations found in whole leaf extracts from Arabidopsis and *Brassica oleracea* (Kim et al., 1999; Awazuhara et al., 2000; Wirtz et al., 2004). This strongly suggests that OAS rather than sulfide limits cysteine synthesis at sulfur sufficient conditions, which is an important feature of the cysteine synthase complex model of regulation (section VI.B).

Total extractable OAS-TL activities determined under substrate saturating conditions will hardly ever constitute a bottleneck for plant growth. However, *in vivo*, lowered availability of substrate concentrations or increased demand may well result in a shortage of cysteine production.

It was suggested that OAS-TL activity can indeed limit rates of cysteine in various stress situation, when high amounts of cysteine are required, e.g. for synthesis phytochelatins in the presence of cadmium (Barroso et al., 1999; Dominguez-Solis et al., 2001; Romero et al., 2001; Dominguez-Solis et al., 2004). Accordingly, OAS-TL overexpression in plastids and cytosol supports the resistance to cadmium and sulfur containing pollutants, which are believed to act toxic via sulfide release (Harada et al., 2001; Noji et al., 2001a).

C. Gene Expression

SAT isoforms in *A. thaliana* show a semi-constitutive transcription pattern during development of the plant according to comprehensive microarray data (see https://www.genevestigator.ethz. ch/ for public database). This is more or less confirmed by individual expression analyses from other plant species, although much less data are available (Saito et al., 1997; Urano et al., 2000). RNAs of cytosolic, mitochondrial and plastidial SAT isoforms have been detected in all analyzed tissues of *A. thaliana*, indicating that cysteine synthesis in all these compartments is required in all tissues during development of higher plants. This also applies to developing seeds, which are capable to assimilate sulfate provided by maternal tissue into cysteine (Tabe and Droux, 2001; Tabe and Droux, 2002). While plastid AtSAT1 has been described as the most prominently transcribed isoform in leaves (Kawashima et al., 2005), this result is challenged by microarray and cell fractionation data. In *Pisum sativum* leaves, approximately 80% of total SAT activity was located in mitochondria and the residual SAT activity was distributed between cytosol and plastids (Ruffet et al., 1995). Species-specific differences in the localization of cysteine synthesis or the fact that the mRNA level of an expressed gene is not necessarily indicative for the amount of protein may be responsible for these discrepancies.

Transcriptional regulation in response to stress as indicated by mRNA contents of SAT isoforms in various species generally shows not more than threefold variation. Only the AtSAT4 gene is reported to be strongly up regulated during sulfur starvation and by treatment with cadmium in roots and leaves (Kawashima et al., 2005). Given the uncertain enzymatic identity of the encoded gene product, it remains currently unclear whether AtSAT4 upregulation is related to cysteine deficiency or is part of a more general response of the plant to cope with stress in general.

Genes encoding OAS-TL-like proteins appear to be more or less ubiquitously expressed in all plant cell types analyzed so far, with little variation in the content of RNA, protein and extractable enzyme activity in response to external factors (Brunold, 1990; Hell et al., 1994; Dominguez-Solis et al., 2001). In fact, mRNA contents of an OAS-TL gene encoding OAS-TL A from Arabidopsis has been used as constitutive control in a gene expression study (Koprivova et al., 2000). However, several experiments strongly suggest the OAS-TL A gene to respond specifically to cadmium or salt exposure with up to a sevenfold increase in mRNA level (Barroso et al., 1999; Dominguez-Solis et al., 2001). Accordingly, overexpression of the corresponding gene conferred enhanced cadmium tolerance (Dominguez-Solis et al., 2004). Remarkably, OAS-TL but also SAT gene expression is elevated in leaf trichomes of Arabidopsis and, according to *in situ* hybridization data, responds to salt and cadmium treatments, presumably to support glutathione synthesis (Gutierrez-Alcala et al., 2000; Howarth et al., 2003).

IV. Structure of Proteins of Cysteine Synthesis in Bacteria and Plants

In their groundbreaking and pioneering work Nicolas Kredich and coworkers revealed in the late 1960s the presence of the CSC in the pathogenic enterobacterium *S. typhimurium*. By using size-exclusion chromatography they determined a total molecular weight of 309 kDa for the hetero-oligomeric CSC, of 160 kDa for the SAT homomer and of 68 kDa for the OAS-TL homomer (Becker et al., 1969; Kredich et al., 1969; Kredich and Becker, 1971). Although the CSC has been known for almost 40 years, the overall structure is still (2007) unresolved. Nonetheless, in the last decades progress has been made to understand the structure/function relationship inside the CSC by analysing separately the structure of the SAT and OAS-TL subunits by X-ray crystallography.

A. Serine Acetyltransferase

The first functional analysis of the SAT domain structure was performed using a eukaryotic SAT from the mitochondria of Arabidopsis. By using the yeast two hybrid system the N-terminal α-helical domain of SAT was identified as the SAT–SAT interaction domain, while the C-terminal domain was revealed to account for both, enzymatic activity and SAT-OAS-TL interaction inside the complex (Bogdanova and Hell, 1997). The idea of a bifunctional C-terminal SAT domain was further strengthened by the modelling of the C-terminus of plant SAT using bacterial acyltransferase structures as templates. These posses an unusual left-handed parallel β-helix (Vuorio et al., 1991; Vaara, 1992). By computational modelling of the SAT C-terminus to these acyltransferases, the C-terminus of SAT could be split into two sections: a left-handed parallel β-helix (LβH) domain, which carries the catalytically active site, and a C-terminal tail that could not be modelled due to low homology to the acyltransferases (Wirtz et al., 2001). Partial deletion of the C-terminal tail of prokaryotic SAT results in loss of interaction with OAS-TL, as shown by co-purification experiments using SAT and OAS-TL of *E. coli* (Mino et al., 1999; Mino et al., 2000b). In addition, several amino acids of the C-terminal tail of plant and bacterial SAT seem to be involved in the feedback inhibition of SAT by cysteine the end product of the pathway (Noji et al., 1998; Inoue et al., 1999; Wirtz and Hell, 2001). Although most features of the C-terminal tail of prokaryotic and eukaryotic SAT seems to be identical, the last C-terminal amino acids are more divergent in prokaryotic and eukaryotic SATs than the LβH structure (Fig. 2). The latter contains an imperfect tandem repeat of a hexapeptide sequence described as [LIV]-[GAED]-X2-[STAV]-X, which is responsible for the β-helical folding (Vaara, 1992). All analysed proteins containing this motive, including SAT, are placed into the superfamily of acyltransferases, whose members are known to act *in vivo* as trimers (Raetz and Roderick, 1995; Beaman et al., 1997). Thus; it was not surprising that bacterial SAT was revealed to be a dimer of trimer in its native uncomplexed form (Fig. 2A, Hindson et al., 2000). In 2004, the three-dimensional structure of SAT homomer from the prokaryotes *E. coli* (Pye et al., 2004) and *Hamophilus influenzea* (Gorman and Shapiro, 2004) were resolved, whereby the predicted overall structural features of SAT could be unequivocally confirmed (Fig. 2B).

Consistently, in *E. coli* and *H. influenzae*, whose SATs share 71% identity at the amino acid sequence level, three SAT monomers form a trimer. The three clefts between the three subunits form a catalytic center (Fig. 2C), which suggests that active site residues from two subunits may contribute to form one joint catalytic center (Pye et al., 2004), see below. Two SAT trimers are arranged in the quaternary structure of a dimer (dimer of trimers) so that the dimer interface is constituted by parts of the N-terminal α-helices of all subunits. The α-helical domain of each subunit is formed by residues 1–140 in the *E. coli* SAT protein and comprises eight α-helices. Helix α1 and α2 as well as α3 and α4 run in an anti-parallel fashion and are connected by 2 β-turns; the residual α-helices are meandering with different angles. The repetition of helices α2–α4 by the molecular threefold axis forms a surface that interacts loosely with the cognate structure in the opposing trimer and thereby stabilizes the dimerization of trimers (Olsen et al., 2004). All of the residues involved in dimerization are found within a N-terminal extension of approximately 80 amino acids that is found only in a subset of SATs (13% of all SAT like proteins). This indicates that SATs fall into two subgroups: Group A, whose members posses the extension and

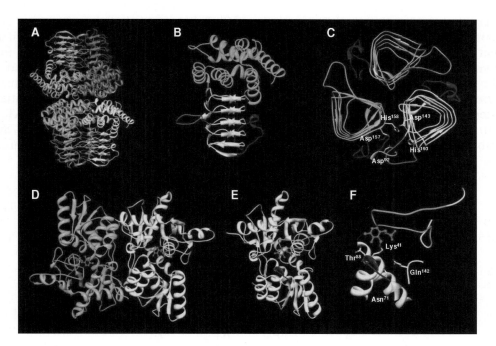

Fig. 2. Three-dimensional structures of prokaryotic SAT and OAS-TL. The figure depicts the overall homomeric quaternary structure of SAT (A) and OAS-TL (D). In both cases the monomers are shown in different colours in order to illustrate the sterical position of the monomers. Each SAT monomer (B) consists of an α-helical domain (green), a β-sheet domain (golden) that included a random coil loop (red) and the C-terminal tail (dark blue), which is responsible for CSC formation. Please note that the last 11 amino acid residues of the C-terminal tail are missing in the structural presentation, since they were not solved by X-ray crystallography. (C) Top view on an SAT trimer, where α-helical domains are deleted for better overview. The active site of SAT lies in between the β-sheet domain of two monomers, thus a SAT 'dimer of trimers' (A) contains six independent active sites. The most important residues for binding of substrates and the enzymatic activity are named and shown in sticks. Each OAS-TL monomer (E) of the native OAS-TL dimer (B) contains one PLP (red) that is bound via an internal Schiff base by Lys[41]. The asparagine loop, which is important for substrate binding and induction of the global conformational change of OAS-TL during catalysis is shown in dark blue. The green colour indicates the β8a–β9a loop that seems to be important for interaction with SAT. Both structural elements are in close proximity to the active site of OAS-TL (F). Important residues for binding of substrates and the enzymatic activity are highlighted in F. The software UCSF Chimera V1.2422 (University of California, San Francisco) was used to visualize SAT (pdb: 1T3d) and OAS-TL (pdb: 1OAS) (*See Color Plates*).

form hexameric SAT, and group B, whose members lacks the extension and are likely to have a trimeric quaternary structure (Gorman and Shapiro, 2004). A proof for this hypothesis has not been provided.

Residues 141–262 of the *E. coli* SAT form the β-helical domain, which is a typical LβH comprising fourteen β-strands forming five coils of the helix. Apart form the final C-terminal tail, there is only one break from the β-helix; this is a loop from residue Gly[184] to His[193] that has a random coil topology. The His[193] of this loop, with His[158] and three aspartic acid residues (Asp[92], Asp[143] and Asp[157]), forms the core of the catalytically active site (Fig. 2C). While Asp[92] and Asp[157] stabilize the positive charge of the amino group of the substrate serine, the imidazole ring of His[158] is supposed to form a classical 'catalytic triad' with the oxygen of the γ-hydroxyl group of serine and the Asp[143] of the adjacent SAT subunit. Thus, six independent active sites exist in the hexameric SAT homomer (Olsen et al., 2004). In the 'catalytic triad', His[158] acts as a base to activate the hydroxyl group of serine for a nucleophilic attack on the carbonyl carbon of the acetyl group of acetyl-CoA. This attack would result in an oxyanion tetrahedral intermediate, whose negative charge would be stabilized by the imidazole of His[193]. The collapse of the oxyanion tetrahedral intermediate would result in the formation of the reaction products: OAS and CoA, whereby the same general base residue (His[158]) acts a general acid and donates a proton of the sulfur atom of CoA. This mechanism would be in good agreement with reaction mechanisms of other acyltransferase, which belongs to the LβH subfamily of acyltransferases (Johnson et al., 2005). Elaborate solvent deuterium kinetic isotope effects measurements indicated that the rate limiting step of this reaction is the nucleophilic attack of the serine hydroxyl group on the thioester of acetyl-CoA (Johnson et al., 2004). Based on initial velocity studies and on the analysis of *H. influenzae* SAT structure in the presence of CoA (1SST) and cysteine, the kinetic mechanisms of SAT is supposed to be sequential equilibrium ordered (Fig. 3A), which would be in agreement with the kinetic mechanism revealed for all other LβH acyltransferases (Sugantino and Roderick, 2002; Hindson and Shaw, 2003; Johnson et al., 2004). The outcome of previous initial velocity studies using the *S. typhimurium* SAT suggested the occurrence of a bi bi ping pong – instead of a sequential – kinetic mechanism (Leu and Cook,

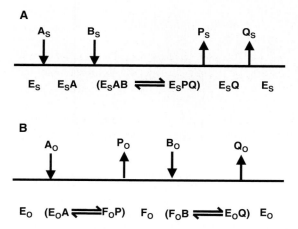

Fig. 3. Kinetics mechanism of serine acetyltransferase and *O*-acetylserine(thiol)lyase. The kinetic mechanisms for SAT (A) and OAS-TL (B) are sequential ordered and bi bi ping pong, respectively. (A) Acetyl-CoA (A_S) binds to SAT (E_S) before serine (B_S) is recruited. Afterwards the catalytic triad is formed, allowing a nucleophilic attack of serine on acetyl-CoA (E_SAB). The collapse of the oxyanion tetrahedral intermediate ($E_SAB \leftrightarrow E_SPQ$) results in the formation of the reaction products: OAS (P_S) and CoA (Q_S). (B) In the first-half reaction OAS-TL binds OAS (A_O) and releases the first product acetate (P_O). E_O represents the internal Schiff-base between PLP and Lys[41] of OAS-TL, while F_O indicates the α-aminoacrylate external Schiff base conformation of OAS-TL. The second substrate sulfide (B_O) attacks in a nucleophilic manner on the α-aminoacrylate intermediate and the final product cysteine (Q_O) is released (second-half reaction).

1994). While almost all results of Leu and Cook (1994) could be confirmed, only the burst kinetic experiment, which gives rise to the hypothesis of a ping pong mechanism for SAT, was not reproducible (Johnson et al., 2004).

The sequential kinetic reaction mechanism of SAT is also consistent with the inhibition mechanism by cysteine (section VI.A), which affects not only one but both substrates. SAT of *E. coli* (1T3D) and *H. influenzae* (1SSQ) were crystallized in the presence of cysteine. In both crystals the cysteine was found in the binding pocket of serine, providing an elegant explanation for the competitive inhibition of cysteine for serine (Olsen et al., 2004; Pye et al., 2004). That the cysteine and the serine binding pockets overlap is also supported by kinetic and microcalorimetric data obtained using serine analogous (Hindson and Shaw, 2003). These results confirm that the cysteine feedback inhibitor binds to the active site of SAT. Beside the direct competition with serine, cysteine also reduced the affinity of SAT for acetyl-CoA, after binding to the enzyme (Kredich and Tomkins, 1966; Leu and Cook, 1994; Johnson et al., 2004). The molecular basis for the reduction of acetyl-CoA affinity is a displacement of the C-terminal tail of SAT in the presence of cysteine. Parts of the C-terminal tail compete with the pantentheinyl arm of acetyl-CoA for binding to the extended LβH loop in the cysteine bound SAT crystals. In the absence of cysteine, acetyl-CoA is able to displace the C-terminal tail in order to enter the binding pocket, while the C-terminal tail closed the acetyl-CoA binding site more efficiently when cysteine is bound to SAT (Olsen et al., 2004). The importance of the C-terminal tail for inhibition of SAT activity by cysteine was also confirmed by mutational analysis of bacterial and plant SATs C-terminal tails (Denk and Böck, 1987; Nakamori et al., 1998; Inoue et al., 1999; Takagi et al., 1999; Wirtz and Hell, 2003a). These results indicate that specific features of SAT domains have the same function in prokaryotic and eukaryotic SAT. Although no eukaryotic SAT has been crystallized up to now, computational modelling of eukaryotic SAT to bacterial SAT suggest that the overall structure of pro- and eukaryotic SATs are identical (Wirtz and Hell, 2006). The basis for this homology modelling is the high sequence identity between the two types of SAT, which is in the order of 34–44%.

B.   O-acetylserine(thiol)lyase

The structures of OAS-TL from the prokaryotes *H. influenzae* and *S. typhimurium* have been determined with and without its substrates (Burkhard et al., 1998; Burkhard et al., 1999). The structure of the first plant OAS-TL was recently solved, demonstrating a high degree of conservation between plant and enterobacterial OAS-TL at the structural level (Bonner et al., 2005). The structure resolved from the plant OAS-TL crystal overall confirms the model of OAS-TL, which was derived from homology modelling of plant to bacterial OAS-TL. The sequence identity between the pro- and eukaryotic OAS-TL proteins is in the same range as for the respective SAT proteins. Taken together these results indicate that the same homology model approach is also suitable for plant SAT (Wirtz and Hell, 2006).

OAS-TL from plants and bacteria form stable homodimers with a molecular weight of 68 to 75 kDa according to their amino acid sequences. Crystallization of OAS-TL from bacteria revealed that the dimers interact only at the dimer interface and are oriented in such a way that the two active sites are facing each other (Fig. 2D). Each OAS-TL subunit carries a tightly bound pyridoxal 5′-phosphate (PLP) at the catalytic center. All active site residues are contributed from one subunit (Rabeh and Cook, 2004). The active site with the PLP is located within a cleft deep in the center of each monomer that is flanked by N- and C-terminal domains (Fig. 2E). This conformation may hamper the standard approach of expression of recombinant proteins with N- or C-terminal tags, resulting in less active or kinetically altered enzymes.

Lys[41] of bacterial OAS-TL is responsible for the linkage to the PLP, whose presence is a prerequisite for an enzymatically active enzyme (Burkhard et al., 1999). Asn[71], which is located on the so-called asparagine loop, and Gln[142] bind OAS and the resulting intermediates along the reaction pathway. Functional analysis of the plant OAS-TL reveals Thr[74] and Ser[75] as important active site residues (Fig. 2F). Most likely both residues are involved in stabilizing the transition state of the second-half reaction (see below). The stabilisation of the transition state has direct impact on the affinity for the second substrate, sulfide (Bonner et al., 2005). The conservation of the active site residues strongly supports the idea of the same kinetic mechanism for OAS-TLs from prokaryotes and eukaryotes.

OAS-TL is one of the most extensively studied enzymes with respect to the reaction mechanism and the kinetic behaviour. The type of reaction kinetic of the enzyme is bi bi ping pong (Cook and Wedding, 1976; Tai et al., 1993): In this type of kinetic, after binding of the first substrate, OAS, the first product, acetate, is released, before the second substrate, sulfide, interacts with the reaction intermediate to give the second product, cysteine (Fig. 3B). Mechanistically, the overall β-substitution of the acetate group of OAS by sulfide can be subdivided into an elimination and an addition reaction, which represent the first-half and the second-half of the reaction, respectively. In short, the first-half reaction includes the binding of OAS and the formation of an α-aminoacrylate intermediate, whereby acetate is released. The second-half reaction is initiated by the attack of sulfide, in the form of HS^-, to the Cβ of α-aminoacrylate intermediate to give a cysteine external Schiff base. Then transimination results in the release of cysteine (Tai et al., 1995; Rabeh and Cook, 2004; Wirtz et al., 2004). The reaction is accompanied by a large conformational change of the protein after the binding of OAS. The binding of OAS in the active site induced a local rearrangement of the asparagine loop (Fig. 2E, F) that shifted the enzyme from the so called 'open' to the 'closed' conformation. Formation of the 'closed' conformation results in expel of bulk solvent from the reaction room and the closure of the active site, which is necessary for the second-half of the reaction, since the α-aminoacrylate is highly reactive (Burkhard et al., 1999). Although the sequence identity of the asparagine loop from prokaryotes and eukaryotes is very high, the triggering function of the asparagine loop for the overall conformational change is doubted in plant OAS-TLs (Bonner et al., 2005). The methods used for identification of chemical mechanisms and different conformational changes of OAS-TL during catalysis included fluorescence (Benci et al., 1997, 1999), phosphorescence (Strambini et al., 1996), and UV visible spectroscopy (Schnackerz et al., 1995) in combination with mutageneic (Rege et al., 1996; Tai et al., 1998) and crystallographic approaches (Burkhard et al., 1998; Burkhard et al., 1999). Most

of this tremendously intensive study is comprehensively summarized by Rabeh and Cook (2004). In agreement with the high K_M of OAS-TL for OAS, the first-half reaction is limiting the overall β-substitution reaction (Woehl et al., 1996), while the second-half of reaction is irreversible (Tai et al., 1995). This biochemical phenotype of OAS-TL is important for cysteine regulation in higher plants. In fact cysteine synthesis must be limited by OAS production for proper regulation of the sulfur assimilation pathway by CSC formation (section VI.C, D) and the transcriptional control of OAS regulated genes. While OAS production by SAT activity is inhibited by cysteine, OAS-TL seems not to be regulated at all. However, sulfide is a competitive substrate inhibitor of bacterial OAS-TL, with a K_I of 50 μM (Cook and Wedding, 1976; Tai et al., 1993). The identification of an allosteric anion-binding site at the dimer interface, which can be occupied by sulfide, led to the hypothesis of an allosterically regulated OAS-TL, due to catabolism of cysteine in bacteria (Burkhard et al., 2000).

C. Cysteine Synthase Complex

The 3D-structure of the CSC is not known. Crystallography of free and complex associated plant SAT was attempted but not successful, due to instability of plant SAT protein (personal communication J. Jez Danforth Center, St. Louis). Moreover, no precise analytical data are available for the molecular weight of both bacterial and plant CSC. The stoichiometry of SAT and OAS-TL subunits in the complex can thus be hypothesized on a theoretical ground. The only estimate of the size of the complex arises from size exclusion chromatography, which revealed a molecular weight of 300 kDa for CSCs from native protein extracts of spinach and tobacco (Droux et al., 1992; Wirtz and Hell, 2007). The size of the bacterial CSC is in the same range (Kredich et al., 1969), indicating an overall similar structure in prokaryotes and eukaryotes. The uncomplexed bacterial SAT is known to be a 'dimer of trimers', while OAS-TL from plants and bacteria form dimeric homomers (see above). On the basis of this knowledge, one hexameric SAT homomer (Fig. 2A) is supposed to form the core of the CSC, which is the completed by the recruitment of two OAS-TL dimers (Fig. 2D). As a consequence,

SAT and OAS-TL subunits would exist in a 6: 4 (1.5: 1) stoichiometric ratio in the CSC. Fluorescence titration assays of CSC yielded an experimentally determined ratio of 1.43: 1 for SAT and OAS-TL, essentially supporting the theoretical stoichiometry (Campanini et al., 2005).

One interaction site of SAT with OAS-TL inside the complex is the C-terminal tail of SAT (Fig. 2B). This is proven by partial deletion of this tail in bacterial SATs, which resulted in the loss of SAT ability to interact with OAS-TL (Mino et al., 1999; Mino et al., 2000b). Co-crystallization of OAS-TL from *H. influenzae* and *A. thaliana* with the last 10 amino acid residues of the SAT C-terminal tail demonstrated that the binding site of this decapeptide is located at the active site of OAS-TL, where also OAS binds (Huang et al., 2005; Francois et al., 2006). The key active site residues Thr^{74}, Ser^{75} and Gln^{147} lock the decapeptide in the binding site via hydrogen bonds (numbering of residues according to AtOAS-TL A, section IV.B), which explains how complex formation downregulates OAS-TL activity (Francois et al., 2006). OAS would compete with the SAT C-terminal tail for binding at the catalytic center of OAS-TL. Assuming this as the only OAS binding site in the complex, such a competition provides also a mechanistic basis for the dissociation of the CSC by OAS. Kinetic characterization of the association of bacterial OAS-TL with the C-terminal tail of SAT showed that the binding affinity of this decapeptide was 250-fold lower than full-length SAT (Mino et al., 2000b). However, the largest part, and thus other potential interaction sites, of SAT were missing in this experiment. Although the C-terminal decapeptide of *H. influenzea* and *S. typhimurium* are highly divergent, the affinity for the C-terminal decapeptide of *H. influenzea* and *S. typhimurium* ($K_D \sim 0.6$ to $1\,\mu M$) were in the same range for OAS-TL of *H. influenzae*, which indicates a general binding mechanism in the γ-proteobacteria (Campanini et al., 2005). In contrast, the plant C-terminal decapeptide appeared to have high affinity for OAS-TL ($K_D = 5$–$100\,nM$) (Kumaran and Jez, 2007), which is even more affine than reported for the full-length SAT from Arabidopsis cytosol and mitochondria. The K_D of both proteins for OAS-TL were independently determined by initial velocity and surface plasmon resonance studies and are in the range of 25 to 40 nM (Droux et al., 1998; Berkowitz et al., 2002). The different structural composition of pro- and eukaryotic decapeptides may be one reason for the variability in their affinity for their respective interaction partner. The only conserved residue in plant and bacterial SAT decapeptides is the very final Ile residue, which, at least in plants, is responsible for the molecular recognition of the decapeptide by OAS-TL (Francois et al., 2006).

The lower affinity of the bacterial decapeptide for OAS-TL compared to the full-length SAT raises the question of whether additional sites for interaction or recognition exist in SAT and OAS-TL. Initial protein–protein interaction analyses using OAS-TL mutants that were genetically modified in the highly conserved $\beta 8A$-$\beta 9A$ surface loop (Fig. 2E) point towards an involvement of this loop in CSC formation (Bonner et al., 2005). The corresponding interaction site of the SAT protein has not been located; its identification would allow a more detailed analysis of the assembly of SAT and OAS-TL subunits inside the CSC.

The association of the SAT C-terminus with the catalytic site of OAS-TL would place the OAS-TL dimers at the distal ends of the SAT 'dimer of trimers'. If symmetry applies here, we may speculate that one of the three SAT C-termini of a trimer interacts with the active site of one subunit of an OAS-TL dimer. In principle, also two C-terminal tails could bind into the two active sites of the OAS-TL homodimer. Both, plant and bacterial C-terminal decapeptides bind to the OAS-TL dimer in a 2: 1 ratio. Nonetheless, one could not conclude from this experiment that two C-terminal tails of SAT bind one OAS-TL dimer, due to possible steric interferences of C-terminal tails in the SAT homohexamer. However, addition of a molar excess of the decapeptide to the CSC results in no detectable binding of the decapeptide, indicating that no active site of complex bound OAS-TL is available (Campanini et al., 2005). The latter fact could be the reason for the strong inactivation of OAS-TL inside the complex, which results in a more than 10-fold decrease of OAS-TL activity in bacterial and plant CSC (Kredich et al., 1969; Droux et al., 1998; Wirtz et al., 2001; Wirtz and Hell, 2006). The effective inactivation of OAS-TL demonstrates the inability of the CSC to channel substrates and is the basis for the regulatory function of the complex *in vivo* (Hell and Hillebrand, 2001).

Beside the effector driven dissociation of the CSC by OAS both plant and bacterial CSC are stabilized by sulfide (Kredich et al., 1969; Wirtz and Hell, 2006). The function of this stabilization *in vivo* could be the increase of OAS levels for enhanced cysteine and glutathione production. Enhanced glutathione production is needed during certain stress conditions like pathogen attack or heavy metal stress, when enough sulfide is present. Without the stabilizing effect of sulfide, OAS would dissociate the CSC, thereby down-regulating its own synthesis. A possible target for sulfide stabilization would be the active site of OAS-TL, where OAS and C-terminal tail of SAT bind. But no stable sulfide binding pocket was found in the active site of OAS-TL, which is consistent with the irreversibility of the second-half reaction (Tai et al., 1995). The identification of an allosteric anion-binding site at the dimer interface of bacterial OAS-TL provides an alternative target site for sulfide (Burkhard et al., 2000). The binding of sulfide to this site causes a conformational shift in the asparagine loop to a new 'inhibited' state of OAS-TL that differs from the 'open' or 'closed' conformation of the enzyme (section IV.B). It is conceivable that the 'inhibited' conformation of OAS-TL stabilizes the C-terminal tail of SAT in the active site of OAS-TL and promotes complex stability.

Due to the inactivity of OAS-TL inside the complex, the complex associated OAS-TL is believed to act as a regulatory subunit of SAT. In theory, association of SAT and OAS-TL activates SAT in order to produce OAS for cysteine synthesis (section VI.E). A trimeric SAT homomer contains three independent active sites, which can all be used to synthesize cysteine. Most likely all active sites can act separately, since in the form III crystal of *H. influenzae* SAT (pbf file: 1SST) all possible binding sites for acetyl-CoA are loaded with CoA in the presence of serine (Olsen et al., 2004). It is difficult to conceive how the binding of one OAS-TL dimer to one or two C-terminal tails at the distal end of a SAT trimer can result in an effective up-regulation of all active sites in the trimer. In agreement with this consideration, neither the affinity for acetyl-CoA and serine nor the maximal activity (V_{max}) of the bacterial SAT is altered by complex formation (Kredich et al., 1969; Cook and Wedding, 1978; Mino et al., 2000a). However, formation of CSC

protects SAT from proteolysis and inactivation by low temperature in *in vitro* experiments (Mino et al., 2000a). Under *in vivo* conditions, both would result in higher SAT activity by CSC formation. Cytosolic SAT of *A. thaliana* was shown to be strongly upregulated by complex formation due to increase of maximal velocity (Bonner et al., 2005) and by increasing the affinity for acetyl-CoA and serine (Droux et al., 1998). The mitochondrial SAT from *A. thaliana* was also characterized in respect to their activation by complex formation. In the latter study only a slight but significant activation of mitochondrial SAT by complex formation was observed. This activation was based on a higher affinity of complex associated SAT for acetyl-CoA (Wirtz et al., 2001). Whether the slight increase of SAT activity by complex formation is sufficient to regulate OAS production under *in vivo* conditions needs to be confirmed. A molecular explanation for the increased acetyl-CoA affinity could be provided by the rearrangement of the C-terminal tail of SAT during binding to OAS-TL. Parts of the same tail were also shown to cover the binding pocket for acetyl-Co after the binding of cysteine, causing a reduction of the affinity of *H. influenzae* SAT for acetyl-CoA (Olsen et al., 2004; Johnson et al., 2005). Taken together, these results put a note of caution to the generally accepted assumption that SAT is activated by complex formation, due to increased affinity or V_{max} of the enzyme in the complex. At least in bacteria, other factors like temperature or proteases are necessary to prevent decreased activity of SAT upon complex formation. The moderate activation of mitochondrial AtSAT upon complex formation in *in vitro* experiments possibly suggests a similar scenario in the mitochondria of plants.

V. S-transfer Reactions and Degradation of Cysteine

Enzymatic degradation of cysteine was hypothesized in the 1940s to take place in bacteria and animals (Fromageot et al., 1940; Smythe, 1945; Fromageot, 1951). Forty years later enzymatically catalyzed degradation of cysteine came to focus also in higher plants as a result of the pioneering work of several research groups (Harrington and Smith, 1980; Rennenberg, 1983; Rennenberg et al.,

1987; Schmidt, 1987). Four major functions for cysteine catabolism in higher plants are discussed in the literature: (a) recycling of cysteine bound sulfur as a result of protein degradation, (b) deposition of elemental sulfur and release of sulfide in response to pathogen defense, (c) transfer of sulfur from cysteine to iron-sulfur clusters and (d) detoxification of cyanide by using cysteine as an acceptor. In contrast to synthesis of cysteine the current knowledge about degradation of cysteine is scarce. Anabolism and catabolism of cysteine must be separated in order to achieve coordination between the two pathways. This separation can be achieved by localization of the two pathways in different subcellular compartments of the plant cell, temporal delimitation of respective protein activities, or the use of different enantiomers of cysteine specifically for degradation (D-cysteine) and synthesis (L-cysteine) (Wirtz and Droux, 2005). Up to now it is unclear if the mandatory separation of cysteine synthesis and degradation is achieved by only one of these mechanisms, several or all of them.

The following reactions (Fig. 4) can be envisaged to occur in the catabolism of cysteine: (A) L-cysteine desulfhydrase reaction or L-cysteine lyase, (B) L-cystine lyase reaction, (C) cysteine desulfurase reaction, (D) D-cysteine desulfhydrase reaction, (E) β-cyanoalanine synthase reaction, (F) decomposition of cysteine by OAS-TL, (G) conversion of cysteine to mercaptopyuvate by transaminases or (H) amino acid oxidases. Since the formation of β-mercaptopyuvate has not been shown in plants so far (Schmidt, 2005) and the kinetic properties of the most abundant OAS-TLs in Arabidopsis does not allow a significant backward reaction (Wirtz et al., 2004), the last three possible reaction mechanisms will not be discussed in this review. Nevertheless, several candidate genes exist in higher plants that may be involved in the degradation of cysteine (Fig. 5).

A.   L-cysteine Desulfhydrase or L-cysteine Lyase

Although other cysteine consuming activities have been shown in higher plants (see below), the enzymes that are the most likely candidate for the degradation of cysteine in higher plants are L-cysteine desulfhydrase (EC 4.4.1.15, L-CDes or L-cysteine lyase, Fig. 4A) or L-cystine lyases. By feeding ^{35}S-labeled L-cysteine (which contains per se cystine) to cultured tobacco cells Harrington and Smith (1980) observed that the initial step in cysteine (or cystine) degradation yielded pyruvate and sulfide in a 1:1 ratio and possibly ammonium. The catabolic reaction was linear with respect to time and amount of protein and had a pH optimum of 8 in crude extracts. A L-CDes with a pH optimum of 8 has also been shown in Streptococcus anginosus (Yoshida et al., 2002). Preliminary kinetic data for the tobacco L-CDes indicated that the K_M is approximately 0.2 mM cysteine. Pre-incubation of the tobacco cells with cysteine results in 15–20-fold increase of the extractable degradative activity, while intracellular cysteine concentrations increases (Harrington and Smith, 1980). In contrast, S fertilization of field grown Brassica napus also significantly increase the contents of total S, glutathione and cysteine, but decreases L-CDes activity. In the same study the activity of L-CDes in B. napus was found to be induced by infection with Pyrenopeziza brassicae that increased cysteine and glutathione contents as well (Bloem et al., 2004). The latter finding gives rises to the hypothesis that L-CDes are involved in defence against pathogens by releasing toxic sulfide. Taken together these results indicate that L-CDes activity is not only regulated by intracellular cysteine levels, but also by unknown endogenous and environmental signals.

So far an enzyme with exclusive L-CDes activity has not been isolated from higher plants, although several candidate genes for L-cysteine desulfhydrases were identified in the genome of Arabidopsis thaliana (Fig. 5). Like in higher plants the activity of the L-CDes in the enterobacterium Salmonella typhimurium is inducible by pre-incubation with cysteine. The L-CDes of S. typhimurium was purified to quasi homogeneity after induction with cysteine. The native protein has a molecular weight of 229 kDa, suggesting a hexa-homoligomeric structure composed of 37 kDa subunits, which contain one PLP per subunit (Kredich et al., 1972). Substrate saturation experiments using the purified enzyme demonstrated positive cooperativity with a Hill-coefficient of 1.9 and a K_M for L-cysteine of 0.17 and 0.21 mM (Kredich et al., 1973). Both the inducibility of the enzymatic activity by cysteine and the biochemical characteristics of the enzyme were similar in S. typhimurium

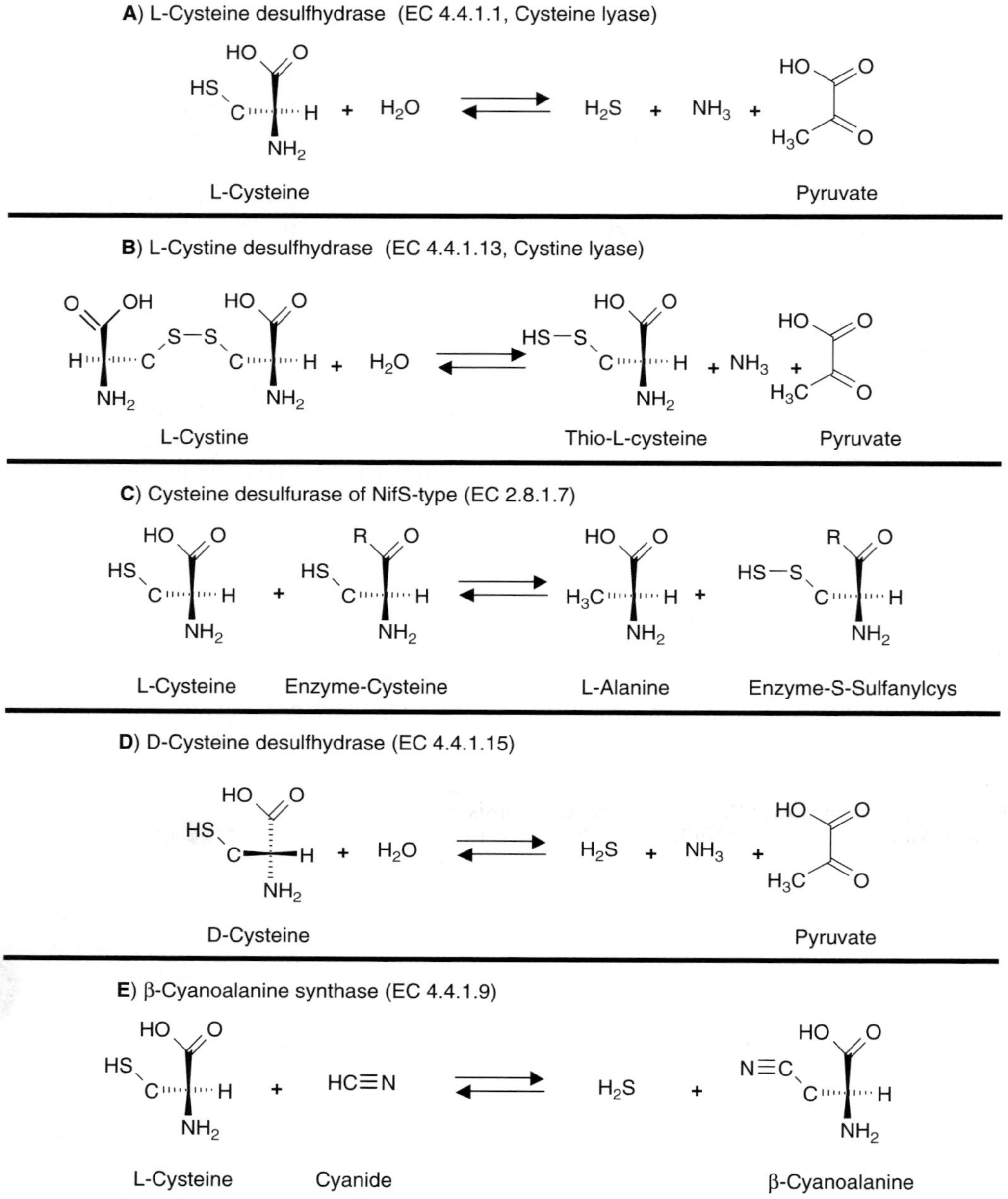

Fig. 4. Cysteine consuming reactions. The figure depicts the candidate reactions for degradation of cysteine. Note that L-cysteine desulfhydrase (A), L-cystine desulfhydrase (B), Cysteine desulfurase (C) and β-cyanoalanine synthase (E) degrade the L-enantiomer of cysteine, while D-cysteine desulfhydrase (D) uses the D-enatiomer of cysteine as substrate. The chemical structures of substrate and products are shown according to IUPAC nomenclature.

and tobacco indicating a conserved regulation of cysteine degradation in pro- and eukaryotes.

B.   L-cystine Desulfhydrase or L-cystine Lyase

L-cystine is the oxidized form of cysteine in which two L-cysteine molecules are linked together via a disulfide bridge. L-cystine was identified in all analyzed higher plants, but the exact degree of oxidized cysteine is hard to quantify, due to oxidation of cysteine during extraction. The L-cysteine pool is kept usually in a reduced state by the reduction/oxidation buffer glutathione, which is present in high excess of L-cysteine. Nevertheless L-cystine can accumulate during oxidative stress

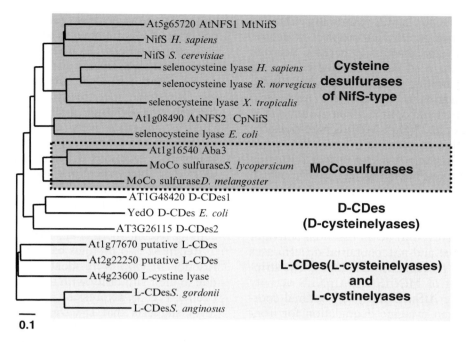

Fig. 5. Sequence relationships of cysteine degrading enzymes. The phylogenetic tree was created with Vector NTI 9 (Invitrogen, Darmstadt) using the full length proteins. Annotation of enzymes is according to their description in the respective GenBank™ accession: AtCpNifS: Q93WX6, AtMtNifS: O49543, NifS *H. sapiens*: Q9Y697, NifS *S. cerevisiae*: NP_009912, selenocysteine lyase *H. sapiens*: NP_057594 *R. norvegicus*: XP_343628, *X. tropicalis*: NP_001011164, *E. coli*: NP_416195, Aba3: Q9C5X8, Molybdenum cofactor (MoCo) sulfurases *S. lycopersicum*: Q8LGM7, *D. melanogaster*: Q9VRA2, D-CDes1:NP_175275, D-CDes2: ABM06014, YedO: NP_416429, At1g77670: ABI49465, At2g22250: AAD23617, At4g23600: BAF01453, L-CDes *S. gordonii*: BAC20222, *S. angionosus*: BAC00815. The Arabidopsis proteins are functionally characterized (section V) or indicated as putative. MoCo sulfurases and NifS proteins from prokaryotes and eukaryotes cluster together and form the group of cysteine desulfurases of NifS type (dark grey). D-cysteine desulfhydrases (white) share higher sequence similarity with cysteine desulfurases than with L-cysteine desulfhydrases (grey).

conditions like high light, cold stress or pathogen attack. Recently, Jones and coworkers revealed that CORI3, which is annotated as a tyrosine aminotransferase-like protein, acts *in vitro* as a L-cystine lyase (EC 4.4.1.13, Fig. 4B, Jones et al., 2003). Interestingly CORI3 is jasmonic acid and salt stress-inducible, which may indicate regulation of cysteine/cystine degradation by stress and hormone signalling (Gong et al., 2001; Sasaki et al., 2001). In animals and bacteria cystathionine γ-lyase is known to act as a L-CDes and L-cystine lyase (Smacchi and Gobbetti, 1998). Usually, cystathionine γ-lyase catalyses the cleavage of cystathionine, which results in the production of cysteine and the byproducts ammonium and α-ketobutyrate. Nevertheless, the enzyme can also use L-cysteine and L-cystine as substrates. For a detailed description of all possible substrates and reaction mechanisms of this multifunctional enzyme look up the enzyme-database BRENDA (http://www.brenda.uni-koeln.de/). Biochemical characterization of the plant cystathionine β-lyase indicates that it have not this wide substrate specificity (Staton and Mazelis, 1991).

C.   Cysteine Desulfurase of NifS-type

The cysteine desulfurases of the NifS-type (EC 2.8.1.7, Fig. 4C) identified in Arabidopsis differ in their subcellular localization and developmental expression patterns. At1g16540 (Aba3) is located in the cytosol (Heidenreich et al., 2005), while At5g65720 (MtNifS, AtNFS1) and At1g08490 (CpNifS, AtNFS2) are transported to mitochondria and plastids, respectively. MtNifS and CpNifC provide elemental sulfur for iron-sulfur cluster formation in the respective organelle, while Aba3 acts as a molybdenum cofactor sulfurase.

Plants with reduced CpNifS expression exhibited chlorosis, a disorganized chloroplast structure, stunted growth and eventually became necrotic and died before seed set (Van Hoewyk et al., 2007), demonstrating that cysteine desulfurase activity in plastids is essential for iron-sulfur cluster formation and optimal plant growth (Ye et al., 2005; Ye et al., 2006). MtNifS is responsible for desulfuration of cysteine in order to provide elemental sulfur for iron-sulfur cluster formation in mitochondria (Frazzon et al., 2007), which is believed to provide iron-sulfur clusters for cytosolic and nuclear iron-sulfur proteins (Balk and Lobreaux, 2005). Very recently the AtSufE protein was resolved to be an essential activator of both plastidic and mitochondrial desulfurases in Arabidopsis (Xu and Moller, 2006). The stringent regulation of MtNifS and CpNifS activity by the presence AtSufE allows an optimal coordination between cysteine degradation for iron-sulfur cluster formation and cysteine synthesis by OAS-TL in mitochondria and plastids. So far it is unknown how the coordination between cysteine degradation for molybdenium cofactor synthesis and cysteine synthesis in the cytosol is achieved. An elegant way would be allosteric inhibition of Aba3 activity by the molybdenum cofactor itself, but a detailed biochemical analysis of Aba3 in this respect is missing to date.

D. D-cysteine Desulfhydrase

In contrast to the former described cysteine degrading activities the D-cysteine desulfhydrase (D-CDes, EC 4.4.1.15, Fig. 4D) uses D-cysteine instead of L-cysteine, which provides an elegant explanation for the presence of cysteine producing and degrading activities in the same sub-cellular compartment. The authors are not aware of any report that unequivocally demonstrate the presence of D-cysteine as a natural occurring metabolite of higher plants. A racemase that converts L-cysteine to D-cysteine has also not been isolated from higher plants (Schmidt, 1986), but protein bound L-cysteine could be converted to D-cysteine non enzymatically by the base-catalysed racemization reaction (Friedman, 1999). In addition, higher plants are suggested to take up D-amino acids from the soil, where they are secreted by microorganisms (Aldag et al., 1971). Although D-cysteine was not isolated from plants

so far, D-amino acids in general are believed to be principle components of gymnosperms as well as of monocotyledonous and dicotyledonous angiosperms (Bruckner and Westhauser, 2003). Green plants and algae are capable of degrading D-cysteine specifically by D-CDes activity (reviewed in Schmidt, 1986). The L-CDes and D-CDes activities were easily distinguishable in cucurbit and tobacco, since the enzymes catalysing both reactions differ in their response to inhibitors, their subcellular localization and had different pH optima. In both plants species the extractable D-CDes activity was more than one order of magnitude higher than the L-CDES activity (Rennenberg et al., 1987).

Recently, two proteins catalysing the D-CDes reaction in vitro were identified in Arabidopsis due to their similarity with the D-CDes of E. coli (YedO, Fig. 5). Expression analysis on transcript level suggests that D-CDes1 (At1g48420) and D-CDes2 (At3g26115) might have specific roles during development of Arabidopsis. While D-CDes1 catalyses specifically the desulfuration of D-cysteine, D-CDes2 accepts L-cysteine beside D-cysteine as substrate (Riemenschneider et al., 2005a; Riemenschneider et al., 2005b). In silico analysis of transit peptides from both proteins indicate a mitochondrial localization, which was proven for D-CDes1 independently by immunological detection and GFP-fusion constructs. Consistent with its sulfur remobilizing function D-CDes activity was identified to be highest in senescent plant leaves. Neither D-CDes1 nor D-CDes2 transcript levels could explain the significant increase of C-Des activity during senescence (Riemenschneider et al., 2005a). The latter results indicate a post-translational control of the D-CDes proteins or the existence of further D-CDes enzymes, which are not identified.

E. β-cyanoalanine Synthase

β-Cyanoalanine synthase (CAS; EC 4.4.1.9, Fig. 4E) catalyzes the formation of the non-protein amino acid β-cyanoalanine from cysteine and cyanide, thereby playing a pivotal role in cyanide detoxification (Floss et al., 1965; Meyers and Ahmad, 1991). CAS activity has been detected in bacteria (Dunnill and Fowden, 1965), insects (Meyers and Ahmad, 1991) and plants (Blumenthal et al., 1968; Hendrickson and Conn, 1969). In plants, ethylene synthesis by 1-aminocyclopropane-1-carboxylic acid oxidase results in the production of cyanide (Peiser et al., 1984). Inhibition of CAS activity by 2-aminoxyacetic acid revealed that detoxification of cyanide is dependant on CAS activity in higher plants. However, several-hundred fold increases of cyanide production during ripening of apple and avocado fruits are not accomplished by a severe induction of extractable CAS activity (Yip and Yang, 1988). Cyanide toxicity is caused by binding of cyanide to the iron atom of the enzyme cytochrome c oxidase in the fourth complex in the mitochondrial membrane. This denatures the enzyme, and the final transport of electrons from cytochrome c oxidase to oxygen cannot be completed. As a result, the electron transport chain is disrupted, meaning that the cell can no longer aerobically produce ATP for energy. Consequently CAS activity is located exclusively in mitochondria of insects (Meyers and Ahmad, 1991). Several lines of evidence indicate that in higher plants CAS is also exclusively localized in mitochondria. Firstly, CAS activity was purified to homogeneity for the first time from mitochondrial fractions of blue lupine (Hendrickson and Conn, 1969), secondly, the identified CAS proteins of spinach and Arabidopsis are both located in the mitochondrial lumen (Warrilow and Hawkesford, 1998; Hatzfeld et al., 2000). Thirdly, the background activity of CAS that was identified in the cytosol of cocklebur (*Xanthium pennsylvanicum*) seeds is attributed to cytosolic OAS-TL (Maruyama et al., 1998). CAS and OAS-TL are structurally very similar proteins, which are able to catalyse both: β-cyanoalanine and cysteine synthesis to a certain extend. For that reason it is hard to distinguish if a protein acts *in vivo* as a CAS or an OAS-TL. Only a careful biochemical characterisation of kinetic constants for both reactions allows a categorization of a certain enzyme to act as CAS or OAS-TL under *in vivo* conditions (Warrilow and Hawkesford, 2000; Yamaguchi et al., 2000). By using these careful biochemical characterization in combination with immunological detection, the putative isoforms of OAS-TL from spinach and potato were identified as CAS (Warrilow and Hawkesford, 1998; Maruyama et al., 2000). Most likely a feature that specifies an authentic OASTL is the ability to interact physically with SAT protein in the CSC. While two-hybrid system

and co-purification experiments show the associated SAT and OAS-TL enzymes (Bogdanova and Hell, 1997; Droux et al., 1998; Wirtz et al., 2001; Hell et al., 2002; Droux, 2003), such an interaction makes no sense for a CAS.

However, in *A. thaliana* both proteins are present in mitochondria (Hatzfeld et al., 2000; Jost et al., 2000; Wirtz et al., 2004). The presence of both enzymatic activities in the mitochondria would allow an efficient detoxification of cyanide via cysteine by consumption of serine, which is produced in mitochondria in high amounts by photorespiration (Fig. 6). In contrast to potato and spinach, Arabidopsis produces cyanogenic glucosinolates, which are part of the defence system against herbivores in all Brassicaceae. It is conceivable that the enhanced cyanide metabolism in the Brassicaceae requires a more efficient mitochondrial cyanide detoxification system consisting of CAS and OAS-TL in mitochondria itself.

In Arabidopsis CAS is 37 kDa protein (trivial name: *At*CysC1), which is encoded by the gene At3g61440. The transcript levels of At3g61440 are almost constant in seeds, rosettes, roots and the inflorescence, with the exception that in pollen only small amounts of transcript could be detected (data from https://www.genevestigator.ethz.ch). In agreement with the finding that enhanced ethylene formation does not induce extractable CAS activity, the transcript level of At3g61440 is kept constant in senescent leaves. Recombinant AtCysC1 has a specific activity of 62 μmol sulfide min^{-1} mg^{-1} protein and K_M values of 2,5 mM and 60 μM for cysteine and cyanide, respectively (Hatzfeld et al., 2000). The spinach CAS has similar biochemical properties (Warrilow and Hawkesford, 2000). It seems doubtful that the low affinities of the recombinant CAS proteins for cyanide allows keeping the cyanide concentration at cellular levels around 0.2 μM, which is far below the K_i of cyanide (10–20 μM) to inhibit respiration (Yip and Yang, 1988).

VI. Regulation of Cysteine Flux in Plants

Regulation in any metabolic pathway is a mechanism to adjust supply and demand. Major triggers for regulation are plant development, environmental factors like stress and nutrient supply.

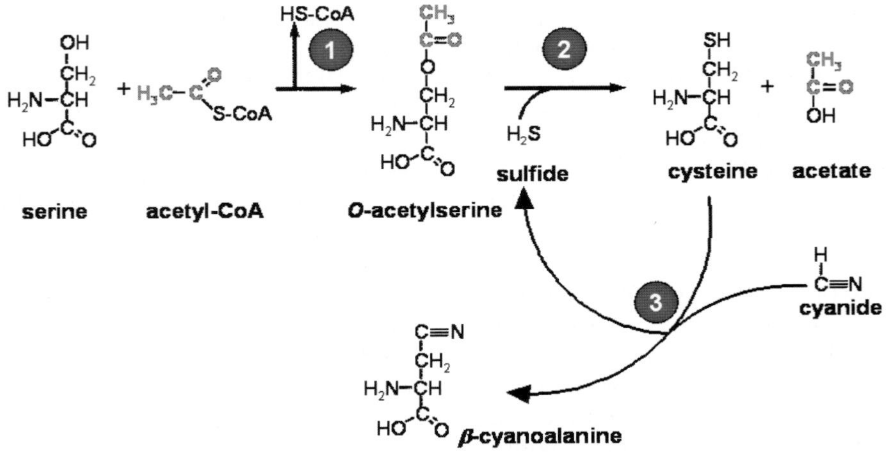

Fig. 6. Detoxification of cyanide in mitochondria of *Arabidopsis thaliana*. Serine acetyltransferase (1) and *O*-acetylserine (thiol)lyase (2) catalyze the synthesis of cysteine in a two step process. The fate of the acetyl-moiety during catalysis is shown in grey. In mitochondria β-cyanoalanine synthase (3) uses cysteine as an acceptor to detoxify cyanide. During the detoxification of cyanide sulfide is released, which can be fixed by OAS-TL to reconstitute the acceptor for cyanide.

The spatial component includes root-to-shoot and source-sink communication, i.e. the supply of sulfate to roots, its transport along the plant axis, the demand by developing seeds or locally stressed organs and transport of reduced sulfur as nutrient or signal within the plant. In all cases the cell, independent of its position in the plant body, has to be able to sense supply and demand of metabolites and to respond adequately. Searches for long-distance signals as well as cellular signal transduction have been in the focus of plant sulfur research. Green algae and in particular *Chlamydomonas* have been and still are model organisms in the latter respect (see chapter by Shibagaki and Grossman, this book). Work on Arabidopsis suggested that cysteine synthesis plays an integral role in the regulation of primary sulfur metabolism. In particular the properties of the cysteine synthase protein complex gave rise to theories about a new metabolite sensing mechanism (Hell and Hillebrand, 2001; Droux, 2003). Moreover, the reaction intermediate of cysteine synthesis, OAS, forms a direct connection of sulfate assimilation with nitrate assimilation and carbon metabolism (Kopriva et al., 2002; Hesse et al., 2004). Network analyses using microarrays confirm these assumptions (see Takahashi and Saito, this book). The regulation of flux into and through cysteine synthesis therefore is of central importance for growth and fitness of the plant. It is noteworthy that this position of cysteine syn-

thesis, although based on totally different regulation, parallels the early regulatory system in enterobacteria.

A. Allosteric Regulation of Cysteine Synthesis by Feedback Inhibition

Feedback inhibition of SAT by cysteine is an important feature for the enterobacteria regulatory system and is also found in several SATs from plants. Unfortunately, no information on feedback sensitivity of SAT from phototrophic bacteria is available. For higher plants an elegant model of flux control of cysteine synthesis was suggested (Saito, 2000), based on the observation that in *A. thaliana* the cytosolic SAT isoform is strongly feedback inhibited by cysteine, whereas the plastid and mitochondrial SAT isoforms are mostly insensitive (Noji et al., 1998). 50% (IC_{50}) inhibition of maximal activity was determined for the cytosolic SATs from watermelon (*Citrullus vulgaris*) at $1.8\,\mu M$ and from *A. thaliana* at $2.9\,\mu M$ cysteine (Saito et al., 1995; Noji et al., 1998). In this model cytosolic SAT activity is considered to particularly important for the control of OAS concentrations, since is potentially involved in the expression control of sulfur metabolism genes in response to sulfate availability. In the organelles feedback-insensitive SAT isoforms would reside to allow independent cysteine formation and this has indeed been reported (Noji et al., 1998; Saito,

2000). An independent analysis of recombinant SAT3 from Arabidopsis mitochondria however, reported an intermediate sensitivity for IC_{50} of 50–70 µM cysteine (Wirtz and Hell, 2003a).

Furthermore, the degree of SAT feedback-sensitivity varies considerably between plant species and subcellular compartment. A plastid SAT from spinach (SAT56) showed an IC_{50} of 7.6 µM cysteine (Noji et al., 2001b). One organelle-localized (SAT1) SAT from tobacco (*Nicotiana tabacum*) showed an IC_{50} value of about 50 µM cysteine and another organelle-localized SAT was not inhibited at all at cysteine concentrations up to 600 µM (SAT4; Wirtz and Hell, 2003a). In contrast, a tobacco cytosolic SAT (SAT7) was indeed inhibited by 50% at about 50 µM cysteine. In pea (*Pisum sativum*) the protein fraction with cytosolic SAT activity was not inhibited by cysteine at all, whereas the SAT activities in isolated chloroplasts and mitochondria displayed IC_{50} values of 33 and 283 µM, respectively (Droux, 2003). A note of caution is added to these data: they were mostly determined using non-purified or recombinant enzymes with N-terminal fusions. Furthermore, these inhibition assays were obviously performed with SAT proteins in the absence of OAS-TL. Whether the degree of feedback inhibition by cysteine changes with the association or dissocitation of the cysteine synthase complex is not known.

Allosteric regulation of cytosolic SAT as primary mechanism of flux control is difficult to interpret in view of a number of *in vivo* experiments. The abiotic defense response of plants that is induced by heavy metals or xenobiotics in order to form glutathione in the cytosol results in an increased flux and accumulation of cysteine which would inhibit SAT (Rüegsegger and Brunold, 1992; Farago and Brunold, 1994). Similarly, fumigation of plants with H_2S results in enhanced levels of cysteine and is even sufficient to cover the sulfur demand of the plant (De Kok et al., 2002). Over-expression of feedback-sensitive SAT from *E. coli* and *A. thaliana* in transgenic plants promotes elevated cysteine levels (Blaszczyk et al., 1999; Harms et al., 2000; Noji and Saito, 2002; Wirtz and Hell, 2003a). Any accumulation of cysteine in a compartment with feedback-sensitive SAT could only be explained, if the subcellular cysteine pools were separated, but the distribution of cysteine within cells is evidently not known to date.

B.   Application of Transgenic Plants

Transcriptional regulation of the sulfate assimilation pathway is most pronounced at the level of sulfate transporter gene expression (Maruyama-Nakashita et al., 2003; Maruyama-Nakashita et al., 2004; Maruyama-Nakashita et al., 2006). Deficiency of sulfate rapidly induces the expression of genes encoding sulfate transporter proteins of the plasmalemma. The genes encoding activation and reduction steps are expressed in most cells at a basal level, but can also be modulated by sulfate supply and demand. However, the mRNA changes are smaller and often even less effected at the protein and activity levels is observed when they are compared to sulfate transport (Logan et al., 1996; Takahashi et al., 1997; Bork and Hell, 2000; Koprivova et al., 2000). As stated above, the response of genes related to cysteine synthesis to sulfate limitation and their corresponding enzyme activity changes are rather small (Hell et al., 1994; Hesse et al., 1999). Thus, at least with respect to sulfate supply, the classical parameter for sulfur research, the importance of transcriptional control seems to decrease from expression of sulfate transporter genes to cysteine synthesis (Hell, 2003). It should be cautioned that other environmental factors such as light and nitrogen nutrition or stress may act differently on the transcription of these genes.

This observation nevertheless suggests that cysteine synthesis is a flux control point for primary sulfur metabolism. If cysteine synthesis is enhanced, a feedback to sulfate uptake and activation would be expected that provides sufficient sulfide and OAS for cysteine synthesis. Understanding of regulation of cysteine synthesis itself is therefore the potential key to the control of assimilatory sulfur metabolism in general.

Modification of the activities of SAT and OAS-TL in the cytosol, plastid and mitochondria are one way to test the regulatory function of cysteine synthesis. Mutants with decreased activities of one of the protein isoforms due to insertional inactivation have not been reported so far. Over-expression approaches in Arabidopsis, tobacco and potato using both proteins basically confirm the limiting role of SAT (Sirko et al., 2004). The expectation of such an approach is an enhanced formation of OAS as far as the substrates serine and acetyl-CoA are available. Total measurable

SAT activity in leaf extracts of wildtype plants of many tested species is about 400 times lower than OAS-TL activity (Droux, 2003). OAS will therefore even in overexpressor lines accumulate only to a moderate extent as long as sufficient sulfide is available, but cysteine and glutathione will show increased steady-state levels. Consequently, the over-expression of OAS-TL in plastids or the cytosol had only minor effects on the contents of cysteine and glutathione, although remarkable effects on stress tolerance were reported (Dominguez-Solis et al., 2001; Harada et al., 2001; Noji et al., 2001a; Youssefian et al., 2001; Kawashima et al., 2004). In line with the above observations isolated chloroplasts from wildtype and transgenic lines with OAS-TL overexpression in this compartment, when supplied with external OAS, showed enhanced cysteine and glutathione formation (Saito et al., 1994). However, assuming strongly increased consumption of cysteine and particularly glutathione under the stress conditions applied in these reports, the improved tolerance of transgenic lines suggests a successfully increased flux rate *in vivo*, even so OAS-TL was already present in the wildtype plants at high activity.

According to the expectation, expression of either bacterial or plant SAT in the cytosol or plastids achieved significantly higher steady-state concentrations of cysteine and glutathione, depending on whether cysteine feedback-sensitive or insensitive SAT proteins were applied. Where determined, OAS contents accumulated too much lower extent than thiols. Unfortunately, these transgenics plants were less intensively tested for their stress tolerance properties (Sirko et al., 2004; Wawrzynski et al., 2006). No reports have included cysteine degrading activity in SAT overexpressors. SAT overexpression studies all indicate that the substrates for OAS-TL, the final step of cysteine synthesis, are rate limiting (Blaszczyk et al., 1999; Harms et al., 2000; Noji and Saito, 2002; Wirtz and Hell, 2003a, 2007).

In addition to these data, an independent experiment using an SAT protein that had been mutated and enzymatically inactivated in its catalytic center, provided astonishing changes in the sulfur composition of the transformed plants (Wirtz and Hell, 2007). This SAT protein from Arabidopsis was still able to interact with OAS-TL, including endogenous OAS-TL, when expressed in the cytosol of tobacco leaves. The tobacco SAT was apparently at least partially out-competed by strongly accumulating mutated SAT, presumably resulting in down-regulation of cysteine synthesis in the cytosol and up-regulation in the chloroplasts and mitochondria. This conclusion by the authors of the study was supported by up to 25 times enhanced cysteine steady-state levels compared to wildtype, twofold increased OAS concentrations and a significant overall accumulation of total sulfur content in the transgenic lines. This was the largest reported enhancement of flux into cysteine so far that strongly suggests that the formation of the CSC together with exchange mechanisms between cellular compartments or lack thereof are of great importance for the regulation of cysteine synthesis (Wirtz and Hell, 2007).

C. Effectors and Properties of the Cysteine Synthase Complex

The availability of substrates is of general importance for any rate of product formation, if the concentrations run below binding constants and Michaelis-Menten constants of the respective protein. In case of cysteine synthesis serine is unlikely to limit, but acetyl-CoA may, at least in some compartments or metabolic situations, become critical for maximal activity of SAT in the CSC (Wirtz and Droux, 2005). As outlined before, the K_M values of OAS-TL for OAS are rather high, giving rise to the assumption that OAS prevents free OAS-TL from maximal rates of cysteine formation under most circumstances. While substrate affinities of bacterial and at least Arabidopsis OAS-TL proteins for sulfide are in the low micromolar range (Wirtz et al., 2004), the true cellular sulfide concentrations are neither known under sufficient nor limiting sulfate supply. Since the integration of sulfide is the decisive step of the entire assimilatory pathway, it seems reasonable to assume that its availability is critical to maintain cysteine synthesis and thus growth. The product cysteine, as discussed before, may control the rate of its formation by feedback inhibition of SAT, but depending on the compartment and the properties of the present SAT isoform.

These considerations suggest OAS and sulfide as potentially most important factors for the flux rate of cysteine synthesis. In fact, both metabolites are not only intermediates of the pathway but have additional functions: OAS seems to be involved in the control of sulfur-related gene expression and OAS as well as sulfide exert allosteric properties on the association of SAT and OAS-TL in the CSC. OAS has long been known to destabilize the CSC in enterobacteria and plants (Kredich et al., 1969; Droux et al., 1998), but only recently its cellular concentrations could be reliably determined. Data from several authors using different tissues from Arabidopsis (leaf, silique, cell culture), soybean (cotyledon) and potato (*in vitro* shoots) revealed overall concentrations of 0.3 to 6 μM after extraction of whole samples, if cells were supplied with sufficient sulfate. Independent of the large differences in the time courses monitored, the typical response to sulfate limitation consisted in an increase of OAS levels to values between 4 and 60 μM. Unfortunately, virtually nothing is known about subcellular differences and possible transport of OAS between subcellular compartments in higher plants (Kim et al., 1999; Awazuhara et al., 2000; Wirtz and Hell, 2003a; Hopkins et al., 2005).

These changes of OAS concentrations in response to sulfate limitation suggest a function in signal transduction that is corroborated by feeding studies. Addition of OAS to barley roots (Smith et al., 1997), *Zea mays* cell cultures (Clarkson et al., 1999) and *Arabidopsis* plants (Koprivova et al., 2000; Kopriva et al., 2002) resulted in the increase of mRNA levels and activities of sulfate transporters, ATP sulfurylase and APS reductase. External concentrations of 1 mM OAS over-ruled the repressing effect of sufficient sulfate in the growth medium. It may be concluded from these findings that OAS is directly or indirectly involved in the control of sulfate uptake and assimilation. This role applies to a large number of genes and not just primary reactions. OAS feeding induced the expression of the nitrilase gene 3 that is involved in glucosinolate degradation in *Arabidopsis* (Kutz et al., 2002). DNA microarray analysis comparing sulfate deficiency and OAS feeding in *Arabidopsis* plants revealed more than hundred up-regulated mRNAs of which about 40% were identical between the sulfate deficiency response and the effect of OAS feeding (Hirai et al., 2003, Takahashi and Saito, this book).

Interesting evidence for the contribution of OAS to sulfur deficiency signalling is provided by identification of a promoter fragment present in the *Arabidopsis Nit3* gene (Kutz et al., 2002) and the soybean β-conglycinin gene (Kim et al., 1999; Ohkama et al., 2002). This element alone in combination with activation elements was sufficient to confer sulfate deficiency and OAS responsiveness to reporter genes. This strongly suggests that OAS is an intermediate in cysteine synthesis and in addition a potential element of signal transduction to regulate sulfur homeostasis in a plant cell. However, detailed analysis of the promoter of sulfate transporter 1;1 (*Sultr1;1*) from Arabidopsis identified a 16 basepair element ('SURE') that was alone sufficient to confer sulfate deficiency response and repression by cysteine and glutathione, but not activation by OAS (Maruyama-Nakashita et al., 2005).

D. Biochemical and Allosteric Properties of the Cysteine Synthase Complex

The CSC is very unusual for a metabolic protein complex because its function in bacteria and plants is obviously different from substrate channeling of OAS (Cook and Wedding, 1978; Droux et al., 1998). It seems now evident that the docking of the C-terminus of SAT to the catalytic cleft of OAS-TL strongly hinders access of OAS and thus results in diffusion of the intermediate into the surrounding solution (Huang et al., 2005; Francois et al., 2006, section III.B). This finding elegantly explains why OAS can dissociate the CSC (via competition with the C-terminus for the catalytic cleft) and why free OAS-TL dimers are required (to catch released OAS and free sulfide). The function of OAS-TL in the CSC was explained as regulatory subunit that acts by stabilisation and activation of SAT (Droux et al., 1998, Fig. 7).

It is indeed remarkable that SAT activity seems to rely on the association with OAS-TL. SAT activity has never been detected in plant or bacterial protein preparations without associated OAS-TL protein. Release of SAT from the complex *in vitro* results in rapid loss of activity (Droux et al., 1998; Wirtz et al., 2001) and to achieve maximal SAT activity for cysteine production a 400-fold excess of OAS-TL activity is required (Ruffet et al., 1994; Droux et al., 1998). This correlates well with the reported 354-fold and 306-fold excess of OAS-TL to SAT activity in chloroplasts of spinach and pea (Ruffet et al., 1994; Ruffet et al., 1995). However,

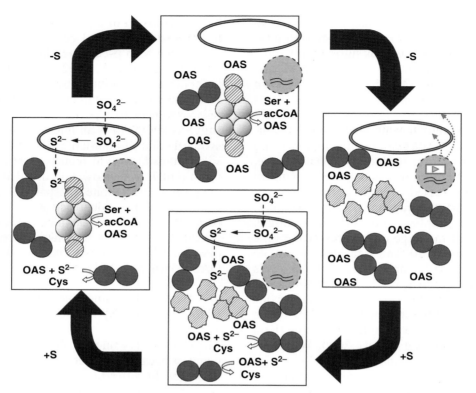

Fig. 7. Hypothesis for the function of the cysteine synthase complex *in planta*. The figure depicts the theoretical model for the sensor and regulatory function of the cytosolic CSC by *O*-acetylserine (OAS) dependant dissociation of the CSC. Under normal sulfur supply (left panel) sulfate (SO_4^{2-}) enters the cell and is reduced to sulfide (S^{2-}) in the plastids (light grey circles). Sulfide leaks as H_2S out or is actively transported across the plastid membranes (dashed arrows) into the cytoplasm. In the cytoplasm sulfide is used by free *O*-acetylserine(thiol)lyase dimers (OAS-TL, dark circles) to synthesize cysteine. The precursor of cysteine is OAS, which is produced by serine acetyltransferase (SAT, light grey circles). Under these conditions SAT is active, since it is associated with OAS-TL in the CSC. OAS releases the CSC since OAS-TL inside the complex (dashed circles) is inactive. At sulfur deficient conditions OAS accumulates, since sulfide is not available for conversion of OAS to cysteine by free OAS-TL (upper panel). Prolonged sulfur deficiency raises OAS levels above a certain threshold and causes the CSC to dissociate (right panel). Dissociation of the CSC results in inactivation of SAT (dented circles) to prevent consumption of acetyl-CoA (acCoA). The high OAS level, the dissociated CSC or a combination of those is the trigger for transcriptional activation of sulfur metabolism related genes in the nucleus (dashed dark circle). The activation of the sulfate transporting and reducing system makes sulfide become available (lower panel). As a result the OAS concentration decreases, since OAS is converted to cysteine by OAS-TL. When the OAS level falls below the threshold for dissociation the CSC, SAT and OAS-TL reassociate to form the CSC. SAT is activated and synthesizes OAS with a rate adapted to the current sulfate supply (left panel).

since the CSC self-assembles even *in vitro* from its constituents in the absence of OAS, it may be suspected that any protein extraction from living cells goes along with a dilution and thus strongly decreases OAS concentrations. This could result in assembly of the CSC from SAT and OAS-TL proteins that under native *in vivo* conditions might have been unbound.

The relevance of OAS-promoted dissociation is supported by precise concentration-dependent dissociation kinetics. While the dissociation of the *S. typhimurium* complex and a plant complex

mixed from *Arabidopsis* and spinach subunits by millimolar concentrations of OAS had long been known (Kredich et al., 1969; Droux et al., 1998), the CSC from *Arabidopsis* mitochondria was quantitatively analyzed using special protein interaction techniques (Berkowitz et al., 2002). An equilibrium dissociation constant of $57\,\mu M$ OAS was observed as well as a dissociation rate constant of $77\,\mu M$ OAS with a cooperative Hill-constant. These findings suggest that the equilibrium of CSC association/dissociation can be effectively shifted almost like a switch (Berkowitz et al., 2002). Since fluctuations of cellular OAS concentrations in this range have been observed (see above) in response to sulfate limitation, the equilibrium of CSC formation could control the rate of cysteine synthesis due to the complex-dependent activity of SAT.

Moreover, stabilization of the complex has already been reported, when 1 mM sulfide was added to the CSC from *S. typhimurium*. This treatment completely prevented the dissociation caused by OAS (Kredich et al., 1969). Again this effect was corroborated for a plant CSC. In the presence of 1 mM OAS and completely dissociated recombinant CSC from *Arabidopsis* mitochondria, the addition of increasing concentrations of sulfide resulted in association of SAT and OAS-TL with about 30 μM sulfide causing 50% decrease of OAS-TL activity due to complex association (Wirtz and Hell, 2006). Although the quantification of sulfide stabilization needs refinement, these *in vitro* data strongly suggest a physiological relevance for the regulation of CSC association and flux control of cysteine synthesis.

E. Metabolite Sensing and Regulation by the Cysteine Synthase Complex

The described evidence was integrated into a model that positions the CSC as a sensor in the regulation of cellular cysteine homeostasis (Droux et al., 1998; Hell, 1998; Hell and Hillebrand, 2001; Hell et al., 2002). The roots are the actual sites of nutrient perception, but decrease of external followed by internal availability of sulfate is important for growth and viability of all cells. Therefore, a basic cellular model might apply to source as well as sink cell. Sulfate deficiency will first result in lack of sulfide and then of cysteine and glutathione until protein biosynthesis stalls. Sulfide is a robust, fully reduced sulfur compound with little toxicity potential due to its very low concentrations. APS and sulfite seem less likely as sensing metabolites because of the lability of APS and high reduction potential of sulfite. In analogy to sulfate assimilation, a reduced nitrogen compound such as ammonia or glutamine has been suspected as sensing metabolite (Crawford and Forde, 2002). This regulatory model is based on the fully associated CSC during sufficient sulfate supply, meaning that all available SAT is bound to OAS-TL but leaving

an excess of free and active OAS-TL dimers (Fig. 7). SAT is active and produces OAS that diffuses into the solution because of blocked catalytic cleft of bound OAS-TL. Instead, free OAS-TL dimers catalyze the formation of cysteine from OAS and sulfide. Under this condition OAS concentrations, including the cytosol, are below the dissociation threshold for complex dissociation. Accordingly, sulfide may stabilize the complex to sustain OAS formation and sulfate transport and assimilation genes are partially repressed.

If sulfate limitation in the environment of the cell results in a limitation of sulfide, OAS concentrations will start to increase, because a critical substrate for cysteine synthesis in missing. The first consequence is OAS accumulation above the dissociation threshold concentration of about 50–80 μM and dissociation of the CSC in the absence of sulfide. Free SAT rapidly loses activity and avoids further consumption of acetyl-CoA. Second, increased OAS concentrations may trigger the de-repression of genes of sulfate transport, *ATP* sufurylase and *APS* reductase. The affinity and capacity for sulfate uptake at the plasmalemma increases and helps to import sulfate if at all available. Sulfide is produced by assimilation and can immediately react with OAS catalyzed by active OAS-TL dimers. When OAS concentrations fall again below the dissociation threshold, the complex can re-assemble including active SAT. The production of OAS starts again at a rate that corresponds to the sulfide status of the cell. Re-adjusted cysteine or glutathione concentrations may repress the uptake and assimilation genes. It should be noted that this model describes the CSC as part of a sensor system that regulates the flux of primary sulfur metabolism in a cell, whereas cysteine is synthesized by free OAS-TL dimers (Fig. 7).

This working model places the CSC in the signal transduction line of sulfate as a macronutrient. Very little is known so far about the upstream events the mediate between sulfate deficiency and transcriptional and growth responses (Schachtman and Shin, 2007). Microarray analyses have revealed numerous downstream events including interaction with nitrogen assimilation (Hirai et al., 2003; Nikiforova et al., 2004; Nikiforova et al., 2005), but only recently the first transcription factor (SLIM1) has been described that seems to control at least part of the genes of the typical sulfate deficiency response (Maruyama-Nakashita et al., 2006). If the CSC model proves to be correct, this excludes by no means the possibility of additional sensors, e.g. for sulfate or glutathione. However, no sensor for any plant nutrient has been described so far (Schachtman and Shin, 2007).

However, a number of open questions put a note of caution to the model. The mechanism of SAT inactivation outside the complex is entirely unknown (section IV.C). So far no post-translational modification has been reported that might account for activity changes, although the phosphorylation of a soybean cytosolic SAT was shown to result in decreased feedback sensitivity for cysteine (Liu et al., 2006). Particularly critical is the lack of correlation between OAS fluctuations, sulfate transport capacity and sulfate transporter kinetics (Hopkins et al., 2005). While OAS concentrations increase early on after transfer of single cells to sulfate deficient medium and approximately in parallel to the expression of a sulfate transporter (Wirtz et al., 2004), the induction of sulfate transporter genes precedes OAS accumulation in long-term experiments (Buchner et al., 2004; Hopkins et al., 2005). The apparently continuous accumulation of OAS can at present only be explained by subcellular compartmentation that removes OAS from the CSC complexes in cytosol, plastids and mitochondria.

On the other hand, the cysteine synthase model agrees with many physiological situations of sulfur metabolism. Optimal sulfur nutrition via H_2S would stabilize the complex to maximize OAS production for cysteine synthesis. This was indeed observed in whole plant experiments (De Kok et al., 2002). Long-term feeding of OAS potentially results in partial complex dissociation, but at the same time the induced sulfate assimilation genes generate more sulfide that stabilizes the complex, allowing for elevated cysteine levels (Neuenschwander et al., 1991). The over-expression of SAT in transgenic plants would produce more complex-bound SAT and consequently more OAS and cysteine formation as was observed (Blaszczyk et al., 1999; Harms et al., 2000; Noji and Saito, 2002; Wirtz and Hell, 2003a). The model is compatible with allosteric regulation by cysteine feedback inhibition of SAT as well (Saito, 2000). This mechanism would act downstream of sulfide and OAS, while the sensing mechanism serves as a trigger for upstream regulation.

References

Aldag R, Young J and Yamamoto M (1971) An enzymatic chromatographic procedure for the determination of D-amino acids in plant and soil extracts. Phytochemistry 10: 267–274

Awazuhara M, Hirai MY, Hayashi H, Chino M, Naito S and Fujiwara T (2000) O-Acetyl-L-serine content in rosette leaves of *Arabidopsis thaliana* affected by S and N nutrition. In: Sulfur nutrition and sulfur assimilation in higher plants, Brunold C, Rennenberg H, De Kok LJ, Stulen I and Davidian J-C (eds), pp 331–333. P. Haupt, Bern

Bae MS, Cho EJ, Choi E-Y and Park OK (2003) Analysis of the *Arabidopsis* nuclear proteome and its response to cold stress. Plant J 36: 652–663

Balk J and Lobreaux S (2005) Biogenesis of iron-sulfur proteins in plants. Trends Plant Sci 10: 324–331

Barroso C, Romero LC, Cejudo FJ, Vega JM and Gotor C (1999) Salt-specific regulation of the cytosolic O-acetylserine(thiol)lyase gene from *Arabidopsis thaliana* is dependent on abscisic acid. Plant Mol Biol 40: 729–736

Beaman TW, Binder DA, Blanchard JS and Roderick SL (1997) Three-dimensional structure of tetrahydrodipicolinate N-succinyltransferase. Biochemistry 36: 489–494

Becker MA, Kredich NM and Tomkins GM (1969) The purification and characterization of O-acetylserine sulfhydrylase-A from *Salmonella typhimurium*. J Biol Chem 244: 2418–2427

Beinert H (2000) A tribute to sulfur. Eur J Biochem 267: 5657–5664

Benci S, Vaccari S, Mozzarelli A and Cook PF (1997) Time-resolved fluorescence of O-acetylserine sulfhydrylase catalytic intermediates. Biochemistry 36: 15419–15427

Benci S, Vaccari S, Mozzarelli A and Cook PF (1999) Time-resolved fluorescence of O-acetylserine sulfhydrylase. Biochim Biophys Acta 1429: 317–330

Berkowitz O, Wirtz M, Wolf A, Kuhlmann J and Hell R (2002) Use of biomolecular interaction analysis to elucidate the regulatory mechanism of the cysteine synthase complex from *Arabidopsis thaliana*. J Biol Chem 277: 30629–30634

Blaszczyk A, Brodzik R and Sirko A (1999) Increased resistance to oxidative stress in transgenic tobacco plants overexpressing bacterial serine acetyltransferase. Plant J 20: 237–243

Bloem E, Riemenschneider A, Volker J, Papenbrock J, Schmidt A, Salac I, Haneklaus S and Schnug E (2004) Sulphur supply and infection with *Pyrenopeziza brassicae* influence L-cysteine desulphydrase activity in *Brassica napus* L. J Exp Bot 55: 2305–2312

Blumenthal SG, Hendrickson HR, Abrol YP and Conn EE (1968) Cyanide metabolism in higher plants. 3. The biosynthesis of β-cyanolanine. J Biol Chem 243: 5302–5307

Bogdanova N and Hell R (1997) Cysteine synthesis in plants: protein–protein interactions of serine acetyltransferase from *Arabidopsis thaliana*. Plant J 11: 251–262

Bonner ER, Cahoon RE, Knapke SM and Jez JM (2005) Molecular basis of cysteine biosynthesis in plants: structural and functional analysis of *O*-acetylserine sulfhydrylase from *Arabidopsis thaliana*. J Biol Chem 280: 38803–38813

Bork C and Hell R (2000) Expression patterns of the sulfite reductase gene from *Arabidopsis thaliana*. In: Sulfur nutrition and sulfur assimilation in higher plants: molecular, biochemical and physiological aspects, Brunold C, Davidian, J-C, De Kok, L, Rennenberg H, Stulen I (eds), pp 299–301. P. Haupt, Bern

Bruckner H and Westhauser T (2003) Chromatographic determination of L- and D-amino acids in plants. Amino Acids 24: 43–55

Brunold C (1990) Reduction of sulfate to sulfide. In: Sulfur nutrition and sulfur assimilation in higher plants, Rennenberg H, Brunold C, De Kok LJ, Stulen E (eds), pp 13–32. SPB Academic

Buchner P, Takahashi H and Hawkesford MJ (2004) Plant sulphate transporters: co-ordination of uptake, intracellular and long-distance transport. J Exp Bot 55: 1765–1773

Burkhard P, Rao GS, Hohenester E, Schnackerz KD, Cook PF and Jansonius JN (1998) Three-dimensional structure of *O*-acetylserine sulfhydrylase from *Salmonella typhimurium*. J Mol Biol 283: 121–133

Burkhard P, Tai CH, Jansonius JN and Cook PF (2000) Identification of an allosteric anion-binding site on *O*-acetylserine sulfhydrylase: structure of the enzyme with chloride bound. J Mol Biol 303: 279–286

Burkhard P, Tai CH, Ristroph CM, Cook PF and Jansonius JN (1999) Ligand binding induces a large conformational change in *O*-acetylserine sulfhydrylase from *Salmonella typhimurium*. J Mol Biol 291: 941–953

Campanini B, Speroni F, Salsi E, Cook PF, Roderick SL, Huang B, Bettati S and Mozzarelli A (2005) Interaction of serine acetyltransferase with *O*-acetylserine sulfhydrylase active site: evidence from fluorescence spectroscopy. Protein Sci 14: 2115–2124

Chen HC, Yokthongwattana K, Newton AJ and Melis A (2003) *SulP*, a nuclear gene encoding a putative chloroplast-targeted sulfate permease in *Chlamydomonas reinhardtii*. Planta 218: 98–106

Clarkson DT, Eugénio D and Sara A (1999) Uptake and assimilation of sulphate by sulphur deficient *Zea mays* cells: the role of *O*-acetyl-L-serine in the interaction between nitrogen and sulphur assimilatory pathways. Plant Physiol Biochem 37: 283–290

Cook PF and Wedding RT (1976) A reaction mechanism from steady state kinetic studies for *O*-acetylserine sulfhydrylase from *Salmonella typhimurium* LT-2. J Biol Chem 251: 2023–2029

Cook PF and Wedding RT (1978) Cysteine synthetase from *Salmonella typhimurium* LT-2. Aggregation, kinetic behavior, and effect of modifiers. J Biol Chem 253: 7874–7879

Crawford NM and Forde BG (2002) Molecular and developmental biology of inorganic nitrogen nutrition. The Arabidopsis Book: 1–25

De Kok LJ, Stuiver CEE, Westerman S and Stulen I (2002) Elevated levels of hydrogen sulfide in the plant environment: nutrient or toxin. In: Air pollution and plant biotechnology. Prospects for phytomonitoring and phytoremediation, Omasa K, Saji H, Youssefian S and Kondo N (eds), pp 201–219. Springer, Tokyo

Denk D and Böck A (1987) L-Cysteine biosynthesis in *Escherichia coli*: nucleotide sequence and expression of the serine acetyltransferase (*cysE*) gene from the wild-type and a cysteine-excreting mutant. J Gen Microbiol 133: 515–525

Dominguez-Solis JR, Gutierrez-Alcala G, Vega JM, Romero LC and Gotor C (2001) The cytosolic *O*-acetylserine(thiol)lyase gene is regulated by heavy metals and can function in cadmium tolerance. J Biol Chem 276: 9297–9302

Dominguez-Solis JR, Lopez-Martin MC, Ager FJ, Ynsa MD, Romero LC and Gotor C (2004) Increased cysteine availability is essential for cadmium tolerance and accumulation in *Arabidopsis thaliana*. Plant Biotech J 2: 469–476

Droux M (2003) Plant serine acetyltransferase: new insights for regulation of sulphur metabolism in plant cells. Plant Physiol Biochem 41: 619–627

Droux M (2004) Sulfur assimilation and the role of sulfur in plant metabolism: a survey. Photosynth Res 79: 331–348

Droux M, Martin J, Sajus P and Douce R (1992) Purification and characterization of *O*-acetylserine-(thiol)-lyase from spinach chloroplasts. Arch Biochem Biophys 295: 379–390

Droux M, Ruffet ML, Douce R and Job D (1998) Interactions between serine acetyltransferase and *O*-acetylserine-(thiol)-lyase in higher plants-structural and kinetic properties of the free and bound enzymes. Eur J Biochem 255: 235–245

Dunnill PM and Fowden L (1965) Enzymatic formation of β-cyanoalanine from cyanide by *Escherichia coli* extracts. Nature 208: 1206–1207

Farago S and Brunold C (1994) Regulation of thiol contents in maize roots by intermediates and effectors of glutathione synthesis. J Plant Physiol 144: 433–437

Floss HG, Hadwiger L and Conn EE (1965) Enzymatic formation of β-cyanoalanine from cyanide. Nature 208: 1207–1208

Francois JA, Kumaran S and Jez JM (2006) Structural basis for interaction of *O*-acetylserine sulfhydrylase and serine acetyltransferase in the *Arabidopsis* cysteine synthase complex. Plant Cell 18: 3647–3655

Frazzon AP, Ramirez MV, Warek U, Balk J, Frazzon J, Dean DR and Winkel BS (2007) Functional analysis of *Arabidopsis* genes involved in mitochondrial iron-sulfur cluster assembly. Plant Mol Biol 64: 225–240

Friedman M (1999) Chemistry, nutrition, and microbiology of D-amino acids. J Agric Food Chem 47: 3457–3479

Fromageot C (1951). In: The enzymes, Sumner J and Myrbäck K (eds), pp 1237–1243. Academic, New York

Fromageot C, Wookey E and Chaix P (1940) Enzymologia 9: 198–214

Gong Z, Koiwa H, Cushman MA, Ray A, Bufford D, Koreeda S, Matsumoto TK, Zhu J, Cushman JC, Bressan RA and Hasegawa PM (2001) Genes that are uniquely stress regulated in salt overly sensitive (sos) mutants. Plant Physiol 126: 363–375

Gorman J and Shapiro L (2004) Structure of serine acetyltransferase from Haemophilus influenzae Rd. Acta Crystallogr D Biol Crystallogr 60: 1600–1605

Gutierrez-Alcala G, Gotor C, Meyer AJ, Fricker M, Vega JM and Romero LC (2000) Glutathione biosynthesis in Arabidopsis trichome cells. Proc Natl Acad Sci U S A 97: 11108–11113

Harada E, Choi EJ, Tsuchisaka A, Obata H and Sano H (2001) Transgenic tobacco plants expressing a rice cysteine synthase gene are tolerant to toxic levels of cadmium. J Plant Physiol 158: 655–661

Harms K, von Ballmoos P, Brunold C, Hofgen R and Hesse H (2000) Expression of a bacterial serine acetyltransferase in transgenic potato plants leads to increased levels of cysteine and glutathione. Plant J 22: 335–343

Harrington H and Smith I (1980) Cysteine metabolism in cultured tobacco cells. Plant Physiol 65: 151–155

Hatzfeld Y, Maruyama A, Schmidt A, Noji M, Ishizawa K and Saito K (2000) β-Cyanoalanine synthase is a mitochondrial cysteine synthase-like protein in spinach and Arabidopsis. Plant Physiol 123: 1163–1172

Heidenreich T, Wollers S, Mendel RR and Bittner F (2005) Characterization of the NifS-like domain of ABA3 from Arabidopsis thaliana provides insight into the mechanism of molybdenum cofactor sulfuration. J Biol Chem 280: 4213–4218

Hell R (1997) Molecular physiology of plant sulfur metabolism. Planta 202: 138–148

Hell R (1998). Molekulare Physiologie des Primärstoffwechsels von Schwefel in Pflanzen. Shaker, Aachen

Hell R (2003) Metabolic regulation of cysteine synthesis and sulfur assimilation. In: Sulfate transport and assimilation in plants, Davidian JC, Grill D, De Kok LJ, Stulen I, Hawkesford MJ, Schnug E and Rennenberg H (eds), pp 21–31. Backhuys Publishers, Leiden

Hell R, Bork C, Bogdanova N, Frolov I and Hauschild R (1994) Isolation and characterization of two cDNAs encoding for compartment specific isoforms of O-acetylserine-(thiol)-lyase from Arabidopsis thaliana. FEBS Lett 351: 257–262

Hell R and Hillebrand H (2001) Plant concepts for mineral acquisition and allocation. Curr Opin Biotechnol 12: 161–168

Hell R, Jost R, Berkowitz O and Wirtz M (2002) Molecular and biochemical analysis of the enzymes of cysteine biosynthesis in the plant Arabidopsis thaliana. Amino Acids 22: 245–257

Hendrickson HR and Conn EE (1969) Cyanide metabolism in higher plants. IV. Purification and properties of the β-cyanolanine synthase of blue lupine. J Biol Chem 244: 2632–2640

Hesse H and Hoefgen R (1998) Isolation of cDNAs encoding cytosolic (Accession No. AF044172) and plastidic (Accession No. AF044173) cysteine synthase isoforms from Solanum tuberosum (PGR98–057). Plant Physiol 116: 1604

Hesse H, Lipke J, Altmann T and Hofgen R (1999) Molecular cloning and expression analyses of mitochondrial and plastidic isoforms of cysteine synthase (O-acetylserine(t hiol)lyase) from Arabidopsis thaliana. Amino Acids 16: 113–131

Hesse H, Nikiforova V, Gakiere B and Hoefgen R (2004) Molecular analysis and control of cysteine biosynthesis: integration of nitrogen and sulphur metabolism. J Exp Bot 55: 1283–1292

Hindson VJ, Moody PC, Rowe AJ and Shaw WV (2000) Serine acetyltransferase from Escherichia coli is a dimer of trimers. J Biol Chem 275: 461–466

Hindson VJ and Shaw WV (2003) Random-order ternary complex reaction mechanism of serine acetyltransferase from Escherichia coli. Biochemistry 42: 3113–3119

Hirai M, Fujiwara T, Awazuhara M, Kimura T, Noji M and Saito K (2003) Global expression profiling of sulfurstarved Arabidopsis by DNA macroarray reveals the role of O-acetyl-L-serine as a general regulator of gene expression in response to sulfur nutrition. Plant J 33: 651–663

Hopkins L, Parmar S, Blaszczyk A, Hesse H, Hoefgen R and Hawkesford MJ (2005) O-acetylserine and the regulation of expression of genes encoding components for sulfate uptake and assimilation in potato. Plant Physiol 138: 433–440

Howarth JR, Dominguez-Solis JR, Gutierrez-Alcala G, Wray JL, Romero LC and Gotor C (2003) The serine acetyltransferase gene family in Arabidopsis thaliana and the regulation of its expression by cadmium. Plant Mol Biol 51: 589–598

Huang B, Vetting MW and Roderick SL (2005) The active site of O-acetylserine sulfhydrylase is the anchor point for bienzyme complex formation with serine acetyltransferase. J Bacteriol 187: 3201–3205

Inoue K, Noji M and Saito K (1999) Determination of the sites required for the allosteric inhibition of serine acetyltransferase by L-cysteine in plants. Eur J Biochem 266: 220–227

Johnson C, Roderick S and Cook P (2005) The serine acetyltransferase reaction: acetyl transfer from an acylpantothenyl donor to an alcohol. Arch Biochem Biophys 433: 85–95

Johnson CM, Huang B, Roderick SL and Cook PF (2004) Chemical mechanism of the serine acetyltransferase from Haemophilus influenzae. Biochemistry 43: 15534–15539

Jones PR, Manabe T, Awazuhara M and Saito K (2003) A new member of plant CS-lyases. A cystine lyase from Arabidopsis thaliana. J Biol Chem 278: 10291–10296

Jost R, Berkowitz O, Wirtz M, Hopkins L, Hawkesford MJ and Hell R (2000) Genomic and functional characterization of the oas gene family encoding O-acetylserine (thiol)

lyases, enzymes catalyzing the final step in cysteine bio-synthesis in *Arabidopsis thaliana*. Gene 253: 237–247

Kawashima CG, Berkowitz O, Hell R, Noji M and Saito K (2005) Characterization and expression analysis of a serine acetyltransferase gene family involved in a key step of the sulfur assimilation pathway in *Arabidopsis*. Plant Physiol 137: 220–230

Kawashima CG, Noji M, Nakamura M, Ogra Y, Suzuki KT and Saito K (2004) Heavy metal tolerance of transgenic tobacco plants over-expressing cysteine synthase. Biotechnol Lett 26: 153–157

Kim H, Hirai MY, Hayashi H, Chino M, Naito S and Fujiwara T (1999) Role of *O*-acetyl-L-serine in the coordinated regulation of the expression of a soybean seed storage-protein gene by sulfur and nitrogen nutrition. Planta 209: 282–289

Kitabatake M, So MW, Tumbula DL and Soll D (2000) Cysteine biosynthesis pathway in the archaeon *Methanosarcina barkeri* encoded by acquired bacterial genes? J Bacteriol 182: 143–145

Kopriva S, Suter M, von Ballmoos P, Hesse H, Krahenbuhl U, Rennenberg H and Brunold C (2002) Interaction of sulfate assimilation with carbon and nitrogen metabolism in *Lemna minor*. Plant Physiol 130: 1406–1413

Koprivova A, Suter M, den Camp RO, Brunold C and Kopriva S (2000) Regulation of sulfate assimilation by nitrogen in *Arabidopsis*. Plant Physiol 122: 737–746

Kredich NM (1996) Biosynthesis of cysteine. In: *Escherichia coli* and *Salmonella typhimurium*. Cellular and molecular biology, Neidhardt FC, Curtiss R, Ingraham JL, Lin ECC, Low KB, Magasanik B, Reznikoff WS, Riley M, Schaechter M and Umberger E (eds), pp 514–527. ASM Press, Washington D.C.

Kredich NM and Becker MA (1971) Cysteine biosynthesis: serine transacetylase and *O*-acetylserine sulfhydrylase (*Salmonella typhimurium*). In: Methods in enzymology, pp 459–471. Academic Press LTD, London, UK

Kredich NM, Becker MA and Tomkins GM (1969) Purification and characterization of cysteine synthetase, a bifunctional protein complex, from *Salmonella typhimurium*. J Biol Chem 244: 2428–2439

Kredich NM, Foote LJ and Keenan BS (1973) The stoichiometry and kinetics of the inducible cysteine desulfhydrase from *Salmonella typhimurium*. J Biol Chem 248: 6187–6196

Kredich NM, Keenan BS and Foote LJ (1972) The purification and subunit structure of cysteine desulfhydrase from *Salmonella typhimurium*. J Biol Chem 247: 7157–7162

Kredich NM and Tomkins GM (1966) The enzymic synthesis of L-cysteine in *Escherichia coli* and *Salmonella typhimurium*. J Biol Chem 241: 4955–4965

Kumaran S and Jez JM (2007) Thermodynamics of the interaction between O-acetylserine sulfhydrylase and the C-terminus of serine acetyltransferase. Biochemistry 46: 5586–5594

Kuske CR, Hill KK, Guzman E and Jackson PJ (1996) Subcellular location of *O*-acetylserine sulfhydrylase isoenzymes in cell cultures and plant tissues of *Datura innoxia* Mill. Plant Physiol 112: 659–667

Kutz A, Muller A, Hennig P, Kaiser WM, Piotrowski M and Weiler EW (2002) A role for nitrilase 3 in the regulation of root morphology in sulphur-starving *Arabidopsis thaliana*. Plant J 30: 95–106

Laudenbach DE, Ehrhardt D, Green L and Grossman A (1991) Isolation and characterization of a sulfur-regulated gene encoding a periplasmically localized protein with sequence similarity to rhodanese. J Bacteriol 173: 2751–2760

Leu LS and Cook PF (1994) Kinetic mechanism of serine transacetylase from *Salmonella typhimurium*. Biochemistry 33: 2667–2671

Liu F, Yoo B-C, Lee J-Y, Pan W and Harmon AC (2006) Calcium-regulated phosphorylation of soybean serine acetyltransferase in response to oxidative stress. J Biol Chem 281: 27405–27415

Logan HM, Cathala N, Grignon C and Davidian JC (1996) Cloning of a cDNA encoded by a member of the *Arabidopsis thaliana* ATP sulfurylase multigene family. Expression studies in yeast and in relation to plant sulfur nutrition. J Biol Chem 271: 12227–12233

Lunn JE, Droux M, Martin J and Douce R (1990) Localization of ATP-sulfurylase and *O*-acetylserine(thiol)lyase in spinach leaves. Plant Physiol 94: 1345–1352

Maier TH (2003) Semisynthetic production of unnatural L-α-amino acids by metabolic engineering of the cysteine-biosynthetic pathway. Nat Biotechnol 21: 422–427

Maruyama A, Ishizawa K and Takagi T (2000) Purification and characterization of β-cyanoalanine synthase and cysteine synthases from potato tubers: are β-cyanoalanine synthase and mitochondrial cysteine synthase same enzyme? Plant Cell Physiol 41: 200–208

Maruyama A, Ishizawa K, Takagi T and Esashi Y (1998) Cytosolic β-cyanoalanine synthase activity attributed to cysteine synthases in cocklebur seeds. Purification and characterization of cytosolic cysteine synthases. Plant Cell Physiol 39: 671–680

Maruyama-Nakashita A, Inoue E, Watanabe-Takahashi A, Yamaya T and Takahashi H (2003) Transcriptome profiling of sulfur-responsive genes in *Arabidopsis* reveals global effects of sulfur nutrition on multiple metabolic pathways. Plant Physiol 132: 597–605

Maruyama-Nakashita A, Nakamura Y, Tohge T, Saito K and Takahashi H (2006) *Arabidopsis* SLIM1 is a central transcriptional regulator of plant sulfur response and metabolism. Plant Cell 18: 3235–3251

Maruyama-Nakashita A, Nakamura Y, Watanabe-Takahashi A, Inoue E, Yamaya T and Takahashi H (2005) Identification of a novel *cis*-acting element conferring sulfur deficiency response in *Arabidopsis* roots. Plant J 42: 305–314

Maruyama-Nakashita A, Nakamura Y, Yamaya T and Takahashi H (2004) A novel regulatory pathway of sulfate uptake in *Arabidopsis* roots: implication of CRE1/WOL/

AHK4-mediated cytokinin-dependent regulation. Plant J 38: 779–789

Mazel D and Marliere P (1989) Adaptive eradication of methionine and cysteine from cyanobacterial light-harvesting proteins. Nature 341: 245–248

Melis A and Chen H-C (2005) Chloroplast sulfate transport in green algae – genes, proteins and effects. Photosynth Res 86: 299–307

Meyers DM and Ahmad S (1991) Link between L-3-cyanoalanine synthase activity and differential cyanide sensitivity of insects. Biochim Biophys Acta 1075: 195–197

Mino K, Yamanoue T, Sakiyama T, Eisaki N, Matsuyama A and Nakanishi K (1999) Purification and characterization of serine acetyltransferase from Escherichia coli partially truncated at the C-terminal region. Biosci Biotechnol Biochem 63: 168–179.

Mino K, Yamanoue T, Sakiyama T, Eisaki N, Matsuyama A and Nakanishi K (2000a) Effects of bienzyme complex formation of cysteine synthetase from Escherichia coli on some properties and kinetics. Biosci Biotechnol Biochem 64: 1628–1640

Mino K, Hiraoka K, Imamura K, Sakiyama T, Eisaki N, Matsuyama A and Nakanishi K (2000b) Characteristics of serine acetyltransferase from Escherichia coli deleting different lengths of amino acid residues from the C-terminus. Biosci Biotechnol Biochem 64: 1874–1880.

Murakoshi I, Kaneko M, Koide C and Ikegami F (1986) Enzymatic synthesis of the neuroexcitatory amino acid quisqualic acid by the cysteine synthase. Phytochemistry 25: 2759–2763

Nakamori S, Kobayashi SI, Kobayashi C and Takagi H (1998) Overproduction of L-cysteine and L-cystine by Escherichia coli strains with a genetically altered serine acetyltransferase. Appl Environ Microbiol 64: 1607–1611

Neuenschwander U, Suter M and Brunold C (1991) Regulation of sulfate assimilation by light and O-acetyl-L-serine in Lemna minor L. Plant Physiol. 97: 253–258

Nicholson ML, Gaasenbeek M and Laudenbach DE (1995) Two enzymes together capable of cysteine biosynthesis are encoded on a cyanobacterial plasmid. Mol Gen Genet 247: 623–632

Nicholson ML and Laudenbach DE (1995) Genes encoded on a cyanobacterial plasmid are transcriptionally regulated by sulfur availability and CysR. J Bacteriol 177: 2143–2150

Niehaus A, Gisselmann G and Schwenn JD (1992) Primary structure of the Synechococcus PCC 7942 PAPS reductase gene. Plant Mol Biol 20: 1179–1183

Nikiforova VJ, Daub CO, Hesse H, Willmitzer L and Hoefgen R (2005) Integrative gene-metabolite network with implemented causality deciphers informational fluxes of sulphur stress response. J Exp Bot 56: 1887–1896

Nikiforova VJ, Gakiere B, Kempa S, Adamik M, Willmitzer L, Hesse H and Hoefgen R (2004) Towards dissecting nutrient metabolism in plants: a systems biology case study on sulphur metabolism. J Exp Bot 55: 1861–1870

Noji M, Inoue K, Kimura N, Gouda A and Saito K (1998) Isoform-dependent differences in feedback regulation and subcellular localization of serine acetyltransferase involved in cysteine biosynthesis from Arabidopsis thaliana. J Biol Chem 273: 32739–32745

Noji M and Saito K (2002) Molecular and biochemical analysis of serine acetyltransferase and cysteine synthase towards sulfur metabolic engineering in plants. Amino Acids 22: 231–243

Noji M, Saito M, Nakamura M, Aono M, Saji H and Saito K (2001a) Cysteine synthase overexpression in tobacco confers tolerance to sulfur-containing environmental pollutants. Plant Physiol 126: 973–980

Noji M, Takagi Y, Kimura N, Inoue K, Saito M, Horikoshi M, Saito F, Takahashi H and Saito K (2001b) Serine acetyltransferase involved in cysteine biosynthesis from spinach: molecular cloning, characterization and expression analysis of cDNA encoding a plastidic isoform. Plant Cell Physiol 42: 627–634

Ohkama N, Goto DB, Fujiwara T and Naito S (2002) Differential tissue-specific response to sulfate and methionine of a soybean seed storage protein promoter region in transgenic Arabidopsis. Plant Cell Physiol 43: 1266–1275

Olsen L, Huang B, Vetting M and Roderick S (2004) Structure of serine acetyltransferase in complexes with CoA and its cysteine feedback inhibitor. Biochemistry 43: 6013–6019

Peiser GD, Wang TT, Hoffman NE, Yang SF, Liu HW and Walsh CT (1984) Formation of cyanide from carbon 1 of 1-aminocyclopropane-1-carboxylic acid during its conversion to ethylene. Proc Natl Acad Sci U S A 81: 3059–3063

Pye VE, Tingey AP, Robson RL and Moody PC (2004) The structure and mechanism of serine acetyltransferase from Escherichia coli. J Biol Chem 279: 40729–40736

Rabeh WM and Cook PF (2004) Structure and mechanism of O-acetylserine sulfhydrylase. J Biol Chem 279: 26803–26806

Raetz CR and Roderick SL (1995) A left-handed parallel beta helix in the structure of UDP-N-acetylglucosamine acyltransferase. Science 270: 997–1000

Ravina CG, Chang CI, Tsakraklides GP, McDermott JP, Vega JM, Leustek T, Gotor C and Davies JP (2002) The sac mutants of Chlamydomonas reinhardtii reveal transcriptional and posttranscriptional control of cysteine biosynthesis. Plant Physiol 130: 2076–2084

Rege VD, Kredich NM, Tai CH, Karsten WE, Schnackerz KD and Cook PF (1996) A change in the internal aldimine lysine (K42) in O-acetylserine sulfhydrylase to alanine indicates its importance in transimination and as a general base catalyst. Biochemistry 35: 13485–13493

Rennenberg H (1983) Cysteine desulfhydrase activity in cucurbit plants: Stimulation by preincubation with L- or D-cysteine. Phytochemistry 22: 1557–1560

Rennenberg H, Arabatzis N and Grundel I (1987) Cysteine desulphydrase activity in higher plants: Evidence for the action of L- and D-cysteine specific enzymes. Phytochemistry 26: 1583–1589

Riemenschneider A, Bonacina E, Schmidt A and Papenbrock J (2005b) Isolation and characterization of a second D-cysteine desulfhydrase-like protein from *Arabidopsis*. In: Sulfur transport and assimilation in plants in the post genomic era, Saito K, De Kok LJ, Stulen I, Hawkesford MJ, Schnug E, Sirko A and Rennenberg H (eds), pp 103–106. Backhuys Publishers, Leiden

Riemenschneider A, Wegele R, Schmidt A and Papenbrock J (2005a) Isolation and characterization of a D-cysteine desulfhydrase protein from *Arabidopsis thaliana*. FEBS J 272: 1291–1304

Rolland N, Droux M and Douce R (1992) Subcellular distribution of O-acetylserine(thiol)lyase in cauliflower (*Brassica oleracea* L.) inflorescence. Plant Physiol 98: 927–935

Romero LC, Dominguez-Solis JR, Gutierrez-Alcala G and Gotor C (2001) Salt regulation of *O*-acetylserine (thiol)lyase in *Arabidopsis thaliana* and increased tolerance in yeast. Plant Physiol Biochem 39: 643–647

Rüegsegger A and Brunold C (1992) Effect of cadmium on γ-glutamylcysteine synthesis in maize seedlings. Plant Physiol 99: 428–433

Ruffet ML, Droux M and Douce R (1994) Purification and kinetic properties of serine acetyltransferase free of *O*-acetylserine(thiol)lyase from spinach chloroplasts. Plant Physiol 104: 597–604

Ruffet ML, Lebrun M, Droux M and Douce R (1995) Subcellular distribution of serine acetyltransferase from *Pisum sativum* and characterization of an *Arabidopsis thaliana* putative cytosolic isoform. Eur J Biochem 227: 500–509

Saidha T, Na SQ, Li JY and Schiff JA (1988) A sulphate metabolizing centre in *Euglena* mitochondria. Biochem J 253: 533–539

Saito K (2000) Regulation of sulfate transport and synthesis of sulfur-containing amino acids. Curr Opin Plant Biol 3: 188–195

Saito K, Inoue K, Fukushima R and Noji M (1997) Genomic structure and expression analyses of serine acetyltransferase gene in *Citrullus vulgaris* (watermelon). Gene 189: 57–63

Saito K, Tatsuguchi K, Takagi Y and Murakoshi I (1994) Isolation and characterization of cDNA that encodes a putative mitochondrion-localizing isoform of cysteine synthase (*O*-acetylserine(thiol)-lyase) from *Spinacia oleracea*. J Biol Chem 269: 28187–28192

Saito K, Yokoyama H, Noji M and Murakoshi I (1995) Molecular cloning and characterization of a plant serine acetyltransferase playing a regulatory role in cysteine biosynthesis from watermelon. J Biol Chem 270: 16321–16326

Sasaki Y, Asamizu E, Shibata D, Nakamura Y, Kaneko T, Awai K, Amagai M, Kuwata C, Tsugane T, Masuda T,

Shimada H, Takamiya K, Ohta H and Tabata S (2001) Monitoring of methyl jasmonate-responsive genes in Arabidopsis by cDNA macroarray: self-activation of jasmonic acid biosynthesis and crosstalk with other phytohormone signaling pathways. DNA Res 8: 153–161

Schachtman DP and Shin R (2007) Nutrient sensing and signaling: NPKS. Annu Rev Plant Biol 58: 47–69

Schiff S, Stern AI, Saidha T and Li J (1993) Some molecular aspects of sulfate metabolism in photosynthetic organisms. In: Sulfur nutrition and assimilation in higher plants. Regulatory, agricultural and environmental aspects, De Kok LJ, Stulen I, Rennenberg H, Brunold C and Rauser WC (eds), SBP Academic, The Hague

Schmidt A (1986) Regulation of sulfur metabolism in plants. Progr. Bot. 48: 133–150

Schmidt A (1987) D-cysteine desulfhydrase from spinach. Methods Enzymol 143: 449–453

Schmidt A (2005) Metabolic background of H_2S release from plants. In: Sino-German workshop on aspects of sulfur nutrition of plants, De Kok LJ and Schnug E (eds), pp 121–129

Schmidt A and Jäger K (1992) Open questions about sulfur metabolism in plants. Annu Rev Plant Physiol Plant Mol Bio 43: 325–349

Schnackerz KD, Tai CH, Simmons JW, 3rd, Jacobson TM, Rao GS and Cook PF (1995) Identification and spectral characterization of the external aldimine of the O-acetylserine sulfhydrylase reaction. Biochemistry 34: 12152–12160

Sirko A, Blaszczyk A and Liszewska F (2004) Overproduction of SAT and/or OASTL in transgenic plants: a survey of effects. J Exp Bot 55: 1881–1888

Smacchi E and Gobbetti M (1998) Purification and characterization of cystathionine γ-lyase from *Lactobacillus fermentum* DT41. FEMS Microbiol Lett 166: 197–202

Smith FW, Hawkesford MJ, Ealing PM, Clarkson DT, Vanden Berg PJ, Belcher AR and Warrilow AG (1997) Regulation of expression of a cDNA from barley roots encoding a high affinity sulphate transporter. Plant J 12: 875–884

Smythe CV (1945). Adv Enzymol 5: 237–247

Staton AL and Mazelis M (1991) The C-S lyases of higher plants: homogenous β-cystathionase of spinach leaves. Arch Biochem Biophys 290: 46–50

Strambini G, Cioni P and Cook P (1996) Tryptophan luminescence as a probe of enzyme conformation along the O-acetylserine sulfhydrylase reaction pathway. Biochemistry 35: 8392–8400

Sugantino M and Roderick SL (2002) Crystal structure of Vat(D): an acetyltransferase that inactivates streptogramin group A antibiotics. Biochemistry 41: 2209–2216

Sun Q, Emanuelsson O and van Wijk KJ (2004) Analysis of curated and predicted plastid subproteomes of Arabidopsis. Subcellular compartmentalization leads to distinctive proteome properties. Plant Physiol 135: 723–734

Tabe LM and Droux M (2001) Sulfur assimilation in developing lupin cotyledons could contribute significantly to the accumulation of organic sulfur reserves in the seed. Plant Physiol 126: 176–187

Tabe LM and Droux M (2002) Limits to sulfur accumulation in transgenic lupin seeds expressing a foreign sulfur-rich protein. Plant Physiol 128: 1137–1148

Tai CH, Nalabolu SR, Jacobson TM, Minter DE and Cook PF (1993) Kinetic mechanisms of the A and B isozymes of O-acetylserine sulfhydrylase from *Salmonella typhimurium* LT-2 using the natural and alternative reactants. Biochemistry 32: 6433–6442

Tai C, Nalabolu S, Simmons J, Jacobson T and Cook P (1995) Acid-base chemical mechanism of O-acetylserine sulfhydrylases-A and -B from pH studies. Biochemistry 34: 12311–12322

Tai CH, Yoon MY, Kim SK, Rege VD, Nalabolu SR, Kredich NM, Schnackerz KD and Cook PF (1998) Cysteine 42 is important for maintaining an integral active site for O-acetylserine sulfhydrylase resulting in the stabilization of the α-aminoacrylate intermediate. Biochemistry 37: 10597–10604

Takagi H, Kobayashi C, Kobayashi S and Nakamori S (1999) PCR random mutagenesis into *Escherichia coli* serine acetyltransferase: isolation of the mutant enzymes that cause overproduction of L-cysteine and L-cystine due to the desensitization to feedback inhibition. FEBS Lett 452: 323–327

Takahashi H, Yamazaki M, Sasakura N, Watanabe A, Leustek T, Engler JA, Engler G, Van Montagu M and Saito K (1997) Regulation of sulfur assimilation in higher plants: a sulfate transporter induced in sulfate-starved roots plays a central role in *Arabidopsis thaliana*. Proc Natl Acad Sci U S A 94: 11102–11107

Urano Y, Manabe T, Noji M and Saito K (2000) Molecular cloning and functional characterization of cDNAs encoding cysteine synthase and serine acetyltransferase that may be responsible for high cellular cysteine content in *Allium tuberosum*. Gene 257: 269–277

Vaara M (1992) Eight bacterial proteins, including UDP-N-acetylglucosamine acyltransferase (LpxA) and three other transferases of *Escherichia coli*, consist of a six-residue periodicity theme. FEMS Microbiol Lett 76: 249–254

Van Hoewyk D, Abdel-Ghany SE, Cohu CM, Herbert SK, Kugrens P, Pilon M and Pilon-Smits EA (2007) Chloroplast iron-sulfur cluster protein maturation requires the essential cysteine desulfurase CpNifS. Proc Natl Acad Sci U S A 104: 5686–5691

Vuorio R, Hirvas L and Vaara M (1991) The Ssc protein of enteric bacteria has significant homology to the acyltransferase Lpxa of lipid A biosynthesis, and to three acetyltransferases. FEBS Lett 292: 90–94

Warrilow A and Hawkesford M (1998) Separation, subcellular location and influence of sulphur nutrition on isoforms of cysteine synthase in spinach. J Exp Bot 49: 1625–1636

Warrilow AG and Hawkesford MJ (2000) Cysteine synthase (O-acetylserine (thiol) lyase) substrate specificities classify the mitochondrial isoform as a cyanoalanine synthase. J Exp Bot 51: 985–993

Wawrzynski A, Kopera E, Wawrzynska A, Kaminska J, Bal W and Sirko A (2006) Effects of simultaneous expression of heterologous genes involved in phytochelatin biosynthesis on thiol content and cadmium accumulation in tobacco plants. J Exp Bot 57: 2173–2182

Wirtz M and Hell R (2001) Recombinant production of cysteine and glutathione by expression of feedback-insensitive serine acetyltransferase in *Escherichia coli*. In Nachwachsende Rohstoffe für die Chemie – 7. Symposium Dresden, pp 722–728. Landwirtschaftsverlag, Münster

Wirtz M and Hell R (2003a) Production of cysteine for bacterial and plant biotechnology: Application of cysteine feedback-insensitive isoforms of serine acetyltransferase. Amino Acids 24: 195–203

Wirtz M and Hell R (2003b) Comparative biochemical characterization of OAS-TL isoforms from *Arabidopsis thaliana*. Backhuys Publishers, Leiden

Wirtz M and Droux M (2005) Synthesis of the sulfur amino acids: cysteine and methionine. Photosynth Res 86: 345–362

Wirtz M and Hell R (2006) Functional analysis of the cysteine synthase protein complex from plants: structural, biochemical and regulatory properties. J Plant Physiol 163: 273–286

Wirtz M and Hell R (2007) Dominant-negative modification reveals the regulatory function of the multimeric cysteine synthase protein complex in transgenic tobacco. Plant Cell 19: 625–639

Wirtz M, Berkowitz O and Hell R (2000) Analysis of plant cysteine synthesis *in vivo* using *E. coli* as a host. In: Sulfur nutrition and sulfur assimilation in higher plants: molecular, biochemical and physiological aspects, Brunold C, Davidian J-C, De Kok L, Rennenberg H, Stulen I (eds), pp 297–298. P. Haupt, Bern

Wirtz M, Droux M and Hell R (2004) O-acetylserine (thiol) lyase: an enigmatic enzyme of plant cysteine biosynthesis revisited in *Arabidopsis thaliana*. J Exp Bot 55: 1785–1798

Wirtz M, Berkowitz O, Droux M and Hell R (2001) The cysteine synthase complex from plants. Mitochondrial serine acetyltransferase from *Arabidopsis thaliana* carries a bifunctional domain for catalysis and protein–protein interaction. Eur J Biochem 268: 686–693

Woehl EU, Tai CH, Dunn MF and Cook PF (1996) Formation of the α-aminoacrylate immediate limits the overall reaction catalyzed by O-acetylserine sulfhydrylase. Biochemistry 35: 4776–4783

Xu XM and Moller SG (2006) AtSufE is an essential activator of plastidic and mitochondrial desulfurases in *Arabidopsis*. EMBO J 25: 900–909

Yamaguchi Y, Nakamura T, Kusano T and Sano H (2000) Three *Arabidopsis* genes encoding proteins with differential activities for cysteine synthase and β-cyanoalanine synthase. Plant Cell Physiol 41: 465–476

Ye H, Abdel-Ghany SE, Anderson TD, Pilon-Smits EA and Pilon M (2006) CpSufE activates the cysteine desulfurase CpNifS for chloroplastic Fe-S cluster formation. J Biol Chem 281: 8958–8969

Ye H, Garifullina GF, Abdel-Ghany SE, Zhang L, Pilon-Smits EA and Pilon M (2005) The chloroplast NifS-like protein of *Arabidopsis thaliana* is required for iron-sulfur cluster formation in ferredoxin. Planta 220: 602–608

Yip WK and Yang SF (1988) Cyanide metabolism in relation to ethylene production in plant tissues. Plant Physiol 88: 473–476

Yoshida Y, Nakano Y, Amano A, Yoshimura M, Fukamachi H, Oho T and Koga T (2002) *lcd* from *Streptococcus angi-* *nosus* encodes a C-S lyase with α,α-elimination activity that degrades L-cysteine. Microbiology 148: 3961–3970

Youssefian S, Nakamura M, Orudgev E and Kondo N (2001) Increased cysteine biosynthesis capacity of transgenic tobacco overexpressing an *O*-acetylserine(thiol) lyase modifies plant responses to oxidative stress. Plant Physiol 126: 1001–1011

Zhang Z, Shrager J, Jain M, Chang C-W, Vallon O and Grossman AR (2004) Insights into the survival of *Chlamydomonas reinhardtii* during sulfur starvation based on microarray analysis of gene expression. Eukaryotic Cell 3: 1331–1348

Chapter 5

Metabolism of Methionine in Plants and Phototrophic Bacteria

Holger Hesse* and Rainer Hoefgen
*Max Planck Institute of Molecular Plant Physiology, Wissenschaftspark Golm,
14424 Potsdam, Germany*

Summary

The sulfur-containing amino acid methionine is a nutritionally important essential amino acid and the precursor of several metabolites that regulate plant growth and responses to the environment. New genetic and molecular data suggest that methionine synthesis and catabolism are coordinately regulated by novel post-transcriptional and post-translational mechanisms. This review focuses on new features reported for the molecular and biochemical aspects of methionine biosynthesis in higher plants with special emphasis on a comparison of the methionine biosynthetic pathway of plants with pathways of phototrophic bacteria (cyanobacteria). Particularly, the impact of the compartmentalization of methionine biosynthesis will be addressed with respect to regulatory aspects.

I. Introduction

Studying the regulation of methionine (Met) homeostasis is crucial for our understanding of physiological and biochemical processes in organisms. As organisms are exposed to environmental changes they have developed response mechanisms to keep metabolites in equilibrium. The process, termed adaptation, reflects the molecular/genetic and biochemical plasticity of a system. Compared to plants cyanobacteria are one of the oldest groups of photoautotrophic organisms on Earth (Schopf, 2000). They have colonized almost every available niche for the last 3.5 billion years, demonstrating an outstanding ability for adaptation to extremely different habitats and play a dominant role in the global nitrogen and carbon cycles. Thus, it was even not surprising to discover abundant cyanobacteria (Johnson and Sieburth, 1979) such as *Synechococcus* (Waterbury et al., 1979) and *Prochlorococcus*

* Corresponding author, Tel.: +49 331 5678247, Fax: +49 331 567898247, E-mail: hesse@mpimp-golm.mpg.de

Rüdiger Hell et al. (eds.), Sulfur Metabolism in Phototrophic Organisms, 93–110.
© 2008 *Springer.*

(Chisholm et al., 1988) thriving in environments with a very poor nutrient supply, such as the vast intertropical gyres of the oceans (Capone, 2000; Karl, 2002; Zehr and Ward, 2002; Delong and Karl, 2005), which were previously considered to be almost empty of living cells. However, their ecological importance was truly unexpected, since it is currently accepted that about half of the global primary production occurs in the oceans (Whitman et al., 1998) and that marine cyanobacteria contribute two-thirds of it (Goericke and Welschmeyer, 1993; Liu et al., 1997). Their metabolism is based on oxygenic photosynthesis, similar to that of eukaryotic algae and plants actually being related to the precursor of plant chloroplasts according to the endosymbiont theory. This process provides ATP and reducing equivalents from the splitting of water, which enables the bacteria to assimilate simple inorganic nutrients for their anabolic demands. Since the discovery of, e.g., *Prochlorococcus* in 1988 (Chisholm et al., 1988), its major importance in the ecology of the oceans has become evident, leading to a large number of genetic studies resulting in sequencing of the genomes of five representative *Prochlorococcus* ecotypes, MED4, MIT9313, MIT9312, and NATL2A (http://img.jgi.doe.gov/cgi-bin/pub/main.cgi) and SS120 (http://www.sb-roscoff.fr/Phyto/ProSS120/), making *Prochlorococcus* one of the most-studied microorganisms from a genomic point of view (Dufresne et al., 2001; Rocap et al., 2003). However, since 2003 the genome of *Synechococcus* is also sequenced (Palenik et al., 2003). The Kazusa Institute in Japan publicly provides sequence information of several cyanobacteria (www.kazusa.or.jp/cyano/). The striking ecological success of cyanobacteria has

been the subject of many studies, focused on some of the most intriguing abilities of this organism to cope with the natural gradients of different parameters occurring along the water column, including light irradiance, which decreases almost four orders of magnitude from the ocean surface to the end of the euphotic zone. Due to their abundance and high metabolic activities, microorganisms may deplete the environment of essential nutrients. A prominent example of an environmental effect caused by the metabolic activity of cyanobacteria is the depletion of carbon dioxide from the atmosphere during the course of evolution, and the concomitant accumulation of the 'waste product' of photosynthesis, oxygen. Deprivation of essential nutrients is frequently the limiting factor in cyanobacterial cell growth, and the need to adapt to periods of nutrient limitation is a major source of selective pressure in diverse natural environments. Research in the past decade has led to fundamental new insights into the molecular mechanisms of these responses. Classically, acclimation responses to nutrient limitation are grouped into specific and general, or common, responses. The specific responses are the acclimation processes that occur as a result of limitation for a particular nutrient, whereas the general responses occur under any starvation condition. Cyanobacterial acclimation to limitation of the macronutrients phosphorus and nitrogen was studied in the past and summarized by Schwarz and Forchhammer (2005). As outlined by the authors little substantial progress has been made in understanding the molecular mechanisms underlying responses to sulfur limitation. This also applies to approaches investigating Met biosynthesis in cyanobacteria.

Met is an essential cellular constituent, an initiator of protein synthesis and a key player in many metabolic activities. Furthermore, Met is the precursor of SAM—S-adenosylmethionine—that is involved in many methylation reactions such as DNA and phospholipid methylation and synthesis of cyclopropane fatty acids. SAM also donates aminopropyl groups to diamines in the synthesis of polyamines such as spermidine, which neutralize the negative charge of nucleic acids in the cells (for reviews see Chiang et al., 1996; Fontecave et al., 2004), is precursor of the vitamin biotin, as well as to synthesize the 'aging' hormone ethylene responsible for fruit ripening. Furthermore, in algae and some bacteria Met is a source of atmospheric sulfur: dimethylsulphide (Amir et al., 2002; see

---

Abbreviations: CDMS – cobalamin-dependent Met synthase; CBL – cystathionine beta-lyase; CBS – cystathionine b-synthase; CGL – cystathionine g-lyase; CGS – cystathionine gamma-synthase; CIMS – cobalamin-independent Met synthase; HCys – homocysteine; HMT – HCys S-methyltransferase; HS – homocysteine synthase; HSK – homoserine kinase; HTA – homoserine transacetylase; HTS – homoserine trans-succinylase; Met, methionine; MMT – Met S-methyltransferase; MS – methionine synthase; MTA – methylthioadenosine; MTR – 5′-methylthioribose; MTR1P – 5′-methyl-thioribose 1-phosphate; OAHS – O-acetyl homoserine sulfhydrylase; OPHS – O-phosphohomoserine; OSHS – O-succinyl homoserine sulfhydrylase; Thr, threonine; SAH – S-adenosylhomocysteine; SAM – S-adenosylmethionine; SAMS – SAM synthetase; SMM – S-methylmethionine; SRH – S-ribosylhomocysteine;

also Chapters 14, 16 and 22, this issue). A derivative of Met, S-methylmethionine (SMM), is used as a major transport molecule for reduced sulfur in some plant species, connecting sink and source organs (Bourgis et al., 1999). Furthermore, Met is an essential amino acid required in the diet of non-ruminant animals. Human and monogastric mammals can synthesize only half of the 20 major proteinogenic amino acids and, therefore, must obtain the others from their diets. Major crops, such as cereals (e.g., corn, rice, wheat, etc.) and legumes, are low in Met (Tabe and Higgins, 1998; Hesse et al., 2001; Galili et al., 2005). Improved nutritional quality may help to solve problems encountered in cases where plant foods are the major or sole source of protein, such as in many developing countries, as well as plant feeds for livestock which are subsequently used as human food.

Recent findings indicate that Met intermediates such as homoserine has yet another important physiological role in microorganisms—it is involved in the synthesis of N-acyl homoserine lactone autoinducer molecules, which constitute quorum-sensing signals and act as cell density-dependent regulators of gene expression (Hanzelka and Greenberg, 1996; Val and Cronan, 1998; Yang et al., 2005). As a process, Met biosynthesis is widely distributed among the three domains of life (bacteria, archaea and eukarya) and could predate their divergence. However, the constituent steps in the pathway, the enzymes catalyzing these steps and the genes encoding these enzymes are not the same among all organisms and there is significant evolutionary plasticity at each of these levels. In plants and microorganisms Met is in general synthesized through three consecutive reactions catalyzed by cystathionine γ-synthase (CGS), cystathionine β-lyase (CBL), and Met synthase (MS) (Fig. 1). Met receives its carbon skeleton from the aspartate-family pathway, while its sulfur moiety is derived from cysteine (Cys). Eventually about 20% of the Met in plants at least is incorporated into proteins while 80% are converted to S-adenosylmethionine (SAM) which thus comprises the actual end product of this biosynthetic pathway as free Met does only occur in marginal concentrations in plants (Giovanelli et al., 1985). This fact indicates the high consumption of Met in different cellular processes. Furthermore, organisms evolved a Met salvage pathway to recover Met without de novo synthesis of the respective precursors. In

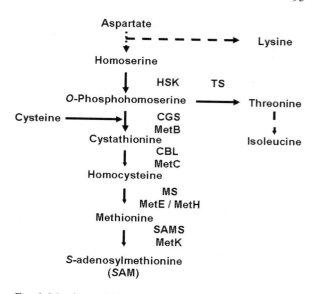

Fig. 1. Met focused biosynthetic pathway of amino acids of the aspartate family in plants. CGS, cystathionine γ-synthase; CBL, cystathionine β-lyase; MS, Met synthase; SAMS, S-AdoMet synthethase. Dashed arrows indicate multiple steps in lysine and isoleucine formation. Corresponding genes in E. coli are given in conventional nomenclature (met…).

plants, CGS, localized in chloroplasts, catalyses the formation of the thioether cystathionine from the substrates Cys and O-phosphohomoserine (OPHS) thus, connecting the aspartate-derived pathway to sulfur assimilation. In yeast, Met is synthesized by direct sulfhydration of O-acetylhomoserine and in bacteria in a different pathway with succinylhomoserine as substrate. It has thus to be emphasized that in microorganisms homoserine is the branch point intermediate leading to the synthesis of Met and threonine (Thr), whereas in plants OPHS is the last common intermediate. This difference of the branch point of the Met and Thr biosynthetic pathway in plants asks for an effective regulation of the respective enzymatic activities. Cyanobacterial orthologs of plant genes have been identified from the Genome Database for Cyanobacteria. Among the listed mutants defective in different pathways none of them is auxotrophic for Met. Genes for the biosynthetic as well as for the salvage pathway can be identified in this database. However, depending on the organism differences in gene annotation are evident and lined out in Gophna et al. (2005). The false annotation has severe consequences for predicting pathways enzymes. As long as the biochemical properties of

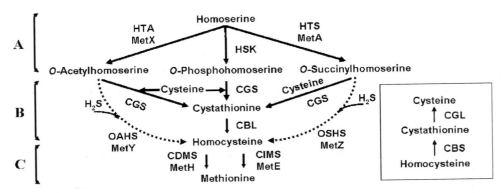

Fig. 2. The biochemical pathway of Met synthesis. Schematic presentation compares mechanistic differences in the synthesis of HCys in plants, bacteria, and fungi. HTA, homoserine trans-acetylase; HSK, homoserine kinase; HTS, homoserine trans-succinylase; CGS, cystathionine γ-synthase; CBL, cystathionine β-lyase; OAHS, O-acetyl homoserine sulfhydrylase (in fungi metY); OSHS, O-succinyl homoserine sulfhydrylase (in bacteria metZ); CDMS, cobalamin-dependent Met synthase (metH); CIMS, cobalamin-independent Met synthase (metE); CBS, cystathionine β-synthase; CGL, cystathionine γ-lyase. A, homoserine activation, B, sulfur incorporation, C, methylation. Dashed lines indicate alternative pathways for direct HCys formation.

the enzymes are not studied, regulatory aspects are difficult to judge. So far one has to assume that due to the relation to bacteria kinetics and regulation are similar (Fig. 1).

It has to be mentioned that several organisms such as yeast or cyanobacteria evolved shunt pathways in which the reduced S-moiety of Cys is directly incorporated to form homocysteine (HCys) via HCys synthase (HS) (Fig. 2). HS is a more general term encompassing two individual enzyme activities, O-acetyl homoserine sulfhydrylase and O-succinyl homoserine sulfhydrylase, starting from succinyl or acetyl homoserine as precursors, respectively (Gophna et al., 2005). However, for plants such an ortholog has not been identified. It is reported that CGS possesses such a side activity (Hesse et al., 2004).

II. Biosynthesis of Methionine Precursors via Alternate Pathways

Intermediates in the synthesis of Met from its precursor homoserine, also a precursor to Thr synthesis, are shown in Fig. 2. There are alternate reactions at each step of the pathway, but the intermediates appear to be conserved for Met synthesis. The first step, homoserine activation, can occur in three ways: the first is acetylation, which is catalyzed by homoserine trans-acetylase (HTA) using acetyl-CoA as a substrate. This reaction is carried out by many bacteria such as *Leptospira*

meyeri (Bourhy et al., 1997) or *Corynebacterium glutamicum* (Park et al., 1998) and fungi such as *Saccharomyces cerevisiae* (Yamagata, 1987) or *Neurospora crassa* (Kerr and Flavin, 1970) or cyanobacteria. Orthologs of the *met2* gene (in analogy to the yeast gene) can be identified for the cyanobacterial strains *Chlorobium tepidum* TLS, strains of *Prochlorococcus marinus* MED4, SS120, and MIT9313, *Synechococcus* sp. WH8102, and *Rhodospseudomonas palustris* CGA009. The second is succinylation, which is catalyzed by homoserine trans-succinylase (HTS) utilizing succinyl-CoA as a substrate. This activity is known in some bacteria such as *Escherichia coli* (Rowbury and Woods, 1964) or *Pseudomonas aeruginosa* (Foglino et al., 1995). Here again orthologs for this gene can be identified in the genomes of *Prochlorococcus marinus* MED4, SS120, and MIT9313, and *Rhodospseudomonas palustris* CGA009. The third alternative reaction is phosphorylation, catalyzed by homoserine kinase (HSK). HSK catalyzes the formation of OPHS from homoserine and is converted in plants in a competing reaction by either CgS or Thr synthase (TS) either by condensation of Cys and OPHS to cystathionine being further converted to Met or directly to Thr (Thr) by TS. Bacteria use OPHS exclusively as precursor for Thr synthesis (Kredich, 1996). In plants HSK is much better characterized than in cyanobacteria. However, due to its low abundance it was not investigated in great detail in the past. The native enzyme from wheat was purified to homogeneity and characterized (Riesmeier et al., 1993). The regulation of the enzyme is variable. Wheat HSK is not inhibited by physiological concentrations of Thr, Met, valine, isoleucine, or SAM (Riesmeier et al., 1993) while older studies revealed that HSK, both from pea and radish, is allosterically inhibited by these amino acids (Thoen et al., 1978; Baum et al., 1983). Greenberg et al. (1988) isolated soybean cell lines with repressed HSK activity resulting in Met accumulation. HSK is localized in plastids of pea probably containing soluble and membrane-associated HSK activities (Muhitch and Wilson, 1983; Wallsgrove et al., 1983). The localization was supported by the isolation of HSK in Arabidopsis. A single gene locus was identified and the predicted enzyme was localized in plastids with a mature mass of 33 kD. The recombinant protein was characterized biochemically to be dependent on Mg^{2+} and K^+ ions (Lee and Leustek, 1999). The respective orthologs can be identified for *Synechocystis* sp. PCC6803, *Anabaena* sp. PCC7120, *Thermosynechococcus elongates* BP-1, *Gloeobacter violaceus* PCC7421, *Chlorobium tepidum* TLS, strains of *Prochlorococcus marinus* MED4, SS120, and MIT9313, *Synechococcus* sp. WH8102, and *Rhodospeudomonas palustris* CGA009.

Although the cognate activities HTA, HSK and HTS in general correspond to one of each of these protein families, evidence from in vitro experiments suggests that membership in a protein family does not unerringly predict substrate specificity. For example, *P. aeruginosa*, an HTS as defined by its activity in vitro is evolutionarily related to the HTA protein found in *Haemophilus influenzae* and not to the *E. coli* enzyme HTS (Foglino et al., 1995). Similarly, *Bacillus subtilis* which is known to possess homoserine transacetylase activity (Brush and Paulus, 1971) is homologous to the *E. coli* HTS (Rodionov et al., 2004). Thus, it is not a priori predictable to which group the respective orthologs from cyanobacteria belong because the sequence homology between both types of transacetylases is relatively high. For example, the annotation predictions of enzymes for *Rhodospeudomonas palustris* CGA009 suggest both, an HTS and an HTA function. This has to be proofed in future by biochemical investigations.

Incorporation of sulfur, the second step of the Met pathway, may also proceed in one of three ways (Fig. 2). Direct incorporation of sulfide (sulfhydrylation) into O-acetyl homoserine by O-acetyl homoserine sulfhydrylase (OAHS) to form HCys is found in several bacteria, fungi, and cyanobacteria (Kerr, 1971; Hwang et al., 2002; CyanoBase). HCys can also be synthesized by a second sulfhydrylation alternative, involving the direct incorporation of sulfide into a different substrate, O-succinyl homoserine, by O-succinyl homoserine sulfhydrylase (OSHS). This activity is found in some bacteria such as *Pseudomonas* species and cyanobacteria (Foglino et al., 1995; Vermeij and Kertesz, 1999; CyanoBase). Contradictory is the fact that cyanobacteria expressing HTS contain also a gene encoding OAHS indicating a still not complete and correct annotation of the genome.

III. Transsulfuration Process to Form Homocysteine

The transfer of the sulfur atom from Cys to HCys through the transsulfuration pathway with cystathionine as intermediate occurs only in plant plastids and has been extensively studied in Arabidopsis (Hesse et al., 2004 and references therein). The first committed step of de novo Met synthesis is a transsulfuration process via a γ-replacement reaction using Cys as the sulfur donor and O-phosphohomoserine as carbon precursor (Figs. 1 and 2). The formation of the thioether cystathionine is catalyzed by cystathionine γ-synthase (CGS) transferring reduced sulfur to a carbon backbone followed by a β-elimination carried out by cystathionine β-lyase (CBL) yielding HCys (Hesse and Höfgen, 2003; Hesse et al., 2004). The carbon/nitrogen precursor of Met synthesis in plants is distinct from that in yeast and bacteria. In yeast, Met is synthesized by direct sulfydration of O-acetylhomoserine; in bacteria, it is through a different pathway with succinyl-homoserine as substrate. In addition, the mature plant CGS exhibits at its N-terminus an additional module of about 120 amino-acids. This particular extension plays a role in its regulation (see section Regulation; Onouchi et al., 2004). Biochemical and kinetic parameters for CGS from diverse sources were extensively analyzed (Hesse et al., 2004 and references therein). The structure, substrate specificity and ping-pong mechanism was recently confirmed through the resolution and analysis of the *Nicotiana tabaccum* CGS (Steegborn et al., 1999). Physiochemical properties of the second enzyme of the transsulfuration pathway (CBL) were relatively similar to those of *Escherichia coli* (Ravanel et al., 1996, 1998). Interestingly, while the bacterial enzyme degrades cystine and cystathionine equally, the reaction with cystine accounted for only 16% with the plant enzyme (Ravanel et al., 1996). Such observations suggest major differences in the amino acids close to the pyridoxal phosphate binding site between the bacterial and plant CBL, as confirmed through resolved structures from both sources (Clausen et al., 1996; Breitinger et al., 2001). The degree of homology CGS is also evident for cyanobacteria as in the genomes of the *Prochlorococcus marinus* strains MED4, SS120, and MIT9313, and the *Synechococcus* strain sp. WH8102 genes encoding CGS can be identified. However, due to rather high homology between CGS and CBL it is not possible to differentiate between both enzymes without having proofed their biochemical properties. Moreover, for *Anabaena* sp. PCC7120, *Chlorobium tepidum* TLS, and *Rhodospeudomonas palustris* CGA009 genes encoding the respective yeast enzymes such as cystathionine β-synthase and cystathionine γ-lyase are present. Here again it is a matter of annotation of these particular genes. All the enzymes for sulfur assimilation described above as well as the reverse transsulfuration enzyme cystathionine γ-lyase (CGL) belong to the same evolutionary family (henceforth the CGS family) and sequence alone may not reliably predict enzymatic activity. It can be assumed that some organisms having an active copy of each CGS, CBL and HS (Hwang et al., 2002; Ruckert et al., 2003), in cases where an organism has both direct incorporation and transsulfuration capacities, the same enzyme can carry out both activities (CGS as well as HS) (Vermeij and Kertesz, 1999; Auger et al., 2002; Hacham et al., 2003). Furthermore, substrate flexibility has also been demonstrated for some of these enzymes—so that a plant enzyme for example, which naturally metabolizes homoserine- phosphate, can also use acetyl-homoserine and succinyl-homoserine to form HCys directly (Thompson et al., 1982a; Kreft et al., 1994; Ravanel et al., 1995, 1998; Hacham et al., 2003).

However, this alternative pathway seems to have only a minor physiological significance in plant cell metabolism regarding the entry of reduced sulfur into Met biosynthesis because HCys formation only takes place in the absence of Cys, and secondly, this direct sulfydration pathway participates in only 3% of total HCys synthesis and has seemingly no physiological significance (Giovanelli et al., 1978; MacNicol et al., 1981; Thompson et al., 1982a, b).

IV. Formation of Methionine by Transmethylation

The last step of Met synthesis is catalyzed by Met synthase (MS), which methylates HCys to form Met using N5-methyltetrahydrofolate as a methylgroup donor (Fig. 2). The function of this enzyme is on the one hand the de novo synthesis of Met and on the other the regeneration of the methyl group of S-adenosylmethionine. Thus, this step is essential even in organisms that do not synthesize Met, for the regeneration of the methyl group of SAM (Ravanel et al., 1998; Eckermann et al., 2000; Droux, 2004; Hesse et al., 2004). Using 5-methyltetrahydropteroyl glutamates as substrates, methylation of HCys can occur using two non-homologous enzymes: the cobalamin-dependent Met synthase (CDMS) in mammals, protists and most bacteria (Krungkrai et al., 1989; Goulding and Matthews, 1997; Evans et al., 2004) or the cobalamin-independent Met synthase (CIMS) as found in Bacteria including cyanobacteria (Whitfield et al., 1970; Cyanobase), *Viridiplantae* (Eichel et al., 1995; Zeh et al., 2002) and fungi (Kacprzak et al., 2003). Orthologs of CDMS have not been identified so far in cyanobacteria. However, it can be assumed that *metE* orthologs are present in cyanobacterial genomes. Since it is known from a survey of 326 algal species that approximately half of them require exogenous vitamin B_{12} because they are cobalamin auxotrophs it has to be assumed that the source of cobalamin seems to be bacteria, indicating an important and unsuspected symbiosis (Croft et al., 2005). Thus, it can be assumed that cyanobacteria have the capability to synthesize cobalamin as bacteria have. So far, the molecular and biochemical characterization of Met synthase from plants is still limited.

One reason for this is the low amount of protein present in plants and another is the substrate specificity of the enzyme. While bacteria are able to use monoglutameric methyltetrahydrofolate, *Catharanthus roseus* (Eichel et al., 1995) accepts only the triglutameric isoform that is not freely available. Neither SAM nor cobalamin is required for activity of the cobalamin-independent Met synthase as demonstrated for MS from *C. roseus* and *S. tuberosum* (Eckermann et al., 2000; Zeh et al., 2002). Intensive studies on MS expression revealed that MS is a low copy gene differentially expressed, at least at potato organs, with elevated levels in flowers, basal levels in sink and source leaves, roots, and stolons, and low levels in stems and tubers. This is in good agreement with protein data except that the protein content in leaves was less than expected from the RNA data. Western blot analysis of subcellular fractions revealed that the protein is located in the cytosol. However, the changing pattern of gene expression during the day/night period implied a light-dependent control of MS-transcription normally seen for enzymes localized in plastids. The expression of MS was shown to be light-inducible with its highest expression at midday, while, during the night, expression dropped to a low, basal mRNA level. These RNA data were not confirmed at the protein level since the protein content remained constant over the whole day. Feeding experiments using detached leaves revealed that sucrose or sucrose-derived products are responsible for the induction of StMS1 gene expression, but with no effect on the protein level in *C. roseus* and *S. tuberosum* (Eckermann et al., 2000; Nikiforova et al., 2002; Zeh et al., 2002). Multiple forms of MS could be identified in plants, and in the genome of *A. thaliana* (Eckermann et al., 2000; Droux, 2004; Hesse et al., 2004; Ravanel et al., 2004). Physico-biochemical investigations were performed to answer the localization of the final synthesizing step. Three *Arabidopsis thaliana* Met synthase (MS) were identified from which two are cytosolically and one plastidial localized (Ravanel et al., 2004). It could be further shown that the plastid isoform exhibited a stronger affinity for the polyglutamate derivatives of folates than the cytosolic isoforms being probably the reason not having detected the plastidial activity earlier (Ravanel et al., 2004). The plastid localization of

MS corroborated earlier observations based on enzyme measurements having been made in pea chloroplasts and mitochondria suggesting that the plastidial de novo synthesis of Met is favored indicating the autonomy of the compartment (Shah and Cossins, 1970; Clandinin and Cossins, 1974). By contrast, the photosynthetic protozoan *E. gracilis* Z. expresses three isoforms of cobalamin-dependent enzymes located in chloroplasts, mitochondria, and cytosol in addition to a cytosolic cobalamin-independent Met synthase (Isegawa et al., 1994).

V. Methionine Recycling

Met is a sulfur-containing amino acid that can be activated by ATP to S-adenosyl-Met (SAM). Radiotracer experiments indicate that more than 80% of the label from ^{14}C-methyllabeled Met was incorporated into lipids, pectines, chlorophyll, and nucleic acids, whereas less than 20% in protein (Giovanelli et al., 1985). Thus, apparently the majority of Met is converted into SAM for transmethylation reactions in plants (Fig. 3). SAM is synthesized from Met and ATP by the enzyme SAM synthetase from which three isoforms exists in plants (Shen et al., 2002). SAM serves as a substrate in many biochemical reactions. The high need for SAM makes it necessary to recycle Met. When SAM is utilized for the synthesis of ethylene, certain polyamines, and siderophores, methylthioadenosine (MTA) is produced as an intermediate which can be recycled to Met and allows high rates of the before mentioned metabolites (Bürstenbinder et al., 2007). This Met salvage pathway was first characterized and was described in plants at the biochemical level and is termed according to the discoverer Yang or MTR cycle (Wang et al., 1982; Yang and Hoffman, 1984; Miyazaki and Yang, 1987). However, this circuit is not unique for plants and was identified also in bacteria, including cyanobacteria, and animals. Although the major progress in detailing enzyme characteristics and in identifying their corresponding genes has come from work on prokaryotes (Sufrin et al., 1995; Cornell et al., 1996; Sekowska et al., 2001; Sekowska and Danchin, 2002), the plant field gain on more and more information (e.g., Sauter et al., 2004, 2005; Bürstenbinder et al., 2007). In bacteria and in plants, MTA is depurinated to 5-methylthioribose (MTR) through

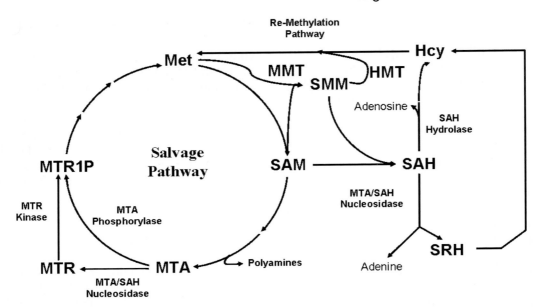

Fig. 3. Met salvage pathway in plants and microorganisms. Met, methionine; MMT, Met S-methyltransferase; HMT, HCys S-methyltransferase; SRH, S-ribosylhomocysteine; MTA, methylthioadenosine; MTR, 5'-methylthioribose; MTR1P, 5'-methyl-thioribose 1-phosphate; SAM, S-adenosylmethionine; SAH, S-adenosylhomocysteine; SMM, S-methylmethionine.

the enzymatic activity of MTA nucleosidase. MTR kinase catalyzes the subsequent phosphorylation of the C-1 hydroxyl group of the Rib moiety of MTR to yield 5-methylthio- Rib-1-P. In animals, MTA phosphorylase carries out both functions in a single ATP-independent step (Schlenk, 1983). Hence, MTR kinase exists in prokaryotes and plant species but not in animals. 5-Methyl-thio- Rib-1-P subsequently undergoes enzymatic isomerization, dehydration, and oxidative decar-boxylation to 2-keto-4-methylthiobutyrate, the immediate precursor of Met (Miyazaki and Yang, 1987). The second possibility to recover Met is the recycling of HCys resulting from hydrolysis of S-adenosylhomocysteine (AdoHCys), pro-duced from cytosolic SAM-dependent methyla-tion (Kloor and Osswald, 2004). It should be noted that AdoHCys is a strong inhibitor of methylation steps and, thus, needs to be quickly transformed into HCys and eventually Met again. Furthermore, accumulation of the hydrolytic product (HCys) of the reaction catalyzed by SAH hydrolase is revers-ible (Kloor et al., 1998; Kloor and Osswald, 2004). Thus, HCys will be eliminated through degrada-tion, transport to other compartments, or through direct conversion into Met, the later step being cat-alyzed by the cytosolic Met synthase (Ravanel et al., 1998, 2004). Furthermore, Met is converted to

S-methylmethionine (SMM) which appears to be a phloem localized transport form of reduced S in plants, as is GSH (Ranocha et al., 2001; Kocsis et al., 2003). The SMM to GSH ratio varies between dif-ferent plants. SMM is synthesized from Met and SAM by SAM:Met S-methyltransferase releasing S-adenosylhomocysteine (MMT). SMM can be reconverted to Met by HCys S-methyltransferase (HMT) methylating HCys yielding two molecules of Met. Essentially this appears to be a shortcut of the SAM methylation cycle and currently it is spec-ulated that its main function is the down regulation of SAM levels in plants. However, as SMM is also transported in the phloem it might well contribute to reduced sulfur supply to sinks. Furthermore, this mechanism presents a second system to remove the excess of HCys produced in the cytosol when folate derivatives are insufficient for Met synthe-sis. SMM is present in millimolar concentration in plants and constituted storage for sulfur and methyl groups, as well a system to maintain the cell SAM level constant as was demonstrated in mutant plants with inactivated Met S-methyltransferase (Kocsis et al., 2003). The above mentioned proc-ess has not been investigated in detail for cyano-bacteria. Analysis of the genome of cyanobacteria revealed that most of the genes are present leading to the assumption that the processes as described above for microorganisms are also functional in cyanobacteria. The major difficulty as already discussed above is the weakness of annotation of open reading frames. SAM synthetase is present in nearly all cyanobacteria indicating, that the initial step, the formation of SAM, is performed. However, somewhat problematic is the classification of genes encoding enzymes of the Yang cycle or of the HCys recycling pathway. Only few genes of the cycle are annotated, however, providing evidence that both recycling pathways are existent. It is also not evident whether cyanobacteria, although living partially in sea water, are able to form dimethylsulfiopropionic acid (DMSP) which derives from SMM as the respective genes were not identified so far. DMSP is an excellent compatible solute widely used for osmoprotection in bacteria and algae and for cryoprotection in algae (Kiene et al., 1996; Malin and Kirst, 1997; Stefels, 2000; Welsh, 2000; see also Chapters 14 and 22, this issue).

VI. Regulatory Aspects of Methionine Homeostasis

The synthesis of Met is subject to a complex regulation (Fig. 4). The whole pathway is regulated by feedback inhibition of aspartate kinase by either lysine or Thr or additionally lysine synergistically together with S-adenosylmethionine (SAM). Earlier studies have shown that the provision of the carbon backbone, e.g., by over-expression of aspartate kinase increased the synthesis of lysine and Thr leading to the assumption that amino acid pools are filled in a controlled order (Galili, 1995; Hesse and Hoefgen, 2003; Hesse et al., 2004; Galili et al., 2005). This result suggests that the carbon flux is limiting the synthesis of Met. Insights into the regulation of the Met network were achieved with a variety of transgenic approaches targeted at the various steps of the transsulfuration pathway, including the cytosolic Met synthase,

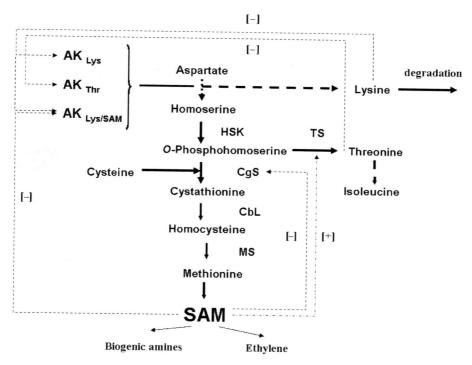

Fig. 4. Schematic diagram of the metabolic networks regulating Lysine, Thr and Met metabolism. Only some of the enzymes and metabolites are specified. Dashed arrows with a 'minus' symbol represent either feedback inhibition loops or repression of gene expression. The dashed and dotted arrow with the 'plus' symbol represents the stimulation of gene expression or enzyme activity. Abbreviations: AK, aspartate kinase (sensitivity to metabolites are indicated); HSK, homoserine kinase; TS, threonine synthase; CGS, cystathionine γ-synthase; CBL, cystathionine β-lyase; MS, Met synthase; SAM, S-adenosyl Met. Dashed lines indicate multiple enzymatic steps.

and also the enzymes synthesizing or competing for the O-phosphohomo- serine substrate, HSK and TS, respectively (Hesse and Hoefgen, 2003; Droux, 2004; Hesse et al., 2004; Lee et al., 2005; Rinder et al., 2007). Conclusion of theses studies revealed that in Arabidopsis CGS was the unique enzyme controlling the flux of sulfur and of the carbon–nitrogen backbone toward Met synthesis either at the level of CgS on RNA level (CgS in Arabidopsis; Chiba et al., 1999) or TS activity level by SAM (TS in Arabidopsis and potato; Curien et al., 1996, 1998; Zeh et al., 2001) depending on the plant species (Hesse and Hoefgen, 2003; Hesse et al., 2004). TS activity is positively regulated by SAM, a direct product of Met, thus favoring carbon flow into Thr biosynthesis in preference to Met synthesis when sufficient SAM is available. Under these conditions the K_M-values of TS for OPHS have been shown to be 250- to 500-fold lower as compared to the competing enzyme, CgS, thus favoring carbon flow into Thr biosynthesis in preference to Met synthesis (Madison and Thompson, 1976; Curien et al., 1996, 1998; Laber et al., 1999). A long standing question was answered recently how SAM is able to activate TS. Now, there is evidence for a counter-exchange transporter of AdoHCys from the plastid to the cytosol with cytosolic synthesized SAM (Ravanel et al., 2004) and SAM transporter (Bouvier et al., 2006) However, it is questionable whether a complete shut down of Met synthesis is reached. The demand of SAM for several processes has to be assumed to be high indicating that most of the synthesized SAM is used. Only a small part is available to be translocated into chloroplasts in order to activate TS. Thus, it can be assumed that carbon provided by aspartate via OPHS is preferentially used to form Met subsequently used for SAM synthesis because in a non-activated condition the K_M value for the substrate is lower compared to TS. Feeding studies with $^{13}CO_2$ revealed that the carbon flux is priorly directed to Met (and, hence, SAM) formation within this pathway and even when compared to other pathways, while lysine and Thr/isoleucine pools provide minor sinks (Hesse, Schauer, Hoefgen, Fernie, unpublished; Fig. 5). From this point of view CGS is the most important enzyme for Met synthesis. As demonstrated by feeding studies, CgS is not an allosteric enzyme since enzyme activity is not inhibited by Met (Thompson et al.,

1982b). Mutants and transgenic plants have been used to display the central role of CgS in the synthesis of Met in Arabidopsis. Most extensive developments on the molecular regulation of the Met network were investigated by characterization of *Arabidopsis thaliana* mutant lines resistant to the toxic analogue of Met, ethionine. These mutant lines called *mto*1(1–7) were characterized by elevated levels of CGS mRNA and of CGS activity as compared to wild type (Onouchi et al., 2004) (Fig. 6). Genetic analysis described a mutation within the conserved exon1 coding region (*mto*1 region) where a single nucleotide changes resulted in one amino acid modification (Chiba et al., 1999, 2003; Amir et al., 2002). This important motif acts in cis to stabilize the CGS mRNA being resistant to degradation (Suzuki et al., 2001; Ominato et al., 2002; Onouchi et al., 2004). Translation of the protein is necessary for the regulation and stability of the mRNA as was confirmed by an experiment using the in vitro translation system of wheat germ extract (Chiba et al., 2003; Lambein et al., 2003). This result is supported by the recent finding for tomato fruits during ripening showing that Met down-regulates CGS transcript levels (Katz et al., 2006). However, this finding is inconsistent to the earlier finding for potato CgS transcript stability which is not influenced by Met (Kreft et al., 2003). Finally, since SAM was revealed as the crucial metabolite involved in this regulation, suggesting that the Met dependent regulation was the result of its transformation into SAM in the cytosol (Onouchi et al., 2004). Interestingly, the mto1 domain showed no features of known SAM binding domains. From this observation, it may be speculated that SAM regulation takes place through a protein intermediate. From the recent investigation using wheat germ extract, one could conclude that this intermediate is constitutively expressed (Chiba et al., 2003; Kreft et al., 2003; Onouchi et al., 2004). A mechanism how SAM might regulate CGS stability is suggested by Onouchi et al. (2005). They suggest a temporary arrest in the translation process at serine-94 (nucleotide 280 from the first ATG of the mRNA) when the SAM level increased which strongly correlates between this arrest and mRNA degradation (Onouchi et al., 2005). Very recently it could be demonstrated that in addition to the full-length CGS transcript, a deleted form exists in Arabidopsis (Hachem et al., 2006). The

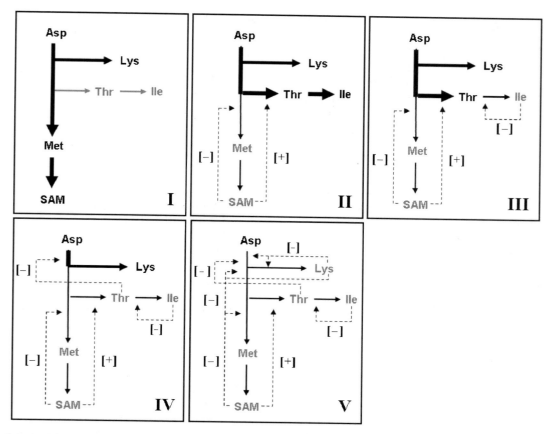

Fig. 5. Schematic presentation of carbon flow and feedback control mechanism in Arabidopsis as model plant. Boxes I–V represents the dynamic regulation of the pathway via product inhibition. Thickness of the arrows indicates the carbon flow in the pathway. Dashed arrows with a 'minus' symbol represent either feedback inhibition loops or repression of gene expression. The dashed arrow with the 'plus' symbol represents the stimulation of gene expression or enzyme activity.

	76		89 99		128/129	
TP	mto1-1-7		Δ 87/90 nt		Catalytic domain	

Fig. 6. The Arabidopsis mRNA of CGS is shown schematically. TP, the chloroplast-targeting transit peptide; mto1 domain indicates the region with point mutations; Δ, indicates the deleted region of 87 or 90 bp; catalytic domain, the catalytic part of CGS that is highly homologous to bacteria CGSs. Nucleotides are numbered from the ATG initiator codon.

deleted transcript of CGS that lacks 90 or 87 nt located internally in the regulatory N-terminal region of CGS localized adjacent to the *mto*1 region maintains the reading frame of the protein (Fig. 6). The deleted region is located within the first exon of the CGS transcript and is not flanked by any known consensus intron/exon boundary sequences, suggesting that the shorter transcript is not a result of consensus alternative splicing.

At the transcript level, it is highly enriched for G-C content (over 70%), suggesting a possible secondary structure (Fig. 7A). When overexpressed in transgenic tobacco plants the transgenic plants accumulated Met to a much higher level than those that expressed the full-length CGS probably because this form of CGS is not subject to feedback regulation by Met, as reported for the full-length transcript (Hachem et al., 2006).

A   B

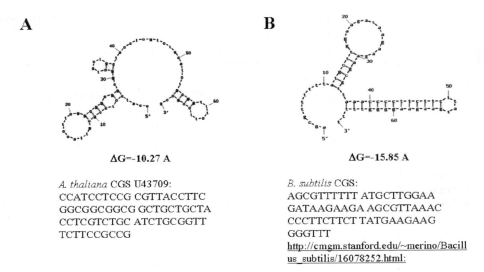

ΔG=-10.27 A   ΔG=-15.85 A

A. thaliana CGS U43709:   *B. subtilis* CGS:
CCATCCTCCG CGTTACCTTC   AGCGTTTTTT ATGCTTGGAA
GGCGGCGGCG GCTGCTGCTA   GATAAGAAGA AGCGTTAAAC
CCTCGTCTGC ATCTGCGGTT   CCCTTCTTCT TATGAAGAAG
TCTTCCGCCG   GGGTTT
                                                http://cmgm.stanford.edu/~merino/Bacill
                                                us_subtilis/16078252.html:

Fig. 7. Prediction of secondary RNA structure for the 90-nt deleted region from Arabidopsis (A) in comparison to *B. subtilis* (B). The predicted calculated ΔG of this stem-loop structure is -10.27 Kcal mol^{-1} for Arabidopsis and -15.85 Kcal mol^{-1} for *B. subtilis*. The prediction was calculated using the software located on the following webpage: www.bioinfo.rpi.edu/applications/hybrid/quikfold.php.

The regulation of Met biosynthesis in cyanobacteria has not been investigated in the past. However, since the genomes of different cyanobacteria ecotypes are known and for most of the enzymatic steps respective genes were identified, one can assume that the controlling mechanisms function nearly identical. Met biosynthesis in *E. coli* has been investigated in details (e.g., for review: Kredich, 1996). Two types of control can be identified in bacteria, the control of gene expression and secondly, the metabolite control mechanism. For *E. coli* transcription nearly all the *Met* genes involved in Met biosynthesis, except *metH*, are under negative control of the MetJ repressor. MetJ interacts with *S*-adenosylmethionine, the pathway's end product, as a corepressor (Saint-Girons et al., 1988). Expression of the *metE, metA, metF, metH*, and *glyA* genes is additionally under positive control of the MetR activator (Maxon et al., 1989; Urbanowski and Stauffer, 1989; Weissbach and Brot, 1991; Mares et al., 1992; Cowan et al., 1993; Lorenz and Stauffer, 1996). HCys may modulate the regulator role of MetR and is required for the *metE* gene activation (Urbanowski and Stauffer, 1987). Furthermore, vitamin B$_{12}$ is involved in *metE* repression, probably by depletion of the coactivator HCys (Wu et al., 1992). MetR is a member of the LysR family of prokaryotic transcriptional regulatory proteins. Common family features are the size (between 300 and 350 amino acids), the formation of either homodimers or homotetramers, the presence of a helix-turn-helix DNA binding motif in the N-terminal region, and the requirement for a small molecule that acts as a co-inducer (Schell, 1993). MetR activates gene expression at one or more loci while negatively regulating the expression of their own genes. In other microorganisms such as *Bacillus subtilis, Clostridium acetobutylicum*, and *Staphylococcus aureus*, several genes involved in the biosynthesis of Met are regulated by a global transcription termination control system called the S box regulon (Grundy and Henkin, 1998). The S box belongs to a group of regulatory elements, so called riboswitches. Riboswitches are RNA elements in the 5' untranslated leaders of bacterial mRNAs that directly sense the levels of specific metabolites with a structurally conserved aptamer domain to regulate expression of downstream genes. Riboswitches use the untranslated leader in an mRNA to form a binding pocket for a metabolite that regulates expression of that gene. Riboswitches are dual function molecules that undergo conformational changes and that communicate metabolite binding typically as either increased transcription termination or reduced translation efficiency via an expression platform (Breaker, 2002; Winkler and Breaker, 2005). So far identified in *B. subtilis* are target genes involved in vitamin B1, B2, and B12 synthesis and transport via *rfn-*, *thi-*, and B_{12}-box, respectively, SAM via the *S*-box, lysine via the *L*-box, and guanine/hypoxanthine via the *G*-box (Nudler and Muronov, 2004). Although not investigated in detail for gram-negative bacteria *S*-boxes play a major role in the regulation of Cys and Met synthesis in gram-positive bacteria such as *Bacillus subtilis*. The S-box regulatory system is used in gram-positive organisms, including members of the *Bacillus/Clostridium/Staphylococcus* group, to regulate expression of genes involved in biosynthesis and transport of Met and *S*-adenosylmethionine (SAM) (Grundy and Henkin, 2004; Rodionov et al., 2004) (Fig. 8). Genes in the S-box family exhibit a pattern of conserved sequence and structural elements in the 5′ region of the mRNA, upstream of the start of the regulated coding sequence(s). These conserved elements include an intrinsic terminator and a competing antiterminator that can sequester sequences that otherwise form the 5′ portion of the terminator helix; residues in the 5′ region of the antiterminator can also pair with sequences located further upstream, and this pairing results in formation of a structure (the anti-antiterminator) that sequesters sequences necessary for formation of the antiterminator. Genetic analyses of S-box leader RNAs supported the model that formation of the anti-antiterminator and tran-

scription termination occur during growth under conditions where Met is abundant, while starvation for Met results in destabilization of the anti-antiterminator, allowing antiterminator formation and read-through of the transcription termination site (Grundy and Henkin, 1998). Overexpression of SAM synthetase in vivo results in increased repression of S-box gene expression, supporting the model that SAM is the effector in vivo (Auger et al., 2002; McDaniel et al., 2003). These results suggested that S-box RNAs directly sense SAM and are therefore members of the family of RNA elements termed 'riboswitches,' which monitor regulatory signals without a requirement for accessory factors such as RNA-binding proteins or ribosomes (Nudler and Mironov, 2004; Tucker and Breaker, 2005). The binding of SAM at the regulatory RNA element could be demonstrated by co-crystallization of an S-adenosylmethionine-responsive riboswitch from *Thermoanaerobacter tengcongensis* with SAM (Montagne and Batey, 2006). Also very recently five novel structured RNA elements were identified by focusing comparative sequence analysis of intergenic regions on genomes of α-proteobacteria (Corbino et al., 2005). One of the five newly found motifs from *Agrobacterium tumefaciens*, termed *metA*, appeared to function as a riboswitch that senses *S*-adenosylmethionine (SAM) (Fig. 8). This SAM-II riboswitch class has a consensus sequence and conserved structure that is distinct from the SAM-I riboswitch reported previously (Epshtein

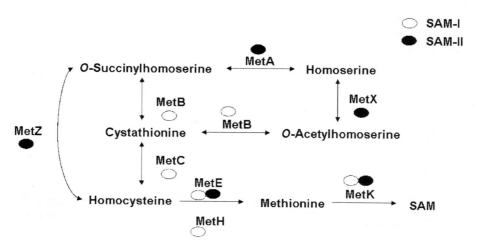

Fig. 8. Comparison of genes in the Met and SAM biosynthetic pathways found downstream of SAM-I and SAM-II riboswitches in gram-negative and positive bacteria.

et al., 2003). Compared with SAM-I aptamers, SAM-II aptamers are smaller and form a simpler secondary structure. However, the SAM-II aptamer exhibits a level of molecular discrimination that is similar to that observed for the SAM-I riboswitch. These findings demonstrate that biological systems use multiple RNA motifs to sense the same chemical compound. Further detailed analyses have to be done in future to investigate this regulatory mechanism in *E. coli* and broaden it to other species such as members of the cyanobacterial family. As two SAM riboswitches systems are identified in bacteria one might argue that this mechanism might be 'RNA World' relics that were selectively retained in certain bacterial lineages or new motifs that have emerged since the divergence of the major bacterial groups. A third regulatory riboswitch was identified very recently in lactic acid bacteria (Fuchs et al., 2006). Here SAM synthetase (MetK) expression is regulated by RNA conformational changes in the 5'UTR due to SAM binding subsequently negatively regulating SAM synthesis. This brings us back to the regulatory mechanism in plants that SAM is able to modulate transcript stability. Perhaps, although CGS does not contain a so far known SAM binding domain, this regulatory mechanism might be a conserved relict. Taking into account that according to the endosymbiontic theory the plastidial targeting developed later than the regulatory mechanism via SAM one might speculate that the mto1 domain and the adjacent 90 nt might have evolved combining both, an open reading frame and a control mechanism via SAM (Fig. 7). Similarities between RNA stem loop structure of CGS from *A. thaliana* and *B. subtilis* are not to neglect. The evolutionary consequences of this finding have to be explored in the future.

References

Amir R, Hacham Y and Galili G (2002) Cystathionine gamma-synthase and threonine synthase operate in concert to regulate carbon flow towards methionine in plants. Trends Plant Sci 7: 153–156

Auger S, Danchin A and Martin-Verstraete I (2002) Global expression profile of *Bacillus subtilis* grown in the presence of sulfate or methionine. J Bacteriol 184: 5179–5186

Baum HJ, Madison JT and Thompson JF (1983) Feedback inhibition of homoserine kinase from radish leaves. Phytochemistry 22: 2409–2412

Bourgis F, Roje S, Nuccio ML, Fisher DB, Tarczynski MC, Li C, Herschbach C, Rennenberg H, Pimenta MJ, Shen TL, Gage DA and Hanson AD (1999) S-methylmethionine plays a major role in phloem sulfur transport and is synthesized by a novel type of methyltransferase. Plant Cell 11: 1485–1498

Bourhy P, Martel A, Margarita D, Saint-Girons I and Belfaiza J (1997) Homoserine O-acetyltransferase, involved in the *Leptospira meyeri* methionine biosynthetic pathway, is not feedback inhibited. J Bacteriol 179: 4396–4398

Bouvier F, Linka N, Isner J-C, Mutterer, Weber APM and Camera B (2006) Arabidopsis SAMT1 defines a plastid transporter regulating plastid biogenesis and plant development. Plant Cell 18: 3088–3105

Breaker RR (2002) Engineered allosteric ribozymes as biosensor components. Curr Opin Biotechnol 13: 31–39

Breitinger U, Clausen T, Ehlert S, Huber R, Laber B, Schmidt F, Pohl E and Messerschmidt A (2001) The three-dimensional structure of cystathionine beta-lyase from Arabidopsis and its substrate specificity. Plant Physiol 126: 631–642

Brush A and Paulus H (1971) The enzymic formation of O-acetylhomoserine in *Bacillus subtilis* and its regulation by methionine and S-adenosylmethionine. Biochem Biophys Res Commun 45: 735–741

Bürstenbinder K, Rzewuski G, Wirtz M, Hell R and Sauter M (2007) The role of methionine recycling for ethylene synthesis in Arabidopsis. Plant J 49: 238–249

Capone DG (2000) The marine microbial nitrogen cycle. In: Kirchman DL (ed) Microbial ecology of the oceans, pp 455–493. Wiley-Liss, Inc, New York

Chiang PK, Gordon RK, Tal J, Zeng GC, Doctor BP, Pardhasaradhi K and McCann PP (1996) S-adenosylmethionine and methylation. FASEB J 10: 471–480

Chiba Y, Ishikawa E, Kijima F, Tyson RH, Kim J, Yamamoto H, Nambara E, Leustek T, Wallsgrove RM and Naito S (1999) Evidence of autoregulation of cystathionine γ-synthase mRNA stability in Arabidopsis. Science 286: 1371–1374

Chiba Y, Sakurai R, Yoshino M, Ominato K, Ishikawa M, Onouchi H and Naito S (2003) S-adenosyl-L-methionine is an effector in the posttranscriptional autoregulation of the cystathionine γ-synthase gene in Arabidopsis. Proc Natl Acad Sci USA 100: 10225–10230

Chisholm SW, Olson RJ, Zettler ER, Goericke R, Waterbury JB and Welschmeyer NA (1988) A novel free-living prochlorophyte abundant in the oceanic euphotic zone. Nature 334: 340–343

Clandinin MT and Cossins EA (1974) Methionine biosynthesis in isolated *Pisum sativum* mitochondria. Phytochemistry 13: 585–591

Clausen T, Wahl MC, Messerschmidt A, Huber R, Fuhrmann JC, Laber B, Streber W and Steegborn C (1999) Cloning, purification and characterization of cystathionine gamma-synthase from *Nicotiana tabacum*. Biol Chem 380: 1237–1242

Corbino KA, Barrick JE, Lim J, Welz R, Tucker BJ, Puskarz I, Mandal M, Rudnick ND and Breaker RR (2005) Evidence for a second class of S-adenosylmethionine riboswitches and other regulatory RNA motifs in alpha-proteobacteria. Genome Biol 6: R70

Cornell KA, Winter RW, Tower PA and Riscoe MK (1996) Affinity purification of 5-methylthioribose kinase and 5-methylthioadenosine/S-adenosylhomocysteine nucleosidase from *Klebsiella pneumoniae*. Biochem J 317: 285–290

Cowan JM, Urbanowski ML, Talmi M and Stauffer GV (1993) Regulation of the *Salmonella typhimurium* metF gene by the MetR protein. J Bacteriol 175: 5862–5866

Croft MT, Lawrence AD, Raux-Deery E, Warren MJ and Smith AG (2005) Algae acquire vitamin B_{12} through a symbiotic relationship with bacteria. Nature 438: 90–93

Curien G, Dumas R, Ravanel S and Douce R (1996) Characterization of an *Arabidopsis thaliana* cDNA encoding an S-adenosylmethionine-sensitive threonine synthase. FEBS Lett 390: 85–90

Curien G, Job D, Douce R and Dumas R (1998) Allosteric activation of *Arabidopsis* threonine synthase by S-adenosylmethionine. Biochemistry 31: 13212–13221

DeLong EF and Karl DM (2005) Genomic perspectives in microbial oceanography. Nature 437: 336–342

Droux M (2004) Sulfur assimilation and the role of sulfur in plant metabolism: a survey. Photosyn Res 79: 331–348

Dufresne A, Salanoubat M, Partensky F, Artiguenave F, Axmann IM, Barbe V, Duprat S, Galperin MY, Koonin EV, Le Gall F, Makarova KS, Ostrowski M, Oztas S, Robert C, Rogozin IB, Scanlan DJ, Tandeau de Marsac N, Weissenbach J, Wincker P, Wolf YI and Hess WR (2003) Genome sequence of the cyanobacterium *Prochlorococcus marinus* SS120, a nearly minimal oxyphototrophic genome. Proc Natl Acad Sci USA 100: 10020–10025

Eckermann C, Eichel J and Schröder J (2000) Plant methionine synthase: new insights into properties and expression. Biol Chem 381: 695–703

Eichel J, González JC, Hotze M, Matthews RG and Schröder J (1995) Vitamin-B_{12}-independent methionine synthase from higher plant (*Catharanthus roseus*). Molecular characterization, regulation, heterologous expression, and enzyme properties. Eur J Biochem 230: 1053–1058

Epshtein V, Mironov AS and Nudler E (2003) The riboswitch-mediated control of sulfur metabolism in bacteria. Proc Natl Acad Sci USA 100: 5052–5056

Evans JC, Huddler DP, Hilgers MT, Romanchuk G, Matthews RG and Ludwig ML (2004) Structures of the N-terminal modules imply large domain motions during catalysis by methionine synthase. Proc Natl Acad Sci USA 101: 3729–3736

Foglino M, Borne F, Bally M, Ball G and Patte JC (1995) A direct sulfhydrylation pathway is used for methionine biosynthesis in *Pseudomonas aeruginosa*. Microbiology 141: 431–439

Fontecave M, Atta M and Mulliez E (2004) S-adenosylmethionine: nothing goes to waste. Trends Biochem Sci 29: 243–249

Fuchs RT, Grundy FJ and Henkin TM (2006) The S_{MK} box is a new SAM-binding RNA for translational regulation of SAM synthetase. Nat Struc Mol Biol 13: 226–233

Galili G (1995) Regulation of lysine and threonine synthesis. Plant Cell 7: 899–906

Galili G, Amir R, Hoefgen R and Hesse H (2005) Improving the levels of essential amino acids and sulfur metabolites in plants. Biol Chem 386: 817–831

Giovanelli J, Mudd SH and Datko AH (1978) Homocysteine biosynthesis in green plants. Physiological importance of the transsulfuration pathway in *Chlorella sorokiniana* growing under steady state conditions with limiting sulfate. J Biol Chem 253: 5665–5677

Giovanelli J, Mudd SH and Datko AH (1985) Quantitative analysis of pathways of methionine metabolism and their regulation in *Lemna*. Plant Physiol 78: 555–560

Goericke R and Welschmeyer NA (1993) The marine prochlorophyte *Prochlorococcus* contributes significantly to phytoplankton biomass and primary production in the Sargasso Sea Deep Sea Res (Part I, Oceanographic Research Papers) 40 Suppl 11–12: 2283–2294

Gophna U, Bapteste E, Doolittle WF, Biran D and Ron EZ (2005) Evolutionary plasticity of methionine biosynthesis. Gene 355: 48–57

Goulding CW and Matthews RG (1997) Cobalamin-dependent methionine synthase from *Escherichia coli*: involvement of zinc in homocysteine activation. Biochemistry 36: 15749–15757

Greenberg JM, Thompson JF and Madison JT (1988) Homoserine kinase and threonine synthase in methionine-overproducing soybean tissue cultures. Plant Cell Rep 7: 477–480

Grundy FJ and Henkin TM (1998) The S box regulon: a new global transcription termination control system for methionine and cysteine biosynthesis genes in Gram-positive bacteria. Mol Microbiol 30: 737–749

Grundy FJ and Henkin TM (2004) Regulation of gene expression by effectors that bind to RNA. Curr Opin Microbiol 7: 126–131

Hacham Y, Gophna U and Amir R (2003). In vivo analysis of various substrates utilized by cystathionine gamma-synthase and O-acetylhomoserine sulfhydrylase in methionine biosynthesis. Mol Biol Evol 20: 1513–1520

Hacham Y, Schuster G and Amir R (2006) An in vivo internal deletion in the N-terminus region of Arabidopsis cystathionine γ-synthase results in CGS expression that is insensitive to methionine. Plant J 45: 955–967

Hanzelka BL and Greenberg EP (1996) Quorum sensing in *Vibrio fischeri*: evidence that S-adenosylmethionine is the amino acid substrate for autoinducer synthesis. J Bacteriol 178: 5291–5294

Hesse H and Hoefgen R (2003) Molecular aspects of methionine biosynthesis in Arabidopsis and potato. Trends Plant Sci 8: 259–262

Hesse H, Kreft O, Maimann S, Zeh M and Hoefgen R (2004) Current understanding of the regulation of methionine biosynthesis in plants. J Exp Bot 55: 1799–1808

Hesse H, Kreft O, Maimann S, Zeh M, Willmitzer L and Hoefgen R (2001) Approaches towards understanding methionine biosynthesis in higher plants. Amino Acids 20: 281–289

Hwang BJ, Yeom HJ, Kim Y and Lee HS (2002) *Corynebacterium glutamicum* utilizes both transsulfuration and direct sulfhydrylation pathways for methionine biosynthesis. J Bacteriol 184: 1277–1286

Isegawa Y, Watanabe F, Kitaoka S and Nakano Y (1994) Subcellular distribution of cobalamin-dependent methionine synthase in *Euglena gracilis*. Z. Phytochem 35: 59–61

Johnson PW and Sieburth JMcN (1979) Chroococcoid cyanobacteria in the sea: a ubiquitous and diverse phototrophic biomass. Limnol Oceanogr 24: 928–935

Kacprzak MM, Lewandowska I, Matthews RG and Paszewski A (2003) Transcriptional regulation of methionine synthase by homocysteine and choline in *Aspergillus nidulans*. Biochem J 376: 517–524

Karl DM (2002) Nutrient dynamics in the deep blue sea. Trends Microbiol 10: 410–418

Katz Y, Galili G and Amir R (2006) Regulatory role of cystathionine-γ-synthase and de novo synthesis of methionine in ethylene production during tomato fruit ripening Plant Mol Biol 61: 255–268

Kerr DS (1971) O-acetylhomoserine sulfhydrylase from *Neurospora*. Purification and consideration of its function in homocysteine and methionine synthesis. J Biol Chem 246: 95–102

Kerr DS and Flavin M (1970) The regulation of methionine synthesis and the nature of cystathionine gamma-synthase in *Neurospora*. J Biol Chem 245: 1842–1855

Kiene RP (1996) Production of methane thiol from dimethylsulfonioproprionate in marine surface waters. Marine Chem 54: 69–83

Kloor D and Osswald H (2004) S-adenosylhomocysteine hydrolase as a target for intracellular adenosine action. Trends Pharmacol Sci J 25: 294–297

Kloor D, Fuchs S, Petroktistis F, Delabar U, Mühlbauer B, Quast U and Osswald H (1998) Effects of ions on adenosine binding and enzyme activity of purified S-adenosylhomocysteine hydrolase from bovine kidney. Biochem Pharmacol 56: 1493–1496

Kocsis MG, Ranocha P, Gage DA, Simon ES, Rhodes D, Peel GJ, Mellema S, Saito K, Awazuhara M, Li CJ, Meeley RB, Tarczynski MC, Wagner C and Hanson AD (2003). Insertional inactivation of the methionine S-methyltransferase gene eliminates the S-methylmethionine cycle and increases the methylation ratio. Plant Physiol 131: 1808–1815

Kredich NM (1996). Biosynthesis of cysteine. In: Neidhardt FC, Curtiss R, Ingraham JL, Lin ECC, Low KB, Magasanik B, Reznikoff WS, Riley M, Schaechter M and Umberger E (eds) *Escherichia coli* and *Salmonella typhimurium*. Cellular and molecular biology, pp 514–527. ASM Press, Washington DC

Kreft BD, Townsend A, Pohlenz HD and Laber B (1994) Purification and properties of cystathionine γ-synthase from wheat (*Triticum aestivum* L.). Plant Physiol 104: 1215–1220

Kreft O, Höfgen R and Hesse H (2003) Functional analysis of cystathionine γ-synthase in genetically engineered potato plants. Plant Physiol 131: 1843–1854

Krungkrai J, Webster HK and Yuthavong Y (1989) Characterization of cobalamin-dependent methionine synthase purified from the human malarial parasite, *Plasmodium falciparum*. Parasitol Res 75: 512–517

Laber B, Maurer W, Hanke C, Grafe S, Ehlert S, Messerschmidt A and Clausen T (1999) Characterization of recombinant *Arabidopsis thaliana* threonine synthase. Eur J Biochem 263: 212–221

Lambein I, Chiba Y, Onouchi H and Naito S (2003) Decay kinetics of autogenously regulated CGS1 mRNA that codes for cystathionine gamma-synthase in *Arabidopsis thaliana*. Plant Cell Physiol 44: 893–900

Lee M and Leustek T (1999) Identification of the gene encoding homoserine kinase from *Arabidopsis thaliana* and characterization of the recombinant enzyme derived from the gene. Arch Biochem Biophys 372: 135–142

Lee M, Martin MN, Hudson AO, Muhitch MJ and Leustek T (2005) Methionine and threonine synthesis are limited by homoserine availability and not the activity of HSK in *A. thaliana*. Plant J 41: 685–696

Liu HB, Nolla HA and Campbell L (1997) *Prochlorococcus* growth rate and contribution to primary production in the Equatorial and Subtropical North Pacific Ocean. Aqua Microb Ecol 12: 39–47

Lorenz E and Stauffer GV (1996) MetR-mediated repression of the *glyA* gene in *Escherichia coli*. FEMS Microbiol Lett 144: 229–233

MacNicol PK, Datko AH, Giovanelli J and Mudd SH (1981) Homocysteine biosynthesis in green plants: physiological importance of the transsulfuration pathway in *Lemna paucicostata*. Plant Physiol 68: 619–625

Madison JT and Thompson JF (1976) Threonine synthetase from higher plants: stimulation by *S*-adenosylmethionine and inhibition by cysteine. Biochem Biophys Res Commun 71: 684–691

Malin G and Kirst GO (1997) Algal production of dimethyl sulfide and its atmospheric role. J Phycol 33: 889–896

Mares R, Urbanowski ML and Stauffer GV (1992) Regulation of the *Salmonella typhimurium* metA gene by the metR protein and homocysteine. J Bacteriol 1174: 390–397

Maxon ME, Redfield B, Cai XY, Shoeman R, Fujita K, Fisher W, Stauffer G, Weissbach H and Brot N (1989) Regulation of methionine synthesis in *Escherichia coli*: effect of the MetR protein on the expression of the metE and metR genes. Proc Natl Acad Sci USA 86: 85–89

McDaniel BAM, Grundy FJ, Artsimovitch I and Henkin TM (2003) Transcription termination control of the S box system: direct measurement of *S*-adenosylmethionine by the leader RNA. Proc Natl Acad Sci USA 100: 3083–3088

Miyazaki JH and Yang SF (1987) The methionine salvage pathway in relation to ethylene and polyamine biosynthesis. Physiol Plant 69: 366–370

Montange RK and Batey RT (2006) Structure of the S-adenosylmethionine riboswitch regulatory mRNA element. Nature 441: 1172–1175

Muhitch MJ and Wilson KG (1983) Chloroplasts are the subcellular location of both soluble and membrane-associated homoserine kinase in pea (Pisum sativum L.) leaves. Z Pflanzenphysiol 110: 39–46

Nikiforova V, Kempa S, Zeh M, Maimann S, Kreft O, Casazza AP, Riedel K, Tauberger E, Hoefgen R and Hesse H (2002) Engineering of cysteine and methionine biosynthesis in potato. Amino Acids 22: 259–278

Nudler E and Mironov AS (2004) The riboswitch control of bacterial metabolism. Trends Biochem Sci 29: 11–17

Ominato K, Akita H, Suzuki A, Kijima F, Yoshino T, Yoshino M, Chiba Y, Onouchi H and Naito S (2002) Identification of a short highly conserved amino-acid sequence as the functional region required for posttranscriptional autoregulation of the cystathionine γ-synthase gene in Arabidopsis. J Biol Chem 277: 36380–36386

Onouchi H, Lambein I, Sakurai R, Suzuki A, Chiba Y and Naito S (2004) Autoregulation of the gene for cystathionine γ-synthase in Arabidopsis: post-transcriptional regulation induced by S-adenosylmethionine. Biochem Soc Trans 32: 597–600

Onouchi H, Nagami Y, Haraguchi Y, Nakamoto M, Nishimura Y, Sakurai R, Nagao N, Kawasaki D, Kadokura Y and Naito S (2005) Nascent peptide-mediated translation elongation arrest coupled with mRNA degradation in the CGS1 gene of Arabidopsis. Genes Dev 19: 1799–1810

Palenik B, Brahamsha F, Larimer M, Land L, Hauser P, Chain J, Lamerdin W, Regala E, Allen E, McCarren J, Paulsen I, Dufresne A, Partensky F, Webb EA and Waterbury JB (2003) The genome of a motile marine Synechococcus. Nature 424: 1037–1042

Park SD, Lee JY, Kim Y, Kim JH and Lee HS (1998) Isolation and analysis of metA, a methionine biosynthetic gene encoding homoserine acetyltransferase in Corynebacterium glutamicum. Mol Cells 8: 286–294

Ranocha P, McNeil SD, Ziemak MJ, Li C, Tarczynski MC and Hanson AD (2001) The S-methylmethionine cycle in angiosperms: ubiquity, antiquity and activity. Plant J 25: 575–584

Ravanel S, Block MA, Rippert P, Jabrin S, Curien G, Rébeillé F and Douce R (2004) Methionine metabolism in plants: chloroplasts are autonomous for de novo methionine synthesis and can import S-adenosylmethionine from the cytosol. J Biol Chem 279: 22548–22557

Ravanel S, Droux M and Douce R. 1995. Methionine biosynthesis in higher plants. I. Purification and characterisation of cystathionine γ-synthase from spinach chloroplasts. Arch Biochem Biophys 316: 572–584

Ravanel S, Gakiere B, Job D and Douce R (1998) The specific features of methionine biosynthesis and metabolism in plants. Proc Natl Acad Sci USA 95: 7805–7812

Ravanel S, Job D and Douce R (1996) Purification and properties of cystathionine β-lyase from Arabidopsis thaliana overexpressed in Escherichia coli. Biochem J 320: 383–392

Riesmeier J, Klonus D and Pohlenz HD (1993) Purification to homogenity and characterisation of homoserin kinase from wheat germ. Phytochemistry 32: 581–584

Rinder J, Casazza AP, Hoefgen R and Hesse H (2007) Regulation of aspartate-derived amino acid homeostasis in potato plants (Solanum tuberosum L.) by expression of E. coli homoserine kinase. Amino Acids, DOI 10.1007/s00726-007-0504-5

Rocap G, Larimer FW, Lamerdin J, Malfatti S, Chain P, Ahlgren NA, Arellano A, Coleman M, Hauser L, Hess WR, Johnson ZI, Land M, Lindell D, Post AF, Regala W, Shah M, Shaw SL, Steglich C, Sullivan MB, Ting CS, Tolonen A, Webb EA, Zinser ER and Chisholm SW (2003) Genome divergence in two Prochlorococcus ecotypes reflects oceanic niche differentiation. Nature 424: 1042–1047

Rodionov DA, Vitreschak AG, Mironov AA and Gelfand MS (2004) Comparative genomics of the methionine metabolism in Gram-positive bacteria: a variety of regulatory systems. Nucl Acids Res 32: 3340–3353

Rowbury RJ and Woods DD (1964) O-Succinylhomoserine as an intermediate in the synthesis of cystathionine by Escherichia coli. J Gen Microbiol 36: 341–358

Ruckert C, Pühler A and Kalinowski J (2003) Genome-wide analysis of the L-methionine biosynthetic pathway in Corynebacterium glutamicum by targeted gene deletion and homologous complementation. J Biotechnol 104: 213–228

Saint-Girons I, Parsot C, Zakin MM, Barzu O and Cohen GN (1988) Methionine biosynthesis in Enterobacteriaceae: biochemical, regulatory, and evolutionary aspects. CRC Crit Rev Biochem 23 Suppl 1: S1–S42

Schell MA (1993) Molecular biology of the LysR family of transcriptional regulators. Annu Rev Microbiol 47: 597–626

Sauter M, Cornell KA, Beszteri S and Rzewuski G (2004) Functional analysis of methylthioribose kinase genes in plants. Plant Physiol 136: 4061–4071

Sauter M, Lorbiecke R, OuYang B, Pochapsky TC and Rzewuski G (2005) The immediate early ethylene response gene OsARD1 encodes an acireductone dioxygenase involved in recycling of the ethylene precursor S-adenosylmethionine. Plant J 44: 718–729

Schlenk F (1983) Methylthioadenosine. Adv Enzymol Relat Areas Mol Biol 54: 195–265

Schopf JW (2000) The fossil record: tracing the roots of the cyanobacterial lineage. In: Whitton BA and Potts M (eds), The ecology of cyanobacteria, pp 13–35. Kluwer Academic Publishers, Dordrecht, The Netherlands

Schwarz R and Forchhammer K (2005) Acclimation of unicellular cyanobacteria to macronutrient deficiency: emergence of a complex network of cellular responses. Microbiology 151: 2503–2514

Sekowska A and Danchin A (2002) The methionine salvage pathway in *Bacillus subtilis*. BMC Microbiology 2: 8

Sekowska A, Mulard L, Krogh S, Tse JKS and Danchin A (2001) MtnK, methylthioribose kinase, is a starvation-induced protein in *Bacillus subtilis*. BMC Microbiology 1: 15

Shah SP and Cossins EA (1970) Pteroylglutamates and methionine biosynthesis in isolated chloroplasts. FEBS Lett 7: 267–270

Shen B, Li CJ and Tarczynski MC (2002) High free-methionine and decreased lignin content result from a mutation in the Arabidopsis S-adenosyl-L-methionine synthetase 3 gene. Plant J 29: 371–380

Steegborn C, Messerschmidt A, Laber B, Streber W, Huber R and Clausen T (1999) The crystal structure of cystathionine gamma-synthase from *Nicotiana tabacum* reveals its substrate and reaction specificity. J Mol Biol 290: 983–996

Stefels J (2000) Physiological aspects of the production and conversion of DMSP in marine algae and higher plants. J Sea Res 43: 183–197

Sufrin JR, Meshnick SR, Spiess AJ, Garofalo-Hannan J, Pan X-Q and Bacchi CJ (1995) Methionine recycling pathways and antimalarial drug design. Antimicrob Agents Chemother 39: 2511–2515

Suzuki A, Shirata Y, Ishida H, Chiba Y, Onouchi H and Naito S (2001) The first exon coding region of cystathionine gamma synthase gene is necessary and sufficient for downregulation of its own mRNA accumulation in transgenic *Arabidopsis thaliana*. Plant Cell Physiol 42: 1174–1180

Tabe LM and Higgins TJV (1998) Engineering plant protein composition for improved nutrition. Trends Plant Sci 3: 282–286

Thoen A, Rognes SE and Aarnes H (1978) Biosynthesis of threonine from homoserine in pea-seedlings. 2. Threonine synthase. Plant Sci Lett 13: 113–119

Thompson GA, Datko AH and Mudd SH (1982a) Methionine synthesis in *Lemna*: inhibition of cystathionine γ-synthase by propargylglycine. Plant Physiol 70: 1347–1352

Thompson GA, Datko AH, Mudd SH and Giovanelli J (1982b) Methionine biosynthesis in *Lemna*. Studies on the regulation of cystathionine γ-synthase, O-phosphohomoserine sulphhydrylase, and O-acetylserine sulphhyrylase. Plant Physiol 69: 1077–1083

Tucker BJ and Breaker RR (2005) Riboswitches as versatile control elements. Curr Opin Struct Biol 15: 342–348

Urbanowski ML and Stauffer GV (1987) Regulation of the metR gene of *Salmonella typhimurium*. J Bacteriol 169: 5841–5844

Urbanowski ML and Stauffer GV (1989) Genetic and biochemical analysis of the MetR activator-binding site in the metE metR control region of *Salmonella typhimurium*. J Bacteriol 1171: 5620–5629

Val DL and Cronan Jr JE (1998) In vivo evidence that S-adenosylmethionine and fatty acid synthesis intermedi-

ates are the substrates for the Lux I family of autoinducer synthases. J Bacteriol 180: 2644–2651

Vermeij P and Kertesz MA (1999) Pathways of assimilative sulphur metabolism in *Pseudomonas putida*. J Bacteriol 181: 5833–5837

Wallsgrove RM, Lea PJ and Miflin BJ (1983) Intracellular localization of aspartate kinase and the enzymes of threonine and methionine biosynthesis in green leaves. Plant Physiol 71: 780–784

Wang SY, Adams DO and Lieberman M (1982) Recycling of 5′-methylthioadenosine-ribose carbon atoms into methionine in tomato tissue in relation to ethylene production. Plant Physiol 70: 117–121

Waterbury JB, Watson SW, Guillard RR and Brand LE (1979) Widespread occurrence of a unicellular marine planktonic cyanobacteria. Nature 277: 293–294

Weissbach H and Brot N (1991) Regulation of methionine synthesis in *Escherichia coli*. Mol Microbiol 5: 1593–1597

Welsh DT (2000) Ecological significance of compatible solute accumulation by micro-organisms: from single cells to global climate. FEMS Microbiol Rev 24: 263–290

Whitfield CD, Steers Jr EJ and Weisbach H (1970) Purification and properties of 5-methyltetrahydropteroyltriglutamate-homocysteine transmethylase. J Biol Chem 245: 390–401

Whitman WB, Coleman DC and Wiebe WJ (1998) Prokaryotes: the unseen majority. Proc Natl Acad Sci USA 95: 6578–6583

Winkler WC and Breaker RR (2005) Regulation of bacterial gene expression by riboswitches. Annu Rev Microbiol 59: 487–517

Wu WF, Urbanowski ML and Stauffer GV (1992) Role of the MetR regulatory system in vitamin B_{12}-mediated repression of the *Salmonella typhimurium* metE gene. J Bacteriol 174: 4833–4837

Yamagata S (1987) Partial purification and some properties of homoserine O-acetyltransferase of a methionine auxotroph of *Saccharomyces cerevisiae*. J Bacteriol 169: 3458–3463

Yang SF and Hoffman NE (1984) Ethylene biosynthesis and its regulation in higher plants. Annu Rev Plant Physiol 35: 155–189

Yang Z, Rogers LM, Song Y, Guo W and Kolattukudy PE (2005) Homoserine and asparagine are host signals that trigger in planta expression of a pathogenesis gene in *Nectria haematococca*. Proc Nat Acad Sci USA 102: 4197–4202

Zeh M, Casazza AP, Kreft O, Rössner U, Biebrich K, Willmitzer L, Höfgen R and Hesse H (2001) Antisense inhibition of threonine synthase leads to high methionine content in transgenic potato plants. Plant Physiol 127: 792–802

Zeh M, Leggewie G, Höfgen R and Hesse H (2002) Cloning and characterization of a cDNA encoding a cobalamin-independent methionine synthase from potato (*Solanum tuberosum* L.). Plant Mol Biol 48: 255–265

Zehr JP and Ward BB (2002) Nitrogen cycling in the ocean: new perspectives on processes and paradigms. Appl Environ Microbiol 68: 1015–1024

Chapter 6

Sulfotransferases from Plants, Algae and Phototrophic Bacteria

Cinta Hernández-Sebastià and Luc Varin*
Centre for Structural and Functional Genomics, Department of Biology, Concordia University, 7141 Sherbrooke West, Montréal, Canada H4A 1R6

Frédéric Marsolais
Agriculture and Agri-Food Canada, Southern Crop Protection and Food Research Centre, 1391 Sandford St., London, Ontario, Canada N5V 4T3

Summary

Sulfotransferases (SULTs) catalyze the transfer of a sulfuryl group (SO_3) from the universal donor 3′-phosphoadenosine 5′-phosphosulfate to a hydroxyl group of various substrates in a process called the sulfonation reaction. These enzymes share highly conserved sequence regions across all kingdoms;

---

* Corresponding author, Fax: 514 848 2881, E-mail: varinl@alcor.concordia.ca

Rüdiger Hell et al. (eds.), Sulfur Metabolism in Phototrophic Organisms, 111–130.
© 2008 *Springer.*

however, their substrates and physiological function are predicted to be very diverse. In mammals, the sulfonation reaction is involved in mechanisms of cellular detoxification and in the modulation of the biological activity of steroid hormones and neurotransmitters. In plants, few SULTs have been characterized despite the large number of sequences found in databases. Understanding of their role in plant biology is still relatively speculative. Since the cloning of the first plant SULT cDNA more than 10 years ago from *Flaveria chloraefolia*, functional genomic approaches, particularly in *Arabidopsis thaliana*, have led to the characterization of brassinosteroid, hydroxyjasmonate, desulfoglucosinolate and flavonoid SULTs. However, these efforts are hindered by a limited knowledge of plant sulfated metabolites. A detailed analysis of plant SULT sequences revealed that at present, phylogeny is of limited value to predict biochemical function. For instance, *A. thaliana* desulfoglucosinolate SULTs form a clade with the *Flaveria* flavonol SULTs. Ten of the 17 *A. thaliana* SULTs belong to a single homogeneous clade, suggesting that most of the divergence occurred after the diversification of this plant lineage. We present here a detailed overview of the molecular phylogeny, characterization, and biological roles of SULTs in plants. We also describe the current state of knowledge of sulfonation in algae, as well as in phototrophic bacteria, where the SULT domain can be present in multidomain proteins.

I. Introduction

Sulfotransferases (SULTs) catalyze the transfer of a sulfuryl group (SO_3) from the universal donor 3′-phosphoadenosine 5′-phosphosulfate (see Chapter 3 in this book) to a hydroxyl group of various substrates in a process called the sulfonation reaction (Fig. 1). There are two classes of SULTs; cytosolic and membrane-associated SULTs. Cytosolic SULTs sulfonate small organic molecules such as steroids, flavonoids, glucosinolates and hydroxyjasmonates, to mention a few. Membrane-associated SULTs sulfonate larger biomolecules such as complex carbohydrates, peptides and proteins. Figure 2 illustrates the structural diversity of selected sulfated metabolites found in plants.

The modification of metabolites by the addition of a sulfonate group can have a profound influence on their biological properties. Initially, the cytosolic SULTs were thought to be primarily involved in detoxification of endogenous and exogenous metabolites. They were considered as part of the detoxification arsenal, which includes hydroxylases and glucuronidases. It is now clear that this is not their only function. For example,

the sulfonation of steroids has been shown to be required for the modulation of their biological activity (Clarke et al., 1982). The presence of the sulfate group on some molecules can also be a prerequisite for their biological activity. For example, it is well known that in mammals the sulfate groups are essential for the molecular interaction between heparan sulfate and antithrombin III accounting for its anticoagulation property (Atha et al., 1985). In another example pertinent to human health, the sulfonation of tyrosine residues on the chemokine receptor CCR5 is a modification that is required for its biological activity, and indirectly for the efficient binding and entry of HIV-1 (Farzan et al., 1999).

The roles of sulfated compounds in bacteria and plants are less clear. However, in a well-documented example, the sulfonation of a glycolipid secondary messenger was shown to modulate the specificity of the interaction between the bacterium *Sinorhizobium meliloti* and alfalfa (Lerouge et al., 1990). *S. meliloti* mutants lacking the SULT responsible for the sulfonation of the glycolipid are unable to induce root nodulation in alfalfa but gain the ability to colonize the roots of vetch (Roche et al., 1991). In this biological system, the sulfonation of a single glycolipid is determinant for the proper cell to cell communication between the bacterium and its natural host plant.

In plants, the sulfonation of brassinosteroids by SULTs from *Brassica napus* was shown to produce a sulfated compound lacking biological activity in a manner similar to the inactivation of estrogens in mammals (Rouleau et al., 1999). The presence of a

---

Abbreviations: DSG – desulfoglucosinolate; PAP – 3′-phosphoadenosine-5′-phosphate; PAPS – 3′-phosphoadenosine-5′-phospho sulfate; PB – 3′-phosphate binding loop; PLMF1 – periodic leaf movement factor 1; PSB – 5′-phosphosulfate binding loop; PSK – phytosulfokine; SULT – sulfotransferase; TPR – tetratricopeptide; TPSULT – tyrosylprotein sulfotransferaseI

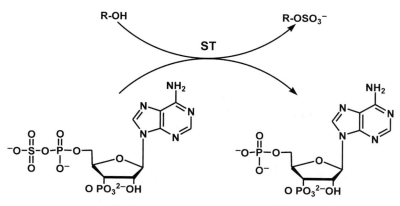

Fig. 1. Enzymatic reaction catalyzed by sulfotransferases.

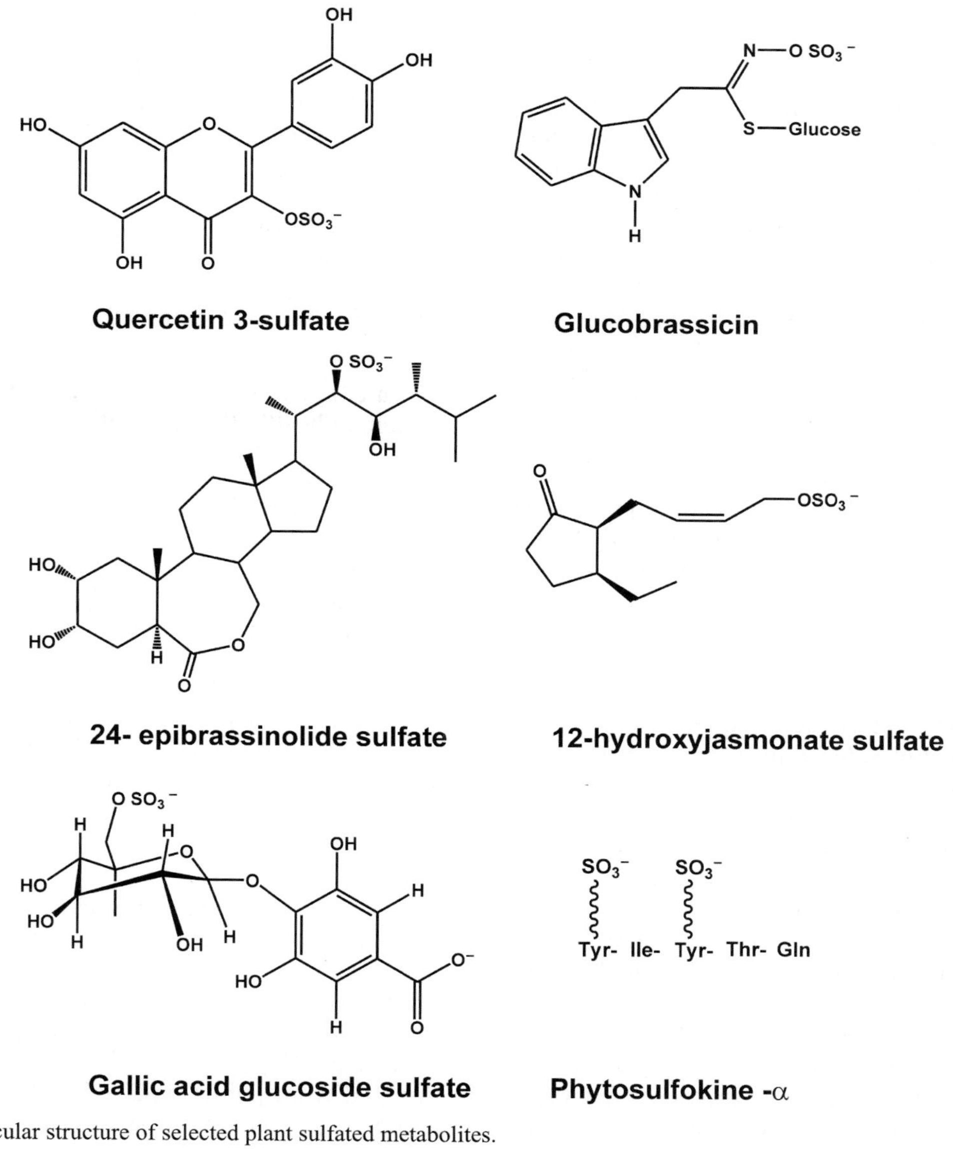

Quercetin 3-sulfate

Glucobrassicin

24- epibrassinolide sulfate

12-hydroxyjasmonate sulfate

Gallic acid glucoside sulfate

Phytosulfokine -α

Fig. 2. Molecular structure of selected plant sulfated metabolites.

sulfate group on some molecules can also be a prerequisite for their function. For example, a sulfated pentapeptide exhibiting mitogenic activity was isolated from the conditioned medium of asparagus cell culture (Matsubayashi and Sakagami, 1996). Structure–activity relationship studies revealed that the presence of two sulfated tyrosine residues was essential for the activity of the peptide hormone. Herein we present the current state of our knowledge on the molecular phylogeny, structure and biological activity of plant, algal and phototrophic bacterial SULTs.

II. Nomenclature

Despite the fact that the literature on plant SULTs is quite limited, the nomenclature of these enzymes is suffering from the same confusion that was initially encountered with their mammalian homologs. Presently, the designations ST and SOT have been used to refer to the same enzymes (Piotrowski et al., 2004; Klein et al., 2006). Furthermore, the numbering of the SULTs in one species is leading to confusion since the same protein can get two different numbers in two different publications (e.g. AtST2a and AtSOT14).

In an effort to rationalize the nomenclature of SULTs across kingdoms, an international committee set up rules to assign names to SULTs (Blanchard et al., 2004). The committee used a system similar to the one applied successfully for the cytochrome P450 monooxygenases. An international curator is responsible to assign final names and can be contacted at rl_blanchard@ fccc.edu. The designation SULT was chosen by the international committee and genes having more than 45% amino acid sequence identity are members of the same family and are assigned the same number. The committee assigned the numbers from 201 to 400 to designate the plant enzymes. SULT401 and higher are reserved for prokaryotes. Members of the same family sharing 60% or more amino acid sequence identity are classified in the same subfamily and are assigned the same letter following the family number. For example, the *Flaveria chloraefolia* flavonol 4'-sulfotransferase was assigned the designation SULT201A2. This enzyme belongs to family 201 and subfamily A, which comprises two other members, the flavonol 3-SULT from the same species and the flavonol 3-SULT from *F. bidentis*. In order to differentiate between orthologs, the family number is preceded by the abbreviation of the species name, for example, (FLACH) for *F. chloraefolia* and (FLABI) for *F. bidentis*. The family number 202 was assigned to the *B. napus* brassinosteroid SULTs, previously called BnST1 to -4. Since the four genes share more than 60% identity at the amino acid level, they were assigned to the same subfamily A. One protein from *A. thaliana* encoded by the locus *At2g03760* (previously called *RaR047* or *AtST1*) also belongs to this subfamily. The hydroxyjasmonate SULT from *A. thaliana* (initially called AtST2a) encoded by the locus *At5g07010* was assigned to family 203. Table 1 shows the result of the application of this nomenclature to the *A. thaliana* SULTs for which the biochemical function is known and to a few related sequences from other plant species.

III. Molecular Phylogeny of Plant Sulfotransferases

The availability of complete or near complete genome sequences for an increasing number of species renders possible to carry out thorough and detailed large-scale analysis of the plant SULT superfamily. The fully sequenced *A. thaliana* genome contains 18 SULT-coding genes including 1 apparent pseudogene (www.arabidopsis.org) (Table 2). For other plants with genomic information (e.g. rice and poplar), SULTs also seem to be encoded by comparatively small gene families. This is contrasting with the size of the glycosyltransferase family, which comprises 120 members in *A. thaliana* including eight apparent pseudogenes (Gachon et al., 2005). These results indicate that the modification of metabolites by sulfonation is a relatively rare event as compared to glycosylation. However, the number of SULTs in plants is significantly higher than in mammals (Table 2).

In spite of the rapid progress in the acquisition of genomic information from plants, only 13 genes encoding plant SULTs have been characterized at the biochemical level and only a handful have been characterized *in planta*. To date, more than 200 plant SULT sequences can be retrieved from public databases. However, their alignment reveals that a smaller number appear

Table 1. Nomenclature of biochemically characterized *Arabidopsis thaliana* sulfotransferases and few selected sequences from other plant species.

Sulfotransferase family	Sulfotransferase subfamily and number	Accession number	Biochemical function
SULT201	A1	Fc_M84135	Flavonol 3-SULT
	A2	Fc_M84136	Flavonol 4′-SULT
	A1	Fb_10277	Flavonol 3-SULT
	B1	At1G74090	Desulfoglucosinolate SULT
	B2	At1G18590	Desulfoglucosinolate SULT
	B3	At1G74100	Desulfoglucosinolate SULT
SULT202	A1	Bn_AF000305	Brassinosteroid SULT
	A2	Bn_AF000306	Brassinosteroid SULT
	A3	Bn_AF000307	Brassinosteroid SULT
	A4	At2G03760	Brassinosteroid SULT
	A5	Bn_AY442306	Brassinosteroid SULT
SULT203	A1	At5G07010	Hydroxyjasmonate SULT
	A2	At5G07000	Unknown
	A3	Pt_LG11001725	Unknown

Table 2. Number of sulfotransferase sequences found in a selected number of species.

Species name	Total number of unique sequences	Putative full-length sequences
Arabidopsis thaliana	18	17
Oryza sativa	35	31
Populus trichocarpa	42	32
Homo sapiens	11	11

to contain a full-length open reading frame. In an attempt to predict biochemical function based on sequence homology, we constructed an unrooted cladogram from 78 plant SULT sequences (Fig. 3). The sequences used to generate the cladogram presented in Fig. 3 were selected according to the following criteria. First, the regions I to IV conserved in all cytosolic SULTs had to be present in the sequence (Varin et al., 1992). Second, only SULTs that could be assembled into full-length sequences were included in the alignment and in the phylogenetic tree. Finally, sequences containing large gaps that might have originated from sequencing errors were eliminated from the dataset. We also generated cladograms from the alignment of the conserved region I or conserved region IV. In all cases, the trees looked similar to the one presented in Fig. 3 (data not shown).

Ten of the 17 *A. thaliana* SULT proteins are present in the same homogeneous clade (Fig. 3). In contrast, the rice and poplar SULTs are distributed in several clades containing sequences from at least two other species. The presence of the majority of the SULT proteins from *A. thaliana* in a homogeneous clade suggests that most of the divergence occurred after the diversification of this plant lineage. The remaining seven sequences are distributed in three clades containing proteins from at least three other species.

As expected, the brassinosteroid SULT enzymes from *B. napus* are present in the clade which also comprises *A. thaliana* At2g03760. The latter has recently been shown to be a brassinosteroid SULT (Marsolais et al., 2007) (see section V.A.4.a). The uncharacterized protein At2g03770 from *A. thaliana* is also present in this clade. The *At2g03770* locus is located adjacent to *At2g03760* on chromosome 2 and probably originates from a recent duplication event. Its position in the brassinosteroid SULT clade suggests that it might also be a brassinosteroid SULT. A number of SULTs from rice and one from barley are present in one branch of this clade. Their involvement in brassinosteroid sulfonation remains to be demonstrated. The absence of a protein from poplar in this clade is surprising considering that its genome is almost fully sequenced and that a conservation of the brassinosteroid SULT function is expected. However, a functional homolog from poplar might be represented in the few genes that had to be excluded following the application of the sequence selection criteria.

To date, all monocotyledonous and dicotyledonous plants that were tested were found to accumulate sulfated hydroxyjasmonates. This

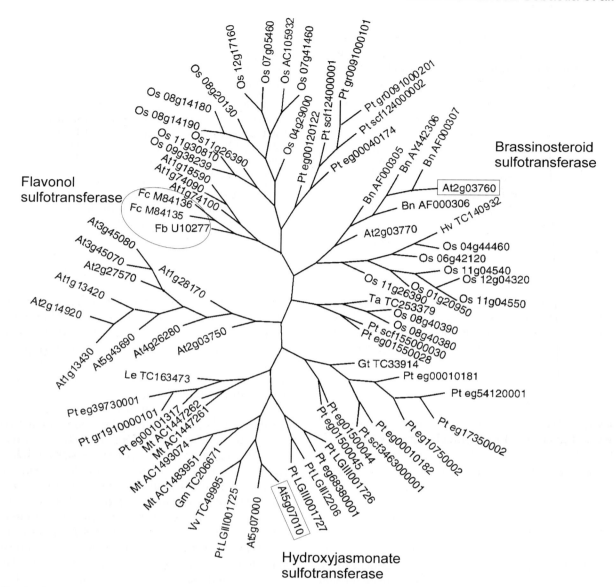

Fig. 3. Cladogram showing the relationship between 78 plant SULT sequences. The accession number is preceded by the initial of the plant species in which they were identified. Abbreviations for species: At, *Arabidopsis thaliana*; Bn, *Brassica napus*; Fb, *Flaveria bidentis*; Fc, *Flaveria chloraefolia*; Gm, *Glycine max*; Gt, *Gentiana triflora*; Hv, *Hordeum vulgare*; Le, *Lycopersicon esculentum*; Mt, *Medicago truncatula*; Os, *Oryza sativa*; Pt, *Populus trichocarpa*; Ta, *Triticum aestivum*; Vv *Vitis vinifera*. The *Arabidopsis* and *Flaveria* genes which were characterized at the biochemical level are highlighted. The sequences were aligned with ClustalW and the cladogram was generated using PAUP 4.0b10 (Ativec) by a neighbor joining method.

suggests that the SULT enzyme involved in their synthesis is conserved among species. Recently, the *A. thaliana* hydroxyjasmonate SULT (AtST2a) was shown to be encoded by the locus *At5g07010* (Gidda et al., 2003). Interestingly, this enzyme is located in a clade comprising members from poplar, grape, alfalfa and soybean (Fig. 3).

Its closest relative is encoded by the adjacent locus *At5g07000* on chromosome 5. Despite an 85% sequence identity with At5g07010, the At5g07000 protein did not accept hydroxy-jasmonates as substrate suggesting that high sequence identity alone is not a good predictor of substrate preference. Alternatively, the absence of activity of At5g07000 with hydroxyjasmonates may have been due to the improper folding of the recombinant enzyme. The second and third closest relatives of At5g07010 are from poplar and grape with 60% and 58% identity, respectively. It will be interesting to test their enzymatic activity with hydroxyjasmonates.

Surprisingly, the closest relatives to the three SULTs involved in glucosinolate biosynthesis in *A. thaliana* are those involved in flavonoid sulfonation in *Flaveria* species (Fig. 3). This result illustrates one of the limitations of molecular phylogeny studies to predict protein function. It is well known that few amino acid changes in SULT sequences can lead to profound changes in substrate specificity (Varin et al., 1995; Sakakibara et al., 1998). However, it is possible that the nature of the desulfoglucosinolate and flavonoid substrates impose similar constraints on the architecture of the protein. It is also possible that a common ancestral protein diverged recently to give rise to the two enzyme groups.

The utility of the cladogram to predict protein function is severely limited by the small number of SULTs that have been thoroughly characterized and by the limited number of sequences that were aligned. The biochemical characterization of more SULTs is required to better understand the evolutionary relationship between SULT genes.

IV. Enzymatic Mechanism and Structural Requirements

The amino acid alignment of the first characterized plant SULT with animal SULTs allowed defining four conserved regions (Varin et al., 1992). The stability of these conserved sequences during evolution suggested that they are essential for the catalytic function of these enzymes. To date, the structures of six SULTs have been solved including one plant enzyme from *A. thaliana* (At2g03760) (Yoshinari et al., 2001; Smith et al., 2004). They share a common globular structure and contain a single α/β fold with a central four- or five-stranded parallel β sheet surrounded by α-helices. The SULT structures are similar to those of nucleotide kinases suggesting a common ancestral protein. The most important structural features of SULTs are briefly described in the following section.

A. 3′-Phosphoadenosine 5′-Phosphosulfate Binding Site

Amino acids from conserved regions I and IV contribute to the PAPS binding site (Fig. 4a and d). Region I forms the 5′-phosphosulfate loop (PSB loop). Hydrogen bonds are formed between the 5′-phosphate and the N6 atom of the strictly conserved lysine residue and the oxygen of the two threonine residues located at the extremity of the PSB loop (Fig. 4a). The 3′-phosphate binding loop (PB loop) is made up of the three amino acids RKG located in region IV (Fig. 4d). Hydrogen bonds are formed between the arginine and glycine residues and the 3′-phosphate of the cosubstrate. The motif GXXGXXK located after the PB loop was shown to be essential for the binding of PAPS and estradiol in the estrogen SULT from mammals (Driscoll et al., 1995). Finally, aromatic amino acids in region II (Trp) and region III (Phe) form a parallel stack with the adenine group of PAPS (Fig. 4b and c).

B. Substrate Binding Site

The flavonol 3- and 4′-SULTs catalyze the transfer of the sulfuryl group of PAPS to position 3 of flavonol aglycones and 4′- of flavonol 3-sulfates (see section V.A.1). To elucidate the structural aspects underlying the difference in substrate specificity of these enzymes, a number of hybrid proteins were constructed by the manipulation of their cloned cDNA sequences (Varin et al., 1995). Analysis of the substrate preference of the chimeric proteins indicated that a segment, designated domain II, contains all the amino acid residues responsible for their substrate and position specificities. Within this domain, two regions of high amino acid divergence corresponding to amino acids 98 to 110 and 153 to 170 were identified. A similar study conducted with the human phenol (SULT1A1) and catecholamine (SULT1A3) SULTs identified the same two variable regions as being responsible for the specificity of the two enzymes (Sakakibara et al., 1998). To refine the analysis of the determinants of specificity, a number of divergent amino acids located in the variable regions of the flavonol 3-SULT were replaced with the corresponding amino acids of the flavonol 4′-SULT (Marsolais and Varin, 1997). No reversal of the specificity was observed after the individual

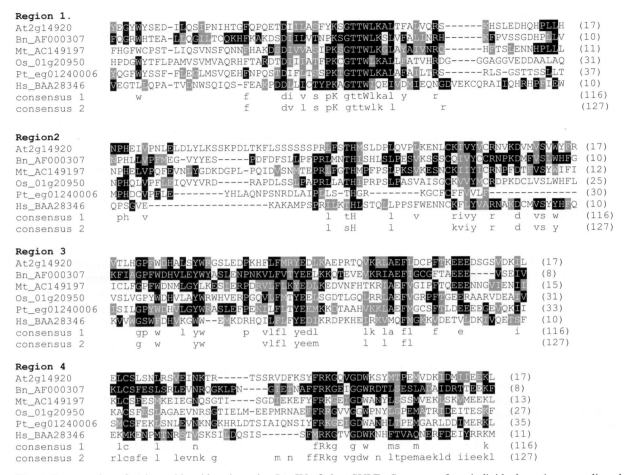

Fig. 4. Conservation of amino acid residues in region I to IV of plant SULTs. Sequences from individual species were aligned separately using ClustalW. The level of conservation is defined by Boxshade 3.21 with a threshold value of 0.9 for black boxes. The number of genes from each species to generate the alignment is shown in the parentheses. Only one gene from each species is illustrated on the figure. Species: *A. thaliana* (At), *Brassica napus* (Bn), *Medicago trunculata* (Mt), *Oryza sativa* (Os) and *Populus trichocarpa* (Pt). Residues conserved in more than 90% of the sequences are shown in black while those conserved in more than 50% of the sequences are shown in gray. Consensus 1 is derived from the alignment of 116 plant sequences. Consensus 2 is derived from the alignment of 116 plant and 11 human SULT sequences.

mutation. However, replacement of leucine 95 of the flavonol 3-SULT by the corresponding tyrosine of the 4′-SULT had different effects on the kinetic constants depending on the flavonoid ring B structure suggesting that the tyrosine side chain may be in direct contact with this part of the molecule. Modeling studies of the flavonol 3- and 4′-SULTs on the structure of the estrogen sulfotransferase showed that the leucine residue of the flavonol 3-SULT and the tyrosine residue of the 4′-SULT are in close proximity with the catalytic site supporting their role in substrate binding (data not shown).

C. Catalytic Mechanism

A great deal of information on the catalytic mechanism of SULTs has been derived from the solved crystal structures and from the results of numerous site-directed mutagenesis studies (Marsolais and Varin, 1998; Chapman et al., 2004). The sulfonation reaction proceeds by an in-line attack of the sulfate group of PAPS. The conserved histidine residue in region II probably assists in the reaction by deprotonating the attacking substrate. Alternatively, the histidine at the active site might act as a nucleophile to form an unstable protein–sulfate complex. The lysine residue located in region I is also clearly involved in catalysis. Initially, this residue is interacting with a conserved serine located in region II. Upon catalysis, the lysine side chain probably interacts with the leaving sulfate group of PAPS (Kakuta et al., 1998). Even a conservative replacement of the lysine with arginine led to a significant reduction in catalytic efficiency without affecting the binding of PAPS supporting its role in catalysis (Marsolais and Varin, 1995).

V. Functional Characterization of Plant Sulfotransferases

A. Cytosolic Plant Sulfotransferases

Cytosolic SULTs sulfonate small organic molecules such as flavonoids, brassinosteroids, glucosinolates and hydroxyjasmonates. The term cytosolic can be misleading since in most cases, the subcellular localization of theses enzymes is unknown. The term cytosolic is in fact referring to the ability of the enzymes to be extracted from cells in a soluble form. Although cytosolic SULTs were initially thought to be primarily involved in detoxification, it is now clear that this is not their main function. The next section describes our current knowledge of the plant SULTs and focuses on the biological significance of the formation of sulfated conjugates.

1. Flavonoid Sulfotransferases from Flaveria Species

The position-specific flavonol SULTs from *F. chloraefolia* were the first plant SULTs to be characterized. These enzymes exhibit strict specificity for position 3 of flavonol aglycones, 3′ and 4′ of flavonol 3-sulfates and 7 of flavonol 3,3′- and 3,4′-disulfates (Fig. 2). Together, they participate in the sequential sulfonation of flavonol polysulfates (Varin and Ibrahim, 1989, 1991). The highly purified enzymes were active as monomers, did not require divalent cations for activity and had a similar molecular mass of 35 kDa. The K_m values of the four enzymes for PAPS and the flavonol acceptors varied between 0.2 and 0.4 μM.

The flavonol 3-SULT was purified to apparent homogeneity and its kinetic properties were extensively studied (Varin and Ibrahim, 1992).

The results of substrate interaction kinetics and product inhibition studies are consistent with an ordered Bi Bi mechanism, where PAPS is the first substrate to bind to the enzyme and PAP the last product to be released. These results are similar to those obtained for several mammalian SULT enzymes (Chapman et al., 2004).

Despite the early discovery and abundance of flavonol sulfates in *Flaveria* species, their biological significance remains to be elucidated. One report suggests that sulfated flavonols might play a role in the regulation of polar auxin transport (Jacobs and Rubery, 1988). Flavonol aglycones, such as quercetin and kaempferol, were found to bind to the naphthylphthalamic acid receptor and thus inhibit polar auxin transport from the basal end of stem cells. In contrast, sulfated flavonols act as antagonists of quercetin and thus allow auxin efflux from tissues where it is produced (Faulkner and Rubery, 1992).

2. Choline Sulfotransferase from Limonium sativum

Choline sulfate accumulates in all the species of the salt stress-tolerant Plumbaginaceae family investigated to date (Hanson et al., 1994) (Fig. 2). It has been proposed that choline sulfate and other betaines act as osmoprotectants in response to salinity or drought stress. *In vivo* feeding experiments with *Limonium* species showed that [^{14}C] choline was converted to the sulfated derivative suggesting the presence of a choline SULT (Hanson et al., 1991). The enzyme was identified and characterized *in vitro* from root extracts of *Limonium sinuatum* (Rivoal and Hanson, 1994). The fact that more than 98% of the choline SULT activity could be recovered in the supernatant following high speed centrifugation suggests that it is cytosolic. The choline SULT activity was found to be strictly dependent on PAPS and showed a pH optimum of 9.0. The apparent K_m values were of 5.5 μM for PAPS and 25 μM for choline. The presence of the choline SULT was investigated in a number of species including non-accumulators of choline sulfate such as barley, maize, sunflower and *Brassica* spp., and was found to be restricted to members of the genus *Limonium*. In spite of the potential of this enzyme for the engineering of plants with increased salt tolerance, no attempts have been made to isolate a cDNA clone encoding this activity.

3. Brassinosteroid Sulfotransferases from Brassica napus

Two members of the plant sulfotransferase superfamily, SULT202A3 and SULT202A5 from *B. napus* (initially called BnST3-4) were found to catalyze the *in vitro* sulfonation of brassinosteroids, as well as mammalian estrogenic steroids and hydroxysteroids (Rouleau et al., 1999; Marsolais et al., 2004) (Fig. 2). SULT202A3 and SULT202A5 are stereospecific for 24-epibrassinosteroids, with a substrate preference for the metabolic precursor 24-epicathasterone. Based on the lack of biological activity of 24-epibrassinolide sulfate, and the induction of *B.napus SULT* genes by salicylic acid, a function in brassinosteroid inactivation was initially hypothesized for these sulfotransferases (Rouleau et al., 1999). However, SULT202A3 and SULT202A5 exhibit a relatively broad specificity towards steroids *in vitro*. Besides salicylic acid, the *B. napus SULT* genes are inducible by ethanol and other xenobiotics, including the herbicide safener naphthalic anhydride (Marsolais et al., 2004), which would be compatible with a general function of the sulfotransferases in detoxication. The observation that constitutive expression of SULT202A3 in transgenic *A. thaliana* did not lead to a brassinosteroid-deficient phenotype suggested that the steroid sulfotransferase is not directly involved in brassinosteroid inactivation (Marsolais et al., 2004).

4. Cytosolic Sulfotransferases from Arabidopsis thaliana

The genome of *A. thaliana* contains a total of 18 sulfotransferase-coding genes based on sequence similarity with previously characterized SULTs from animals and plants. Of these, one (*At3g51210*) is probably a pseudogene since it encodes a truncated protein lacking the strictly conserved region I. To date, seven *A. thalina* SULTs have been characterized at the biochemical level.

a. Brassinosteroid SULTs (At2g03760 and At2g14920)

At2g03760

SULT202A4 (*At2g03760, RaR047*) was first characterized as a gene up-regulated in response to pathogens and the pathogen-related signals, methyl jasmonate and salicylic acid (Lacomme and Roby, 1996). As can be seen in Fig. 3, *At2g03760* is an ortholog of the *B. napus SULT* genes in *A. thaliana*. *At2g03760* shares 87% of amino acid sequence identity with its closest relative SULT202A5. Together, At2g03760 and the *B. napus* SULT proteins belong to the SULT202 family of plant soluble SULTs, according to the proposed guidelines for SULT nomenclature (Table 1) (http://www.fccc.edu/research/labs/blanchard/sult/) (Marsolais et al., 2000; Blanchard et al., 2004). *At2g03760* is located on chromosome 1 and clustered with two other SULT genes of unknown function, *At2g03770*, which is also part of the SULT202 family (53% identity with *At2g03760* in amino acid sequence), and the more distant *At2g03750* (Fig. 3).

Recombinant SULT202A4 displayed similar substrate specificity towards brassinosteroids as the previously characterized SULT202A3 and SULT202A5 (Rouleau et al., 1999; Marsolais et al., 2007). Among naturally occurring brassinosteroids, the enzyme was stereospecific for 24-epibrassinosteroids, since no activity was observed with the 24-epimers castasterone and brassinolide. The preferred substrates of SULT202A4 were 24-epicathasterone (V_{max}/K_m 8.3 pkatal mg^{-1} μM^{-1}), followed by 6-deoxo-24-epicathasterone (V_{max}/K_m 2.2 pkatal mg^{-1} μM^{-1}) (Marsolais et al., 2007). SULT202A3 was previously hypothesized to transfer the sulfonate group at position 22 of 24-epibrassinosteroids, based on its lack of catalytic activity with the synthetic 22-deoxy-24-epiteasterone, and the fact that estrogens are sulfonated at position 17, since 17β-estradiol 3-methyl ether was conjugated but not estrone (Rouleau et al., 1999). Like SULT202A5 (Marsolais et al., 2004), SULT202A4 displayed a significant but low catalytic activity with the synthetic 22-deoxy-24-epiteasterone. Like SULT202A3, SULT202A4 catalyzed the sulfonation of both hydroxysteroids and estrogens (Rouleau et al., 1999), whereas SULT202A5 only accepted hydroxysteroids (Marsolais et al., 2004). The apparent K_m for PAPS of SULT202A4 was 3 μM.

At2g14920

At2g14920 was recently identified and characterized as a novel enzyme which sulfonates brassinosteroids *in vitro* (Marsolais et al., 2007). Recombinant At2g14920 was screened against a wide array of potential substrates, including phenolic acids, desulfoglucosinolates, flavonoids, steroids, gibberellic acids, cytokinins, phenylpropanoids, hydroxyjasmonates and coumarins. Catalytic activity was detected exclusively with brassinosteroids. In contrast with SULT202A4, no enzymatic activity was observed towards hydroxysteroids or estrogens. Among brassinosteroids tested, At2g14920 was specific for biologically active end-products of the biosynthetic pathway, including castasterone, brassinolide, related 24-epimers, and the naturally occurring (22R, 23R)-28-homobrassinosteroids. The preferred substrates of the enzyme were (22R, 23R)-28-homocastasterone (V_{max}/K_m 2.7 pkatal mg^{-1} μM^{-1}), followed by (22R, 23R)-28-homobrassinolide (V_{max}/K_m 1.8 pkatal mg^{-1} μM^{-1}). The apparent K_m of the enzyme for PAPS was 0.4 μM (Marsolais et al., 2007).

The *A. thaliana* genome contains two closely related genes, *At1g13420* and *At1g13430*, located in tandem on chromosome 1. At1g13430 is more closely related to At2g14920 (84% identical in amino acid sequence) than At1g13420 (69%). It is possible that *At1g13420* and *At1g13430* also encode brassinosteroid SULTs, given the high degree of sequence relatedness.

b. Hydroxyjasmonate SULT (At5g07010)

12-Hydroxyjasmonate, also known as tuberonic acid, was first isolated from *Solanum tuberosum* and was shown to have tuber-inducing properties (Yoshihara et al., 1989) (Fig. 2). It is derived from the ubiquitously occurring jasmonic acid, an important signaling molecule mediating diverse developmental processes and plant defense responses. It has recently been shown that 12-hydroxyjasmonate and its sulfated derivative occur naturally in *A. thaliana* (Gidda et al., 2003). Treatments with methyl jasmonate led to an increase in the amount of 12-hydroxyjasmonate and 12-hydroxyjasmonate sulfate in this plant, and it has been proposed that the metabolism of jasmonic acid to 12-hydroxyjasmonate sulfate might be a route leading to its inactivation.

The enzyme catalyzing the sulfonation of 12-hydroxyjasmonate is encoded by the gene *SULT203A1* (At5g07010, previously *AtST2a*) (Gidda et al., 2003) (Table 1). The recombinant SULT203A1 protein was found to exhibit strict specificity for 11- and 12-hydroxyjasmonate with

K_m values of 50 and 10 μM, respectively. The K_m value for PAPS was found to be 1 μM. *SULT203A1* expression was induced following treatment with methyljasmonate and 12-hydroxyjasmonate. In contrast, the expression of the methyljasmonate-responsive gene *Thi2.1*, a marker gene in plant defense responses, is not induced upon treatment with 12-hydroxyjasmonate indicating the existence of independent signaling pathways. The presence of two independent response pathways suggests that the function of SULT203A1 might be to directly control the biological activity of 12-hydroxyjasmonate.

At5g07010 is localized on chromosome 5 and is clustered with the locus *At5g07000*. The latter encodes SULT203A2 that shares 85% amino acid identity with SULT203A1. SULT203A2 was also expressed in *Escherichia coli* and could be recovered in a soluble form (Gidda et al., 2003). However, no activity was observed with any of the substrates tested including 11- and 12-hydroxyjasmonate. Other substrates structurally related to 11- and 12-hydroxyjasmonate will need to be tested to elucidate the biochemical function of this enzyme.

c. Desulfoglucosinolate SULTs (At1g74100, At1g74090 and At1g18590)

Glucosinolates are sulfated, non-volatile thioglucosides derived from aliphatic, indolyl or aromatic amino acids (Fig. 2). They are found throughout the order Capparales (Cronquist, 1988), particularly in the Brassicaceae family, which includes important crop plants such as *B. napus* and the model species *A. thaliana*. Upon mechanical damage, infection or pest attack, cellular breakdown exposes the stored glucosinolates to catabolic enzymes (e.g. myrosinases), yielding a variety of reactive products such as isothiocyanates, organic nitriles, thiocyanates and oxazolidine-2-thiones (Bones and Rossiter, 1996). These catabolites contribute to the distinctive flavor and aroma of cruciferous plants and are believed to play an important role in plant protection against herbivores and pathogen attack (Chew, 1988). In addition, the glucosinolate degradation products have toxic effects on animals and humans (Fenwick et al., 1983). The last step in glucosinolate biosynthesis is catalyzed by a desulfoglucosinolate SULT (DSG-SULT). A DSG-SULT has been partially purified from

Lepidium sativum (Glendening and Poulton, 1990) and *Brassica juncea* (Jain et al., 1990), and its activity was detected in cell-free extracts of several crucifers, including *A. thaliana* and other *Brassica* species (Glendening and Poulton, 1990). The *L. sativum* desulfoglucosinolate SULT had apparent K_m for PAPS and desulfo-benzylglucosinolate of 60 and 82 µM, respectively (Glendening and Poulton, 1990).

The presence of glucosinolates in *A. thaliana* and the availability of its genome sequence facilitated the identification of the three DSG-SULTs present in this plant (Varin and Spertini, 2003; Piotrowski et al., 2004; Klein et al., 2006). Different names have been used to describe these proteins (AtST5a-c and AtSOT16-18) and in the following section, the nomenclature described in Table 1 will be used. Two different approaches were used to identify the genes encoding DSG-SULTs in *A. thaliana*. Varin and Spertini (2003) used a systematic approach of cloning all SULT-coding genes from *A. thaliana* and assay the recombinant proteins with desulfoallyl-, desulfobenzyl- and desulfoindolylglucosinolates. Three DSG-SULTS (SULT201B1, SULT201B2, and SULT201B3) with different substrate specificities were identified using this approach. Piotrowski et al. (2004) isolated a cDNA clone induced by the phytotoxin coronatine, a structural homolog of jasmonic acid, using differential mRNA display. The cDNA was found to correspond to *SULT201B1*. Upon wounding, an immediate and transient increase in *SULT201B1* transcript was observed both locally and systemically. The coronatine-induced SULT was shown to prefer desulfoindolylglucosinolates, whereas long chain desulfoglucosinolates derived from methionine are the preferred substrates of the two close homologs, SULT201B2 and SULT201B3. The substrate preference of SULT201B1 and its induction by coronatine and methyl jasmonate correlates with a previous report demonstrating that *B. napus* responds to methyl jasmonate treatment by increasing the accumulation of indolylglucosinolates (Doughty et al., 1995). The K_m values ranged from 50 to 100 µM for desulfoglucosinolates and 25 to 100 µM for PAPS (Klein et al., 2006). Green fluorescent protein fusions were used to study the sub-cellular localization of the DSG-SULTs of *A. thaliana*. The results indicated that the three SULTs have a cytosolic localization (Klein et al., 2006).

d. Flavonoid SULT (At3g45070)

To characterize the biochemical function of At3g45070, a large number of substrates were tested including desulfo-derivatives of most of the known plant sulfated metabolites as well as a collection of metabolites for which no sulfated derivatives have been reported in the literature (Gidda and Varin, 2006). When expressed in *E. coli*, At3g45070 exhibited specificity for position 7 of flavones, flavonols and their monosulfate derivatives. The substrate specificity studies clearly indicate that At3g45070 prefers flavonols over flavones. Furthermore, among the flavonols tested, At3g45070 was shown to have a higher affinity for the monosulfate derivatives as compared with the corresponding aglycones, suggesting the existence in *A. thaliana* of another flavonol SULT that may sulfonate the 3-hydroxyl group. The fact that At3g45070 could sulfonate flavones (apigenin, chrysin, 7-hydroxyflavone), flavonols with different substitution patterns on ring B (kaempferol, isorhamnetin, quercetin, myricetin 3, 5, 7, 3′, 4′, 5′ hexahydroxyflavone) and an isoflavone (genistein) albeit with different catalytic efficiencies suggested that the enzyme does not exhibit strict structural requirements for defined ring B and C structures. However, the position specificity exhibited by At3g45070, and the low enzyme activity observed with 6- or 8-hydroxylated flavonols indicated that the enzyme exhibits strict structural requirements for the flavonoid A ring. In contrast, the two isoforms of the previously characterized flavonol 7-SULT from *F. bidentis* were found to exhibit strict structural requirements for the three rings of the flavonol skeleton (Varin and Ibrahim, 1991). The fact that At3g45070 exhibits a broad substrate specificity suggests that sulfonation of flavonols in *A. thaliana* does not follow a stepwise or sequential order as observed in *Flaveria* species (Varin and Ibrahim, 1989). The K_m values for kaempferol 3-sulfate and PAPS were found to be 2 and 1 µM, respectively with a V_{max} of 285 pkatal/mg^{-1}.

The substrate specificity of At3g45070 is consistent with the reported occurrence of kaempferol, quercetin and myricetin in *A. thaliana* (Shirley et al., 1995; Burbulis et al., 1996). However, the absence of any reports on the presence of flavonol sulfates in *A. thaliana* suggests that these compounds might be present in very low quantities and possibly in specific tissues or at specific developmental stages or desulfated during isolation because of their labile nature. RT- PCR experiments indicated that *At3g45070* is expressed only at the early stage of seedling development and in siliques and inflorescence stem in mature plants. Using biochemical and visualization techniques it has been shown that flavonoid accumulation in *A. thaliana* seedlings is developmentally regulated, and in the mature plants, flavonoid accumulation is restricted to flowers, immature siliques and upper inflorescence stems (Peer et al., 2001). The pattern of expression of *At3g45070* therefore parallels flavonoid accumulation in *A. thaliana*. The discovery of a flavonol sulfotransferase from *A. thaliana* suggests that flavonol sulfates are more widely distributed in plants than once thought and this model plant could be used to study their biological significance.

The At3g45070 and At3g45080 enzymes exhibit 86% amino acid sequence identity, suggesting that they might represent isoenzymes sharing similar substrates. Attempts to characterize the biochemical function of At3g45080 were unsuccessful (Gidda and Varin, unpublished results). Therefore, it remains to be determined if this enzyme exhibits overlapping or different substrate and position preferences as compared with those of At3g45070.

B.   Membrane-Bound Plant Sulfotransferases

In mammals, 12 membrane-associated SULTs have been characterized extensively (Chapman et al., 2004). These enzymes are known to play critical functions in anticoagulation, angiogenesis, leucocyte adhesion, cartilage development, corneal transparency, neuronal function, lymphocyte binding, T-cell response, and HSV-1 entry. They catalyze the sulfonation of sugar residues in heparan, keratan, dermatan and chondroitin to mention a few, and of tyrosyl residues in peptides and proteins. Because of their biological importance and medical relevance, there is intense interest in understanding their properties and mode of action. In contrast, very little is known about membrane-associated plant SULTs. To date, only two have been partially characterized. Interestingly, both are involved in the sulfonation of regulatory or signalling molecules.

1.   Gallic Acid Glucoside Sulfotransferase from *Mimosa pudica*

Mimosa pudica has the ability to close its leaves at night (nyctinasty) or in response to a mechanical stimulus (seismonasty). Following mechanical stimulation, the leaves close in 1 to 2 s, and in the absence of further stimulation, they recover their original position during a recuperation phase lasting 2 to 5 min. The movements are taking place at the motor organs (pulvini) localized at the base of the petiole and leaflets. A combination of electrical and chemical signals has been shown to be involved in this movement (Satter, 1990).

Early in the 20th century, scientists reported that an extract from *Mimosa*, or other plants exhibiting nyctinastic movement was able to induce leaf closure when applied to the cut stem of the seisomonastic plant *M. pudica*. The substance that could induce the movement was later found to be gallic acid 4-*O*-(β-D-glucopyranosyl-6'-sulfate) and was named the periodic leaf movement factor 1 (PLMF-1) (Schildknecht and Schumacher, 1981). Structure–activity relationship studies of PLMF-1 revealed that the presence of the sulfate group was required for biological activity.

A SULT that catalyzes the transfer of the sulfuryl group from PAPS to gallic acid glucoside was characterized from *M. pudica* plasma membrane protein preparations (Varin et al., 1997). The enzyme was found to exhibit strict specificity and high affinity for the substrate gallic acid glucoside (K_m 3.0 μM) and cosubstrate PAPS (K_m 0.5 μM). The PLMF-1 SULT activity was detected only in plasma membrane protein extracts from pulvini, suggesting that the site of synthesis of this compound is restricted to the motor organs. Although the PLMF-1 SULT has never been purified to apparent homogeneity, the availability of antibodies that reacted with this protein allowed detecting a 42 kDa band in plasma membrane fractions exhibiting PLMF-1 SULT activity. The 7 to 12 kDa difference between the molecular mass of the *Mimosa* SULT and the 30–35 kDa reported for the plant and animal cytosolic SULTs might reflect the presence of an additional transmembrane domain anchoring the protein in the plasma membrane of the pulvini. Indirect immunogold labeling of sections from primary and secondary pulvini indicated the presence of the gold particles on the plasma membrane of the sieve tubes. The localization of the PLMF-1 SULT in the phloem cells of the motor organs coincides with the site of transport of the electrical signal that is triggered following stimulation and supports the model that has been proposed for the mode of action of PLMF-1 (Schildknecht and Meir-Augenstein, 1990).

2.   Tyrosylprotein Sulfotransferase from Oryza sativa

Protein tyrosine O-sulfonation is one of the most frequent posttranslational modifications occurring in many secretory and membrane bound proteins of eukaryotes (Niehrs et al., 1994). In mammalian cells, this sulfonation reaction is catalyzed by tyrosylprotein sulfotransferase (TPSULT), a membrane-bound enzyme localized in the *trans*-Golgi network (Lee and Huttner, 1983; Baeuerle and Huttner, 1987). Human and mouse TPSULT cDNAs have been cloned and shown to encode type II transmembrane proteins of 370 amino acids with apparent molecular masses of 54 kDa (Ouyang et al., 1998).

In plants, the first report on the characterization of a TPSULT had to await the isolation of a disulfated pentapeptide [Tyr(SO$_3$H)-Ile-Tyr(SO$_3$H)-Thr-Gln] named phytosulfokine-α (PSK-α) which was purified from a conditioned medium derived from asparagus cell culture (Matsubayashi and Sakagami, 1996). PSK-α triggers cell proliferation at nanomolar concentrations synergistically with other plant hormones. Structure–activity relationship studies demonstrated that the sulfonation of the tyrosine residues is essential for the mitogenic activity of PSK-α. Subsequently, PSK-α was shown to be present in monocotyledonous and dicotyledonous cell cultures, and PSK-α encoding genes from rice and *A. thaliana* have been cloned and characterized (Yang et al., 1999, 2001). Putative receptor proteins for this autocrine-type growth factor were identified by photoaffinity labelling of plasma membrane fractions derived from rice, carrot and tobacco cells suggesting the widespread occurrence of the binding proteins (Matsubayashi and Sakagami, 1999, 2000). A PSK receptor was purified and cloned from carrot microsomal fractions, belonging to the leucine-rich repeat receptor-like kinases (Matsubayashi et al., 2002).

An *in vitro* enzyme assay to detect TPSULT activity was developed using a 14 amino acid synthetic oligopeptide precursor derived from the rice PSK-α cDNA sequence (Hanai et al., 2000). The precursor contained the mature YIYTQ sequence and the acidic amino acid residues at the N-terminal of the first tyrosine residue of mature PSK-α. TPSULT activity was found in microsomal membrane preparations from rice, asparagus and carrot cells (Hanai et al., 2000). The widespread distribution of TPSULT activity in higher plants suggests that tyrosine O-sulfonation is a ubiquitous posttranslational process involved in the modification of proteins and peptides. The asparagus enzyme exhibited a broad pH optimum of 7.0–8.5, required manganese ions for optimal activity and appeared to be membrane-localized in the Golgi apparatus. The apparent K_m of the rice enzyme for the peptide precursor was 71 µM, with a V_{max} of 1.0 pmol min^{-1} mg^{-1}. Substrate specificity studies revealed that acidic amino acid residues adjacent to the tyrosine residues of the acceptor peptide were essential for activity, indicating that, as in mammals, tyrosine sulfonation takes place prior to the proteolytic processing and maturation of the peptide hormone. Structural requirements for the peptide substrate and the properties of the rice enzyme are similar to those reported for mammalian TPSULT enzymes. So far, the cloning of a plant TPSULT has not been reported, and no candidate sequences are available. Cloning of a plant TPSULT will allow studying the regulation of its expression and its tissue distribution to better understand its function. A cloned plant TPSULT may also assist the identification of additional substrates beside the PSK precursor. An improved understanding of the biology and substrate specificity of the plant TPSULT is also relevant to the production of pharmaceutical proteins in plants by molecular farming, since a large number of therapeutic proteins and peptides are modified by tyrosine sulfonation in their native form (Niehrs et al., 1994).

VI.   Sulfotransferases in Algae and Phototrophic Bacteria

A.   Algae

Several genera of marine macroalgae synthesize sulfated polysaccharides which constitute the major compound of their cell wall (see Chapters

15 and 22 in this book). These polyanionic molecules chelate metallic ions, and provide a hydration shell to the organism. The commercially valuable sulfated polysaccharides are the carageenans from red algae (Rhopophyta) and sulfated fucans from brown algae (Phaeophyceae). The sulfated fucans have anticoagulant properties, and present a possible alternative to medical treatment with heparin (Mourao, 2004). Carageenans are used in various products due to their properties as hydrocolloids, including as food thickeners, but also in cosmetic and pharmaceutical products, and in various industrial applications (McHugh, 2003). Carageenans are sulfated galactans composed of alternating 1,3-α-1,4-β D-galactose units substituted with one (κ-), two (ι-) or three sulfate residues (λ-) per disaccharide monomer. Sulfated fucans are polysaccharides mainly constituted of sulfated L-fucose, with substitutions at positions 2, 3 and/or −4, depending on the species (Berteau and Mulloy, 2003).

Sulfated fucans are also present in echinoderms, such as sea urchins and sea cucumbers, while sulfated galactans are found in marine invertebrates ascidians. Aquino et al. (2005) reported that marine angiosperms, the seagrassses, contain sulfated polysaccharides in their cell walls, whereas they are absent from land plants. The structure of the polysaccharide from *Ruppia maritima* was determined to be a sulfated galactan containing regular tetrasaccharide units. This discovery suggested that biosynthesis of sulfated polysaccharides may be a result of physiological adaptation to marine environments.

In mammals, numerous genes have been cloned coding for Golgi-localized, membrane-bound SULTs involved in the biosynthesis of sulfated oligosaccharides and glycosaminoglycans (Fukuda et al., 2001). Carbohydrate SULTs have also been characterized from symbiotic rhizobacteria, and participate in the biosynthesis of nodulation (Nod) factors acting as signals eliciting developmental programs leading to nodule formation in the plant root. NodH SULT from *Sinorhizobium meliloti* catalyzes the sulfonation of the 6-reducing end of *N*-acetylglucosamine in an oligosaccharide (Ehrhardt et al., 1995), while NoeE from *Rhizobium* sp. NGR234 sulfates a 2-*O*-methyl fucose residue substituting the reducing end of the *N*-acetylglucosamine oligomer (Hanin et al., 1997). The mammalian and rhizo-

bacterial carbohydrate SULTs share consensus sequences which are the hallmark of all SULTs characterized to date (Marsolais and Varin, 1995; Kakuta et al., 1998). Yet, at present, no algal candidate genes are available for the SULTs involved in the biosynthesis of sulfated polysaccharides. This may be due to the scarcity of genomic information available from algae, although a few expressed sequence tag sequencing projects have been initiated (Grossman, 2005). Alternatively, algal carbohydrate SULT genes may have diverged to the extent that they cannot be readily identified using bioinformatic search tools. Algal carbohydrate SULT genes may constitute a highly attractive target for future research, to produce existing or new sulfated carbohydrate polymers, through enzymatic synthesis or metabolic engineering. However, carrageenans and sulfated fucans can act as elicitors of plant defence responses (Mercier et al., 2001; Klarzynski et al., 2003), and therefore, their production in plants may have deleterious effects. This is indeed a yet unexplored area.

B. Phototrophic Bacteria

Recent sequencing of bacterial genomes has uncovered a multitude of SULT genes of unknown function. They are particularly well represented in cyanobacteria, but also in other phototrophic bacteria. A search of the Entrez conserved protein domain database (http://www.ncbi.nlm.nih.gov) revealed that cyanobacterial sequences are represented in proteins defined by five different Pfam SULT conserved domains. As compared with plant and mammalian SULTs, the unique feature of bacterial sequences is that the SULT domain may be associated with other functional domains in a single protein. The SULTs present in phototrophic bacteria can be classified into four structural types, based on their domain architecture (visualized in the Pfam database, http://www.sanger.ac.uk/Software/Pfam/) (Table 3, Fig. 5). For each type illustrated in Fig. 5, a cyanobacterial SULT was arbitrarily chosen as a model. Information about possible function of the SULTs may sometimes be inferred from their domain structure and genomic context. In the first structural type, the SULT domain is located C-terminal to a series of tetratricopeptide (TPR) repeats (Table 3, Fig. 5). The number, subtype and arrangement of the TPR repeats can be variable. Among pro-

Table 3. Sulfotransferase sequences from phototrophic bacteria. Sequences are listed with their NCBI protein database annotation, species of origin, accession no. and length. aa: amino acid.

Sulfotransferase structural type	Protein model	Other selected proteins from cyanobacteria	Presence in other phototrophic bacteria	Taxonomic distribution
TPR: sulfotransferase	Hypothetical protein, *Gloeobacter violaceus* (NP_924330.1) (631 aa)	Sulfotransferase TPR repeat, *Prochlorococcus marinus* (NP_893237.1) (584 aa); Probable TPR domain protein, *Synechococcus* sp. (ZP_0108118) (606 aa)	Purple non-sulfur bacteria, for example sulfotransferase, *Rhodospirillum rubrum* (YP_427300.1) (656 aa)	Present in other proteobacteria; sulfotransferase, *Chromohalobacter salexigens* (ABE59338) (1415 aa) has fused TPR, *O*-GlcNAc- and sulfo- transferase domains
Sulfotransferase	Hypothetical protein, *Synechocystis* sp. (NP_942202.1) (316 aa)	Sulfotransferase, *Crocosphera watsonii* (ZP_00516056.1) (273 aa); unknown, *Prochloron didemni* (AAT37520.1) (277 aa); hypothetical protein, *G. violaceus* (NP_924845) (320 aa)	Purple non-sulfur bacteria, for example sulfotransferase, *Rhodospirillum rubrum* (ABC21658) (289 aa); green sulfur bacteria, for example sulfotransferase, *Chlorobium phaeobacteroides* (ZP_00533423) (343 aa)	Present in other proteobacteria and Gram-positive bacteria
Polyketide synthase: sulfotransferase	CurM, *Lyngbya majuscula* (AAT70108) (2147 aa)	None	Absent	Unique; Polyketide synthase, *Pseudomonas entomophila* (YP_610919.1) (1217 aa) has a related sulfotransferase domain
GAF: sulfotransferase	GAF, Synechococcus sp. (ABB34328) (529 aa)	None	Absent	Unique

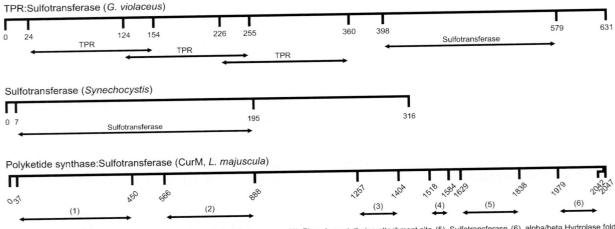

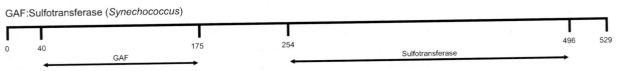

Fig. 5. Protein domain architecture of SULTs from phototrophic bacteria. Schematic representation of the four structural types of SULT present in phototrophic bacteria, based on their domain structure. The protein models represented are listed in Table 1.

teins of known function, the TPR repeat domains present in the *Gloeobacter* and *Rhodospirillum* SULTs listed in Table 3 are most similar to that of eukaryotic *O*-linked *N*-acetylglucosamine (*O*-GlcNAc) transferases (data not shown), which catalyze posttranslational modification of proteins on serine and threonine residues. In *O*-GlcNAc transferases, the TPR repeat domain is involved in protein-protein interaction, and assists substrate recognition (Iyer and Hart, 2003). Interestingly, a related hybrid protein from *Chromohalobacter salexigens* has a fusion of the TPR, *O*-GlcNAc transferase and SULT domains (Table 3). Proteins from the second structural type have no additional domains beside the SULT domain (Table 3, Fig. 5). The *Synechocystis* SULT used as a protein model is encoded on a plasmid, pSYSM, and its gene located within a cluster comprised of eight glycosyl transferases and two homologues of polysaccharide transporters, which may participate in the biosynthesis and secretion of unknown exopolysaccharides (Kaneko et al., 2003). Within the same structural type, the SULT gene from the cyanobacterial symbiont *Prochloron didemni* (Table 3) is located in a cluster with a nonribosomal peptide synthetase, which may be involved in the biosynthesis of bioactive nonribosomal cyclic peptides (Schmidt et al., 2004). The third structural type is characterized by a fusion of polyketide synthase and SULT domains, represented by CurM from *Lyngbya majuscula* (Table 3, Fig. 5). The *curM* locus is part of large gene cluster involved in the biosynthesis of curacin A, comprised of a nonribosomal peptide synthetase and multiple polyketide synthases (Chang et al., 2004). CurM is the only example of its type within the phototrophic bacteria. However, a protein harbouring a polyketide synthase:sulfotransferase fusion has also been reported from *Pseudomonas entomophila* (Table 3). In the fourth structural type, represented by the GAF protein from *Synechococcus* sp. (Table 3), the SULT domain is fused with a GAF domain, involved in ligand binding in phytochromes and cGMP-dependent 3′,5′-cyclic phosphodiesterase (Fischer et al., 2005; Gross-Langenhoff et al., 2006). The multitude of SULT sequences and multiplicity of protein domain architectures suggest the presence of a large variety of sulfated macromolecules and secondary metabolites in cyanobacteria and phototrophic bacteria, which are yet completely uncharacterized.

VII. Future Prospects in Sulfotransferase Research

Structural and functional genomic approaches have demonstrated great potential for identifying enzymes catalyzing the sulfonation reaction. While only four plant SULT sequences were known 10 years ago, there are now more than 200. The acquisition of genomic information allowed developing powerful tools to study the biological function of the SULT-coding genes. For example, a centralized database of *A. thaliana* microarray data provides researchers with valuable information to understand how SULT-coding genes are regulated during development, in adaptation to stress, as well as in various mutant backgrounds. In addition, international efforts to produce collections of T-DNA insertion mutants may allow an assessment of *in vivo* gene function. Despite the availability of functional genomic tools, and although many plant SULTs have been expressed heterologously in the past few years, the catalytic activity of only a few has been demonstrated *in vivo*. The low abundance of functional studies conducted with SULTs from phototrophic bacteria is even more dramatic. Studies conducted with *A. thaliana* so far illustrate the problems encountered when trying to assess the biochemical function of plant SULTs. In the case of DSG-SULTs, substrates could be predicted based on prior knowledge of glucosinolates accumulating in *A. thaliana*. However, no other sulfated metabolites were known from this plant, making difficult the selection of potential substrates that could be assayed with the remaining 14 uncharacterized SULTs. Although the approach of testing a collection of potential substrates proved to be successful for the characterization of the hydroxyjasmonate SULT, it did not allow demonstrating unambiguously the function of other SULTs present in *A. thaliana*. In future, unbiased metabolite profiling experiments will need to be conducted to resolve this problem. Specific protocols allowing the isolation and characterization of labile sulfated compounds by mass spectrometry will need to be developed, and systematically applied to have a more advanced understanding of the sulfonated compounds accumulating in plants. This knowledge will greatly facilitate future studies to elucidate the biochemical function of the SULTs, and the biological significance of the accumulation of their sulfonated enzymatic products.

Acknowledgements

We thank the Natural Sciences and Engineering Research Council of Canada for its support. We thank Alex Molnar for expert graphical assistance in the preparation of a figure.

References

Aquino RS, Landeira-Fernandez AM, Valente AP, Andrade LR, Mourao PA (2005) Occurrence of sulfated galactans in marine angiosperms: evolutionary implications. Glycobiology 15: 11–20

Atha DH, Lormeau JC, Petitou M, Rosenberg RD, Choay J (1985) Contribution of monosaccharide residues in heparin binding to antithrombin III. Biochemistry 24: 6723–6729

Baeuerle PA, Huttner WB (1987) Tyrosine sulfation is a trans-Golgi-specific protein modification. J Cell Biol 105: 2655–2664

Berteau O, Mulloy B (2003) Sulfated fucans, fresh perspectives: structures, functions, and biological properties of sulfated fucans and an overview of enzymes active toward this class of polysaccharide. Glycobiology 13: 29R–40R

Blanchard RL, Freimuth RR, Buck J, Weinshilboum RM, Coughtrie MW (2004) A proposed nomenclature system for the cytosolic sulfotransferase (SULT) superfamily. Pharmacogenetics 14: 199–211

Bones AMR, Rossiter JT (1996) The myrosinase-glucosinolate system, its organisation and biochemistry. Physiol Plant 97: 194–208

Burbulis IE, Iacobucci M, Shirley BW (1996) A null mutation in the first enzyme of flavonoid biosynthesis does not affect male fertility in Arabidopsis. Plant Cell 8: 1013–1025

Chang Z, Sitachitta N, Rossi JV, Roberts MA, Flatt PM, Jia J, Sherman DH, Gerwick WH (2004) Biosynthetic pathway and gene cluster analysis of curacin A, an antitubulin natural product from the tropical marine cyanobacterium Lyngbya majuscula. J Nat Prod 67: 1356–1367

Chapman E, Best MD, Hanson SR, Wong CH (2004) Sulfotransferases: structure, mechanism, biological activity, inhibition, and synthetic utility. Angew Chem Int Ed Engl 43: 3526–3548

Chew FS (1988) Biologically Active Natural Products. American Chemical Society, Washington, D.C.

Clarke C, Thorburn P, McDonald D, Adams JB (1982) Enzymic synthesis of steroid sulphates. XV. Structural domains of oestrogen sulphotransferase. Biochim Biophys Acta 707: 28–37

Cronquist A (1988) Evolution and Classification of Flowering Plants. New York Botanical Garden, New York

Doughty KJ, Kiddle GA, Pye BJ, Wallsgrove RM, Pickett JA (1995) Selective induction of glucosinolates in oilseed rape leaves by methyl jasmonate. Phytochemistry 38: 347–350

Driscoll WJ, Komatsu K, Strott CA (1995) Proposed active site domain in estrogen sulfotransferase as determined by mutational analysis. Proc Natl Acad Sci USA 92: 12328–12332

Ehrhardt DW, Atkinson EM, Faull KF, Freedberg DI, Sutherlin DP, Armstrong R, Long SR (1995) In vitro sulfotransferase activity of NodH, a nodulation protein of Rhizobium meliloti required for host-specific nodulation. J Bacteriol 177: 6237–6245

Farzan M, Mirzabekov T, Kolchinsky P, Wyatt R, Cayabyab M, Gerard NP, Gerard C, Sodroski J, Choe H (1999) Tyrosine sulfation of the amino terminus of CCR5 facilitates HIV-1 entry. Cell 96: 667–676

Faulkner IJ, Rubery PH (1992) Flavonoids and flavonol sulfates as probes of auxin transport regulation in Cucurbita pepo hypocotyl segments and vesicles. Planta 186: 618–625

Fenwick GR, Heaney RK, Mullin WJ (1983) Glucosinolates and their breakdown products in food and food plants. Crit Rev Food Sci Nutr 18: 123–201

Fischer AJ, Rockwell NC, Jang AY, Ernst LA, Waggoner AS, Duan Y, Lei H, Lagarias JC (2005) Multiple roles of a conserved GAF domain tyrosine residue in cyanobacterial and plant phytochromes. Biochemistry 44: 15203–15215

Fukuda M, Hiraoka N, Akama TO, Fukuda MN (2001) Carbohydrate-modifying sulfotransferases: structure, function, and pathophysiology. J Biol Chem 276: 47747–47750

Gachon CM, Langlois-Meurinne M, Saindrenan P (2005) Plant secondary metabolism glycosyltransferases: the emerging functional analysis. Trends Plant Sci 10: 542–549

Gidda SK, Miersch O, Levitin A, Schmidt J, Wasternack C, Varin L (2003) Biochemical and molecular characterization of a hydroxyjasmonate sulfotransferase from Arabidopsis thaliana. J Biol Chem 278: 17895–17900

Gidda SK, Varin L (2006) Biochemical and molecular characterization of flavonoid 7-sulfotransferase from Arabidopsis thaliana. Plant Physiol Biochem 44: 628–636

Glendening TM, Poulton JE (1990) Partial purification and characterization of a 3′-phosphoadenosine 5′-phosphosulfate: desulfoglucosinolate sulfotransferase from cress (Lepidium sativum). Plant Physiol 94: 811–818

Gross-Langenhoff M, Hofbauer K, Weber J, Schultz A, Schultz JE (2006) cAMP is a ligand for the tandem GAF domain of human phosphodiesterase 10 and cGMP for the tandem GAF domain of phosphodiesterase 11. J Biol Chem 281: 2841–2846

Grossman AR (2005) Paths toward algal genomics. Plant Physiol 137: 410–427

Hanai H, Nakayama D, Yang H, Matsubayashi Y, Hirota Y, Sakagami Y (2000) Existence of a plant tyrosylprotein sulfotransferase: novel plant enzyme catalyzing tyrosine O-sulfation of preprophytosulfokine variants in vitro. FEBS Lett 470: 97–101

Hanin M, Jabbouri S, Quesada-Vincens D, Freiberg C, Perret X, Prome JC, Broughton WJ, Fellay R (1997) Sulphation of Rhizobium sp. NGR234 Nod factors is dependent on noeE, a new host-specificity gene. Mol Microbiol 24: 1119–1129

Hanson AD, Rathinasabapathi B, Chamberlin B, Gage DA (1991) Comparative physiological evidence that beta-alanine betaine and choline-O-sulfate act as compatible osmolytes in halophytic Limonium species. Plant Physiol 97: 1199–1205

Hanson AD, Rathinasabapathi B, Rivoal J, Burnet M, Dillon MO, Gage DA (1994) Osmoprotective compounds in the Plumbaginaceae: a natural experiment in metabolic engineering of stress tolerance. Proc Natl Acad Sci USA 91: 306–310

Iyer SP, Hart GW (2003) Roles of the tetratricopeptide repeat domain in O-GlcNAc transferase targeting and protein substrate specificity. J Biol Chem 278: 24608–24616

Jacobs M, Rubery PH (1988) Naturally occurring auxin transport regulators. Science 241: 346–349

Jain JC, GrootWassink JWD, Kolenovsky AD, Underhill EW (1990) Purification and properties of 3′-phosphoadenosine 5′-phosphosulfate: desulphoglucosinolate sulphotransferase from Brassica juncea cell cultures. Phytochemistry 29: 1425–1428

Kakuta Y, Pedersen LG, Pedersen LC, Negishi M (1998) Conserved structural motifs in the sulfotransferase family. Trends Biochem Sci 23: 129–130

Kaneko T, Nakamura Y, Sasamoto S, Watanabe A, Kohara M, Matsumoto M, Shimpo S, Yamada M, Tabata S (2003) Structural analysis of four large plasmids harboring in a unicellular cyanobacterium, Synechocystis sp. PCC 6803. DNA Res 10: 221–228

Klarzynski O, Descamps V, Plesse B, Yvin JC, Kloareg B, Fritig B (2003) Sulfated fucan oligosaccharides elicit defense responses in tobacco and local and systemic resistance against tobacco mosaic virus. Mol Plant Microbe Interact 16: 115–122

Klein M, Reichelt M, Gershenzon J, Papenbrock J (2006) The three desulfoglucosinolate sulfotransferase proteins in Arabidopsis have different substrate specificities and are differentially expressed. FEBS J 273: 122–136

Lacomme C, Roby D (1996) Molecular cloning of a sulfotransferase in Arabidopsis thaliana and regulation during development and in response to infection with pathogenic bacteria. Plant Mol Biol 30: 995–1008

Lee RW, Huttner WB (1983) Tyrosine-O-sulfated proteins of PC12 pheochromocytoma cells and their sulfation by a tyrosylprotein sulfotransferase. J Biol Chem 258: 11326–1334

Lerouge P, Roche P, Faucher C, Maillet F, Truchet G, Prome JC, Denarie J (1990) Symbiotic host-specificity of Rhizobium meliloti is determined by a sulphated and acylated glucosamine oligosaccharide signal. Nature 344: 781–784

Marsolais F, Boyd J, Paredes Y, Schinas AM, Garcia M, Elzein S, Varin L (2007) Molecular and biochemical characterization of two brassinosteroid sulfotransferases from Arabidopsis, AtST4a (At2g14920) and AtST1 (At2g03760). Planta 225: 1233–1244

Marsolais F, Gidda SK, Boyd J, Varin L (2000) Plant soluble sulfotransferases: Structural and functional similarity with mammalian enzymes. Rec Adv Phytochem 34: 433–456

Marsolais F, Hernández Sebastià CH, Rousseau A, Varin L (2004) Molecular and biochemical characterization of BNST4, an ethanol-inducible steroid sulfotransferase from Brassica napus, and regulation of BNST genes by chemical stress and during development. Plant Sci 166: 1359–1370

Marsolais F, Varin L (1995) Identification of amino acid residues critical for catalysis and cosubstrate binding in the flavonol 3-sulfotransferase. J Biol Chem 270: 30458–30463

Marsolais F, Varin L (1997) Mutational analysis of domain II of flavonol 3-sulfotransferase. Eur J Biochem 247: 1056–1062

Marsolais F, Varin L (1998) Recent developments in the study of the structure–function relationship of flavonol sulfotransferases. Chem Biol Interact 109: 117–122

Matsubayashi Y, Ogawa M, Morita A, Sakagami Y (2002) An LRR receptor kinase involved in perception of a peptide plant hormone, phytosulfokine. Science 296: 1470–1472

Matsubayashi Y, Sakagami Y (1996) Phytosulfokine, sulfated peptides that induce the proliferation of single mesophyll cells of Asparagus officinalis L. Proc Natl Acad Sci USA 93: 7623–7627

Matsubayashi Y, Sakagami Y (1999) Characterization of specific binding sites for a mitogenic sulfated peptide, phytosulfokine-alpha, in the plasma-embrane fraction derived from Oryza sativa L. Eur J Biochem 262: 661–671

Matsubayashi Y, Sakagami Y (2000) 120- and 160-kDa receptors for endogenous mitogenic peptide, phytosulfokine-alpha, in rice plasma membranes. J Biol Chem 275: 15520–15525

McHugh D (2003) A guide to the seaweed industry. FAO fisheries technical paper no 441. FAO, Rome

Mercier L, Lafitte C, Borderies G, Briand X, Esquerré-Tugayé M-T, Fournier J (2001) The algal polysaccharide carrageenans can act as an elicitor of plant defence. New Phytol 149: 43–52

Mourao PA (2004) Use of sulfated fucans as anticoagulant and antithrombotic agents: future perspectives. Curr Pharm Des 10: 967–981

Niehrs C, Beisswanger R, Huttner WB (1994) Protein tyrosine sulfation, 1993 – an update. Chem Biol Interact 92: 257–271

Ouyang Y, Lane WS, Moore KL (1998) Tyrosylprotein sulfotransferase: purification and molecular cloning of an enzyme that catalyzes tyrosine O-sulfation, a common posttranslational modification of eukaryotic proteins. Proc Natl Acad Sci USA 95: 2896–2901

Peer WA, Brown DE, Tague BW, Muday GK, Taiz L, Murphy AS (2001) Flavonoid accumulation patterns of transparent testa mutants of Arabidopsis. Plant Physiol 126: 536–548

Piotrowski M, Schemenewitz A, Lopukhina A, Muller A, Janowitz T, Weiler EW, Oecking C (2004) Desulfoglucosinolate sulfotransferases from *Arabidopsis thaliana* catalyze the final step in the biosynthesis of the glucosinolate core structure. J Biol Chem 279: 50717–50725

Rivoal J, Hanson AD (1994) Choline-O-sulfate biosynthesis in plants: identification and partial characterization of a salinity-inducible choline sulfotransferase from species of *Limonium* (Plumbaginaceae). Plant Physiol 106: 1187–1193

Roche P, Debelle F, Maillet F, Lerouge P, Faucher C, Truchet G, Denarie J, Prome JC (1991) Molecular basis of symbiotic host specificity in Rhizobium meliloti: nodH and nodPQ genes encode the sulfation of lipo-oligosaccharide signals. Cell 67: 1131–1143

Rouleau M, Marsolais F, Richard M, Nicolle L, Voigt B, Adam G, Varin L (1999) Inactivation of brassinosteroid biological activity by a salicylate-inducible steroid sulfotransferase from *Brassica napus*. J Biol Chem 274: 20925–20930

Sakakibara Y, Takami Y, Nakayama T, Suiko M, Liu MC (1998) Localization and functional analysis of the substrate specificity/catalytic domains of human M-form and P-form phenol sulfotransferases. J Biol Chem 273: 6242–6247

Satter RH (1990) Leaf movement: an overview of the field. In: Satter RL, Gorton, HL and Vogelman TC (eds) The Pulvinus: Motor Organ For Leaf Movement, pp 1–9. American Society of Plant Physiologists, Rockville, MD

Schildknecht H, Schumacher K (1981) Ein och wirksamer leaf movement factor aus *Acacia karroo*. Chem-Ztg 105: 287–290

Schildknecht H, Meier-Augenstein W (1990) Role of turgorins in leaf movement. In: Satter RL, Gorton HL and Vogelman TC (eds) The Pulvinus: Motor Organ For Leaf Movement, pp 205–213. American Society of Plant Physiologists, Rockville, MD

Schmidt EW, Sudek S, Haygood MG (2004) Genetic evidence supports secondary metabolic diversity in *Prochloron* spp., the cyanobacterial symbiont of a tropical ascidian. J Nat Prod 67: 1341–1345

Shirley BW, Kubasek WL, Storz G, Bruggemann E, Koornneef M, Ausubel FM, Goodman HM (1995) Analysis of Arabidopsis mutants deficient in flavonoid biosynthesis. Plant J 8: 659–671

Smith DW, Johnson KA, Bingman CA, Aceti DJ, Blommel PG, Wrobel RL, Frederick RO, Zhao Q, Sreenath H, Fox BG, Volkman BF, Jeon WB, Newman CS, Ulrich EL, Hegeman AD, Kimball T, Thao S, Sussman MR, Markley JL, Phillips GN, Jr. (2004) Crystal structure of At2g03760, a putative steroid sulfotransferase from *Arabidopsis thaliana*. Proteins 57: 854–857

Varin L, Chamberland H, Lafontaine JG, Richard M (1997) The enzyme involved in sulfation of the turgorin, gallic acid 4-*O*-(β-D-glucopyranosyl-6'-sulfate) is pulvinilocalized in *Mimosa pudica*. Plant J 12: 831–837

Varin L, DeLuca V, Ibrahim RK, Brisson N (1992) Molecular characterization of two plant flavonol sulfotransferases. Proc Natl Acad Sci USA 89: 1286–1290

Varin L, Ibrahim RK (1989) Partial purification and characterization of three flavonol-specific sulfotransferases from *Flaveria chloraefolia*. Plant Physiol 90: 977–981

Varin L, Ibrahim RK (1991) Partial purification and some properties of flavonol 7-sulfotransferase from *Flaveria bidentis*. Plant Physiol 95:1254–1258

Varin L, Ibrahim RK (1992) Novel flavonol 3-sulfotransferase. purification, kinetic properties, and partial amino acid sequence. J Biol Chem 267:1858–1863

Varin L, Marsolais F, Brisson N (1995) Chimeric flavonol sulfotransferases define a domain responsible for substrate and position specificities. J Biol Chem 270: 12498–12502

Varin L, Spertini D (2003) Desulfoglucosinolate sulfotransferases, sequences coding the same and uses thereof for modulating glucosinolate biosynthesis in plants. Patent WO 03/010318-A1 Concordia University, Canada

Yang H, Matsubayashi Y, Nakamura K, Sakagami Y (1999) Oryza sativa PSK gene encodes a precursor of phytosulfokine-alpha, a sulfated peptide growth factor found in plants. Proc Natl Acad Sci USA 96: 13560–13565

Yang H, Matsubayashi Y, Nakamura K, Sakagami Y (2001) Diversity of Arabidopsis genes encoding precursors for phytosulfokine, a peptide growth factor. Plant Physiol 127: 842–851

Yoshihara T, Omer E-SA, Koshino H, Sakamura S, Kikuta Y, Koda Y (1989) Structure of tuber inducing stimulus from potato leaves. Agric Biol Chem: 2835–2837

Yoshinari K, Petrotchenko EV, Pedersen LC, Negishi M (2001) Crystal structure-based studies of cytosolic sulfotransferase. J Biochem Mol Toxicol 15: 67–75

Chapter 7

Cysteine Desulfurase-Mediated Sulfur Donation Pathways in Plants and Phototrophic Bacteria

Lolla Padmavathi, Hong Ye, Elizabeth AH Pilon-Smits, and Marinus Pilon*
Department of Biology, Colorado State University, Fort Collins, CO 80523, USA

Summary

Cysteine is the sulfur donor for a number of important cofactor biosynthetic pathways including the synthesis of iron–sulfur clusters, thiamine, biotin and molybdenum cofactor. NifS-like cysteine desulfurase enzymes are key components in these pathways, catalyzing the initial release of S from cysteine. NifS-like enzymes do not work alone but are the first component of a sulfur transfer pathway from cysteine to cofactor. *In vivo*, NifS-like cysteine desulfurases work in concert with assembly factor proteins to which they transfer the released S and which serve to regulate the cysteine desulfurase activity and orchestrate the delivery of S to downstream targets. In plants, the chloroplast localized iron–sulfur assembly machinery resembles at least in part a machinery that in bacteria is responsible for the synthesis of iron–sulfur clusters under oxidative stress and iron limitation. A similar system operates

---

*Corresponding author, Phone: 970 491 0803, Fax: 970 491 0649, E-mail: pilon@lamar.colostate.edu

in photosynthetic bacteria. While we are just beginning to unravel the mechanisms of S-dependent cofactor assembly systems it is already evident that these pathways play pivotal roles in cellular metabolism, and particularly are important to the function of plant plastids.

I. Introduction

The amino acid cysteine plays an important role as a component of proteins and glutathione. In addition, cysteine is required for the synthesis of essential sulfur-containing cofactors. The sulfur in Iron–Sulfur [Fe–S] clusters, thiamine, molybdenum cofactor (Moco), lipoic acid and biotin is derived from cysteine, as is the S in some modified tRNA species. A key enzymatic activity in these reactions is provided by NifS-like cysteine desulfurase enzymes, which decompose cysteine to alanine and sulfide (see Fig. 1). The first of these pyridoxal 5′-phosphate (PLP) dependent enzymes identified was the NifS enzyme of *Azotobacter vinelandii* which is required for the Fe–S cofactor assembly in dinitrogenase. It is now evident that NifS-like enzymes are ubiquitous and play key roles in the synthesis of a variety of S containing compounds and cofactors. Because free sulfur is potentially toxic, cells need to couple the biosynthesis of S-containing cofactors to the activation of S from cysteine (see Hell and Wirtz, chapter 4). Therefore, NifS-like enzymes pair up with downstream-acting protein partners that may serve to activate the cysteine desulfurase and that form sulfur transfer pathways mediating the transfer of S from cysteine into the various target compounds. We are just beginning to see how these cysteine desulfurase dependent systems are regulated. This chapter aims to review our current knowledge of cysteine desulfurase mediated sulfur activation pathways and its role in cofactor assembly, with emphasis on photosynthetic organisms.

II. Iron–Sulfur Cluster Assembly

A. Overview of Iron–Sulfur Cluster Function

Iron occurs in a large variety of cofactors, but we can distinguish three main groups: iron sulfur,

Abbreviations: CysD – cysteine desulfurase; [Fe–S] – iron–sulfur cluster; Moco – molybdenum cofactor; NiR – nitrite reductase; PLP – pyridoxal 5′-phosphate; SiR – sulfite reductase

heme plus siroheme, and finally non-heme iron. Iron–sulfur ([Fe–S]) clusters are an ancient class of prosthetic groups, consisting of iron and sulfur atoms. The architecture of various types of [Fe–S] clusters is described by Beinert and coworkers (1997, 2000). As a protein cofactor, [Fe–S] clusters are usually bound to polypeptides by covalent bonding between iron atoms of the [Fe–S] cluster and sulfur of cysteine residues in the polypeptide. An exception is the Rieske-type [2Fe-2S] cluster, in which one iron atom of the [Fe–S] cluster is coordinated to two histidines. Since the iron in [Fe–S] clusters can easily gain or lose an electron, switching between Fe^{2+} and Fe^{3+}, [Fe–S] clusters are ideal cofactors for proteins that function in electron transport chains (photosynthesis and respiration) and catalyze redox reactions. Iron–sulfur clusters also act as catalytic centers, as sensors of iron and oxygen, and as regulators of gene expression (for a review see Beinert and Kiley 1999).

B. Iron–Sulfur Cluster Assembly Systems in Microbes

Iron–sulfur cluster assembly can be divided into three steps: (i) mobilization of S and Fe, (ii) cluster assembly, and (iii) insertion in apo-proteins. Many aspects of [Fe–S] cluster assembly are conserved from bacteria to eukaryotes. Much of the nomenclature for [Fe–S] assembly components is derived from bacterial systems where components were first discovered. To put the plant machinery in perspective we start here with a brief overview of microbial [Fe–S] assembly systems. For more details about Fe–S assembly in microbes see reviews by Frazzon et al. (2002), Johnson et al. (2005) and Lill and Muhlenhoff (2005).

In bacteria, four machineries have been found to assemble [Fe–S] clusters, each of which is encoded by a gene cluster. The first Fe–S assembly machinery studied in detail was the *nif* system of *Azotobacter vinelandii*, which is responsible for the formation of Fe–S clusters for dinitrogenase, required under nitrogen fixation conditions (Zheng et al., 1993). The *A. vinelandii nif* gene cluster includes a cysteine desulfurase (CysD) encoding gene, *nifS*, as well as the other

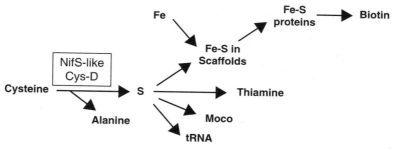

Fig. 1. Overview of cysteine desulfurase dependent biosynthetic pathways.

genes *nifU, iscA^nif, nifV* and *cysE*, all thought to be involved in Fe–S cluster formation. NifS-like proteins are pyridoxal 5′-phosphate (PLP)-dependent, enzymes that produce elemental sulfur from cysteine or selenium from selenocysteine, leaving alanine (Mihara et al., 1997; Mihara and Esaki 2002). Some enzymes also show activity with L-cystine as substrate and produce pyruvate, ammonia and elemental sulfur, via the formation of an enzyme complexed cysteine persulfide intermediate (Clausen et al., 2000). *In vitro*, most NifS-like enzymes show equal or higher activity on selenocysteine compared to cysteine.

On the basis of sequence similarity (Mihara et al., 1997) the NifS-like proteins are divided into group I (NifS/IscS-like) and group II (SufS-like) proteins. Purified group I enzymes efficiently use L-cysteine as a substrate and produce alanine and elemental sulfur as products via the formation of an enzyme bound cysteinyl persulfide intermediate (Zheng et al., 1994). Enzymes acting by this mechanism thus have high endogenous cysteine desulfurase activity. Purified group II enzymes have much lower cysteine desulfurase activity compared to selenocysteine lyase activity. The cysteine desulfurase activity of many of these enzymes is now known to be activated *in vivo* by other proteins.

A second NifS-like protein, IscS, occurs in *A. vinelandii*, and has a housekeeping function in the formation of cellular Fe–S proteins other than dinitrogenase (Zheng et al., 1998). The *iscS* gene is present in a gene cluster that contains paralogs of some of the *nif* genes (*iscU*, similar to the N-terminus of *nifU*, and *iscA*); thus the *nif* and *isc* clusters share a similar organization (Zheng et al., 1998). The IscU- and perhaps IscA-like proteins are thought to serve a scaffold function for the [Fe–S] cluster during its synthesis and before

its transfer to the target protein, and conserved cysteines play a pivotal role in this process (Agar et al., 2000; Krebs et al., 2001). IscA may be an alternative scaffold for IscU because it can carry a transient [2Fe–2S] cluster, which subsequently can be transferred to ferredoxin or biotin synthase (Ollagnier-de-Choudens et al., 2004). However, alternative functions for IscA have been proposed. Some studies supported a role of IscA as an iron-binding protein, which subsequently donates iron for the [Fe–S] cluster assembly on IscU (Ding et al., 2004a, 2004b, 2005). Other data support a role of IscA type proteins in iron sensing (see below). The Isc gene cluster further includes an Hsp70 and Hsp40 and a ferredoxin type protein. Based on work done in yeast mitochondria (Muhlenhoff et al., 2003), the Hsp type proteins may be involved in the transfer of clusters from IscU scaffold proteins with which they interact (for a review on eukaryotic [Fe–S] assembly with a focus on yeast, see Lill and Muhlenhoff, 2005). Homologues of the *nif/isc* genes have been discovered in several other bacteria including *E. coli* (Zheng et al., 1998) and are also present in the mitochondria of eukaryotes (Lill and Kispal, 2000), as described below.

A third set of genes involved in Fe–S cluster formation that was first described for *E. coli* and *Erwinia chrysanthemi* is present in the *suf* operon (Takahashi and Tokumoto, 2002). The *suf* operon of *E.coli* is upregulated in response to oxidative stress and iron-limitation (Zheng et al., 2001, Outten et al., 2004) whereas expression of the *isc*-operon requires iron sufficient conditions. A major function of the *suf*-genes may be in protecting the cell from oxidative stress and iron starvation (Nachin et al., 2003; Outten et al., 2004). Besides a NifS-like protein (SufS/CsdB) the Suf operon encodes SufA, SufB, SufC, SufD and SufE. SufA is related in sequence to IscA^{nif} and IscA and may have a scaffold function (Ollagnier-de Choudens et al., 2003), while SufE was shown to activate SufS (Loiseau et al., 2003; Outten et al., 2003). SufC constitutes a non-intrinsic cytosolic member of the ABC domain transporter superfamily. SufC forms a complex with SufB and SufD but the precise biochemical role of this complex is not yet clear (Nachin et al., 2003; Outten et al., 2003).

Finally, a fourth bacterial [Fe–S] machinery may be present in *Escherichia coli* (Loiseau et al., 2005). This simple gene cluster called *csd* is composed of *csdA* and *csdE*. The *csdA* gene encodes a cysteine desulfurase (Mihara et al., 2000; Loiseau et al., 2005) and *csdE* encodes a SufE-like protein that activates the cysteine desulfurase. The Csd system is proposed to supply [Fe–S] clusters for quinolinate synthase, NadA (Loiseau et al., 2005).

C. Functions of [Fe–S] Proteins in Chloroplasts

Chloroplasts use [Fe–S] proteins in photosynthetic electron transport, nitrogen and sulfur assimilation, and various other plastidic processes. In chloroplasts five cluster types have been found thus far. These [Fe–S] cluster types include [2Fe–2S] (found in for instance ferredoxin), Rieske-type [2Fe-2S] (found in the cytochrome-b/f complex and TIC55, a protein involved in precursor import), [3Fe–4S] (found in FD-GOGAT or glutamate synthase), [4Fe–4S] (found in PSI and in ferredoxin dependent thioredoxin reductase), and finally siroheme-[4Fe–4S], a unique cofactor in which the [4Fe–4S] is covalently bound to a siroheme (found in nitrite reductase and sulfite reductase). A more extensive overview of the various functions of iron–sulfur proteins in plants is given by Ye et al., (2006b). The function and synthesis of [Fe–S] proteins in plants (mitochondria, plastids and cytosol) were reviewed recently by Balk and Lobreaux (2005).

D. [Fe–S] Cluster Biogenesis in Chloroplasts and Photosynthetic Bacteria

1. Overview

Iron–sulfur cluster insertion into apoferredoxin was observed in isolated spinach chloroplasts and chloroplast fractions (Takahashi et al., 1986).

Moreover, *in vitro* synthesized ferredoxin acquires a [2Fe–2S] cofactor after it is imported into isolated pea chloroplasts (Li et al., 1990; Pilon et al., 1995), suggesting that [Fe–S] clusters can be synthesized within this compartment independent of cytosol or other sub-cellular organelles. Cysteine was identified as the source of sulfur for the *in vitro* insertion of a [Fe–S] cluster in ferredoxin (Takahashi et al., 1986, 1990). As a specialized system, the chloroplast [Fe–S] biosynthetic machinery has more recently been investigated using *Arabidopsis thaliana* as a model. With the availability of complete genome sequences of cyanobacterial species, the existence of NifS-like proteins, present in suf-type operons and possibly isc operons have been discovered largely by comparing the genomes of *Synechocystes*, with those of *E.coli*, *A.vinelandii* and *A. thaliana*.

2. Cysteine Desulfurase

Because cysteine is required as a source of sulfur, a cysteine desulfurase is an essential component of any [Fe–S] biosynthetic machinery. In *Arabidopsis*, *CpNifS* (At1g08490) encodes a plastidic cysteine desulfurase (Leon et al., 2002; Pilon-Smits et al., 2002). CpNifS converts cysteine to alanine and provides sulfur for [Fe–S] assembly. Like all other NifS-like proteins, CpNifS has also selenocysteine lyase activity. Its selenocysteine lyase activity is much higher than its cysteine desulfurase activity. CpNifS is a class II NifS-like protein and most similar to SufS among all NifS-like proteins in *Escherichia coli*. CpNifS is essential for the [Fe–S] cluster formation activity of chloroplast stroma (Ye et al., 2005). In stroma, CpNifS is detected both in a 600 kDa complex and in dimeric form as indicated by gelfiltration (Ye et al., 2005). Repression of *CpNifS* expression by RNAi is lethal and causes a defect in the maturation of plastidic but not mitochondrial FeS proteins (Van Hoewyk et al., 2007).

CpSufE, a SufE-like protein encoded by At4g26500, is a cysteine desulfurase activator for CpNifS (Ye et al., 2006a). CpSufE forms a heterotetrameric complex with CpNifS, stimulating cysteine desulfurase activity 40–60-fold and increasing the substrate affinity of CpNifS toward cysteine. *In vitro* reconstitution experiments show that the CpNifS [2Fe-2S] cluster assembly activity in ferredoxin is enhanced 20-fold by CpSufE (Ye et al., 2006a). These activities need an essential cysteine residue in CpSufE, which is likely the acceptor site for intermediate sulfur. This cysteine is not required for binding to CpNifS. Therefore, excess cysteine-mutated SufE displays a dominant negative effect over the wild-type protein. CpSufE function is essential for seedling viability (Xu and Moller, 2006; Ye et al., 2006a). The subcellular location of SufE is a matter of interest. Whereas Ye et al. (2006a) detected the protein in chloroplasts by analyzing GFP-fusions as well as by immunoblotting, data presented by Xu and Moller (2006) suggest that in *Arabidopsis* CpSufE may have a dual localization in both plastids and mitochondria. The dual localization reported by Xu and Moller (2006) is supported by GFP-fusions, complementation of a T-DNA KO only by intact precursor SufE or by both a mitochondrial targeted and plastid targeted SufE and by interaction of SufE with both CpNifS and the mitochondrial NifS, Nfs2. If correct, this would be the first reported case of a type-I, IscS-like, cysteine desulfurase acting with a SufE-like protein.

The cyanobacterium *Synechocystis* sp. PCC 6803 has four NifS-like proteins. The genes slr0387 and sll0704 encode enzymes (Sscsd 1 & 2) similar to NifS of *A.vinelandii* and IscS of *E. coli.* (Kato et al., 2000). They catalyze the desulfuration of L-cysteine, producing alanine and elemental sulfur, and likely play an important role in [Fe–S] synthesis in this cyanobacterium. Typical for NifS-like proteins, the two enzymes also show selenocysteine lyase activity. Sscsd1 has a higher substrate specificity for L-selenocysteine and a much higher specific activity towards the selenium substrate. Sscsd1 and 2 also produced [2Fe-2S] *in vitro* that facilitated the formation of holoferredoxin (Kato et al., 2000). These cyanobacterial IscS related proteins most likely work in concert with an Isc-type assembly machinery.

In addition to the two iscS-like genes described above, *Synechocystis* sp. PCC 6803 also has genes encoding type-II NifS-like proteins. The *slr2143* gene encodes a cystine C-S lyase (Clausen et al., 2000) and *slr0077*, encodes a protein which acts as both a cysteine desulfurase and cystine lyase (Kessler, 2004). Slr0077 is essential (Seidler et al., 2001) but the IscS-like type-I cysteine desulfurases are not. Also called SufS, the Slr0077 enzyme is most related in predicted protein sequence to the *Arabidopsis* chloroplast NifS-like enzyme CpNifS (Pilon-Smits et al., 2002). Slr0077 shows cysteine desulfurase activity under reducing conditions and cystine lyase activity in a partially oxidizing environment (Kessler, 2004). This twin reaction mechanism of Slr0077 might be a significant feature of [Fe–S] synthesis in the cyanobacterium that responds to the redox status of the cell.

3. IscA and Nfu Scaffolds Proteins

CpIscA (At1g10500) may serve as a scaffold protein for [Fe–S] synthesis in plastids (Abdel-Ghany et al., 2005). After acquiring sulfur from cysteine via CpNifS and ferrous iron from media, CpIscA is able to assemble a [2Fe–2S] cluster resulting in dimeric holo-CpIscA *in vitro*. This holo-CpIscA can be isolated by gelfiltration holding its transient cluster. Upon incubation with apoferredoxin CpIscA can transfer its cluster, resulting in holo-ferredoxin, which is then active when tested for electron transfer (Abdel-Ghany et al., 2005). The presence of the CpIscA scaffold improves the [2Fe–2S] reconstitution in ferredoxin. Interestingly, CpIscA is mostly present in a Ð600 Kda chloroplast stromal complex, similar to CpNifS (Abdel-Ghany et al., 2005). Recent work by Yabe and Nakai (2006) reported that the chloroplast IscA in *A.thaliana* is a non-essential or redundant scaffold. The authors showed that a chloroplast Nfu-like protein called Nfu2 acts as an essential scaffold whose deficiency affected the production or accumulation of CpIscA. Thus CpIscA seems to operate downstream of NfU (see below).

In the cyanobacterium *Synechocystis* sp. PCC 6803, *sufA* and *sufE* exist in a separate operon that is not contiguous with the other *suf* genes as is the case in *E. coli* (Wang et al., 2004). A recent and interesting report proposes a model for the function of the *suf* and *isc* regulons and the regulatory role of *sufA* and *iscA* in cyanobacteria (Balasubramanian et al., 2006). An *iscA* null mutant had upregulated expression of the *suf* and *isc* genes and *isiA* (a marker for iron limitation), meaning the iscA mutant mistakenly senses iron-limitation under normal conditions and responds to it. Therefore iscA may be part of iron sensing and homeostasis in cyanobacteria (Balasubramanian et al., 2006). The mRNA levels of *suf* genes were increased in a sufA null mutant during oxidative stress (Balasubramanian et al., 2006). Therefore the Suf system has been proposed to be involved in [Fe–S] synthesis/repair under oxidative stress. The cyanobacterial and *Arabidopsis* sufA homologs contain two additional highly conserved cysteine residues (absent in non-photosynthetic bacteria) which could possibly act as a redox sensitive regulatory switch by reversible disulfide bond formation, and set in motion the suf, isc and possibly other factors of cyanobacterial [Fe–S] biosynthesis machinery. This recent work by Balasubramanian et al. (2006) proposes new roles for *iscA* and *sufA* in iron homeostasis and oxidative stress response, in addition to the role of alternative scaffolds earlier assigned to them (Ollagnier de Choudens et al., 2004; Abdel-Ghany et al., 2005). In view of these recent reports and the cyanobacterial origin of chloroplasts, it would be interesting to look into the roles of IscA and/or SufA in iron sensing/homeostasis and oxidative stress responses.

In addition to CpIscA, Nfu1–3 are scaffold proteins active in chloroplasts (Leon et al., 2003; Touraine et al., 2004; Yabe et al., 2004). Nfu1 (At4g01940) and Nfu2 (At5g49940) are able to restore the growth of a scaffold-mutated yeast, strain $\Delta isu1\Delta nfu1$, suggesting a role as a scaffold. Recombinant Nfu2 contains a labile [2Fe–2S] cluster, which can be transferred to apo-ferredoxin resulting in holo-ferredoxin formation (Leon et al., 2003; Yabe et al., 2004). The analysis of T-DNA insertion lines revealed that Nfu2 is required for assembling [4Fe–4S] clusters of photosystem I and the [2Fe-2S] cluster of ferredoxin in chloroplasts (Touraine et al., 2004; Yabe et al., 2004). Interestingly, a subset but not all chloroplast [Fe–S] proteins are affected in the *Nfu2* mutant, suggesting alternative scaffolds are functional in chloroplasts. Two more Nfu-like proteins, Nfu4 and Nfu5, are located in mitochondria (Leon et al., 2003). Notably, proteins with similarity to the N-terminus of NifU (IscU-like proteins) are all localized in mitochondria in *Arabidopsis* and are absent in plastids (Leon et al., 2005).

In cyanobacteria there is a gene denoted nfu encoding a protein similar to the C-terminal domain of NifU of *A.vinelandii* (Nishio and Nakai 2000). It was shown to be an essential scaffold protein (Seidler et al., 2001, Balasubramanian et al., 2006). On the other hand, null mutants of *iscA, sufA* and the *iscA-sufA* double mutant expressed wild-type levels of Nfu and growth rates comparable to wild-type under normal conditions, iron-limited conditions, or oxidative stress (Balasubramanian et al., 2006).

4. Other Factors

Components of the SufBCD complex have been found in plants and are active in the plastids. In Arabidopsis, *SufB* (At4g04770) encodes a protein with ATPase activity (Xu et al., 2005). This gene can complement sufB deficiency in *Escherichia coli*. *Arabidopsis SufC* (At3g10670) is an ABC type ATPase (Xu and Moller, 2004), which can partially rescue growth defects in an *Escherichia coli sufC* mutant. Finally, the *Arabidopsis SufD* (At1g32500) encodes a protein with homology to the bacterial SufD, mutation of which results in impaired embryogenesis and abnormal growth of *Arabidopsis* (Hjorth et al., 2005). Like their bacterial homologues, the chloroplast SufBCD proteins form a complex (Xu and Moller, 2004; Xu et al., 2005), displaying ATPase activity. Although it is likely that SufBCD play some role in [Fe–S] cluster biogenesis, the exact role in the process remains to be characterized not just in plants but in any organism.

Two additional components were identified in screens for mutants that are pleiotropically affected in photosynthesis. One such protein is HCF101 for High Chlorophyll Fluorescence (Stockel and Oelmuller, 2004). *HCF101* (At3g24430) encodes a protein with sequence similarity to P-loop ATPases (Lezhneva et al., 2004). It is required for the biogenesis of [4Fe–4S] clusters for photosystem I (PSI) and ferredoxin-thioredoxin reductase (FTR) in chloroplasts. The exact biochemical function of HCF101 remains to be elucidated. Another factor which was originally identified in a photosystem I mutant screen is *APO1* (Accumulation of Photosystem One1). *APO1* (At1g64810) is a member of a novel gene family so far found only in vascular plants, (Amann et al., 2004). It is involved in the assembly of [4Fe–4S] cluster-containing complexes of chloroplasts, e.g. PSI. However, a direct connection between APO1 and [Fe–S] assembly needs to be shown because the protein could also affect PS1 maturation without acting directly on [Fe–S] assembly.

The existence of a complete *suf* operon has been reported in the cyanobacteria *Synechocystis, Synechococcus* and *Anabaena* (Wang et al., 2004). DNA sequence comparison shows that the cyanobacterial *sufBCDS* cluster is homologous to the suf operon of *E. coli* (Wang et al., 2004). The *sufBCDS* genes are cotranscribed in *Synechococcus* sp. PCC 7002. A gene named *sufR* is located upstream of the *sufBCDS* cluster in cyanobacteria and has been proposed to be a transcriptional repressor of the operon (Wang et al., 2004). SufR has a DNA binding domain at its N-terminus and a metal binding region with conserved cysteines at the C-terminus, which could be acting as a sensor of oxidative stress or iron limitation. *sufR* null mutants of *Synechococcus* sp. PCC 7002 had elevated transcript levels of *sufBCDS* compared to wild type and also showed higher growth rates under iron limitation (Wang et al., 2004).

Little is known about [Fe–S] biosynthesis mechanisms in other photosynthetic prokaryotes, though the existence of [Fe–S] clusters and hence their biosynthetic machineries can be predicted. The occurrence of *nifS, nifU* and ferredoxin as part of the *nif* gene cluster for maturation of [Fe–S] and Mo cofactor containing nitrogenase was reported in the photosynthetic bacterium *Rhodopseudomonas palustris* (Oda et al., 2005). Microarray data indicate upregulation of [Fe–S] assembly/repair genes in a facultative phototrophic bacterium *Rhodobacter sphaeroides* upon exposure to hydrogen peroxide (Zeller et al., 2005).

5.   Summary of Plastidic and Cyanobacterial [Fe–S] Assembly Systems

In summary, chloroplasts and photosynthetic bacteria conserve a complete SUF-type machinery, including homologs to all components encoded by the *E. coli sufABCDSE* gene cluster for [Fe–S] biogenesis. Because green chloroplasts produce saturated oxygen levels in the light, it is not unexpected that a SUF-like system operates in chloroplasts. However, the complete chloroplast [Fe–S] machinery is more complex than bacterial [FeS] assembly systems. The chloroplast system also includes three NifU-like proteins and HCF101, a protein that has no homologue in bacterial [Fe–S] machineries. Furthermore APO1, a protein that is unique in vascular plants is required

for the accumulation of a subset of [Fe–S] proteins in plastids. The severe phenotypes caused by mutations of individual machinery components point to the general importance and complexity of the chloroplast [Fe–S] machinery. For instance, loss of function mutants for *CpNifS* are inviable (Van Hoewyk et al., 2007). An insertion in *SufE* causes embryonic lethality (Xu and Moller, 2006; Ye et al, 2006a). *Nfu2* T-DNA insertion mutants are dwarfed and yellowish (Touraine et al., 2004; Yabe et al., 2004). Abnormal plastid structure and impaired embryogenesis is observed for *SufC* or *SufD* mutants (Xu and Moller, 2004; Hjorth et al., 2005). Finally, seedling lethality and high chlorophyll fluorescence is observed for *HCF101* mutants (Lezhneva et al., 2004).

6.   The Integration of Iron and Sulfur Metabolism in Plastids

Because the CpNifS-dependent machinery supplies [Fe–S] clusters for a range of chloroplast activities, plastid [Fe–S] biogenesis is essential for photosynthetic carbon fixation, nitrogen assimilation, sulfur assimilation, and other biosynthetic processes. Because of the importance of iron and sulfur assimilation for agricultural crops, the effects of the plastid [Fe–S] machinery on the homeostasis of iron and sulfur in plants is of particular interest.

In *Arabidopsis* leaf tissue, about 70% of the iron is located in chloroplasts and at least half of that iron is associated with thylakoids (Shikanai et al., 2003). Fe^{2+} transport across the chloroplast inner envelope membrane is measurable, and determined to be light-dependent. Furthermore, Fe^{2+} transport across inner envelope membrane is stimulated by an electrochemical proton gradient, and reduced by negating the potential gradient, suggesting that iron transport into chloroplasts is via a Fe^{2+}/H^+ symport mechanism (Shingles et al., 2002). In thylakoid membranes, iron is predominantly used for electron transfer. Approximately half of the thylakoid iron is in the photosystem I in form of [4Fe–4S] clusters. The remaining half is present as non-heme iron, iron in cytochromes in the form of heme and Rieske-type [2Fe–2S] clusters (Raven et al., 1999). In the stroma, iron is used in [Fe–S] clusters, hemes, and non-heme iron in for instance Fe-SOD. Because excess free iron would be toxic, surplus iron is stored in stromal ferritin (Petit et al., 2001). Because iron is one of the most limiting micronutrients to plants yet potentially toxic in excess, we expect a sophisticated homeostasis mechanism to control cellular iron, particularly in plastids. Would there be cross-talk between the various biosynthetic pathways that involve Fe cofactors in plastids? We consider this more than likely and possible interactions are indicated in Fig. 2. The pathways for inserting iron in non-heme Fe

proteins such as FeSOD are largely unknown. However, much is known about the synthesis of heme and siroheme cofactors, which are produced through the tetrapyrrole pathway starting from glutamate. This pathway also serves to synthesize chlorophyll A and B (for review see Cornah et al., 2003). Feedback regulation in the tetrapyrrole pathway is well established. Heme accumulation may be an important factor in this process. Furthermore, the accumulation of the intermediate Mg-protoporphyrin IX is in

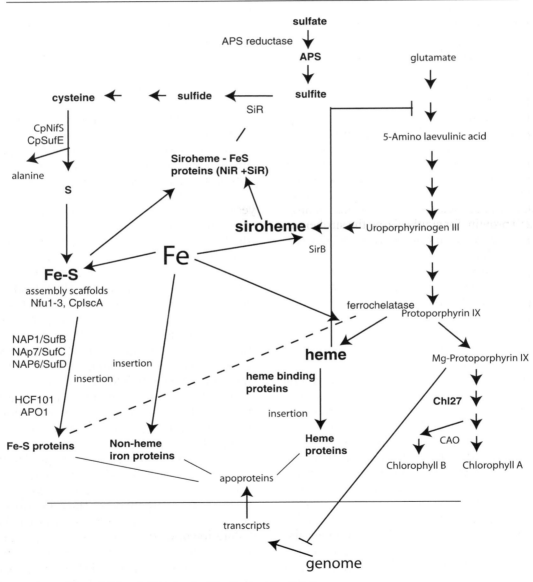

Fig. 2. Interactions of Fe and S metabolism in plastids. See text for details.

all likelihood an important signal affecting the expression of nuclear genes for chloroplast precursor proteins (Cornah et al., 2003). Finally, there may be cross-talk between [Fe–S] assembly and heme synthesis because ferrochelatase activity is downregulated in [Fe–S] assembly mutants, at least in mitochondria of yeast (Lange et al., 2004); a similar regulation may occur in plastids. Cross-talk is also suggested by the phenotype of the *laf6* (= *SufB*) mutant of *Arabidopsis*, which accumulates protoporphyrin IX an intermediate in chlorophyll and heme synthesis (Moller et al., 2001).

The plastid is also a key compartment for sulfur assimilation from sulfate to cysteine (for reviews see: Hawkesford, 2007; Hell and Wirtz, 2007). Notably, out of four enzymes involved in S assimilation in plastids, two are [Fe–S] proteins: APS reductase and sulfite reductase (SiR) (Fig. 2). Thus, sulfur assimilation is likely dependent on the plastid [Fe–S] machinery. The step catalyzed by sulfite reductase is even more dependent on the [Fe–S] assembly machinery. This enzyme employs a unique siroheme-[4Fe–4S] as its prosthetic group, requiring not only a direct incorporation of a [4Fe–4S] cluster but also a sirohydrochlorin ferrochelatase (SirB), which is also a [2Fe–2S] protein, for synthesizing the siroheme. Moreover, the six electron reduction catalyzed by SiR needs ferredoxin, a [2Fe–2S] protein, for providing electrons. Thus, the SiR-catalyzed step in sulfur assimilation is completely dependent on the plastid [Fe–S] machinery, strongly suggesting a central role of the [Fe–S] machinery in plastid sulfur homeostasis.

E. Function and Assembly of Fe–S Clusters in Plant Mitochondria and Cytosol

Iron sulfur clusters play crucial roles in mitochondrial function. For instance, aconitase, a key enzyme in the Krebs cycle in the matrix, is an Fe–S protein. Furthermore respiratory electron transport chain complexes I, II and III all have Fe–S cofactors (for review see Balk and Lobreaux, 2005). In Arabidopsis, a second cysteine desulfurase, Nfs2, is present in mitochondria (Kushnir et al., 2001), where another [Fe–S] biogenesis machinery is present. In addition, a third cysteine desulfurase activity is found in the cytosol (Heidenreich et al., 2005). This activity is due to

the NifS-like domain of the ABA3 protein, which functions in Moco synthesis and is most likely not relevant to [Fe–S] synthesis. Work in yeast, which can grow anaerobically using fermentation, suggested that Fe–S cluster formation is the only essential function of mitochondria (Lill and Kispal, 2000) and cytosolic Fe–S clusters depend on the mitochondrial Isc machinery involving homologues of the genes encoded by the *nif/isc* clusters of bacteria. The yeast mitochondrial NifS-like protein, IscS, is essential for this reason (Kispal et al., 1999). A similar mitochondrial machinery, dependent on Nfs2, may be present in plants (Kushnir et al., 2001).

In *Arabidopsis* the mitochondrial ABC transporter Sta1 could be involved in the transport of [Fe–S], or an unknown precursor of these clusters, from mitochondria to the cytosol (Kushnir et al., 2001). The *sta1* mutants were dwarfed, chlorotic, had distorted nuclei, and accumulated higher amounts of free (non heme, non protein) iron in mitochondria and showed increased expression of mitochondrial cysteine desulfurase: effects collectively known as the '*starik*' phenotype (Kushnir et al., 2001). Sta1 is a functional ortholog of Atm1p, the mitochondrial ABC transporter in yeast (Kushnir et al., 2001). Complementation of mutant *atm1* yeast cells with *Arabidopsis* Sta1 restored maturation of cytosolic [Fe–S] protein Leu1p (Kushnir et al., 2001). However, in *Arabidopsis sta1* mutants, the activities of cytosolic and mitochondrial isoforms of aconitase were similar to wild type, probably due to the presence of redundant mitochondrial ABC transporters such as Sta2 and Sta3 that perhaps compensated the loss of Sta1. The *starik* phenotype is thought to be a pleiotropic effect resulting from an imbalance in the intracellular iron homeostasis (Kushnir et al., 2001). This report indicates that there is a mechanism of iron movement between mitochondria and cytosol and that [Fe–S] generated in mitochondria could be exported to cytosol for the maturation of cytosolic Fe–S proteins as seen in *atm1* yeast cells (Kushnir et al., 2001). In that case, there could be scaffold proteins in the cytosol that help transfer mitochondrial [Fe–S] to cytosolic apoproteins. The mechanisms for the maturation of cytosolic [Fe–S] proteins remain to be elucidated.

The existence of Fe–S proteins in the cytosol (e.g. an aconitase) raises the question if there is a separate machinery for the synthesis of [Fe–S] clusters and their incorporation into cytosolic apo-proteins. We do not yet know of a cytosolic [Fe–S] assembly system in plants. Recently, components of such a machinery have been reported in mammalian (Pondarre et al., 2006; Tong and Rouault 2006) and yeast cells (Balk et al., 2004, 2005; Hausmann et al., 2005). A matter of debate is the presence of an IscS activity in the cytosol or nucleus. The observation in yeast that IscS is required for the thio-modification of cytoplasmic tRNA *in vivo*, strongly supports the presence of IscS outside the mitochondria (Nakai et al., 2004).

III.   Iron–Sulfur Cluster Dependent Cofactor Assembly Pathways

A.   *Radical SAM Enzymes*

Radical SAM enzymes are [4Fe–4S] containing proteins that use S-adenosyl methionine (SAM) as a cofactor (Layer et al., 2004). These enzymes form a superfamily of proteins that now includes more than 600 members, present in a diverse group of organisms from bacteria to plants and humans (Sofia et al., 2001). The functions supported by radical SAM proteins so far identified include a diverse set of reactions such as sulfur transfer reactions, heme and chlorophyll biosynthesis, ring forming reactions like thiazole formation, antibiotic and herbicide biosynthesis and DNA repair (Sofia et al., 2001). The activities of most SAM radical enzymes are extremely oxygen sensitive. All these enzymes contain a [4Fe–4S] cluster which is thought to be the main catalytic site (Walsby et al., 2005). Some enzymes like biotin synthase and lipoyl synthase have an additional [2Fe–2S] cluster. Other examples of the radical SAM enzymes identified are HemN (oxygen independent coproporphyrinogen synthase, involved in tetrapyrrole biosynthesis), Biotin synthase, MoaA (involved in molybdenum cofactor biosynthesis), littorine mutase (an alkaloid generating enzyme in *Datura*) and HydE/G, two enzymes required for the maturation of an Fe-hydrogenase in the chloroplast of *Chlamydomonas reinhardtii* (Layer et al., 2005). This large class of iron–sulfur proteins is probably very important for plant metabolism, but thus far only few radical SAM enzymes have been characterized in plants and other photosynthetic organisms.

A characteristic feature of the radical SAM superfamily members is the presence of a conserved three-cysteine motif binding the [4Fe–4S] cluster. Three of four iron atoms bind to the three cysteines. The fourth iron, bound to a non-cysteine ligand, provides a unique site for coordination with the cofactor S-adenosyl methionine. Electron transfer from the iron–sulfur cluster to SAM results in SAM cleavage and generates a highly reactive 5′-deoxyadenosyl radical that abstracts a H atom from the substrate and initiates the reaction, a feature common to all SAM radical enzyme catalysed reactions (Walsby et al., 2005).

It is obvious that this important class of enzymes must depend on the iron–sulfur synthesis machinery for their [Fe–S] clusters. However, to date the processes involved in formation and transfer of [Fe–S] to SAM radical proteins have not been studied in detail. In one report the SAM radical enzyme lipoyl synthase from *E.coli* when coexpressed with the *A.vinelandii* Isc operon is active without *in vitro* reconstitution (Cicchillo et al., 2004). It would be interesting to investigate if a specialized mechanism exists for the [Fe–S] supply to SAM radical proteins.

B.   *Biotin Synthesis*

Biotin, a member of the vitamin B family is a water soluble cofactor for certain enzymes involved in fatty acid and carbohydrate metabolism. Biotin could also be involved in modulating gene expression (Che et al., 2003). Plants and bacteria possess a similar biosynthetic pathway (Schneider and Lindqvist, 2001) for the synthesis of this essential nutrient. The mechanism of biotin synthase reaction has been well studied in *E. coli*. Biotin synthase, also known as BioB contains a [4Fe–4S] and a [2Fe–2S]. It is a radical SAM enzyme requiring S-adenosyl methionine as a cofactor. The [2Fe–2S] is proposed to be the immediate sulfur donor to dethiobiotin (Jameson et al., 2004), resulting in the loss of sulfur from the [2Fe–2S] and consequent inactivation of the enzyme. *In vitro*, NifS from *A.vinelandii* and C-DES from *Synechocystis* could mobilize sulfur from cysteine for reconstitution of the [2Fe–2S] into the apoprotein of *E. coli* biotin synthase and restore its activity (Bui et al., 2000).

In plants, biotin synthesis occurs in the cytosol and mitochondria. The last step involves the conversion of dethiobiotin to biotin and involves the insertion of a sulfur atom into dethiobiotin. The enzyme catalyzing this reaction, biotin synthase, occurs in the mitochondrial matrix (Baldet et al., 1997). The identity of the sulfur donor in plants remains unknown but based on analogy with bacterial systems this may involve S from a 2Fe–2S cluster, ultimately derived from cysteine and activated via the mitochondrial NifS-like enzyme. Indeed it has been shown that in *Arabidopsis* mitochondria, biotin synthase requires other mitochondrial components for its activity (Picciocchi et al., 2003). It is highly probable that mitochondrial cysteine desulfurase, using cysteine as the substrate could replenish sulfur for [2Fe–2S] of biotin synthase.

C. NAD Synthesis

NAD and NADP are important coenzymes in biological redox reactions. In all NAD biosynthetic pathways known thus far, quinolinate is a precursor of nicotinic acid, which is a component of NAD. Quinolinate is synthesized from aspartate or tryptophan. In the aspartate pathway, aspartate is oxidized by L-aspartate oxidase (product of the NadB gene) to iminoaspartate, which is converted to quinolinate by the enzyme quinolinate synthase (NadA). NadA has been characterized in *E. coli*. It is a [4Fe-4S] containing enzyme. The iron–sulfur cluster is sensitive to oxygen and is essential for the quinolinate synthase activity (Ollagnier-de Choudens et al., 2005). An *E. coli* strain lacking IscS was unable to synthesize NAD, and required nicotinic acid for growth (Lauhon and Kambampati 2000). A recent report (Katoh et al., 2006) shows that in *Arabidopsis* NadA, along with two other enzymes of the aspartate to quinolinate pathway (NadB and quinolinate phosphoribosyl transferase) is essential. T-DNA disruption of the corresponding genes was embryonic lethal. The three proteins were found to be located in the plastid (Katoh et al., 2006). Most likely, the *Arabidopsis* plastid NadA is also an iron–sulfur protein like its *E. coli* counterpart, and could be sensitive to oxidative stress and plastid iron status. Therefore this enzyme could be a link between NAD synthesis, redox status and photosynthesis in chloroplasts. The *Arabidopsis* NadA protein has

an extra amino terminal SufE-like domain that may provide SufE activity to recruit and enhance the cysteine desulfurase CpNifS specifically for the activation of quinolinate synthase.

IV. Synthesis of Thiamine

Thiamine or vitamin B1 is an essential cofactor in all cells. Thiamine consists of pyrimidine and thiazole moieties. The mechanism of biosynthesis of this essential vitamin, especially its thiazole ring has been a matter of interest over the past 15 years (Julliard and Douce, 1991; Belanger et al., 1995; Park et al., 2003; Dorrestein et al., 2004). In higher plants, the chloroplast is a site of thiamine synthesis. When stromal proteins from spinach chloroplasts were incubated with substrates such as glyceraldehyde-3-phosphate, pyruvate, tyrosine, cysteine and MgATP, thiazole was synthesized *in vitro* (Julliard and Douce, 1991). Stromal proteins formed thiamine when provided with thiazole and pyrimidine moieties, showing that chloroplast stroma has all the enzymes and substrates required for thiazole synthesis and condensation of thiazole and pyrimidine, to form thiamine (Julliard and Douce, 1991).

Cysteine is an essential substrate for thiazole formation. It is quite likely that the chloroplast NifS is involved in the mobilization of sulfur from cysteine, for thiazole synthesis. It has been shown in bacterial systems that in addition to the components of a multi-enzyme complex involved in the biosynthesis of the thiazole ring (Leonardi et al., 2003), a NifS-like protein is essential, along with cysteine and other substrates for *in vitro* thiazole synthesis (Park et al., 2003, Leonardi and Roach, 2004). An *E. coli* IscS-deletion strain required thiazole in the medium for growth (Lauhon and Kambampati, 2000). Recently, a thiazole biosynthetic enzyme Thi1 has been identified in *A. thaliana*, that seems to be targeted to both chloroplasts and mitochondria (Ribeiro et al., 2005). It would be interesting to know if thiamine is synthesized in both chloroplasts and mitochondria since both organelles contain cysteine desulfurases.

V. Synthesis of Molybdenum Cofactor

Plants have only four Mo requiring proteins: nitrate reductase, xanthin dehydrogenase, and

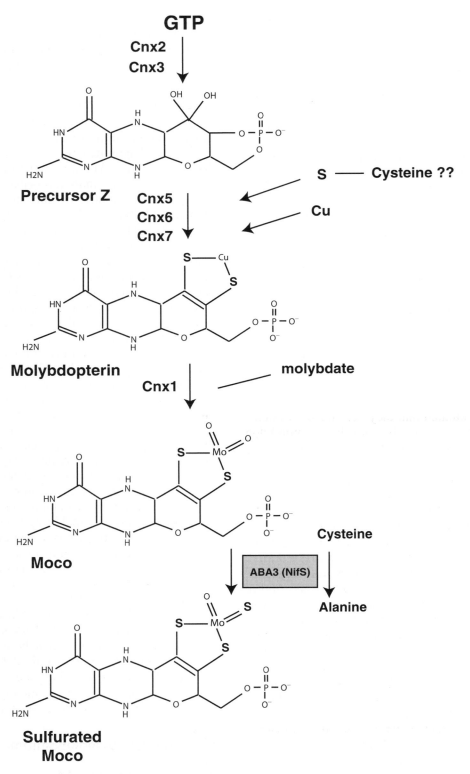

Fig. 3. Molybdenum cofactor biosynthesis. See text for details.

aldehyde oxidase are active in the cytosol while sulfite oxidase is most likely active in peroxisomes (Mendel and Hansch, 2002). With the exception of bacterial nitrogenase, all biologically active Mo occurs in a special pterin-derived cofactor termed molybdenum cofactor or Moco, which is assembled in the cytosol (Mendel, 2005). In plants, MoCo deficiency leads to reduced nitrate reductase activity and N depletion, as well as reduced phytohormone synthesis (Mendel and Hansch, 2002). The studies performed in plants have significantly contributed to what we know about Moco synthesis and this is now one of the best-understood cofactor assembly systems (Mendel, 2005). Seven enzymes are required for Moco synthesis in plants (see Fig. 3). The bacterial homologue of the first of these enzymes, Cnx2, is called MoaA and is a member of the radical SAM enzyme family in bacteria. Therefore, since radical SAM enzymes are [4Fe–4S] enzymes, we can expect that the activity of the Fe–S assembly machinery is required for Moco synthesis. The conversion of precursor-Z to molybdopterin involves the insertion of two S atoms, that will later hold Mo in Moco. Presently the source of these two S atoms is unknown, but it can be speculated that they derive from cysteine. The insertion of molybdenum in the molybdopterin skeleton requires the CNX1 protein. The structure of CNX1 with the molybdopterin bound revealed that the enzyme makes an adenylated intermediate, molydopterin-AMP, which was found to be bound to the enzyme. Furthermore, Cu was found to be bound to the sulfhydryls of the molybdopterin moiety and it was suggested that the presence of Cu served to protect the thiols and facilitate Mo insertion (Kuper et al., 2004). The activities of xanthin dehydrogenase and aldehyde oxidase require a special modification of Moco: the replacement of O by S to form sulfurated Moco (see Fig. 3). The S for this modification is derived from cysteine. In *Arabidopsis*, a specialized cytosolic enzyme, ABA3 catalyzes this reaction (Xiong et al., 2001; Heidenreich et al., 2005). ABA3 contains a PLP containing NifS-like domain required specifically for this activation. ABA3 is the 3rd NifS-like enzyme in *Arabidopsis*, next to CpNifS and mitochondrial Nfs2. The activity of ABA3, which is induced by drought stress and cold treatment, may regulate the activities of xantine dehydrogenase and aldehyde oxidase and therefore cellular phytohormone levels through this sulfurylation step.

VI.   Conclusions and Outlook

Cysteine is the sulfur donor in a wide spectrum of biosynthetic reactions mediated by NifS-like and PLP-dependent cysteine desulfurase enzymes. This is a critical role of cysteine next to its role as an amino acid subunit of peptides and proteins. A central role of cysteine desulfurases is in the synthesis of [FeS] clusters and other S containing cofactors. Because [FeS] containing enzymes play such pivotal roles in assimilation reactions (including those for sulfur) and in the biosynthesis of other cofactors acting both as catalysts but also as non-catalytic donors, using transient [FeS] clusters, the machinery for [FeS] synthesis is essential. Iron–sulfur clusters are ancient cofactors that probably evolved very early in the evolution of life or even before that time in a reducing atmosphere where these clusters may have formed spontaneously. The accumulation of oxygen, due to oxygenic photosynthesis, presented cells with a challenge with respect to their [FeS] assembly because the clusters are very sensitive to oxygen. Several types of NifS-like protein dependent [FeS] machinery, perhaps with different sensitivities to oxygen have evolved. We still have not identified all the components needed for [FeS] and other S containing cofactors but we are getting closer. Once all components are known we can try to understand the regulation of cysteine desulfurase dependent pathways and how cofactor assembly and cysteine availability for protein synthesis are balanced.

Acknowledgements

Work in the authors' laboratories was supported by grants from the National Science Foundation (Grant # NSF-MCB-9982432 to EAHPS and grant # NSF IBN-0418993 to MP) and by a grant from the United States Department of Agriculture NRI program (grant # USDA-NRI, 2005-35318-16212 to EAHPS and MP).

References

Abdel-Ghany SE, Ye H, Garifullina GF, Zhang L, Pilon-Smits EAH and Pilon M (2005) Iron–sulfur cluster biogenesis in chloroplasts. Involvement of the scaffold protein CpIscA. Plant Physiol 138: 161–172

Agar JN, Krebs C, Frazzon J, Huynh BH, Dean DR and Johnson MK (2000) IscU as a scaffold for iron–sulfur cluster biosynthesis: sequential assembly of [2Fe–2S] and [4Fe–4S] clusters in IscU. Biochemistry 39: 7856–7862

Amann K, Lezhneva L, Wanner G, Herrmann R and Meurer J. (2004) Accumulation of Photosystem One1, a member of a novel gene family, is required for accumulation of [4Fe–4S] cluster-containing chloroplast complexes and antenna proteins. Plant Cell 16: 3084–3097.

Balasubramanian R, Shen G, Bryant DA and Golbeck JH (2006) Regulatory roles for IscA and SufA in iron homeostasis and redox stress responses in the cyanobacterium *Synechococcus sp.* strain PCC 7002. J Bacteriol 188: 3182–3191

Baldet P, Alban C and Douce R (1997) Biotin synthesis in higher plants: purification and characterization of bioB gene product equivalent from *Arabidopsis thaliana* overexpressed in *Escherichia coli* and its subcellular localization in pea leaf cells. FEBS Lett 419: 206–210

Balk J and Lobreaux S (2005) Biogenesis of iron–sulfur proteins in plants. Trends in Plant Sci 10: 324–331.

Balk J, Aguilar Netz DJ, Tepper K, Pierik AJ and Lill R (2005) The essential WD40 protein Cia1 is involved in a late step of cytosolic and nuclear iron–sulfur protein assembly. Mol Cell Biol 25: 10833–10841

Balk J, Pierik AJ, Netz DJ, Muhlenhoff U, Lill R (2004) The hydrogenase-like Nar1p is essential for maturation of cytosolic and nuclear iron–sulphur proteins. EMBO J 23: 2105–2115

Beinert H (2000) Iron–sulfur proteins: ancient structures, still full of surprises. J Biol Inorg Chem 5: 2–15

Beinert H and Kiley PJ (1999) [Fe–S] proteins in sensing and regulatory functions. Curr Opin Chem Biol 3: 152–157

Beinert H, Holm RH, and Munck E (1997) Iron–sulfur clusters: nature's modular, multipurpose structures. Science 277: 653–659

Belanger FC, Leustek T, Chu B and Kriz AL (1995) Evidence for the thiamine biosynthetic pathway in higher-plant plastids and its developmental regulation. Plant Mol Biol 29: 809–821

Bui BT, Escalettes F, Chottard G, Florentin D and Marquet A (2000) Enzyme-mediated sulfide production for the reconstitution of [2Fe–2S] clusters into apo-biotin synthase of Escherichia coli. Sulfide transfer from cysteine to biotin. Eur J Biochem 267: 2688–2694

Che P, Weaver LM, Wurtele ES and Nikolau BJ (2003) The role of biotin in regulating 3-methylcrotonyl-coenzyme a carboxylase expression in *Arabidopsis*. Plant Physiol 131: 1479–1486

Cicchillo RM, Lee KH, Baleanu-Gogonea C, Nesbitt NM, Krebs C and Booker SJ (2004) *Escherichia coli* lipoyl synthase binds two distinct [4Fe–4S] clusters per polypeptide. Biochemistry 43: 11770–11781

Clausen T, Kaiser JT, Steegborn C, Huber R and Kessler D (2000) Crystal structure of the cystine C-S lyase from *Synechocystis*: stabilization of cysteine persulfide for FeS cluster biosynthesis. Proc Natl Acad Sci USA 97: 3856–3861

Cornah JE, Terry MJ and Smith AG. (2003) Green or red: what stops the traffic in the tetrapyrrole pathway? Trends Plant Sci 8:224–230.

Ding B, Smith ES and Ding H (2005) Mobilization of the iron centre in IscA for the iron–sulphur cluster assembly in IscU. Biochem J 389: 797–802

Ding H and Clark RJ (2004a) Characterization of iron binding in IscA, an ancient iron–sulphur cluster assembly protein. Biochem J 15: 433–440

Ding H, Clark RJ and Ding B (2004b) IscA mediates iron delivery for assembly of iron–sulfur clusters in IscU under the limited accessible free iron conditions. J Biol Chem 279: 37499–37504

Dorrestein PC, Zhai H, McLafferty FW and Begley TP (2004) The biosynthesis of the thiazole phosphate moiety of thiamin: the sulfur transfer mediated by the sulfur carrier protein ThiS. Chem Biol 11: 1373–1381

Frazzon J, Fick JR and Dean DR (2002) Biosynthesis of iron–sulphur clusters is a complex and highly conserved process. Biochem Soc Trans 30: 680–685

Hausmann A, Aguilar Netz DDJ, Balk J, Pierik AJ, Muhlenhoff U and Lill R (2005) The eukaryotic P loop NTPase Nbp35: an essential component of the cytosolic and nuclear iron–sulfur protein assembly machinery. Proc Natl Acad Sci USA 102: 3266–3271

Hawkesford MJ (2007) Uptake, allocation and subcellular transport of sulfate (this book).

Heidenreich T, Wollers S, Mendel RR and Bittner F (2005) Characterization of the NifS-like domain of ABA3 from *Arabidopsis thaliana* provides insight into the mechanism ofmolybdenum cofactor sulfuration. J. Biol. Chem. 280: 4213–4218

Hell and Wirtz (2007) Metabolism of cysteine in plants and phototrophic bacteria (this book).

Hjorth E, Hadfi K, Zauner S and Maier UG.(2005) Unique genetic compartmentalization of the SUF system in cryptophytes and characterization of a SufD mutant in *Arabidopsis thaliana*. FEBS Lett. 579: 1129–1135

Jameson GN, Cosper MM, Hernandez HL, Johnson MK and Huynh BH (2004) Role of the [2Fe–2S] cluster in recombinant *Escherichia coli* biotin synthase. Biochemistry 43: 2022–2031

Johnson D, Dean, D Smith AD and Johnson MK (2005) Structure, function and formation of biological iron–sulfur clusters. Annu Rev Biochem 74: 247–81.

Julliard J-H and Douce R (1991) Biosynthesis of the thiazole moiety of thiamin (vitamin B1) in higher plant chloroplasts. Proc Natl Acad Sci USA 88: 2042–2045

Kato S, Mihara H, Kurihara T, Yoshimura T and Esaki N (2000) Gene cloning, purification, and characterization of two cyanobacterial NifS homologs driving iron–sulfur cluster formation. Biosci Biotechnol Biochem 64: 2412–2419

Katoh A, Uenohara K, Akita M and Hashimoto T (2006) Early steps in the biosynthesis of NAD in *Arabidopsis thaliana* start with aspartate and occur in the plastid. Plant Physiol 141: 851–857

Kessler D (2004) Slr0077 of *Synechocystis* has cysteine desulfurase as well as cystine lyase activity. Biochem Biophys Res Commun 320: 571–577

Kispal G, Csere P, Prohl C and Lill R (1999) The mitochondrial proteins Atm1p and Nfs1p are essential for biogenesis of cytosolic Fe/S proteins. EMBO J 18: 3981–3989.

Krebs C, Agar JN, Smith AD, Frazzon J, Dean DR, Huynh BH and Johnson MK (2001) IscA, an alternate scaffold for Fe–S cluster biosynthesis. Biochemistry 40:14069–14080

Kuper J, Llamas A, Hecht HJ, Mendel RR and Schwarz G. (2004) Structure of the molybdopterin-bound Cnx1G domain links molybdenum and copper metabolism. Nature 430: 803–806.

Kushnir S, Babiychuk E, Storozhenko S, Davey MW, Papenbrock J, De Rycke R, Engler G, Stephan UW, Lange H, Kispal G, Lill R and Van Montagu M (2001) A Mutation of the mitochondrial ABC transporter Sta1 leads to dwarfism and chlorosis in the *Arabidopsis* mutant *starik*. Plant Cell 13: 89–100

Lange H, Mühlenhoff U, Denzel M, Kispal G and Lill R (2004) The heme synthesis defect of mutants impaired in mitochondrial iron-sulfur protein biogenesis is caused by reversible inhibition of ferrochelatase. J Biol Chem 279: 29101–29108

Lauhon CT and Kambampati R (2000) The *iscS* gene in *Escherichia coli* is required for the biosynthesis of 4-thiouridine, thiamin, and NAD. J Biol Chem 275: 20096–20103

Layer G, Heinz DW, Jahn D and Schubert WD. (2004) Structure and function of radical SAM enzymes. Curr Opin Chem Biol 8:468–476

Layer G, Kervio E, Morlock G, Heinz DW, Jahn D, Retey J and Schubert WD (2005) Structural and functional comparison of HemN to other radical SAM enzymes. Biol Chem 386: 971–980

Leon S, Touraine B, Briat J-F and Lobreaux S (2002) The *AtNFS2* gene from *Arabidopsis thaliana* encodes a NifS-like plastidial cysteine desulphurase. Biochem J 366:557–564

Leon S, Touraine B, Ribot C, Briat JF and Loberaux S (2003) Iron–sulphur cluster assembly in plants: distinct NFU proteins in mitochondria and plastids from *Arabidopsis thaliana*. Biochem J 371: 823–830

Leon S, Touraine B, Briat JF and Lobreaux S. (2005) Mitochondrial localization of *Arabidopsis thaliana* Isu Fe–S scaffold proteins. FEBS Lett. 579:1930–1934.

Leonardi R and Roach PL (2004) Thiamine biosynthesis in *Escherichia coli: in vitro* reconstitution of the thiazole synthase activity. J Biol Chem 279: 17054–17062

Leonardi R, Fairhurst SA, Kriek M, Lowe DJ and Roach PL (2003) Thiamine biosynthesis in *Escherichia coli*: isolation and initial characterisation of the ThiGH complex. FEBS Lett 539: 95–99

Lezhneva L, Amann K and Meurer J (2004) The universally conserved HCF101 protein is involved in assembly of [4Fe–4S]-cluster-containing complexes in *Arabidopsis thaliana* chloroplasts. Plant J 37: 174–85.

Li H-M, Theg SM, Bauerle CM and Keegstra K (1990) Metal-ion-center assembly of ferredoxin and plastocyanin in isolated chloroplasts. Proc Natl Acad Sci USA 87: 6748–6752.

Lill R and Kispal G (2000) Maturation of cellular Fe-S proteins: an essential function of mitochondria. Trends Biochem Sci 25: 352–356

Lill R and Muhlenhoff U (2005) Iron–sulfur protein biogenesis in eukaryotes. Trends Biochem Sci 30: 133–141.

Loiseau L, Ollagnier-de-Choudens S, Nachin L, Fontecave M and Barras F (2003) Biogenesis of Fe–S cluster by the bacterial Suf system: SufS and SufE form a new type of cysteine desulfurase. J Biol Chem 278: 38352–38359

Loiseau L, Ollagnier-de Choudens S, Lascoux D, Forest E, Fontecave M and Barras F (2005) Analysis of the heteromeric CsdA-CsdE cysteine desulfurase, assisting [Fe–S] cluster biogenesis in *Escherichia coli*. J Biol Chem 280: 26760–26769

Mendel RR (2005) Molybdenum: biological activity and metabolism. Dalton Trans 21: 3404–3409.

Mendel RR and Hansch R. (2002) Molybdoenzymes and molybdenum cofactor in plants. J Exp Bot 53:1689–1698

Mihara H and Esaki N (2002) Bacterial cysteine desulfurases: their function and mechanisms. Appl Microbiol Biotechnol 60: 12–23

Mihara H, Kurihara T, Yoshimura T, Soda K and Esaki N (1997) Cysteine sulfinate desulfinase, a NIFS-like protein of *Escherichia coli* with selenocysteine lyase and cysteine desulfurase activities. Gene cloning, purification, and characterization of a novel pyridoxal enzyme. J Biol Chem 272: 22417–22424

Mihara H, Kurihara T, Yoshimura T and Esaki N (2000) Kinetic and mutational studies of three NifS homologs from *Escherichia coli*: mechanistic difference between L-cysteine desulfurase and L-selenocysteine lyase reactions. J Biochem 127: 559–567

Moller GM, Kunkel T and Chua N-H (2001) A plastidic ABC protein involved in intercompartmental communication of light signaling. Genes Dev 15: 90–103.

Muhlenhoff U, Gerber J, Richhardt N and Lill R (2003) Components involved in assembly and dislocation of iron–sulfur clusters on the scaffold protein Isu1p. EMBO J 22: 4815–4825

Nachin L, Loiseau L, Expert D and Barras F (2003) SufC: an unorthodox cytoplasmic ABC/ATPase required for [Fe–S] biogenesis under oxidative stress. EMBO J 22: 427–437

Nakai Y, Umeda N, Suzuki T, Nakai M, Hayashi H, Watanabe K and Kagamiyama H. (2004) Yeast Nfs1p is involved in thio-modification of both mitochondrial and cytoplasmic tRNAs. J Biol Chem 279:12363–12368.

Nishio K and Nakai M (2000) Transfer of iron–sulfur cluster from NifU to apoferredoxin. J Biol Chem 275: 22615–22618

Oda Y, Samanta SK, Rey FE, Wu L, Liu X, Yan T, Zhou J and Harwood CS (2005) Functional genomic analysis of three nitrogenase isozymes in the photosynthetic bacterium *Rhodopseudomonas palustris*. J Bacteriol 187: 7784–7794

Ollagnier-de Choudens S, Nachin L, Sanakis Y, Loiseau L, Barras F and Fontecave M (2003) SufA from *Erwinia chrysanthemi*. Characterization of a scaffold protein required for iron–sulfur cluster assembly. J Biol Chem 278:17993–18001

Ollagnier-de-Choudens S, Sanakis Y and Fontecave M (2004) SufA/IscA: reactivity studies of a class of scaffold proteins involved in [Fe–S] cluster assembly. J Biol Inorg Chem 9: 828–838

Ollagnier de Choudens S, Loiseau L, Sanakis Y, Barras F and Fontecave M (2005) Quinolinate synthetase, an iron–sulfur enzyme in NAD biosynthesis. FEBS Lett 579: 3737–3743

Outten FW, Wood MJ, Munoz FM and Storz G (2003) The SufE protein and the SufBCD complex enhance SufS cysteine desulfurase activity as part of a sulfur transfer pathway for Fe–S cluster assembly in *Escherichia coli*. J Biol Chem 278: 45713–45719

Outten FW, Djaman O and Storz G (2004) A *suf* operon requirement for Fe–S cluster assembly during iron starvation in *Escherichia coli*. Mol Microbiol 52: 861–872

Park J-H, Dorrestein PC, Zhai H, Kinsland C, McLafferty FW and Begley TP (2003) Biosynthesis of the thiazole moiety of thiamin pyrophosphate (vitamin B1). Biochemistry 42: 12430–12438

Petit J-M, Briat J-F and Lobreaux S (2001) Structure and differential expression of the four members of the *Arabidopsis thaliana* ferritin gene family. Biochem J 359: 575–582

Picciocchi A, Douce R and Alban C (2003) The plant biotin synthase reaction. Identification and characterization of essential mitochondrial accessory protein components. J Biol Chem 278: 24966–24975

Pilon M, America T, van 't Hof R, de Kruijff B and Weisbeek P (1995) Protein translocation into chloroplasts. In: Advances in molecular and cell biology (Rothman SS Ed.) Membrane protein transport. JAI Press, Greenwich. Vol 4, pp 229–255

Pilon-Smits EA, Garifullina GF, Abdel-Ghany S, Kato S, Mihara H, Hale KL, Burkhead JL, Esaki N, Kurihara T and Pilon M (2002) Characterization of a NifS-like chloroplast protein from *Arabidopsis*. Implications for its role in sulfur and selenium metabolism. Plant Physiol 130: 1309–1318

Pondarre C, Antiochos BB, Campagna DR, Clarke SSL, Greer EL, Deck KM, McDonald A, Han AP, Medlock A, Kutok JL, Anderson SA, Eisenstein RS and Fleming MD (2006) The mitochondrial ATP-binding cassette transporter Abcb7 is essential in mice and participates in cytosolic iron–sulfur cluster biogenesis. Hum Mol Genet 15: 953–964

Raven JA, Evans MC and Korb RE (1999) The role of trace metals in photosynthetic electron transport in O_2-evolving organisms. Photosynthesis Res 60: 111–149

Ribeiro DT, Farias LP, de Almeida JD, Kashiwabara PM, Ribeiro AF, Silva-Filho MC, Menck CF and Van Sluys MA (2005) Functional characterization of the *thi1* promoter region from *Arabidopsis thaliana*. J Exp Bot 56: 1797–1804

Schneider G and Lindqvist Y (2001) Structural enzymology of biotin biosynthesis. FEBS Lett 495: 7–11

Seidler A, Jaschkowitz K and Wollenberg M (2001) Incorporation of iron–sulphur clusters in membrane-bound proteins. Biochem Soc Trans 29: 418–421

Shikanai T, Müller-Moulé P, Munekage Y, Niyogi K and Pilon M (2003) *PAA1*, a P-type ATPase of *Arabidopsis*, Functions in Copper Transport in Chloroplasts. Plant Cell 15: 1333–1346.

Shingles R, North M and McCarty RE (2002) Ferrous Ion Transport across Chloroplast Inner Envelope Membranes. Plant Physiol 128: 1022–1030

Sofia HJ, Chen G, Hetzler BG, Reyes-Spindola JF and Miller NE (2001) Radical SAM, a novel protein superfamily linking unresolved steps in familiar biosynthetic pathways with radical mechanisms: functional characterization using new analysis and information visualization methods. Nucleic Acids Res 29: 1097—1106

Stockel J and Oelmuller R (2004) A novel protein for photosystem I biogenesis. J. Biol. Chem. 279: 10243–10251

Takahashi Y and Tokumoto U (2002) A third bacterial system for the assembly of iron–sulfur clusters with homologs in archaea and plastids. J Biol Chem 277: 28380–28383

Takahashi Y, Mitsui A, Hase T and Matsubara H (1986) Formation of the iron sulfur cluster of ferredoxin in isolated chloroplasts. Proc Natl Acad Sci USA 83: 2434–2437

Takahashi Y, Mitsui A and Matsubara H (1990) Formation of the Fe–S cluster of ferredoxin in lysed spinach chloroplasts. Plant Physiol 95: 97–103

Tong WH and Rouault TA (2006) Functions of mitochondrial ISCU and cytosolic ISCU in mammalian iron–sulfur cluster biogenesis and iron homeostasis. Cell Metab 3: 199–210

Touraine B, Boutin J, Marion-Poll A, Briat J, Peltier G and Lobreaux S (2004) Nfu2: a scaffold protein required for [4Fe–4S] and ferredoxin iron–sulfur cluster assembly in *Arabidopsis* chloroplasts. Plant J 40:101–111

Van Hoewyk D, Abdel-Ghany SE, Cohu C, Herbert S, Kugrens P, Pilon M and Pilon-Smits EAH (2007) Chloroplast iron-sulfur cluster protein maturation requires the essential cysteine desulfurylase CpNifS. Proc Natl Acad Sci USA 104:5686–5691

Walsby CJ, Ortillo D, Yang J, Nnyepi MR, Broderick WE, Hoffman BM and Broderick JB (2005) Spectroscopic approaches to elucidating novel iron–sulfur chemistry in the "radical-SAM" protein superfamily. Inorg Chem 44: 727–741

Wang T, Shen G, Balasubramanian R, McIntosh L, Bryant DA and Golbeck JH (2004) The *sufR* gene (*sll0088* in *Synechocystis sp.* strain PCC 6803) functions as a repressor of the *sufBCDS* operon in iron–sulfur cluster biogenesis in *cyanobacteria*. J Bacteriol 186: 956–967

Xiong L, Ishitani M, Lee H and Zhu JK (2001) The *Arabidopsis LOS5/ABA3* locus encodes a molybdenum cofactor sulfurase and modulates cold stress- and osmotic stress-responsive gene expression. Plant Cell. 13:2063–83

Xu XM and Moller SG. (2004) AtNAP7 is a plastidic SufC-like ATP-binding

Xu XM and Moller SG (2006) AtSufE is an essential activator of plastidic and mitochondrial desulfurases in *Arabidopsis*. EMBO J 25: 900–909

Xu XM, Adams S, Chua NH and Moller SG (2005) AtNAP1 represents an atypical SufB protein in *Arabidopsis* plastids. J Biol Chem 280: 6648–6654

Yabe T and Nakai M (2006) *Arabidopsis* AtIscA-I is affected by deficiency of Fe-S cluster biosynthetic scaffold AtCnfU-V. Biochem Biophys Res Commun 340: 1047–1052

Yabe T, Morimoto K, Kikuchi S, Nishio K, Terashima I and Nakai M (2004) The *Arabidopsis* chloroplastic NifU-like protein CnfU, which can act as an iron–sulfur cluster scaffold protein, is required for biogenesis of ferredoxin and photosystem I. Plant Cell 16: 993–1007

Ye H, Garifullina GF, Abdel-Ghany S, Zhang L, Pilon-Smits EAH and Pilon M (2005) AtCpNifS is required for iron–sulfur cluster formation in ferredoxin in vitro. Planta 220:602–608

Ye H, Abdel-Ghany SE, Anderson TD, Pilon-Smits EA and Pilon M (2006a) CpSufE activates the cysteine desulfurase CpNifS for chloroplastic Fe–S cluster formation. J Biol Chem 281: 8958–8969

Ye H, Pilon M and Pilon-Smits EAH (2006b) Iron Sulfur Cluster Biogenesis in Chloroplasts. New Phytologist 171: 285–292.

Zeller T, Moskvin OV, Li K, Klug G and Gomelsky M (2005) Transcriptome and physiological responses to hydrogen peroxide of the facultatively phototrophic bacterium *Rhodobacter sphaeroides*. J Bacteriol 187: 7232–7242

Zheng L, White RH, Cash VL, Jack RF and Dean DR (1993) Cysteine desulfurase activity indicates a role for NIFS in metallocluster biosynthesis. Proc Natl Acad Sci USA 90: 2754–2758

Zheng L, White RH, Cash VL and Dean DR (1994) Mechanism for the desulfurization of L-cysteine catalyzed by the *nifS* gene product. Biochemistry 33: 4714–4720

Zheng L, Cash VL, Flint DH and Dean DR (1998) Assembly of Iron–Sulfur clusters. Identification of an iscSUA-hscBA-fdx gene cluster from *Azotobacter vinelandii*. J Biol Chem 273: 13264–13272

Zheng M, Wang X, Templeton LJ, Smulski DR, LaRossa RA and Storz G (2001) DNA microarray-mediated transcriptional profiling of the *Escherichia coli* response to hydrogen peroxide. J Bacteriol 183: 4562–4570

Chapter 8

Molecular Biology and Functional Genomics for Identification of Regulatory Networks of Plant Sulfate Uptake and Assimilatory Metabolism

Hideki Takahashi*
RIKEN Plant Science Center, 1-7-22 Suehiro-cho, Tsurumi-ku, Yokohama 230-0045, Japan

Kazuki Saito
*RIKEN Plant Science Center, 1-7-22 Suehiro-cho, Tsurumi-ku, Yokohama 230-0045, Japan
Graduate School of Pharmaceutical Sciences, Chiba University, 1-33 Yayoi-cho, Inage-ku, Chiba 263-8522, Japan*

Summary

Uptake of sulfate from the environment is critical for sulfur metabolism as it controls the quantity of sulfur to be distributed through the metabolic pathways. Similar to the uptake systems for other nutrient ions, transport of sulfate can be resolved into high- and low-affinity phases. Sulfur limitation stimulates the high-affinity sulfate transport system that essentially facilitates the uptake of sulfate in roots. Apparently, the induction of sulfate uptake by sulfur limitation is driven by demands of sulfur. Recent molecular biological studies have unveiled some important aspects behind the regulatory cascades of plant sulfur response and sulfate assimilation. In this review, we describe the regulatory steps controlling

* Corresponding author, Phone: +81 45 503 9577, Fax: +81 45 503 9650, E-mail: hideki@riken.jp

Rüdiger Hell et al. (eds.), Sulfur Metabolism in Phototrophic Organisms, 149–159.
© 2008 Springer.

sulfate uptake and assimilatory metabolism in *Arabidopsis thaliana*, and further discuss on the networks of metabolic and hormonal regulation based on recent findings that deal with transcriptome and metabolome data analyses.

I. Sulfate Transport Systems in Plants

A. Sulfate Transporter Gene Family

As described in recent reviews, *Arabidopsis* genome encodes 12 distinct sulfate transporter genes that are classified into 4 functional groups (Buchner et al., 2004b; Takahashi et al., 2006). All 12 isoforms of sulfate transporters from *Arabidopsis* are structurally similar to the one first identified from *Stylosanthes hamata* by functional complementation of a yeast sulfate transporter mutant (Smith et al., 1995). Two additional homologues representing the group 5 members (Buchner et al., 2004b) appear to have low sequence similarity to sulfate transporters, though the exact functions have been remained unverified. Likewise the case of other nutrient transporters, multiplicities of isoforms may reflect complexities of the whole-plant sulfate transport systems in vascular plants. In fact, the functionality and localization of sulfate transporters are diversified among the isoforms, which lead us to set out for physiological characterization and dissection of the functions of individual transport components using T-DNA insertion mutants of *Arabidopsis* (Takahashi et al., 2006). The following two sections summarize the roles of individual transporters that correspond to the uptake and internal transport systems that have been verified from the analysis of T-DNA insertion mutants. In addition to this review article, general features of plant sulfate transport systems have been described by Hawkesford in Chapter 2.

B. Uptake of Sulfate

The initial uptake of sulfate occurs at the root surface. Earlier physiological studies suggested the high-affinity kinetics (phase I) of sulfate uptake system predominates under sulfur limited condition (Clarkson et al., 1983; Deane-Drum-

Abbreviations: OAS – O-acetyl-L-serine; SULTR – sulfate transporter; SURE – sulfur responsive element

mond, 1987). The high-affinity sulfate transporters that correspond to this sulfur-limitation inducible transport system are represented by the Group 1 members (Takahashi et al., 2006). The first isolated cDNAs, *SHST1* and *SHST2*, from *Stylosanthes hamata* were able to complement the yeast sulfate transporter mutant, and exhibited saturable kinetics of sulfate uptake with micromolar K_m values (Smith et al., 1995). In addition, the *SHST1* and *SHST2* mRNAs accumulated in root tissues when plants were starved for sulfate. The *Arabidopsis SULTR1;1* and *SULTR1;2* showed similar characteristics (Takahashi, et al., 2000; Shibagaki et al., 2002; Yoshimoto et al., 2002). They were localized in the root hairs, epidermis and cortex of roots, and their transcripts were significantly accumulated during sulfur limitation. The uptake of sulfate and the overall sulfur status was significantly affected by deletion of *SULTR1;2*, suggesting this transporter plays a major role in facilitating the uptake of sulfate in *Arabidopsis* roots (Maruyama-Nakashita et al., 2003). In addition, induction of *SULTR1;1* mRNA in the *sultr1;2* mutant indicates compensatory and demand-driven regulation of this isoform. Presence of the inducible isoforms of high-affinity sulfate transporters appears to be common in various plant species (Howarth et al., 2003; Buchner et al., 2004a, 2004b), which may allow flexible adaptation to fluctuating sulfur conditions in the environment.

C. Internal Transport of Sulfate

Pathways that mediate internal translocation of sulfate are complicated and are not completely resolved. However, studies using the knockouts of *Arabidopsis* sulfate transporters suggested several potential components that facilitate distribution of sulfate through the vasculature. In *Arabidopsis*, a low-affinity sulfate transporter, SULTR2;1, is suggested to play a pivotal role in controlling transport of sulfate at xylem parenchyma cells (Takahashi et al., 1997, 2000;

Awazuhara et al., 2005). *SULTR2;1* was strictly regulated by sulfur status, exhibiting drastic induction of its mRNA in the roots of sulfur-starved plants. In addition to regulation of its transcript levels, the activity of SULTR2;1 was modulated by the presence of a co-localizing component, SULTR3;5, in the pericycle and xylem parenchyma cells of roots (Kataoka et al., 2004a). For sufficient transport of sulfate from root tissues to shoots, release of sulfate pools from vacuoles was an additional rate-limiting step under sulfur-limited conditions. The tonoplast-localizing sulfate transporters that serve for this essential step were encoded by *SULTR4;1* and *SULTR4;2* in *Arabidopsis* (Kataoka et al., 2004b). Both were inducible by sulfur limitation, and T-DNA insertion mutants showed accumulation of sulfate in the vacuoles, leading to a significant decrease in distribution of the incorporated sulfate to shoots. Besides root-to-shoot transport, source-to-sink transport is suggested to be important for allocation of sulfur storage from old to young tissues (Bourgis et al., 1999; Herschbach et al., 2000). In *Arabidopsis*, the function of a phloem-localizing component of sulfate transport system has been evidenced from the analysis of knockout mutant. SULTR1;3, the third member of high-affinity sulfate transporters, was localized in the companion cells of transport phloem, and was essentially required for the movement of sulfate from cotyledons to shoot meristems and roots in *Arabidopsis* (Yoshimoto et al., 2003).

II.   Regulation by Sulfur

A.   Demand-Driven Regulation

The external supply of sulfate is primarily important for regulation of sulfate uptake and assimilation (Leustek et al., 2000; Saito, 2004; Takahashi et sh1., 2006). Particularly, high-affinity sulfate transporters that facilitate the initial uptake of sulfate in roots are strictly regulated by sulfate availabilities (Takahashi et al., 2000; Yoshimoto et al., 2002; Shibagaki et al., 2002; Maruyama-Nakashita et al., 2004a, 2004c). In addition to sulfate, metabolites of sulfur assimilatory pathways are known to have major impacts on gene expression of sulfate transporters. When sulfate is adequately or excessively supplied from the environment, plants may synthesize cysteine, methionine and glutathione, causing feedback regulation of the uptake of sulfate. On feeding experiments, excessive application of cysteine and glutathione led to a drastic decrease of sulfate uptake activity in roots (Smith et al., 1997; Vidmar et al., 2000). In *Arabidopsis*, decrease of sulfate uptake is attributable to repression of *SULTR1;1* and *SULTR1;2*, both being down-regulated significantly in the presence of thiols in the medium (Maruyama-Nakashita et al., 2004c, 2005). In addition to the high-affinity sulfate transporters that facilitate the initial sulfate uptake, transcripts of *SULTR2;1* low-affinity sulfate transporter was regulated under a similar scheme (Vidmar et al., 2000). A split-root experiment suggests repressive signals can be translocated from distant organs through phloem (Lappartient et al., 1999). These results suggest that sulfate uptake and vascular transport are both controlled in a demand-driven manner not only by a local signal but also by systemic requirement of sulfur (Lappartient et al., 1999; Herschbach et al., 2000).

B.   The Action of a Cysteine Precursor

O-acetyl-L-serine (OAS), the precursor of cysteine synthesis, positively affects the expression of high-affinity sulfate transporters (Smith et al., 1997; Maruyama-Nakashita et al., 2004c). In barley, induction of *HVST1* sulfate transporter by OAS was accompanied with increase in sulfate uptake activities (Smith et al., 1997). In addition, transcripts for adenosine 5'-phophosulfate reductase, the pivotal enzyme that controls the flux of sulfur through the sulfate assimilation pathway (Vauclare et al., 2002), were induced by the addition of OAS to the medium after nitrogen starvation (Koprivova et al., 2000). From these findings, one may postulate that OAS could provide signals of sulfur limitation response. Identification of an OAS accumulating *Arabidopsis* mutant, *osh1*, which stimulates the activity of sulfur limitation-responsive promoter of β-conglycinin β-subunit, supports this regulatory model (Ohkama-Ohtsu et al., 2004). Furthermore, microarray studies and other expression analysis indicate that feeding of OAS gives a pattern of gene expression profile that mimics sulfur limitation (Hirai et al., 2003, 2004). However, a time-course measurements of tissue OAS contents indicates that induction of sulfate transporter precedes accumulation of OAS under sulfur limitation (Hopkins et al., 2005). This contrasts with the hypothesis placing OAS as a signal mediator that triggers up-regulation of sulfur assimilatory pathways, questioning when and how the accumulation of OAS serves for regulation of gene expression in sulfur-starved plants. Excessive accumulation of OAS simply may imply disturbance of cysteine synthesis by limited supply of sulfide under prolonged sulfur starvation (Hopkins et al., 2005). Biochemical studies indicate that OAS has an ability to control the activity and structure of cysteine synthase complex (Wirtz et al., 2004). The enzyme complex is dissociated and associated reversibly by OAS and sulfide, respectively. The two different states provide production of cysteine by free OAS(thiol)lyase or synthesis of OAS by the cysteine synthase complex that is reconstituted under an excess of sulfide over OAS. Apparently, a switching mechanism exists for the control of cysteine synthesis. Verification of these conceptual models hypothesizing the OAS actions awaits further investigation.

III. The *cis*-Acting Element of Sulfur Response

Cumulative information from the molecular biological works on plant sulfur response provided us with a notion that numbers of sulfur-responsive genes are coordinately regulated for metabolisms and probably for stress mitigation. A basic question arose here whether all these genes are regulated by sulfur under the same mechanisms? Transcriptional regulation of yeast *MET* genes is a well characterized example of sulfur response (Thomas and Surdin-Kerjan, 1997). *MET* genes that encode the enzymes for assimilatory sulfur metabolism have common *cis*-acting elements within their 5'-regions that are capable of binding the components of transcription factors regulated by availability of *S*-adenosylmethionine. Unlike this highly organized regulatory system in yeast, higher plants must have developed more complicated mechanisms, as has been anticipated from several modes of metabolic and plant hormone-mediated regulation discussed further in this article.

For the plant sulfur response, a sulfur-responsive *cis*-acting element, SURE, is reported from *Arabidopsis*. SURE is a 7-bp sulfur-responsive element identified in the 5'-region of *Arabidopsis* *SULTR1;1* sulfate transporter gene (Maruyama-Nakashita et al., 2005). Measurements of reporter activities of a series of promoter deletion constructs in transgenic *Arabidopsis* plants indicated that the sulfur-responsive region can be delimited to a specific GGAGACA sequence between the -2773 and -2767 region of *SULTR1;1*. Under this element, the reporter activity was induced by sulfur limitation or by OAS treatment, but was repressed by cysteine and glutathione. The SURE sequence contained the core sequence of the auxin response factor (ARF) binding site (Ulmasov et al., 1999; Hagen and Guilfoyle, 2002). *SULTR1;1* is slightly induced by auxin treatment, but SURE itself exhibited no response to auxin. These observations suggest that transcriptional activation of *SULTR1;1* in response to sulfur deficiency could involve a transcription factor which may structurally resemble to ARF in DNA-binding mechanisms; however the specific SURE-binding candidate has not yet been identified. The data of microarray analysis of a time course series of sulfate deprivation response indicated that SUREs are present in numbers of sulfur-responsive genes that are co-activated with *SULTR1;1* (Maruyama-Nakashita et al., 2005). Similar sequences were found in the sulfur-responsive 5'-regions of NIT3 nitrilase (Kutz et al., 2002) and β-conglycinin β-subunit (Awazuhara et al., 2002), suggesting generality of SURE-mediated regulation in plant sulfur response. However, interestingly, *SULTR1;2* showed no consensus of SURE sequences within its 5'-region. Although it is induced by sulfur limitation and its significance in facilitating sulfate uptake is remarkable (Shibagaki et al., 2002; Yoshimoto et al., 2002; Maruyama-Nakashita et al., 2003), this sulfate transporter is suggested to be regulated in a different manner independent of the SURE-mediated regulation.

IV. Regulation by Nitrogen and Carbon

Sulfate is metabolized to a sulfur-containing amino acid, cysteine, using sulfide and OAS. Reduction of sulfate primarily occurs in the presence of ATP and reducing cofactors that can be provided by photosynthesis or from the pentose phosphate cycle (Leustek et al., 2000; Saito, 2004). OAS, the other substrate for cysteine synthesis, originates from serine. In chloroplasts, OAS is synthesized from 3-phosphoglycerate through the enzymes catalyzing dephosphorylation and trans-amination reactions (Ho and Saito, 2001). This indicates supply of carbon skeletons of cysteine is tightly linked with carbon and nitrogen metabolisms. Reminiscent of these metabolic linkages, transcripts for sulfate transporters and adenosine 5′-phophosulfate reductases, the key steps facilitating the sulfate assimilatory pathway, were both regulated by supply of nitrogen and carbon (Kopriva et al., 1999; Vidmar et al., 1999; Koprivova et al., 2000; Hesse et al., 2003; Wang et al., 2003; Maruyama-Nakashita et al., 2004b). In the case of high-affinity sulfate transporters, the sulfur-deficiency response of *SULTR1;1* and *SULTR1;2* was strongly attenuated in *Arabidopsis* roots by depletion of nitrogen and carbon sources from the medium (Maruyama-Nakashita et al., 2004b). In barley roots, sulfate uptake activities were transiently induced by replenishment of nitrate or ammonium to nitrogen-starved plants, which correlated with the increase of *HVST1* transcripts (Vidmar et al., 1999). In addition to sulfate transporters, various nutrient transporters facilitating the uptake of essential elements in *Arabidopsis* roots were positively and coordinately regulated during the day time and by supply of carbon source (Lejay et al., 2003). Metabolites of glycolytic pathway are suggested to be involved in this general regulation, although the molecular mechanisms are unknown.

V. Plant Hormone Signals

A. Cytokinin in Plant Sulfur Response

In addition to metabolic regulation, recent studies suggested that sulfur assimilation may involve plant hormone signals for its own regulation. The most evident case indicating a linkage between the component of signal perception and downstream gene expression is the cytokinin-dependent signaling cascade that participates in negative regulation of sulfate uptake in *Arabidopsis* roots (Maruyama-Nakashita et al., 2004c). This regulatory

pathway was initially found by screening of plant hormones that affect the expression of *SULTR1;2* promoter-GFP. Among the plant hormones tested, cytokinin treatment specifically repressed the accumulation of GFP under sulfur-limited conditions. Commensurate with repression of the signals of GFP reporter, *SULTR1;1* and *SULTR1;2* mRNAs were coordinately down-regulated by the addition of cytokinin. This regulatory pathway involved a two-component phospho-relay of a cytokinin receptor histidine kinase, CRE1/WOL/ AHK4 (Inoue et al., 2001; Kakimoto, 2003). Accordingly, the *Arabidopsis cre1–1* mutant was unable to regulate the high-affinity sulfate transporters in response to cytokinin (Maruyama-Nakashita et al., 2004c). Currently, we consider that cytokinin and sulfur may work independently for the control of sulfate uptake. When sulfate is limiting, sulfur-specific signals activate the expression of high-affinity sulfate transporters for the acquisition of sulfur source. By contrast, cytokinin provides a negative signal attenuating the influx of sulfate, similar to the case in phosphate uptake that also involves CRE1 cytokinin receptor for its regulation (Martin et al., 2000; Franco-Zorrilla et al., 2002). It is reported that a reporter gene fusion construct of a sulfur limitation-responsive promoter of β-conglycinin β-subunit responds positively to cytokinin in parallel with the induction of adenosine 5′-phophosulfate reductase (Ohkama et al., 2002), which is opposite to the response of sulfate transporters (Maruyama-Nakashita et al., 2004c). Under cytokinin treatment, repression of *SULTR1;1* and *SULTR1;2* sulfate transporters may lead to a substantial decline of internal sulfur status, consequently stimulating the expression of sulfur-responsive genes. Alternatively, the regulatory mechanisms of cytokinin response may differ between these two groups that apparently exhibit similar responses to sulfur limitation.

B. Interaction of Auxin and Glucosinolate Biosynthesis

Plants alter their root architectures during nutrient deficient conditions. The mechanisms and the sites of root growth differ among the nutrients absorbed by plant roots; sulfur deficiency generally stimulates the growth of lateral roots both in length and numbers (Hell and Hillebrand, 2001).

In *Arabidopsis*, degradation of indole glucosinolates is suggested to be involved in this growth regulation (Kutz et al., 2002). Under sulfur deficiency, thioglucosidase releases the aglycon from indole glucosinolates, and indole acetonitrile formed spontaneously after this reaction will be catalyzed by nitrilase, generating indole acetic acid (IAA) as a product that may stimulate the root growth. The transcript of *NIT3* nitrilase was inducible by sulfur deprivation in *Arabidopsis* roots, providing a mechanism for root growth under sulfur deficiency (Kutz et al., 2002). In addition to these findings, microarray studies indicated that some auxin-responsive genes were positively regulated in sulfate-starved plants (Hirai et al., 2003; Maruyama-Nakashita et al., 2003; Nikiforova et al., 2003). However, direct evidence for increase of IAA content has not been shown for sulfate-starved plants, suggesting IAA derived from degradation of glucosinolates might locally affect the growth of lateral roots.

C. Jasmonic Acid Signaling

Jasmonic acid (JA) is a plant hormone synthesized by oxidation of linolenic acid constituting the membrane lipids. Several lines of evidence suggest significance of JA-signaling in mitigation of oxidative stresses and synthesis of defense chemicals, both being related with sulfur metabolism. Shortage of sulfur supply leads to a significant decrease of cellular glutathione contents, and therefore may cause oxidative stresses. Corollary of this, the expression of genes for JA synthesis was induced by sulfur limitation or by knockout of *SULTR1;2* sulfate transporter (Hirai et al., 2003; Maruyama-Nakashita et al., 2003; Nikiforova et al., 2003). Under sulfur deficiency, synthesis of glucosinolates can be down-regulated for recycling of sulfur, and induction of JA synthesis may occur in parallel. When provided with ample supply of sulfur, JA treatment caused induction of metabolic genes for glutathione synthesis (Xiang and Oliver, 1998). Recent microarray studies additionally indicate that both reduction of sulfate to thiols and synthesis of glucosinolates are induced by JA treatment (Jost et al., 2005; Sasaki-Sekimoto et al., 2005). These findings suggest that JA may act positively both for active synthesis of antioxidants and glucosinolate production, which appears to be contradictory to the case in sulfur-starved plants. Under sulfur deficiency, nutritional demand of sulfur recycling may presumably override the JA-derived signals that are required for the synthesis of glucosinolates acting as defense chemicals against pathogenic attack. In fact, a JA-deficient mutant lacking allene oxide synthase gene showed normal sulfur-limitation responses, as represented by induction of sulfate transporters and adenosine 5′-phophosulfate reductase, and repression of genes for glucosinolate synthesis under sulfur-deficient conditions (Yano et al., 2005). This supports the hypothesis suggesting divergence of sulfur limitation signals from the JA-mediated regulatory cascades. Findings of regulatory complexes of JA with sulfur metabolisms and antioxidant recycling systems suggest significance of sulfur fertilization on stressed environments that plants may encounter in nature.

VI. Prospects of Transcriptome and Metabolome Analyses for Novel Gene Findings

A. Co-Regulated Genes

In the functional genomics era, the use of microarrays has become a powerful tool to characterize the changes of transcriptomes at given conditions or genetic varieties. Most recently, the use of Affymetrix GeneChip arrays provided us a holistic view of transcript profiles that covers more than 70% of the coding genes in the *Arabidopsis* genome. Studies using this pre-manufactured array or other microarrays with equivalent qualities provided us unprecedented images of plant sulfur response that occurs on transcript regulation (Hirai et al., 2003, 2004, 2005; Maruyama-Nakashita et al., 2003, 2005; Nikiforova et al., 2003). The whole dataset of transcriptome appears to be a promiscuous mass, but with an appropriate experimental design and proper handling of data clusters, typical features of sulfur response can arise extractable.

In general, sulfur-deficiency responsive genes are co-regulated with high-affinity sulfate transporters in *Arabidopsis* roots (Maruyama-Nakashita et al., 2005, 2006). Appearance of their transcripts on a time course of sulfate deprivation allowed us to identify the presence of *cis*-acting element, SURE, within their promoter regions, although some exceptions such as *SULTR1;2* and *SULTR2;1* were present in the same category. More recently, genetic analysis of sulfur limitation response-less *Arabidopsis* mutants identified a key transcription factor, SLIM1, which is necessary for the coordinate expression of genes for sulfur assimilation and secondary sulfur metabolisms (Maruyama-Nakashita et al., 2006). Transcriptome analysis of *slim1* mutant revealed that many of the sulfur metabolic pathways are regulated by SLIM1, suggesting its hub-like function within the regulatory cascade (Maruyama-Nakashita et al., 2006) (Fig. 1).

In addition to direct or specific effects of sulfate or related sulfur metabolites on transcript regulation of sulfate transport and metabolism, the array data indicated that supply of nitrogen, carbon and plant hormones may additionally influence sulfur metabolism (Wang et al., 2003; Hirai et al., 2004; Jost et al., 2005; Sasaki-Sekimoto et al., 2005). The effects of nitrogen and carbon on the array of sulfur-responsive genes are suggested to be the global response (Hirai et al., 2004). Under such conditions, sulfur-responsive genes may fluctuate with the entire metabolic systems, as represented by the changes of energy metabolism and/or photosynthesis, rather than with a specific cue of sulfur demand that directly controls the uptake of sulfate and assimilatory metabolism.

B. Approaches from Metabolomics

As for the metabolome analysis, comprehensiveness is rather obscure as compared with the transcriptome data, and the numbers of annotated metabolites are still limited (Hirai et al., 2004, 2005; Nikiforova et al., 2005). However, challenges to the untargeted metabolite analysis have made substantial progresses in determining the functions of metabolic enzymes and genetic diversities of glucosinolate synthesis in *Arabidopsis*. Data processing thorough a batch-learning self-organized map analysis, genes and metabolites of glucosinolate synthesis were co-clustered within the same group or in its vicinities, forming gene-to-gene or metabolite-to-metabolite networks (Hirai et al., 2004, 2005). This method has been shown to be useful for novel gene finding that eventually assisted functional identification of desulfoglucosinolate sulfotransferase (Hirai et al., 2005). Genetic diversities of glucosinolate compositions have been well characterized in *Arabidopsis*, as have been exemplified by polymorphisms of key metabolic enzymes between several different ecotypes (Halkier and Gerschenzon, 2006). More

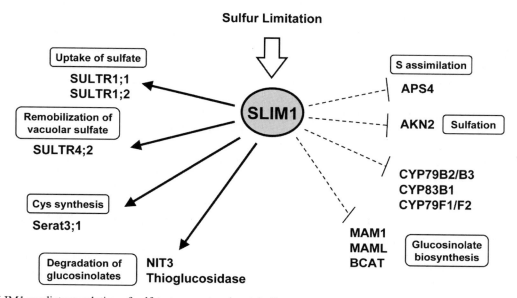

Fig. 1. SLIM1-mediate regulation of sulfate transport and metabolism.
AKN2, 5′adenylylsulfate kinase; APS4, ATP sulfurylase; BCAT, branched-chain amino acid aminotransferase; CYP, cytochrome P450; MAM, methyl(thio)alkylmalate synthase; NIT3, nirilase; Serat, serine acetyltransferase; SULTR, sulfate transporter.

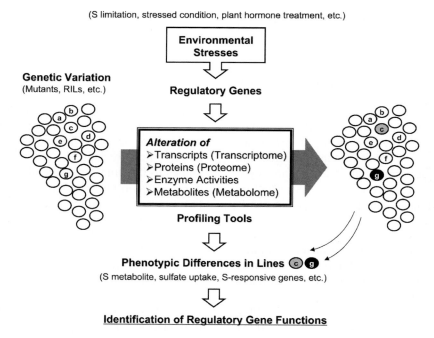

Fig. 2. Omics-based profiling of a large-scale genetic variation for functional identification of regulatory genes in plant metabolisms.

recently, quantitative trait loci representing these known differences of glucosinolate compositions among the *Arabidopsis* ecotypes were resolved by liquid chromatography mass spectrometry-based metabolic profiling of recombinant inbred line populations, suggesting a potential of this untargeted approach for gene finding (Keurentjes et al., 2006).

Approaches from the untargeted metabolomics were successful for identification of the enzymes for synthesis of secondary metabolites through quantitative detection of the accumulating final products. However, these would not be the case with nutrient assimilation and primary metabolisms where the metabolites can easily be distributed or metabolized in reversible pathway networks. Redundancies of the enzymes may also hamper identification of their specific functionalities. In addition, unlike the metabolic enzymes that may have direct impacts on modulating the contents of metabolites, the functions of regulatory genes are suggested to be rather conditional. In other words, positive or negative effects of regulatory genes on fluctuation of tissue metabolite levels might only be detectable during certain stressed conditions where these metabolites are necessarily accumu-

lated. The untargeted or -omics-based approaches are still under way for identification of regulatory gene functions (Fig. 2). The use of genetic variations, comparative analysis of multiple environmental setups, throughputness of metabolite measurements or gene expression analyses, and statistics for interpretation of holistic data outputs, may become the prerequisites for findings of novel regulatory genes modulating the profiles of plant metabolic systems.

References

Awazuhara M, Fujiwara T, Hayashi H, Watanabe-Takahashi A, Takahashi H and Saito K (2005) The function of SULTR2;1 sulfate transporter during seed development in *Arabidopsis thaliana*. Physiol Plant 125: 95–105

Awazuhara M, Kim H, Goto DB, Matsu A, Hayashi H, Chino M, Kim S-G, Naito S and Fujiwara T (2002) A 235-bp region from a nutritionally regulated soybean seed-specific gene promoter can confer its sulfur and nitrogen response to a constitutive promoter in aerial tissues of *Arabidopsis thaliana*. Plant Sci 163: 75–82

Bourgis F, Roje S, Nuccio ML, Fisher DB., Tarczynski MC, Li C, Herschbach C, Rennenberg H, Pimenta MJ, Shen TL, Gage DA and Hanson AD (1999) *S*-methylmethionine plays a major role in phloem sulfur transport and is synthesized by a novel type of methyltransferase. Plant Cell 11: 1485–1498

Buchner P, Stuiver CE, Westerman S, Wirtz M, Hell R, Hawkesford MJ and De Kok LJ (2004a) Regulation of sulfate uptake and expression of sulfate transporter genes in *Brassica oleracea* as affected by atmospheric H_2S and pedospheric sulfate nutrition. Plant Physiol 136: 3396–3408

Buchner P, Takahashi H and Hawkesford MJ (2004b) Plant sulphate transporters: co-ordination of uptake, intracellular and long-distance transport. J Exp Bot 55: 1765–1773

Clarkson DT, Smith FW and Vanden Berg PJ (1983) Regulation of sulfate transport in a tropical legume, *Macroptilium atropurpureum* cv. Siratro. J Exp Bot 34: 1463–1483

Deane-Drummond CE (1987) The regulation of sulfate uptake following growth of *Pisum sativum* L. seedlings in S nutrient limiting conditions. Interaction between nitrate and sulphate transport. Plant Sci 50: 27–35

Franco-Zorrilla JM, Martin AC, Solano R, Rubio V, Leyva A and Paz-Ares J (2002) Mutations at *CRE1* impair cytokinin-induced repression of phosphate starvation responses in *Arabidopsis*. Plant J 32: 353–360

Hagen G and Guilfoyle TJ (2002) Auxin-responsive gene expression: genes, promoters and regulatory factors. Plant Mol Biol 49: 373–385

Halkier BA and Gershenzon J (2006) Biology and biochemistry of glucosinolates. Annu Rev Plant Biol 57: 303–333

Hell R and Hillebrand H (2001) Plant concepts for mineral acquisition and allocation. Curr Opin Biotechnol 12: 161–168

Herschbach C, van Der Zalm E, Schneider A, Jouanin L, De Kok LJ and Rennenberg H (2000) Regulation of sulfur nutrition in wild-type and transgenic poplar over-expressing γ-glutamylcysteine synthetase in the cytosol as affected by atmospheric H_2S. Plant Physiol 124: 461–473

Hesse H, Trachsel N, Suter M, Kopriva S, von Ballmoos P, Rennenberg H and Brunold C (2003) Effect of glucose on assimilatory sulphate reduction in *Arabidopsis thaliana* roots. J Exp Bot 54: 1701–1709

Hirai MY, Fujiwara T, Awazuhara M, Kimura T, Noji M and Saito K (2003) Global expression profiling of sulfur-starved Arabidopsis by DNA macroarray reveals the role of O-acetyl-L-serine as a general regulator of gene expression in response to sulfur nutrition. Plant J 33: 651–663

Hirai MY, Klein M, Fujikawa Y, Yano M, Goodenowe DB,. Yamazaki Y, Kanaya S, Nakamura Y, Kitayama M, Suzuki H, Sakurai N, Shibata D, Tokuhisa J, Reichelt M, Gershenzon J, Papenbrock J and Saito K (2005) Elucidation of gene-to-gene and metabolite-to-gene networks in arabidopsis by integration of metabolomics and transcriptomics. J Biol Chem 280: 25590–25595

Hirai MY, Yano M, Goodenowe DB, Kanaya S, Kimura T, Awazuhara M, Arita M, Fujiwara T and Saito K (2004) Integration of transcriptomics and metabolomics for understanding of global responses to nutritional stresses in *Arabidopsis thaliana*. Proc Natl Acad Sci USA 101: 10205–10210

Hopkins L, Parmar S, Blaszczyk A, Hesse H, Hoefgen R and Hawkesford MJ (2005) O-acetylserine and the regulation of expression of genes encoding components for sulfate uptake and assimilation in potato. Plant Physiol 138: 433–440

Ho C-L and Saito K (2001) Molecular biology of the plastidic phosphorylated serine biosynthetic pathway in *Arabidopsis thaliana*. Amino Acids 20: 243–259

Howarth JR, Fourcroy P, Davidian J-C, Smith FW and Hawkesford MJ (2003) Cloning of two contrasting high-affinity sulphate transporters from tomato induced by low sulphate and infection by the vascular pathogen *Verticillium dahlia*. Planta 218: 58–64

Inoue T, Higuchi M, Hashimoto Y, Seki M, Kobayashi M, Kato T, Tabata S, Shinozaki K and Kakimoto T (2001) Identification of CRE1 as a cytokinin receptor from *Arabidopsis*. Nature 409: 1060–1063

Jost R, Altschmied L, Bloem E, Bogs J, Gershenzon J, Hahnel U, Hansch R, Hartmann T, Kopriva S, Kruse C, Mendel RR, Papenbrock J, Reichelt M, Rennenberg H, Schnug E, Schmidt A, Textor S, Tokuhisa J, Wachter A, Wirtz M, Rausch T and Hell R (2005) Expression profiling of metabolic genes in response to methyl jasmonate reveals regulation of genes of primary and secondary sulfur-related pathways in *Arabidopsis thaliana*. Photosynth Res 86: 491–508

Kakimoto T (2003) Perception and signal transduction of cytokinins. Annu Rev Plant Biol 54: 605–627

Kataoka T, Hayashi N, Yamaya T and Takahashi H (2004a) Root-to-shoot transport of sulfate in Arabidopsis: evidence for the role of SULTR3;5 as a component of low-affinity sulfate transport system in the root vasculature. Plant Physiol 136: 4198–4204

Kataoka T, Watanabe-Takahashi A, Hayashi N, Ohnishi M, Mimura T, Buchner P, Hawkesford MJ, Yamaya T and Takahashi H (2004b) Vacuolar sulfate transporters are essential determinants controlling internal distribution of sulfate in Arabidopsis. Plant Cell 16: 2693–2704

Keurentjes JJ, Fu J, de Vos CH, Lommen A, Hall RD, Bino RJ, van der Plas LH, Jansen RC, Vreugdenhi D and Koornneef M (2006) The genetics of plant metabolism. Nat Genet 38: 842–849

Kopriva S, Muheim R, Koprivova A, Trachsel N, Catalano C, Suter M and Brunold C (1999). Light regulation of assimilatory sulphate reduction in *Arabidopsis thaliana*. Plant J 20: 37–44

Koprivova A, Suter M, Op den Camp R, Brunold C and Kopriva S (2000) Regulation of sulfur assimilation by nitrogen in Arabidopsis. Plant Physiol 122: 737–746

Kutz A, Muller A, Hennig P, Kaiser WM, Piotrowski M, Weiler EW (2002) A role for nitrilase 3 in the regulation of root morphology in sulphur-starving *Arabidopsis thaliana*. Plant J 30: 95–106

Lejay L, Gansel X, Cerezo M, Tillard P, Müller C, Krapp A, von Wirén N, Daniel-Vedele F and Gojon A (2003) Regulation of root ion transporters by photosynthesis: functional importance and relation with hexokinase. Plant Cell 15: 2218–2232

Lappartient AG, Vidmar JJ, Leustek T, Glass AD and Touraine B (1999) Inter-organ signaling in plants: regulation of ATP sulfurylase and sulfate transporter genes expression in roots mediated by phloem-translocated compound. Plant J 18: 89–95

Leustek T, Martin MN, Bick JA and Davies JP (2000) Pathways and regulation of sulfur metabolism revealed through molecular and genetic studies. Annu Rev Plant Physiol Plant Mol Biol 51: 141–165

Martin AC, del Pozo JC, Iglesias J, Rubio V, Solano R, de la Pena A, Leyva A and Paz-Ares J (2000) Influence of cytokinins on the expression of phosphate starvation responsive genes in Arabidopsis. Plant J 24: 559–567

Maruyama-Nakashita A, Inoue E, Watanabe-Takahashi A, Yamaya T and Takahashi H (2003) Transcriptome profiling of sulfur-responsive genes in Arabidopsis reveals global effect on sulfur nutrition on multiple metabolic pathways. Plant Physiol 132: 597–605

Maruyama-Nakashita A, Nakamura Y, Tohge T, Saito K and Takahashi H (2006) Arabidopsis SLIM1 is a central transcriptional regulator of plant sulfur response and metabolism. Plant Cell 18: 3235–3251

Maruyama-Nakashita A, Nakamura Y, Watanabe-Takahashi A, Inoue E, Yamaya T and Takahashi H (2005) Identification of a novel cis-acting element conferring sulfur deficiency response in Arabidopsis roots. Plant J 42: 305–314

Maruyama-Nakashita A, Nakamura Y, Watanabe-Takahashi A, Yamaya T and Takahashi H (2004a) Induction of SULTR1;1 sulfate transporter in Arabidopsis roots involves protein phosphorylation/dephosphorylation circuit for transcriptional regulation. Plant Cell Physiol 45: 340–345

Maruyama-Nakashita A, Nakamura Y, Yamaya T and Takahashi H (2004b) Regulation of high-affinity sulphate transporters in plants: towards systematic analysis of sulphur signalling and regulation. J Exp Bot 55: 1843–1849

Maruyama-Nakashita A, Nakamura Y, Yamaya T and Takahashi H (2004c) A novel regulatory pathway of sulfate uptake in Arabidopsis roots: Implication of CRE1/WOL/AHK4-mediated cytokinin-dependent regulation. Plant J 38: 779–789

Nikiforova V, Freitag J, Kempa S, Adamik M, Hesse H and Hoefgen R (2003) Transcriptome analysis of sulfur depletion in Arabidopsis thaliana: interlacing of biosynthetic pathways provides response specificity. Plant J 33: 633–650

Nikiforova VJ, Kopka J, Tolstikov V, Fiehn O, Hopkins L, Hawkesford MJ, Hesse H and Hoefgen R (2005) Systems rebalancing of metabolism in response to sulfur deprivation, as revealed by metabolome analysis of Arabidopsis plants. Plant Physiol 138: 304–318

Ohkama N, Takei K, Sakakibara H, Hayashi H, Yoneyama T and Fujiwara T (2002) Regulation of sulfur-responsive gene expression by exogenously applied cytokinins in Arabidopsis thaliana. Plant Cell Physiol 43: 1493–1501

Ohkama-Ohtsu N, Kasajima I, Fujiwara T and Naito S (2004) Isolation and characterization of an Arabidopsis mutant that overaccumulates O-acetyl-L-Ser. Plant Physiol 136: 3209–3222

Saito K (2004) Sulfur assimilatory metabolism. The long and smelling road. Plant Physiol 136: 2443–2450

Sasaki-Sekimoto Y, Taki N, Obayashi T, Aono M, Matsumoto F, Sakura N, Suzuki H, Hirai MY, Noji M, Saito K, Masuda T, Takamiya K, Shibata D and Ohta H (2005) Coordinated activation of metabolic pathways for antioxidants and defence compounds by jasmonates and their roles in stress tolerance in Arabidopsis. Plant J 44: 653–668

Shibagaki N, Rose A, Mcdermott JP, Fujiwara T, Hayashi H, Yoneyama T and Davies JP (2002) Selenate-resistant mutants of Arabidopsis thaliana identify Sultr1;2, a sulfate transporter required for efficient transport of sulfate into roots. Plant J 29: 475–486

Smith FW, Ealing PM, Hawkesford MJ and Clarkson DT (1995) Plant members of a family of sulfate transporters reveal functional subtypes. Proc Natl Acad Sci USA 92: 9373–9377

Smith FW, Hawkesford MJ, Ealing PM, Clarkson DT, Vanden Berg PJ, Belcher AR and Warrilow AGS (1997) Regulation of expression of a cDNA from barley roots encoding a high affinity sulfate transporter. Plant J 12: 875–884

Takahashi H, Watanabe-Takahashi A, Smith FW, Blake-Kalff M, Hawkesford MJ and Saito K (2000) The roles of three functional sulphate transporters involved in uptake and translocation of sulphate in Arabidopsis thaliana. Plant J 23: 171–182

Takahashi H, Yamazaki M, Sasakura N, Watanabe A, Leustek T, de Almeida Engler J, Engler G, Van Montagu M and Saito K (1997) Regulation of sulfur assimilation in higher plants: a sulfate transporter induced in sulfate starved roots plays a central role in Arabidopsis thaliana. Proc Natl Acad Sci USA 94: 11102–11107

Takahashi H, Yoshimoto N and Saito K (2006) Anionic nutrient transport in plants: the molecular bases of sulfate transporter gene family. In: Setlow JK (ed) Genetic Engineering, Principles and Methods, Vol 27, pp 67–80. Springer, New York

Thomas D and Surdin-Kerjan Y (1997) Metabolism of sulfur amino acids in Saccharomyces cerevisiae. Microbiol Mol Biol Rev 61: 503–532

Ulmasov T, Hagen G and Guilfoyle TJ (1999) Dimerization and DNA binding of auxin response factors. Plant J 19: 309–319

Vauclare P, Kopriva S, Fell D, Suter M, Sticher L, von Ballmoos P, Krahenbuhl U, den Camp RO and Brunold C (2002) Flux control of sulphate assimilation in Arabidopsis thaliana: adenosine 5′-phosphosulphate reductase is more susceptible than ATP sulphurylase to negative control by thiols. Plant J 31: 729–740

Vidmar JJ, Schjoerring JK, Touraine B and Glass ADM (1999) Regulation of the *hvst1* gene encoding a high-affinity sulfate transporter from *Hordeum vulgare*. Plant Mol Biol 40: 883–892

Vidmar JJ, Tagmount A, Cathala N, Touraine B and Davidian J-CE (2000) Cloning and characterization of a root specific high-affinity sulfate transporter from *Arabidopsis thaliana*. FEBS Lett 475: 65–69

Wang R, Okamoto M, Xing X and Crawford NM (2003) Microarray analysis of the nitrate response in Arabidopsis roots and shoots reveals over 1,000 rapidly responding genes and new linkages to glucose, trehalose-6-phosphate, iron, and sulfate metabolism. Plant Physiol 132: 556–567

Wirtz M, Droux M and Hell R (2004) *O*-acetylserine (thiol) lyase: an enigmatic enzyme of plant cysteine biosynthesis revisited in *Arabidopsis thaliana*. J Exp Bot 55: 1785–1798

Xiang C and Oliver DJ (1998) Glutathione metabolic genes coordinately respond to heavy metals and jasmonic acid in Arabidopsis. Plant Cell 10: 1539–1550

Yano M, Hirai MY, Kusano M, Kitayama M, Kanaya S and Saito K (2005) Studies for elucidation of jasmonate signaling in the response of Arabidopsis to sulfur deficiency. In: Saito K, De Kok LJ, Stulen I, Hawkesford MJ, Schnug E, Sirko A, Rennenberg H (eds) Sulfur Transport and Assimilation in Plants in the Postgenomic Era, pp 165–168. Backhuys Publishers, Leiden, The Netherlands

Yoshimoto N, Inoue E, Saito K, Yamaya T and Takahashi H (2003) Phloem-localizing sulfate transporter, Sultr1;3, mediates re-distribution of sulfur from source to sink organs in Arabidopsis. Plant Physiol 131: 1511–1517

Yoshimoto N, Takahashi H, Smith FW, Yamaya T and Saito K (2002) Two distinct high-affinity sulfate transporters with different inducibilities mediate uptake of sulfate in *Arabidopsis* roots. Plant J 29: 465–473

Chapter 9

Biosynthesis, Compartmentation and Cellular Functions of Glutathione in Plant Cells

Andreas J. Meyer* and Thomas Rausch
Heidelberg Institute of Plant Sciences, University of Heidelberg, Im Neuenheimer Feld 360, 69120 Heidelberg, Germany

Summary

Glutathione is the most abundant low molecular weight thiol in all plant cells with the only exception of some plant species that produce and accumulate homologous tripeptides to similar levels. The broad range of functions of glutathione in terms of detoxification of heavy metals, xenobiotics and reactive oxygen species (ROS) has been highlighted in numerous reviews before. Glutathione S-conjugates formed during detoxification of electrophilic xenobiotics are immediately sequestered to the vacuole for degradation. This degradation is initiated by cleavage of the two terminal amino acids of glutathione. The cleavage of the γ-peptide bond between glutamate and cysteine involves a specific γ-glutamyl transpeptidase. Other members of this gene family are suggested to be involved in glutathione catabolism in the apoplast and linked to long-distance transport of glutathione. Recent findings on the biosynthesis and compartmentation now begin to illuminate how the biosynthesis of glutathione is regulated at the molecular level and how different subcellular pools of glutathione are interconnected. Glutamate-cysteine ligase (GSH1) is

* Corresponding author, Fax: 0049 6221 545859, E-mail: ameyer@hip.uni-hd.de

Rüdiger Hell et al. (eds.), Sulfur Metabolism in Phototrophic Organisms, 161–184.
© 2008 *Springer.*

the key regulatory enzyme of glutathione biosynthesis. Redox-dependent modulation of GSH1 activity also makes GSH1 a key factor in cellular redox homeostasis. Current work indicates that the redox state of the cellular glutathione redox buffer can be read out and directly transferred to target proteins by glutaredoxins. In this way glutathione is both, a scavenger for toxic compounds and a sensor for environmental signals which impact on the cellular redox state. This review aims at describing the important recent results on the cellular glutathione homeostasis in plant cells and highlighting the implications for glutathione-based redox sensing and signaling.

I. Introduction

Thiol-containing compounds are key players in a broad range of significant metabolic reactions and are essential for redox signaling. The tripeptide glutathione (reduced form: GSH; oxidized form: GSSG) is the most abundant low molecular weight thiol in almost all eukaryotic and many prokaryotic cells. Glutathione is essential for detoxification of xenobiotics and, being the precursor of heavy metal binding phytochelatins, also for the detoxification of heavy metals. Furthermore glutathione is an essential part of the glutathione–ascorbate cycle (GAC) and thus also involved in the detoxification of reactive oxygen species (ROS). Detoxification of ROS through GAC goes along with reversible oxidation and reduction of glutathione and thus immediate effects on the glutathione redox potential. Because glutathione is present in low millimolar concentrations in plant cells and thus the dominating redox buffer besides ascorbate, changes of the redox potential of the glutathione pool will have important effects on other cellular redox systems and thiol containing proteins in particular.

To understand the broad range of functions of glutathione and the integration of glutathione in the complex cellular signaling network it is essential to investigate the biosynthesis and the compartmentation of glutathione metabolism. The focus of this work is thus to review recent

progress on different factors affecting glutathione homeostasis in plants and thereby outlining the foundations of glutathione-dependent signaling events during stress reactions.

II. Biosynthesis of Glutathione

A. Evolution of GSH Biosynthesis

Glutathione and its homologues are the most prominent low molecular weight thiols in virtually all eukaryotic cells (with the exception of cells that lack mitochondria; Fahey et al., 1984; Newton et al., 1996) and in most Gram-negative bacteria, including cyanobacteria and purple bacteria (Fahey, 2001). In Gram-positive bacteria glutathione is less frequently found and in many cases replaced by other redox active thiol compounds (Fahey, 2001). Glutathione is generally synthesized in two ATP-dependent steps from its constituent amino acids glutamate, cysteine and glycine. In some plant species glycine is replaced by other residues (see below). The two enzymes involved in this biosynthetic pathway are glutamate-cysteine ligase (GSH1; GSHA in bacteria) and glutathione synthetase (GSH2; GSHB in bacteria). GSH1 catalyzes the formation of the atypical peptide bond between the γ-carboxylic group of glutamate and the amino group of cysteine. GSH2 subsequently catalyzes the formation of the peptide bond between the carboxylic group of cysteine and the amino group of glycine. Both enzymes are highly conserved between different plants, but surprisingly plant GSH1 was shown to be highly divergent from other eukaryotic organisms (May and Leaver, 1994).

Generally it is assumed that the need for a stable cellular redox buffer and compounds keeping ROS under control arose with the evolution of oxygenic photosynthesis about 2.6 billion years ago (Des Marais, 2000). From

---

Abbreviations: APS – adenosine 5′-phosphosulfate; BSO-L-buthionine-(S,R)-sulfoximine; ER – endoplasmic reticulum; DHAR – dehydroascorbate reductase; GAC – glutathione–ascorbate cycle; γ-EC – γ-glutamylcysteine; GGT – γ-glutamyl transpeptidase; GR – glutathione reductase; GRX – glutaredoxin; GSB – glutathione *S*-bimane; GSH – reduced glutathione; GSSG – oxidized glutathione; GST – glutathione *S*-transferase; MCB – monochlorobimane; OPT – oligopeptide transporter; PC – phytochelatin; PCS – phytochelatin synthase; ROS – reactive oxygen species; TRX – thioredoxin

cyanobacteria in which GSHA originally evolved the gene was transmitted to other species including archaea and proteobacteria. In proteobacteria that made use of the increasing atmospheric oxygen concentration by developing aerobic metabolism the presence of GSH would have been advantageous because of increasing amounts of ROS.

Direct comparison of sequences for GSH1 from a broad range of organisms from all kingdoms showed that the sequences cluster in three distinct groups (Copley and Dhillon, 2002). Group 1 comprises primarily γ-proteobacteria including *Escherichia coli*, group 2 comprises most eukaryotic organisms including human and *Saccharomyces cerevisiae* but excluding plants. Plant GSH1 sequences form a third group together with α-proteobacteria and archaea. Despite the highly divergent forms of the same enzymatic function, careful analysis of small blocks of conserved sequence motifs indicated that the three groups of sequences are indeed distantly related (Copley and Dhillon, 2002). While non-plant eukaryotic organisms might have received their GSH1 genes from α-proteobacterial progenitors of mitochondria (Fahey et al., 1984; Fahey and Sundquist, 1991; Fahey, 2001), plants were assumed to have received their GSH1 genes from the cyanobacterial progenitor of chloroplasts. It is puzzling, however, that the plant GSH1 sequences are very similar to a number of α-proteobacterial sequences (see below). In this context it remains unknown whether the α-proteobacterial progenitor gene in plants replaced an already present gene or whether the ancestral plastidic genes were lost first, the α-proteobacterial genes subsequently filling the functional gap (Copley and Dhillon, 2002).

Evolutionary relationships are even less clear for GSH2. In this case all known eukaryotic sequences are related, but none of them show any significant homology with bacterial GSHB (Copley and Dhillon, 2002). Polekhina et al. (1999) suggested that eukaryotic GSH2 did not evolve from a bacterial ancestor. Instead, both bacterial GSHB and eukaryotic GSH2 might have evolved independently from ancestors that had the characteristic fold of the ATP-grasp superfamily. The members of this family exhibit a distinct carboxylate-amine/thiol ligase activity (Galperin and Koonin, 1997).

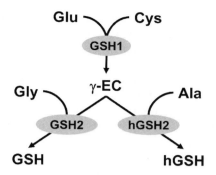

Fig. 1. Pathways for biosynthesis of GSH and hGSH. All depicted biosynthetic steps consume one ATP per synthesized molecule.

Many members of the Fabaceae contain homoglutathione (γ-Glu-Cys-β-Ala, hGSH) besides glutathione (Fig. 1) (Price, 1957; Klapheck, 1988). Analysis of the molecular basis of homoglutathione biosynthesis in *Medicago trunculata* showed that homoglutathione synthetase (hGSH2) is closely related to glutathione synthetase. A gene duplication event for GSH2 after divergence of Fabales from Solanales and Brassicales and subsequent substitution of two highly conserved alanines by Leu-534 and Pro-535 in one of the two copies gave rise to synthesis of hGSH (Frendo et al., 2001).

B. Biochemistry of GSH1 and GSH2 Enzymes: From Structure to Regulation

Although plant GSH1 proteins and GSHI from E. Coli share no statistically significant similarities in their sequences (May and Leaver, 1994; Copley and Dhillon, 2002), the recent elucidation of the GSH1 protein structures from *E. coli* (Hibi et al., 2004) and *Brassica juncea* (Hothorn et al., 2006) has revealed overall similarity in the protein fold with some additional plant GSH1-specific properties. The primary plant GSH1 sequence folds into a six-stranded antiparallel β-sheet forming a bowl-like structure, which is flanked by helical regions. While the catalytic residues were conserved between the plant and *E. coli* enzymes, an unexpected β-hairpin was discovered in the plant enzyme, which is absent in the *E. coli* enzyme. While a detailed description of the structure of plant GSH1 is beyond the scope of this review, it is important to note that for the *B. juncea* enzyme, the protein structure has revealed the presence of two intramolecular disulfide bridges (Hothorn et al., 2006). One of these disulfide bridges,

CC1, has been proposed to position the β-hairpin (see above) in a way that the access of cysteine to its binding site is allowed, whereas reduction of CC1 possibly shields the Cys binding site. In agreement with this assumption, mutating these Cys residues resulted in a 10-fold reduced enzyme activity of the mutant recombinant protein. Using the mutant *Bj*GSH1 protein, the possible role of the second disulfide bridge (CC2) could be addressed. Again, reduction of this disulfide bridge resulted in a further decrease of enzyme activity (about fourfold; Hothorn et al., 2006).

How exactly the previously reported redox control of the plant GSH1 enzyme activity (Hell and Bergmann, 1990; Jez et al., 2004) relates to the roles of CC1 and CC2 for regulating GSH1 enzyme activity in vivo remains to be convincingly shown. In their seminal paper on the plant GSH1 enzyme, Hell and Bergmann (1990) revealed that upon addition of DTT the tobacco enzyme apparently underwent a profound structural change, resulting in inactivation and changed mobility in gel filtration chromatography (i.e. from approx. 60 to 30 kDa). In their recent study on recombinant Arabidopsis GSH1 protein, Jez et al. (2004) could basically confirm this observation. As under reduced conditions, the tobacco GSH1 enzyme (Hell and Bergmann, 1990) and the Arabidopsis enzyme (Jez et al., 2004) showed similar behavior, it can be excluded that CC1 is relevant for this structural change, as the presence of CC1 Cys residues is not conserved in plant GSH1 enzymes (i.e. absent in tobacco). While Hothorn et al. (2006) confirmed for the enzyme from *B. juncea* the occurrence of a profound structural change in response to the redox environment, these authors advocate a redox-regulated monomer-homodimer switch, based on their results from size exclusion chromatography and analysis of *Bj*GSH1 crystal structure. Interestingly, this redox-induced monomer-homodimer switch is not observed for recombinant GSH1 proteins from the structurally closely related α-proteobacteria (R. Gromes, M. Hothorn and T. Rausch, unpublished). Also, the amino acid residues forming the interface of the homodimer in the *Bj*GSH1 enzyme are conserved in GSH1 proteins across the plant kingdom but not in α-proteobacteria. It is intriguing, that the enzyme catalyzing the rate-limiting step in the biosynthesis of one of the cells major antioxidants should be itself under tight redox control. In fact, the results discussed above suggest that, under normal conditions (i.e. reducing milieu in the chloroplast stroma) the plant GSH1 enzyme would operate "with brakes on", being fully activated only under oxidizing conditions as encountered upon stress exposure. While definite proof for this attractive hypothesis remains to be provided, Jez and colleagues have recently presented first supportive evidence for in vivo operation of the proposed redox switch (Jez et al., 2006; Hicks et al., 2007).

In contrast to GSH1, plant GSH2 enzymes belong to one large family with all other eukaryotic GSH2 enzymes, which shows no similarity with bacterial GSH2 enzymes (Wang and Oliver, 1996; Copley and Dhillon, 2002). While GSH2 from *E. coli* is a functional tetramer of approximately 300-residue subunits (Hara et al., 1996), the larger eukaryotic GSH2 subunits from mammals and yeast (approx. 470 residues) form a homodimer (Polekhina et al., 1999; Gogos and Shapiro, 2002). Likewise, the analysis of recombinant *At*GSH2 protein indicated that the enzyme operates as a homodimer (Jez and Cahoon, 2004). While no plant GSH2 structure has as yet been reported, the fairly high conservation of eukaryotic GSH2 proteins (approx. 40% sequence identity) supports the assumption that the kinetic mechanism is conserved among eukaryotic GSH2 enzymes. Despite the lack of sequence conservation, their difference in oligomer structure and molecular masses of subunits, GSH2 enzymes from bacteria and eukaryotes share a common protein fold and belong to the ATP-grasp structural family (Hara et al., 1996; Galperin and Koonin, 1997; Polekhina et al., 1999; Gogos and Shapiro, 2002).

Only recently have the catalytic properties of plant GSH1 and GSH2 enzymes been studied with recombinant enzymes from *Arabidopsis thaliana* after expression in *E. coli* (Jez and Cahoon, 2004; Jez et al., 2004). In both enzymes, the three substrates form ternary complexes in the active site, with all substrates mutually affecting their binding. Detailed kinetic analysis of GSH1 and GSH2 has revealed that both enzymes appear to operate via a random ter-reaction mechanism with a preferred order of substrate addition. Thus, in *At*GSH1, binding of the substrates ATP or glutamate increases the affinity for the other substrate

2.5-fold, whereas the positive interaction between cysteine and glutamate results in an even higher reciprocal increase of binding affinities (16-fold; Jez et al., 2004). Similarly, in AtGSH2 binding of γ-glutamylcysteine or ATP increase the affinity of the enzyme for the other substrate 10-fold (Jez and Cahoon, 2004), whereas binding of either glycine or γ-glutamylcysteine decrease binding of the other substrate by almost sevenfold.

Since the cloning of the first plant cDNAs encoding GSH1 and GSH2 (May and Leaver, 1994; Ullmann et al., 1996; Wang and Oliver, 1996), numerous studies have addressed the control of their expression at the transcript level. The results of these studies have largely confirmed that under conditions where GSH biosynthesis is upgraded in response to various developmental or environmental cues, GSH1 expression appears to be more affected, corroborating its proposed role as catalyzing the limiting step (Schäfer et al., 1998; Xiang and Oliver, 1998; Xiang et al., 2001). However, some reports also document a coordinate increase of GSH1 and GSH2 mRNAs. Thus, in B. juncea transcript amounts were increased for both genes (Schäfer et al., 1998). Likewise, in A. thaliana the infection with Phytophthora brassicae caused a more than twofold coordinate increase of GSH1 and GSH2 transcripts (Parisy et al., 2007).

Recently, Wachter et al. (2005) reported on different transcript populations for GSH1 in A. thaliana and B. juncea. In both species, two TATA-boxes located in proximity give rise to two distinct transcript populations, differing in their length of 5′UTR sequence. The quantitative ratio of long to short 5′UTR mRNAs was dependent on the developmental stage (Wachter et al., 2005) and was also affected by several stress-related cues (Wachter, 2004). The biological significance of different 5′UTRs is not yet known, however, it may be speculated that transcript stability and/or binding of protein factors to the 5′UTR could be affected (Xiang and Bertrand, 2000; Wachter, 2004). Mapping of transcript start sites revealed that mRNAs were initiated about 30 bp downstream of both TATA-boxes, indicating that both were operative. Thus it can be assumed that two closely spaced, overlapping promoters differentially regulate the formation of short and long 5′UTR transcripts in these species. Both transcript classes code for the same GSH1

protein with functional transit peptide, rendering an effect of different 5′UTR structure on GSH1 targeting unlikely (Wachter et al., 2005).

Transcriptional control of GSH1 expression in response to hormonal (e.g. jasmonic acid) and stress-related cues (e.g. heavy metal exposure) undoubtedly contributes to the observed changes in GSH1 activity, however, earlier work has already indicated that other regulatory mechanisms are likely to be involved. Thus, in Arabidopsis suspension-cultured cells, a stress-induced increase of GSH1 enzyme activity was observed in the absence of transcript increase (May et al., 1998). While several independent studies support the existence of a post-transcriptional control of GSH1 expression (May et al., 1998; Xiang and Bertrand, 2000), our factual knowledge about the underlying molecular mechanism(s) is fragmentary at best. Clearly, there is an urgent need for further detailed analysis of the observed interactions of GSH1 transcripts with 5′UTR-binding proteins, as work has not progressed beyond initial observations (Xiang and Bertrand, 2000; Wachter, 2004). The postulated redox-control of such an interaction would provide a feedback mechanism to assure cellular glutathione homeostasis. It is noteworthy that such a control would have to operate in the cytosol and would therefore respond to the cytosolic redox poise, whereas the GSH1 enzyme is exclusively targeted to the plastids (Wachter et al., 2005). Therefore, such a mechanism would constitute a conduit for redox communication between both compartments. In addition to the postulated redox-mediated translational control, the post-translational control of GSH1 activity via a redox-mediated monomer-homodimer switch, as observed in vitro with recombinant GSH1 enzyme (see above; Hothorn et al., 2006), may account for an additional level of regulation.

The Cd-sensitive Arabidopsis cad2-1 mutant, which shows a 2 amino acid deletion in the GSH1 protein, provided the first genetic evidence for a causal link between GSH biosynthesis and a specific GSH function in planta (Cobbett et al., 1998). The recent elucidation of plant GSH1 structure (Hothorn et al., 2006) indicated that this deletion most likely affects the position of residues involved in glutamate binding. As a result of reduced GSH1 activity, the cad2-1 mutant has a significantly decreased

GSH content (45% as compared to wild-type). Conversely, and as expected, its cysteine content is increased about twofold as compared to wild-type plants (Cobbett et al., 1998). The second GSH1 mutant, *rml1* (*r*oot-*m*eristem-*l*ess), depicts a strong developmental phenotype, its root meristem being nonfunctional (Vernoux et al., 2000). This mutant documented for the first time a direct link between GSH content and GSH function during plant development, and revealed its essential role in initiating and maintaining cell division during post-embryonic root development. The *rml1* mutation, in which an aspartate is exchanged for an asparagine (D250N), is most likely affected in adenine nucleotide binding (Hothorn et al., 2006). This mutant has only about 3% GSH as compared to the wild-type and shows a threefold increased cysteine content. The *rax1-1* mutant was initially identified as showing a constitutive expression of an otherwise stress-inducible ascorbate peroxidase (Ball et al., 2004). While this GSH1 mutant showed a reduction of GSH content (about 30% of the wild-type) similar to *cad2-1*, its cysteine content was unaffected, in marked contrast to *cad2-1* and *pad2-1* (see below). Surprisingly, the molecular basis of the *rax1-1* mutation is a single conservative amino acid exchange (R229K). The functional analysis of recombinant *rax1-1* protein revealed an about fivefold higher K_m towards cysteine (with V_{max} reduced by about 50%), in agreement with the position of R^{229} being proximal to the cysteine-binding pocket (Hothorn et al., 2006). Thus, this mutant also supports the widely held assumption that cysteine availability may affect GSH biosynthesis. Recently, a fourth GSH1 mutant, *pad2-1*, has been shown to be impaired in resistance towards *P. brassicae* and *Pseudomonas syringae* (Parisy et al., 2007). The Arabidopsis *pad2-1* mutant was originally shown to exhibit a reduced accumulation of the phytoalexin camalexin (Glazebrook et al., 1997); however, later work revealed that this was not the cause for the observed phenotype (Zhou et al., 1999). The *pad2-1* mutant shows a S298N substitution, located close to the cysteine binding site. While its GSH content is only about 20%, this mutant exhibits an about fivefold increased cysteine content. While it is tempting to speculate that the decreased GSH content is causally related to the reduced resistance towards pathogens, it cannot be excluded that the strong increase in cysteine content may also contribute to the observed phenotype. In summary, the analysis of several GSH1 mutants has provided proof of the (direct or indirect) role of GSH in several vital plant functions, including plant development and tolerance against abiotic and biotic stress.

III. Compartmentation of Glutathione Metabolism in Plants

A. Transport of GSH and Precursors Across the Chloroplast Envelope

A key parameter in cellular glutathione homeostasis is efficient transport of glutathione and/or its precursors between different organelles. For a long time, the abundant low molecular weight thiols cysteine and especially GSH have been assumed to fulfill the function as a transport metabolite. In terms of glutathione biosynthesis the differential subcellular localization of GSH1 and GSH2 in Arabidopsis highlights the importance for consideration of subcellular compartments in GSH metabolism. At the same time exclusive plastidic localization of GSH1 and dual targeting of GSH2 to both cytosol and plastids (Wachter et al., 2005) points away from cysteine and GSH transport and rather implies export of γ-EC from plastids to provide cytosolic GSH2 with its substrate. Export of γ-EC from plastids was suggested by Meyer and Fricker (2002) after in situ labeling of Arabidopsis cell cultures with monochlorobimane (MCB) for several hours. This fluorescent dye has been shown to predominantly label glutathione in live cells (Meyer et al., 2001; Cairns et al., 2006). Long-term labeling of Arabidopsis cells for several hours resulted in a steady increase in fluorescence which could be blocked by the GSH biosynthesis inhibitor L-buthionine-(S,R)-sulfoximine (BSO) indicating demand-driven de novo synthesis of GSH und conditions of extended exposure to MCB as an electrophilic xenobiotic. Because this synthesis phase could also be abolished by removal of sulfate from the external medium this implied that the steady increase in fluorescence was not only due to GSH biosynthesis but in fact mirrored flux through the entire sulfur assimilation pathway starting from external sulfate and running down to GSH. The first step of sulfate reduction is catalyzed by adenosine 5′phosphosulfate (APS) reductase, an enzyme that contains a sub-domain homologous to glutaredoxins (GRXs) at its carboxy-terminus. Therefore APS reductase is assumed to use electrons delivered by GSH for reduction of APS (Bick et al., 1998). This demand for plastidic GSH as electron donor for continued sulfur assimilation indicates that the plastidic GSH pool is not depleted during the incubation with MCB (Meyer and Fricker, 2002). Conversely, this implies that not GSH but rather its precursor γ-EC is exported from the plastids to cover cytosolic demand for reduced sulfur. This hypothesis was further supported through the isolation of null-mutants for *GSH2* from Arabidopsis T-DNA insertion collections. Homozygous *gsh2* null-mutants hyperaccumulate γ-EC to levels 200-fold greater than wild-type GSH and 5,000-fold greater than wild-type γ-EC (Pasternak et al., 2008). In situ labeling with MCB showed that in this case the extreme concentration of γ-EC led to partial labeling of this pool. The label was predominantly in the cytosol and thus indicated that γ-EC was exported. This result is also corroborated by biochemical in vitro assays on GSH1 showing that this enzyme would quickly be inhibited by accumulating γ-EC albeit with 50% efficiency compared with the normal feedback inhibitor GSH. Furthermore, cytosol specific complementation of the *gsh2* knockouts with wild-type *GSH2* restored the wild-type phenotype and low molecular weight thiol levels almost similar to wild-type.

The cytosolic complementation of *gsh2* knockouts strongly suggests efficient transport of GSH from the cytosol into the plastid. Preliminary data indicating uptake of radioactive GSH into isolated wheat chloroplasts have been presented (Noctor et al., 2000; Noctor et al., 2002). From their data, Noctor and colleagues concluded that the rate of GSH uptake is sufficient to play a significant role in determining the GSH pool on either side of the membrane. Further analysis of the cytosolic complementation of *gsh2* knockouts and quantitative analysis of plastidic GSH in these mutants is expected to provide final evidence for the efficiency of this transport. Despite the increasing evidence

for γ-EC as the major metabolite for export of reduced sulfur from the plastids, export of GSH from the plastids can still not be ruled out. The fact that plastid-specific complementations of a *gsh2* knockout are fully viable indicates that GSH can also be transported from plastids to the cytosol (M. Pasternak and A.J. Meyer, unpublished).

The accumulating evidence for efficient export of γ-EC from the plastids now immediately raises the question for a γ-EC transporter on the chloroplast envelope. Because such a transporter when knocked out should severely affect cellular GSH homeostasis and thus result in clear phenotypes candidate proteins should eventually appear in genetic screens. However, none of the solute transporters on the plastid envelope membrane characterized to date has been linked to GSH metabolism (Weber et al., 2005).

B.   Glutathione in Other Organelles

Besides photosynthesis mitochondrial respiration is the second main source of ROS in plant cells and at the same time mitochondria house a range of different redox pathways, utilized for protection from oxidative damage and assembly of the organelle (Koehler et al., 2006; Logan, 2006). Thus, glutathione is required within mitochondria to buffer against ROS production and to avoid oxidative damage. Similar to chloroplasts, the most important pathway for ROS detoxification is the GAC (see below), which has been shown to be present in mitochondria with all participating enzymes (Chew et al., 2003b). Given that GSH is not synthesized in mitochondria it has to be imported from the cytosol. Putative transporters for GSH have been identified based on homology with a high affinity glutathione transporter (hgt1p) from *S. cerevisiae* (Bourbouloux et al., 2000). The identified family of nine genes in Arabidopsis is designated as *OPT1–OPT9* (Koh et al., 2002) and at least *AtOPT6* as well as some homologues form other species have been shown to complement the yeast *hgt1* knockout mutant (Bogs et al., 2003; Cagnac et al., 2004; Zhang et al., 2004). Both, OPT3 and OPT6 are predicted to be targeted to the mitochondria by different bioinformatics tools. However, restricted tissue-specific expression of these genes (Stacey et al., 2006) suggests that these proteins are not the essential GSH-transporters for uptake of GSH into the mitochondria. Transport studies with isolated mitochondria from rats suggest that GSH is taken up into mitochondria via dicarboxylate and 2-oxoglutarate carriers (Chen and Lash, 1998). The fact that the glutathione reductase 1 (GR1, AT3G54660) is dually targeted to both chloroplasts and mitochondria (Chew et al., 2003a) together with the fact that a knockout of this mutant is embryo lethal (Tzafrir et al., 2004) might indicate that GSSG cannot be exported from mitochondria for reduction in the cytosol.

The first reaction of the conversion of glycolate to glycine during photorespiration produces the toxic intermediate glyoxalate and the toxic by-product H_2O_2. Containment of these reactions in the peroxisomes avoid that the toxic products can harm other reaction in the chloroplasts. The GAC was also shown to be present in peroxisomes of pea (*Pisum sativum* L.) (Jimenez et al., 1997). As a result of senescence, the pools of GSH and GSSG were considerably increased in peroxisomes while the oxidative damage in mitochondria was significantly accelerated already. This observation lead to the suggestion that peroxisomes may function longer than mitochondria in the oxidative mechanisms of senescence (Jimenez et al., 1998). The presence of the GAC in peroxisomes implies the presence of glutathione reductase(s) (GR) in this compartment. While ascorbate peroxidase has bee found in proteomic studies of Arabidopsis peroxisomes (Fukao et al., 2002) there is currently no genetic nor proteomic evidence for a GR in peroxisomes (Heazlewood et al., 2007). Nevertheless, glutathione is required as a redox buffer and is used by a number of enzymes as a substrate. Glutathione-depending enzymes shown to be present in peroxisomes of non-plant organisms include glutathione peroxidase in rat peroxisomes (Singh et al., 1994) and glutathione S-transferase in *S. cerevisiae* (Barreto et al., 2006). Expression of a novel glutathione specific redox sensor in tobacco peroxisomes suggests that the redox potential of the glutathione redox buffer in peroxisomes is highly reducing and thus similar to that in the cytosol (M. Schwarzländer et al., submitted)[1]

The endoplasmic reticulum (ER) is the compartment in which proteins destined for the secretory pathway are folded. Attainment of their native structure often includes the formation of intramolecular disulfide bridges. Glutathione is known to be the principle redox buffer in the ER, but contrary to the highly reduced state of cytosolic glutathione the ratio of GSH to GSSG in the ER is between 1:1 and 3:1 (Hwang et al., 1992). Targeting of the glutathione dependent redox-sensitive GFP (roGFP) to the ER recently allowed to directly showing the highly oxidized state of the luminal glutathione pool in tobacco (Meyer et al., 2007). Maintenance of such steep gradients across the ER membrane for both GSH and GSSG can only be achieved through active transport mechanisms, but to date no such transporter has been identified. A large fraction of the luminal glutathione pool was shown to be present in the form of mixed disulfides with proteins, which may play a role as a glutathione reserve and a component of the luminal redox buffering system, but may also play a more active role in the process of native protein disulfide bond formation (Bass et al., 2004). Due to its high degree of oxidation glutathione in the ER lumen has long been suspected the prime source of oxidative power for protein folding, but this hypothesis was refuted by the observation that in yeast oxidizing equivalents are provided by the protein Ero1 (Cuozzo and Kaiser, 1999). Despite being present mainly in the oxidized form, glutathione still acts as a source of reducing equivalents by playing a direct role in the isomerization of luminal oxidoreductases and maintenance of theses enzymes in their reduced state (Jessop and Bulleid, 2004; Sevier et al., 2007).

Nuclear compartmentalization of glutathione with ATP-dependent maintenance of the nucleoplasm/cytosol concentration gradient has been shown for hepatocytes based on specific labeling of GSH with MCB (Bellomo et al., 1992). This observation is highly surprising as the only gateway for exchange of macromolecules between cytosol and nucleoplasm is the nuclear pore complex (NPC). This complex has a 9 nm aqueous pore and allows passive diffusion of molecules up to about 60 kDa, molecules over 50 kDa at a very slow rate (Meier, 2007). It was also shown that MCB adducts injected into cells immediately accumulated in the nucleus (Briviba et al., 1993). Labeling of GSH in living plant cells with MCB also very quickly gives rise to very strong nuclear labeling, but even under conditions where normal vacuolar sequestration of glutathione-bimane conjugates is prevented no indication of differential labeling between cytoplasm and nucleus could be observed (Gutiérrez-Alcalá et al., 2000;

Meyer et al., 2001; Hartmann et al., 2003). Tight redox control within the nucleus is essential for normal functioning and it has been shown that the members of the TGA family of transcription factors need to be in the reduced state to allow efficient DNA binding (Despres et al., 2003). A specific role of the glutathione redox buffer in this control is strongly supported by the observation that TGA transcription factors interact with a subclass of GRXs in Arabidopsis (Ndamukong et al., 2007).

C.   Compartmentation Between 'Sink' and 'Source' Tissues

Glutathione is the major form of systemically transported reduced sulfur (Foyer et al., 2001). Glutathione is thought to be synthesized in vegetative shoot tissues and transported to generative tissues and developing seeds in particular as well as in the opposite direction to the roots. Long-distance transport of glutathione and/or the precursor γ-EC along the phloem has been shown (Lappartient and Touraine, 1996; Herschbach and Rennenberg, 2001; Li et al., 2006). Split-root experiments in which one half of the root system was exposed to low sulfate conditions indicate that a systemic signal, most likely glutathione, controls sulfate uptake in the roots (Lappartient et al., 1999). High glutathione levels are correlated with decreased sulfate uptake and thus glutathione might contribute to a well-balanced feedback control mechanism for sulfate uptake. Conversely, artificially reduced GSH levels during growth on BSO relieved the expression of APS reductase, the key enzyme for sulfate reduction (Vauclare et al., 2002). It is not clear, however, whether this control of sulfate uptake in the roots and further assimilation is solely controlled by glutathione or whether other components of the glutathione biosynthesis pathway upstream of GSH are also involved. Other possible control mechanisms might include O-acetylserine, sucrose and different phytohormones (Kopriva, 2006).

Homozygous *gsh1* knockouts are not capable of synthesizing γ-EC and hence GSH. Due to this defect these mutants show an embryo-lethal phenotype (Cairns et al., 2006). Despite not being able to synthesize GSH the homozygous knockout embryos develop to their normal size indicating that cell division during seed development can progress without any GSH or, alternatively, supply with GSH from maternal tissues. In situ

labeling of GSH with MCB failed to detect significant amounts of GSH (Cairns et al., 2006), but it can not be excluded that limited amounts of GSH are indeed supplied by the mother plant. Very intense fluorescent labeling at the chalaza suggested release of GSH from the phloem at this point. Because embryonic tissues are symplastically isolated from maternal tissues, organic and inorganic compounds supplied to the embryo have to cross three apoplastic borders on the way from the phloem to the embryo (Stadler et al., 2005). Transport of GSH would thus require highly efficient transport systems on membranes along this pathway. Proteins of the OPT family have been suggested to facilitate the membrane transport of GSH (Bourbouloux et al., 2000; Cagnac et al., 2004), but no information is available on the involvement of OPTs in seed supply. Alternatively, GSH supplied to the seed via the phloem might be largely degraded to the amino acids, which are then taken up by the embryo. In animals, GSH present in the extracellular space is degraded by subsequent action of γ-glutamyl transpeptidases (GGTs) and dipeptidases to the respective amino acids, which can then be taken up (Meister, 1988; Lieberman et al., 1996). Arabidopsis also contains a family of 4 GGT genes with homology to human and mouse GGTs (Storozhenko et al., 2002). Two of the GGTs have recently been assigned to the apoplast (Storozhenko et al., 2002; Martin et al., 2007; Ohkama-Ohtsu et al., 2007a). The fact that the apoplast does not contain significant amounts of glutathione (Foyer and Noctor, 2005) might be indicative of high catabolic activity of GGTs towards GSH and GSSG.

D.   Catabolism of Glutathione

Inhibition of GSH biosynthesis by BSO leads to almost complete depletion of the entire GSH pool within 3 days indicating a turnover of the GSH pool in normal metabolism (A.J. Meyer, unpublished). It is less clear, however, which metabolic pathways are responsible for this turnover or in which cellular processes GSH is being used up. GSH is well known to form mixed disulfides with a large number of proteins (Shelton et al., 2005) and it might be possible that a certain amount of GSH is continuously lost during protein turnover. To date, however, the effect of protein degradation on the GSH pool has not been studied.

Direct metabolic turnover of the GSH pool in the cytosol is unlikely because the distinct γ-amide bond between glutamate and cysteine can be released by only very few enzymes and thus protects GSH from catabolism by intracellular aminopeptidases. Lack of efficient degradation capabilities for the γ-peptide bond also contributes to hyperaccumulation of γ-EC in *gsh2* knockout mutants (Pasternak et al., 2008). The only enzymes known to be capable of cleaving the γ-peptide bond are the GGTs (see above), which are located in either the apoplast or the vacuole (Grzam et al., 2007, Martin et al., 2007; Ohkama-Ohtsu et al., 2007a; Ohkama-Ohtsu et al., 2007b). Similar to glutathione *S*-conjugates, GSSG might be exported to the vacuole for degradation. Such sequestration has been discussed as an overspill valve for extremely high GSSG concentrations in the cytosol under conditions of extreme stress (Foyer et al., 2001). The only cellular compartment for which high concentrations of GSSG have been shown is the ER (Hwang et al., 1992) and it can be assumed that with each vesicle transported to the plasma membrane or the vacuole some glutathione is lost. In both cases this glutathione would become accessible to degradation by GGTs. In analogy to animals other enzyme activities of the γ-glutamyl cycle have been described. These enzymes include carboxypeptidases, Cys-Gly dipeptidases, γ-glutamyl cyclotransferase and 5-oxo-prolinase (Martin, 2003 and refs. cited therein). However, so far none of these enzymes have been characterized at the molecular level.

IV. Cellular Functions of Glutathione in Plants

A. Detoxification of Xenobiotics and Endogenous Compounds

Plants are generally taking up many toxic compounds from their natural environment with little indiscrimination (Coleman et al., 1997). There are, however, efficient detoxifying systems in place to avoid long-term damage. The most important pathway for detoxification of electrophilic compounds is based on conjugation of these compounds to glutathione through reactions catalyzed by glutathione *S*-transferases (GSTs). The Arabidopsis nuclear genome contains 54 genes with high homology to GSTs (Edwards and Dixon, 2005). Most plant GSTs are predicted to be present in the cytosol, but there are also reports of microsomal GSTs, nuclear- or apoplast-localized enzymes and of gene products bearing plastid-targeting signal peptides (Frova, 2003 and refs. cited therein). The GSTs are a group of homo- or heterodimeric enzymes and due to multiple heterodimer formation a large number of different combinations is possible, which might contribute to the broad range of different substrates for conjugation to glutathione (Edwards and Dixon, 2005). After conjugation the glutathione-moiety acts as an efficient tag marking the conjugate for vacuolar sequestration. This sequestration is achieved by multidrug-resistance associated proteins (MRPs), a subfamily of ATP-binding cassette proteins (Rea et al., 1998; Martinoia et al., 2000; Rea, 2007). A number of vacuolar ABC-transporters have been shown to transport glutathione *S*-conjugates in vitro (Martinoia et al., 1993; Li et al., 1995), but even multiple knockouts of ABC-transporters did not significantly affect vacuolar sequestration in vivo (A.J. Meyer and M.D. Fricker, unpublished results). The latter observation is supported by the identification of at least 10 MRPs on the tonoplast (Jaquinod et al., 2007). In vitro ABC-transporters on the tonoplast membrane have also been shown to transport GSSG with Km values of 73 to 400 µM (Foyer et al., 2001). It has been discussed whether vacuolar sequestration of GSSG excessively formed under conditions of oxidative stress might contribute to maintenance of a reduced cytosol (Foyer et al., 2001), but this hypothesis has not been tested experimentally in vivo.

After their formation glutathione *S*-conjugates are further processed. Using MCB as a fluorescent probe for GSH it was shown that vacuolar sequestration of glutathione *S*-bimane adducts is very fast and completed within 30 min (Fricker et al., 2000; Meyer and Fricker, 2000; Meyer et al., 2001; Meyer and Fricker, 2002). Using the in vivo labeling of GSH with MCB in combination with HPLC analysis it was recently shown that degradation of glutathione *S*-conjugates in Arabidopsis is much slower than vacuolar sequestration (Grzam et al., 2006; Ohkama-Ohtsu et al., 2007b). Vacuolar degradation of conjugates in Arabidopsis is initiated by a rate-limiting γ-glutamyl transpeptidase cleaving the γ-amide bond and releasing glutamate (GGT4, At4g29210; Grzam et al., 2007; Ohkama-Ohtsu et al., 2007b). The remaining cysteinylgylcine conjugate is quickly undergoing further degradation to cysteine conjugates (Grzam et al., 2006, 2007). The latter reaction might be catalyzed by a vacuolar dipeptidase, but no enzymatic activity has been ascribed yet. Despite very fast vacuolar sequestration of glutathione S-conjugates the cytosolic enzyme phytochelatin synthase (PCS) has been suggested to play an important role in catabolism of glutathione S-conjugates by cleaving the glycine residue in the initial degradation step (Beck et al., 2003; Blum et al., 2007). The minor activity of PCS towards glutathione S-conjugates and the concomitant release of γ-EC-conjugates is, however, much slower than sequestration of the glutathione S-conjugates even in the presence of PCS-activating heavy metals (Grzam et al., 2006).

Besides exogenous toxic compounds normal metabolism also generates a range of toxic by-products, which need to be detoxified. One example for endogenous toxic compounds that are detoxified via conjugation with GSH is methylglyoxal. Glycolysis leads to formation of highly toxic methylglyoxal through non-enzymatic β-elimination of the phosphate group of triose phosphates. Methylglyoxal is reacting non-enzymatically with GSH to form a hemithioacetal, which is then further degraded by the enzymes glyoxalase I and II (Singla-Pareek et al., 2003). During this reaction GSH is released again. Similarly, GSH is required as a cofactor in isomerization reactions catalyzed by zeta- and phi-isoforms of GSTs (Edwards et al., 2000; Edwards and Dixon, 2005). Such an isomerase activity has been exploited to bioactivate a thiadiazolidine herbicide through GST-mediated isomerization with a GSH conjugate as an intermediate (Edwards et al., 2000).

B. Detoxification of Heavy Metals via Phytochelatins: the Glutathione-Based First Line of Defense

Higher plants (but also *Schizosaccharomyces pombe* and *Caenorhabditis elegans*; Rea et al., 2004) may detoxify heavy metals via enzymatic biosynthesis of metal-binding thiolpeptides, the so-called phytochelatins (Clemens et al., 2002; Tong et al., 2004; Clemens, 2006). The enzyme PCS is associated with the papain superfamily of cysteine proteases (Rea, 2006; Romanyuk et al., 2006). It operates as a γ-glutamylcysteine dipeptidyl transferase, transferring a γ-glutamylcysteine unit from one GSH molecule to another to form PC_2, which may be further extended by repeated transfer of additional γ-glutamylcysteine units. Depending on the species and tissue, phytochelatins may assume different lengths, but in general PC_{2-4} are the most abundant ones. The reaction mechanism of the PCS enzyme has attracted much interest, in particular its activation by heavy metal ions. Initially, PCS was thought to be directly activated by binding of heavy metal ions to Cys residues located in its highly conserved N-terminal half which has been shown to provide the fold for core catalysis. However, closer inspection later revealed that PCS catalyzes a bisubstrate transpeptidation reaction in which both free GSH and its corresponding metal thiolate are co-substrates (Vatamaniuk et al., 2000; Rea et al., 2004). During catalysis, the PCS enzyme forms a covalent γ-glutamylcysteine acyl intermediate. In fact, transient acylation of PCS occurs at two sites, however, with different ligand requirements (Vatamaniuk et al., 2004). Recently, the existence of prokaryotic PCS homologs has been discovered. These enzymes are only half the size of eukaryotic PCS proteins, and contain only the N-terminal catalytic domain (Tsuji et al., 2004; Vivares et al., 2005). For the enzyme from *Nostoc* sp., *Ns*PCS, an enzymatic function has been demonstrated; however, this enzyme shows both GSH hydrolase and PCS activities, being more peptidase- than transpeptidase in its mode of action (Tsuji et al., 2004; Vivares et al., 2005; Rea, 2006). The protein structure of the *Ns*PCS enzyme has recently been resolved (Vivares et al., 2005).

PCS expression appears to be constitutive in roots and shoots of higher plants, but may be further up-regulated in response to heavy metal exposure at the transcript and protein level, respectively (Lee et al., 2002; Heiss et al., 2003). In plants, PCS activation provides the first line of defense against toxic levels of heavy metal ions, positioning its precursor glutathione in the center of heavy metal tolerance. In fact, the first

Arabidopsis mutations showing increased Cd sensitivity could be assigned to defects in PCS (*cad1-1*; Howden et al., 1995b) and GSH1 (*cad2-1*; Howden et al., 1995a; Cobbett et al., 1998), respectively (see above). When PCS activity is up-regulated in response to Cd exposure, phytochelatins (expressed as GSH equivalents) may accumulate 10-fold the tissues GSH concentration (Schäfer et al., 1998; Haag-Kerwer et al., 1999; Heiss et al., 1999; Heiss et al., 2003 and refs. cited therein), causing a transient decrease of GSH content (Ducruix et al., 2006; Herbette et al., 2006; Nocito et al., 2006). In response to the strong metabolic sink for GSH generated during the rapid accumulation of phytochelatins, the biosynthesis of GSH and cysteine and the entire sulfate assimilation pathway all become activated (Schäfer et al., 1998; Heiss et al., 1999). This activation is based on a coordinate transcriptional up-regulation (Schäfer et al., 1998; Xiang and Oliver, 1998; Heiss et al., 1999; Xiang et al., 2001; Herbette et al., 2006; Nocito et al., 2006), and extends to the transcriptional activation of high affinity sulfate transporters (Nocito et al., 2002; Nocito et al., 2006). During early biosynthesis of PCs, γ-glutamylcysteine content may, at least transiently, increase while cysteine levels remain unaffected (Ducruix et al., 2006; Nocito et al., 2006), pointing to a limitation of glycine and/or GSH2 activity. As feeding glycine to Cd-exposed Arabidopsis suspension-culture cells prevented the accumulation of γ-glutamylcysteine, it is likely that glycine may turn into a limiting factor when GSH synthesis is upgraded during PC synthesis; however, this may largely depend on the tissue and be less relevant in leaves with active photorespiration. This interpretation is supported by earlier studies on poplar transformed with γ-glutamylcysteine synthetase from *E. coli* (Noctor et al., 1999), where γ-glutamylcysteine was shown to accumulate in leaves only in the dark phase.

The observation of a (at least transiently) reduced GSH content as early as 2 h after onset of Cd exposure indicates that the cellular redox poise will be significantly affected (Herbette et al., 2006; Nocito et al., 2006). In their recent study on Cd-exposed maize seedlings, Nocito et al. (2006) reported a threefold decrease of reduced GSH 3 h after Cd exposure, with the content of γ-glutamylcysteine remaining almost unaffected. Such a shift in the glutathione-based cellular redox potential is expected to act as a strong signal, and indeed, genes involved in the response to reactive oxygen species were among the Cd-responsive genes in Arabidopsis (Suzuki et al., 2001) and *B. juncea* (Minglin et al., 2005). Recently, Gillet et al. (2006) have shown in a proteomic approach that in the unicellular photosynthetic algae *Chlamydomonas reinhardtii* several proteins related to oxidative stress response are indeed up-regulated in response to Cd exposure, and that most of the Cd-sensitive proteins were also regulated via thioredoxin (TRX) and/or GRX.

Based on the concept of phytoremediation, several attempts have been made to upgrade the potential of plants to accumulate heavy metals (Pilon-Smits, 2005 and refs. cited therein). Since detoxification of heavy metals via the GSH-based formation of PCs is pivotal to both heavy metal tolerance and accumulation, genes of the cysteine biosynthesis pathway, genes encoding GSH biosynthesis (GSH1 and GSH2), and PCS, have all been overexpressed in plants, individually or in combination (Zhu et al., 1999a,b; Xiang et al., 2001; Dominguez-Solis et al., 2004; Bittsanszky et al., 2005; Pomponi et al., 2006; Wawrzynski et al., 2006). While, as expected, an increased Cd tolerance and accumulation have been achieved in several plant species, including tobacco, poplar and *B. juncea*, the overall potential of engineering PC biosynthesis, directly or indirectly, appears to be rather moderate. In particular, the phytoremediation-relevant root-shoot transfer of heavy metals proved to remain a major bottleneck. While for Arabidopsis, transfer of PCs and Cd from root to shoot and vice versa has been demonstrated (Gong et al., 2003; Chen et al., 2006), the overexpression of PCS from Arabidopsis in tobacco enhanced only Cd tolerance but not Cd transfer from root to shoot (Pomponi et al., 2006). Whether at the whole plant level, redistribution of the PC precursor GSH via modulation of its long-distance transport plays a role in the potential of roots (or shoots) to mount their PC defense line is not known, however, the observed changes in expression of a putative GSH transporter in *B. juncea* at both transcript and protein level is indicative of such an adaptation (Bogs et al., 2003). To what extent the GSH-degrading γ-glutamyl transpeptidases are involved in long-distance transport of GSH (i.e. by catalyzing phloem loading and/or unloading of GSH or GSSG) remains to be conclusively shown (Ohkama-Ohtsu et al., 2007a); if their role in GSH transport can be confirmed it would be interesting to study their expression in response to Cd exposure.

In summary, GSH plays its pivotal role in cellular detoxification of heavy metals most likely in two ways, i.e. being the precursor for PC biosynthesis, and acting as a cellular redox sensor due to the transient but significant decrease of reduced GSH in response to heavy metal exposure.

C.   GSH as a Reductant in Normal Metabolism

GSH is now well established as an important factor in reversible protein modification catalyzed by GRXs (Fernandes and Holmgren, 2004; Shelton et al., 2005). The original discovery of GRXs as important mediators of GSH-derived electrons, however, was related to a metabolic process. The GRX system was first discovered in thioredoxin deficient *E. coli*, which were still able to reduce ribonucleotides (Holmgren, 1976). In this case electrons are transferred from NADPH to glutathione catalyzed by GR and finally to the target protein ribonucleotide reductase in a reaction catalyzed by GRXs. Besides a large number of diverse GRXs plants also contain GRX-motifs fused to other proteins. The best described example is the APS reductase which contains a GRX-like domain at its C-terminus and uses GSH as hydrogen donor (Bick et al., 1998).

While the main function of plant GSTs is the conjugation of xenobiotics (see section IV.A.) some subgroups of this highly diverse gene family have other catalytic functions that do not lead to formation of glutathione S-conjugates. Several isoforms belonging to the tau-, phi- and theta-classes of GSTs have been shown to exhibit glutathione-dependent peroxidase activity with reductive activity towards organic hydroperoxides. In these reactions GSH is oxidized to the respective sulfenic acid, which spontaneously reacts with a second GSH molecule to form GSSG (Edwards et al., 2000).

The most prominent use of GSH as an electron donor in normal metabolism, however, is the GAC

in which electrons are transferred from GSH to dehydroascorbate (DHA) to regenerate ascorbate. The enzyme responsible for this electron transfer is dehydroascorbate reductase (DHAR), which has also been shown to belong to the GST superfamily (Dixon et al., 2002). These enzymes, in contrast to most other GSTs, are monomeric. The active site serine residue in this case is replaced by a single cysteine, a catalytically essential residue that has been proposed to form mixed disulfides with GSH during the catalytic cycle (Dixon et al., 2002). Arabidopsis contains four different DHARs which are predicted to be targeted to the cytosol, the mitochondria and chloroplasts. The presence of DHAR in both mitochondria and chloroplasts has been confirmed by proteomic studies (Chew et al., 2003b).

D.   The Glutathione–Ascorbate Cycle (GAC): Removing Reactive Oxygen Species (ROS)

Depending on their in situ concentration and on the metabolic context, ROS may act as cellular stressors, or, alternatively, as primary signals which initiate specific developmental or defense programs, either directly or via affecting the redox state of cellular antioxidant molecules (Foyer and Noctor, 2005 and refs. cited therein). Consequently, plants have developed mechanisms to induce, to increase or to quench ROS accumulation. Thus, under pathogen attack, plasma membrane bound NADPH oxidase is activated and initiates the "oxidative burst" in the apoplast, which in turn plays a major role in orchestrating the cellular defense reaction, culminating in programmed cell death (hypersensitive response) (Vanacker et al., 2000; Dangl and Jones, 2001; Ameisen, 2002; Mou et al., 2003). Also, it was recently shown that NADPH oxidase-mediated formation of ROS in the apoplast provides an important signal for developmental processes like root hair formation (Foreman et al., 2003). Conversely, to cope with excessive and unregulated accumulation of stress-induced ROS in different compartments, e.g. chloroplasts, mitochondria, peroxisomes and the cell wall space, a set of small antioxidant molecules provides a regulated network present in different cellular compartments (Foyer and Noctor, 2005; Kanwischer et al., 2005). Included are glutathione, ascorbic acid, α-tocopherol and several secondary plant products with antioxidant activities (e.g. flavonoids). While α-tocopherol is a membrane-soluble, lipophilic antioxidant, glutathione and ascorbic acid are soluble antioxidants present in up to millimolar concentrations in most cellular compartments. Functionally, the latter are linked in the GAC (Fig. 2). In principle, electrons from NADPH are used to detoxify ROS (namely H_2O_2) via a sequence of enzymatic steps, which involve the sequential oxidation/reduction of glutathione and ascorbic acid, reflecting the redox potentials of the named antioxidants. The GAC has been suggested to operate in the cytosol, chloroplasts, mitochondria and peroxisomes (Chew et al., 2003b; Kuzniak and Sklodowska, 2005a; Leterrier et al., 2005). While for chloroplasts and mitochondria, GAC function is certainly required for ROS detoxification in these organelles, its operation in peroxisomes might also be required to regenerate NADP for metabolic function (del Rio et al., 2002; Kuzniak and Sklodowska, 2005b). Interestingly, for GR and several other GAC enzymes dual targeting to chloroplasts and mitochondria has been demonstrated (Creissen et al., 1995; Chew et al., 2003b). As the biosynthesis of glutathione and ascorbic acid are strictly compartmentalized (Smirnoff et al., 2001; Wachter et al., 2005), transport systems must exist to shuttle these antioxidants between different compartments. As yet, little is known about the involved transporters and their mechanism for antioxidant exchange between different compartments. While in the apoplast, ascorbic acid appears to provide the only redox buffer (Horemans et al., 2000; Sanmartin et al., 2003; Foyer and Noctor, 2005; Pignocchi et al., 2006), in other organelles glutathione and ascorbic acid always operate together in the GAC. The role of the central vacuole in cellular redox poise has not attracted much attention. However, clearly ROS, including H_2O_2, may equilibrate with the vacuolar compartment which often makes up 90% of the cellular volume. Several secondary plant products, including flavonoids, which exhibit antioxidant activity, are localized in the vacuole and may possibly act as redox buffer. Recently, Bienert et al. (2007) could show that H_2O_2 is transported via certain types of aquaporins (Arabidopsis TIP1;1 and TIP1;2), providing a route for regulated movement of H_2O_2 across the tonoplast membrane. Interestingly, Arabidopsis lines mutated in TIP1;1 exhibit a strong growth phenotype (cell and plant death) with a signature reminiscent of oxidative stress (Ma et al., 2004).

The above considerations assign a central role to the antioxidant pair glutathione–ascorbate for

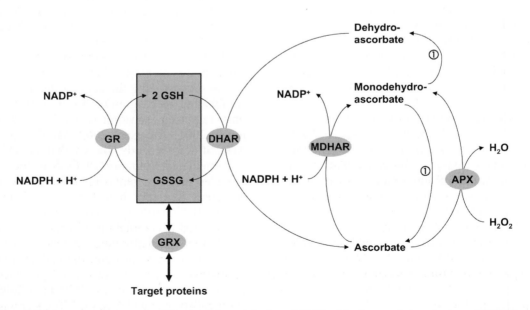

Fig. 2. The glutathione–ascorbate cycle. GR: glutathione reductase; DHAR: dehydroascorbate reductase; MDHAR: monodehydroascorbate reductase; APX: ascorbate peroxidase; GRX: glutaredoxin; ①: spontaneous disproportionation of monodehydroascorbate. The ascorbate part of the cycle is not shown stoichiometrically.

maintaining cellular redox poise. In particular, regulation and compartmentation of glutathione and ascorbate biosynthesis, degradation, and the exchange of glutathione and ascorbate between different compartments via specific transporters all impact on GAC function by affecting the pool sizes of both antioxidants. As both compounds are also largely, but not exclusively synthesized in the leaves, at the whole plant level source-sink relationships are superimposed on the local distributions. Thus, ascorbate levels in the root are more than 10-fold lower than in photosynthetic tissues (C. Kiefer and T. Rausch, unpublished). Within GAC itself, individual components of this complex enzyme network, including multiple isoforms for SOD, APX, MDHAR, DHAR, and GR for different cellular compartment, have all be shown to be regulated in their expression and/or activity in response to increased ROS exposure. Mutant analysis and attempts to engineer the expression of single enzymes of GAC, or of enzymes involved in the biosynthesis of one of the major cellular antioxidants have revealed substantial "cross talk" between GAC components. Prominent examples are the GSH1 mutant *rax1* in Arabidopsis (Ball et al., 2004), the effect of ectopic overexpression of DHAR (Chen et al., 2003), the tocopherol cyclase mutant *vte1*, and the low ascorbate *vtc1* mutant (Kanwischer et al., 2005). In all these cases, changing the level of one antioxidant significantly affected the concentration of other antioxidants. In the *rax1* mutant (Ball et al., 2004), a significantly decreased affinity of GSH1 for its substrate cysteine results in a lowered GSH content (Hothorn et al., 2006), causing a constitutive up-regulation of a specific APX isoform. In the *vte1* mutant, increased contents of GSH and AA were observed, whereas in the *vtc1* mutant the content of α-tocopherol was enhanced (Kanwischer et al., 2005). Interestingly, the ectopic expression of a wheat DHAR in tobacco and in maize caused not only an increase of total AA content, but also led to an up-regulation of total GSH content, and in both cases the reduced forms were even more affected (Chen et al., 2003). The molecular mechanisms of these various types of "cross-talk" between the different biosynthetic pathways and/or redox states of individual antioxidants are still a matter of speculation. Obviously, a comprehensive and dynamic analysis of all enzymes (including isoforms) and

antioxidant metabolites at compartmental and high temporal resolution, respectively, would be required to develop a model for the observed interdependencies. While microarray analysis allows to simultaneously follow the expression of all contributing enzyme isoforms, at least at the transcript level, the recent demonstration of post-transcriptional and post-translational controls (Hothorn et al., 2006; Jez et al., 2006) as well as the dynamic changes of subcellular compartmentation preclude any simply predictions based on transcriptomics alone.

Finally, another important aspect of GAC-mediated redox control in different cellular compartments is the continuous requirement for NADPH to reduce GSSG via GR. At present, no experimental studies or model calculations are available that would address the metabolic costs of GAC operation in different tissues and under different stress conditions. While in light-exposed leaves, the required NADPH is continuously regenerated via photosynthetic electron transport, the situation is different in the dark and in heterotrophic tissues. Here, GAC operation is most likely fueled by NADPH originating from the oxidative pentose phosphate (OPP) cycle. It would be interesting to determine whether oxidative stress results in significant changes of OPP metabolites and what the metabolic costs are in terms of glucose oxidized as compared to cellular respiration.

E. Glutathione-Dependent Redox Signaling

The most important function of glutathione is its role in redox buffering of the intracellular milieu and, related to this, sensing and transmission of deviations form the steady state redox poise. Evolution of GSH is thought to be related to the evolution of oxygenic photosynthesis and it is assumed that the two step biosynthesis pathway has evolved in forward direction (see section II.A). Due to rapid autocatalytic oxidation of cysteine in the presence of transition metals and concomitant formation of hydroxyl radicals through a Fenton reaction cysteine does not meet the requirements of a suitable redox sensor (Meyer and Hell, 2005). Formation of the distinct γ-amide bond between glutamate and cysteine leads to blocking of the α-amino group of cysteine and an α-amino acid like domain at the glutamate residue. This feature already greatly reduces metal-catalyzed thiol oxidation of the cysteine residue, which can be even further reduced by adding the glycine residue to the C-terminus of cysteine (Fahey and Sundquist, 1991). Despite these important features in terms of interactions with transition metals, the redox properties of GSH and its precursors are highly similar. The standard reduction potentials range from −216 mV for the cysteine/cystine couple and −240 mV for the GSH/GSSG couple (pH 7.0, 25 °C) (Schafer and Buettner, 2001; Jones, 2002). The reduction potential of thiols is largely dependent on the degree of protonation of the thiol group and thus for quantitative comparisons an adjustment of −5.9 mV per 0.1 increase in pH needs to be considered for quantitative calculations. More important, however, is the fact that two GSH molecules form a single GSSG molecule during oxidation. Thus, the concentration of GSH enters the Nernst equation as a squared term resulting in dependence of the reduction potential on both, the total concentration of GSH equivalents ([GSH] + 2[GSSG]) and the degree of oxidation of the glutathione pool (Equ. 1). The GSH concentration in the cytosol is considered to be in the low millimolar range (Meyer et al., 2001) and thus 10–50 times higher than the cysteine concentration. With the assumption of a similar degree of oxidation the reduction potential of glutathione would be about 200 mV more negative than that of cysteine. Replacement of the glycine residue by other amino acids like serine or glutamate does not significantly change the redox properties of the thiol group (Krezel and Bal, 2003).

$$E_{hc} = -240 \text{ mV} - (59.1/2) \text{ mV} * \log ([GSH]^2/[GSSG]) \qquad (1)$$

The reduction state of the glutathione pool is frequently described as > 90% (Noctor, 2006). Assuming a total glutathione concentration of 1 mM and a degree of oxidation of only 1% at pH 7.0 would, according to Equ. 1, result in a reduction potential of −219 mV. The recently developed roGFP enables direct redox measurements in living cells (Dooley et al., 2004; Hanson et al., 2004; Jiang et al., 2006). This sensor has a midpoint reduction potential of −280 mM at pH 7.0 (Hanson et al., 2004) and equilibrates specifically with the cellular glutathione redox buffer (Meyer et al., 2007).

Imaging of roGFP expressed in the cytosol of Arabidopsis and tobacco showed that the sensor is almost completely reduced. This observation strongly suggests that the actual reduction potential of the cellular glutathione redox buffer is far more negative than formerly assumed on the basis of biochemical analysis of cell extracts. With total glutathione concentrations in the low millimolar range GSSG would thus be present in only sub-micromolar concentrations.

The reduction potential of glutathione can equilibrate with other redox buffers. Unless enzymatically catalyzed such an equilibration would be far too slow to have any physiological significance. A particular family of disulfide-oxidoreductases, the GRXs, are capable of reversibly transferring electrons between glutathione and thiol groups on target proteins (Fig. 3). It is now apparent that GRXs are involved in a large number of different cellular processes and that GRXs play a crucial role in response to oxidative stress in all organisms (Fernandes and Holmgren, 2004; Shelton et al., 2005; Xing et al., 2006). The ability of GRXs to rapidly equilibrate the glutathione reduction potential with the reduction potential of protein thiols can be exploited for signaling purposes. Analogous to O-phosphorylation reversible post-translational modification of specific thiols can also efficiently regulate protein functions. Depending on the protein structure this redox-dependent regulation can occur in different forms. Single cysteine residues might be glutathionylated and in other cases formation of disulfide bridges between two cysteines might occur. Based on these two possibilities target proteins for redox-dependent modification have been classified in Type I- and Type II-nanoswitches (Schafer and Buettner, 2001; Meyer and Hell, 2005).

The Arabidopsis genome encodes at least 31 GRXs (Lemaire, 2004; Xing et al., 2006), which are predicted to be present in all subcellular compartments apart the vacuole. Based on the cysteine motif of the active site plant GRXs cluster in three distinct groups, denominated CPYC-, CGFS- and CC-type (Xing et al., 2006). The CPYC group is structurally equivalent to the classical dithiol GRXs from other organisms, while the CC-type GRXs are specific for plants and most prominent in higher plants (Xing et al., 2006). The high number of different GRXs implicates a large number of diverse target proteins.

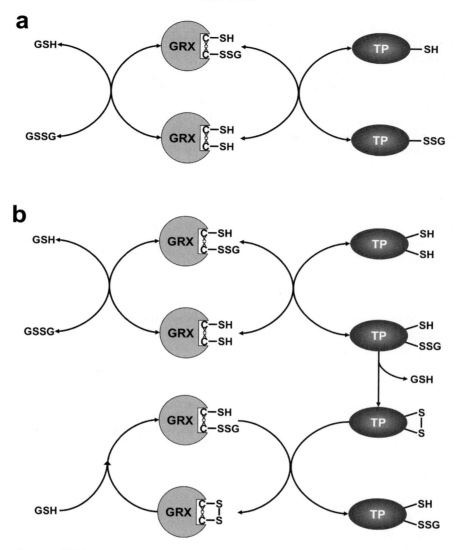

Fig. 3. Mechanism for reversible interaction of glutaredoxins with target proteins. (a) Glutathionylation of a type I target protein (TP). (b) Oxidation and reduction of a type II target protein. Cartoons shown for the oxidation reactions (top) are for a dithiol GRX, but reactions are exactly the same for monothiol GRXs, because the catalytic reaction leading to oxidation of a target protein involves only the N-terminal cysteine of the GRX motif. The glutathionylated intermediate of the target protein is instable and a second nearby cysteine can substitute the glutathione moiety thus forming an intramolecular disulfide bridge. In contrast to the oxidation reaction, the reduction of target protein disulfides by GRX might involve both active site cysteines (bottom).

Because the catalytic mechanism involves only the N-terminal cysteine and because GRXs only transfer GSH rather than forming mixed disulfides with target proteins (Peltoniemi et al., 2006), most approaches for trapping of mixed disulfides between GRXs and their targets similar to successful approaches for TRXs have failed. The only exception is a report by Rouhier et al. (2005) in which 94 putative targets of poplar GRX C4 were trapped after mutating the second cysteine from the active site. Applying a different approach based on labeling of proteins with radioactive GSH and mass-spectrometry, different studies have recently shown that plant cells, like other eukaryotic cells, contain several proteins that undergo *S*-glutathionylation (Ito et al., 2003; Dixon et al., 2005; Michelet et al., 2005). *S*-glutathionylation can protect protein thiols from irreversible oxidation by ROS (Ghezzi, 2005). This protective mechanism is important for retaining protein structure during desiccation of resurrection plants (Kranner et al., 2002). In addition, reversible protein *S*-glutathionylation as well as GRX-dependent formation of disulfide bridges are means for regulation of signal transduction and are currently emerging as novel mechanisms involved in cellular regulation. To be effective as a regulatory mechanism a number of different criteria need to be met: First, the function of the modified protein must change. Second, the redox-dependent change must occur within intact cells in response to a physiological stimulus. Third, redox-dependent changes must occur at relatively low degree of oxidation of the cellular glutathione pool, i.e. very low concentrations of GSSG. Fourth, there must be rapid and efficient mechanisms for specific protein modifications in place. And finally, there must be rapid and efficient mechanisms for reversing the redox-dependent change. While the first criterion, change of function, is fulfilled by the described glutathionylated proteins TRXf, aldolase and triose-phosphate isomerase (Ito et al., 2003; Michelet et al., 2005), all other criteria remain largely unresolved, mainly because few target proteins have been established so far.

The best evidence for full reversibility of GRX-dependent post-translational protein modification results from expression of redox-sensitive fluorescent proteins (rxYFP, (Østergaard et al., 2001); roGFP (Hanson et al., 2004; Jiang et al., 2006). These proteins act as artificial targets for GRXs (Østergaard et al., 2004; Meyer et al., 2007) and redox-dependent changes in fluorescence can directly be observed in living cells. Changes of fluorescence occur in response to the cytosolic glutathione levels and different stimuli affecting the glutathione pool. Due to its redox-potential of −280 mV roGFP is extremely sensitive to minute changes of GSSG indicating that the cytosolic glutathione pool in non-stressed cells is almost completely reduced with only sub-micromolar concentrations of GSSG (Meyer et al., 2007). Kinetic analysis of fluorescence showed that the redox-dependent alteration of roGFP in living cells occurs much faster than in vitro indicating the involvement of interacting proteins, which are likely to be GRXs. GRXs at the same time guarantee the full reversibility of the reaction.

In the light of the very small number of described glutathionylated proteins the major challenge for the future thus is the identification of target proteins interacting with GRXs and exploring the specificity in these interactions. In this context it will also be necessary to study the exact localization of the entire set of GRXs at subcellular level and the expression during development and under stress conditions. Identification of target proteins will likely highlight the potential for cross-talk between different signal transduction pathways, which has been shown for the cross-talk between plastidic GRX and TRX systems already (Michelet et al., 2005). Due to the emerging complexity of interacting signaling pathways the entire signaling network cannot be described solely by biochemical approaches. Instead, quantitative descriptions of glutathione-dependent signaling and its integration into the cellular signaling system, will also rely on the use of computer-based modeling tools.

Acknowledgements

Research in our laboratories has been supported by the Deutsche Forschungsgemeinschaft (ME1567/3-2, RA415/10-1, RA8-3/4 and RA9-3/4) and the University of Heidelberg. Further support to TR was provided by the KWS SAAT AG, Einbeck, and the Südzucker AG, Mannheim.

References

Ameisen J (2002) On the origin, evolution, and nature of programmed cell death: A timeline of four billion years. Cell Death Differ 9: 367–393

Ball L, Accotto GP, Bechtold U, Creissen G, Funck D, Jimenez A, Kular B, Leyland N, Mejia-Carranza J, Reynolds H, Karpinski S and Mullineaux PM (2004) Evidence for a direct link between glutathione biosynthesis and stress defense gene expression in Arabidopsis. Plant Cell 16: 2448–2462

Barreto L, Garcera A, Jansson K, Sunnerhagen P and Herrero E (2006) A peroxisomal glutathione transferase of *Saccharomyces cerevisiae* is functionally related to sulfur amino acid metabolism. Eukaryot Cell 5: 1748–1759

Bass R, Ruddock LW, Klappa P and Freedman RB (2004) A major fraction of endoplasmic reticulum-located glutathione is present as mixed disulfides with protein. J Biol Chem 279: 5257–5262

Beck A, Lendzian K, Oven M, Christmann A and Grill E (2003) Phytochelatin synthase catalyzes key step in turnover of glutathione conjugates. Phytochemistry 62: 423–431

Bellomo G, Vairetti M, Stivala L, Mirabelli F, Richelmi P and Orrenius S (1992) Demonstration of nuclear compartmentalization of glutathione in hepatocytes. Proc Natl Acad Sci U S A 89: 4412–4416

Bick JA, Aslund F, Chen Y and Leustek T (1998) Glutaredoxin function for the carboxyl-terminal domain of the plant-type 5′-adenylylsulfate reductase. Proc Natl Acad Sci U S A 95: 8404–8409

Bienert GP, Moller ALB, Kristiansen KA, Schulz A, Moller IM, Schjoerring JK and Jahn TP (2007) Specific aquaporins facilitate the diffusion of hydrogen peroxide across membranes. J Biol Chem 282: 1183–1192

Bittsanszky A, Komives T, Gullner G, Gyulai G, Kiss J, Heszky L, Radimszky L and Rennenberg H (2005) Ability of transgenic poplars with elevated glutathione content to tolerate zinc(2+) stress. Environ Int 31: 251–254

Blum R, Beck A, Korte A, Stengel A, Letzel T, Lendzian K and Grill E (2007) Function of phytochelatin synthase in catabolism of glutathione-conjugates. Plant J 49: 740–749

Bogs J, Bourbouloux A, Cagnac O, Wachter A, Rausch T and Delrot S (2003) Functional characterization and expression analysis of a glutathione transporter, BjGT1, from Brassica juncea: Evidence for regulation by heavy metal exposure. Plant Cell Environ 26: 1703–1711

Bourbouloux A, Shahi P, Chakladar A, Delrot S and Bachhawat AK (2000) Hgt1p, a high affinity glutathione transporter from the yeast Saccharomyces cerevisiae. J Biol Chem 275: 13259–13265

Briviba K, Fraser G, Sies H and Ketterer B (1993) Distribution of the monochlorobimane-glutathione conjugate between nucleus and cytosol in isolated hepatocytes. Biochem J 294: 631–633

Cagnac O, Bourbouloux A, Chakrabarty D, Zhang M-Y and Delrot S (2004) AtOPT6 transports glutathione derivatives and is induced by primisulfuron. Plant Physiol 135: 1378–1387

Cairns NG, Pasternak M, Wachter A, Cobbett CS and Meyer AJ (2006) Maturation of Arabidopsis seeds is dependent on glutathione biosynthesis within the embryo. Plant Physiol 141: 446–455

Chen A, Komives EA and Schroeder JI (2006) An improved grafting technique for mature Arabidopsis plants demonstrates long-distance shoot-to-root transport of phytochelatins in Arabidopsis. Plant Physiol 141: 108–120

Chen Z and Lash LH (1998) Evidence for mitochondrial uptake of glutathione by dicarboxylate and 2-oxoglutarate carriers. J Pharmacol Exp Ther 285: 608–618

Chen Z, Young TE, Ling J, Chang S-C and Gallie DR (2003) Increasing vitamin C content of plants through enhanced ascorbate recycling. Proc Natl Acad Sci U S A 100: 3525–3530

Chew O, Rudhe C, Glaser E and Whelan J (2003a) Characterization of the targeting signal of dual-targeted pea glutathione reductase. Plant Mol Biol 53: 341–356

Chew O, Whelan J and Millar AH (2003b) Molecular definition of the ascorbate-glutathione cycle in Arabidop-

sis mitochondria reveals dual targeting of antioxidant defenses in plants. J Biol Chem 278: 46869–46877

Clemens S (2006) Evolution and function of phytochelatin synthases. J Plant Physiol 163: 319–332

Clemens S, Palmgren MG and Kramer U (2002) A long way ahead: Understanding and engineering plant metal accumulation. Trends Plant Sci 7: 309–315

Cobbett CS, May MJ, Howden R and Rolls B (1998) The glutathione-deficient, cadmium-sensitive mutant, cad2-1, of Arabidopsis thaliana is deficient in γ-glutamylcysteine synthetase. Plant J 16: 73–78

Coleman JOD, Blake-Kalff MMA and Davies TGE (1997) Detoxification of xenobiotics by plants: Chemical modification and vacuolar compartmentation. Trends Plant Sci 2: 144–151

Copley SD and Dhillon JK (2002) Lateral gene transfer and parallel evolution in the history of glutathione biosynthesis genes. Genome Biol 3: 1–16

Creissen G, Reynolds H, Xue Y and Mullineaux P (1995) Simultaneous targeting of pea glutathione reductase and of a bacterial fusion protein to chloroplasts and mitochondria in transgenic tobacco. Plant J 8: 167–175

Cuozzo JW and Kaiser CA (1999) Competition between glutathione and protein thiols for disulphide-bond formation. Nat Cell Biol 1: 130–135

Dangl J and Jones J (2001) Plant pathogens and integrated defence responses to infection. Nature 411: 826–833

del Rio LA, Corpas FJ, Sandalio LM, Palma JM, Gomez M and Barroso JB (2002) Reactive oxygen species, antioxidant systems and nitric oxide in peroxisomes. J Exp Bot 53: 1255–1272

Des Marais DJ (2000) Evolution: When did photosynthesis emerge on earth? Science 289: 1703–1705

Despres C, Chubak C, Rochon A, Clark R, Bethune T, Desveaux D and Fobert PR (2003) The Arabidopsis NPR1 disease resistance protein is a novel cofactor that confers redox regulation of DNA binding activity to the basic domain/leucine zipper transcription factor TGA1. Plant Cell 15: 2181–2191

Dixon DP, Davis BG and Edwards R (2002) Functional divergence in the glutathione transferase superfamily in plants. Identification of two classes with putative functions in redox homeostasis in Arabidopsis thaliana. J Biol Chem 277: 30859–30869

Dixon DP, Skipsey M, Grundy NM and Edwards R (2005) Stress-induced protein S-glutathionylation in Arabidopsis. Plant Physiol 138: 2233–2244

Dominguez-Solis J, Lopez-Martin M, Ager F, Ynsa M, Romero L and Gotor C (2004) Increased cysteine availability is essential for cadmium tolerance and accumulation in Arabidopsis thaliana. Plant Biotechnol J 2: 469–476

Dooley CT, Dore TM, Hanson GT, Jackson WC, Remington SJ and Tsien RY (2004) Imaging dynamic redox changes in mammalian cells with green fluorescent protein indicators. J Biol Chem 279: 22284–22293

Ducruix C, Junot C, Fievet J, Villiers F, Ezan E and Bourguignon J (2006) New insights into the regulation of phytochelatin biosynthesis in *A. thaliana* cells from metabolite profiling analyses. Biochimie 88: 1733–1742

Edwards R and Dixon D (2005) Plant glutathione transferases. Methods Enzymol 401: 169–186

Edwards R, Dixon DP and Walbot V (2000) Plant glutathione S-transferases: Enzymes with multiple functions in sickness and in health. Trends Plant Sci 5: 193–198

Fahey RC (2001) Novel thiols of prokaryotes. Annu Rev Microbiol 55: 333–356

Fahey RC and Sundquist AR (1991) Evolution of glutathione metabolism. Adv Enzymol RAMB 64: 1–53

Fahey R, Newton G, Arrick B, Overdank-Bogart T and Aley S (1984) *Entamoeba histolytica*: A eukaryote without glutathione metabolism. Science 224: 70–72

Fernandes A and Holmgren A (2004) Glutaredoxins: Glutathione-dependent redox enzymes with functions far beyond a simple thioredoxin backup system. Antioxid Redox Signal 6: 63–74

Foreman J, Demidchik V, Bothwell J, Mylona P, Miedema H, Torres M, Linstead P, Costa S, Brownlee C, Jones J, Davies J and Dolan L (2003) Reactive oxygen species produced by NADPH oxidase regulate plant cell growth. Nature 422: 442–446

Foyer CH and Noctor G (2005) Redox homeostasis and antioxidant signaling: A metabolic interface between stress perception and physiological responses. Plant Cell 17: 1866–1875

Foyer CH, Theodoulou FL and Delrot S (2001) The functions of inter- and intracellular glutathione transport systems in plants. Trends Plant Sci 6: 486–492

Frendo P, Jimenez MJ, Mathieu C, Duret L, Gallesi D, Van de Sype G, Herouart D and Puppo A (2001) A *Medicago truncatula* homoglutathione synthetase is derived from glutathione synthetase by gene duplication. Plant Physiol 126: 1706–1715

Fricker MD, May M, Meyer AJ, Sheard N and White NS (2000) Measurement of glutathione levels in intact roots of Arabidopsis. J Microsc (Oxf) 198: 162–173

Frova C (2003) The plant glutathione transferase gene family: Genomic structure, functions, expression and evolution. Physiol Plant 119: 469–479

Fukao Y, Hayashi M and Nishimura M (2002) Proteomic analysis of leaf peroxisomal proteins in greening cotyledons of *Arabidopsis thaliana*. Plant Cell Physiol 43: 689–696

Galperin MY and Koonin EV (1997) A diverse superfamily of enzymes with ATP-dependent carboxylate-amine/thiol ligase activity. Protein Sci 6: 2639–2643

Ghezzi P (2005) Regulation of protein function by glutathionylation. Free Radic Res 39: 573–580

Gillet S, Decottignies P, Chardonnet S and Le Marechal P (2006) Cadmium response and redoxin targets in *Chlamydomonas reinhardtii*: A proteomic approach. Photosynth Res 89: 201–211

Glazebrook J, Zook M, Mert F, Kagan I, Rogers EE, Crute IR, Holub EB, Hammerschmidt R and Ausubel FM (1997) Phytoalexin-deficient mutants of Arabidopsis reveal that *pad4* encodes a regulatory factor and that four *pad* genes

contribute to downy mildew resistance. Genetics 146: 381–392

Gogos A and Shapiro L (2002) Large conformational changes in the catalytic cycle of glutathione synthase. Structure 10: 1669–1676

Gong JM, Lee DA and Schroeder JI (2003) Long-distance root-to-shoot transport of phytochelatins and cadmium in Arabidopsis. Proc Natl Acad Sci U S A 100: 10118–10123

Grzam A, Tennstedt P, Clemens S, Hell R and Meyer AJ (2006) Vacuolar sequestration of glutathione S-conjugates outcompetes a possible degradation of the glutathione moiety by phytochelatin synthase. FEBS Lett 580: 6384–6390

Grzam A, Martin MN, Hell R and Meyer AJ (2007) γ-Glutamyl transpeptidase GGT4 initiates vacuolar degradation of glutathione S-conjugates in Arabidopsis. FEBS Lett 581: 3131–3138

Gutiérrez-Alcalá G, Gotor C, Meyer AJ, Fricker M, Vega JM and Romero LC (2000) Glutathione biosynthesis in Arabidopsis trichome cells. Proc Natl Acad Sci U S A 97: 11108–11113

Haag-Kerwer A, Schafer H, Heiss S, Walter C and Rausch T (1999) Cadmium exposure in *Brassica juncea* causes a decline in transpiration rate and leaf expansion without effect on photosynthesis. J Exp Bot 50: 1827–1835

Hanson GT, Aggeler R, Oglesbee D, Cannon M, Capaldi RA, Tsien RY and Remington SJ (2004) Investigating mitochondrial redox potential with redox-sensitive green fluorescent protein indicators. J Biol Chem 279: 13044–13053

Hara T, Kato H, Katsube Y and Oda J (1996) A pseudo-michaelis quaternary complex in the reverse reaction of a ligase: Structure of *Escherichia coli* B glutathione synthetase complexed with ADP, glutathione, and sulfate at 2.0 A resolution. Biochemistry 35: 11967–11974

Hartmann TN, Fricker MD, Rennenberg H and Meyer AJ (2003) Cell-specific measurement of cytosolic glutathione in poplar leaves. Plant Cell Environ 26: 965–975

Heazlewood JL, Verboom RE, Tonti-Filippini J, Small I and Millar AH (2007) SUBA: The Arabidopsis subcellular database. Nucleic Acids Res 35: D213–D218

Heiss S, Schafer HJ, Haag-Kerwer A and Rausch T (1999) Cloning sulfur assimilation genes of *Brassica juncea* L.: Cadmium differentially affects the expression of a putative low-affinity sulfate transporter and isoforms of ATP sulfurylase and APS reductase. Plant Mol Biol 39: 847–857

Heiss S, Wachter A, Bogs J, Cobbett C and Rausch T (2003) Phytochelatin synthase (PCS) protein is induced in *Brassica juncea* leaves after prolonged Cd exposure. J Exp Bot 54: 1833–1839

Hell R and Bergmann L (1990) γ-glutamylcysteine synthetase in higher plants: Catalytic properties and subcellular localization. Planta 180: 603–612.

Herbette S, Taconnat L, Hugouvieux V, Piette L, Magniette M, Cuine S, Auroy P, Richaud P, Forestier C, Bourguignon J, Renou J, Vavasseur A and Leonhardt N (2006) Genome-wide transcriptome profiling of the early cadmium response of Arabidopsis roots and shoots. Biochimie 88: 1751–1765

Herschbach C and Rennenberg H (2001) Significance of phloem-translocated organic sulfur compounds for the regulation of sulfur nutrition. Progr Bot 62: 177–193

Hibi T, Nii H, Nakatsu T, Kimura A, Kato H, Hiratake J and Oda Ji (2004) Crystal structure of γ-glutamylcysteine synthetase: Insights into the mechanism of catalysis by a key enzyme for glutathione homeostasis. Proc Natl Acad Sci U S A 101: 15052–15057

Hicks LM, Cahoon RE, Bonner ER, Rivard RS, Sheffield J and Jez JM (2007) Thiol-based regulation of redox-active glutamate-cysteine ligase from *Arabidopsis thaliana*. Plant Cell 19: 2653–2661

Holmgren A (1976) Hydrogen donor system for *Escherichia coli* ribonucleoside-diphosphate reductase dependent upon glutathione. Proc Natl Acad Sci U S A 73: 2275–2279

Horemans N, Foyer CH and Asard H (2000) Transport and action of ascorbate at the plant plasma membrane. Trends Plant Sci 5: 263–267

Hothorn M, Wachter A, Gromes R, Stuwe T, Rausch T and Scheffzek K (2006) Structural basis for the redox control of plant glutamate cysteine ligase. J Biol Chem 281: 27557–27565

Howden R, Andersen CR, Goldsbrough PB and Cobbett CS (1995a) A cadmium-sensitive, glutathione-deficient mutant of *Arabidopsis thaliana*. Plant Physiol 107: 1067–1073

Howden R, Goldsbrough PB, Andersen CR and Cobbett CS (1995b) Cadmium-sensitive, *cad1* mutants of *Arabidopsis thaliana* are phytochelatin deficient. Plant Physiol 107: 1059–1066

Hwang CC, Sinskey AJ and Lodish HF (1992) Oxidized redox state of glutathione in the endoplasmic reticulum. Science 257: 1496–1502

Ito H, Iwabuchi M and Ogawa K (2003) The sugar-metabolic enzymes aldolase and triose-phosphate isomerase are targets of glutathionylation in *Arabidopsis thaliana*: Detection using biotinylated glutathione. Plant Cell Physiol 44: 655–660

Jaquinod M, Villiers F, Kieffer-Jaquinod S, Hugouvieux V, Bruley C, Garin J and Bourguignon J (2007) A proteomics dissection of *Arabidopsis thaliana* vacuoles isolated from cell culture. Mol Cell Proteomics 6: 394–412

Jessop CE and Bulleid NJ (2004) Glutathione directly reduces an oxidoreductase in the endoplasmic reticulum of mammalian cells. J Biol Chem 279: 55341–55347

Jez JM and Cahoon RE (2004) Kinetic mechanism of glutathione synthetase from *Arabidopsis thaliana*. J Biol Chem 279: 42726–42731

Jez JM, Cahoon RE and Chen S (2004) *Arabidopsis thaliana* glutamate-cysteine ligase: Functional properties, kinetic mechanism, and regulation of activity. J Biol Chem 279: 33463–33470

Jez JM, Cahoon RE, Bonner ER and Chen S (2006) Redox-regulation of glutathione synthesis in plants. FASEB J 20: A41–A42

Jiang K, Schwarzer C, Lally E, Zhang S, Ruzin S, Machen T, Remington SJ and Feldman L (2006) Expression and characterization of a redox-sensing green fluorescent protein (reduction-oxidation-sensitive green fluorescent protein) in Arabidopsis. Plant Physiol 141: 397–403

Jimenez A, Hernandez JA, Del Rio LA and Sevilla F (1997) Evidence for the presence of the ascorbate–glutathione cycle in mitochondria and peroxisomes of pea leaves. Plant Physiol 114: 275–284

Jimenez A, Hernandez JA, Pastori G, del Rio LA and Sevilla F (1998) Role of the ascorbate-glutathione cycle of mitochondria and peroxisomes in the senescence of pea leaves. Plant Physiol 118: 1327–1335

Jones D (2002) Redox potential of GSH/GSSG couple: Assay and biological significance. Methods Enzymol 348: 93–112

Kanwischer M, Porfirova S, Bergmüller E and Dörmann P (2005) Alterations in tocopherol cyclase activity in transgenic and mutant plants of Arabidopsis affect tocopherol content, tocopherol composition, and oxidative stress. Plant Physiol 137: 713–723

Klapheck S (1988) Homoglutathione: Isolation, quantification and occurrence in legumes. Physiol Plant 74: 727–732

Koehler C, Beverly K and Leverich E (2006) Redox pathways of the mitochondrion. Antioxid Redox Signal 8: 813–822

Koh S, Wiles AM, Sharp JS, Naider FR, Becker JM and Stacey G (2002) An oligopeptide transporter gene family in Arabidopsis. Plant Physiol 128: 21–29

Kopriva S (2006) Regulation of sulfate assimilation in Arabidopsis and beyond. Ann Bot 97: 479–495

Kranner I, Beckett RP, Wornik S, Zorn M and Pfeifhofer HW (2002) Revival of a resurrection plant correlates with its antioxidant status. Plant J 31: 13–24

Krezel A and Bal W (2003) Structure–function relationships in glutathione and its analogues. Org Biomol Chem 1: 3885–3890

Kuzniak E and Sklodowska M (2005a) Compartment-specific role of the ascorbate–glutathione cycle in the response of tomato leaf cells to *Botrytis cinerea* infection. J Exp Bot 56: 921–933

Kuzniak E and Sklodowska M (2005b) Fungal pathogen-induced changes in the antioxidant systems of leaf peroxisomes from infected tomato plants. Planta 222: 192–200

Lappartient AG and Touraine B (1996) Demand-driven control of root ATP sulfurylase activity and SO$_4^{2-}$ uptake in intact canola. Plant Physiol 111: 147–157

Lappartient AG, Vidmar JJ, Leustek T, Glass AD and Touraine B (1999) Inter-organ signaling in plants: Regulation of ATP sulfurylase and sulfate transporter genes expression in roots mediated by phloem-translocated compound. Plant J 18: 89–95

Lee SM, Moon JS, Domier LL and Korban SS (2002) Molecular characterization of phytochelatin synthase expression in transgenic Arabidopsis. Plant Physiol Biochem 40: 727–733

Lemaire SD (2004) The glutaredoxin family in oxygenic photosynthetic organisms. Photosynth Res 79: 305–318

Leterrier M, Corpas FJ, Barroso JB, Sandalio LM and del Rio LA (2005) Peroxisomal monodehydroascorbate reductase. Genomic clone characterization and functional analysis under environmental stress conditions. Plant Physiol 138: 2111–2123

Li Y, Dankher OP, Carreira L, Smith AP and Meagher RB (2006) The shoot-specific expression of gamma-glutamyl-cysteine synthetase directs the long-distance transport of thiol-peptides to roots conferring tolerance to mercury and arsenic. Plant Physiol 141: 288–298

Li ZS, Zhao Y and Rea PA (1995) Magnesium adenosine 5'-triphosphate-energized transport of glutathione-S-conjugates by plant vacuolar membrane vesicles. Plant Physiol 107: 1257–1268

Lieberman MW, Wiseman AL, Shi Z-Z, Carter BZ, Barrios R, Ou C-N, Chevez-Barrios P, Wang Y, Habib GM, Goodman JC, Huang SL, Lebovitz RM and Matzuk MM (1996) Growth retardation and cysteine deficiency in gamma-glutamyl transpeptidase-deficient mice. Proc Natl Acad Sci U S A 93: 7923–7926

Logan DC (2006) The mitochondrial compartment. J Exp Bot 57: 1225–1243

Ma S, Quist T, Ulanov A, Joly R and Bohnert H (2004) Loss of TIP1;1 aquaporin in Arabidopsis leads to cell and plant death. Plant J 40: 845–859

Martin MN (2003) Biosynthesis and metabolism of glutathione in plants. In: Setlow JK (ed) Genetic engineering, Vol 25, pp 163–188. Kluwer Academic/Plenum, London

Martin MN, Saladores PH, Lambert E, Hudson AO and Leustek T (2007) Localization of members of the γ-glutamyl transpeptidase family identifies sites of glutathione and glutathione S-conjugate hydrolysis. Plant Physiol 140: 1715–1732

Martinoia E, Erwin G, Roberto T, Kreuz K and Amrhein N (1993) ATP-dependent glutathione S-conjugate 'export' pump in the vacuolar membrane of plants. Nature 364: 247–249

Martinoia E, Massonneau A and Frangne N (2000) Transport processes of solutes across the vacuolar membrane of higher plants. Plant Cell Physiol 41: 1175–1186

May MJ and Leaver CJ (1994) Arabidopsis thaliana γ-glutamylcysteine synthetase is structurally unrelated to mammalian, yeast, and Escherichia coli homologs. Proc Natl Acad Sci U S A 91: 10059–10063

May MJ, Vernoux T, Sanchez-Fernandez R, Van Montagu M and Inze D (1998) Evidence for posttranscriptional activation of γ-glutamylcysteine synthetase during plant stress responses. Proc Natl Acad Sci U S A 95: 12049–12054

Meier I (2007) Composition of the plant nuclear envelope: Theme and variations. J Exp Bot 58: 27–34

Meister A (1988) Glutathione metabolism and its selective modification. J Biol Chem 263: 17205–17208

Meyer AJ and Fricker MD (2000) Direct measurement of glutathione in epidermal cells of intact Arabidopsis roots by two-photon laser scanning microscopy. J Microsc (Oxf) 198: 174–181

Meyer AJ and Fricker MD (2002) Control of demand-driven biosynthesis of glutathione in green Arabidopsis suspension culture cells. Plant Physiol 130: 1927–1937

Meyer A and Hell R (2005) Glutathione homeostasis and redox-regulation by sulfhydryl groups. Photosynth Res 86: 435–457

Meyer AJ, May MJ and Fricker M (2001) Quantitative in vivo measurement of glutathione in Arabidopsis cells. Plant J 27: 67–78

Meyer AJ, Brach T, Marty L, Kreye S, Rouhier N, Jacquot J-P and Hell R (2007) Redox-sensitive GFP in Arabidopsis thaliana is a quantitative biosensor for the redox potential of the cellular glutathione redox buffer. Plant J 52: 973–986

Michelet L, Zaffagnini M, Marchand C, Collin V, Decottignies P, Tsan P, Lancelin J-M, Trost P, Miginiac-Maslow M, Noctor G and Lemaire SD (2005) Glutathionylation of chloroplast thioredoxin f is a redox signaling mechanism in plants. Proc Natl Acad Sci U S A 102: 16478–16483

Minglin L, Yuxiu Z and Tuanyao C (2005) Identification of genes up-regulated in response to Cd exposure in Brassica juncea L. Gene 363: 151–158

Mou Z, Fan W and Dong X (2003) Inducers of plant systemic acquired resistance regulate NPR1 function through redox changes. Cell 113: 935–944

Ndamukong I, Abdallat A, Thurow C, Fode B, Zander M, Weigel R and Gatz C (2007) SA-inducible Arabidopsis glutaredoxin interacts with TGA factors and suppresses JA-responsive PDF1.2 transcription. Plant J 50: 128–139

Newton G, Arnold K, Price M, Sherrill C, Delcardayre S, Aharonowitz Y, Cohen G, Davies J, Fahey R and Davis C (1996) Distribution of thiols in microorganisms: Mycothiol is a major thiol in most actinomycetes. J Bacteriol 178: 1990–1995

Nocito FF, Pirovano L, Cocucci M and Sacchi GA (2002) Cadmium-induced sulfate uptake in maize roots. Plant Physiol 129: 1872–1879

Nocito FF, Lancilli C, Crema B, Fourcroy P, Davidian J-C and Sacchi GA (2006) Heavy metal stress and sulfate uptake in maize roots. Plant Physiol 141: 1138–1148

Noctor G (2006) Metabolic signalling in defence and stress: The central roles of soluble redox couples. Plant Cell Environ 29: 409–425

Noctor G, Arisi A, Jouanin L and Foyer C (1999) Photorespiratory glycine enhances glutathione accumulation in both the chloroplastic and cytosolic compartments. J Exp Bot 50: 1157–1167

Noctor G, Veljovic-Jovanovic S and Foyer CH (2000) Peroxide processing in photosynthesis: Antioxidant coupling and redox signalling. Philos Trans R Soc Lond B Biol Sci 355: 1465–1475

Noctor G, Gomez L, Vanacker H and Foyer CH (2002) Interactions between biosynthesis, compartmentation and transport in the control of glutathione homeostasis and signalling. J Exp Bot 53: 1283–1304

Ohkama-Ohtsu N, Radwan S, Peterson A, Zhao P, Badr A, Xiang C and Oliver D (2007a) Characterization of the extracellular gamma-glutamyl transpeptidases, GGT1 and GGT2, in Arabidopsis. Plant J 49: 865–877

Ohkama-Ohtsu N, Zhao P, Xiang C and Oliver D (2007b) Glutathione conjugates in the vacuole are degraded by gamma-glutamyl transpeptidase GGT3 in Arabidopsis. Plant J 49: 878–888

Østergaard H, Henriksen A, Hansen FG and Winther JR (2001) Shedding light on disulfide bond formation: Engineering a redox switch in green fluorescent protein. EMBO J 20: 5853–5862

Østergaard H, Tachibana C and Winther JR (2004) Monitoring disulfide bond formation in the eukaryotic cytosol. J Cell Biol 166: 337–345

Parisy V, Poinssot B, Owsianowski L, Buchala A, Glazebrook J and Mauch F (2007) Identification of PAD2 as a γ-glutamylcysteine synthetase highlights the importance of glutathione in disease resistance of Arabidopsis. Plant J 49: 159–172

Pasternak M, Lim B, Wirtz M, Hell R, Cobbett CS and Meyer AJ (2008) Restricting glutathione biosynthesis to the cytosol is sufficient for normal plant development. Plant J, in press

Peltoniemi MJ, Karala A-R, Jurvansuu JK, Kinnula VL and Ruddock LW (2006) Insights into deglutathionylation reactions: Different intermediates in the glutaredoxin and protein disulfide isomerase catalyzed reactions are defined by the γ-linkage present in glutathione. J Biol Chem 281: 33107–33114

Pignocchi C, Kiddle G, Hernandez I, Foster SJ, Asensi A, Taybi T, Barnes J and Foyer CH (2006) Ascorbate oxidase-dependent changes in the redox state of the apoplast modulate gene transcript accumulation leading to modified hormone signaling and orchestration of defense processes in tobacco. Plant Physiol 141: 423–435

Pilon-Smits E (2005) Phytoremediation. Annu Rev Plant Biol 56: 15–39

Polekhina G, Board P, Gali R, Rossjohn J and Parker M (1999) Molecular basis of glutathione synthetase deficiency and a rare gene permutation event. EMBO J 18: 3204–3213

Pomponi M, Censi V, Di Girolamo V, De Paolis A, di Toppi L, Aromolo R, Costantino P and Cardarelli M (2006) Overexpression of Arabidopsis phytochelatin synthase in tobacco plants enhances Cd(2+) tolerance and accumulation but not translocation to the shoot. Planta 223: 180–190

Price CA (1957) A new thiol in legumes. Nature 180: 148–149

Rea PA (2006) Phytochelatin synthase, papain's cousin, in stereo. Proc Natl Acad Sci U S A 103: 507–508

Rea PA (2007) Plant ATP-binding cassette transporters. Annu Rev Plant Biol 58: 347–375

Rea PA, Li ZS, Lu YP, Drozdowicz YM and Martinoia E (1998) From vacuolar GS-X pumps to multispecific ABC transporters. Annu Rev Plant Physiol Plant Mol Biol 49: 727–760

Rea PA, Vatamaniuk OK and Rigden DJ (2004) Weeds, worms, and more. Papain's long-lost cousin, phytochelatin synthase. Plant Physiol 136: 2463–2474

Romanyuk ND, Rigden DJ, Vatamaniuk OK, Lang A, Cahoon RE, Jez JM and Rea PA (2006) Mutagenic definition of a papain-like catalytic triad, sufficiency of the N-terminal domain for single-site core catalytic enzyme acylation, and C-terminal domain for augmentative metal activation of a eukaryotic phytochelatin synthase. Plant Physiol 141: 858–869

Rouhier N, Villarejo A, Srivastava M, Gelhaye E, Keech O, Droux M, Finkemeier I, Samuelsson G, Dietz K, Jacquot J and Wingsle G (2005) Identification of plant glutaredoxin targets. Antioxid Redox Signal 7: 919–929

Sanmartin M, Pavlina D. Drogoudi, Tom Lyons IP, Jeremy Barnes and Kanellis AK (2003) Over-expression of ascorbate oxidase in the apoplast of transgenic tobacco results in altered ascorbate and glutathione redox states and increased sensitivity to ozone. Planta 216: 918–928

Schafer FQ and Buettner GR (2001) Redox environment of the cell as viewed through the redox state of the glutathione disulfide/glutathione couple. Free Radic Biol Med 30: 1191–1212

Schäfer HJ, Haag-Kerwer A and Rausch T (1998) cDNA cloning and expression analysis of genes encoding GSH synthesis in roots of the heavy-metal accumulator Brassica juncea L.: Evidence for Cd-induction of a putative mitochondrial γ-glutamylcysteine synthetase isoform. Plant Mol Biol 37: 87–97

Sevier C, Qu H, Heldman N, Gross E, Fass D and Kaiser C (2007) Modulation of cellular disulfide-bond formation and the ER redox environment by feedback regulation of Ero1. Cell 129: 333–344

Shelton M, Chock P and Mieyal J (2005) Glutaredoxin: Role in reversible protein S-glutathionylation and regulation of redox signal transduction and protein translocation. Antioxid Redox Signal 7: 348–366

Singh AK, Dhaunsi GS, Gupta MP, Orak JK, Asayama K and Singh I (1994) Demonstration of glutathione peroxidase in rat liver peroxisomes and its intraorganellar distribution. Arch Biochem Biophys 315: 331–338

Singla-Pareek SL, Reddy MK and Sopory SK (2003) Genetic engineering of the glyoxalase pathway in tobacco leads to enhanced salinity tolerance. Proc Natl Acad Sci U S A 100: 14672–14677

Smirnoff N, Conklin P and Loewus F (2001) Biosynthesis of ascorbic acid in plants: A renaissance. Annu Rev Plant Physiol Plant Mol Biol 52: 437–467

Stacey M, Osawa H, Patel A, Gassmann W and Stacey G (2006) Expression analyses of Arabidopsis oligopeptide transporters during seed germination, vegetative growth and reproduction. Planta 223: 291–305

Stadler R, Lauterbach C and Sauer N (2005) Cell-to-cell movement of green fluorescent protein reveals postphloem transport in the outer integument and identifies symplastic domains in Arabidopsis seeds and embryos. Plant Physiol 139: 701–712

Storozhenko S, Belles-Boix E, Babiychuk E, Herouart D, Davey MW, Slooten L, Van Montagu M, Inze D and Kushnir S (2002) γ-glutamyl transpeptidase in transgenic tobacco plants. Cellular localization, processing, and biochemical properties. Plant Physiol 128: 1109–1119

Suzuki N, Koizumi N and Sano H (2001) Screening of cadmium-responsive genes in *Arabidopsis thaliana*. Plant Cell Environ 24: 1177–1188

Tong Y, Kneer R and Zhu Y (2004) Vacuolar compartmentalization: A second-generation approach to engineering plants for phytoremediation. Trends Plant Sci 9: 7–9

Tsuji N, Nishikori S, Iwabe O, Shiraki K, Miyasaka H, Takagi M, Hirata K and Miyamoto K (2004) Characterization of phytochelatin synthase-like protein encoded by alr0975 from a prokaryote, Nostoc sp. Pcc 7120. Biochem Biophys Res Commun 315: 751–755

Tzafrir I, Pena-Muralla R, Dickerman A, Berg M, Rogers R, Hutchens S, Sweeney TC, McElver J, Aux G, Patton D and Meinke D (2004) Identification of genes required for embryo development in Arabidopsis. Plant Physiol 135: 1206–1220

Ullmann P, Gondet L, Potier S and Bach TJ (1996) Cloning of *Arabidopsis thaliana* glutathione synthetase (*GSH2*) by functional complementation of a yeast *gsh2* mutant. Eur J Biochem 236: 662–669

Vanacker H, Carver TL and Foyer CH (2000) Early H_2O_2 accumulation in mesophyll cells leads to induction of glutathione during the hyper-sensitive response in the barley-powdery mildew interaction. Plant Physiol 123: 1289–1300

Vatamaniuk OK, Mari S, Lu Y-P and Rea PA (2000) Mechanism of heavy metal ion activation of phytochelatin (PC) synthase. J Biol Chem 275: 31451–31459

Vatamaniuk OK, Mari S, Lang A, Chalasani S, Demkiv LO and Rea PA (2004) Phytochelatin synthase, a dipeptidyltransferase that undergoes multisite acylation with γ-glutamylcysteine during catalysis: Stoichiometric and site-directed mutagenic analysis of *Arabidopsis thaliana* PCS1-catalyzed phytochelatin synthesis. J Biol Chem 279: 22449–22460

Vauclare P, Kopriva S, Fell D, Suter M, Sticher L, von Ballmoos P, Krahenbuhl U, den Camp RO and Brunold C (2002) Flux control of sulphate assimilation in *Arabidopsis thaliana*: Adenosine 5′-phosphosulphate reductase is more susceptible than ATP sulphurylase to negative control by thiols. Plant J 31: 729–740

Vernoux T, Wilson RC, Seeley KA, Reichheld JP, Muroy S, Brown S, Maughan SC, Cobbett CS, Van Montagu M, Inze D, May MJ and Sung ZR (2000) The ROOT MERISTEMLESS1/CADMIUM SENSITIVE2 gene defines a glutathione-dependent pathway involved in initiation and maintenance of cell division during postembryonic root development. Plant Cell 12: 97–110

Vivares D, Arnoux P and Pignol D (2005) A papain-like enzyme at work: Native and acyl-enzyme intermediate structures in phytochelatin synthesis. Proc Natl Acad Sci U S A 102: 18848–18853

Wachter A (2004) Glutathion-Synthese und -kompartimentierung in der Pflanze: Nachweis komplexer Regulationsmechanismen. PhD thesis, University of Heidelberg.

Wachter A, Wolf S, Steininger H, Bogs J and Rausch T (2005) Differential targeting of GSH1 and GSH2 is achieved by multiple transcription initiation: Implications for the compartmentation of glutathione biosynthesis in the Brassicaceae. Plant J 41: 15–30

Wang CL and Oliver DJ (1996) Cloning of the cDNA and genomic clones for glutathione synthetase from *Arabidopsis thaliana* and complementation of a *gsh2* mutant in fission yeast. Plant Mol Biol 31: 1093–1104

Wawrzynski A, Kopera E, Wawrzynska A, Kaminska J, Bal W and Sirko A (2006) Effects of simultaneous expression of heterologous genes involved in phytochelatin biosynthesis on thiol content and cadmium accumulation in tobacco plants. J Exp Bot 57: 2173–2182

Weber A, Schwacke R and Flügge U (2005) Solute transporters of the plastid envelope membrane. Annu Rev Plant Biol 56: 133–164

Xiang C and Bertrand D (2000) Glutathione synthesis in Arabidopsis: Multilevel controls coordinate responses to stress. In: Brunold C, Rennenberg H, De Kok LJ, Stulen I and Davidian J-C (eds) Sulfur nutrition and sulfur assimilation in higher plants, pp 409–412. Paul Haupt Publishers, Berne

Xiang C and Oliver DJ (1998) Glutathione metabolic genes coordinately respond to heavy metals and jasmonic acid in Arabidopsis. Plant Cell 10: 1539–1550

Xiang C, Werner BL, Christensen EM and Oliver DJ (2001) The biological functions of glutathione revisited in Arabidopsis transgenic plants with altered glutathione levels. Plant Physiol 126: 564–574

Xing S, Lauri A and Zachgo S (2006) Redox regulation and flower development: A novel function for glutaredoxins. Plant Biol 8: 547–555

Zhang M-Y, Bourbouloux A, Cagnac O, Srikanth CV, Rentsch D, Bachhawat AK and Delrot S (2004) A novel family of transporters mediating the transport of glutathione derivatives in plants. Plant Physiol 134: 482–491

Zhou N, Tootle TL and Glazebrook J (1999) Arabidopsis *pad3*, a gene required for camalexin biosynthesis, encodes a putative cytochrome p450 monooxygenase. Plant Cell 11: 2419–2428

Zhu YL, Pilon-Smits EAH, Jouanin L and Terry N (1999a) Overexpression of glutathione synthetase in Indian mustard enhances cadmium accumulation and tolerance. Plant Physiol 119: 73–80

Zhu YL, Pilon-Smits EAH, Tarun AS, Weber SU, Jouanin L and Terry N (1999b) Cadmium tolerance and accumulation in Indian mustard is enhanced by overexpressing gamma-glutamylcysteine synthetase. Plant Physiol 121: 1169–1177

Chapter 10

Sulfolipid Biosynthesis and Function in Plants

Christoph Benning* and R. Michael Garavito
Department of Biochemistry and Molecular Biology, Michigan State University, East Lansing, MI 48824, USA

Mie Shimojima
Graduate School of Bioscience and Biotechnology, Tokyo Institute of Technology, Nagatsuta-cho, Midori-ku. Yokohama 226-8501, Japan

Summary

The plant sulfolipid sulfoquinovosyldiacylglycerol accounts for a large fraction of organic sulfur in the biosphere. Aside from sulfur amino acids, sulfolipid represents a considerable sink for sulfate in plants. Plant sulfolipid is found in the photosynthetic membranes of plastids and provides negative charge in the thylakoid membrane where it is thought to stabilize photosynthetic complexes. As the plant sulfolipid is a non-phosphorous glycolipid, its synthesis does not impinge on the supply of phosphate, which is a macronutrient limiting plant growth in many natural environments. Indeed, plants evolved homeostatic mechanisms to balance the amount of sulfolipid with anionic phospholipids maintaining a proper level of anionic charge in the photosynthetic membrane. The strong anionic nature of the sugar sulfonate head group of sulfolipid also makes this lipid an interesting compound for biotechnological applications. As bacterial and plant genes encoding sulfolipid enzymes are now available, biotechnological approaches can be developed to produce the plant sulfolipid in sufficient amounts to pursue the development of practical applications.

Corresponding author, Phone: 517 355 1609, Fax: 517 353 9334, E-mail: benning@msu.edu

Rüdiger Hell et al. (eds.), Sulfur Metabolism in Phototrophic Organisms, 185–200.
© 2008 *Springer.*

I. Introduction

The plant sulfolipid, sulfoquinovosyldiacylglycerol (SQDG), is a glycoglycerolipid characterized by a 6-deoxy-6-sulfoglucose head group
at the *sn*-3 position of the glycerol backbone of
diacylglycerol (DAG) as shown in Fig. 1. Over
the past 50 years, much has been written about
the presumed biosynthetic pathway of plant sulfolipid and its possible function in plants. With
recent advances in molecular and biochemical analysis, data are now available for a more
informed discussion of the biosynthesis and
function of SQDG in bacteria, algae and plants
and its biotechnological potential. Bacteria are
considered here, because the first genes encoding the enzymes directly involved in SQDG
biosynthesis were discovered in the purple bacterium *Rhodobacter sphaeroides* (Benning and
Somerville, 1992a; Benning and Somerville,
1992b) and the cyanobacterium *Synechococcus*
PCC7942 (Güler et al., 1996; Güler et al., 2000).
These studies prepared the way for the final elucidation of SQDG biosynthesis in plants and
algae. Aside from a brief historic overview, we
will focus on the more recent literature providing
the current view of sulfolipid biosynthesis and
function. For additional background information
the reader is referred to previous reviews on the
subject (Benson, 1963; Haines, 1973; Harwood,
1980; Barber and Gounaris, 1986; Mudd and
Kleppinger-Sparace, 1987; Kleppinger-Sparace
et al., 1990; Heinz, 1993; Marechal et al., 1997;
Benning, 1998; Okanenko, 2000; Frentzen,
2004).

A. Discovery and Historic Perspective

Sulfolipid was discovered by A.A. Benson and
coworkers in alcohol extracts of [^{35}S]-sulfate-
labeled algae, purple bacteria and plants (Benson
et al., 1959). The structure of SQDG was
subsequently elucidated as summarized by A.A.
Benson (Benson, 1963) and later confirmed by
mass spectrometric analysis (Budzikiewicz et al.,
1973; Gage et al., 1992; Kim et al., 1997). Ana-

lyzing water soluble sulfur-labeled compounds,
a sulfoquinovose nucleotide was tentatively
identified in *Chlorella* extracts, as well as small
sulfur-labeled compounds (Shibuya et al., 1963)
interpreted as biosynthetic intermediates. As we
now know, A.A. Benson correctly proposed that
a sulfoquinovose nucleotide is the direct head
group donor for sulfolipid biosynthesis (Benson,
1963). Moreover, G.A. Barber was the first to
suggest that the sulfoquinovose nucleotide could
be formed from a 4-keto-6-deoxy-α-D-glucose
nucleotide or similar molecule (Barber, 1963)
and Lehmann and Benson demonstrated that
sulfoquinovose can be chemically generated by
addition of sulfite to methyl 5,6-glucoseenide
(Lehmann and Benson, 1964). A more detailed
hypothesis of the nucleotide pathway was provided by Pugh and colleagues (Pugh et al., 1995b)
following the genetic demonstration that a sugar
nucleotide modifying enzyme was critical for
sulfolipid biosynthesis in the purple bacterium
Rhodobacter sphaeroides (Benning and Somerville, 1992a). Based on the isolation of the genes
encoding the enzymes of sulfolipid biosynthesis,
their biochemical analysis *in vitro*, and the structure determination of the UDP-sulfoquinovose
(UDP-SQ) forming enzyme as described in detail
below, it is now generally accepted that sulfolipid is synthesized in plants by the nucleotide
pathway shown in Fig. 2. Sulfonated 3-carbon or
2-carbon compounds found in plants or bacteria
are thought to be degradation products of SQDG
rather than biosynthetic intermediates (Roy
et al., 2003).

B. Occurrence in Plants and Bacteria

The plant sulfolipid SQDG represents an
important component of the global sulfur cycle
(Harwood and Nicholls, 1979) as it is found in
most photosynthetic and a few non-photosynthetic
organisms. It has been estimated that sulfolipid is
approximately equal to glutathione in abundance,
but presumably an order of magnitude less abundant than sulfur amino acids bound in proteins of
plant tissues (Heinz, 1993). The relative sulfolipid content in plant and algae tissues has been
reported to be as low as 2% and as high as 50% of
total polar lipids in marine plants and algae (Harwood, 1980; Dembitsky et al., 1990; Dembitsky
et al., 1991; Heinz, 1993). Because sulfolipid is

Abbreviations: DAG – diacylglycerol; ER – endoplasmic
reticulum; PG – phosphatidylglycerol; SQDG – sulfoquinovosyldiacylglycerol; TLC – thin-layer chromatography;
UDP-SQ – UDP-sulfoquinovose

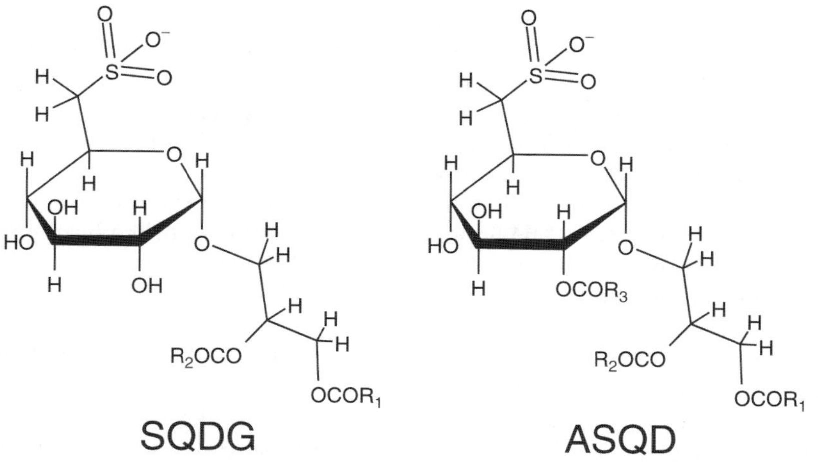

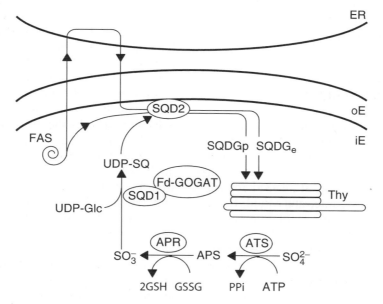

Fig. 1. Structures of sulfoquinovosyldiacylglycerol (SQDG) and 2-acyl- sulfoquinovosyldiacylglycerol (ASQD).

Fig. 2. Sulfolipid biosynthesis in plant chloroplasts. The inner envelope membrane (iE), the outer envelope membrane (oE), the endoplasmic reticulum (ER), and the thylakoid membranes (Thy) are shown. With the exception of the fatty acid synthase complex (FAS) depicted as a spiral, enzymes are shown as ovals. They include the UDP-SQ synthase (SQD1), the SQDG synthase (SQD2), ferredoxin-dependent glutamate synthase (Fd-GOGAT), ATP-sulfurylase (ATS) and adenosylphosphosulfate reductase (APR). Sulfolipid (SQDG) can be derived from the plastid pathway of thylakoid lipid biosynthesis ($SQDG_p$), or the ER pathway of thylakoid lipid biosynthesis ($SQDG_e$). Arrows indicate the direction of net flux in the pathways. Substrates and intermediates are: APS, adenosylphosphosulfate; ATP, adenosine triphosphate; GSSG, oxidized glutathione; GSH, reduced glutathione; PPi, orthophosphate; SO_3^-, sulfite; SO_4^{2-}, sulfate; UDP-Glc; UDP-glucose; UDP-SQ, UDP-sulfoquinovose.

found in photosynthetic membranes of bacteria and plants where it is associated with photosynthetic complexes, and because chlorophyll content seems to correlate with sulfolipid content, it was concluded that sulfolipid must be important for photosynthesis (Barber and Gounaris, 1986). However, there is no strict correlation between the competence of an organism to conduct photosynthesis and the presence of the SQDG. For example, *Sinorhizobium meliloti* (Cedergren and Hollingsworth, 1994; Weissenmayer et al., 2000), different species of *Caulobacter* and *Brevundimonas* (Abraham et al., 1997), and even Gram-positive bacteria (Langworthy et al., 1976; Sprott et al., 2006) have been reported to contain SQDG. The sulfolipid SQDG as well as other sulfolipids are present in the non-photosynthetic diatom *Nitzschia alba* (Anderson et al., 1978). The opposite has been observed as well, for example the lack of sulfolipid genes and, therefore, SQDG in a cyanobacterium (Selstam and Campbell, 1996; Nakamura et al., 2003). It should be cautioned, though, that SQDG content of bacteria and plants can change depending on growth conditions (Gage et al., 1992; Benning et al., 1993). As full-genome sequences for many bacteria are now available, a search for genes encoding putative sulfolipid biosynthetic enzymes provides a broad picture of the distribution and evolution of sulfolipid biosynthesis in different bacteria and plants, as will be discussed below.

II. Biosynthesis of Sulfoquinovosyldiacylglycerol

During the early phase of research exploring the biosynthetic pathway for SQDG in seed plants, isolated chloroplasts served as a facile model and were employed in numerous studies (Haas et al., 1980; Kleppinger-Sparace et al., 1985; Joyard et al., 1986; Kleppinger-Sparace and Mudd, 1987; Kleppinger-Sparace and Mudd, 1990; Pugh et al., 1995a; Pugh et al., 1995b; Roy and Harwood, 1999). The general conclusion from these experiments was that chloroplasts are fully capable of synthesizing SQDG from labeled sulfate when energy requirements were met either by photosynthesis or by the supply of nucleotides. Therefore, in seed plants chloroplasts must contain the biosynthetic machinery to provide the sulfur and carbon precursors for sulfolipid biosynthesis. By synthesizing the proposed head group donor for SQDG biosynthesis, the sugar nucleotide UDP-SQ, Heinz and colleagues were able to assay the SQDG synthase in chloroplast envelopes (Heinz et al., 1989). This assay was used to characterize the SQDG synthase and determine its localization on the inside of the inner envelope membrane of chloroplasts (Seifert and Heinz, 1992; Tietje and Heinz, 1998). However, the definitive determination

of the pathway of sulfolipid biosynthesis as shown in Fig. 2 was made possible by the identification of the enzymes at the molecular level and the biochemical characterization of the recombinant proteins. Sulfolipid biosynthesis in plants requires at minimum two specific enzymes: (1) the UDP-SQ synthase (SQD1), which is responsible for the biosynthesis of the head group donor, and (2) the SQDG synthase (SQD2) catalyzing the final assembly of sulfolipid.

A. Biosynthesis of UDP-Sulfoquinovose

Aside from the sulfoquinovosylated oligosaccharide side chain of a cytochrome *b* protein from an archaeon (Zähringer et al., 2000), sulfoquinovose has only been reported to be a substituent of sulfolipid or its precursor UDP-SQ. The existence of UDP-SQ in biological materials was originally suggested by A.A. Benson based on the tentative identification of a sulfur labeled sugar nucleotide in extracts of *Chlorella* (Shibuya et al., 1963). A more certain identification of UDP-SQ in an organism was accomplished in extracts of a sulfolipid-deficient *sqdD* mutant of *R. sphaeroides* (Rossak et al., 1995). This mutant lacks the presumed SQDG synthase encoded by *sqdD* and accumulated the UDP-SQ precursor to the extent that it could be readily analyzed. Subsequently, UDP-SQ was identified in extracts from different plants (Tietje and Heinz, 1998). Using genetic analysis, Benning and Somerville identified four genes, *sqdA, sqdB, sqdC,* and *sqdD* of *Rhodobacter sphaeroides* essential for SQDG biosynthesis (Benning and Somerville, 1992a; Benning and Somerville, 1992b). The first gene, *sqdA*, encodes an acyltransferase–like protein (Benning and Somerville, 1992b) and its precise role in SQDG biosynthesis remains to be shown. The *sqdB* gene forms an operon with *sqdC* and *sqdD* (Benning and Somerville, 1992a), with the *sqdD* gene encoding a predicted glycosyltransferase, the presumed SQDG synthase in this bacterium (Rossak et al., 1995). The *sqdC* gene encodes a small reductase-like protein and its targeted disruption led to the accumulation of sulfoquinovosyl-1-*O*-dihydroxyacetone in the respective mutant (Rossak et al., 1997). The original interpretation was that the SqdC and SqdD proteins form a functional SQDG synthase with SqdC providing substrate-specificity to the enzyme (Rossak et al., 1997). However, this hypothesis needs to be revisited as it seems possible that in bacteria SQDG is assembled in a different way than shown in Fig. 2 for plants.

Neither *sqdA*, nor *sqdC* and *sqdD* of *R. sphaeroides* seem to be involved in the biosynthesis of UDP-SQ and are also not found to be conserved in cyanobacteria and plants. However, the *sqdB* encoded protein of *R. sphaeroides* resembles sugar nucleotide modifying enzymes and has its apparent orthologue in every SQDG-producing organism studied thus far (see evolution of SQDG biosynthesis below). Its discovery in *R. sphaeroides* (Benning and Somerville, 1992a) led to the identification of the respective *sqdB* orthologue in the cyanobacterium *Synechococcus* sp. PCC7942 (Güler et al., 1996), and the orthologues in plants (Essigmann et al., 1998; Shimojima and Benning, 2003) and algae (Riekhof et al., 2003; Sato et al., 2003b) designated *SQD1*, and laid the foundation for our current understanding of UDP-SQ biosynthesis. The Arabidopsis SQD1 protein was localized in the stroma of the chloroplast (Essigmann et al., 1998). The structure of SQD1 from Arabidopsis was predicted based on its similarity to other sugar-nucleotide modifying enzymes and a detailed reaction mechanism was proposed (Essigmann et al., 1999) based on a 4-keto-5,6-glucoseene intermediate as suggested by Pugh and coworkers (Pugh et al., 1995b). The SQD1 protein contains a tightly-bound NAD^+ residue participating in the formation of a 4-keto hexosyl group. This 4-keto intermediate is converted to a 4-keto-5,6-glucoseene intermediate to which sulfite is added. Subsequently, the 4-keto group reverts back to the hydroxyl after accepting a hydride from the NADH bound in the active site, which thereby releases UDP-SQ and regenerates NAD^+. Detailed structural analysis by X-ray crystallography of the Arabidopsis SQD1 dimeric protein with NAD^+ and UDP-glucose (Mulichak et al., 1999) supported this reaction mechanism and identified the active site residues as shown in Fig. 3.

The recombinant SQD1 protein was subsequently shown to convert UDP-glucose and sulfite to UDP-SQ *in vitro* (Sanda et al., 2001). However, the rate at which the recombinant enzyme catalyzed this reaction was very low. This left the question whether ancillary proteins and additional cofactors might be involved *in vivo*, or whether sulfite might not be the native substrate for this enzyme. When the SQD1 protein was isolated from spinach chloroplasts, it purified as a 250 kDa complex suggesting the presence of additional proteins in the native complex (Shimojima and Benning, 2003). The K_M of this native complex for sulfite was at least fourfold reduced compared to recombinant SQD1 suggesting an increased affinity for this substrate. Further analysis of the native complex revealed that ferredoxin-dependent glutamate synthase (FdGOGAT) co-purifies and tightly associates with SQD1 (Shimojima et al., 2005). The physiological/biochemical relevance of this interaction is not yet fully understood. However, as FdGOGAT is a flavin-containing protein, it could intermittently and covalently bind sulfite and deliver it to the active site of SQD1. Modeling of the cyanobacterial FdGOGAT and Arabidopsis SQD1 into a plausible complex supports this suggestion (Shimojima et al., 2005). It was therefore proposed that FdGOGAT moonlights in the synthesis of UDP-SQ by providing sulfite to the SQD1 active site. Because sulfite is a highly cytotoxic compound that is maintained at very low level in plants, the interaction of FdGOGAT and SQD1 could overcome this impasse by channeling sulfite to SQD1. However, further experiments will be required to support this current working hypothesis.

B. Assembly of Sulfoquinovosyldiacylglycerol

For the final assembly of SQDG, *R. sphaeroides* could not serve as a universal model because plants have neither orthologues of *sqdC* nor *sqdD*. When the *sqdB* gene of the cyanobacterium *Synechococcus* sp. PCC7942 was first isolated, it was found in an operon next to a gene now called *sqdX* (Güler et al., 2000). This gene encodes a protein with similarity to other lipid glycosyltransferases (Berg et al., 2001), but different from that encoded by the *R. sphaeroides sqdD* gene. Disruption of the *sqdX* gene of *Synechococcus* sp. PCC7942 led to SQDG deficiency implicating this protein in SQDG biosynthesis in this bacterium (Güler et al., 2000). Using this cyanobacterial gene, it was subsequently possible to isolate the plant orthologue, *SQD2*, encoding the plant sulfolipid

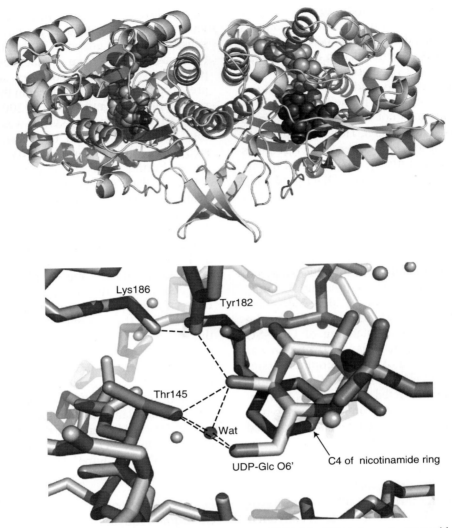

Fig. 3. Crystal structure of the Arabidopsis SQD1 protein (PDB entry 1QRR). The SQD1 dimer structure with the NAD$^+$ cofactors (light gray solid spheres) and the UDP-Glucose substrates (dark grey solid spheres) is shown in the top panel. The bottom panel depicts the active site of an SQD1 monomer with three critical amino acid residues that hydrogen-bond with the substrate UDP-glucose O6′ hydroxyl group, and one water molecule (Wat). The carbon 4 of the nicotine amide ring of NAD$^+$ cofactor is indicated.

synthase (Yu et al., 2002). Disruption of this gene by T-DNA insertion resulted in complete loss of SQDG in Arabidopsis. Moreover, co-expression of the *SQD1* and *SQD2* cDNAs from Arabidopsis in *E. coli* led to sulfolipid biosynthesis in this bacterium normally lacking this lipid (Yu et al., 2002). Thus, these two plant genes were sufficient to reconstitute SQDG biosynthesis demonstrating that the ultimate sulfur donor for SQDG biosynthesis must be a common intermediate of the sulfur assimilation pathway, such as sulfite,

present in *E. coli*. The SQD2 protein was localized inside the plastid (Yu et al., 2002) consistent with biochemical data showing that the sulfolipid synthase is associated with the inside of the inner envelope membrane (Seifert and Heinz, 1992; Tietje and Heinz, 1998). *In vitro* data or crystallographic data on this membrane-bound enzyme are not yet available. Based on these data, the nucleotide pathway of sulfolipid biosynthesis in plants inside chloroplast and the proteins involved are summarized as shown in Fig. 2.

III. Evolution of Sulfolipid Biosynthesis

In the era of genomics it is now possible to scan for the presence of putative sulfolipid genes in a large number of genomes. This approach provides independent means to examine the distribution of sulfolipid biosynthetic competence in the natural world and suggests clues towards the evolution of sulfolipid biosynthesis. Taking this approach, it is immediately clear that the two plant genes *SQD1* and *SQD2* evolved at different times. While many genomes show readily identifiable *SQD1/sqdB* orthologues suggesting that the capability to produce UDP-SQ evolved early and presumably only once, SQDG synthases were recruited differently in cyanobacteria/plants versus non-cyanobacteria and archaea. The alpha-proteobacteria *R. sphaeroides* and *S. meliloti*, in which the *sqd* genes have been well studied, harbor an SQDG synthase encoded by the *sqdD*

gene (Rossak et al., 1995; Weissenmayer et al., 2000). This gene is not related to the cyanobacterial *sqdX* gene (Güler et al., 2000) and its plant orthologue *SQD2* (Yu et al., 2002).

Figure 4 shows the approximate relatedness of predicted SQD1/sqdB-like protein sequences in bacteria, archaea and plants. Overall UDP-SQ synthases are closely related to UDP-glucose epimerases and similar sugar nucleotide-modifying enzymes. Therefore, many of the presumed UDP-SQ synthases are not correctly annotated in the public databases. On the other hand, without biochemical analysis, one cannot be certain that the respective protein indeed represents a UDP-SQ synthase. At this time, seven UDP-SQ synthases have been experimentally verified as indicated in Fig. 4. There is a clear cutoff in sequence similarity between presumed and known UDP-SQ synthases and other enzyme classes and it is apparent that animals do not have an UDP-SQ synthase

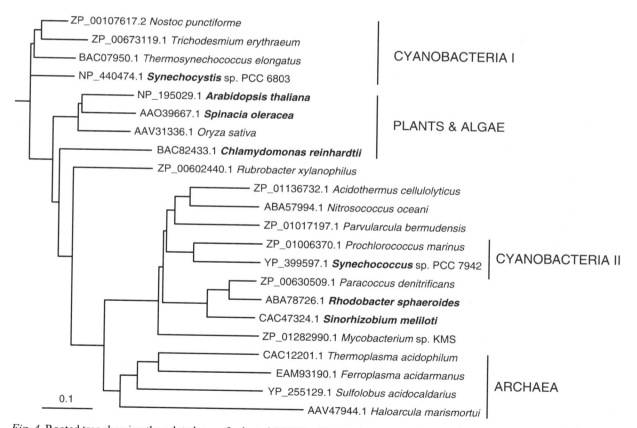

Fig. 4. Rooted tree showing the relatedness of selected SQD1/sqdB protein sequences. Sequences were aligned and the tree was built using Clustal W software (Thompson et al., 1994). Genbank accession numbers and species names are indicated. Species in which the respective UDP-SQ synthase was experimentally verified are shown in bold.

related to SQD1/sqdB consistent with their inability to produce SQDG. Only sea urchins have been reported to contain SQDG and its derivatives in their gut (Sahara et al., 1997), but it seems that these lipids are derived from their algal diet rather than being synthesized by this organism. Overall, SQD1/sqdB encoding sequences are present in a wide range of archaea, Gram + and Gram- bacteria, cyanobacteria, algae and plants. Whether all these organisms produce sulfolipid is not clear. It is also possible that UDP-SQ serves as sulfoquinovosyl donor for syntheses other than sulfolipid, exemplified by the presence of sulfoquinovose in a glycoprotein of *Sulfolobus acidocaldarius*, an archaeon (Zähringer et al., 2000). Thus, the capability to synthesize UDP-SQ evolved very early on, but SQDG biosynthesis itself arose independently at least twice, in bacteria other than cyanobacteria, and in plants and cyanobacteria.

IV. Biological Functions of Sulfolipid

A. In Vitro *and Correlative Studies*

Early inferences regarding the function of sulfolipid were mostly based on considerations of chemical and molecular properties of SQDG (Haines, 1983), the prevalence of SQDG in photosynthetic organisms, and the analysis of SQDG in typical phototrophs. For example, SQDG content increases during chloroplast development (Shibuya and Hase, 1965; Leech et al., 1973) and SQDG-specific antibodies inhibit the biochemical activity of photosynthetic membranes (Radunz and Schmid, 1992). Moreover, SQDG is associated with pigment protein complexes of photosynthetic membranes (Menke et al., 1976; Gounaris and Barber, 1985; Pick et al., 1985; Sigrist et al., 1988) and SQDG is required for the functional reconstitution of membrane-bound enzymes *in vitro* (Pick et al., 1987; Vishwanath et al., 1996). Sulfolipid specifically interacts with signal peptides of chloroplast targeted proteins suggesting a role during the import of nuclear encoded proteins into plastids (van't Hof et al., 1993; Inoue et al., 2001). Sulfolipid has also been shown to bind to an annexin in a calcium-dependent manner presumably at the cytosolic face of the outer envelope (Seigneurin-Berny et al., 2000). However, the biological significance of this interaction seen in *in vitro* experiments is not clear.

Effects of environmental factors on membrane lipid composition have led to suggestions about the conditional importance of plant sulfolipid. For example, a positive correlation has been observed between sulfolipid content and salt exposure, particularly in halophytes (Kuiper et al., 1974; Müller and Santarius, 1978; Ramani et al., 2004; Hamed et al., 2005), or drought exposure (Quartacci et al., 1995; Taran et al., 2000). Increases in sulfolipid have also been observed in cold-hardened plants (Kuiper, 1970) or during the winter season in pine needles (Oquist, 1982). However, one of the most striking environmental factors affecting sulfolipid content in bacteria and plants has been phosphate deprivation (Gage et al., 1992; Benning et al., 1993; Güler et al., 1996; Härtel et al., 1998; Essigmann et al., 1998; Sato et al., 2000; Yu et al., 2002; Yu and Benning, 2003; Sato, 2004) suggesting that sulfolipid can substitute to some extent for phosphatidylglycerol to maintain the anionic surface charge of thylakoid membranes.

B. Analysis of Sulfolipid-Deficient Mutants

The knowledge of the genes required for the biosynthesis of SQDG has provided genetic tools to study the function of the sulfolipid *in vivo*. An interesting picture has emerged from the study of SQDG-deficient mutants of bacteria, Arabidopsis and the unicellular alga *Chlamydomonas reinhardtii* as recently summarized by others (Sato, 2004; Frentzen, 2004). Bacterial mutants of the photosynthetic purple bacterium *Rhodobacter sphaeroides* (Benning et al., 1993), the cyanobacterium *Synechococcus* sp. PCC7942 (Güler et al., 1996) or the non-photosynthetic bacterium *Sinorhizobium meliloti* (Weissenmayer et al., 2000) disrupted in the *sqdB* gene essential for SQDG biosynthesis and completely lacking sulfolipid did not show any growth deficiencies under optimal growth conditions. However, it should be noted that for *Synechococcus* sp. PCC7942, subtle effects of SQDG deficiency on photosystem II chemistry were observed, but were apparently not limiting to growth (Güler et al., 1996). Lack of sulfolipid did not affect nodule formation induced by *S. meliloti* (Weissenmayer et al., 2000). However, *R. sphaeroides* and *Synechoc-*

occus sp. PCC7942 *sqdB* mutants ceased growth much earlier than the wild type under phosphate-limited growth conditions (Benning et al., 1993; Güler et al., 1996). In the respective wild-type strains, phosphatidylglycerol content decreased while SQDG content increased during phosphate deprivation suggesting that SQDG can substitute for the anionic phospholipid under phosphate-limited conditions. The respective *sqdB* mutants maintained their level of phosphatidylglycerol following the onset of phosphate limitation and therefore depleted in this essential macro nutrient sooner than did wild type. Therefore, in these bacteria sulfolipid appears to be of conditional importance.

Similar observations were made with an SQDG-deficient *sqd2* mutant of Arabidopsis which showed a growth-impairment only after severe phosphate depletion (Yu et al., 2002). Phosphatidylglycerol content remained high in the mutant following phosphate limitation as was observed for the bacteria. However, from the analysis of phosphatidylglycerol-deficient mutants it became clear that this phospholipid has roles beyond that of SQDG in Arabidopsis as *pgp1* mutants impaired in phosphatidylglycerolphosphate synthase activity showed considerable growth and photosynthesis defects depending on the severity of the allele (Xu et al., 2002; Hagio et al., 2002; Babiychuk et al., 2003). When a weak *pgp1-1* allele was combined with an insertion disruption allele of *sqd2* in Arabidopsis, total anionic lipids were reduced to one third of wild-type level leading to a more severe impairment of growth and photosynthesis than in either mutant alone (Yu and Benning, 2003). This result suggests that anionic lipids such as SQDG or phosphatidylglycerol are essential for a functional photosynthetic membrane. This result also shows that under certain circumstances, in this case when the overall anionic lipid content is low, SQDG-deficiency can cause a visible phenotype even under normal growth conditions, unlike the three bacterial examples described above. Indeed, an SQDG-deficient mutant of *Chlamydomonas reinhardtii* (Sato et al., 1995; Minoda et al., 2002; Minoda et al., 2003; Sato et al., 2003a) has shown growth defects and impairment of photosynthesis without additional phosphate-limitation. The stability of photosystem II was impaired, in particular at higher temperatures, and the mutant was more

sensitive to inhibitors of photosystem II activity. However, it is difficult to conclude from these results whether the observed defects were specifically due to the lack of SQDG or due to the overall decrease in anionic lipids. Moreover, the exact molecular defect in the UV-induced *hf-2 C. reinhardtii* mutant used in these studies is not known, although the mutant has been backcrossed repeatedly to the wild-type in the more recent studies to minimize the chance of interfering background mutations. When the *SQD1* orthologue of *C. reinhardtii* was specifically disrupted by plasmid insertion, complete loss of SQDG as well as of an acylated derivative of SQDG (Fig. 2) was observed and the mutant was more sensitive to a photosystem II inhibitor confirming previous results observed for the *hf-2* mutant (Riekhof et al., 2003). However, the loss of two lipids in this *sqd1* mutant still left ambiguity with respect to a causal link between a specific SQDG-defect and the observed increased sensitivity against a photosystem II inhibitor.

A cleaner picture emerged in a direct comparison of *sqdB*-disrupted mutants of *Synechococcus* sp. PCC6803 and *Synechococcus* sp. PCC7642 (Aoki et al., 2004), where in the former strain SQDG seemed to be essential for growth under optimal conditions as was observed in the *C. reinhardtii* studies, and in the latter it was dispensable. Therefore, from the limited analysis of SQDG-deficient mutants in different organisms one must conclude that SQDG is required, either conditionally or outright, to maintain a functional photosynthetic membrane, in particularly photosystem II. One function of SQDG seems to be the maintenance of anionic lipid content in the thylakoid membrane under phosphate-limited conditions and possible other conditions yet to be discovered. It seems even possible that photosynthetic complexes have specific binding sites for anionic lipids as was observed in the crystal structure of photosystem I for phosphatidylglycerol (Jordan et al., 2001). Recently, an SQDG molecule was located close to a regulatory lysine residue of cytochrome f in the crystal structure of the cytochrome b_6f complex of *C. reinhardtii* (Stroebel et al., 2003), corroborating the specific interaction of this lipid with complexes of the photosynthetic membrane. As more refined and higher resolution structures become available, it is possible that more lipid-binding sites in these complexes will be discovered which can be tested for functionality by directed mutational analysis.

V. Biotechnological Applications and Production of Sulfolipids

The discoverer of the plant sulfolipid, A.A. Benson, considers it "Nature's Finest Surfactant Molecule" (Benson, 2002). However, few detailed studies of biophysical properties of sulfolipid are available, (Shipley et al., 1973; Webb and Green, 1991; Howard and Prestegard, 1996; Matsumoto et al., 2005a). One reason might be that the plant sulfolipid is not readily available in larger quantities and with high purity to broadly study its applications as a natural detergent molecule. However, beginning with the discovery that SQDG from a marine cyanobacterium potentially has antiviral properties (Gustafson et al., 1989), attention has turned to this lipid class, and an exponentially increasing number of papers has been published documenting the biological effects of natural or synthetic SQDG or its derivatives, e.g. the monoacylated or beta-linked forms, in a number of systems. While it is outside the authors' expertise to critically evaluate the medical or physiological properties of SQDG and its derivatives in the employed model systems, the summary of the different reports in Table 1 might enable the reader to obtain an initial assessment of the biotechnological potential of this compound class.

Chemical synthesis of sulfolipid, in particular its chiral synthesis, is complex and presumably expensive (Gordon and Danishefsky, 1992; Hanashima et al., 2000a; Hanashima et al., 2000b; Hanashima et al., 2001) but has led to the limited availability of stereoisomers and different derivatives of the naturally occurring SQDG used in the studies described above. The alternative production of sulfolipids from natural resources is currently impeded by two factors: 1. While basic protocols for the isolation and purification of SQDG from plant tissues are available (O'Brien and Benson, 1964; Norman et al., 1996), they are not practical for a large scale commercial setting because they are based on halogenated solvents, which require expensive measures to contain their environmental hazards. There is a clear need for the development of an economic and environmentally friendly extraction and purification procedure for SQDG if this natural compound class should become commercially valuable. 2. The yield of SQDG from readily available agricultural materials is relatively low. Attempts at cultivation of cyanobacteria for the optimization of SQDG yield have been made (Archer et al., 1997), but what is needed is an even richer natural resource for SQDG. The availability of sulfolipid biosynthetic genes and our increasing knowledge about sulfur and sulfolipid metabolism of plants might enable us one day to engineer plants that are sufficiently rich in SQDG content to allow a commercial harvest of this valuable natural compound.

Table 1. Summary of reported biological effects of the plant sulfolipid and its derivatives.

Biological effect	Sulfolipid[a]	Citation
Anti AIDS/inhibition of reverse transcriptase	SQDG, SQMG, acylated forms of SQDG	Gustafson et al., 1989; Gordon and Danishefsky, 1992; Reshef et al., 1997; Ohta et al., 1998; Loya et al., 1998
Inhibition of mammalian polymerases	SQDG, SQMG, βSQDG	Mizushina et al., 1998; Ohta et al., 1999; Ohta et al., 2000; Murakami et al., 2002; Mizushina et al., 2003a; Murakami et al., 2003a; Murakami et al., 2003b; Mizushina et al., 2003b; Kuriyama et al., 2005; Matsumoto et al., 2005b
Anti-tumor effects	SQDG, SQMG	Shirahashi et al., 1993; Sahara et al., 1997; Ohta et al., 2001; Quasney et al., 2001; Sahara et al., 2002; Murakami et al., 2004; Hossain et al., 2005; Matsubara et al., 2005; Maeda et al., 2005
Inhibition of Telomerase	SQDG	Eitsuka et al., 2004
Immunosuppressant	βSQDG	Matsumoto et al., 2000; Matsumoto et al., 2004; Shima et al., 2005
Anti-inflammatory effects	SQDG	Vasange et al., 1997; Golik et al., 1997

[a]SQDG, sulfoquinovosyldiacylglycerol (naturally occurring α-anomeric form); βSQDG, synthetic β-anomeric form; SQMG, sulfoquinovosylmonoacylglycerol.

VI.   Concluding Remarks

Research towards the biosynthesis and function of the plant sulfolipid has come a long way since the discovery of SQDG approximately 50 years ago. Based on genetic and biochemical analyses we are reasonably certain now that two proteins, SQD1 and SQD2, are crucial for SQDG biosynthesis in plants. The pathway depicted in Fig. 2 summarizes our current knowledge about SQDG biosynthesis in plants and cyanobacteria. What remains uncertain is the involvement and role of ancillary proteins that make the process sufficiently efficient *in vivo*. Although the first SQDG genes were identified in purple bacteria, their exact roles are less clear and it seems possible that sulfolipid is assembled differently in alpha-proteobacteria than in plants. The UDP-SQ synthase, on the other hand, is well conserved in many diverse organisms. This raises the question whether sulfoquinovose can be a moiety of biomolecules other than SQDG as has been reported for a bacterial glycoprotein (Zähringer et al., 2000).

The function of sulfolipid *in vivo* can be traced to its anionic non-phosphorous properties, which apparently make this lipid a highly suitable component of photosynthetic membranes. As new high-resolution structures of photosynthetic and other biosynthetic membrane complexes in the chloroplast become available, it seems likely that new specific binding sites for SQDG in these complexes might be discovered. These anticipated findings will enable more direct approaches to test specific roles of SQDG.

With the availability of the genes required for the biosynthesis of SQDG, biotechnological approaches towards the production of sulfolipid can now be devised. The number of publications reporting potentially beneficial health effects and applications for SQDG and its derivates has increased exponentially during the past three years. It seems likely that as SQDG becomes more widely available, necessary follow up studies will become more facile.

Acknowledgements

Work on sulfolipid biosynthesis and the regulation of thylakoid lipid biosynthesis in the Benning lab has been supported in part by grants from the U.S. National Science Foundation (MCB-0109912) and the U.S. Department of Energy (DE-FG02-98ER20305).

References

Abraham WR, Meyer H, Lindholst S, Vancanneyt M, Smit J (1997) Phospho- and sulfolipids as biomarkers of *Caulobacter sensu lato, Brvundimaonas* and *Hyphomonas.* System Appl Microbiol 20: 522–539

Anderson R, Livermore BP, Kates M, Volcani BE (1978) The lipid composition of the non-photosynthetic diatom Nitzschia alba. Biochim Biophys Acta 528: 77–88

Aoki M, Sato N, Meguro A, Tsuzuki M (2004) Differing involvement of sulfoquinovosyl diacylglycerol in photosystem II in two species of unicellular cyanobacteria. Eur J Biochem 271: 685–693

Archer SD, McDonald KA, Jackman AP (1997) Effect of light irradiance on the production of sulfolipids from *Anabaena* 7120 in a fed-batch photobioreactor. Appl Biochem Biotechnol 67: 139–152

Babiychuk E, Müller F, Eubel H, Braun HP, Frentzen M, Kushnir S (2003) Arabidopsis phosphatidylglycerophosphate synthase 1 is essential for chloroplast differentiation, but is dispensable for mitochondrial function. Plant J 33: 899–909

Barber GA (1963) The formation of uridine diphosphate L-Rhamnose by enzymes of the tobacco leaf. Arch Biochem Biophys 103: 276–282

Barber J, Gounaris K (1986) What role does sulfolipid play within the thylakoid membrane? Photosynthes Res 9: 239–249

Benning C (1998) Biosynthesis and function of the sulfolipid sulfoquinovosyl diacylglycerol. Annu Rev Plant Physiol Plant Mol Biol 49: 53–75

Benning C, Beatty JT, Prince RC, Somerville CR (1993) The sulfolipid sulfoquinovosyldiacylglycerol is not required for photosynthetic electron transport in *Rhodobacter sphaeroides* but enhances growth under phosphate limitation. Proc Natl Acad Sci U S A 90: 1561–1565

Benning C, Somerville CR (1992a) Identification of an operon involved in sulfolipid biosynthesis in *Rhodobacter sphaeroides.* J Bacteriol 174: 6479–6487

Benning C, Somerville CR (1992b) Isolation and genetic complementation of a sulfolipid-deficient mutant of *Rhodobacter sphaeroides.* J Bacteriol 174: 2352–2360

Benson AA (1963) The plant sulfolipid. Adv Lipid Res 1: 387–94

Benson AA (2002) Paving the path. Annu Rev Plant Biol 53: 1–25

Benson AA, Daniel H, Wiser R (1959) A sulfolipid in plants. Proc Natl Acad Sci U S A 45: 1582–1587

Berg S, Edman M, Li L, Wikstrom M, Wieslander A (2001) Sequence properties of the 1,2-diacylglycerol 3-glucosyltransferase from *Acholeplasma laidlawii* membranes. Recognition of a large group of lipid glycosyltransferases in eubacteria and archaea. J Biol Chem 276: 22056–22063

Budzikiewicz H, Rullkötter J, Heinz E (1973) Massenspetroskopische Untersuchungen an Glycosylglyceriden. Z Naturforsch [C] 28: 499–504

Cedergren RA, Hollingsworth RI (1994) Occurrence of sulfoquinovosyl diacylglycerol in some members of the family Rhizobiaceae. J Lipid Res 35: 1452–1461

Dembitsky VM, Pechenkina-Shubina EE, Rozentsvet OA (1991) Glycolipids and fatty acids of some seaweeds and marine grasses from the Black Sea. Phytochemistry 30: 2279–2283

Dembitsky VM, Rozentsvet OA, Pechenkina EE (1990) Glycolipids, phospholipids and fatty acids of brown algae species. Phytochemistry 29: 3417–3421

Eitsuka T, Nakagawa K, Igarashi M, Miyazawa T (2004) Telomerase inhibition by sulfoquinovosyldiacylglycerol from edible purple laver (*Porphyra yezoensis*). Cancer Lett 212: 15–20

Essigmann B, Güler S, Narang RA, Linke D, Benning C (1998) Phosphate availability affects the thylakoid lipid composition and the expression of SQD1, a gene required for sulfolipid biosynthesis in Arabidopsis thaliana. Proc Natl Acad Sci U S A 95: 1950–1955

Essigmann B, Hespenheide BM, Kuhn LA, Benning C (1999) Prediction of the active-site structure and NAD(+) binding in SQD1, a protein essential for sulfolipid biosynthesis in Arabidopsis. Arch Biochem Biophys 369: 30–41

Frentzen M (2004) Phosphatidylglycerol and sulfoquinovosyldiacylglycerol: anionic membrane lipids and phosphate regulation. Curr Opin Plant Biol 7: 270–276

Gage DA, Huang ZH, Benning C (1992) Comparison of sulfoquinovosyl diacylglycerol from spinach and the purple bacterium *Rhodobacter spaeroides* by fast atom bombardment tandem mass spectrometry. Lipids 27: 632–636

Golik J, Dickey JK, Todderud G, Lee D, Alford J, Huang S, Klohr S, Eustice D, Aruffo A, Agler ML (1997) Isolation and structure determination of sulfonoquinovosyl dipalmitoyl glyceride, a P-selectin receptor inhibitor from the alga *Dictyochloris fragrans*. J Nat Prod 60: 387–389

Gordon DM, Danishefsky SJ (1992) Synthesis of a cyanobacterial sulfolipid: confirmation of its structure, stereochemistry and anti-HIV-1 activity. J Am Chem Soc 114: 659–663

Gounaris K, Barber J (1985) Isolation and characterisation of a photosystem II reaction center lipoprotein complex. FEBS Lett 188: 68–72

Güler S, Essigmann B, Benning C (2000) A cyanobacterial gene, *sqdX*, required for biosynthesis of the sulfolipid sulfoquinovosyldiacylglycerol. J Bacteriol 182: 543–545

Güler S, Seeliger A, Härtel H, Renger G, Benning C (1996) A null mutant of *Synechococcus* sp. PCC7942 deficient in the sulfolipid sulfoquinovosyl diacylglycerol. J Biol Chem 271: 7501–7507

Gustafson KR, Cardellina JH, Fuller RW, Weislow OS, Kiser RF, Snader KM, Patterson GM, Boyd MR (1989) AIDS-antiviral sulfolipids from cyanobacteria (blue-green algae). J Natl Cancer Inst 81: 1254–1258

Haas R, Siebertz HP, Wrage K, Heinz E (1980) Localization of sulfolipid labeling within cells and chloroplasts. Planta 148: 238–244

Hagio M, Sakurai I, Sato S, Kato T, Tabata S, Wada H (2002) Phosphatidylglycerol is essential for the development of thylakoid membranes in *Arabidopsis thaliana*. Plant Cell Physiol 43: 1456–1464

Haines TH (1973) Sulfolipids and halosulfolipids. In JA Erwin, ed, Lipids and Biomembranes of Eucryotic Organisms. Academic, New York, pp 197–232

Haines TH (1983) Anionic lipid headgroups as a proton-conducting pathway along the surface of membranes: a hypothesis. Proc Natl Acad Sci U S A 80: 160–164

Hamed LB, Youssef NB, Ranieri A, Zarrouk M, Abdelly C (2005) Changes in content and fatty acid profiles of total lipids and sulfolipids in the halophyte *Crithmum maritimum* under salt stress. J Plant Physiol 162: 599–602

Hanashima S, Mizushina Y, Ohta K, Yamazaki T, Sugawara F, Sakaguchi K (2000a) Structure–activity relationship of a novel group of mammalian DNA polymerase inhibitors, synthetic sulfoquinovosylacylglycerols. Jpn J Cancer Res 91: 1073–1083

Hanashima S, Mizushina Y, Yamazaki T, Ohta K, Takahashi H, Koshino H, Sahara H, Sakaguchi K, Sugawara F (2000b) Structural determination of sulfoquinovosyldiacylglycerol by chiral syntheses. Tetrahedron Lett 41: 4403–4407

Hanashima S, Mizushina Y, Yamazaki T, Ohta K, Takahashi S, Sahara H, Sakaguchi K, Sugawar F (2001) Synthesis of sulfoquinovosylacylglycerols, inhibitors of eukaryotic DNA polymerase alpha and beta. Bioorg Med Chem 9: 367–376

Härtel H, Essigmann B, Lokstein H, Hoffmann-Benning S, Peters-Kottig M, Benning C (1998) The phospholipid-deficient *pho1* mutant of *Arabidopsis thaliana* is affected in the organization, but not in the light acclimation, of the thylakoid membrane. Biochim Biophys Acta 1415: 205–218

Harwood JL (1980) Sulfolipids. In PK Stumpf, ed, The Biosynthesis of Plants, Vol. 4. Academic, New York, pp 301–320

Harwood JL, Nicholls RG (1979) The plant sulpholipid—a major component of the sulphur cycle. Biochem Soc Trans 7: 440–447

Heinz E (1993) Recent investigations on the biosynthesis of the plant sulfolipid. In LJ De Kok, ed, Sulfur Nutrition and Assimilation in Higher Plants. SPB Academic, The Hague, The Netherlands, pp 163–178

Heinz E, Schmidt H, Hoch M, Jung KH, Binder H, Schmidt RR (1989) Synthesis of different nucleoside 5′-diphospho-sulfoquinovoses and their use for studies on sulfolipid biosynthesis in chloroplasts. Eur J Biochem 184: 445–453

Hossain Z, Kurihara H, Hosokawa M, Takahashi K (2005) Growth inhibition and induction of differentiation and apoptosis mediated by sodium butyrate in Caco-2 cells with algal glycolipids. In Vitro Cell Dev Biol Anim 41: 154–159

Howard KP, Prestegard JH (1996) Conformation of sulfoquinovosyldiacylglycerol bound to a magnetically oriented membrane system. Biophys J 71: 2573–2582

Inoue K, Demel R, de Kruijff B, Keegstra K (2001) The N-terminal portion of the preToc75 transit peptide interacts with membrane lipids and inhibits binding and import of precursor proteins into isolated chloroplasts. Eur J Biochem 268: 4036–4043

Jordan P, Fromme P, Witt H, Klukas O, Saenger W, Krauss N (2001) Three-dimensional structure of cyanobacterial photosystem I at 2.5 A resolution. Nature 411: 909–917

Joyard J, Blee E, Douce R (1986) Sulfolipid synthesis from 35SO4 and [1–14C]acetate in isolated intact spinach chloroplasts. Bichim Biophys Acta 879: 78–87

Kim YH, Yoo JS, Kim MS (1997) Structural characterization of sulfoquinvosyl, monogalactosyl and digalactosyl diacylglcyerols by FAB-CID-MS/MS. J Mass Spec 32: 968–977

Kleppinger-Sparace KF, Mudd JB (1987) Biosynthesis of sulfoquinovosyldiacylglycerol in higher plants: the incorporation of 35SO4 by intact chloroplasts in darkness. Plant Physiol 84: 682–687

Kleppinger-Sparace KF, Mudd JB (1990) Biosynthesis of sulfoquinovosyldiacylglycerol in higher plants: use of adenosine-5′-phosphosulfate and adenosine-3′-phosphate 5′-phosphosulfate as precursors. Plant Physiol 93: 256–263

Kleppinger-Sparace KF, Mudd JB, Bishop DG (1985) Biosynthesis of sulfoquinovosyldiacylglycerol in higher plants: the incorporation of 35SO4 by intact chloroplasts. Arch Biochem Biophys 240: 859–865

Kleppinger-Sparace KF, Sparace SA, Mudd JB (1990) Plant Sulfolipids. In H Rennenberg, C Brunold, LJ De Kok, I Stulen, eds, Sulfur Nutrition and Sulfur Assimilation in Higher Plants. SPB Academic, The Hague, pp 77–88

Kuiper PJC (1970) Lipids in alfalfa leaves in relation to cold hardiness. Plant Physiol 45: 684–686

Kuiper PJC, Kähr M, Stuiver CEE, Kylin A (1974) Lipid composition of whole roots and of Ca2+ and Mg2+-activated adenosine triphosphatases from wheat and oat as related to mineral nutrition. Physiol Plant 32: 33–36

Kuriyama I, Musumi K, Yonezawa Y, Takemura M, Maeda N, Iijima H, Hada T, Yoshida H, Mizushina Y (2005) Inhibitory effects of glycolipids fraction from spinach on mammalian DNA polymerase activity and human cancer cell proliferation. J Nutr Biochem 16: 594–601

Langworthy TA, Mayberry WR, Smith PF (1976) A sulfonolipid and novel glucosamidyl glycolipids from the extreme thermoacidophile Bacillus acidocaldarius. Biochim Biophys Acta 431: 550–569

Leech RM, Rumsby MG, Thomson WW (1973) Plastid differentiation, acyl lipid, and fatty acid changes in developing green maize leaves. Plant Physiol 52: 240–245

Lehmann J, Benson AA (1964) The plant sulfolipid. IX. Sulfosugar syntheses from methyl hexoseenides. J Am Chem Soc 86: 4469–4472

Loya S, Reshef V, Mizrachi E, Silberstein C, Rachamim Y, Carmeli S, Hizi A (1998) The inhibition of the reverse transcriptase of HIV-1 by the natural sulfoglycolipids from cyanobacteria: contribution of different moieties to their high potency. J Nat Prod 61: 891–895

Maeda N, Hada T, Murakami-Nakai C, Kuriyama I, Ichikawa H, Fukumori Y, Hiratsuka J, Yoshida H, Sakaguchi K, Mizushina Y (2005) Effects of DNA polymerase inhibitory and antitumor activities of lipase-hydrolyzed glycolipid fractions from spinach. J Nutr Biochem 16: 121–128

Marechal E, Block MA, Dorne A-J, Joyard J (1997) Lipid synthesis and metabolism in the plastid envelope. Physiol Plant 100: 65–77

Matsubara K, Matsumoto H, Mizushina Y, Mori M, Nakajima N, Fuchigami M, Yoshida H, Hada T (2005) Inhibitory effect of glycolipids from spinach on in vitro and ex vivo angiogenesis. Oncol Rep 14: 157–160

Matsumoto K, Sakai H, Ohta K, Kameda H, Sugawara F, Abe M, Sakaguchi K (2005a) Monolayer membranes and bilayer vesicles characterized by alpha- and beta-anomer of sulfoquinovosyldiacylglycerol (SQDG). Chem Phys Lipids 133: 203–214

Matsumoto K, Sakai H, Takeuchi R, Tsuchiya K, Ohta K, Sugawara F, Abe M, Sakaguchi K (2005b) Effective form of sulfoquinovosyldiacylglycerol (SQDG) vesicles for DNA polymerase inhibition. Colloids Surf B Biointerfaces 46: 175–181

Matsumoto K, Takenouchi M, Ohta K, Ohta Y, Imura T, Oshige M, Yamamoto Y, Sahara H, Sakai H, Abe M, Sugawara F, Sato N, Sakaguchi K (2004) Design of vesicles of 1,2-di-O-acyl-3-O-(beta-D-sulfoquinovosyl)-glyceride bearing two stearic acids (beta-SQDG-C18), a novel immunosuppressive drug. Biochem Pharmacol 68: 2379–2386

Matsumoto Y, Sahara H, Fujita T, Hanashima S, Yamazaki T, Takahashi S, Sugawara F, Mizushina Y, Ohta K, Takahashi N, Jimbow K, Sakaguchi K, Sato N (2000) A novel immunosuppressive agent, SQDG, derived from sea urchin. Transplant Proc 32: 2051–2053

Menke W, Radunz A, Schmid GH, Koenig F, Hirtz RD (1976) Intermolecular interactions of polypeptides and lipids in the thylakoid membrane. Z Naturforsch [C] 31: 436–444

Minoda A, Sato N, Nozaki H, Okada K, Takahashi H, Sonoike K, Tsuzuki M (2002) Role of sulfoquinovosyl diacylglycerol for the maintenance of photosystem II in Chlamydomonas reinhardtii. Eur J Biochem 269: 2353–2358

Minoda A, Sonoike K, Okada K, Sato N, Tsuzuki M (2003) Decrease in the efficiency of the electron donation to tyrosine Z of photosystem II in an SQDG-deficient mutant of Chlamydomonas. FEBS Lett 553: 109–112

Mizushina Y, Maeda N, Kawasaki M, Ichikawa H, Murakami C, Takemura M, Xu X, Sugawara F, Fukumori Y, Yoshida H, Sakaguchi K (2003a) Inhibitory action of emulsified sulfoquinovosyl acylglycerol on mammalian DNA polymerases. Lipids 38: 1065–1074

Mizushina Y, Watanabe I, Ohta K, Takemura M, Sahara H, Takahashi N, Gasa S, Sugawara F, Matsukage A, Yoshida S, Sakaguchi K (1998) Studies on inhibitors of mammalian DNA polymerase alpha and beta: sulfolipids from a pteridophyte, *Athyrium niponicum*. Biochem Pharmacol 55: 537–541

Mizushina Y, Xu X, Asahara H, Takeuchi R, Oshige M, Shimazaki N, Takemura M, Yamaguchi T, Kuroda K, Linn S, Yoshida H, Koiwai O, Saneyoshi M, Sugawara F, Sakaguchi K (2003b) A sulphoquinovosyl diacylglycerol is a DNA polymerase epsilon inhibitor. Biochem J 370: 299–305

Mudd JB, Kleppinger-Sparace KF (1987) Sulfolipids. In PK Stumpf, ed, The Biochemistry of Plants, Vol. 9. Lipids: Structure and Function. Academic, New York, pp 275–288

Mulichak AM, Theisen MJ, Essigmann B, Benning C, Garavito RM (1999) Crystal structure of SQD1, an enzyme involved in the biosynthesis of the plant sulfolipid headgroup donor UDP-sulfoquinovose. Proc Natl Acad Sci U S A 96: 13097–13102

Müller M, Santarius KA (1978) Changes in chloroplast membrane lipids during adaptation of barely to extreme salinity. Plant Physiol 62: 326–329

Murakami C, Miuzno T, Hanaoka F, Yoshida H, Sakaguchi K, Mizushina Y (2004) Mechanism of cell cycle arrest by sulfoquinovosyl monoacylglycerol with a C18-saturated fatty acid (C18-SQMG). Biochem Pharmacol 67: 1373–1380

Murakami C, Takemura M, Yoshida H, Sugawara F, Sakaguchi K, Mizushina Y (2003a) Analysis of cell cycle regulation by 1-mono-O-acyl-3-O-(alpha-D-sulfoquinovosyl)-glyceride (SQMG), an inhibitor of eukaryotic DNA polymerases. Biochem Pharmacol 66: 541–550

Murakami C, Yamazaki T, Hanashima S, Takahashi S, Ohta K, Yoshida H, Sugawara F, Sakaguchi K, Mizushina Y (2002) Structure-function relationship of synthetic sulfoquinovosyl-acylglycerols as mammalian DNA polymerase inhibitors. Arch Biochem Biophys 403: 229–236

Murakami C, Yamazaki T, Hanashima S, Takahashi S, Takemura M, Yoshida S, Ohta K, Yoshida H, Sugawara F, Sakaguchi K, Mizushina Y (2003b) A novel DNA polymerase inhibitor and a potent apoptosis inducer: 2-mono-O-acyl-3-O-(alpha-D-sulfoquinovosyl)-glyceride with stearic acid. Biochim Biophys Acta 1645: 72–80

Nakamura Y, Kaneko T, Sato S, Mimuro M, Miyashita H, Tsuchiya T, Sasamoto S, Watanabe A, Kawashima K, Kishida Y, Kiyokawa C, Kohara M, Matsumoto M, Matsuno A, Nakazaki N, Shimpo S, Takeuchi C, Yamada M, Tabata S (2003) Complete genome structure of *Gloeobacter violaceus* PCC 7421, a cyanobacterium that lacks thylakoids. DNA Res 10: 137–145

Norman HA, Mischke CF, Allen B, Vincent JS (1996) Semipreparative isolation of plant sulfoquinovosyldiacylglyc-erols by solid phase extraction and HPLC procedures. J Lipid Res 37: 1372–1376

O'Brien JS, Benson AA (1964) Isolation and fatty acid composition of the plant sulfolipid and galactolipids. J Lipid Res 5: 432–434

Ohta K, Hanashima S, Mizushina Y, Yamazaki T, Saneyoshi M, Sugawara F, Sakaguchi K (2000) Studies on a novel DNA polymerase inhibitor group, synthetic sulfoquinovosylacylglycerols: inhibitory action on cell proliferation. Mutat Res 467: 139–152

Ohta K, Mizushina Y, Hirata N, Takemura M, Sugawara F, Matsukage A, Yoshida S, Sakaguchi K (1998) Sulfoquinovosyldiacylglycerol, KM043, a new potent inhibitor of eukaryotic DNA polymerases and HIV-reverse transcriptase type 1 from a marine red alga, *Gigartina tenella*. Chem Pharm Bull (Tokyo) 46: 684–686

Ohta K, Mizushina Y, Hirata N, Takemura M, Sugawara F, Matsukage A, Yoshida S, Sakaguchi K (1999) Action of a new mammalian DNA polymerase inhibitor, sulfoquinovosyldiacylglycerol. Biol Pharm Bull 22: 111–116

Ohta K, Mizushina Y, Yamazaki T, Hanashima S, Sugawara F, Sakaguchi K (2001) Specific interaction between an oligosaccharide on the tumor cell surface and the novel antitumor agents, sulfoquinovosylacylglycerols. Biochem Biophys Res Commun 288: 893–900

Okanenko A (2000) Sulfoquinovosyldiacylglycerol in higher plants: Biosynthesis and physiological function. In C Brunold, ed, Sulfur Nutrition and Sulfur Assimilation in Higher Plants. Paul Haupt, Bern, CH, pp 203–216

Oquist G (1982) Seasonally induced changes in acyl lipids and fatty acids of chloroplast thylakoids of *Pinus silvestris*. Plant Physiol 69: 869–875

Pick U, Gounaris K, Weiss M, Barber J (1985) Tightly bound sulfolipids in chloroplast CF0-CF1. Biochim Biophys Acta 808: 415–420

Pick U, Weiss M, Gounaris K, Barber J (1987) The role of different thylakoid glycolipids in the function of reconstituted chloroplast ATP synthase. Biochim Biophys Acta 891: 28–39

Pugh CE, Hawkes T, Harwood JL (1995a) Biosynthesis of sulphoquinovosyldiacylglycerol by chloroplast fractions from pea and lettuce. Phytochemistry 39: 1071–1075

Pugh CE, Roy AB, Hawkes T, Harwood JL (1995b) A new pathway for the synthesis of the plant sulpholipid, sulphoquinovosyldiacylglycerol. Biochem J 309 (Pt 2): 513–519

Quartacci MF, Pinzino C, Sgherri C, Navari-Izzo F (1995) Lipid composition and protein dynamics in thylakoids of two wheat cultivars differently sensitive to drought. Plant Physiol 108: 191–197

Quasney ME, Carter LC, Oxford C, Watkins SM, Gershwin ME, German JB (2001) Inhibition of proliferation and induction of apoptosis in SNU-1 human gastric cancer cells by the plant sulfolipid, sulfoquinovosyldiacylglyc-erol. J Nutr Biochem 12: 310–315

Radunz A, Schmid GH (1992) Binding of lipids onto polypeptides of the thylakoid membrane I. Galactolipids and sulpholipid as prosthetic groups of core peptides of the photosystem II complex. Z Naturforsch [C] 47: 406–415

Ramani B, Zorn H, Papenbrock J (2004) Quantification and fatty acid profiles of sulfolipids in two halophytes and a glycophyte grown under different salt concentrations. Z Naturforsch [C] 59: 835–842

Reshef V, Mizrachi E, Maretzki T, Silberstein C, Loya S, Hizi A, Carmeli S (1997) New acylated sulfoglycolipids and digalactolipids and related known glycolipids from cyanobacteria with a potential to inhibit the reverse transcriptase of HIV-1. J Nat Prod 60: 1251–1260

Riekhof WR, Ruckle ME, Lydic TA, Sears BB, Benning C (2003) The sulfolipids 2′-O-acyl-sulfoquinovosyldiacylglycerol and sulfoquinovosyldiacylglycerol are absent from a Chlamydomonas reinhardtii mutant deleted in SQD1. Plant Physiol 133: 864–874

Rossak M, Schäfer A, Xu N, Gage DA, Benning C (1997) Accumulation of sulfoquinovosyl-1-O-dihydroxyacetone in a sulfolipid-deficient mutant of Rhodobacter sphaeroides inactivated in sqdC. Arch Biochem Biophys 340: 219–230

Rossak M, Tietje C, Heinz E, Benning C (1995) Accumulation of UDP-sulfoquinovose in a sulfolipid-deficient mutant of Rhodobacter sphaeroides. J Biol Chem 270: 25792–25797

Roy AB, Harwood JL (1999) Re-evaluation of plant sulpholipid labelling from UDP-[14C]glucose in pea chloroplasts. Biochem J 344 Pt 1: 185–187

Roy AB, Hewlins MJ, Ellis AJ, Harwood JL, White GF (2003) Glycolytic breakdown of sulfoquinovose in bacteria: a missing link in the sulfur cycle. Appl Environ Microbiol 69: 6434–6441

Sahara H, Hanashima S, Yamazaki T, Takahashi S, Sugawara F, Ohtani S, Ishikawa M, Mizushina Y, Ohta K, Shimozawa K, Gasa S, Jimbow K, Sakaguchi K, Sato N, Takahashi N (2002) Anti-tumor effect of chemically synthesized sulfolipids based on sea urchin's natural sulfonoquinovosylmonoacylglycerols. Jpn J Cancer Res 93: 85–92

Sahara H, Ishikawa M, Takahashi N, Ohtani S, Sato N, Gasa S, Akino T, Kikuchi K (1997) In vivo anti-tumour effect of 3′-sulphonoquinovosyl 1′-monoacylglyceride isolated from sea urchin (Strongylocentrotus intermedius) intestine. Br J Cancer 75: 324–332

Sanda S, Leustek T, Theisen MJ, Garavito RM, Benning C (2001) Recombinant Arabidopsis SQD1 converts UDP-glucose and sulfite to the sulfolipid head group precursor UDP-sulfoquinovose in vitro. J Biol Chem 276: 3941–3946

Sato N (2004) Roles of the acidic lipids sulfoquinovosyl diacylglycerol and phosphatidylglycerol in photosynthesis: their specificity and evolution. J Plant Res 117: 495–505

Sato N, Aoki M, Maru Y, Sonoike K, Minoda A, Tsuzuki M (2003a) Involvement of sulfoquinovosyl diacylglycerol in the structural integrity and heat-tolerance of photosystem II. Planta 217: 245–251

Sato N, Hagio M, Wada H, Tsuzuki M (2000) Environmental effects on acidic lipids of thylakoid membranes. Biochem Soc Trans 28: 912–914

Sato N, Sonoike K, Tsuzuki M, Kawaguchi A (1995) Impaired photosystem II in a mutant of Chlamydomonas reinhardtii defective in sulfoquinovosyl diacylglycerol. Eur J Biochem 234: 16–23

Sato N, Sugimoto K, Meguro A, Tsuzuki M (2003b) Identification of a gene for UDP-sulfoquinovose synthase of a green alga, Chlamydomonas reinhardtii, and its phylogeny. DNA Res 10: 229–237

Seifert U, Heinz E (1992) Enzymatic characteristics of UDP-sulfoquinovose:diacylglycerol sulfoquinovosyltransferase from chloroplast envelopes. Bot Acta 105: 197–205

Seigneurin-Berny D, Rolland N, Dorne AJ, Joyard J (2000) Sulfolipid is a potential candidate for annexin binding to the outer surface of chloroplast. Biochem Biophys Res Commun 272: 519–524

Selstam E, Campbell D (1996) Membrane lipid composition of the unusual cyanobacterium Gloeobacter violaceus sp. PCC7421, which lacks sulfoquinovosyl diacylglycerol. Arch Microbiol 166: 132–135

Shibuya I, Hase E (1965) Degradation and formation of sulfolipid occurring concurrently with de- and re-generation of chloroplasts in the cells of Chlorella protothecoides. Plant Cell Phyisol 6: 267–283

Shibuya I, Yagi T, Benson AA (1963) Sulfonic acids in algae. Microalgae and Photosynthetic bacteria, Jpn. Soc. Plant Physiol., Tokyo University Press, Tokyo, pp 627–636

Shima H, Tsuruma T, Sahara H, Takenouchi M, Takahashi N, Iwayama Y, Yagihashi A, Watanabe N, Sato N, Hirata K (2005) Treatment with beta-SQAG9 prevents rat hepatic ischemia-reperfusion injury. Transplant Proc 37: 417–421

Shimojima M, Benning C (2003) Native uridine 5′-diphosphate-sulfoquinovose synthase, SQD1, from spinach purifies as a 250-kDa complex. Arch Biochem Biophys 413: 123–130

Shimojima M, Hoffmann-Benning S, Garavito RM, Benning C (2005) Ferredoxin-dependent glutamate synthase moonlights in plant sulfolipid biosynthesis by forming a complex with SQD1. Arch Biochem Biophys 436: 206–214

Shipley GG, Green JP, Nichols BW (1973) The phase behavior of monogalactosyl, digalactosyl, and sulphoquinovosyl diglycerides. Biochim Biophys Acta 311: 531–544

Shirahashi H, Murakami N, Watanabe M, Nagatsu A, Sakakibara J, Tokuda H, Nishino H, Iwashima A (1993) Isolation and identification of anti-tumor-promoting principles from the fresh-water cyanobacterium Phormidium tenue. Chem Pharm Bull (Tokyo) 41: 1664–1666

Sigrist M, Zwillenberg C, Giroud CH, Eichenberger W, Boschetti A (1988) Sulfolipid associated with the light harvesting complex associated with photosystem II apoproteins of *Chlamydomonas reinhardtii*. Plant Sci 58: 15–23

Sprott GD, Bakouche L, Rajagopal K (2006) Identification of sulfoquinovosyl diacylglycerol as a major polar lipid in *Marinococcus halophilus* and *Salinicoccus hispanicus* and substitution with phosphatidylglycerol. Can J Microbiol 52: 209–219

Stroebel D, Choquet Y, Popot JL, Picot D (2003) An atypical haem in the cytochrome b(6)f complex. Nature 426: 413–418

Taran N, Okanenko A, Musienko N (2000) Sulpholipid reflects plant resistance to stress-factor action. Biochem Soc Trans 28: 922–924

Thompson JD, Higgins DG, Gibson TJ (1994) CLUSTAL W: improving the sensitivity of progressive multiple sequence alignment through sequence weighting, position-specific gap penalties and weight matrix choice. Nucleic Acids Res 22: 4673–4680

Tietje C, Heinz E (1998) Uridine-diphospho-sulfoquinovose:diacylglycerol sulfoquinovosyltransferase activity is concentrated in the inner membrane of chloroplast envelopes. Planta 206: 72–78

van't Hof R, van Klompenburg W, Pilon M, Kozubek A, Korte-Kool G, Demel RA, Weisbeek PJ, de Kruijff B (1993) The transit sequence mediates the specific interaction of the precursor of ferredoxin with chloroplast envelope membrane lipids. J Biol Chem 268: 4037–4042

Vasange M, Rolfsen W, Bohlin L (1997) A sulphonoglycolipid from the fern *Polypodium decumanum* and its effect on the platelet activating-factor receptor in human neutrophils. J Pharm Pharmacol 49: 562–566

Vishwanath BS, Eichenberger W, Frey FJ, Frey BM (1996) Interaction of plant lipids with 14 kDa phospholipase A2 enzymes. Biochem J 320 (Pt 1): 93–99

Webb MS, Green BR (1991) Biochemical and biophysical properties of thyalkoid acyl lipids. Bichim Biophys Acta 1060: 133–158

Weissenmayer B, Geiger O, Benning C (2000) Disruption of a gene essential for sulfoquinovosyldiacylglycerol biosynthesis in *Sinorhizobium meliloti* has no detectable effect on root nodule symbiosis. Mol Plant Microbe Interact 13: 666–672

Xu C, Härtel H, Wada H, Hagio M, Yu B, Eakin C, Benning C (2002) The *pgp1* locus of Arabidopsis encodes a phosphatidylglycerol synthase with impaired activity. Plant Physiol 129: 594–604

Yu B, Benning C (2003) Anionic lipids are required for chloroplast structure and function in Arabidopsis. Plant J 36: 762–770

Yu B, Xu C, Benning C (2002) Arabidopsis disrupted in SQD2 encoding sulfolipid synthase is impaired in phosphate-limited growth. Proc Natl Acad Sci U S A 99: 5732–5737

Zähringer U, Moll H, Hettmann T, Knirel YA, Schäfer G (2000) Cytochrome b558/566 from the archaeon *Sulfolobus acidocaldarius* has a unique Asn-linked highly branched hexasaccharide chain containing 6-sulfoquinovose. Eur J Biochem 267: 4144–4149

Chapter 11

Sulfur-Containing Secondary Metabolites and Their Role in Plant Defense

Meike Burow and Ute Wittstock*

Institut für Pharmazeutische Biologie, Technische Universität Braunschweig, Mendelssohnstr. 1, D-38106 Braunschweig, Germany

Jonathan Gershenzon

Max Planck Institute for Chemical Ecology, Department of Biochemistry, Beutenberg Campus, Hans-Knöll-Str. 8, D-07745 Jena, Germany

Summary

Although sulfur-containing products of secondary metabolism are rather unusual plant constituents, they play an important role in plant–pest interactions in a variety of different plant families and constitute major chemical defenses in the Brassicaceae, Alliaceae and Asteraceae. Besides their role as key players in two activated plant defense systems, the glucosinolate–myrosinase system of the Brassicaceae and the alliin–alliinase system of the Alliaceae, sulfur-containing compounds include prominent phytoalexins

---

* Corresponding author, Phone: +49 (0) 531 391 5681, Fax: +49 (0) 531 391 8104, E-mail: u.wittstock@tu-bs.de

Rüdiger Hell et al. (eds.), Sulfur Metabolism in Phototrophic Organisms, 201–222.
© 2008 *Springer.*

with structures ranging from elemental sulfur to the alkaloid-like camalexin and complex polypeptides like the defensins. Due to the diversity in chemical structures and modes of action, sulfur-containing natural products provide plants with a versatile array of chemical defenses against a broad range of potential enemies. The use of the model plant *Arabidopsis thaliana* has considerably advanced our understanding of the biochemistry of several classes of these compounds including the glucosinolates, camalexin and the defensins at the molecular level. This has paved the way for metabolic engineering projects aimed at elucidating the ecological roles of sulfur-containing defenses as well as at improving the nutritional value and pest resistance of crop plants.

I. Introduction

Compared to the huge classes of most ubiquitously distributed secondary metabolites such as terpenoids and alkaloids, the sulfur-containing products of secondary metabolism are rather unusual plant constituents. Consequently, their role in plant defense seems to be rather limited at first glance. However, a closer look reveals that sulfur-containing secondary metabolites are involved in several different types of chemical defenses, including constitutive, induced and activated defenses in a broad range of higher plant species as well as mosses and algae. Among these sulfur-containing defenses, one finds both compounds that occur only in a few related plant families and constitute major chemical defenses in these families (e.g. camalexin and related compounds from the Brassicaceae, the glucosinolates from the Brassicales, the alliines from the Alliaceae, and the thiophenes from the Asteraceae) as well as compounds which are distributed rather widely (e.g. elemental sulfur and the defensins). Only a few of these chemical defenses, namely the glucosinolates, camalexin and the defensins, have been studied extensively on a molecular level. In this review, we survey the principal sulfur-containing secondary metabolites in plants summarizing what is known about their structures, distribution and biosynthesis. Special emphasis is placed on the

roles of these compounds in defense against herbivores and pathogens.

II. Sulfur and Activated Plant Defense

Among the sulfur-containing secondary metabolites, two groups of compounds are involved in so-called activated plant defense systems, the glucosinolates and the alliins. In the intact plant tissue, these compounds, which are relatively physiologically inert by themselves, are spatially separated from their hydrolyzing enzymes, the myrosinases and alliinases, respectively. When the tissue is damaged, for example upon herbivore attack, the parent compounds are converted to biologically active products by the action of the hydrolyzing enzymes. The precursors as well as the activating enzymes are usually stored in very high amounts so that a fast release of sufficient amounts of hydrolysis products is ensured. Besides their role in plant defense, many studies on both glucosinolates and alliins have also been motivated by the health-promoting effects of these compounds in the human diet. During the past decade, research on glucosinolates has been benefited enormously from the resources that have become available through the *Arabidopsis* genome and transcriptome projects.

A. Glucosinolates

1. Structure, Occurrence and Biological Significance

The core structure of glucosinolates consists of a β-D-thioglucose group linked to a (Z)-N-hydroxyiminosulfate ester and a variable side chain that is derived from an amino acid and can be aliphatic, aromatic or indolic (Fig. 1a; Ettlinger and Lundeen, 1957). To date, about 140 glucosi-

---

Abbreviations: ESP – epithiospecifier protein; IAOx – indole-3-acetaldoxime; MCSO – (+)-*S*-methyl-L-cysteine sulfoxide; PeCSO – (+)-*S*-trans-1-propenyl-L-cysteine sulfoxide; PCSO – (+)-*S*-propyl-L-cysteine sulfoxide; S⁰ – elemental sulfur; S-cells – sulfur-rich cells; SEM–EDX – scanning electron microscopy–energy dispersive x-ray microanalysis; TFP – thiocyanate-forming protein

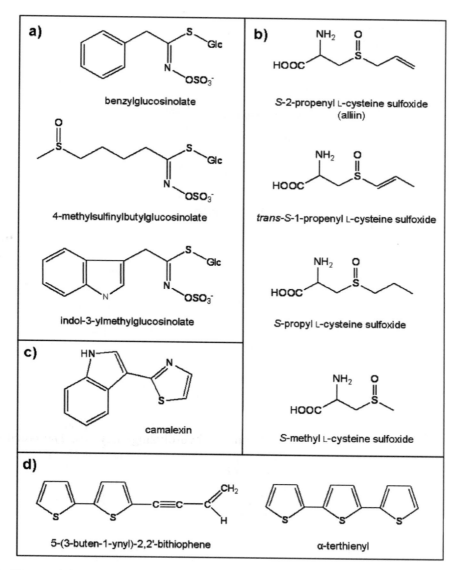

Fig. 1. Examples of sulfur-containing defense compounds. (a) Chemical structures of an aromatic, an aliphatic and an indole glucosinolate. (b) *S*-2-propenyl L-cysteine sulfoxide (alliin), the precursor of the characteristic flavor compounds of garlic, *trans-S*-1-propenyl L-cysteine sulfoxide and *S*-propyl L-cysteine sulfoxide, both found in onion, and *S*-methyl L-cysteine sulfoxide, present in most *Allium* species and in some Brassicaceae. (c) Camalexin (3-thiazol-2'-yl-indole), the major phytoalexin in *Arabidopsis thaliana*, (d) 5-(3-buten-1-ynyl)-2,2'-bithiophene, a predominant dithiophene in roots of *Tagetes* species, and the trithiophene α-terthienyl.

nolate structures have been identified (D'Auria and Gershenzon, 2005; Halkier and Gershenzon, 2006). Their occurrence seems to be restricted to the order Brassicales which includes the agriculturally important Brassicaceae as well as the Capparaceae, and the Caricaceae. The identification of glucosinolates in representatives of the genus *Drypetes* (Euphorbiaceae) is the only

report of their occurrence outside the Brassicales (Fahey et al., 2001).

While intact glucosinolates play a role as cues for host identification, oviposition and feeding of insects specialized on glucosinolate-containing plants (see below), their defensive potential arises mainly from the hydrolysis products formed by the action of myrosinases upon tissue damage

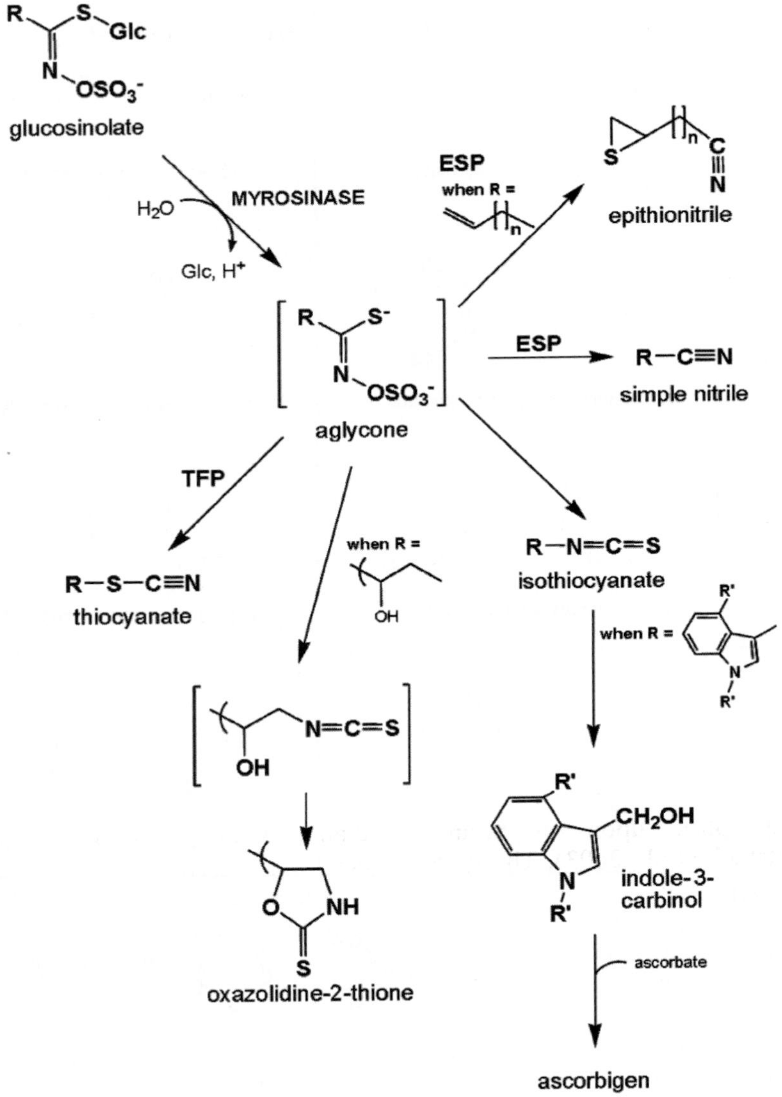

Fig. 2. Glucosinolate hydrolysis. Upon damage of plant tissue, myrosinases catalyze the hydrolysis of glucosinolates to glucose and unstable aglyca. These aglyca can rapidly rearrange to form isothiocyanates or, in case of glucosinolates with a 2-hydroxy-lated side chain, oxazolidine-2-thiones. Spontaneous rearrangement of the unstable isothiocyanates derived from indole glu-cosinolates leads to the generation of indole-3-carbinols which subsequently react with free ascorbate to form ascorbigens. In the presence of an epithiospecifier protein (ESP), simple nitriles or epithionitriles are formed at the expense of isothiocyanates. Organic thiocyanates are formed in the presence of a thiocyanate-forming protein (TFP). (R, R', variable side chains).

(Matile, 1980; Rask et al., 2000). Myrosinases (EC 3.2.1.147) catalyze the hydrolysis of the thi-oglycosidic linkage in the glucosinolate skeleton leading to the formation of glucose and an unsta-ble aglycone which can spontaneously undergo a Lossen rearrangement to an isothiocyanate (Fig. 2; Ettlinger and Lundeen, 1957; Benn, 1977; Bur-meister et al., 2000). If the glucosinolate side chain bears a hydroxyl group in the C-2 posi-

tion, the corresponding isothiocyanate spontane-ously cyclizes to form an oxazolidine-2-thione. Isothiocyanates derived from indole glucosi-nolates are also unstable and rapidly rearrange to indole-3-carbinols which subsequently react with free ascorbate to form ascorbigens (Chevolleau et al., 1997; Agerbirk et al., 1998). In many plant species, glucosinolates can be hydrolyzed to alternative products, such as simple nitriles, epithionitriles, and thiocyanates due the presence of additional protein factors (Tookey, 1973; Hasapis and MacLeod, 1982; Petroski and Kwolek, 1985; Lambrix et al., 2001; Burow et al., 2007). Epithiospecifier proteins (ESPs) have been identified in *Brassica napus*, *B. oleracea*, and *Arabidopsis thaliana* and redirect the hydrolysis of glucosinolates from isothiocyanate to nitrile formation without having hydrolytic activity on glucosinolates themselves (Bernardi et al., 2000; Lambrix et al., 2001; Burow et al., 2006a; Matusheski et al., 2006). In *Lepidium sativum*, benzyl thiocyanate formation is promoted by a thiocyanate-forming protein with high sequence similarity to ESPs (Burow et al., 2007).

Many glucosinolate hydrolysis products have been demonstrated to act as toxins, growth inhibitors, or feeding deterrents on a wide range of potential enemies, including mammals, birds, insects, mollusks, aquatic invertebrates, nematodes, bacteria, and fungi (Chew, 1988; Louda and Mole, 1991; Li et al., 2000; Tierens et al., 2001; Buskov et al., 2002; Wittstock et al., 2003; Lazzeri et al., 2004; Noret et al., 2005). When tested for their activities against insects, glucosinolate breakdown products have been reported to be effective in the gas phase, upon contact, and after ingestion (Wittstock et al., 2003). The type of hydrolysis products which have been most frequently demonstrated to be toxic to both generalist and specialist insects are the isothiocyanates (Lichtenstein et al., 1962; Seo and Tang, 1982; Li et al., 2000; Agrawal and Kurashige, 2003). Although isothiocyanates are known to react with amino and sulfhydryl groups of proteins *in vitro* (Kawakishi and Kaneko, 1987), their mode of action remains to be investigated *in vivo*. Thiocyanates and simple nitriles have also been reported to be toxic to some insect species, but no information is available on the biological activities of epithionitriles (Wittstock et al., 2003). Although only few studies have been carried out to compare the effects of different hydrolysis products derived from the same parent glucosinolate, nitriles are generally considered to be less toxic to insects than isothiocyanates and may therefore serve other functions than direct chemical defense (Lambrix et al., 2001; Burow et al., 2006b).

Although the glucosinolate–myrosinase system has been shown to defend plants against generalist herbivores (Li et al., 2000), a number of insect species has developed sensory and biochemical adaptations that allow them to colonize glucosinolate-containing plants without negative effects. Besides the intact glucosinolates (Chun and Schoonhoven, 1973; Nielsen, 1978; Städler, 1978; Louda and Mole, 1991), such specialized feeders also use glucosinolate hydrolysis products as feeding and oviposition cues (Rojas, 1999; Gabrys and Tjallingii, 2002; Miles et al., 2005). For example, both isothiocyanates and simple nitriles are involved in the long distance attraction of some specialized beetles (Pivnick et al., 1992; Bartlet et al., 1997). Allyl isothiocyanate has been reported to attract wasps that parasitize insect herbivores on brassicaceous plants (Titayavan and Altieri, 1990; Pivnick, 1993; Murchie et al., 1997). Biochemical mechanisms used by specialist herbivores on the Brassicaceae to overcome the glucosinolate–myrosinase system have recently been studied on a molecular level. Larvae of the small cabbage white butterfly *Pieris rapae* can circumvent poisoning by the glucosinolate–myrosinase system by redirecting glucosinolate hydrolysis away from isothiocyanate formation towards the formation of the corresponding nitriles which can then be excreted with the feces (Wittstock et al., 2004). A contrasting mechanism was found in the diamondback moth *Plutella xylostella*. The larvae accumulate a sulfatase in their guts which converts the glucosinolates ingested with the plant material into desulfoglucosinolates and thereby prevents the hydrolysis of glucosinolates into the toxic isothiocyanates (Ratzka et al., 2002). Other insect herbivores are able to sequester intact glucosinolates from their host plants (Aliabadi et al., 2002; Müller et al., 2002). The cabbage aphid, *Brevicoryne brassicae*, has been shown to produce its own myrosinase, which most probably aids in the amplification of an alarm signal within an aphid colony upon attack by predators (Bridges et al., 2002).

2.   Compartmentalization of the Glucosinolate–Myrosinase System

As a fundamental principle of activated plant defenses, the compartmentalization of glucosinolates, the biologically inactive substrates, separately from myrosinases, the activating enzymes, has been investigated in several species of the Brassicaceae on the cellular and subcellular level using histological, immunocytochemical, autoradiographic, promoter-GUS fusion and *in situ* hydridization techniques. While the compartmentalization of the glucosinolate–myrosinase system appears to be achieved on the subcellular level in seeds and seedlings of *B. napus* and *B. juncea* because myrosinase and glucosinolates were detected in the same cells (Kelly et al., 1998; Thangstad et al., 2001), the two components are stored in separate cells of other tissues. High concentrations of sulfur detected by x-ray analyses indicate the accumulation of glucosinolates in specific cells between the vascular bundles and the endodermis (S-cells) and in epidermis cells of *A. thaliana* flower stalks as well as in the root cap and the cortex of *Sinapis alba* roots (Wei et al., 1981; Koroleva et al., 2000). Immunocytochemical studies on different organs of *B. napus* and *A. thaliana* revealed that the myrosinases are localized in the vacuoles of idioblasts in the phloem parenchyma (Andreasson et al., 2001; Ueda et al., 2006). The presence of myrosinases in phloem cells which are presumably adjacent to the glucosinolate-rich S-cells (Koroleva et al., 2000) was also confirmed by investigations on the activities of myrosinase promoters using promoter-GUS and promoter-GFP fusion constructs in *A. thaliana* (Husebye et al., 2002; Thangstad et al., 2004). In the latter studies, myrosinase expression was also shown to be directed to guard cells. As the compartmentalization on the cellular level is destroyed upon attack by chewing herbivores, it enables the plant to release the biologically active compounds within a short period of time from preformed precursors that can be stored safely. In contrast, sucking herbivores may be able to circumvent the disruption of multiple cells and thereby the release of toxic products (Moran et al., 2002).

3. Glucosinolate Biosynthesis

As major secondary metabolites of the agriculturally important crops of the Brassicaceae family such as oil seed rape and cabbage, the nutritional value of glucosinolates as constituents of animal feed and the human diet has been studied extensively (Verhoeven et al., 1997; Griffiths et al., 1998). Both anti-nutritional and health-promoting compounds have been identified among the glucosinolate hydrolysis products (Bradshaw et al., 1984; Zhang et al., 1992; Fahey et al., 1997). Together with the known defensive role of the glucosinolate–myrosinase system, this has prompted the desire to specifically manipu-

late glucosinolate profiles in agricultural crops to improve their nutritional value as well as their pest resistance and has been one major driving force of biosynthetic studies. To date, most steps of the pathway leading from few protein amino acids (alanine, leucine, isoleucine, methionine, valine, phenylalanine, tyrosine, tryptophan) to the variety of glucosinolate structures have been identified. In the first phase of glucosinolate biosynthesis, certain protein amino acids that serve as precursors for aliphatic or aromatic glucosinolates are elongated by sequential insertions of one to nine methylene groups into the side chain (Fahey et al., 2001). In the second phase, a protein amino acid or its chain-elongated derivative is converted to the core structure common to all glucosinolates. The first step in this reaction sequence, the conversion of the amino acid into an aldoxime, is catalyzed by cytochrome P450-dependent monooxygenases of the CYP79 family (Wittstock and Halkier, 2002). Aldoxime formation from tryptophan seems to be critical not only for the biosynthesis of indole glucosinolates, but also for camalexin biosynthesis and auxin homeostasis (Fig. 3; Hull et al., 2000). After the biosynthesis of the core structure, the glucosinolate side chain can undergo various secondary modifications, such as hydroxylation, *O*-methylation, desaturation, acylation, and glycosylation in the third phase of the pathway (Fahey et al., 2001; Halkier and Gershenzon, 2006). With the identification of most of the genes involved in glucosinolate biosynthesis in *Arabidopsis* (Grubb and Abel, 2006), metabolic engineering of the pathway in crop plants has become a realistic possibility (Halkier and Du, 1997).

B. S-Alkyl-Cysteine Sulfoxides (Alliins)

1. Structure, Bioactivation and Ecological Roles

Like glucosinolates, the *S*-alkyl-cysteine sulfoxides (alliins) found in the Alliaceae, which includes widely used vegetable and spice plants such as onion (*Allium cepa* L.), garlic (*Allium sativum* L.), leek (*Allium porrum* L.), and chives (*Allium schoenoprasum* L.), are preformed secondary metabolites that are activated by hydrolyzing enzymes upon tissue damage. Various pharmacological effects have been ascribed especially

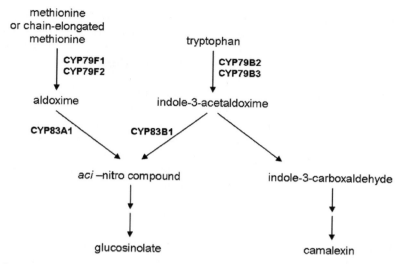

Fig. 3. Metabolic link between glucosinolate and camalexin biosynthesis. The central intermediate in the biosynthesis of both indole glucosinolates and camalexin is indole-3-acetaldoxime which is formed from tryptophan by the cytochrome P450-dependent monooxygenases CYP79B2 and CYP79B3 and is subsequently directed into indole glucosinolate or camalexin biosynthesis. Aldoxime formation from methionine and its chain-elongated derivatives in the biosynthesis of aliphatic glucosinolates is catalyzed by two other members of the CYP79 family, CYP79F1 and CYP79F2.

to garlic or its ingredients, e.g. inhibition of blood clotting, lowering of blood lipid level and blood pressure as well as antiviral, antifungal, antimicrobial, and cancerostatic properties (Tsai et al., 1985; Block, 1992; Agarwal, 1996; Griffiths et al., 2002). The compounds that are responsible for the typical flavor and odor of garlic, onions, and related species have been intensely studied since the 1940s when it was recognized for the first time that odorless precursors present in the intact plant tissue are cleaved to organosulfur volatiles upon tissue damage (Cavallito et al., 1945; Stoll and Seebeck, 1947). In *Allium* species, the diversity of flavor compounds produced can mainly be attributed to four (+)-*S*-alkyl-L-cysteine sulfoxides (Fig. 1b), but more structures of flavor precursors have been identified (Kubec et al., 2000; Kubec et al., 2002a; Kubec et al., 2002b; Jones et al., 2004). (+)-*S*-alkyl-L-cysteine sulfoxides are also present in several members of the Brassicaceae, in tropical plants of the Phytolaccaceae and Olacaceae as well as in fruiting bodies of Basidiomycetes (Gmelin et al., 1976; Stoewsand, 1995; Kubota et al., 1998; Kubec and Musah, 2001; Kubec et al., 2001).

The bioactivation of (+)-*S*-alkyl-L-cysteine sulfoxides upon tissue disruption is accomplished by a group of *C-S*-lyases known as alliinases (EC 4.4.1.4). Alliinases catalyze the cleavage of the $C_\beta–S_\gamma$ bond of (+)-*S*-alkyl-L-cysteine sulfoxides which results in the generation of pyruvate, ammonia, and thiosulphinates (Fig. 4). The latter are very reactive compounds and undergo subsequent chemical rearrangements to form dialk(en)yl sulfides, dithiines, ajoenes, and various other volatile and non-volatile organosulfur compounds (for review see Whitaker, 1976; Block, 1992). In onion, the relative contribution of the different flavor precursors to the blend of sulfur-containing compounds may be influenced by the distinct substrate specificity of onion alliinase which prefers *S-trans*-1-propenyl-L-cysteine sulfoxide (PeCSO) over *S*-methyl-L-cysteine sulfoxide (MCSO) and *S*-propyl-L-cysteine sulfoxide (PCSO) (Nock and Mazelis, 1987; Coolong and Randle, 2003). The formation of the lachrymatory volatile released by chopped onions, propanthial-*S*-oxide, from PeCSO requires not only alliinase activity but also the presence of propanthial-*S*-oxide synthase (Imai et al., 2002). Spatial separation of (+)-*S*-alkyl-L-cysteine sulfoxides and alliinases in intact plants is achieved on the subcellular and the cellular level, respectively, depending on the plant species. The (+)-*S*-alkyl-L-cysteine sulfoxides are cytosolic compounds as demonstrated by cell fractionation after tracer feeding studies in onion (Lancaster and Collin, 1981; Lancaster et al., 1989). Onion alliinase is sequestered in the vacuoles throughout the leaf tissue (Lancaster and Collin, 1981). In contrast, garlic alliinase is accumulated only in the vacuoles of the bundle sheath cells as shown by staining of alliinase activity and immunolocalization in sections of garlic cloves (Ellmore and Feldberg, 1994).

The sulfur-containing compounds formed from (+)-S-alkyl-L-cysteine sulfoxides upon tissue disruption are thought to serve a protective function against pathogens and herbivores (Ellmore and Feldberg, 1994; Ankri and Mirelman, 1999). However, certain specialized fungi and insect herbivores of *Allium* species have adapted to these chemicals and use them for host finding and identification. For example, germination of sclerotia of *Sclerotium cepivorum* causing the white rot disease in onion and garlic depends on the presence of the (+)-S-alkyl-L-cysteine sulfoxide hydrolysis products under field conditions (Coley-Smith, 1986). As an example of insects specialized on onion and closely related *Allium* species, females of the onion fly (*Delia antiqua*) respond to *n*-dipropyl disulfide in a dose-dependent manner during host plant finding and oviposition as demonstrated by both behavioural assays and electroantennogram recordings (Romeis et al., 2003). Larvae of the leek moth (*Acrolepiopsis assectella*) even sequester (+)-S-alkyl-L-cysteine sulfoxides from leek and other *Allium* species to protect themselves from predatory ants (Le Roux et al., 2002).

2. Biosynthesis of Alliins

The positive effects of (+)-S-alkyl-L-cysteine sulfoxides and their hydrolysis products in numerous pharmacological test systems have motivated a large number of studies on their biosynthesis. Two pathways for (+)-S-alkyl-L-cysteine sulfoxide formation have been proposed based on tracer feeding studies. Central intermediates in the first pathway identified in leaves of *A. cepa*, *A. sativum*, and *A. siculum* are the so-called γ-glutamyl peptides (glutamyl-cysteine derivatives) formed from glutathione (Fig. 4; Granroth, 1970; Lancaster and Shaw, 1989). Apart from their role as intermediate in (+)-S-alkyl-L-cysteine sulfoxide biosynthesis, γ-glutamyl peptides have been suggested to represent a storage form for nitrogen and sulfur (Lancaster and Shaw, 1991; Randle et al.,

2002). This assumption is supported by the finding that several species of the Fabaceae family accumulate γ-glutamyl peptides in their seeds although they do not synthesize any (+)-S-alkyl-L-cysteine sulfoxide (Ellis and Salt, 2003). A role for these peptides in selenium tolerance has also been proposed (Ellis and Salt, 2003). The alk(en)yl side chains of these peptides and the corresponding (+)-S-alkyl-L-cysteine sulfoxides (except methyl) are thought to derive from the chemical reaction of cysteine with methacrylic acid, a transient intermediate in valine catabolism (Suzuki et al., 1962; Granroth, 1970; Lancaster and Shaw, 1989; Parry and Lii, 1991). Whether this reaction with cysteine occurs before or after its incorporation into glutathione is not known. As an alternative route, (+)-S-alkyl-L-cysteine sulfoxides can be synthesized via direct alk(en)ylation of cysteine or thioalk(en)ylation of O-acetyl serine followed by enzymatic oxidation to a cysteine sulfoxide derivative (Granroth, 1970; Lancaster and Shaw, 1989; Ohsumi et al., 1993). It remains to be elucidated to which extent the two pathways contribute to (+)-S-alkyl-L-cysteine sulfoxide biosynthesis in different plant species and throughout plant development (Jones et al., 2004). The site of (+)-S-alkyl-L-cysteine sulfoxide biosynthesis is photosynthetically active leaves. The (+)-S-alkyl-L-cysteine sulfoxides as well as γ-glutamyl peptides are subsequently transferred to the bulbs where non-protein cysteine and glutathione derivatives can account for up to 1–5% of the dry weight (Lancaster and Kelly, 1983; Lancaster et al., 1986).

Interestingly, the hydrolytic enzymes that mediate the bioactivation of alliins in the Alliaceae and of glucosinolates in the Brassicales share a number of common features which may therefore be crucial for the functioning of activated plant defense systems. Both alliinases and myrosinases are homodimeric glycoproteins which have been reported to form stable complexes with lectins (Palmieri et al., 1986; Rabinkov et al., 1995; Smeets et al., 1997; Burmeister et al., 2000; Rask et al., 2000; Kuettner et al., 2002). High quantities of the enzymes activating alliins or glucosinolates upon tissue disruption are stored in the vacuoles of idioblastic cells or throughout the tissue to facilitate the rapid production of bioactive compounds (Lancaster and Collin, 1981; Ellmore and Feldberg, 1994; Andreasson and Jorgensen, 2003).

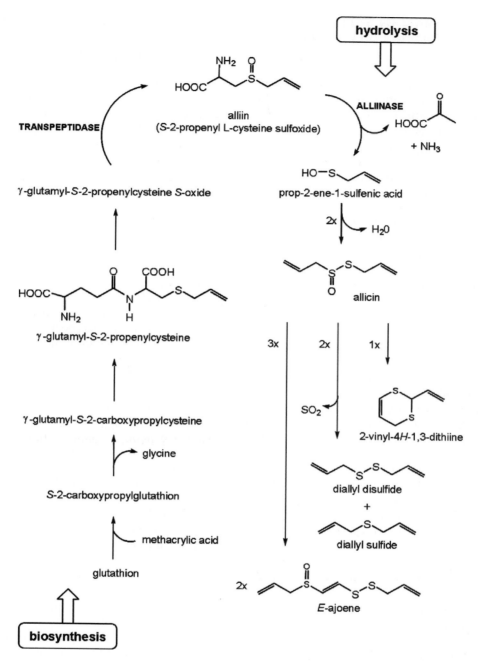

Fig. 4. Biosynthesis and hydrolysis of alliin (*S*-2-propenyl L-cysteine sulfoxide). Alliin biosynthesis has been proposed to be initiated by *S*-alkylation of glutathione with methacrylic acid as alkyl donor. The γ-glutamyl peptide obtained by removal of the glycine group is decarboxylated and then oxidized to γ-glutamyl-*S*-2- propenylcysteine *S*-oxide. Final loss of the γ-glutamyl group by a transpeptidase leads to alliin. Upon tissue disruption, alliinase hydrolyzes the C_β–S_γ bond of alliin which results in the formation of pyruvate, ammonia, and prop-2-ene sulfenic acid which spontaneously reacts to form the thiosulphinate allicin (=*S*-allyl prop-2-ene-1-sulfinothioate). Allicin can undergo further chemical rearrangement, e.g. to *E*-ajoene (=1-allyl-2-((*E*)-2-(allylsulfinyl)vinyl) disulfane), diallyl disulfide (=1,2-diallyldisulfane), diallyl sulfide (=diallylsulfane), or 2-vinyl-4H-1,3-dithiine.

III. Induced Defense by Sulfur and Sulfur-Containing Compounds

While activated plant defenses require the storage of preformed components to enable rapid activation upon attack, other defenses are only biosynthesized after attack by a herbivore or pathogen. Biosynthesis of such defenses is induced by signals from the herbivore or pathogen that trigger a local or systemic signal transduction cascade in the plant resulting in the expression of defense responses (Kessler and Baldwin, 2002). One group of induced chemical defenses are the phytoalexins, compounds of diverse chemical structures, that are only produced in response to pathogen attack. Many of the most prominent phytoalexins are sulfur-containing metabolites with structures ranging from elemental sulfur to the alkaloid-like camalexin and complex polypeptides like the defensins.

A. Elemental Sulfur

Elemental sulfur has been demonstrated to inhibit the growth or spore germination of many fungal pathogens representing ascomycetes, basidiomycetes, and deuteromycetes in both field and laboratory experiments, whereas bacteria and the oomycete *Phytophtera palmivora* appear to be generally less affected (for review see Williams and Cooper, 2004). Fungal cells can take up S^0 into their cytoplasm where it is thought to cause a multiple site inhibition of the mitochondrial respiratory chain. However, further research is needed to completely unravel the mode of toxicity of elemental sulfur to fungal pathogens. Although elemental sulfur (S^0) is probably man's oldest fungicide and is still a common component in integrated pest control programs (Williams and Cooper, 2004), it was recognized as the only inorganic phytoalexin only ten years ago (Cooper et al., 1996; Resende et al., 1996; Dixon, 2001). After infestation with the soil-borne vascular pathogen *Verticillium dahliae*, disease-resistant genotypes of *Theobroma cacao* (cacao, Sterculiaceae) deposited up to $116 \mu g$ S^0 (g fresh weight)$^{-1}$ in the stem, whereas no S^0 was recovered from plants of the susceptible genotype or from intact or wounded plants (Resende et al., 1996). The sulfur persisted in stems of resistant *T. cacao* plants for more than 60 days after inoculation indicating that it was unavailable to living cells and that the levels of S^0 were too high to enable fungal growth (Cooper et al., 1996). Accumulation of elemental sulfur has been observed also in disease-resistant genotypes of tomato (*Lycopersicon esculentum*, Solanaceae) after infection with *V. dahliae* (Williams et al., 2002). Although the concentration of S^0 in whole tissue extracts was an order of magnitude lower as compared to *T. cacao*, the quantities were still sufficient to inhibit colonization of the vascular system by the pathogen. The accumulation of elemental sulfur in tomato coincided with higher concentrations of sulfate and a transient increase in L-cysteine and glutathione (Williams et al., 2002). Much smaller amounts of S^0 were detected in the vascular tissues of resistant plants after inoculation with *Ralstonia solanacearum* (Southern bacterial wilt) as compared to those infected with *V. dahliae* (Williams and Cooper, 2003).

More recently, fungal or bacterial pathogen-induced S^0 production has been reported for several different plant species from distantly related families with a stronger and more rapid response in resistant genotypes as compared to susceptible genotypes, e.g. in *Gossypium hirsutum* (cotton, Malvaceae) infected with *V. dahliae* and in *Nicotiana tabacum* (tobacco, Solanaceae) and *Phaseolus vulgaris* (French bean, Fabaceae) infected with *Fusarium oxysporum* (Fusarium wilt) (Williams and Cooper, 2003). No S^0 was recovered from *Fragaria vesca* (strawberry, Rosaceae) or *Zea mays* (Maize, Poaceae) challenged with *V. dahliae* or the bacterial pathogen *Erwinia stewartii*, respectively. Interestingly, high constitutive levels of S^0 were present in the cotyledons of *Brassica oleracea* and leaves of *Arabidopsis thaliana* (both Brassicaceae) (Williams and Cooper, 2003). Finally, elemental sulfur was also found in cuticular waxes of several gymnosperms and angiosperms where it might play a role in plant defense (Kylin et al., 1994). Although the metabolic capacity to produce elemental sulfur appears to be widespread among higher plants, the biosynthetic pathway leading to S^0 formation has remained unresolved. Proposed pathways include the formation from sulfide via oxidation by sulfide oxidase (Joyard et al., 1988) or by cytochromes (Krauss et al., 1984), the degradation of glutathione or cysteine by cysteine desulfhydrase (Rennenberg et al., 1987, Schmidt, 1987), and the hydrolysis of glucosinolates to simple nitriles (Bones and Rossiter, 1996; Foo et al., 2000).

Coupled scanning electron microscopy–energy dispersive x-ray microanalysis (SEM–EDX) of pathogen-infected tomato and cacao stems revealed high concentrations of S^0 scattered in xylem parenchyma cells adjacent to xylem vessels, within vessel walls, and in gels occluding vessels (Cooper et al., 1996; Williams et al., 2002). This accumulation in the vascular tissue supports the idea that S^0 serves as a barrier to vertical and lateral spreading of the pathogen. In contrast, the high constitutive amounts of elemental sulfur present in *A. thaliana* are not restricted to the xylem and may therefore serve a different biological function (Williams and Cooper, 2003).

B.   Camalexin and Related Phytoalexins

1.   Structure and Biological Significance

The presence of a specific class of alkaloid-like phytoalexins in members of the Brassicaceae that has a core structure consisting of an indole ring system coupled to a sulfur-containing moiety at the C-3 position is a relatively recent discovery (Fig. 1c; Conn et al., 1988; Browne et al., 1991; Tsuji et al., 1992; Zook et al., 1998). While the actual role of these compounds in plant defense and their mode of action has remained unresolved, numerous investigations have analyzed the signaling cascades that are associated with their induction. Camalexin (3-thiazol-2'-yl-indole) has become the best studied example of this class of phytoalexins. In *Arabidopsis thaliana*, camalexin production was found to be induced in response to a variety of pathogens, *e.g. Pseudomonas syringae* and *Alternaria bassicicola,* as well as elicitors that generate reactive oxygen species (Tsuji et al., 1992; Tsuji et al., 1993; Reuber et al., 1998; Zhao et al., 1998; Thomma et al., 1999; Roetschi et al., 2001). The induction of camalexin biosynthesis appears to be controlled by a complex network of signaling pathways (Denby et al., 2004; Hansen and Halkier, 2005). Both the upregulation of the key biosynthetic proteins and camalexin accumulation are restricted to the site of infection (Fig. 5; Schuhegger et al., 2006b).

Among the tools that have frequently been used to study the role of camalexin in plant defense are the PAD3 knock-out mutants of *A. thaliana* that completely lack the capability of producing camalexin (Glazebrook et al., 1997; Zhou et al.,

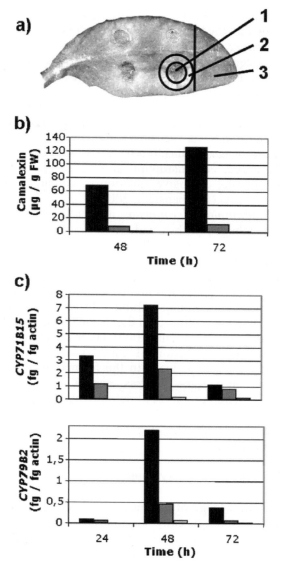

Fig. 5. Camalexin induction is strictly localized to the site of pathogen infection. (a) After droplet infection of *Arabidopsis thaliana* with *Alternaria alternata*, leaves were separated into infection site (1), surrounding ring (2), and the rest of the leaf (3). Camalexin accumulation (b) and transcript induction of the biosynthetic genes *CYP79B2* and *CYP71B15* (c) were highly localized to the infection site (black bars). Gray bars, ring sorrounding infection site; white bars, rest of leaf. This figure was kindly provided by Erich Glawischnig.

1999). Interestingly, only necrotrophic fungi that cause cell disruption, e.g. *Botrytis cinerea* and *Alternaria brassicicola,* show increased growth on the *pad3* knock-out mutant as compared to wild-type *A. thaliana* (Thomma et al., 1999; Ferrari et al., 2003; van Wees et al., 2003). By contrast, the growth of the biotrophic fungi *Hyaloperonospora parasitica* and *Erysiphe orontii* does not seem to be affected by the induction of camalexin (Reuber et al., 1998; Mert-Türk et al., 2003). Similarly, the bacterial pathogen *Pseudomonas syringae* showed the same growth rates on *A. thaliana* wild-type plants and *pad3* mutants, although *P. syringae* was sensitive to camalexin when tested *in vitro* (Glazebrook and Ausubel, 1994). Therefore, cell disruption appears to be essential for the defensive function of camalexin in *A. thaliana* (Kliebenstein, 2004).

Sclerotinia sclerotiorum, a necrotrophic fungal pathogen with a wide host range, has been demonstrated to detoxify camalexin and related indole phytoalexins as well as synthetic analogues by glycosylating the C-6 or the N-1 atom in the indole moiety (Pedras and Ahiahonu, 2002). The resulting compounds exhibit substantially lower antifungal activities than the original compounds (Pedras and Ahiahonu, 2002). Hydroxylation and subsequent oxidation of camalexin in *Rhizoctonia solani* also leads to the formation of less toxic derivatives (Pedras and Khan, 1997). These different modes of detoxification may also explain observed differences in the sensitivity to camalexins among fungal pathogens (Kliebenstein, 2004).

2. Camalexin Biosynthesis

Our knowledge of camalexin biosynthesis is still limited. Only recently, the first genes involved in the biosynthetic pathway have been identified in *A. thaliana*. In feeding experiments using labeled indole, tryptophan and indol-3-acetaldoxime (IAOx), all three compounds were shown to be precursors for camalexin biosynthesis (Glawischnig et al., 2004) contrasting with previous studies that had suggested a tryptophan-independent pathway (Tsuji et al., 1993; Zook and Hammerschmidt, 1997). The conversion of tryptophan to IAOx, a reaction that is common to the biosynthesis of glucosinolates and camalexin (Fig. 3), is catalyzed by two substrate-specific cytochrome P450 monooxygenases, CYP79B2 and CYP79B3 (Hull et al., 2000; Mikkelsen et al., 2000; Glawischnig et al., 2004). Consistent with this pathway being the only one operating in camalexin biosynthesis in *A. thaliana*, the *cyp79B2/cyp79B3* double knock-out mutant (Zhao et al., 2002) was found to be completely devoid of camalexin production, whereas the *cyp79B2* and *cyp79B3* single mutants still produced about 50% and about 100% of wild-type level, respectively (Glawischnig et al., 2004). Since *cyp79B2* but not *cyp79B3* has been reported to be upregulated upon induction of camalexin production after inoculation with the bacterial pathogen *Erwinia carotovora* or after treatment with $AgNO_3$ (Brader et al., 2001; Glawischnig et al., 2004), the two genes may have different roles in camalexin and glucosinolate biosynthesis (Hansen and Halkier, 2005). The formation of the thiazole ring in the camalexin structure has been proposed to be accomplished by condensation of indole-3-carboxaldehyde with L-cysteine and subsequent cyclization (Zook and Hammerschmidt, 1997). The final decarboxylation of 2-(indol-3-yl)-4,5-dihydro-1, 3-thiazole-4-carboxylic acid (dihydrocamalexic acid) is catalyzed by the cytochrome P450 CYP71B15, also named PAD3 (Schuhegger et al., 2006a).

C. Defensin and Related Peptides

In addition to the low molecular weight compounds that are typically involved in plant defense, cysteine-rich peptides with antimicrobial properties have been identified in mammals, amphibians, insects, plants and even in microorganisms themselves (Rao, 1995). Defensins and thionins are two groups of such peptides that are found in a broad variety of plant families (Bohlmann, 1994; Terras et al., 1995; Harrison et al., 1997). Thionins accumulate in the seed endosperm of many plant species where they are believed to serve as sulfur storage compounds as well as chemical defenses (Bohlmann et al., 1988; Schrader-Fischer and Apel, 1993). The structurally related plant defensins (originally termed γ-thionins) were first discovered in grains of wheat and barley (Colilla et al., 1990; Mendez et al., 1990) and since then in various monocot and dicot species, e.g. Poaceae, Asteraceae, Fabaceae, Brassicaceae, Hippocastanaceae and Saxifragaceae (Bloch and Richardson, 1991; Terras et al., 1992; Terras et al., 1993; Moreno et al., 1994; Osborne et al., 1995). The most frequently studied effect of defensins is their antifungal activity. With a typical length of 45 to 54 amino acids, plant defensins are larger than those from insects and mammals (Broekaert et al., 1995). The characteristic three-dimensional structure of plant defensins consists of a triple stranded antiparallel β-sheet and an α-helix motif stabilized by disulfide bridges formed by eight highly conserved cysteine residues (for review see Boman, 1995). Antifungal peptides from insects and mammals most commonly possess only three cysteine residues.

Based on amino acid sequence identities, plant defensins can be further divided into different subfamilies (Thevissen et al., 2000). The sequence-based classification mirrors to a large extend the grouping of defensins according to the plant families they originate from as well as classifications that have been made based on the most obvious symptoms of defensin action on fungi (Fig. 6). Antifungal defensins from the Brassicaceae and Saxifragaceae impair hyphal growth by inducing tip ballooning and branch formation of susceptible fungi and have therefore been designated morphogenic defensins (Broekaert et al., 1995; Osborne et al., 1995). The non-morphogenic defensins present in the Asteraceae, Hippocastanaceae and Fabaceae retard hyphal growth of a different spectrum of pathogenic fungi without causing any morphological changes and they also differ from the morphogenic defensins (Broekaert et al., 1995; Osborne et al., 1995). The antifungal activities of both morphogenic and non-morphogenic defensins depend on their binding to highly specific binding sites on the fungal plasma membrane (Thevissen et al., 2000). In contrast to insect and mammalian defensins (Kagan et al., 1990; Cociancich et al., 1993), plant defensins lack the ability to form voltage-gated ion channels in biomembranes (Thevissen et al., 2000). A third subgroup of plant defensins that are found in the Poaceae has been reported to inhibit α-amylase and/or eukaryotic cell free translation (Mendez et al., 1990; Bloch and Richardson, 1991). Defensins of the α-amylase inhibitor type appear to play a role in plant defense against herbivores rather than against pathogenic fungi (Shade et al., 1994).

Apart from their expression as a response to pathogen attack, plant defensins are also constitutively expressed in seeds of various plant species where they mainly accumulate in the outer cell wall of the seed coat (Terras et al., 1995). The rather small amounts of defensins released from germinating seeds of *Raphanus sativus* (about 1 μg per seed) proved sufficient to inhibit the growth of several pathogenic fungi and to protect the emerging seedling (Terras et al., 1995). In the leaves of *R. sativus*, expression of the antifungal peptide Rs-AFP1 was found to be locally and systemically induced on the transcript level in response to infection with the fungal pathogens *Alternaria brassicicola* and *Botrytis cinerea* (Terras et al., 1995). Consistent with its presumed defensive function against fungi, overexpression of the Rs-AFP1 cDNA in tobacco reduced the lesion size upon infection with the foliar fungal pathogen *Alternaria longipes*. The existence of constitutive as well as pathogen-induced expression of defensins within the same plant has also been reported for other plant species. As an example, inoculation of pea pods (*Pisum sativum*, Fabaceae) with compatible and incompatible strains of *Fusarium solani* leads to an increased defensin transcript level, whereas the corresponding gene appears to be constitutively expressed in leaf epidermis cells (Broekaert et al., 1995).

The use of molecular techniques has facilitated the identification of cDNAs encoding presumed defensin homologues in different families, e.g. in the Fabaceae, Solanaceae and Brassicaceae (Stiekema et al., 1988; Ishibashi et al., 1990; Chiang and Hadwiger, 1991; Gu et al., 1992; Karunanandaa et al., 1994; Doughty et al., 1998; Silverstein et al., 2005). Although the actual roles of the corresponding peptides in plant defense await further investigation, their identification will benefit studies on structure-function relationships and the evolutionary origin of plant defensins.

IV. Constitutive Sulfur-Containing Defenses

Although most recent investigations have focused on induced plant defenses, constitutive defenses play an important role in the protection of plants from herbivores and pathogens. They are often deployed by plants that are likely to suffer frequent or serious damage or in plant organs that possess a high fitness value. Constitutive expression of chemical defenses enables the plant to protect itself even during the initial phase of attack before induced responses can act. As initial damage by a herbivore or pathogen may be too rapid or too severe to deploy induced defenses successfully, the use of constitutive defenses is regarded as the less risky defense strategy but is likely associated with higher costs (Gershenzon, 1994; Wittstock and Gershenzon, 2002). The constitutive or induced formation of sulfur-containing secondary metabolites requires the diversion of sulfur from primary metabolism and so is often accompanied by the regulation of sulfur uptake and assimilation (Rausch and Wachter, 2005; Hawkesford, this volume; Hell and Wirtz, this volume). The metabolic importance of sulfur may explain why only a very small proportion of constitutive defenses in plants is made up of sulfur-containing metabolites. Glucosinolates and alliins represent a special type of constitutive defenses. They are permanently stored as inactive precursors but are only activated upon attack (see above). Apart from these activated defenses, the thiophenes found in the Asteraceae and several other families are the only group of constitutive sulfur defenses that is relatively well characterized and whose defensive function has been studied in some detail (Kagan, 1991).

A. Thiophenes

Since the first report on the isolation of α-terthiophene from *Tagetes erecta* (Zehmeister and Sease, 1947) more than 150 thiophene structures comprised of one to three thiophene rings coupled in α-position and a side chain with a variable number of double and triple bonds have been described (Fig. 1d; Kagan, 1991). Upon irradiation with long wavelength ultraviolet light (UV-A, 320–400 nm), thiophenes exhibit substantial antiviral, antibacterial, antifungal, nematocidal, and insecticidal properties, whereas their toxicity is much lower in the absence of light (Gommers and Geerlings, 1973; Camm et al., 1975; Chan et al., 1975; Bakker et al., 1979; Champagne et al., 1986; Hudson et al., 1986; Downum et al., 1991).

Thiophenes are type II photodynamic photosensitizers that produce singlet oxygen and thereby damage biomembranes (Arnason et al.,

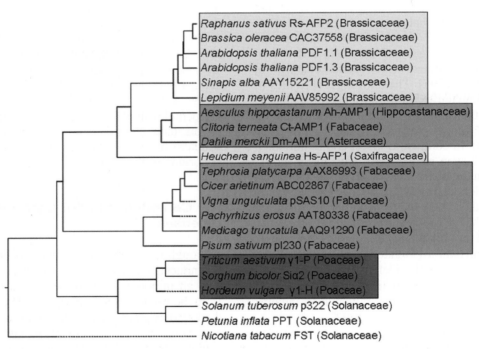

Fig. 6. Phenogram of plant defensins constructed from an alignment of amino acid sequences. Based on sequence similarities, the morphogenic defensins present in the Brassicaceae and Saxifragaceae (light gray), the non-morphogenic defensins from the Asteraceae, Hippocastanaceae and Fabaceae (gray), and the α-amylase-inhibiting defensins from the Poaceae (dark gray) form distinct groups. GenBank accession numbers (if not indicated in the figure): *R. sativus* RS-AFP2, P30230; *A. thaliana* PDF1.1, AAY27736; *A. thaliana* PDF1.3, NP 180171; *A. hippocastanum* Ah-AMP1, 1BK8; *C. terneata* Ct-AMP1, AAB34971; *D. merckii* Dm-AMP1, S66221; *H. sanguinea* Hs-AFP1, AAB34974; *V. unguiculata* pSAS10, P18646; *P. sativum* pI230, CAA36474; *T. aestivum* γ1-P, P20158; *S. bicolor* Siα2, P21924; *H. vulgare* γ1-H, P20230; *S. tuberosum* p322, P20346; *P. inflata* PPT, Q40901; *N. tabacum* FST, P32026. The sequence alignment by Clustal W and the tree construction were carried out using the program MegAlign.

1981; Downum et al., 1982; McLachlan et al., 1984). When the effects of isolated thiophenes on the larval development of *Ostrinia nubilalis*, *Euxoa messoria*, and *Manduca sexta* were tested in artificial diet studies, mortality rates of larvae reared in the presence of UV-light were higher compared to those of larvae kept without UV-light (Champagne et al., 1986). In the same study, thiophene toxicity was shown to depend on the thiophene structure as well as on the insect species. However, the action of some thiophenes that have been demonstrated to be highly toxic to blood-feeding and phytophagous insect larvae is not strictly dependent on light-mediated activation (Wat et al., 1981; Champagne et al., 1984; Downum et al., 1984; Philogéne et al., 1985). While the light-independent mechanism of action against insect herbivores is poorly understood, nematocidal activity of thiophenes in the absence of light has been proposed to depend on enzymatic activation in the root in response to endoparasitic nematode attack (Gommers, 1982; Hudson and Towers, 1991).

Some phytophagous insects evolved biochemical or behavioural adaptations to protect themselves from the toxic effects of the thiophenes in their host plants. Higher levels of cytochrome P450 monooxygenases in the generalist insect herbivore *Ostrinia nubilalis* as compared to the Solanaceae-specialist *Manduca sexta* resulted in more rapid metabolism of α-terthienyl and thereby conferred higher tolerance to this thiophene (Iyengar et al., 1990). Larvae of *Argyrotaenia velutinana* (Lepidoptera: Tortricidae) are specialized feeders on *Chrysanthemum leucanthemum* (Asteraceae), a plant species that contains both polyacetylenes and thiophenes, although direct exposure to light increases larval mortality (Guillet et al., 1995). The induced hiding behaviour of the larvae in combination with the preference of the females to oviposit in the shade reduce photo-activation of the plant defensive compounds. Another specialist insect herbivore on *C. leucanthemum*, larvae of *Chlorochlamys chloroleucaria* (Lepidoptera: Geometridae), preferentially feed on the non-phototoxic pollen of their host plant despite the increased risk of predation (Guillet et al., 1995).

Thiophene biosynthesis has been most extensively investigated in the genus *Tagetes* (marigolds) known as a rich source of bi- and trithiophenes (Bohlmann and Zdero, 1985). Although thiophenes are also formed in above-ground organs (Croes et al., 1994), root cultures or roots of wild-type plants or chemically mutagenized lines have been most frequently used for labeling experiments with $[^{35}S]SO_4^{2-}$. The polyacetylene tridecapentaynene, the immediate precursor of thiophene biosynthesis, has been proposed to be formed from oleic acid via crepenyic acid by sequential chain shortening and desaturation (Sörensen et al., 1954; Gunstone et al., 1967; Bohlmann et al., 1976; Jacobs et al., 1995). Thiophene formation involves the addition of reduced sulfur (presumably from L-cysteine) to conjugated triple bonds and subsequent ring formation, the removal of the terminal methyl group, and the modification of a vinyl group. The elucidation of the order in which these steps occur has been hampered by difficulties in detecting the proposed intermediates. Moreover, none of the enzymes involved has been biochemically characterized to date (Jacobs et al., 1995).

B.   Sulfur-Containing Metabolites in Algae and Mosses

Brown algae of the genus *Dictyopteris* use fatty acid-derived C_{11} sulfur compounds to deter feeding by herbivorous amphipods whereas sea urchins appaer to be unaffected (Moore, 1977; Hay et al., 1998; Schnitzler et al., 1998). Analogous to C_{11} hydrocarbons present in the thalli of many algae, the sulfur-containing C_{11} compounds in *Dictyopteris* are thought to be biosynthesized via oxidative degradation of highly saturated eicosanoids (Pohnert and Boland, 1996; Hombeck and Boland, 1998; Schnitzler et al., 1998). Sulfur-containing secondary metabolites have also been isolated from the liverwort *Corsinia coriandrina* (Corsiniaceae) including the isothiocyanate coriandrin [(Z)-2-(4-methoxyphenyl)ethenyl isothiocyanate]. The ecological role of coriandrin and related compounds, however, remains to be elucidated (von Reuß and König, 2005).

Acknowledgements

We thank Erich Glawischnig (Lehrstuhl für Genetik, Technische Universität München, Freising, Germany) for providing Fig. 5 and Katharina

Schramm (Max Planck Institute for Chemical Ecology, Jena, Germany) for help with the procurement of literature.

References

Agarwal KC (1996) Therapeutic actions of garlic constituents. Med Res Rev 16: 111–124

Agerbirk N, Olsen CE and Sorensen H (1998) Initial and final products, nitriles, and ascorbigens produced in myrosinase-catalyzed hydrolysis of indole glucosinolates. J Agric Food Chem 46: 1563–1571

Agrawal AA and Kurashige NS (2003) A role for isothiocyanates in plant resistance against the specialist herbivore *Pieris rapae*. J Chem Ecol 29: 1403–1415

Aliabadi A, Renwick JAA and Whitman DW (2002) Sequestration of glucosinolates by harlequin bug *Murgantia histrionica*. J Chem Ecol 28: 1749–1762

Andreasson E and Jorgensen LB (2003) Localization of Plant Myrosinases and Glucosinolates. In: Romeo J (ed) Integrative Phytochemistry: from Ethnobotany to Molecular Ecology, Recent Advances in Phytochemistry, Vol. 37, pp 79–99. Elsevier Science, Amsterdam

Andreasson E, Jorgensen LB, Hoglund AS, Rask L and Meijer J (2001) Different myrosinase and idioblast distribution in *Arabidopsis* and *Brassica napus*. Plant Physiol 127: 1750–1763

Ankri S and Mirelman D (1999) Antimicrobial properties of alliicin from garlic. Microbes Infect 1: 125–129

Arnason T, Chan GFQ, Downum C-K and Towers GHN (1981) Oxygen requirement for near-UV mediated cytotoxicity of α-terthienyl to *Escherichia coli* and *Saccharomyces cerevisia*. Photochem Photobiol 33: 821–824

Bakker J, Gommer FJ, Nieuwenhuis I and Wynberg H (1979) Photoactivation of the nematocidal compound α-terthienyl from roots of marigolds (*Tagetes* species). J Biol Chem 254: 1841–1844

Bartlet E, Blight MM, Lane P and Williams IH (1997) The responses of the cabbage seed weevil *Ceutorhynchus assimilis* to volatile compounds from oilseed rape in a linear track olfactometer. Entomol Exp Appl 85: 257–262

Benn M (1977) Glucosinolates. Pure Appl Chem 49: 197–210

Bernardi R, Negri A, Ronchi S and Palmieri S (2000) Isolation of the epithiospecifier protein from oil-rape (*Brassica napus* ssp *oleifera*) seed and its characterization. FEBS Lett 467: 296–298

Bloch C, Jr. and Richardson M (1991) A new family of small (5 kDa) protein inhibitors of insect alpha-amylases from seeds or sorghum (*Sorghum bicolor* (L) Moench) have sequence homologies with wheat gamma-purothionins. FEBS Lett 279: 101–104

Block E (1992) The organosulfur chemistry of the genus *Allium* - Implications for the organic chemistry of sulfur. Angew Chem Int Ed Engl 31: 1135–1178

Bohlmann F and Zdero C (1985) Naturally Occurring Thiophenes. In: Weissberger A and Taylor EC (eds) Thiophene and its Derivatives, pp 261–323. Wiley, New York

Bohlmann F, Zdero C and Grenz M (1976) Polyacetylene compounds. Part 241. Constituents of some genera of the tribes Helenieae and Senecioneae. Phytochemistry 15: 1309–1310

Bohlmann H (1994) The role of thionins in plant protection. Crit Rev Plant Sci 13: 1–16

Bohlmann H, Clausen S, Behnke S, Giese H, Hiller C, Reimann PU, Schrader G, Barkholt V and Apel K (1988) Leaf-specific thionins of barley: a novel class of cell wall proteins toxic to plant pathogenic fungi and possibly involved in the defence mechanism of plants. EMBO J 7: 1559–1566

Boman HG (1995) Peptide antibiotics and their role in innate immunity. Annu Rev Immunol 13: 61–92

Bones AM and Rossiter JT (1996) The myrosinase-glucosinolate system, its organisation and biochemistry. Physiol Plant 97: 194–208

Brader G, Tas E and Palva ET (2001) Jasmonate-dependent induction of indole glucosinolates in *Arabidopsis* by culture filtrates of the nonspecific pathogen *Erwinia carotovora*. Plant Physiol 126: 849–860

Bradshaw JE, Heaney RK, Smith WH, Gowers S, Gemmell DJ and Fenwick GR (1984) The glucosinolate content of some fodder Brassicas. J Sci Food Agric 35: 977–981

Bridges M, Jones AME, Bones AM, Hodgson C, Cole R, Bartlet E, Wallsgrove R, Karapapa VK, Watts N and Rossiter JT (2002) Spatial organization of the glucosinolate–myrosinase system in *Brassica* specialist aphids is similar to that of the host plant. Proc R Soc Lond Ser B Biol Sci 269: 187–191

Broekaert WF, Terras FRG, Cammue BPA and Osborne RW (1995) Plant defensins: novel antimicrobial peptides as components of the host defense system. Plant Physiol 108: 1353–1358

Browne LM, Conn KL, Ayer WA and Tewari JP (1991) The camalexins: new phytoalexins produced in leaves of *Camalexin sativa* (Cruciferae). Tetrahedron 47: 3909–3914

Burmeister WP, Cottaz S, Rollin P, Vasella A and Henrissat B (2000) High resolution x-ray crystallography shows that ascorbate is a cofactor for myrosinase and substitutes for the function of the catalytic base. J Biol Chem 275: 39385–39393

Burow M, Bergner A, Gershenzon J and Wittstock U (2007) Glucosinolate hydrolysis in *Lepidium sativum* - identification of the thiocyanate-forming protein. Plant Mol Biol 63: 49–61

Burow M, Markert J, Gershenzon J and Wittstock U (2006a) Comparative biochemical characterization of nitrile-forming proteins from plants and insects that alter myrosinase-catalyzed hydrolysis of glucosinolates. FEBS J 273: 2432–2446

Burow M, Müller R, Gershenzon J and Wittstock U (2006b) Altered glucosinolate hydrolysis in genetically engineered

Arabidopsis thaliana and its influence on the larval development of *Spodoptera littoralis*. J Chem Ecol 32: 2333–2349

Buskov S, Serra B, Rosa E, Sørensen H and Sørensen JC (2002) Effects of intact glucosinolates and products produced from glucosinolates in myrosinase-catalyzed hydrolysis on the potato cyst nematode (*Globodera rostochiensis* cv. Woll). J Agric Food Chem 50: 690–695

Camm EL, Towers GHL and Mitchell JC (1975) UV-mediated antibiotic activity of some Compositae species. Phytochemistry 14: 2007–2011

Cavallito CJ, Bailey JH and Buck JS (1945) The antibacterial principle of *Allium satvum*. III. Its precursor and "essential oil of garlic". J Amer Chem Soc 67: 1032–1033

Champagne DE, Arnason JT, Philogéne BJR, Campbell G and McLachlan D (1984) Photosensitization and feeding deterrence of *Euxoa messoria* by α-terthienyl, a naturally occurring thiophene from the Asteraceae. Experientia 40: 577–578

Champagne DE, Arnason JT, Philogéne BJR, Morand P and Lam J (1986) Light-mediated allelochemical effects of naturally occurring polyacetylenes and thiophenes from Asteraceae on herbivorous insects. J Chem Ecol 12: 835–858

Chan GFQ, Towers GHL and Mitchell JC (1975) Ultraviolet-mediated antibiotic activity of thiophene compounds of *Tagetes*. Phytochemistry 14: 2295–2296

Chevolleau S, Gasc N, Rollin P and Tulliez J (1997) Enzymatic, chemical, and thermal breakdown of H-3-labeled glucobrassicin, the parent indole glucosinolate. J Agric Food Chem 45: 4290–4296

Chew FS (1988) Biological Effects of Glucosinolates. In: Cutler HG (ed) Biologically Active Natural Products. pp 155–181. American Chemical Society, Washington, DC

Chiang CC and Hadwiger LA (1991) The *Fusarium solani*-induced expression of a pea gene family encoding high cysteine content proteins. Mol Plant Microbe Interact 4: 324–331

Chun MW and Schoonhoven LM (1973) Tarsal contact chemosensory hairs of large white butterfly *Pieris brassicae* and their possible role in oviposition behavior. Entomol Exp Appl 16: 343–357

Cociancich S, Ghazi A, Hetru C, Hoffmann JA and Letellier L (1993) Insect defensin, an inducible antibacterial peptide, forms voltage- dependent channels in *Micrococcus luteus*. J Biol Chem 268: 19239–19245

Coley-Smith JR (1986) Interactions between *Sclerotium cepivorum* and cultivars of onion, leek, garlic, and *Allium fistulosum*. Plant Pathol 35: 362–369

Colilla FJ, Rocher A and Mendez E (1990) Gamma-purothionins: amino acid sequences of two polypeptides of a new family of thionins from wheat endosperm. FEBS Lett 270: 191–194

Conn KL, Tewari JP and Dahiya JS (1988) Resistance to *Alternaria brassicae* and phytoalexin-elicitation in rapeseed and other crucifers. Plant Sci 56: 21–25

Coolong TW and Randle WM (2003) Sulfur and nitrogen availability interact to affect the flavor biosynthetic pathway in onion. J Amer Soc Hort Sci 128: 776–783

Cooper RM, Resende MLV, Flood J, Rowan MG, Beale MH and Potter U (1996) Detection and cellular localization of elemental sulphur in disease-resistant genotypes of *Theobroma cacao*. Nature 379: 159–162

Croes AF, Jacobs JJMR, Arroo RRJ and Wullems GJ (1994) Thiophene biosynthesis in *Tagestes* roots: molecular versus metabolic regulation. Plant Cell Tiss Org Cult 38: 159–164

D'Auria JC and Gershenzon J (2005) The secondary metabolism of *Arabidopsis thaliana*: growing like a weed. Curr Opin Plant Biol 8: 308–316

Denby KJ, Kumar P and Kliebenstein DJ (2004) Identification of *Botrytis cinerea* susceptibility loci in *Arabidopsis thaliana*. Plant J 38: 473–486

Dixon RA (2001) Natural products and plant disease resistance. Nature 411: 843–847

Doughty J, Dixon S, Hiscock SJ, Willis AC, Parkin IAP and Dickinson HG (1998) PCP-A1, a defensin-like *Brassica* pollen coat protein that binds the *S* locus glycoprotein, is the product of gametophytic gene expression. Plant Cell 10: 1333–1347

Downum KR, Hanncock REW and Towers GHN (1982) Mode of action of α-terthienyl on *Escherichia coli*: Evidence for a photodynamic effect on membranes. Photochem Photobiol 36: 517–532

Downum KR, Rosenthal GA and Towers GHN (1984) Photoxicity of the allelocheical, α-terthienyl, to larvae of *Manduca sexta* (L.) (Sphingidae). Pest Biochem Physiol 22: 104–109

Downum KR, Swain LA and Faleiro LJ (1991) Influence of light on plant allelochemicals: A synergistic defense in higher plants. Arch Insect Biochem Physiol 17: 201–211

Ellis DR and Salt DE (2003) Plants, selenium and human health. Curr Opin Plant Biol 6: 273–279

Ellmore GS and Feldberg RS (1994) Alliin lyase localization in bundle sheaths of the garlic clove (*Allium sativum*). Amer J Bot 81: 89–94

Ettlinger MG and Lundeen AJ (1957) 1st synthesis of a mustard oil glucoside – the enzymatic Lossen rearrangement. J Amer Chem Soc 79: 1764–1765

Fahey JW, Zalcmann AT and Talalay P (2001) The chemical diversity and distribution of glucosinolates and isothiocyanates among plants. Phytochemistry 56: 5–51

Fahey JW, Zhang YS and Talalay P (1997) Broccoli sprouts: An exceptionally rich source of inducers of enzymes that protect against chemical carcinogens. PNAS 10367–10372

Ferrari S, Plotnikova JM, De Lorenzo G and Ausubel FM (2003) *Arabidopsis* local resistance to *Botrytis cinerea* involves salicylic acid and camalexin and requires EDS4 and PAD2, but not SID2, EDS5 or PAD4. Plant J 35: 193–205

Foo HL, Gronning LM, Goodenough L, Bones AM, Danielsen BE, Whiting DA and Rossiter JT (2000) Purification and characterisation of epithiospecifier protein from *Brassica napus*: enzymic intramolecular sulphur addition within alkenyl thiohydroximates derived from alkenyl glucosinolate hydrolysis. FEBS Lett 468: 243–246

Gabrys B and Tjallingii WF (2002) The role of sinigrin in host plant recognition by aphids during initial plant penetration. Entomol Exp Appl 104: 89–93

Gershenzon J (1994) Metabolic costs of terpenoid accumulation in higher plants. J Chem Ecol 20: 1281–1328

Glawischnig E, Hansen BG, Olsen CE and Halkier BA (2004) Camalexin is synthesized from indole-3-acetaldoxime, a key branching point between primary and secondary metabolism in *Arabidopsis*. PNAS 101: 8245–8250

Glazebrook J and Ausubel FM (1994) Isolation of phyto-alexin-deficient mutants of *Arabidopsis thaliana* and characterization of their interactions with bacterial pathogens. PNAS 91: 8955–8959

Glazebrook J, Rogers EE and Ausubel FM (1997) Use of *Arabidopsis* for genetic dissection of plant defence responses. Ann Rev Genet 31: 547–569

Gmelin R, Luxa H-H, Roth K and Höfle G (1976) Dipeptide precursor of garlic odour in *Marasmius* species. Phytochemistry 15: 1717–1721

Gommers FJ (1982) Dithiophenes as singlet oxygen sensitisers. Photochem Photobiol 35: 615–619

Gommers FJ and Geerlings JWG (1973) Lethal effect of near-ultraviolet light on *Pratylenchus penetrans* from roots of *Tagetes*. Nematologica 19: 389–393

Granroth B (1970) Biosynthesis and decomposition of cysteine derivatives in onion and other *Allium* species. Ann Acad Sci Fenn A 154: 1–71

Griffiths DW, Birch ANE and Hillman JR (1998) Antinutritional compounds in the Brassicaceae - analysis, biosynthesis, chemistry and dietary effects. J Hort Sci Biotechnol 73: 1–18

Griffiths G, Trueman L, Crowther T and Smith B (2002) Onions – a global benefit to health. Phytother Res 19: 603–615

Grubb CD and Abel S (2006) Glucosinolate metabolism and its control. Trends Plant Sci 11: 89–100

Gu Q, Kawata EE, Morse MJ, Wu HM and Cheung AY (1992) A flower-specific cDNA encoding a novel thionin in tobacco. Mol Gen Genet 234: 89–96

Guillet G, Lavigne ME, Philogéne BJR and Arnason JT (1995) Behavioral adaptations of two phytophagous insects feeding on two species of Asteraceae. J Insect Behav 8: 533

Gunstone FD, Kilcast D, Powell RG and Taylor GM (1967) *Afzelia cuanzensis* seed oil: a source of crepenynic and 14,15-dehydrocrepenynic acid. Chem Commun 143: 295–296

Halkier BA and Du LC (1997) The biosynthesis of glucosinolates. Trends Plant Sci 2: 425–431

Halkier BA and Gershenzon J (2006) Biology and biochemistry of glucosinolates. Ann Rev Plant Biol 57: 303–333

Hansen BG and Halkier BA (2005) New insight into the biosynthesis and regulation of indole compounds in *Arabidopsis thaliana*. Planta 221: 603–605

Harrison SJ, Marcus JP, Goulter KC, Green JL, Maclean DJ and Manners JM (1997) An antimicrobial peptide from the Australian native *Hardenbergia violacea* provides the first functionally characterised member of a subfamily of plant defensins. Aust J Plant Physiol 24: 571–578

Hasapis X and MacLeod AJ (1982) Benzylglucosinolate degradation in heat-treated *Lepidium sativum* seeds and detection of a thiocyanate-forming factor. Phytochemistry 21: 1009–1013

Hay ME, Piel J, Boland W and Schnitzler I (1998) Seaweed sex pheromones and their degradation products frequently suppress amphipod feeding but rarely suppress sea urchin feeding. Chemoecology 8: 91–98

Hombeck M and Boland W (1998) Biosynthesis of the algal pheromon fucoserratene by the freshwater diatom *Asterionella formasa* (Bacillariophyceae). Tetrahedron 54: 11033–11042

Hudson JB, Graham EA, Micki N, Hudson L and Towers GHN (1986) Antiviral activity of the photoactive thiophene α-terthienyl. Photochem Photobiol 44: 477–482

Hudson JB and Towers GHN (1991) Therapeutic potential of plant photosensitisers. Pharmacol Ther 49: 181–222

Hull AK, Vij R and Celenza JL (2000) *Arabidopsis* cytochrome P450s that catalyze the first step of tryptophan-dependent indole-3-acetic acid biosynthesis. PNAS 97: 2379–2384

Husebye H, Chadchawan S, Winge P, Thangstad OP and Bones AM (2002) Guard cell- and phloem idioblast-specific expression of thioglucoside glucohydrolase 1 (myrosinase) in *Arabidopsis*. Plant Physiol 128: 1180–1188

Imai S, Tsuge N, Tomotake M, Nagatome Y, Sawada H, Nagata T and Kumagai H (2002) An onion enzyme that makes the eye water. Nature 419: 685

Ishibashi N, Yamauchi D and Minamikawa T (1990) Stored mRNA in cotyledons of *Vigna unguiculata* seeds: nucleotide sequence of cloned cDNA for a stored mRNA and induction of its synthesis by precocious germination. Plant Mol Biol 15: 59–64

Iyengar S, Arnason JT, Philogéne BJR, Werstiuk NH and Moran P (1990) Comparative metabolism of the phototoxic allelochemical alpha-terthienylin in three species of lepidopterans. Pest Biochem Physiol 37: 154–164

Jacobs JJMR, Arroo RRJ, De Koning EA, Klunder J, Croes AF and Wullems GJ (1995) Isolation and characterization of mutants of thiophene synthesis in *Tagetes erecta*. Plant Physiol 107: 807–814

Jones MG, Hughes J, Tregova A, Milne J, Tomsett AB and Collin HA (2004) Biosynthesis of the flavour precursors of onion and garlic. J Exp Bot 55: 1903–1918

Joyard J, Forest E, Blée E and Douce R (1988) Characterization of elemental sulfur in isolated intact spinach chloroplasts. Plant Physiol 88: 961–964

Kagan BL, Selsted ME, Ganz T and Lehrer RI (1990) Antimicrobial defensin peptides form voltage-dependent ion-permeable channels in planar lipid bilayer membranes. PNAS 87: 210–214

Kagan J (1991) Naturally occurring di- and trithiophenes. Progr Chem Org Nat Prod 56: 87–169

Karunanandaa B, Singh A and Kao TH (1994) Characterization of a predominantly pistil-expressed gene encoding a gamma-thionin-like protein of *Petunia inflata*. Plant Mol Biol 26: 459–64

Kawakishi S and Kaneko T (1987) Interactions of proteins with allyl isothiocyanate. J Agric Food Chem 35: 85–88

Kelly PJ, Bones AM and Rossiter JT (1998) Sub-cellular immunolocalization of the glucosinolate sinigrin in seedlings of *Brassica juncea*. Planta 206: 370–377

Kessler A and Baldwin IT (2002) Plant responses to insect herbivory: The emerging molecular analysis. Ann Rev Plant Biol 53: 299–328

Kliebenstein DJ (2004) Secondary metabolites and plant/ environment interactions: a view through *Arabidopsis thaliana* tinged glasses. Plant Cell Environ 27: 675–684

Koroleva OA, Davies A, Deeken R, Thorpe MR, Tomos AD and Hedrich R (2000) Identification of a new glucosinolate-rich cell type in *Arabidopsis* flower stalk. Plant Physiol 124: 599–608

Krauss F, Schäfer W and Schmidt A (1984) Formation of elemental sulfur by *Chlorella fusca* during growth on L-cysteine ethylester. Plant Physiol 74: 176–182

Kubec R, Kim S, McKeon DM and Musah RA (2002a) Isolation of *S-n*-butylcysteine sulfoxide and six *n*-butyl-containing thiosulfinates from *Allium siculum*. J Nat Prod 65: 960–964

Kubec R and Musah RA (2001) Cysteine sulfoxide derivatives in *Petiveria alliacea*. Phytochemistry 58: 981–985

Kubec R, Svobodova M and Velisek J (2000) Distribution of *S*-alk(en)ylcysteine sulfoxide in some *Allium* species. Identification of a new flavor precursor: *S*-ethylcysteine sulfoxide (ethiin). J Agric Food Chem 48: 428–433

Kubec R, Svobodova M and Velisek J (2001) Gas-chromatographic determination of *S*-methylcysteine sulfoxide in cruciferous vegetables. Eur Food Res Technol 213: 386–388

Kubec R, Velisek J and Musah RA (2002b) The amino acid precursors and odor formation in society garlic (*Tulbaghia violacea* Harv.). Phytochemistry 60: 21–25

Kubota K, Hirayama H, Sato Y, Kobayashi A and Sugawara F (1998) Amino acid precursors of the garlic-like odour in *Scorodocarpus borneensis*. Phytochemistry 49: 99–102

Kuettner EB, Hilgenfeld R and Weiss MS (2002) The active principle of garlic at atomic resolution. J Biol Chem 227: 46402–46407

Kylin H, Atuma S, Hovander L and Jensen S (1994) Elemental sulphur (S_8) in higher plants: biogenic or anthropogenic origin? Experientia 50: 80–85

Lambrix VM, Reichelt M, Mitchell-Olds T, Kliebenstein DJ and Gershenzon J (2001) The *Arabidopsis* epithiospecifier protein promotes the hydrolysis of glucosinolates to nitriles and influences *Trichoplusia ni* herbivory. Plant Cell 13: 2793–2807

Lancaster JE and Collin HA (1981) Presence of alliinase in isolated vacuoles and of alkyl cysteine sulphoxides in the cytoplasm of bulbs of onion (*Allium cepa*). Plant Sci Lett 22: 169–176

Lancaster JE and Kelly KE (1983) Quantitative analysis of the *S*-alk(en)yl-L-cysteine sulfoxides on onion (*Allium cepa* L.). J Sci Food Agric 34: 1229–1235

Lancaster JE, McCallion J and Shaw ML (1986) The dynamics of the flavour precursors, the *S*-alk(en)yl-L-cysteine sulphoxides, during leaf blade and scale development in the onion (*Allium cepa*). Physiol Plant 66: 293–297

Lancaster JE, Reynolds PHS, Shaw ML, Dommisse EM and Munro J (1989) Intra-cellular localization of the biosynthetic pathway to flavour precursors in onion. Phytochemistry 38: 461–464

Lancaster JE and Shaw ML (1989) γ-glutamyl peptides in the biosynthesis of *S*-alk(en)yl-L-cysteine sulphoxides (flavour precursors) in *Allium*. Phytochemistry 28: 455–460

Lancaster JE and Shaw ML (1991) Metabolism of γ-glutamyl peptides during development, storage and sprouting of onion bulbs. Phytochemistry 30: 2857–2859

Lazzeri L, Curto G, Leoni O and Dallavalle E (2004) Effects of glucosinolates and their enzymatic hydrolysis products via myrosinase on the root-knot nematode *Meloidogyne incognita* (Kofoid et White) Chitw. J Agric Food Chem 52: 6703–6707

Le Roux AM, Le Roux G and Thibout E (2002) Food experience on the predatory behavior of the ant *Myrmica rubra* towards a specialist moth, *Acrolepiopsis assectella*. J Chem Ecol 28: 2307–2314

Li Q, Eigenbrode SD, Stringham GR and Thiagarajah MR (2000) Feeding and growth of *Plutella xylostella* and *Spodoptera eridania* on *Brassica juncea* with varying glucosinolate concentrations and myrosinase activities. J Chem Ecol 26: 2401–2419

Lichtenstein EP, Morgan DG and Strong FM (1962) Naturally occurring insecticides – Identification of 2-phenylethyli-sothiocyanate as an insecticide occuring naturally in edible part of turnips. J Agric Food Chem 10: 30–33

Louda S and Mole S (1991) Glucosinolates: Chemistry and Ecology. In: Rosenthal GA and Berenbaum MR (eds) Herbivores: Their Interaction with Secondary Plant Metabolites, pp 123–164. Academic., San Diego

Matile P (1980) The mustard oil bomb – Compartmentation of the myrosinase system. Biochem Physiol Pflanzen 175: 722–731

Matusheski NV, Swarup R, Juvik JA, Mithen R, Bennett M and Jeffery EH (2006) Epithiospecifier protein from broccoli (*Brassica oleracea* L. ssp *italica*) inhibits formation of the anticancer agent sulforaphane. J Agric Food Chem 54: 2069–2076

McLachlan D, Arnason JT and Lam J (1984) The role of oxygen in photosensitization with polyacetylenes and thiophene derivatives. Photochem Photobiol 39: 177–182

Mendez E, Moreno A, Colilla F, Pelaez F, Limas GG, Mendez R, Soriano F, Salinas M and de Haro C (1990) Primary structure and inhibition of protein synthesis in eukaryotic cell-free system of a novel thionin, gamma-hordothionin, from barley endosperm. Eur J Biochem 194: 533–539

Mert-Türk F, Bennett MH, Mansfield JW and Holub EB (2003) Camalexin accumulation in *Arabidopsis thaliana* following abiotic elicitation or inoculation with virulent or avirulent *Hyaloperonospora parasitica*. Physiol Mol Plant Pathol 62: 137–145

Mikkelsen MD, Hansen CH, Wittstock U and Halkier BA (2000) Cytochrome P450 CYP79B2 from *Arabidopsis* catalyzes the conversion of tryptophan to indole-3-acetaldoxime, a precursor of indole glucosinolates and indole-3-acetic acid. J Biol Chem 275: 33712–33717

Miles CI, del Campo ML and Renwick JAA (2005) Behavioral and chemosensory responses to a host recognition cue by larvae of *Pieris rapae*. J Comp Physiol A – Neuroethol Sens Neural Behav Physiol 191: 147–155

Moore RE (1977) Volatile compounds from marine algae. Acc Chem Res 10: 40–47

Moran PJ, Cheng Y, Cassell JL and Thompson GA (2002) Gene expression profiling of *Arabidopsis thaliana* in compatible plant–aphid interactions. Arch Insect Biochem Physiol 51: 182–203

Moreno M, Segura A and Garcia-Olmedo F (1994) Pseudothionin-St1, a potato peptide active against potato pathogens. Eur J Biochem 223: 135–139

Müller C, Boeve JL and Brakefield P (2002) Host plant derived feeding deterrence towards ants in the turnip sawfly *Athalia rosae*. Entomol Exp Appl 104: 153–157

Murchie AK, Smart LE and Williams IH (1997) Responses of *Dasineura brassicae* and its parasitoids *Platygaster subuliformis* and *Omphale clypealis* to field traps baited with organic isothiocyanates. J Chem Ecol 23: 917–926

Nielsen JK (1978) Host plant discrimination within Cruciferae – Feeding responses of four leaf beetles (Coleoptera-Chrysomelidae) to glucosinolates, cucurbitacins and cardenolides. Entomol Exp Appl 24: 41–54

Nock LP and Mazelis M (1987) The *C-S* lyases of higher plants. Plant Physiol 85: 1079–1083

Noret N, Meerts P, Poschenrieder C, Barcelo J and Escarre J (2005) Palatability of *Thlaspi caerulescens* for snails: influence of zinc and glucosinolates. New Phytol 165: 763–772

Ohsumi C, Hayashi T and Sano K (1993) Formation of alliin in the culture tissues of *Allium sativum*. Oxidation of *S*-allyl-L-cysteine. Phytochemistry 33: 107–111

Osborne RW, De Samblanx GW, Thevissen K, Goderis I, Torrekens S, Van Leuven F, Attenborough S, Rees SB and Broekaert WF (1995) Isolation and characterisation of plant defensins from seeds of Asteraceae, Fabaceae, Hippocastanaceae and Saxifragaceae. FEBS Lett 368: 257–262

Palmieri S, Iori R and Leoni O (1986) Myrosinase from *Sinapis alba* L. – a new method of purification for glucosinolate analyses. J Agric Food Chem 34: 138–140

Parry RJ and Lii F-L (1991) Investigations of the biosynthesis of *trans*-(+)-*S*-1-propenyl-L-cysteine sulfoxide. Elucidation of the stereochemistry of the oxidative decarboxylation process. J Amer Chem Soc 113: 4704–4706

Pedras MSC and Ahiahonu PWK (2002) Probing the phytopathogenic stem root fungus with phytoalaexins and analogues: unprecedented glycosylation of camalexin and 6-methoxycamalexin. Bioorg Med Chem 10: 3307–3312

Pedras MSC and Khan AQ (1997) Unprecedented detoxication of the cruciferous phytoalexin camalexin by a root phytopathogen. Bioorg Med Chem Lett 7: 2255–2260

Petroski RJ and Kwolek WF (1985) Interaction of a fungal thioglucoside glucohydrolase and cruciferous plant epithiospecifier protein to form 1-cyanoepithioalkanes: implications of an allosteric mechanism. Phytochemistry 24: 213–216

Philogéne BJR, Arnason JT, Berg CW, Duval F, Champagne D, Taylor RG, Leitch L and Morand P (1985) Synthesis and evaluation of the naturally occurring phototoxin, alpha-terthienyl, as a control agent for larvae of *Aedes intrudens*, *Aedes atropalpus* (Diptera: Culicidae) and of *Simulium verecundum* (Diptera: Simuliidae). J Econ Entomol 78: 121–126

Pivnick KA (1993) Response of *Meteorus leviventris*, (Hymenoptera, Braconidae) to mustard oils in field trapping experiments. J Chem Ecol 19: 2075–2079

Pivnick KA, Lamb RJ and Reed D (1992) Response of flea beetles, *Phyllotreta* spp., to mustard oils and nitriles in field trapping experiments. J Chem Ecol 18: 863–873

Pohnert G and Boland W (1996) Biosynthesis of the algal pheromone hormosirene by the freshwater diatom *Gomphonema parvulum* (Bacillariophyceae). Tetrahedron 52: 10073–10082

Rabinkov A, Wilchek M and Mirelman D (1995) Alliinase (alliin lyase) from garlic (*Alium sativum*) is glycosylated at ASN[146] and forms a complex with a garlic mannose-specific lectin. Glycoconjugate J 12: 690–698

Randle WM, Kopsell DE and Kopsell DA (2002) Sequentially reducing sulfate fertility during onion growth and development affects bulb flavor at harvest. HortScience 37: 118–121

Rao AG (1995) Antimicrobial peptides. Mol Plant Microbe Interact 8: 6–13

Rask L, Andreasson E, Ekbom B, Eriksson S, Pontoppidan B and Meijer J (2000) Myrosinase: gene family evolution and herbivore defense in Brassicaceae. Plant Mol Biol 42: 93–113

Ratzka A, Vogel H, Kliebenstein DJ, Mitchell-Olds T and Kroymann J (2002) Disarming the mustard oil bomb. PNAS 99: 11223–11228

Rausch T and Wachter A (2005) Sulfur metabolism: a versatile platform for launching defence operations. Trends Plant Sci 10: 503–509

Rennenberg H, Arabatzis N and Grundel I (1987) Cysteine desulphhydrase activity in higher plants: evidence for the action of L- and D-cysteine specific enzymes. Phytochemistry 26: 1583–1589

Resende MLV, Flood J, Ramsden JD, Rowan MG, Beale MH and Cooper RM (1996) Novel phytoalexins including elemental sulphur in the resistance of cocoa (*Theobroma cacao* L.) to Verticillium wilt (*Verticillium dahliae* Kleb.). Physiol Mol Plant Pathol 48: 347–359

Reuber TL, Plotnikova JM, Dewdney J, Rogers EE, Wood W and Ausubel FM (1998) Correlation of defense gene induction defects with powdery mildew susceptibility in *Arabidopsis* enhanced disease susceptibility mutants. Plant J 16: 473–485

Roetschi A, Si-Ammour A, Belbahri L, Mauch F and Mauch-Mani B (2001) Characterization of an Arabidopsis-Phytophthora pathosystem: resistance requires a functional PAD2 gene and is independent of salicylic acid, ethylene and jasmonic acid signalling. Plant Journal 28: 293–305

Rojas JC (1999) Electrophysiological and behavioral responses of the cabbage moth to plant volatiles. J Chem Ecol 25: 1867–1883

Romeis J, Ebbinghaus D and Scherkenbeck J (2003) Factors accounting for the variability in the behavioral response of the onion fly (*Delia antiqua*) to n-dipropyl disulfide. J Chem Ecol 29: 2131–2142

Schmidt A (1987) D-cysteine desulfhydrase from spinach. Meth Enzymol 143: 449–453

Schnitzler I, Boland W and Hay M (1998) Organic sulfur compounds from *Dictyopteris ssp.* deter feeding by an herbivorous amphipod (*Ampithoe longimana*) but not by an herbivorous sea urchin (*Arbacia punctulata*). J Chem Ecol 24: 1715–1732

Schrader-Fischer G and Apel K (1993) The anticyclic timing of leaf senescence in the parasitic plant *Viscum album* is closely correlated with the selective degradation of sulfur-rich viscotoxins. Plant Physiol 101: 745–749

Schuhegger R, Nafisi M, Mansourova M, Petersen BL, Olsen CE, Svatos A, Halkier BA and Glawischnig E (2006a) CYP71B15 (PAD3) catalyzes the final step in camalexin biosynthesis. Plant Physiol 141: 1248–1254

Schuhegger R, Rauhut T and Glawischnig E (2006b) Regulatory variability of camalexin biosynthesis. J Plant Physiol 10: 10

Seo ST and Tang CS (1982) Hawaiian fruit flies (Diptera, Tephritidae) - Toxicity of benzyl isothiocyanate against eggs or 1st instars of three species. J Econ Entomol 75: 1132–1135

Shade RE, Schroeder HE, Pueyo JJ, Tabe LM, Murdock LL, Higgins TJV and Chrispeels MJ (1994) Transgenic pea seeds expressing the α-amylase inhibitor of the common bean are resistant to bruchid beetles. Biotechnol Prog 12: 793–796

Silverstein KAT, Graham MA, Paape TD and VandenBosh KA (2005) Genome organization of more than 300 defensin-like genes in *Arabidopsis*. Plant Physiol 138: 600–610

Smeets K, Van Damme EJM, Van Leuven F and Peumans WJ (1997) Isolation and characterization of lectins and lectin-alliinase complexes from bulbs of garlic (*Allium sativum*) and ramsons (*Allium ursinum*). Glycocon J 14: 331–343

Sörensen JS, Holme D, Borlaug ET and Sorensen NA (1954) Studies related to naturally occurring acetylene compounds. XX. A preliminary communication on some polyacetylenic pigments from compositae plants. Acta Chem Scand 8: 1769–1778

Städler E (1978) Chemoreception of host plant-chemicals by ovipositing females of *Delia (Hylemya) brassicae*. Entomol Exp Appl 24: 711–720

Stiekema WJ, Heidekamp F, Dirkse WG, van Beckum J, de Haan P, ten Bosch C and Louwerse JD (1988) Molecular cloning and analysis of four potato tuber mRNAs. Plant Mol Biol 11: 255–269

Stoewsand GS (1995) Bioactive organosulfur phytochemicals in *Brassica oleracea* vegetables – a review. Food Chem Toxicol 33: 537–543

Stoll A and Seebeck E (1947) Über Allin, die genuine Muttersubstanz des Knoblauchöls. Experientia 3: 114–115

Suzuki T, Sugii M and Kakimoto T (1962) Metabolic incorporation of L-valine-[14C] into S-(2-carboxypropyl)glutathione and S-(2-carboxypropyl)cysteine in garlic. J Amer Chem Soc 69: 328–331

Terras FRG, Eggermont K, Kovaleva V, Raikhel NV, Osborne RW, Kester A, Rees SB, Torrekens S, Van LF, Vanderleyden J, Cammune BPA and Broekaert WF (1995) Small cysteine-rich antifungal proteins from radish: their role in host defence. Plant Cell 7: 573–588

Terras FRG, Schoofs HME, De BMFC, Van LF, Rees SB, Vanderleyden J, Cammune BPA and Broekaert WF (1992) Analysis of two novel classes of plant antifugal proteins from radish (*Raphanus sativus* L.) seeds. J Biol Chem 267: 15301–15309

Terras FRG, Schoofs H, Thevissen K, Osborn RW, Vanderleyden J, Cammue B and Broekaert WF (1993) Synergistic enhancement of the antifungal activity of wheat and barley thionins by radish and oilseed rape 2S albumins and by barley trypsin inhibitors. Plant Physiol 103: 1311–1319

Thangstad OP, Bones AM, Holton S, Moen L and Rossiter JT (2001) Microautoradiographic localisation of a glucosinolate precursor to specific cells in *Brassica napus* L. embryos indicates a separate transport pathway into myrosin cells. Planta 213: 207–213

Thangstad OP, Gilde B, Chadchawan S, Seem M, Husebye H, Bradley D and Bones AM (2004) Cell specific, cross-species expression of myrosinases in *Brassica napus*, *Arabidopsis thaliana* and *Nicotiana tabacum*. Plant Mol Biol 54: 597–611

Thevissen K, Osborne RW, Acland DP and Broekaert WF (2000) Specific binding sites for an antifungal plant defensin from dahlia (*Dahlia merckii*) on fungal cells are required for antifungal activity. Mol Plant Microbe Interact 13: 54–61

Thomma BPHJ, Nelissen I, Eggermont K and Broekaert WF (1999) Deficiency in phytoalexin production causes enhanced susceptibility of *Arabidopsis thaliana* to the fungus *Alternaria brassicicola*. Plant J 19: 163–171

Tierens K, Thomma BPH, Brouwer M, Schmidt J, Kistner K, Porzel A, Mauch-Mani B, Cammue BPA and Broekaert WF (2001) Study of the role of antimicrobial glucosinolate-derived isothiocyanates in resistance of *Arabidopsis* to microbial pathogens. Plant Physiol 125: 1688–1699

Titayavan M and Altieri MA (1990) Synomone-mediated interactions between the parasitoid *Diaeretiella rapae* and *Brevicoryne brassicae* under field conditions. Entomophaga 35: 499–507

Tookey HL (1973) Crambe thioglucoside glucohydrolase (EC 3.2.3.1) – Separation of a protein required for epithiobutane formation. Can J Biochem 51: 1654–1660

Tsai Y, Cole LL, Davis LE, Lockwood SJ, Simmons V and Wild GC (1985) Antiviral properties of garlic: *in vitro* effects on influenza B, *Herpes simplex* and Coxsackie viruses. Planta Med 51: 460–461

Tsuji J, Jackson E, Gage D, Hammerschmidt R and Somerville SC (1992) Phytoalexin accumulation in *Arabidopsis thaliana* during the hypersensitive response to *Pseudomonas syringae* pv *syringae*. Plant Physiol 98: 1304–1309

Tsuji J, Zook M, Somerville SC, Last RL and Hammerschmidt R (1993) Evidence that tryptophan is not a direct biosynthetic intermediate of camalexin in *Arabidopsis thaliana*. Physiol Mol Plant Pathol 43: 221–229

Ueda H, Nishiyama C, Shimada T, Koumoto Y, Hayashi Y, Kondo M, Takahashi T, Ohtomo I, Nishimura M and Hara-Nishimura I (2006) AtVAM3 is required for normal specification of idioblasts, myrosin cells. Plant Cell Physiol 47: 164–175

van Wees SCM, Chang HS, Zhu T and Glazebrook J (2003) Characterization of the early response of *Arabidopsis* to *Alternaria brassicicola* infection using expression profiling. Plant Physiol 132: 606–617

Verhoeven DTH, Verhagen H, Goldbohm RA, Vandenbrandt PA and Vanpoppel G (1997) A review of mechanisms underlying anticarcinogenicity by *Brassica* vegetables. Chem Biol Interact 103: 79–129

von Reuß SH and König WA (2005) Olefinic isothiocyanates and iminodithiocarbonates from the liverwort *Corsinia coriandrina*. Eur J Org Chem 6: 1184–1188

Wat C-K, Prasad SK, Graham EA, Partingtn S, Arnason T and Towers GHN (1981) Photosentization of invertebrates by natural polyacetylenes. Biochem Syst Ecol 9: 59–62

Wei X, Roomans GM, Seveus L and Pihakaski K (1981) Localization of glucosinolates in roots of *Sinapis alba* using x-ray-microanalysis. Scanning Electron Microscopy: 481–488

Whitaker JR (1976) Development of flavor, odor, and pungency in onion and garlic. Adv Food Res 22: 73–133

Williams JS and Cooper RM (2003) Elemental sulphur is produced by diverse plant families as a component of defence against fungal and bacterial pathogens. Physiol Mol Plant Pathol 63: 3–16

Williams JS and Cooper RM (2004) The oldest fungicide and newest phytoalexin – a reappraisal of the fungitoxicity of elemental sulphur. Plant Pathol 53: 263–279

Williams JS, Hall SA, Hawkesford MJ, Beale MH and Cooper RM (2002) Elemental sulfur and thiol accumulation in tomato and defense against a fungal vascular pathogen. Plant Physiol 128: 150–159

Wittstock U, Agerbirk N, Stauber EJ, Olsen CE, Hippler M, Mitchell-Olds T, Gershenzon J and Vogel H (2004) Successful herbivore attack due to metabolic diversion of a plant chemical defense. PNAS 101: 4859–4864

Wittstock U and Gershenzon J (2002) Constitutive plant toxins and their role in defense against herbivores and pathogens. Curr Opin Plant Biol 5: 300–307

Wittstock U and Halkier BA (2002) Glucosinolate research in the *Arabidopsis* era. Trends Plant Sci 7: 263–270

Wittstock U, Kliebenstein D, Lambrix VM, Reichelt M and Gershenzon J (2003) Glucosinolate hydrolysis and its impact on generalist and specialist insect herbivores. In: Romeo JT (ed) Integrative Phytochemistry: from Ethnobotany to Molecular Ecology, Recent Advances in Phytochemistry, pp 101–126. Elsevier, Amsterdam

Zehmeister L and Sease JW (1947) A blue-fluorescing compound, terthienyl, isolated from marigold. J Amer Chem Soc 69: 273

Zhang GY, Talalay P, Cho CG and Posner GH (1992) A major inducer of anticarcinogenic protective enzymes from broccoli: isolation and elucidation of structure. PNAS 89: 2399–2403

Zhao JM, Williams CC and Last RL (1998) Induction of *Arabidopsis* tryptophan pathway enzymes and camalexin by amino acid starvation, oxidative stress, and an abiotic elicitor. Plant Cell 10: 359–370

Zhao Y, Hull AK, Gupta NR, Goss KA, Alonso J, Ecker JR, Normanly J, Chory J and Celenza JL (2002) Trp-dependent auxin biosynthesis in *Arabidopsis*: involvement of cytochrome P450s CYP79B2 and CYP79B3. Genes Dev 16: 3100–3112

Zhou N, Tootle TL and Glazebrook J (1999) *Arabidopsis* PAD3, a gene required for camalexin biosynthesis, encodes a putative cytochrome P450 monooxygenase. Plant Cell 11: 2419–2428

Zook M and Hammerschmidt R (1997) Origin of the thiozole ring of camalexin, a phytoalexin from *Arabidopsis thaliana*. Plant Physiol 113: 463–468

Zook M, Leege L, Jacobson D and Hammerschmidt R (1998) Camalexin accumulation in *Arabis lyrata*. Phytochemistry 49: 2287–2289

Chapter 12

Sulfite Oxidation in Plants

Robert Hänsch and Ralf R. Mendel*
*Department of Plant Biology, Technical University of Braunschweig,
Humboldtstrasse 1, D-38106 Braunschweig, Germany*

Summary

Sulfite oxidation in plants was a matter of controversial discussion for a long time and still is not finally understood. There is no doubt anymore about the occurrence of sulfite oxidation besides primary sulfate assimilation that takes place in the chloroplast. Sulfate is reduced via sulfite to organic sulfide which is essential for the biosynthesis of S-containing amino acids and other compounds like glutathione. However, it has also been reported that sulfite can be oxidized back to sulfate, e.g. when plants were subjected to SO_2 gas. Work from our laboratory has identified sulfite oxidase as a member of molybdenum-containing enzymes in plants, which seems to be the most important way to detoxify excess of sulfite. In this paper we show how plant cells separate the two counteracting pathways – sulfate assimilation and sulfite detoxification – into different cell organelles. We discuss how these two processes are (co-)regulated and what kind of other sulfite oxidase activities occur in the plant.

I. Sulfur Cycling in Nature

Sulfur is an essential macronutrient for plants, animals and microorganism and plays a critical role in the catalytic and electrochemical functions of biomolecules in the cell. Sulfur is found in the two amino acids cysteine and methionine, in oligopeptides (glutathione and phytochelatins), vitamins and cofactors (biotin, molybdenum cofactor [Moco], thiamine, CoenzymeA, and S-adenosyl-Methionine), in phytosulfokin hormones (Matsubayashi, Sakagami 1996) and a variety of secondary products (see

---

*Corresponding author, Phone: +49 (0) 531 391 5870, Fax: +49 (0) 531 391 8128, E-mail: r.mendel@tu-bs.de

Leustek 2002). Disulfide bonds between polypeptides mediated by cysteine are very important in protein assembly and structure. Sulfur itself belongs to the chalcogen family; other members of the family are oxygen, selenium, tellurium, and polonium. Because of the electronic status of sulfur, this element can undergo four different oxidation states: $\beta \leftarrow 6$ (sulfate, SO_4^{2-}), $\beta \leftarrow 4$ (sulfite, SO_3^{2-}), $\leftarrow'$ (elemental sulfur S^0) and -2 (sulfide, H_2S), which is important for their biological activity and allows to recycle it in a biogeochemical way including (i) assimilative sulfate reduction, (ii) desulfuration, (iii) oxidation of organic sulfur compounds, and (iv) mineralization of organic sulfur to the inorganic form. Human impact on the sulfur cycle is exerted mainly by producing toxic sulfur dioxide in industry and by motor cars. Sulfur dioxide can be further reduced to sulfide or re-oxidized to sulfate in different enzymatic and non-enzymatic reactions.

II. Sulfate Reduction in Plants

Plants take up sulfate from soil into the roots and translocate it via the xylem to the green parts of the plant where it is stored as the major anionic component of vacuolar sap (Kaiser et al. 1989; Leustek and Saito 1999). Plastids are the limiting organelles where assimilatory sulfate reduction takes place, only cysteine synthesis enzymes are localized in the following three compartments: plastids, cytosol and mitochondria. Sulfate assimilation starts with the activation by ATP to adenosine-5'-phosphosulfate (5'-adenylylsulfate [APS]) and further conversion via sulfite to the final sulfide and requires one ATP and eight electrons coming from reduced glutathione (Hell 1997; Bick and Leustek 1998). Sulfide is later coupled to O-acetyl-Ser to form cysteine (see recent reviews Leustek 2002; Droux 2004; Saito 2004). Details are given in chapters I.1 and I.2 in this book.

Sulfate itself can be also covalently bound to a variety of compounds, a process termed sulfation which also begins with APS synthesis. APS is then phosphorylated by APS-kinase to form 3'-phosphoadenosine-5'-phosphosulfate (PAPS) which is used as a sulfuryl donor by a variety of sulfotransferases forming a sulfate ester bond. The most prominent example for this class of compounds formed by sulfation are glucosinolates that function as insect feeding deterrents produced by different species in the *Brassicaceae* family. Here, glucosinolates contain two forms of sulfur in different oxidation states: the reduced form is a thioether derived from cysteine, whereas the oxidized form is a sulfamate which comes from the sulfation pathway (Leustek 2002). For further details see chapter I.6 in this book.

III. The Ambivalent Nature of Sulfite: an Important but Toxic Intermediate

Sulfite plays an important role in the reductive and oxidative sulfur metabolism in pro- and eukaryotes. In plants, sulfite (SO_3^{2-}) with an oxidation state of $\beta \leftarrow 4$ is the first important intermediate in the reduction pathway of sulfur originating from sulfate. Here, sulfite is also the starting point for the formation of sulfur lipids (Yu et al. 2002). Some microorganisms use sulfite as sole electron source. Sulfite is the key intermediate in the oxidation of reduced sulfur compounds to sulfate and the major product of most dissimilatory sulfur-oxidizing prokaryotes (Kappler and Dahl 2001). In animals, it is known since 1953 (Heimberg et al. 1953) that sulfate is produced from sulfite in an enzymatic reaction. Sulfite oxidases (SO) catalyzes the reaction $SO_3^{2-} + H_2O \rightarrow SO_4^{2-} + 2H^+ + 2e^-$, which is the terminal step in the oxidative degradation of cysteine and methionine.

Deficiencies of SO lead to major neurological abnormalities and early death in the studied animals and humans (Calabrese et al. 1981; Kisker et al. 1997; Garrett et al. 1998). In the mammalian system, SO is localized in the intermembrane space of mitochondria (Cohen et al. 1972) where electrons derived from sulfite can passes via the enzyme's heme domain on to cytochrome c, the physiological electron acceptor.

Sulfite is also known for many years to damage plants (Hill and Thomas 1933; Moyer and Geo 1935), for review see Peiser and Yang (1985) and Heber and Hüve (1998). As nucleophilic agent, sulfite is able to attack diverse substrates (Peiser and Yang 1985), where it opens S-S bridges. This reaction – so-called sulfitolysis – can cause inactivation of proteins when incubated with sulfite

---

Abbreviations: Moco – Molybdenum cofactor; GFP – green fluorescent protein; PTS – peroxisomal targeting sequence; SO – sulfite oxidase

(Ziegler 1974) or even when plants are exposed to high concentrations of SO_2 gas (Tanaka et al. 1982). Sulfitolysis of oxidized thioredoxin can interfere with the regulation of enzymes of the Calvin cycle (Würfel et al. 1990). These effects cause severe reduction in plant growth. The susceptibility to SO_2 may vary considerably between the different species and depends on combinations of duration and dosage of SO_2, but also on physiological and environmental factors (Rennenberg 1984). There are two systems discussed to control the internal SO_2 concentration: (i) control of uptake of the gas by the laminar boundary layer, the cuticle or the guard cells, and (ii) the rate of its metabolic conversion and translocation. In plants, theoretically there are two possibilities to handle SO_2: (i) to feed it into the assimilation stream of sulfur for producing of cysteine, methionine or other reduced sulfur compounds, or (ii) to re-oxidize it to sulfate.

As gaseous substance, SO_2 enters plant tissues mainly via their stomata (Rennenberg and Polle 1994; Rennenberg and Herschbach 1996) and is transformed into sulfite and/or bisulfite ions on the wet surface of guard cells and in the cytoplasmic fluid, which results in a proton generation:

$$SO_2 + H_2O \rightarrow [SO_2 \cdot H_2O] \rightarrow HSO_3^- + H^+ \leftrightarrow SO_3^{2-} + 2H^+$$

The guard cells are able to respond to different levels of SO_2 with stomatal closing or opening (Rao and Anderson 1983). In acidic environments, HSO_3^- is prevailing, while under alkaline conditions in the chloroplasts, SO_2 is chemically converted into SO_3^{2-} (Heber et al. 1987). The flow of SO_2 between the gaseous phase of the intercellular space and the liquid phase of the apoplast and/or cytosol seems to be continuous.

When applied as the major source of sulfur at non-toxic dosages, i.e. in soils that do not fulfill the needs for sulfate, biomass production depends on the supply of SO_2 or other volatile sulfur-compounds in the air (summarized in Rennenberg 1984). Furthermore, the amount of SO_2 taken up by the leaves can regulate the sulfur uptake by the roots (Herschbach and Rennenberg 2001). However, when the uptake of SO_2 exceeds a certain threshold that differs from species to species, toxicity effects occur that lead to growth problems of the plant (Linzon 1978). Here, active detoxification of sulfite is necessary for the survival of the whole plant. Until

recently, metabolic conversion was interpreted to mean reductive detoxification leading to sulfide that is used to produce cysteine (Heber and Hüve 1998). This process is well understood because it forms part of the sulfate assimilation pathway. But is has been also reported that sulfite can be oxidized to sulfate. Upon foliar application of labeled $^{35}SO_2$ this gas is rapidly metabolized in the light and also in dark, with sulfate as the main end product (Garsed and Read 1977; Van der Kooij et al. 1997). The possibility to explain these results by oxidative conversion of sulfite to sulfate was largely neglected because this step would counteract the assimilatory pathway. Yet, experimental data were accumulating that needed further explanations (Rennenberg et al. 1982).

IV.   Sulfite Oxidase Activities in Plants

For decades, occurrence and nature of a sulfite oxidizing activity in higher plants were controversially discussed as shown in the following history.

- Already in 1944, Thomas and coworkers showed high concentrations of sulfate in SO_2-treated alfalfa and sugar beet (Thomas et al. 1944).
- Fromageot described sulfite oxidation by oat roots (Fromageot et al. 1960).
- Apoplastic peroxidases of barley leaves can efficiently detoxify sulfite: after infiltration of sulfite-containing buffer through the stomata, sulfate could be extracted in increasing amounts over time from the apoplastic washing solution (Pfanz et al. 1990).
- Later, apoplastic peroxidases were discussed to oxidize sulfite using H_2O_2 and different phenolic compounds (Pfanz and Oppmann 1991).
- Miszalski and Ziegler suggested a non-enzymatic oxidation of sulfite, initiated (i) by superoxide anions formed on the reduction site of the electron transport system in chloroplasts, (ii) by free radicals such as OH^-, or (iii) by H_2O_2 (Miszalski and Ziegler 1992).
- Intact chloroplasts isolated from spinach (*Spinacia oleracea*) fed with radioactively labeled sulfite showed a sulfite oxidation activity (Dittrich et al. 1992). This reaction was discussed to proceed via a radical chain reaction involving light-dependent photosynthetic electron transport which was found to be enhanced by light and to be sensitive to inhibitors of the photosynthetic electron transport (Dittrich et al. 1992). However, in addition to this

non-enzymatic light-dependent sulfite oxidation there should also be an enzymatic reaction because sulfite oxidation could also be detected in the dark.

- Jolivet et al. (1995a, 1995b) described a sulfite oxidizing activity associated with isolated thylakoid membranes that was not induced through the photosynthetic radical-dependent oxidation chain reaction. A protein preparation gave activities 50 times higher than in crude extract of spinach leaves, and SDS-PAGE analysis showed four major protein bands (65, 53, 36 and 33 kDa), that were discussed to represent either different subunits of an even more complex enzyme or to be contaminating bands because an *in gel*-staining assay was not successful (Jolivet et al. 1995a).

All publications presented describe sulfite oxidizing activities as non-enzymatic or enzymatic reactions in the apoplastic space or in steps associated with the light-dependent photosynthetic electron transport or other unknown reactions in chloroplasts. Yet, the main problem was still unsolved: a chloroplast-localized sulfite oxidizing activity would counteract sulfate assimilation residing in the same organelle. How could a plant cell regulate these two conflicting pathways in one and the same compartment? The discussion of this obvious problem became an unexpected turn when we viewed sulfite oxidation from the point of eukaryotic molybdenum metabolism.

In mammals, SO is well studied: it is an enzyme containing molybdenum in the active site and is localized in the intermembrane space of mitochondria (Cohen et al. 1972). It is a two-domain protein consisting of a molybdenum domain and a heme domain and it is responsible for detoxifying sulfite in the course of amino acid decomposition. By screening an *A. thaliana* cDNA library using the amino acid sequence of human (XP_006727) or chicken SO (P07850), we identified plant SO as the fourth plant enzyme containing molybdenum (Eilers et al. 2001). The isolated full-length cDNA of *Arabidopsis*-SO has a single open reading frame of 1182 bp encoding a protein of 393 amino acids (43.3 kDa) with 47% identity to the primary sequence of the molybdenum cofactor-domain of chicken SO. However, the sequence for the heme domain known from animal SO was lacking in this plant clone and was also absent in the genomic region. The genomic sequence showed a single open reading frame with 11 introns located on chromosome III. High strin-

gency hybridization of *Arabidopsis* genomic DNA with the isolated cDNA clone as probe demonstrated that the gene encoding for the *Arabidopsis*-SO (At-*so*) is single copy gene.

The alignment of molybdenum cofactor-domains of SOs from different sources with *Arabidopsis*-SO demonstrated considerable overall homology, identifying these enzymes as members of a common family (Eilers et al. 2001). Plant SO turned out to be conserved among higher plants because antibodies raised against *Arabidopsis*-SO detected a dominantly cross-reacting protein of about 45 kDa in a wide range of species belonging to a variety of both herbaceous (dicots and monocots) and woody (e.g. poplar) plants (Eilers et al. 2001). In *Arabidopsis*, SO shows a constitutive expression in all tissues tested and also over the day without any pronounced diurnal rhythm (Hänsch et al. 2006). Hence one can conclude that plant SOs are widely distributed among higher plants and are expressed as a housekeeping gene. Recently, a SO-specific sequence was also detected in the genome of the green alga *Chlamydomonas reinhardtii* (Emilio Fernandez, personal communication) and in the moss *Physcomitrella patens* (Ralf Reski, personal communication).

V. Biochemical Properties of Plant Sulfite Oxidase (EC 1.8.3.1)

For recombinant expression, the isolated *Arabidopsis*-cDNA was cloned into an expression vector allowing the expression and purification as His-tagged protein from *E. coli*. This protein exhibited a sulfite-dependent SO activity when using ferricyanide as artificial (Eilers et al. 2001) or oxygen as natural electron acceptor (Hänsch et al. 2006). No activity was found with cytochrome c as electron acceptor as expected, since the heme domain known to mediate electron transfer between the molybdenum cofactor-domain and cytochrome c in rat hepatic SO is missing in the plant enzyme. HPLC analysis of the oxidation product of the molybdenum cofactor confirmed its pterin nature as found in animals. And also the spectroscopic properties of recombinant plant SO identified it as member of the general SO family: on the basis of the UV-visible absorption and the EPR signature it was evident that the molybdenum centre of *Arabidopsis*-SO is fundamentally similar to that of the vertebrate proteins (Eilers et al. 2001; Hemann et al. 2005).

The Km-value of 22.6 µM for sulfite using oxygen as electron acceptor was in the same range as shown for the artificial acceptor ferricyanide determined to be 33.8 µM which is in the range as found for rat SO (Eilers et al. 2001; Hänsch et al. 2006). When plant SO uses molecular oxygen as terminal electron acceptor, the question arises what could be the second end product besides of sulfate? This second reaction product turned out to be hydrogen peroxide (H_2O_2). We showed this by two different methods: (i) Nag et al. (2000) described the specific formation of a yellow-orange peroxo-disulfatotitanate(IV)-complex $[Ti(O_2)(SO_4)_2]^{2-}$ from the hydroxylcation $[Ti(OH)_2(H_2O)_4]^{2+}$ in the presence of H_2O_2 and its detection at 405 nm, and (ii) the fluorescent dye lucigenin is known from Rost et al. (1998) to react specifically with H_2O_2 but not with other reactive oxygen species. Both assays were positive for the plant SO (Hänsch et al. 2006). However, adding low amounts of catalase to both the titanate-complex assay and the fluorescent-dye assay abolished H_2O_2 accumulation completely.

VI. Plant Sulfite Oxidase is a Peroxisomal Enzyme

Analysis of SO in 17 plant species *in silico* revealed that all plant SO-proteins possess a C-terminal peroxisomal targeting sequence (Nakamura et al. 2002). The C-terminal SNL-tripeptide of the *Arabidopsis*-protein (Eilers et al. 2001) is very similar to the C-terminal amino acid motif serine-lysine-leucine (SKL) which is the consensus peroxisomal targeting sequence 1 (PTS1) and which is sufficient to direct polypeptides to peroxisomes *in vivo* in plants, animals and yeast. This non-cleaved tripeptide motif, consisting of a small, a basic and a hydrophobic residue or a variant thereof, resides at the extreme C-terminus and occurs in the majority of peroxisomal matrix proteins (Hayashi et al. 1996; Mullen et al. 1997). Plant PTS1 motifs apparently exhibit more sequence variability as compared to accepted signals in animals (Mullen et al. 1997).

Having the *Arabidopsis*-SO clone at hands (Eilers et al. 2001) and making use of antiserum that we generated against plant SO we finally answered the question of its subcellular localization. Antibodies directed against plant SO were applied for histochemical studies by transmission electron microscopy. Immunogold experiments performed on ultrathin sections of *Arabidopsis thaliana* leaves and of protoplast-derived micro-colonies of *Nicotiana plumbaginifolia* demonstrated for both species that gold labels were exclusively located in peroxisomes, and only a few were observed in other organelles or the cytoplasm (Nowak et al. 2004). To validate these results, we generated GFP::SO fusion constructs, transferred the genes via particle gun into tobacco leaves and monitored transient expression by confocal laser scanning microscopy. A punctuate fluorescence pattern was observed. The overlay of chlorophyll autofluorescence demonstrates that GFP was clearly excluded from the chloroplasts. To distinguish between peroxisomes and mitochondria we performed double transformation experiments with different fluorescent proteins and excluded mitochondria as targets by counterstaining with MitoTracker-Red (Nowak et al. 2004). Thus, independent lines of experimental evidence unequivocally demonstrate that plant SO is a peroxisomal enzyme.

A shared feature of all peroxisomes is their ability to metabolize hydrogen peroxide (H_2O_2), consequently protecting the rest of the cell from this toxic byproduct (Johnson and Olsen 2001). Our studies identified oxygen as the new final electron acceptor thereby generating H_2O_2 as reaction product in addition to sulfate which might explain why plant SO is localized in peroxisomes while animal SO occurs in mitochondria where it uses cytochrome c as electron acceptor. H_2O_2 is a highly reactive molecule that can be decomposed by peroxisomal catalase. Yet, there is another possible way for removing H_2O_2: In clouds and rain droplets, H_2O_2 was identified as one of the most effective non-enzymatic oxidants for HSO_3^- (Clegg and Abbatt 2001). Under our experimental conditions, sulfite did not spontaneously oxidize to sulfate, however the addition of physiological concentrations of H_2O_2 in the micromolar range led to the conversion of sulfite into sulfate (Hänsch et al. 2006). So we suggest that in the case of high sulfite concentrations in the plant cell, the production of H_2O_2 by SO can help to detoxify further sulfite molecules by a non-enzymatic reaction subsequent to enzymatic sulfite oxidation, thus increasing sulfite

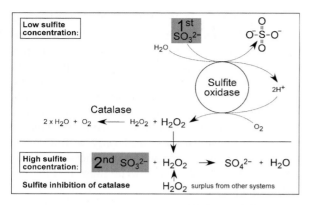

Fig. 1. Proposed interaction of plant SO and catalase. Plant SO oxidizes of sulfite and generates equimolar amount of H_2O_2. At low sulfite concentrations, all H_2O_2 formed will be immediately degraded by peroxisomal catalase. But at high sulfite concentrations, however, catalase will be inhibited by sulfite. The H_2O_2 molecule generated by the plant SO reaction can non-enzymatically oxidize a second molecule of sulfite (according to Hänsch et al. 2006).

removal (Fig. 1). And this makes sense because it has been shown previously that peroxisomal catalase is inhibited when leaves were treated with sulfite (Veljović-Jovanović et al. 1998) – the half-maximal inhibition was below of 500 μM sulfite. Here, on one hand the plant SO could play a role for protecting this important enzyme from sulfite damage and on the other hand: excess of sulfite will inhibit the catalase and the increasing H_2O_2 can help to reduce toxic sulfite. Hence we assume that SO could possibly serve as "safety valve" to detoxify excess amounts of sulfite and protect the cell from sulfitolysis.

VII. Compartmentalization of Sulfur Metabolism

Cells solved the problem of having two important conflicting pathways by separating them into different compartments. This rule holds also true for sulfur metabolism: sulfate assimilation takes place in the chloroplasts whereas sulfite detoxification by the SO is peroxisomally localized. However, peroxisomes seem to be not the only known sulfite-oxidizing organelles. Although the peroxisomal molybdoenzyme SO (EC 1.8.3.1) is the only biochemically and genetically characterized SO, there is still sulfite oxidation going on

in the cell. In non-green suspension cultures of mutants lacking the molybdenum cofactor and therefore also peroxisomal SO, the sulfite oxidizing capacity of the cell extract does not go down to zero but to 40% of the wildtype-level (Eilers et al. 2001). The origin of this residual activity remains unclear.

How do the two pathways of chloroplast-based sulfate assimilation and peroxisomal sulfite oxidation interact and how are they co-regulated? Chloroplasts and peroxisomes are closely associated within the plant cell which is the basis for photorespiration where intermediates are crossing back and forth between these two organelles and mitochondria (Buchanan et al. 2000). Obviously, this association forms the basis for the rapid and efficient metabolic channeling of the two toxic metabolites: sulfite and H_2O_2. Finally, another cell-compartment seems to be involved in this metabolic process as well: the end product of sulfite oxidation – sulfate – is stored in the vacuole or could be transported out of the cell. The internal sulfate reserve in the vacuoles may buffer the flux of sulfate through the plant. While the nature of a tonoplast sulfate influx transporter is still unsolved (Buchner et al. 2004), recently Kataoka et al. (2004) could demonstrate SULTR4-type vacuole transporters to facilitate the efflux of sulfate. Sulfate uptake into chloroplast is described to be mediated by the same group 4 transporter family (for review see Leustek 2002 and chapter I.2 in this book). But for peroxisomes only one porin is known as transport system for a variety of different inorganic and organic anions (Reumann et al. 1998), which could principally assist sulfate or sulfite transport.

In the future, more information on the subcellular transport of SO_2, sulfite and sulfate will sharpen our view of the complex regulatory interaction between chloroplasts and peroxisomes and thus will shed more light on the fate of sulfite during assimilatory or dissimilatory processes.

Acknowledgements

We are grateful to Christina Lang for critical comments on the manuscript. Our work was financially supported by the Deutsche Forschungsgemeinschaft.

References

Bick JA, Leustek T (1998) Plant sulfur metabolism – the reduction of sulfate to sulfite. Cur Opin Plant Biol 1: 240–244

Buchanan BB, Gruissem W, Russell JL (2000) Biochemistry & Molecular Biology of Plants. American Society of Plant Physiologists, Rockville, Maryland

Buchner P, Stuiver CE, Westerman S, Wirtz M, Hell R, Hawkesford MJ, De Kok LJ (2004) Regulation of sulfate uptake and expression of sulfate transporter genes in *Brassica oleracea* as affected by atmospheric H_2S and pedospheric sulfate nutrition. Plant Physiol 136: 3396–3408

Calabrese E, Sacco C, Moore G, DiNardi S (1981) Sulfite oxidase deficiency: a high risk factor in SO_2, sulfite, and bisulfite toxicity? Med Hypotheses 7: 133–145

Clegg SM, Abbatt JPD (2001) Oxidation of SO_2 by H_2O_2 on ice surfaces at 228 K: a sink for SO_2 in ice clouds. Atmos Chem Phy Discussions 1: 77–92

Cohen HJ, Betcher-Lange S, Kessler DL, Rajagopalan KV (1972) Hepatic sulfite oxidase. Congruency in mitochondria of prosthetic groups and activity. J Biol Chem 247: 7759–7766

Dittrich AP, Pfanz H, Heber U (1992) Oxidation and reduction of sulfite by chloroplasts and formation of sulfite addition compounds. Plant Physiol 98: 738–744

Droux M (2004) Sulfur assimilation and the role of sulfur in plant metabolism: a survey. Photosynth Res 79: 331–348

Eilers T, Schwarz G, Brinkmann H, Witt C, Richter T, Nieder J, Koch B, Hille R, Hänsch R, Mendel RR (2001) Identification and biochemical characterization of *Arabidopsis thaliana* sulfite oxidase. A new player in plant sulfur metabolism. J Biol Chem 276: 46989–46994

Fromageot P, Vaillant R, Perez-Milan H (1960) Oxidation of sulfite to sulfate by oat roots. Biochim Biophys Acta 44: 77–85

Garrett RM, Johnson JL, Graf TN, Feigenbaum A, Rajagopalan KV (1998) Human sulfite oxidase R160Q: identification of the mutation in a sulfite oxidase-deficient patient and expression and characterization of the mutant enzyme. Proc Natl Acad Sci U S A 95: 6394–6398

Garsed SG, Read DJ (1977) Sulphur dioxide metabolism in soy-bean, *Glycine max* var. Biloxi. I. The effects of light and dark on the uptake and translocation of $^{35}SO_2$. New Phytol 78: 111–119

Hänsch R, Lang C, Riebeseel E, Lindigkeit R, Gessler A, Rennenberg H, Mendel RR (2006) Plant sulfite oxidase as novel producer of H_2O_2: combination of enzyme catalysis with a subsequent non-enzymatic reaction step. J Biol Chem 281: 6884–6888

Hayashi M, Aoki M, Kato A, Kondo M, Nishimura M (1996) Transport of chimeric proteins that contain a carboxy-terminal targeting signal into plant microbodies. Plant J 10: 225–234

Heber U, Hüve K (1998) Action of SO_2 on plants and metabolic detoxification of SO_2. Int Rev Cytochem 177: 255–286

Heber U, Laisk A, Pfanz H, Lange OL (1987) Wann ist SO_2 Nährstoff und wann Schadstoff? Ein Beitrag zum Waldschadensproblem. AFZ 27/28/29: 700–705

Heimberg M, Fridovich I, Handler P (1953) The enzymatic oxidation of sulfite. J Biol Chem 204: 913–926

Hell R (1997) Molecular physiology of plant sulfur metabolism. Planta 202: 138–148

Hemann C, Hood BL, Fulton M, Hänsch R, Schwarz G, Mendel RR, Kirk ML, Hille R (2005) Spectroscopic and kinetic studies of *Arabidopsis thaliana* sulfite oxidase: nature of the redox-active orbital and electronic structure contributions to catalysis. J Am Chem Soc 127: 16567–16577

Herschbach C, Rennenberg H (2001) Significance of phloem-translocated organic sulfur compounds for the regulation of sulfur nutrition. Progr Bot 62: 177–192

Hill GR, Thomas MD (1933) Influence of leaf destruction by sulphur dioxide and by clipping on yield of alfalfa. Plant Physiol 8: 223–245

Johnson TL, Olsen LJ (2001) Building new models for peroxisome biogenesis. Plant Physiol 127: 731–739

Jolivet P, Bergeron E, Meunier JC (1995a) Evidence for sulfite oxidase activity in spinach leaves. Phytochemistry 40: 667–672

Jolivet P, Bergeron E, Zimierski A, Meunier JC (1995b) Metabolism of elemental sulfur and oxidation of sulfite by wheat and spinach chloroplasts. Phytochemistry 38: 9–14

Kaiser G, Martinoia E, Schroppelmeier G, Heber U (1989) Active-transport of sulfate into the vacuole of plant cells provides halotolerance and can detoxify SO_2. J Plant Physiol 133: 756–763

Kappler U, Dahl C (2001) Enzymology and molecular biology of prokaryotic sulfite oxidation. FEMS Microbiol Lett 203: 1–9

Kataoka T, Watanabe-Takahashi A, Hayashi N, Ohnishi M, Mimura T, Buchner P, Hawkesford MJ, Yamaya T, Takahashi H (2004) Vacuolar sulfate transporters are essential determinants controlling internal distribution of sulfate in *Arabidopsis*. Plant Cell 16: 2693–2704

Kisker C, Schindelin H, Pacheco A, Wehbi WA, Garrett RM, Rajagopalan KV, Enemark JH, Rees DC (1997) Molecular basis of sulfite oxidase deficiency from the structure of sulfite oxidase. Cell 91: 973–983

Leustek T (2002) Sulfate Metabolism. In: The Arabidopsis Book, CR Somerville and EM Meyerowitz (eds.), American Society of Plant Biologists, Rockville

Leustek T, Saito K (1999) Sulfate transport and assimilation in plants. Plant Physiol 120: 637–644

Linzon SN (1978) Effects of airborne sulfur pollutants on plants. In: Sulfur in the Environment Part II. Ecological Impacts, JO Nriagu (ed.), Wiley, New York: 109–162

Matsubayashi Y, Sakagami Y (1996) Phytosulfokine, sulfated peptides that induce the proliferation of single mesophyll cells of *Asparagus officinalis* L. Proc Natl Acad Sci U S A 93: 7623–7627

Miszalski Z, Ziegler H (1992) Superoxide dismutase and sulfite oxidation. Z Naturforsch 47c: 360–364

Moyer DT, Geo RH (1935) Absorption of sulphus dioxide by Alfalfa and its relation to leaf injury. Plant Physiol 10: 291–307

Mullen RT, Lee MS, Flynn CR, Trelease RN (1997) Diverse amino acid residues function within the type 1 peroxisomal targeting signal. Implications for the role of accessory residues upstream of the type 1 peroxisomal targeting signal. Plant Physiol 115: 881–889

Nag S, Saha K, Choudhuri MA (2000) A rapid and sensitive assay method for measuring amine oxidase based on hydrogen peroxide-titanium complex formation. Plant Sci 157: 157–163

Nakamura T, Meyer C, Sano H (2002) Molecular cloning and characterization of plant genes encoding novel peroxisomal molybdoenzymes of the sulphite oxidase family. J Exp Bot 53: 1833–1836

Nowak K, Luniak N, Witt C, Wustefeld Y, Wachter A, Mendel RR, Hänsch R (2004) Peroxisomal localization of sulfite oxidase separates it from chloroplast-based sulfur assimilation. Plant Cell Physiol 45: 1889–1894

Peiser G, Yang SF (1985) Biochemical and physiological effects of SO_2 on nonphotosynthetic processes in plants. In: WE Winner, HA Mooney and RA Goldstein (eds.), Sulfur Dioxide and Vegetation, Stanford University Press, Stanford, California: 148–161

Pfanz H, Dietz K-J, Weinerth I, Oppmann B (1990) Detoxification of sulfur dioxide by apoplastic peroxidases. In: Sulfur Nutrition and Sulfur Assimilation in Higher Plants, H Rennenberg, C Brunold, LJ De Kok and I Stulen (eds.), SPB Academic: 229–233

Pfanz H, Oppmann B (1991) The possible role of apoplastic peroxidases in detoxifying the air pollutant sulfur dioxide. In: Biochemical, Molecular and Physiological Aspects of Plant Peroxidases, J Lobarzewski, H Greppin, C Penel and T Gaspar (eds.), University of Geneva, Geneva: 401–417

Rao IM, Anderson LE (1983) Light and stomatal metabolism: II. Effects of sulfite and arsenite on stomatal opening and light modulation of enzymes in epidermis. Plant Physiol 71: 456–459

Rennenberg H (1984) The fate of excess sulfur in higher plants. Annu Rev Plant Physiol 35: 121–153

Rennenberg H, Herschbach C (1996) Responses of plants to atmospheric sulphur. In: Plant Response to Air Pollution, M Yunus and M Iqbal (eds.), Wiley, Chichester: 285–294

Rennenberg H, Polle A (1994) Metabolic consequences of atmospheric sulphur influx into plants. In: Plant Responses to the Gaseous Environment, R Alscher and A Wellburn A (eds.), Chapman & Hall, London: 165–181

Rennenberg H, Sekiya J, Wilson LG, Filner P (1982) Evidence for an intracellular sulfur cycle in cucumber leaves. Planta 154: 516–524

Reumann S, Maier E, Heldt HW, Benz R (1998) Permeability properties of the porin of spinach leaf peroxisomes. Eur J Biochem 251: 359–366

Rost M, Karge E, Klinger W (1998) What do we measure with luminol-, lucigenin- and penicillin-amplified chemiluminescence? 1. Investigations with hydrogen peroxide and sodium hypochlorite. J Biolumin Chemilumin 13: 355–363

Saito K (2004) Sulfur assimilatory metabolism. The long and smelling road. Plant Physiol 136: 2443–2450

Tanaka K, Otsubo T, Kondo N (1982) Participation of hydrogen peroxide in the inactivation of Calvin cycle SH enzymes in SO_2-fumigated spinach leaves. Plant Cell Physiol 23: 1009–1018

Thomas MD, Hendricks RH, Hill GR (1944) Some chemical reactions of sulfur dioxide after absorption by Alfalfa and sugar beets. Plant Physiol 19: 212–226

Van der Kooij TAW, de Kok LJ, Haneklaus S, Schnug E (1997) Uptake and metabolism of sulphur dioxide by *Arabidopsis thaliana*. New Phytol 135: 101–107

Veljović-Jovanović S, Oniki T, Takahama U (1998) Detection of monodehydro ascorbic acid radical in sulfite-treated leaves and mechanism of its formation. Plant Cell Physiol 39: 1203–1208

Wurfel M, Haberlein I, Follmann H (1990) Inactivation of thioredoxin by sulfite ions. FEBS Lett 268: 146–148

Yu B, Xu C, Benning C (2002) *Arabidopsis* disrupted in SQD2 encoding sulfolipid synthase is impaired in phosphate-limited growth. Proc Natl Acad Sci U S A 99: 5732–5737

Ziegler I (1974) Action of sulphite on plant malate dehydrogenase. Phytochemistry 13: 2411–2416

Chapter 13

The State of Sulfur Metabolism in Algae: From Ecology to Genomics

Nakako Shibagaki* and Arthur Grossman

Carnegie Institution, Department of Plant Biology, 260 Panama Street, Stanford, CA 94305, USA

*Corresponding author, Phone: +1 650 325 1521 X 241, Fax: + 1 650 325 6857, E-mail: snakako@stanford.edu

Rüdiger Hell et al. (eds.), Sulfur Metabolism in Phototrophic Organisms, 231–267.
© 2008 *Springer.*

Summary

Sulfur, primarily in the form of sulfate, is transported into algal and plant cells and reduced to sulfide in the chloroplast. Both sulfate and sulfide can be incorporated into a variety of sulfur-containing compounds critical for protein, lipid and polysaccharide synthesis, as well as signaling molecules. Most of our current knowledge about sulfur metabolism and the acclimation of photosynthetic organisms to conditions of sulfur deprivation, especially at the molecular level, are from studies that have exploited the model plant *Arabidopsis thaliana*, or the freshwater alga *Chlamydomonas reinhardtii*. However, there are also novel aspects of the biosynthesis and function of sulfur metabolites, especially volatile metabolites synthesized by marine algae that might have antifreeze and antioxidant functions and also influence global features of climate. This chapter describes sulfate assimilation in algae and the adaptation and acclimation of algae to changing sulfur environments.

I. Algae and the Global Sulfur Cycle

Sulfur (S) is a macroelement that is incorporated into proteins (as cysteine and methionine), lipids and polysaccharides. It is also required for the production of molecules that help organisms cope with reactive oxygen species (ROS) and heavy metals, and may be integral to key cellular regulators. Generally, the dominant, fully oxidized form

Abbreviations: APR – APS reductase; APS – adenosine 5′-phosphosulfate; APSK – APS (or AKN)_APS kinase; APS (or AKN)_APS kinase; ARS – arylsulfatase; ATS – ATP sulfurylase; CCN – cloud condensation nuclei; DMS – dimethylsulfide; Cp – chloroplast; DMSHB – 4-dimethylsulphonio-2-hydroxybutyrate; DMSO – dimethyl sulfoxide; DMSP – dimethylsulfoniopropionate; ER – endoplasmic reticulum; γ-ECS – γ-glutamylcysteine synthetase; GPX – glutathione peroxidase; GRX – glutaredoxin; GSH – glutathione; GST – glutathione S-transferase; LSU – the large subunit; MSA – methane sulfonic acid; MTOB – 4-methylthio-2-oxobutylate; OAS – O-acetylserine; OASTL – O-acetylserine(thiol)lyase; OPH – O-phosphohomoserine; PAPS – 3′-phosphoadenosine 5′-phosphosulfate; PC – phytochelatin; PCS – phytochelatin synthase; PDI – protein disulfide isomerase; PG – phosphatidyl glycerol; Pi – phosphate; PS – the photosystem; ROS – reactive oxygen species; Rubisco – ribulose-1,5-bisphosphate carboxylase; SAC – sulfur-acclimation mutants; SAT – serine acetyltransferase; SIR – sulfite reductase; SMT – selenocysteine methyltransferase; SQDG – sulfoquinovosyl diacylglycerol; TMD – transmembrane domain; TRX – thioredoxin

of S, sulfate (SO_4^{2-}), is the most stable S form in the oxidizing environment of the Earth. SO_4^{2-} is rapidly taken up by microbes and plants, activated by conjugation to ATP, and can be used immediately for the sulfation of various compounds, primarily polysaccharides, or reduced and then incorporated into the amino acids cysteine and methionine and the antioxidant, tripeptide glutathione. High concentrations of SO_4^{2-} have been measured in numerous terrestrial and aquatic environments. However, since animals do not have enzymes that reduce SO_4^{2-} to sulfide, and the latter is critical for cysteine and methionine synthesis, the S-containing amino acids are essential in mammalian diets (Tabe and Higgins, 1998).

S compounds are introduced into the biosphere from natural sources (e.g. volcano activities) as well as from activities of humans and other organisms. Organic, S-containing compounds released into the environment as a consequence of biotic activity may be recycled into the available S pool through the action of hydrolytic and redox active enzymes produced by soil and aquatic microbes. It is important to note that algae play a pivotal role in S cycling and are responsible for the majority of SO_4^{2-} reduction in seawater. One reduced S compound produced by algae in the marine environment is dimethyl sulfide (DMS) (Andreae and Raemdonch, 1983), which arises primarily from the enzymatic cleavage of dimethylsulfoniopropionate (DMSP). In the 1970s, DMS was proposed to be a volatile compound and to be involved in cycling of S from the oceanic to terrestrial environments (Lovelock, 1972). Indeed, DMS has been estimated to be responsible for as much as 50% of the global, biogenic S input into the atmosphere. The oxidation products of DMS serve to nucleate the formation of clouds over the oceans (Habicht et al., 2002; Lomans et al., 2002; Lovelock, 1972), which is likely to strongly influence the Earth's climate, as proposed by Charlson et al. in 1987 ("CLAW hypothesis" after the initials of authors). Atmospheric DMS would be converted mainly to SO_2 via oxidation; the oxidation is triggered by interactions with OH radicals, which can be formed by UV-driven photodissociation of ozone (Andreae and Crutzen, 1997; Charlson et al., 1987; Kieber et al., 1996). SO_4^{2-} particles formed from atmospheric SO_2 behave as cloud condensation nuclei (CCN). Marine algae may accumulate extremely high intracellular levels of DMSP as light levels increase and/or ammonium levels decline, causing elevated DMS emission, elevated acidic S aerosols and CCN. The resulting increased cloud cover will lower the light intensities that reach the ocean surface. Furthermore, rainfall from clouds formed as a consequence of the accumulation of S aerosols will return certain elements to the oceans, including nitrogen (N), which is often lacking in marine environments. This feedback system demonstrates how light levels may indirectly alter the metabolism of marine algae, and the changing spectrum of metabolites produced may result in changes in atmospheric conditions that influence the number of quanta that reach the ocean surfaces. A diagram depicting the major transformations of S that occur on the Earth is given in Fig. 1. A more complete discussion of the global S cycle is presented by Giordano et al. in Chapter V-23 of this book.

II. Sulfur Metabolism

SO_4^{2-} is the major source of S for the growth of bacteria and plants. The SO_4^{2-} assimilation pathway was elucidated in early studies using the organisms *Escherichia coli*, *Neurospora crassae*, *Saccharomyces cerevisiae* and the alga *Chlorella pyrenoidosa* (Schiff, 1959) and

Euglena gracilis (Buetow and Buchanan, 1964; Goodman and Schiff, 1964). Collections of algal mutants impaired in SO_4^{2-} assimilation have been invaluable for elucidating S metabolism (Hodson and Schiff, 1971a; Hodson and Schiff, 1971b; Hodson et al., 1971). Furthermore, prior to the development of sophisticated molecular technologies, algae and bacteria were the subject of extensive biochemical analyses directed toward the isolation and characterization of enzymes catalyzing each step in the pathway required for the assimilation of SO_4^{2-}. Much of our knowledge concerning the acquisition and assimilation of SO_4^{2-} in photosynthetic organisms has come from studies of vascular plants (Grossman and Takahashi, 2001; Maruyama-Nakashita et al., 2003; Maruyama-Nakashita et al., 2004; Takahashi et al., 2003), with considerable information gleaned from examination of the algae (Grossman and Takahashi, 2001; Pollock et al., 2005) and cyanobacteria (Green et al., 1989; Grossman and Takahashi, 2001; Schiff, 1979). In this chapter we will primarily summarize work that has allowed for elucidation of SO_4^{2-} acquisition and reduction processes in algae, and how these photosynthetic organisms respond when growth becomes limited by lowS availability (see section V).

A. SO_4^{2-} Transport Systems

Algae have several SO_4^{2-} transport systems residing on the cellular membranes. The transport process is energy dependent, and oxide anions of group IV elements (e.g. chromate, selenate and molybdate) directly compete with SO_4^{2-} at the site of transport (Ramus, 1974; Ryan et al., 1987). Early studies of SO_4^{2-} uptake in the fresh water cyanobacterium *Synechococcus* sp. PC7942 (formerly *Anacystis nidulans*) demonstrated that the process is light-dependent and energy-requiring (Jeanjean and Broda, 1977; Utkilen, 1976). Genes encoding a cyanobacterial SO_4^{2-} transporter (ABC transporter type) were first identified in *Synechococcus* sp. PCC7942 (Green et al., 1989; Laudenbach and Grossman, 1991); the subunits of this transporter are designated CysA, CysT, CysW and SbpA. The genes for these transporter subunits plus CysR are clustered on a *Synechococcus* sp. PCC7942 plasmid; the *cysR* gene encodes a transcription factor. Since a deletion of CysA, CysT

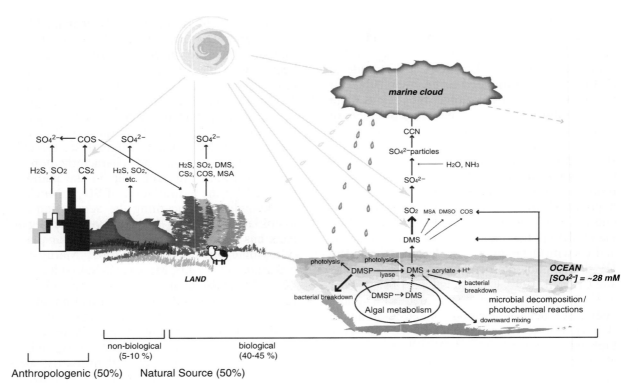

Fig. 1. *The sulfur cycle.* DMSP, a prominent S metabolite synthesized by marine algae (and plants) can reach an intra-cellular concentration of 50–400 mM (Stefels, 2000). Excess DMSP is exported from the cell, and most is degraded by bacteria or by the activity of extra-cellular, algal DMSP lyases. Products of the lyase reaction are DMS, acrylate and H+. Acrylate may serve as a grazer repellent (Wolfe et al., 1997). DMS is volatile and can be released, to some extent, from the ocean surface into the atmosphere (Groene, 1995) at the rate of 0.6–1.7 TmolS/year (13, 4), although it is subject to photolysis (7–40%) or biological comsumption (129). DMSP and DMS concentrations in sea water are in the nM range (Barnes, 1994; Kieber, 1996). ROS in the atmosphere, generated by photo-dissociation of ozone by UV light, causes oxidation of DMS and the formation of SO_2, with the production of lesser amounts of MSA (20%) and DMSO (14%) (Andreae and Crutzen, 1997; Charlson et al., 1987; Kieber et al., 1996). SO_4^{2-} particles can form from the SO_2; this process is affected by various atmospheric factors including the concentration of H_2O and NH_3. SO_4^{2-} particles of a certain size behave as CCN. COS is the most abundant S compound in the atmosphere. It is produced by burning coal and oil, which leads to oxidation of CS_2 present in coal. It is also a product of photooxidation of DMS (Barnes et al., 1994). COS is taken up by plants or oxidized to form SO_4^{2-} leading to SO_4^{2-} aerosols and CCN formation.

or CysW resulted in decreased SO_4^{2-} uptake and no growth when SO_4^{2-} was provided as the only S source, this transport complex is likely the sole or major SO_4^{2-} transporter in *Synechococcus* sp. PCC7942. Interestingly, disruption of the *sbpA* gene did not result in a growth phenotype when cells were maintained on SO_4^{2-}, although the mutant could not synthesize a high capacity uptake system during S starvation (Laudenbach and Grossman, 1991).

Once SO_4^{2-} is transported into a photosynthetic eukaryotic cell, it must be routed into the plastids where it is reduced. Recently, a bacte-

rial-type SO_4^{2-} transporter was discovered in *C. reinhardtii* (but not in vascular plants), which was found to be associated with the chloroplast envelope (Fig. 2). Initially, a single gene encoding a subunit of this transporter was isolated and designated *SulP* (GenBank accession AF467891) (Chen and Melis, 2004; Chen et al., 2003). SulP is predicted to be a transmembrane polypeptide with strong similarity to the ABC transporter subunit CysT. *SulP* mRNA and protein increase when *C. reinhardtii* is starved for S, and a *sulP* antisense strain exhibited reduced SO_4^{2-} uptake capacity, lower photosynthesis (as measured by

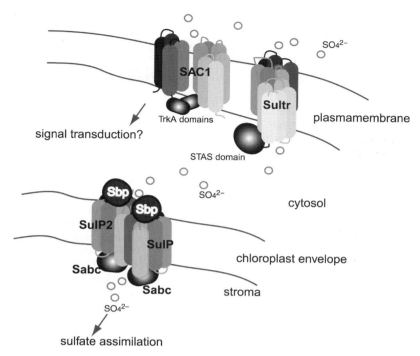

Fig. 2. *Sulfate transport systems*. SAC1, anticipated to function on the plasmamembrane where is may sense intra- and/or extra-cellular SO_4^{2-} conditions and initiate a signaling cascade that regulates expression of genes involved in S metabolism. H^+/SO_4^{2-} transporters and SAC1-like proteins, which show a high degree of sequence similarity to Na^+/SO_4^{2-} transporters, may also be located on the plasmamembrane, while bacterial type SO_4^{2-} transporters, complexes containing SulP, Sbp and Sabc, function on the chloroplast envelope to import SO_4^{2-} into the stroma (where the SO_4^{2-} can be reduced by photosynthetic electron transport).

light-saturated rates of O_2 evolution), low levels of ribulose-1,5-bisphosphate carboxylase (Rubisco) and the photosystem (PS) II reaction center polypeptide D1, and signs of S deficiency, even in the presence of moderate concentrations of SO_4^{2-} (the cells cannot efficiently acquire S and begin to exhibit responses associated with S starvation) (Chen and Melis, 2004; Chen et al., 2005). Failure to isolate *SulP* antisense transformants with a large reduction in the *SulP* transcript level suggests that a complete loss of gene function would be lethal and that this transporter defines the major path for SO_4^{2-} import into *C. reinhardtii* plastids.

Recently, additional components of the chloroplast, bacterial-type SO_4^{2-} transporter have been identified (Melis and Chen, 2005), including another transmembrane protein and the nucleotide and substrate binding proteins. Of the four nuclear genes encoding this SO_4^{2-} permease holocomplex (Table 1), there are two encoding chloroplast envelop-targeted transmembrane proteins,

SulP and SulP2 (AY536251), a stromal-targeted ATP-binding protein (Sabc, AY536252) and a substrate-binding protein (Sbp, AY536253). The mature SulP and SulP2 polypeptides are in the inner chloroplast membrane and contain seven transmembrane domains and two relatively large hydrophilic loops.

The nuclear genome of the primitive red alga *Cyanidioschyzon merolae* also contains genes for all components of a bacterial-type SO_4^{2-} transporter; CysA and Sbp are each probably encoded by a single gene while two CysT-like genes (which also resemble CysW) are in tandem and just downstream of a gene encoding a LysR-like transcription regulator (http://merolae.biol.s.u-tokyo.ac.jp/) (Matsuzaki et al., 2004). The plastid genome sequence of *C. merolae* (AY286123 for the genome of strain DBV201, and AB002583 for the genome of strain 10D) (Clowney and Suzuki, direct submission) (Ohta et al., 2003) also contains a single *cysT*-like gene. In contrast, the nuclear genome of the diatom *Thalassiosira*

Table 1. C. reinhardtii genes encoding enzymes of the SO_4^{2-} assimilation pathway and GSH/PC synthesis.

GenBank accession number or Gene model in ver3	Name	Description
AF467891	*SulP*	ABC transporter type-sulfate transporter
AY536251	*SulP2*	
AY536252	*Sabc*	
AY536253	*Sbp*	
e_gwW.4.24.1	*Sultr1*	putative H+/sulfate transporter
Chlre2_kg.scaffold_32000067	*Sultr2*	putative H+/sulfate transporter
Chlre2_kg.scaffold_35000027	*Sultr3*	putative H+/sulfate transporter
e_gwW.10.232.1	*SAC1-LIKE*	putative Na+/sulfate transporter
gwH.10.31.1	*SAC1-LIKE*	putative Na+/sulfate transporter
gwH.11.15.1	*SAC1-LIKE*	putative Na+/sulfate transporter
U57088	*ATS1*	ATP sulfurylase 1
estExt_gwp_1W.C_50037	*ATS2*	ATP sulfurylase2
estExt_fgenesh2_kg.C_280057	*AKN2*	Adenylylsulfate kinase 2
AF498290	*APR1*	APS reductase 1
estExt_GenewiseH_1.C_460003	*SIR1*	ferredoxin dependent sulfite reductase 1
estExt_fgenesh2_pg.C_290083	*SIR2*	ferredoxin dependent sulfite reductase 2
estExt_fgenesh2_pg.C_90006 or estExt_fgenesh1_pg.C_90006	*SIR3*	NADH/NADPH dependent sulfite reductase 3
AAM23309	*SAT1*	Serine acetyltransferase 1
estExt_fgenesh2_pg.C_540058	*SAT2*	Serine acetyltransferase 2
e_gwW.4.121.1	*SAT3*	Serine acetyltransferase 3
estExt_gwp_1W.C_290041	*OASTL1*	O-acetylserine(thiol)lyase 1
SKA_Chlre2_kg.scaffold_67000049	*OASTL2*	O-acetylserine(thiol)lyase 2
MHS_estExt_GenewiseW_1.C_90074	*OASTL3*	O-acetylserine(thiol)lyase 3
AF078693	*OASTL4 (Cys1ACr)*	O-acetylserine(thiol)lyase 4
estExt_fgenesh1_pm.C_270009	*Cystathionine gamma-synthase*	
estExt_gwp_1H.C_450020	*Cystathionine beta-lyase*	
e_gwW.795.8.1	*METH*	cobalamin-dependent methionine synthase
Chlre2_kg.scaffold_82000007	*METE*	cobalamin-independent methionine synthase
estExt_fgenesh2_kg.C_70040	*GSH1*	g-glutamylcysteine synthetase
e_gwH.13.43.1	*GSH2*	glutathione synthethase
e_gwH.6.128.1	*PCS*	phytochelatin synthase

pseudonana encodes a CysA-like polypeptide, which may be part of a SO_4^{2-} transport system (or another ABC transporter), but there are no genes encoding CysT- or Sbp-like polypeptides on the nuclear genome (http://genome.jgi-psf.org/thaps1/thaps1.home.html). The plastid genome of this organism has not been sequenced, but it may harbor genes for the other SO_4^{2-} transport components. Indeed, genes encoding potential CysT and CysA homologs (but not Sbp) are encoded on chloroplast genomes of various algae including *Nephroselmis olivacea* (AF137379) (Turmel et al., 1999), *Mesostigma viride* (AF166114) (Lemieux et al., 2000), *Zygnema circumcarinatum* (Ay958086) (Turmel et al., 2005), *Pro-*

totheca wickerhamii (AJ245645, Knauf,U direct submission) and *Chlorella vulgaris* C-27 (AB001684) (Wakasugi et al., 1997). Interestingly, there are also *cysA*- and *cysT*-like genes on the chloroplast genomes of the lower vascular plants *Marchantia polymorpha* (liverwort, X04465) (Ohyama et al., 1988) and *Anthoceros formosae* (hornwort, AB086179) (Kugita et al., 2003), but not on the nuclear or plastid genomes of vascular plants. These findings suggest that genes encoding polypeptides of the SO_4^{2-} transport system that function in chloroplasts of algae and lower eukaryotic photosynthetic organisms are similar to prokaryotic SO_4^{2-} transport systems, and that some or all of these genes were transferred from the original endosymbiont genome to the nuclear genome of the host organism. The distribution of genes for various component of this bacterial-type SO_4^{2-} transporter between nuclear and plastid genomes appears to differ among different organisms, and further analysis of this distribution and the way in which the synthesis of nuclear- and chloroplast-encoded components of the transporter are coordinated will be of considerable evolutionary interest.

In vascular plants, SO_4^{2-} transport across plasmamembrane and tonoplast membranes is performed by H^+/SO_4^{2-} co-transporters (SLC26 family proteins) (Buchner et al., 2004; Maruyama-Nakashita et al., 2004; Saito, 2000). Recently, genes encoding the putative plasmamembrane SLC26 type H^+/SO_4^{2-} transporters of *C. reinhardtii* have been isolated and characterized, and have been designated *SULTR1*, *SULTR2*, and *SULTR3* (Fig. 2). These findings suggest that similar SO_4^{2-} transport systems may function on the plasmamembranes of vascular plants and *C. reinhardtii*. In addition, the *C. reinhardtii* genome contains three *SAC1*-LIKE genes that exhibit strong sequence similarity to mammalian Na^+/SO_4^{2-} transporters (SLC13 family). Two of these *SAC1*-LIKE genes are arranged in tandem, and in the same orientation, on the *C. reinhardtii* genome (Pootakham and Grossman, unpublished). Both H^+- and Na^+-dependent type SO_4^{2-} transporters have distinctive small domains in the region of the polypeptide predicted to be in cytosol (STAS and TrkA_C domain, respectively), whose functions are of great interest but remain unknown (2). All genes associated with SO_4^{2-}-assimilation, are given in Table 1.

Some cyanobacterial genomes (e.g. *Synechocystis* sp. PCC6803 and *Synechococcus* sp. PCC7942) also contain putative SLC26 family type H^+/SO_4^{2-} transporters (BAA16785 and ABB57410, respectively). It is not clear whether or not these putative transporters really transport SO_4^{2-} since some SLC26 family members are responsible for transporting other anions. This will only be revealed with additional genetic and biochemical studies.

B.   ATP Sulfurylase

The first enzyme in the pathway leading to the reduction and assimilation of SO_4^{2-} is ATP sulfurylase (*ATS* and ATS for gene and protein desig-

nations, respectively). ATS catalyzes activation of SO_4^{2-} to form adenosine 5'-phosphosulfate (APS). Two independent proteins with ATS activity were purified from *Euglena gracilis*; one was associated with chloroplasts and the other with the cytosol (Li et al., 1991). The *C. reinhardtii* (http://genome.jgi-psf.org/Chlre3/Chlre3.home.html), *T. pseudonana* and *C. merolae* nuclear genomes all appear to contain two genes encoding *ATS*. The polypeptide encoded by *C. reinhardtii ATS1* (accession U57088) has a plastidic transit peptide (Yildiz et al., 1996), suggesting that this gene product functions in the chloroplast. There is no obvious signal peptide associated with the ATS2 polypeptide.

C.   APS Kinase

APS generated in the ATS reaction can function directly in sulfation reactions, undergo a second phosphorylation to yield 3'-phosphoadenosine 5'-phosphosulfate (PAPS) (Arz et al., 1994; Lee and Leustek, 1998), or the SO_4^{2-} ligand of APS can be further reduced to generate sulfide. APS kinase, designated APSK or AKN (adenylsulfate kinase) catalyzes the phosphorylation of APS to yield PAPS. AKN appears to be encoded by a single gene in *C. reinhardtii*. PAPS, like APS, can also be used by sulfotransferases to catalyze the sulfation of various metabolites including flavanols, choline and glucosides. This non-reductive incorporation of SO_4^{2-} into various molecules occurs in the cytoplasm. In most eukaryotic algae, PAPS is not the major substrate used for the reduction SO_4^{2-} (Gao et al., 2000); the SO_4^{2-} ligand on APS generally serves as the substrate for reduction (Bick et al., 2000; Kopriva et al., 2001; Kopriva and Koprivova, 2004).

D.   APS and PAPS Reductases

The SO_4^{2-} of APS is reduced to sulfite via a reaction catalyzed by APS reductase (APS sulfotransferase; gene and protein designations *APR* and APR, respectively). Studies focused on this reaction were recently reviewed (Kopriva and Koprivova, 2004). The green, unicellular alga *Chlorella* (Schmidt, 1972), *Euglena gracilis* (Li and Schiff, 1991) and the marine macrophytic red alga *Porphyra yezoensis* (Kanno et al., 1996) were used for purifying APR. The purified proteins were tested for their enzymatic properties, including their requirement for thiols. A cDNA encoding APR was cloned from the marine chlorophyte *Enteromorpha intestinalis* (AF069951) by complementation of the *E. coli cysH* mutant (Gao et al., 2000) and also from *C. reinhardtii* (Ravina et al., 2002), which appears to have a single *APR1* gene (AF498290). The proteins encoded by *E. intestinalis* and *C. reinhardtii* cDNAs have the same structure as analogous proteins of vascular plants; the protein has a predicted transit peptide, a PAPS reductase-like domain and a carboxy-terminal thioredoxin domain, which is thought to be required for releasing sulfite, the reduction intermediate, from the protein (Weber et al., 2000).

Kopriva et al. (2002) observed that the APR proteins have specific cysteine residues within the PAPS reductase-like domain that bind an [4Fe–4S]$^{2+}$ cluster (Kopriva et al., 2001). In contrast, the PAPS reductase does not have these [4Fe–4S]$^{2+}$ binding residues, suggesting that PAPS reductase may be the reductase of choice in environments with low iron (Fe) availability. Kopriva et al. (2002) suggested, based on phylogenetic analysis, that the pathway leading to SO_4^{2-} reduction via the APR reaction represents the ancestral pathway for SO_4^{2-} assimilation and that PAPS reductase evolved in response to either Fe or S poor environments.

APR is likely to be the primary route for SO_4^{2-} reduction in most eukaryotic algae since little PAPS reductase activity was detected in three chlorophytes (*E. intestinalis*, *Tetraselmis* sp., and *Dunaliella salina*), two diatoms (*Thalassiosira weissflogii* and *Thalassiosira oceanica*), two prymnesiophytes (*Isochrysis galbana* and *Emiliania huxleyi*), and a dinoflagellate (*Heterocapsa triquetra*) (Gao et al., 2000). In contrast, most marine and freshwater cyanobacteria have evolved PAPS reductase (http:// cyano.genome.jp) (Kopriva et al., 2002), with some exceptions (Schmidt and Trüper, 1977), one of which is *Plectonema* sp. PCC73110 (Kopriva and Koprivova, 2004). *APR* genes in algae (*E. intestinalis, C. merolae* and *C. reinhardtii*) encode proteins with the APR signature cysteines. However, there are two genes on the nuclear genome of the marine centric diatom *T. pseudonana* (gene models grail.75.38.1 and new V2.0.genewise.6.577.1) that encode a protein like APR, but with some unique features. While these polypeptides have a thioredoxin-

like domain, they do not appear to contain the APR-specific signature cysteines, suggesting that these APR-like polypeptides may not have the [4Fe–4S]$^{2+}$ cluster and that they might use PAPS as their substrate Hence, in the oceans, where Fe can be severely limiting, *T. pseudonana* might have developed a means of SO_4^{2-} reduction that avoids the use of the [4Fe–4S]$^{2+}$ cluster and likely uses PAPS as the substrate for reduction; this conclusion requires biochemical confirmation.

E. Sulfite Reductase

SO_3^{2-} generated by the APS reductase reaction is reduced to sulfide by plastidic sulfite reductase (*SIR* and SIR for the gene and protein, respectively). This ferredoxin-dependent activity was detected in cell-free extracts from the green alga *Chlorella* (Schmidt, 1973). In *C. reinhardtii* there are two genes encoding a eukaryotic-type sulfite reductase (*SIR1* and *SIR2*), which use reduced ferredoxin as reductant, and a single gene for a putative bacterial-type enzyme (*SIR3*). The latter has both flavin and pyridine nucleotide binding domains and probably uses NADPH or NADH as reductant. The *T. pseudonana* and *C. merolae* nuclear genomes also contain genes encoding both eukaryotic and prokaryotic SIR, probably two of each type.

F. Serine Acetyltransferase

The sulfide generated by the SIR reaction is combined with O-acetylserine (OAS) to form cysteine in a reaction catalyzed by O-acetylserine(thiol)lyase (OASTL). OAS, the immediate substrate for incorporation of sulfide, is formed from serine and acetylCoA in a reaction catalyzed by serine acetyltransferase (SAT). Interestingly, *Synechococcus* sp. PCC7942 harbors a plasmid containing genes associated with S metabolism. The transcription factor CysR, encoded by a gene within the SO_4^{2-} transporter gene cluster, is activated along with the transporter genes. CysR appears to control expression of genes on a *Synechococcus* sp. PCC7942 plasmid; among these plasmid genes is *srpH*, which codes for SAT (Nicholson et al., 1995). These results suggest that there is sequential activation of genes associated with S-deprivation responses in *Synechococcus* sp. PCC7942. The *C. reinhardtii* genome contains at least three *SAT* genes, based on our analysis of the genome sequence. The *SAT1* gene (AY095344) has been described by Ravina et al. (2002). While SAT1 and SAT2 polypeptides are predicted to be targeted to the chloroplast, there is no reliable gene model, at this point, for the third *SAT* gene.

G. OASTL

In vascular plants, OASTL is encoded by multiple genes and the different gene products are present in the cytosol, plastids, and mitochondria. The *C. reinhardtii* genome also harbors multiple *OASTL* genes (there are probably four genes; *OASTL1–4*). The polypeptides encoded by these genes are highly homologous to OASTL from bacteria and vascular plants. OASTL encoded by the *C. reinhardtii Cys1ACr* gene (Ravina et al., 1999) *OASTL4* has a putative plastid transit peptide and is predicted, based on the program PSORT, to be localized to the chloroplast stroma. The *T. pseudonana* genome also contains multiple genes encoding SAT and OASTL. The existence of multiple SAT and OASTL isoforms suggests that these activities may be required in different subcellular compartments. The *Synechococcus* sp. PCC7942 *srpG* gene, which encodes OASTL, is present on the same plasmid as *srpH* (Nicholson et al., 1995; Nicholson and Laudenbach, 1995), and is also regulated by S availability and the regulatory element CysR.

H. Methionine Synthesis

Methionine is synthesized from cysteine and O-phosphohomoserine (OPH) by three consecutive reactions catalyzed by cystathionine γ-synthase, cystathionine β-lyase and methionine synthase; intermediates in this pathway are cystathionine and homocysteine. Cystathionine γ-synthase catalyzes the formation of cystathionine. Its activity is controlled by OPH and S-adenosylmethionine availability. OPH, which is generated from aspartate, can serve as the substrates for both cystathionine γ-synthase and threonine synthase, and the relative activities of these enzymes are controlled by the S status of cells; as S levels decline the activity of cystathionine γ-synthase increases relative to threonine synthase. Cystathionine β-lyase generates homocysteine from cystathionine, while methionine synthase catalyzes

the methylation of homocysteine to form methionine. The *C. reinhardtii* genome appears to have a single copy of the cystathionine β-lyase gene and two genes encoding methionine synthase, METE and METH. The METE methionine synthase is a vitamin B_{12} (cobalamin)-independent enzyme with no predicted transit peptide sequence (Kurvari et al., 1995). Expression of *METE* appears to increase in activated *C. reinhardtii* gametes. This finding raised the possibility of methionine being involved in signal transduction or cellular responses associated with fertilization. The METH methionine synthase is a vitamin B_{12}-dependent enzyme. It appears that *C. reinhardtii* preferentially uses the vitamin B_{12}-dependent form of methionine synthase, which has a higher rate of catalysis, but in the absence of vitamin B_{12}, the alga survives by activating expression of *METE*. The controlled synthesis of the two different methionine synthase proteins is through a riboswitch mechanism. Vitamin B_{12} appears to bind to the *METE* mRNA, thereby blocking its translation; this promotes the synthesis of the *METH* transcript during vitamin B_{12}-sufficient growth. Work is currently being done to more clearly define the binding of vitamin B_{12} to the mRNA and the structural and regulatory consequences of that binding (Croft et al., 2005; Croft, The 12th International Conference on Cell and Molecular Biology of Chlamydomonas, May 9–14, 2006, Portland, Oregon).

I. Glutathione Synthesis

Glutathione (GSH) is a small thiol compound (γ-Glu-Cys-Gly) that plays a key role in helping cells control damage that may be caused by the accumulation of ROS species. Thus, GSH synthesis is likely to have co-evolved with the evolution of oxygenic photosynthesis, and GSH accumulation has been noted in most cyanobacteria and eukaryotic algae (Fahey et al., 1987). GSH is synthesized by the ATP-dependent addition of cysteine to glutamate by γ-glutamylcysteine synthetase (γ-ECS or GSH1) to form γ-glutamylcysteine, followed by the ATP-dependent addition of glycine to γ-glutamylcysteine by GSH synthetase (GSH2). The synthesis of GSH in *C. reinhardtii* is enhanced by exposure of the cells to heavy metals (Howe and Merchant, 1992). The *C. reinhardtii* genome contains a single gene encoding each of these enzymes, and the polypeptides encoded by both have putative plastid transit sequences. Amino acid sequences of γ-ECS among various organisms exhibit considerable divergence, while the GSH2 sequences are more highly conserved (Copley and Dhillon, 2002). γ-ECSs of cyanobacteria form an independent phylogenetic group relative to eukaryotes. Furthermore, the vascular plant and cyanobacterial γ-ECSs appear evolutionarily more distant than the vascular plant and α-proteobacteria γ-ECS (Ashida et al., 2005), suggesting that the plant gene may have evolved from a α-protepbacterial gene following horizontal gene transfer. Plant type γ-ECSs differ from those of bacteria or non-plant eukaryotes both in structure and regulation (Copley and Dhillon, 2002; Jez et al., 2004). Vascular plant γ-ECS functions as a monomer and is subject to inhibition by GSH. Based on sequence similarity, the *C. reinhardtii* γ-ECS resembles those of vascular plants. In contrast, the nuclear genomes of *T. pseudonana* and *C. merolae* contain a single γ-*ECS* gene encoding a product that clusters into independent phylogenetic groups that are distant from those in other organisms (the encoded proteins are most similar those of *Homo sapiens*). The diversity of GSH1 sequences suggests marked evolutionary tailoring of the gene product, a product critical for the fitness of organisms under extreme environmental conditions (see section III.D.).

J. Phytochelatin Synthase

Phytochelatin (PC) is a small peptide (γ-Glu-Cys) n-Gly (n = 2–11) involved in the detoxification of heavy metals. PC is synthesized non-ribosomally from GSH by the enzyme PC synthase (PCS). Genes for PCS have been identified in eukaryotic algae, cyanobacteria, vascular plants, fungi and nematodes (Clemens, 2006).

Cyanobacterial PCS genes have been cloned and the encoded polypeptide studied at both structural and catalytic levels (Tsuji et al., 2005; Tsuji et al., 2004; Vivares et al., 2005). The cyanobacterial PCS is distantly related to PCS of other organisms, but it resembles the amino terminal domain of eukaryotic PCS. The *Nostoc* (*Anabaena*) sp. PCC7120 enzyme has been shown to cleave glycine from GSH (first step of the reaction), but exhibited low activity for the second reaction step, the step that results in PCS formation (Harada et al., 2004; Tsuji

et al., 2004). Mutagenesis of *A. thaliana* PCS (Rea et al., 2004) and the solved crystal structure of the *Nostoc* enzyme (Vivares et al., 2005) have led to a proposed mechanism for PCS action (Clemens, 2006). However detailed comparisons of the sequence and activity of *A. thaliana* and *Nostoc* PCS using ion-pair HPLC has led to a mechanistic model (Romanyuk et al., 2006) that is not congruent with previous models (Clemens, 2006).

Aspects of PCS structure and function have not been extensively studied in eukaryotic algae other than *Dunaliella tertiolecta*; this organism has been used to determine how heavy metals influence PC accumulation in algae (Hirata et al., 2001; Tsuji et al., 2003). Based on nuclear genome sequences, *C. reinhardtii* and *C. merolae* have one putative *PCS* gene each; the PCS proteins in these algae appear to be phylogenetically distant. In contrast, the *T. pseudonana* nuclear genome appears to contain three putative *PCS* genes.

K. DMSP Synthesis

Marine algae grow in S-rich environments and use SO_4^{2-} and other S compounds in the synthesis of unique algal products including agar, alginates and carageenans; all of these compounds are highly sulfated polysaccharides. The sulfonium compound DMSP is another unique S-containing molecule that is synthesized by marine algae, but rarely by vascular plants. As discussed below, the synthesis of DMSP appears to be important for adaptation/acclimation to changes in osmolarity and salinity (Stefels, 2000).

Most small flagellates, coccolithophores and dinoflagellates possess high levels of activities required for DMSP synthesis, while these activities are generally low in diatoms (Malin and Kirst, 1997; Matrai and Keller, 1994). Some algae accumulate extremely high DMSP levels (>1 M in some dinoflagellates, but more typically 50–400mM), which can be a major S sink; DMSP biosynthesis can divert a significant proportion of S from the SO_4^{2-} assimilation pathway in marine algae. Marine algae may increase DMSP synthesis upon exposure to high intensity light and by increasing the daylength; both conditions generate elevated DMS emission (Karsten et al., 1990; Stefels, 2000). Furthermore, the flux of S through the DMSP pathway may be enhanced under N-limitation conditions, where protein synthesis ceases and the S that would normally be allocated to the synthesis of proteins becomes available for making more DMSP (Gröne T and O., 1992; Keller et al., 1999).

The biosynthesis of DMSP occurs via a unique pathway, as shown in Fig. 3, which was elucidated using in vivo isotope labeling of the green macro-alga *E. intestinalis* (Gage et al., 1997). This pathway has little in common with the pathway for DMSP synthesis in vascular plants, which has been studied in *Wollastonia biflora* (James et al., 1995) and *Spartina alterniflora* (Kocsis et al., 1998). Methionine serves as the precursor to DMSP in all known biosynthetic pathways, and the first step in the pathway is a transamination reaction that removes NH$_3$ from methionine resulting in the formation of 4-methylthio-2-oxobutylate (MTOB). The higher levels of DMSP in N-starved relative to unstarved cells (Groene, 1995) may reflect elevated transamination activity as proteins in the cell are recycled and more free methionine becomes more abundant. Reduction and S-methylation of MTOB produces 4-dimethyl-sulphonio-2-hydroxybutyrate (DMSHB), a novel compound identified in diverse phytoplankton groups, suggesting that the same pathway for DMSP synthesis as in *E. intestinalis* is present in a range of different algal species. Finally, DMSHB is converted to DMSP by an oxidative decarboxylation reaction.

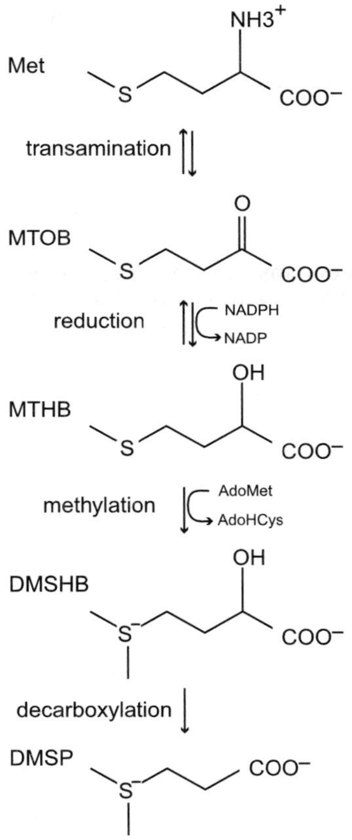

L.   DMSP Degradation and Its Consequences

DMSP that is secreted or leaks into the extracellular environment may be rapidly degraded. One degradation pathways involves DMSP lyase activity, which may be produced by the alga itself, by marine bacteria (de Souza and Yoch, 1995a; de Souza and Yoch, 1995b; Yoch et al., 1997) or by fungi (Bacic and Yoch, 1998). DMSP lyases catalyze the cleavage of DMSP, generating the volatile gas DMS, plus acrylate and H$^+$. The acrylate can potentially serve as a repellent to grazers (Wolfe et al., 1997) while the DMS is released into the atmosphere and influences the formation of clouds (see below). DMSP lyase activity has been measured in various algae (benthic macroalgae and pelagic phytoplankton) (Cantoni and Anderson, 1956; Challenger and Simpson, 1948; de Souza and Yoch, 1996; Karsten et al., 1990; Nishiguchi and Somero, 1992; Stefels and van Boekel, 1993), and often accumulates in the extracellular space of algal tissue (Stefels and van Boekel, 1993). DMSP may also be consumed by bacteria which can degrade the compound by demethylation and demethionylation, producing first methanethiol and then methionine. However, some algae may also accumulate high intracellular DMSP levels, and after they die and lyse, the released DMSP may be degraded by extracellular algal, bacterial and fungal DMSP lyases (Bacic and Yoch, 1998; de Souza and Yoch, 1995a; de Souza and Yoch, 1995b; Yoch et al., 1997), releasing significant amounts of DMS into the atmosphere. It has been difficult to evaluate how much of the DMSP that is generated is completely consumed

Fig. 3. DMSP synthesis. The pathway presented below shows the biosynthesis of DMSP from methionine in marine algae (Gage et al., 1997). Abbreviations are; Met, methionine; MTOB, 4-methylthio-2-oxobutyrate; MTHB, 4-methylthio-2-hydroxybutyrate; DMSHB, 4-dimethilsulphonio-2-hydroxy-butyrate; DMSP3-, dimethylsulphoniopropionate; AdoMet, S-adenosylmethionine; AdoHCys, S-adenosylhomocysteine.

and how much is converted to DMS and venti-lated into the atmosphere. It is also a challenge to quantify the relationship between phytoplank-ton populations or measured DMSP lyase activity with DMS fluxes at the ocean surface (Charlson et al., 1987). The situation is complicated by the fact that there is significant variation in the lev-els of DMSP synthesis among different marine algal species (Karsten et al., 1992; Malin and Kirst, 1997), which may also be significantly affected by environmental conditions, as dis-cussed above. Simo and Pedros-Alio (Simo and Pedros-Alio, 1999) demonstrated a correlation between DMS production and the mixed layer depth, but this finding cannot be solely explained by enhancement of DMSP production at elevated light intensities; the explanation must also involve degradation of DMSP by marine bacteria, which would result in diminished DMS production.

The interest in DMS formation and release into the atmosphere stems from the fact that this volatile compound can generate SO_4^{2-} aerosols that can nucleate the formation of clouds (Bates et al., 1987), as discussed in sections I and III.E.

III.   Non-Protein S Compounds

The thiol groups present on cysteine and methio-nine provide the substrate for disulfide bond formation in and between polypeptides. These bonds or bridges may stabilize proteins and pro-tein complexes and serve as redox-active sites that can modulate catalytic activity. The thiol groups of cysteine often bind specific ligands such as Fe, which is critical for electron trans-port/redox-driven processes in the cell. Cysteines in proteins may also undergo glutathionylation and S-nitrosylation; these reversible modifica-tions may unveil different activity states of a pro-tein and be intergral to the regulation of cellular metabolism. However, S is also present in several other biologically active molecules that serve a variety of functions.

A.   Sulpholipids

The sulfolipid sulfoquinovosyl diacylglycerol (SQDG) has been estimated to be one of the most abundant S-containing organic compounds in bio-logical systems (Harwood and Nicholls, 1979),

and its presence has been established in a variety of algae after first being discovered in vascular plants and cyanobacteria. SQDG is a prominent molecule in photosynthetic membranes of green algae and vascular plants. The unique cyanobac-terium *Gloeobacter violaceus* sp. PCC7421, has a cytoplasmic membrane that houses the photosyn-thetic machinery, but lacks internal thylakoids (the internal photosynthetic membranes in most cells) and is completely devoid of SQDG (Selstam and Campbell, 1996). The relative amount of SQDG in thylakoids is 11–14%; it is significantly lower in the cyanobacterial cytoplasmic membranes (Murata and Sato, 1983; Omata and Murata, 1983). Sulfolipid synthesis and function have been reviewed by Benning (Benning, 1998), and are also discussed by Benning et al. in Chapter II-11 in this book.

A number of factors may influence SQDG levels in the membranes. During phosphorus (P) -limited growth, the ratio of phospholipids to non-phospholipids in cellular membranes declines, and a greater proportion of the lipid is in the form of SQDG; SQDG synthesis may increase during P-limitation to help maintain a constant level of anionic lipid in the membranes. For example, in *Synechococcus* sp. PCC7942, phosphatidyl glyc-erol (PG) and SQDG accumulate to 7.2 and 22.3 mol%, respectively, under growth conditions in which P is limiting, while those values are 16.6 and 10.3 mol% under P-replete condition (Guler et al., 1996).

The *C. reinhardtii* genome contains a single gene for UDP-sulfoquinovose synthase, *SQD1* (AB116936, homolog of *sqdB*), which attaches SO_4^{2-} to UDP-glucose to generate the sulfoqui-novose moiety. The SQD1 protein has a putative plastid transit peptide at its amino-teminus and a phylogenetic analysis places the proteins into two distinct groups (Sato et al., 2003). One group is represented by the *Synechococcus* sp. PCC7942 protein, which shows strong similarity to the pro-tein in anoxygenic photosynthetic bacteria such as *R. sphaeroides*. The second goup contains SQD1 of *Synechocystis* sp. PCC6803, *T. elongates*, *C. reinhardtii* and vascular plants. For the second group of organisms, association of SQDG with PS II has been proven experimentaly; SQDG was shown to be associated with the PS II core and light-harvesting complexes of *C. reinhardtii* (Sato et al., 1995; Sigrist et al., 1988), and also PS II of both *T. elongatus* (Loll et al., 2005) and vascular plants (Gounaris et al., 1985; Kruse et al., 2000; Murata et al., 1990; Trémoliéres et al., 1994). Thus, the molecular structure of SQD1 appears to correlate with the dependency of PS II activity on SQDG (Sato, 2004). The results suggest two lineages of PS II evolution, one in which the photosynthetic apparatus required SQDG for activity and the other in which the sulfolipid could be replaced by other lipids without a marked loss of PS II activity.

The second enzyme involved in sulfolipid biosynthesis, SQD2 (homolog of SqdX), adds the sulfoquinovose to diacylglycerol to form SQDG. In *C. reinhardtii* this activity appears to be encoded by *SQD2a* (e_gwW.1.590.1) and *SQD2b* (e_gwW.45.101.1). The levels of transcripts from the *SQD1* and *SQD2a* genes increase during S starvation (Zhang et al., 2004), which probably helps sustain SQDG synthesis even when S levels are declining (Sugimoto et al., 2005), even though in *C. reinhardtii* there is an accelerated rate of SQDG degradation upon S deprivation (Sugimoto et al., 2005), which may correlate with the decline in photosynthetic electron transport.

Finally, there is a minor sulfolipid species, 2'-O-acyl-sulfoquinovosyldiacylglycerol (ASQD), that has been detected in *C. reinhardtii* cells (Riekhof et al., 2003). The fatty acids in ASQD are predominantly unsaturated, while those in SQDG are mostly saturated. Since the deletion of *SQD1* in *C. reinhardtii* eliminates the synthesis of both SQDG and ASQD, ASQD is likely synthesized by acylation of UDP-sulfoquinovose or SQDG (Riekhof et al., 2003).

B. Sulfated Oligosaccharides

Algae as well as vascular plants are producers of neutral polysaccharides such as starch and cellulose. Most marine algae also accumulate sulfated polysaccharides, which may function to maintain thallus hydration and/or to capture and store specific ions. The character of the algal polysaccharides may typify certain algal groups (Kloareg and Quatrano, 1988). Sulfated polysaccharides, such as agar and carrageenan cover the surfaces of red algal cells (Rhodophyta), filling spaces between the cell wall microfibrils (Craigie, 1990). Agars and carrageenans are 1,3-α-1,4-β-galactans substituted by zero (agarose), one (κ-carrageenan),

two (ι-carrageenan), or three (λ-carrageenan) SO_4^{2-} groups per disaccharidic monomer. These high molecular mass compounds have been used as stabilizer and thickeners in foods, the base for cosmetics and polyelectrolyte films (Schoeler et al., 2006), and for centuries have also been exploited for their anti-infective properties (Fabregas et al., 1999; Hetland, 2003; Huleihel et al., 2002). The unicellular red algae *Porphyridium aerugineum* secretes a gel-like polysaccharide that encapsulates the organism. Pulse-chase experiments have revealed that during the logarithmic phase of growth, the alga actively synthesizes and accumulates sulfated oligosaccharides within the cell body (probably in Golgi bodies). As much as 50% of exogenously supplied $^{35}SO_4^{2-}$ can be incorporated into *P. aeruginoseum* oligosaccharide (Ramus and Groves, 1972). Incorporation of radioactive SO_4^{2-} begins almost immediately after feeding labeled SO_4^{2-} to S-starved cells (Ramus and Groves, 1972), which is congruent with activation of the SO_4^{2-} assimilation pathway during S deprivation and the rapid synthesis of APS and PAPS, the immediate substrates for the sulfation reactions. Pulse-labeling experiments have also been performed to examine carrageenan synthesis in another red algal species, *Chondrus crispus* (Irish moss) (Loewus et al., 1971). Additionally, sulfated oligosaccharides are known to function as signaling molecule that are involved in symbiotic interactions between organisms, as exemplified by the interactions between *Rhizobium meliloti* and legumes (vascular plants) that lead to the production of nodRM1 in the bacterial cells and the subsequent nodulation of the legume roots (Schultze et al., 1992). An example of an algal signaling process that employs S-containing signaling molecules occurs in the interactions between the endophytic green alga *Acrochaete operculata* and *C. crispus*, its red algal host. *A. operculata* secrets enzymes that degrade *C. crispus* cell wall material, releasing carageenan oligosaccharides. These sulfated oligosaccharides lead to further stimulation of *A. operculata* pathogenicity (Bouarab et al., 1999).

C. Thioredoxin and Glutaredoxin

Thioredoxin (TRX) is a family of small heat stable oxidoreductases. TRXs have a conserved active site -Cys-Gly(Ala/Pro)-Pro-Cys- that contains two highly reactive cysteine molecules that can form a disulfide bond with each other, and also catalyze reversible disulfide bond formation in target proteins in a redox sensitive manner; the modulation of protein disulfide bond formation can dramatically alter catalytic activity (Vlamis-Gardikas and Holmgren, 2002). For example, in chloroplasts, excitation of the photosynthetic electron transport system can promote the reduction of TRX by reduced ferredoxin, which in turn can activate the Calvin Cycle by reducing regulatory disulfides present in enzymes required for Calvin Cycle activity; these enzymes include NADP-malate dehydrogenase and fructose 1,6-bisphosphatase (Geigenberger et al., 2005; Lemaire et al., 2005). In *C. reinhardtii*, expression of the TRX was shown to be regulated by heavy metal exposure, light and the circadian clock (Lemaire et al., 1999a; Lemaire and Miginiac-Maslow, 2004; Lemaire et al., 1999b), which reflects the redox function of the molecule.

TRX exists in all-free living organisms and in all of the organelles; TRXf and TRXm are in chloroplasts, TRXh in the cytosol, and TRXo in mitochondria. There are also relatively newly identified members of the TRX family of proteins, including TRXx, TRXy and the *C. reinhardtii* flageller type TRX (Lemaire and Miginiac-Maslow, 2004). Each TRX isoform interacts with a specific reductase that uses a specific electron doner for the reduction reaction; shown in studies with both *C. reinhardtii* (Huppe et al., 1990; Huppe et al., 1991) and vascular plants (Florencio et al., 1988) there is a NADPH-thioredoxin reductase in the cytosol and mitochondria, and a ferredoxin-thioredoxin reductase in chloroplasts.

For the algae, a TRX-like activity was first isolated from *Euglena gracilis* in 1975 (Munavalli et al., 1975), with subsequent isolations from the green algae *Scenedesmus obliguus* (Wagner and Follmann, 1977) and *Chlorella* (Tsang, 1981) and from cyanobacteria (Gleason et al., 1985; Whittaker and Gleason, 1984). As reviewed by Lemaire et al. (Lemaire and Miginiac-Maslow, 2004), the genome sequences of *C. reinhardtii* and *Synechocystis* sp. PCC6803 have revealed eight and four genes encoding TRX, respectively, while the *A. thaliana* genome encodes nineteen TRX polypetides. Generally, the genomes of photosynthetic organisms contain more TRX genes than those of non-photosynthetic organisms. Two of the TRX isoforms in *C. reinhardtii* function to control movement of the alga by modulating flagella action in response to redox conditions, which involves controlling the conformational states of dynein within the flagella (Harrison et al., 2002).

Recent proteomic approaches in which a mutated TRX (the mutant protein lacks one of the redox-active cysteines, allowing the second redox-active cysteine to bind target polypeptides without the subsequent disulfide bond reduction; therefore the TRX does not release the target protein) linked to an affinity column has facilitated identification of TRX target proteins (Balmer et al., 2003). This method has also been applied to studies of cyanobacteria (Lindahl and Florencio, 2003), in which disruption of the *trxA* gene is lethal under both photoautrophic and heterotrophic growth conditions (Navarro and Florencio, 1996). TRXa of *Synechocystis* sp. PCC6803 was found to interact with enzymes involved in various metabolic processes, while similar studies with vascular plants identified different processes; in cyanobacteria both ATP sulfurylase and ferredoxin-dependent sulfite reductase were found to interact with TRXa. In contrast, application of the same methods to *C. reinhardtii* identified targets that were similar to those identified in vascular plants, but also revealed the Ran protein as a TRX target (Lemaire et al., 2004). the RAN protein is a small GTPase involved in the biogenesis of the nucleus and cell cycle control.

In addition to TRX, the superfamily of TRX proteins includes glutaredoxin (GRX), protein disulfide isomerase (PDI), glutathione S-transferase (GST) and GSH peroxidase. These proteins all have the TRX motif and are involved in regulating cell processes in response to cellur redox conditions. There are three and six GRX family genes on the *Synechocystis* sp. PCC6803 and *C. reinhardtii* genomes, respectively, while the *A. thaliana* genome encodes 30 GRX polypeptides (Lemaire, 2004; Lemaire and Miginiac-Maslow, 2004). Like TRX, GRX also appears to be present in all subcellular compartments, although the full scope of its activities is not known. The reduction of GRX by GSH can trigger glutathionylation of target proteins, which in turn may alter their activities (see below).

PDIs, proteins unique to eukaryotes, catalyze the reduction–oxidation or isomerization of protein disulfides in the endoplasmic reticulum (ER), as mostly studied in mammian cells (Buchanan and Balmer, 2005; Ferrari and Soling, 1999; Jordan and Gibbins, 2006; Wilkinson and Gilbert, 2004). The *C. reinhardtii* RB60 protein is a unique PDI that functions in plastids (Trebitsh et al., 2001) to regulate translation of the *psbA* transcript, which encodes the PS II reaction center protein D1 (Danon and Mayfield, 1991). This protein may also localize and function in the ER, which is suggested by the finding that the polypeptide has both a signal sequence and an ER retention sequence, which has been further substantiated by experimentation (Levitan et al., 2005). RB60, which contains two TRX domains (Kim and Mayfield, 2002), regulates disulfide bond formation in the RNA binding domains of RB47 in response to redox conditions, which are probably reported by PS I-dependent reduction of the ferredoxin-TRX system (Trebitsh et al., 2000; Trebitsh et al., 2001) or ADP-dependent phosphorylation in the dark (Danon and Mayfield, 1994a; Danon and Mayfield, 1994b). Reduction of the RB47 disulfide bond changes the binding of RB47 to the 5'UTR of the *psbA* mRNA (Fong et al., 2000; Kim and Mayfield, 1997), providing a light-dependent switch for the translation of *psbA* mRNA (elevated light levels promote translation).

D.   Glutathione/Phytochelatin

GSH (γ-Glu-Cys-Gly), which is extremely sensitive to the oxidative state of the cell (Meyer and Hell, 2005), may represent a family of functionally distinct peptides. It is known to accumulate to relatively high concentrations (mM range) in the cytoplasm of many photosynthetic cells, and may function as a non-protein, storage form of reduced S. Glutathione S-transferases (GSTs) catalyze the conjugation of GSH to target molecules, either xenobiotics or endogenous molecules, promoting deposition of these molecules in the vacuole (or in animals, it can cause expulsion of the molecule from the cell) (Marrs, 1996). GSH is known to serve as an electron donor in SO_4^{2-} assimilation (two electron reduction of APS to generate SO_3^{2-}) and in the ROS scavenging system, which is dependent upon reduced ascorbate (Noctor and Foyer, 1998). It also serves as a redox buffer, stabilizing redox homeostasis, based on its interconversion

between the reduced GSH and oxidized GSSG forms. Changes in the redox state of the cell can also elicit oxidation/reduction of thiol groups of the cysteine residues of proteins, which can lead to conformation changes in those proteins that alter their activity. Hence, GSH can serve as a redox sensor that modulates protein activity through redox buffering or by directly promoting the reduction of cysteine thiols.

One example of a critical and novel GSH/GSSG controlled process in photosynthetic organisms concerns the arrest of chloroplast translation, discovered for the large subunit (LSU) of Rubisco in *C. reinhardtii*, during exposure of the cells to high light conditions. Decreased LSU synthesis was observed within a few minutes of shifting low-light acclimated cells to high light (Shapira et al., 1997). Photo-oxidative stress through the generation of ROS and a resulting higher ratio of GSSG to GSH in the chloroplast was found to be the cause of the transient translation arrest (Irihimovitch and Shapira, 2000). Structural changes resulting from the interactions between LSU and GSH cause exposure of an RNA recognition motif at the amino-terminus of LSU; this motif is buried in the holoenzyme (or not accessible to RNA in the LSU homodimer) under low light condition. The LSU surface-exposed amino-terminal region, which shows homology to an RNA binding domain, binds to the 5'UTR of *LSU* mRNA to induce translational stalling (Cohen et al., 2005; Cohen et al., 2006; Yosef et al., 2004). Recently it was shown that this mechanism is conserved in LSU in a range of photosynthetic organisms from the purple bacteria to land plants (Cohen et al., 2006).

As many soils and bodies of water are contaminated with heavy metals such as cadmium (Cd) and lead (Pb), the rapid development of systems for bioremediation has become imperative. Plants and yeast are known to enhance synthesis of GSH and its derivative phytochelatin (PC) in response to elevated heavy metal levels; these molecules help detoxify the heavy metals through chelation followed by sequestration or expulsion (Gekeler et al., 1988; Howe and Merchant, 1992). Relationships between resistance to heavy metals, mainly Cd, and the accumulation of GSH/PC has been studied in several fresh water alga including *Chlorella* (El-Naggar and El-Sheekh, 1998; Kaplan et al., 1995), *Euglena gracilis* (Aviles et al., 2005; Watanabe and Suzuki, 2004), *Scenedesmus vacuolatus* and *S. acutus* (Le Faucheur et al., 2005; Torricelli et al., 2004), the marine algae *D. tertiolecta* (Tsuji et al., 2003) and *Tetraselmis suecica* (Perez-Rama et al., 2006), the marine macroalgae *Enteromorpha prolifera* and *E. linza* (Malea et al., 2006), the marine diatoms *Ditylum brightwellii* and *T. pseudonana* (Rijstenbil et al., 1994) and the model alga *C. reinhardtii* (Nagel et al., 1996; Nagel and Voigt, 1995).

The toxic heavy metal, Cd, has been closely associated with studies of PCs; this metal has been shown to accumulate in chloroplasts of *C. reinhardtii* (Nagel et al., 1996; Nagel and Voigt, 1995), and to inhibit photosynthesis (Faller et al., 2005). GSH and PC were found to be chelators of Cd in *C. reinhardtii* (Hu et al., 2001). The initial response of the diatom *Phaeodactylum tricornutum* to Cd exposure was to induce the synthesis of PC (Morelli and Scarano, 2001) and to generate Cd-GSH complexes; the Cd in these complexes was transferred from GSH to PC to form stable, high molecular mass Cd-PC complexes (Morelli et al., 2002). Strong heavy metal tolerance was observed for *Chlamydomonas acidophila*. Interestingly, the levels of PC in the cells did not correlate with the level of Cd accumulation, while the levels of GSH and γ-ECS activity appeared more important to both accumulation and resistance to Cd. Furthermore, a *C. acidophila* strain that died following exposure to Cd exhibited starch granule accumulation and disruption of cellular membranes; the more resistant *C. acidophila* strain accumulated little starch (with little observed disruption of cellular membranes) (Nishikawa et al., 2006). Hence, mechanisms that control photosynthetic activity, the sensitivity of photosynthesis to heavy metals, and the structure of the photosynthetic machinery are probably all critical for heavy metal tolerance.

In the marine green *alga D. tertiolecta*, more pronounced synthesis of PC was elicited by Zn than by Cd, which contrasts with results from vascular plants in which Cd elicits a stronger response than Zn. This response was a consequence of Zn-triggered induction of γ-ECS and GS, but not of PCS (Tsuji et al., 2003). Furthermore, a higher GSH/GSSG ratio in *C. reinhardtii* cells was achieved by decreasing levels of free radicals through the introduction of the gene for $\Delta(1)$-pyrroline-5-carboxylate synthetase (stimulates proline synthesis); the resulting Cd toler-

ance was through enhanced Cd chelation by PC (Siripornadulsil et al., 2002). These results suggest that the mechanism for Cd torelance depends on increased synthesis and accumulation of GSH rather than PC, even though PC would still be required for detoxification. Furthermore, it was recently shown that Cd treatment of *C. reinhardtii* cells enabled them to better cope with arsenate toxicity (Kobayashi et al., 2006), probably because of elevated GSH/PC accumulation resulting from increased SO_4^{2-} uptake and cysteine synthase activity that led to a 7.5% increase in the S content of cells and an enhanced supply of glutamate, which is essential for PC synthesis (Domínguez et al., 2003).

Glutathionylation of proteins, mostly studied in mammalian cells, controls the activities of a number of specific target proteins. The reaction involves the formation of a disulfide bond between GSH and the cysteinyl thiol group in proteins. Over the last three years targets for glutathionylation have been identified in *A thaliana* (Ito et al., 2003) as well as *C. reinhardtii* (Michelet et al., 2005). TRX had been known to be a target of glutathionylation in humans (Casagrande et al., 2002). Recently, TRXfs, isoforms of TRX that function in chloroplasts, were shown to undergo glutathionylation, which in turn prevents reduction of ferredoxin-TRX reductase and the subsequent activation of TRXf target proteins (Michelet et al., 2005). These findings demonstrate crosstalk between two redox systems.

Glutathione peroxidase (GPX), an enzyme critical for alleviating detrimental effects of oxidative stress, and especially the peroxidation of lipids, catalyzes the oxidation of GSH to GSSG coupled to the reduction of hydrogen peroxide to water. Steroid and lipid hydroperoxides can also act as substrates for the enzyme. *Synechocystis* sp. PCC6803 possesses GPXs (Gpx-1, Gpx-2) which use NADPH and not glutathione as the electron donor, and unsaturated fatty acid hydroperoxides and alkyl hydroperoxides as preferred electron acceptors (Gaber et al., 2001). Accumulation of the *Gpx-1* and *Gpx-2* transcripts and polypeptides occurs in response to oxidative or salt stress, and the disruption of each gene results in elevated accumulation of lipid hydroperoxide and a decrease in O_2 evolution, suggesting that these GPX proteins help protect membrane integrity and the stability of the photosynthetic apparatus when the cells are experiencing oxidative stresses (Gaber et al., 2004).

The *C. reinhardtii* genome encodes at least six GPX polypeptides, some of which contain selenocysteine in their catalytic site (see below). GPX activity was only elevated in cells grown in medium containing selenium, and the rise in this GPX activity correlated with a loss of ascorbic acid peroxidase activity (Yokota et al., 1988). The *GPXH* gene (AF014927) of *C. reinhardtii* has been shown to be transcriptionally-activated in cells exposed to singlet oxygen (Leisinger et al., 2001); transcripts from this gene also accumulate in cells exposed to S-deprivation (Zhang et al., 2004). Furthermore, *GPX* was identified as a gene involved in halotolerance in the marine green algae *C. reinhardtii* W80 (a halotolerant strain). This GPX does not contain selenocysteine and could use fatty acid hydroperoxides, but not H_2O_2 or phospholipid hydroperoxide, as substrates (Takeda et al., 2003). Introduction of the *GPX* gene of *C. reinhardtii* W80 into either the nuclear or chloroplast genomes of tobacco resulted in transgenic plants with increased tolerance to conditions that would cause oxidative stress. The membrane structure in these transgenic plants is probably stabilized since, upon chilling or exposure to elevated salt conditions, these transgenic lines exhibited reduced lipid hydroperoxidation and elevated photosynthetic capacity relative to nontransgenic lines (Yoshimura et al., 2004).

E.   DMSP and DMS

Marine algae accumulate concentrations of DMSP, probably in the cytoplasm of the cell, that range from 1 to 460 mM (Reed, 1983). DMSP was demonstrated to accumulate in chloroplast of the plant *Wollastonia biflora* (Trossat et al., 1998). The functions of DMSP have not been clearly established, although it, as well as its breakdown products, probably participates in a variety of processes. Levels and patterns of intracellular DMSP accumulation over the life cycle of algae can be very different in different species (Matrai and Keller, 1994), suggesting potential differences in the role of DMSP among algal species. DMSP accumulates in polar algae, where it has been proposed to serve as a cryoprotectant (Kirst et al., 1991). It has been suggested that DMSP has osmoprotectant activity (Dickson and Kirst, 1987) and can replace the N-containing osmolyte glycine betaine in N-limited marine environments (Stefels, 2000). This hypothesis is supported by the finding that supplementation of growth media with DMSP rescued the high osmotic pressure-resistant growth of an *E. coli* mutant with increased sensitivity to osmotic conditions (Summers et al., 1998), and also by reports demonstrating a negative correlation between available N levels and DMSP accumulation in various marine algae (Keller et al., 1999). In laboratory experiments, DMSP was found to function as an antioxidant, scavenging harmful hydroxyl radicals generated as a consequence of photosynthetic electron transport (Sunda et al., 2002). Furthermore, DMSP accumulation in five different antarctic green algae was positively correlated with both daylength and light intensity (Karsten et al., 1990), and its accumulation in *T. pseudonana* and *E. huxleyi* (coccolithophore) was enhanced when these organisms experienced CO_2 or Fe deprivation (Sunda et al., 2002). The actual data for seasonal changes of DMS emission from the oceans (Dacey et al., 1998; Simo and Pedros-Alio, 1999) also supports the idea that DMSP functions as an antioxidant. The breakdown products of DMSP, including acrylate, DMS, DMSO (oxidized form DMS), and methane sulfuric acid (MSA) were found to efficiently scavenge hydroxyl radicals (as efficient as GSH) (Sunda et al., 2002). DMS may also have a protective role; it is an uncharged molecule that can readily penetrate cellular membranes and may be responsible for protecting membrane complexes/proteins from the potential damaging consequences of oxidative reactions.

Acrylate has been proposed to have antimicrobial activity (Groene, 1995; Sieburth, 1960; Stefels, 2000). Enzymatic cleavage of secreted DMSP into DMS and acrylate by algal DMSP lyase occurs mostly when the enzyme and substrate leak from the cell due to cell breakage. This finding has raised the possibility that DMSP and DMSP lyase in marine algae may function as part of a grazing-activated defence system (Wolfe et al., 1997).

F.   S-Containing Cofactors

Two very important S-containing cofactors are thiamine, which is required for the activity of various enzymes including pyruvate dehydrogenase, α-ketoglutarate dehydrogenase and transketolase, and S-adenosylmethionine, which is the methyl donor for numerous methylation reactions.

IV. Selenocysteine Metabolism

Selenium (Se) has properties that are similar to that of S, both in the forms in which it exists and in its molecular interactions; like S, Se can have oxidation states of $+6$, $+4$ and -2. While Se is considered an essential trace element in animals, excess accumulation of Se and Se compounds (e.g. H_2Se) can be highly toxic. The biological activities of Se are also similar to that of S. Selenate and selenite can be imported into cells via SO_4^{2-} transporters and incorporated into organic molecules by enzymes of the SO_4^{2-} assimilation pathway; selenocysteine, which can be incorporated into protein, is a major end product of Se assimilation (Lauchli, 1993). Both free Se and Se incorporated into proteins can be toxic to vascular plants at relatively low levels. Some plants exhibit high Se tolerance and have the capacity for hyeraccumulation of Se. A key enzyme involved in Se tolerance and hyperaccumulation is selenocysteine methyltransferase (SMT) (LeDuc et al., 2004; Lyi et al., 2005). SMT converts selenocysteine to methylselenocysteine, and the latter molecule is probably less toxic than the former since it cannot be incorporated into proteins. Furthermore, methylselenocysteine is a substrate for reactions that produce volatile seleno-compound. The accumulation of methylselenocysteine in *Allium* and *Brassica* may be beneficial to the human diet since this compound has anticarcinogenic activity (Dashwood, 1998).

Se-containing proteins have been identified in bacteria, archaea and eukaryotes (Gladyshev and Kryukov, 2001), although there are some interesting divergences in selenoprotein distribution in eukaryote lineages. Some eukaryotic genomes, such as those of *A. thaliana* and *S. cerevisiae*, lack genes encoding selenoprotein, while a number of selenoproteins are encoded on the genomes of animals and lower eukaryotes, including *C. reinhardtii*. Phylogenetic analyses of selenoproteins suggest that they evolved before the separation of the animal and plant kingdoms, but were lost in some lineages, including those that developed into vascular plants (Novoselov et al., 2002). Ten selenoproteins have been identified in *C. rein-*

hardtii (Novoselov et al., 2002), many of which have animal homologs. At least four of these selenoproteins, a TRX reductase, two GPXs and a methionine S-sulfoxide reductase, have antioxidant function. The *C. reinhardtii* thioredoxin reductase, designated TR1 (AAN32903), is probably located in the cytosol and has a selenocysteine-containing active site in its carboxy-terminal flexible tail (this has been observed in homologs from other organisms). Maintaining a selenocysteine active site may confer a broader pH optimim and substrate specificity to the enzyme than is characteristic of cysteine-type thioredoxin reductases (Gromer et al., 2003). A selenocysteine is also part of the catalytic site of GPX; the enzyme with a selenocyteine-based active site, typical of animal GPX, has a much higher activity than the enzyme that has cysteine-based active site, typical of plant GPX (Beeor-Tzahar et al., 1995). The *C. reinhardtii* genome appears to contain at least five genes encoding GPX, two of which contain selenocysteine, PHGPx1 (AY051144) (Fu et al., 2002) and PHGPx1 (EST clone P83564); in the other GPX proteins the selenocysteine has been replaced by cysteine. However, the development of resistance to oxidative stress in *C. reinhardtii* was not observed to be dependent upon Se, possibly because this alga has four cysteine-type GPX polypeptides (Novoselov et al., 2002).

V. Adaptation and Acclimation to S Deficiency

Freshwater and terrestrial habitats can be limiting for S (lake sediment may have a SO_4^{2-} concentration of 200 μM), while S is generally abundant in oceans (~28 mM). Furthermore, there is little storage of S in most organisms and, unlike the situation for plants, unicellular algae cannot cope with S deprivation by redistributing S between organs or tissues. However, algae have developed diverse mechanisms for optimizing S utilization and for tuning cellular metabolism to S availability; these processes help alga sustain viability even when nutrients may be severely limiting. A number of algal and cyanobacterial cells have been shown to be very sensitive to S levels in the environment, exhibiting rapid acclimation responses that can be observed at the levels of pigmentation (Collier and Grossman, 1992), cell wall ultrastructure (Jensen and Rachlin, 1984), the production of specific enzymes (de Hostos et al., 1988; Quisel et al., 1996), and changes in transport characteristics (Yildiz et al., 1994), metabolic profiles (Bolling and Fiehn, 2005) and metabolic activities (Wykoff et al., 1998).

Algae, and especially single-celled algae, offer a number of advantages as subjects for studies of nutrient deprivation. These organisms can often be grown on defined medium, and transferred to medium with a different nutrient composition within minutes. Cultures of single-celled algae can be synchronized and rapidly assayed for biological activities. They do not have the complexity of macrophytic organisms, which are composed of different cell, tissue and organ types, with the potential to redistribute metabolites among the various organs. Furthermore, genetic studies using algae such as *C. reinhardtii* have demonstrated the ease with which mutant strains abberant for their responses to nutrient deprivation can be isolated and analyzed. Several *C. reinhardtii* mutants that do not properly acclimate to S-deprivation have already been identified, and characterizations of these mutants have facilitated identification of novel regulatory genes. This section describes the adaptations of algae to the S environment, and the ways in which they can acclimate when S levels decline.

A. Cell Growth and Division

D. salina grown under limiting S conditions appears to have a smaller volume, by a factor of −2.4, than cells maintained on S-replete medium (Giordano et al., 2000). In contrast, *C. reinhardtii* cells stop dividing soon after exposure to S deprivation but increase in size (the cells fill with starch) (Zhang et al., 2002). The cells remain viable in this 'low metabolic state' for as long as two weeks (Grossman, unpublished).

B. Changes Extracellular Proteins

Profiles of *C. reinhardtii* extracellular polypeptides change dramatically upon either S or P starvation (de Hostos et al., 1989; Takahashi et al., 2001); some polypeptides disappear while others accumulate. Two prominent Extra-Cellular Polypeptides, ECP76 (76 kDa; AF359251) and ECP88 (88 kDa; AF359252) are synthesized specifically during S starvation (Takahashi et al., 2001). The genes encoding ECP76 and ECP88 are regulated at the level of transcription; they appear to be rapidly activated when S is eliminated from the growth medium. Furthermore, the ECP76 and ECP88 mRNAs are rapidly degraded once S-deprived cells are administered adequate levels of SO_4^{2-} (the half-lives of these mRNAs under both S-replete and S-deprivation conditions was < 10 min). The ECP76 and ECP88 polypeptides are 39% identical and have features of cell wall proteins, but contain zero and one S-containing amino acid (in the mature protein), respectively. These results suggest that highly regulated processes tailors the protein-rich cell wall of *C. reinhardtii* for S deprivation, recycling S in cell wall proteins for other cellular activities and perhaps also modifying the cell wall for limited expansion/division. Scavenging S-amino acids from proteins is also a strategy used by other organisms including *Saccharomyces cerevisiae* (Baudouin-Cornu et al., 2001) and vascular plants (soybean seed storage protein β-conglycinin, β-subunit) (Coates et al., 1985; Gayler and Sykes, 1985).

Interestingly, metabolic profiling has revealed that during S-deprivation, *C. reinhardtii* accumulates more than 50 times more 4-hydroxyproline and five times more glycerol than unstarved cells, while the levels of these compounds do not change during Fe, P or N deprivation (Bolling and Fiehn, 2005). The increased hydroxyproline level may reflect the restructuring of the cell wall.

C. Arylsulfatase and Hydrolytic Activity

The SO_4^{2-} anion in the soil is often complexed with cations as insoluble salts and may also tightly adhere to the surface of soil particles. Furthermore, a large proportion of soil SO_4^{2-} may be covalently bonded to organic molecules in the form of SO_4^{2-} esters and sulfonates. *C. reinhardtii*, along with other soil microbes, has developed a number of strategies to scavenge SO_4^{2-} from both internal and external S pools as environmental S concentrations become limiting. One of the extracellular polypeptides secreted by *C. reinhardtii* in response to S starvation is an arylsulfatase (ARS) (de Hostos et al., 1989; Lien and Schreiner, 1975; Schreiner et al., 1975); the production of ARS is analogous to the production of periplasmic carbonic anhydrase or periplasmic phosphatases when algal cells are starved for inorganic carbon or P, respectively. The predicted amino acid sequences of ARS1

(X52304) (de Hostos et al., 1988; de Hostos et al., 1989) and ARS2 (AF333184) (Davies et al., 1992; Ravina et al., 2002) are 99% identical and transcripts from both genes appear to accumulate with similar kinetics upon exposure of cells to S-limiting conditions (Eberhard et al., 2006; Ravina et al., 2002; Zhang et al., 2004). The induction of *ARS1* is sensitive to general kinase inhibitors, to some extent (Irihimovitch and Stern, 2006), suggesting that ARS regulation involves a phosphorelay (Pollock et al., 2005) (Gonzalez-Ballester and Grossman, unpublished). The *ARS1* and *ARS2* genes are arranged in tandem on the *C. reinhardtii* genome, although, based on sequence similarity, there are as many as 16 other *ARS* gene family members.

D. SO₄²⁻ Transport

As in a number of other organisms (bacteria, fungi and vascular plants), algae develop high affinity and/or high capacity SO_4^{2-} transport systems that enable them to better cope with conditions of S starvation. The capacity for SO_4^{2-} uptake by *Synechococcus* sp. PCC7942 increases 10–20-fold during the first 24 h of S-deprivation, but there is no increase in the affinity of the cell for SO_4^{2-}. These results suggest that there is no induction of a higher affinity SO_4^{2-} transporter when the cells are deprived of S, but that the number of active transporter molecules increases (Green and Grossman, 1988; Jeanjean and Broda, 1977). More recent work has demonstrated that the increase in SO_4^{2-} transport was a consequence of increased transcription from a cluster of genes encoding a bacterial type SO_4^{2-} transporter (Laudenbach and Grossman, 1991). In the case of *C. reinhardtii*, Yildiz et al (Yildiz et al., 1994) showed that S starvation elicited changes in SO_4^{2-} transport activity; there was an increase in the V_{max} and a decrease in the K_m (increase in affinity) associated with whole cell SO_4^{2-} transport. We have recently demonstrated that the nuclear genome of *C. reinhardtii* encodes six putative SO_4^{2-} transporters, three resembling the H^+/SO_4^{2-} transporters of vascular plants and three resembling the Na^+/SO_4^{2-} transporters of animals. The transcripts for three of the putative transporters accumulate to high levels during S deprivation (Pootakham and Grossman, unpublished).

E. SO₄²⁻ Assimilation

In many organisms the activity of the SO_4^{2-} assimilation pathway increases upon S deprivation, and the abundance of many S compounds in cells may drop to below detectable levels (Bolling and Fiehn, 2005). For example, the activity of ATS, which catalyzes the activation of SO_4^{2-} for subsequent assimilation, increased fourfold in *D. salina salina* when the alga were deprived of S (Giordano et al., 2000). This increase is partly a consequence of elevated transcript accumulation, as was previously demonstrated for *C. reinhardtii* (Ravina et al., 2002; Yildiz et al., 1996; Zhang et al., 2004). Both *SAT1* and *Cys1ACr* (*OASTL4*) mRNAs accumulate in *C. reinhardtii* in response to S starvation (Zhang et al., 2004); a corresponding increase in total OASTL activity was also demonstrated (Ravina et al., 1999). The levels of transcripts encoding APR and SIR also increased during S deprivation, although the changes appear to be less pronounced than those observed for *ATS, SAT, OASTL* and *SQD* transcripts (Zhang et al., 2004). Furthermore, the pattern of *APR1* transcript abundance appears different from the other assimilation genes in that it peaks between 2 and 4 h following the imposition of S deprivation conditions, and then begins to decline. In contrast, there is a sustained increase in APR activity in cellular extracts derived from S-deprived cells (Ravina et al., 2002). These results raise the possibility that the level of APR1 activity may be controlled, at least in part, at the posttranslational level. There are even larger changes in levels of transcripts encoding SAC1-like proteins (resemble $Na^+/$ SO_4^{2-} transporters), ARS and ECP76 in response to S limitation. There are also some temporal distinctions in the kinetics of mRNA accumulation for the different S-regulated genes; *ECP76* and *ECP88* mRNAs begin to accumulate at a later time following removal of SO_4^{2-} from the medium than transcripts for ARS and the putative SO_4^{2-} transporters (Takahashi et al., 2001; Zhang et al., 2004). Figure 4 shows the kinetics of accumulation of some transcripts during acclimation of *C. reinhardtii* to S deprivation (Zhang et al., 2004). Interestingly, mRNAs for ARS2, ATS1, ECP76 and ECP88 (Takahashi et al., 2001) all have very short half-lives, suggesting stringent control of transcript levels, especially during periods in which the S concentrations in the environment are

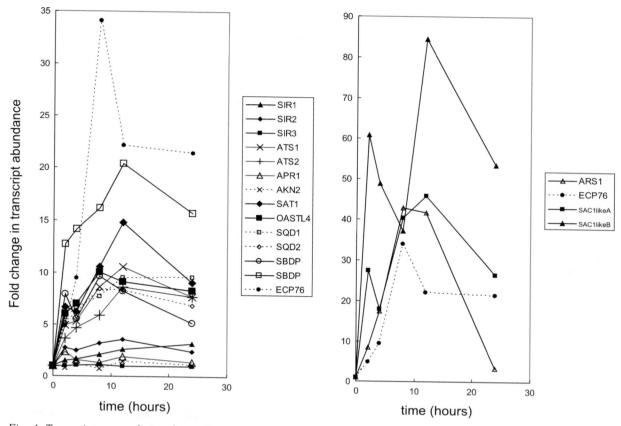

Fig. 4. Transcript accumulation during S starvation. All of the data shown in this figure are from microarray experiments reported by Zhang et al (2004). The experiments were performed with total RNA from *C. reinhardtii* cells (strain CC425) grown in medium lacking S for the times indicated (x-axis; hours). Transcript abundance relative to the control (time point zero) is shown for selected genes related to S assimilation. The number values obtained for the *SAC1*-like gene A is a mixture of two *SAC1*-like genes located in tandem (e_gwW.10.232.1 and gwH.10.31.1) as well as *SAC1*-like gene B (gwH.11.15.1).

fluctuating. Generally, the findings discussed above demonstrate that the responses of *C. reinhardtii* to S-deprivation involve a temporal program of gene activations, and that transcript turnover rates are also critical for rapidly eliminating activities once S is introduced back into the environment.

F. Metabolic Changes Elicited by S Deprivation

Algal cells experiencing S limited growth, slow their anabolic processes and achieve a new ion/metabolite homeostasis. For example, when *D. salina* is starved for S, nitrate reductase and phosphoenolpyruvate carboxylase activities declined by 4- and 11-fold, respectively (Giordano et al., 2000). There is also a marked decrease in photosynthesis (Wykoff et

al., 1998), a loss of pigmentation and a reduction in the levels of transcripts encoding proteins of the photosynthetic apparatus during both N- and S-deprivation (Plumley and Schmidt, 1989; Zhang et al., 2004).

1. Down-Regulating Photosynthesis

Nutrient deprivation often leads to a reduction in the pigments associated with photosynthetic function. *Synechococcus* sp. PCC 7942 loses most of its photosynthetic pigments, which include phycocyanin, allophycocyanin and chlorophyll, during N or S deprivation; this is accompanied by a loss of thylakoid membranes (Collier and Grossman, 1992; Collier and Grossman, 1994; van Waasbergen et al., 2002; Yamanaka et al., 1980). When *D. salina* is limited for S, the maximal photosynthetic rate and the affinity of the cell for CO_2 declined to one-third of that of cells grown in S-replete medium (Giordano et al., 2000). In *C. reinhardtii*, S deprivation inhibits chloroplast protein synthesis, which in turn prevents the repair of PS II by blocking the turnover of damaged D1 reaction-center polypeptides (Wykoff et al., 1998). The decline in PS II activity has been shown to be essential for survival of cells during S deficiency since a specific mutant (*sac1*, see section VI) is unable to inactivate PS II and dies much more quickly than S-starved wild-type cells (Davies et al., 1994). This result indicates that it is absolutely critical to regulate photosynthetic activity during S limitation.

2. Accumulation of Starch

S-starved *C. reinhardtii* cells accumulate high levels of starch (Ball et al., 1990; Zhang et al., 2002). The accumulation of transcripts encoding granule-bound starch synthase I (STA2, AAL28128) increases rapidly, by up to sixfold, upon S deprivation; this increase is less in the *sac1* mutant (Zhang et al., 2004), suggesting that starch accumulation during S deprivation may be regulated, at least in part, by transcript accumulation that is either a direct or indirect response to S deprivation. Starch accumulation in *C. reinhardtii* is not exclusively an S deprivation response; it has been observed to occur under many other conditions, including during N and P starvation (Ball et al., 1990; Klein, 1987; Libessart et al., 1995; Matagne et al., 1976), following exposure to Cd (Aguilera and Amils, 2005; Visviki and Rachlin, 1994), and in flagellar mutants (Hamilton et al., 1992). A uniting feature of these results is the inverse correlation between the storage of chemical bond energy in a form of starch and energy utilization for growth and/or motility.

3. H_2 Production in S-Starved Cells

Because of an increased demand for energy and the need for society to move away from fossil fuel-based sources (since they have been a direct cause of elevated atmospheric CO_2 levels and global warming), several researchers have focused on the bio-production of H_2, a phenomenon observed over one-half century ago by Gaffron and Rubin (Gaffron and Rubin, 1942). Recently, there has

been considerable examination of the economic feasibility of building a bio-factory for the production of H_2 (Ghirardi et al., 2000; Melis and Happe, 2001; Melis and Happe, 2004; Melis et al., 2004). Light-dependent H_2 production involves the excitation of the photosystems and the transfer of electrons either from H_2O or polysaccharides to ferredoxin; reduced ferredoxin is used by the Fe-hydrogenase to reduce protons to H_2. Both hydrogenase activity and the expression of the hydrogenase genes (*HYD*) are inhibited by O_2 (Ghirardi et al., 2000).

C. reinhardtii has two hydrogenase genes, *HYD* (AAL23572) and *HYD* (AAL23573). When *C. reinhardtii* was starved for S, the level of PS II-dependent O_2 evolution declined, resulting in the development of anoxia in sealed culture vessels. These conditions triggered the degradation of starch that had accumulated during oxic growth, which was able to serve as a source of electrons for PS II-independent H_2 production (Fouchard et al., 2005; Posewitz et al., 2004b). Furthermore, screening *C. reinhardtii* for strains with a reduced capacity for H_2 production resulted in identification of the novel, radical S-methionine proteins HYDEF (AAS92601) (fusion protein) and HYDG (AAS92602), both of which are essential for hydrogenase biogenesis (Posewitz et al., 2004a). Studies directed toward defining factors controlling starch accumulation and utilization, and modulation of PS II activity as cells experiencing S deprivation may provide directions for genetically tailoring PS II-independent and PS II-dependent H_2 production. Controlling the activities of S acquisition and assimilation may also help modulate the metabolism of the cell for more efficient H_2 production. For example, a strain of *C. reinhardtii* producing less SulP, generated by antisense technology (Chen et al., 2005), appears to be S-deprived when exposed to SO_4^{2-} levels that normally do not elicit S-deprivation responses; the cells have decreased O_2 production as a consequence of a moderate decrease in PS II activity (they may not be able to rapidly repair damaged PS II because low levels of SO_4^{2-} in the chloroplast would lead to lower rates of cysteine/methionine production which would retard translation). This characteristic may be leveraged in a multipronged approach directed toward engineering high level H_2 production (Fouchard et al., 2005; Jo et al., 2006; Zhang and Melis, 2002).

S-starvation conditions have been used to elicit H_2 production in other photosynthetic microbes, including the unicellular cyanobacteria *Gloeocapsa alpicola* and *Synechocystis* sp. PCC6803 (Antal and Lindblad, 2005). Finally, the production of starch during S-deprivation of algal cells can be used to drive H_2 production in other organisms. For example, algae can synthesize lactate from starch during fermentation and excrete the lactate into the medium. The excreted lactate can be used as an electron donor by certain photosynthetic bacteria for light-dependent H_2 production (Kawaguchi et al., 2001).

VI. Regulatory Mechanisms Controlling Responses to Sulfur Deficiency

C. reinhardtii is a haploid alga that has been used for numerous genetic screens. Since it is a unicell, molecular analyses are not complicated by the presence of specific cell and tissue types, and patterns of gene expression that might be associated with developmental processes. Exploiting specific biochemical and molecular tools has led to the establishment of a powerful screen, based on the induction of extracellular ARS, to identify *C. reinhardtii* mutants with aberrant responses to S deficiency (Davies et al., 1994; Pollock et al., 2005). In this screen, the colorimetric reagent X-SO_4^{2-} is sprayed directly onto *C. reinhardtii* colonies growing under S-deprivation conditions to identify strains unable to synthesize the extracellular ARS (Davies et al., 1994). While several sulfur-acclimation (sac) mutants have been identified, the *sac1* mutant has been most thoroughly characterized so far. A scheme showing some of the major aspects of regulation during S deprivation are shown in Fig. 5.

A. SAC1

SAC1 is a regulatory protein that is required for accumulation of several transcripts encoding proteins involved in S acquisition and assimilation, and for restructuring of both the cell wall and photosynthetic apparatus in response to S-deprivation conditions. Transcripts from *ECP76*, *ECP88*, *ARS1* and *ARS2*, which all increase during S deprivation, show little change in abundance in *sac1* mutants, suggesting that SAC1

controls transcription of these, and many other genes (Ravina et al., 2002; Takahashi et al., 2001; Zhang et al., 2004). However, some transcripts, still accumulate to high levels in the *sac1* mutant during S deprivation (Zhang et al., 2004), indicating that factors in addition to SAC1 control transcript levels during S deprivation.

Another mutant designated *sac3* exhibits constitutive low level ARS activity during S-replete growth (see below). Interestingly, ARS activity during S-replete growth is lower in the *sac1 sac3* double mutant than in the *sac3* strain (Davies et al., 1999), raising the possibility that SAC1 may also influence ARS production/activity even under S-replete conditions. Furthermore, even though APR1 mRNA does not accumulate in a *sac1* mutant during S starvation, APR activity in cell extracts derived from both the mutant and wild-type strains increased following removal of S from the medium (Ravina et al., 2002). These results suggest that post-transcriptional processes participate in the control of APR1 enzymatic activity, and that these processes are not under the control of SAC1. The *sac1* mutant also develops high affinity SO_4^{2-} transport in response to S-deprivation, although the levels of these transporters may be significantly lower than in wild-type cells since the Vmax for SO_4^{2-} uptake increases to a level that is one-third of that of wild-type cells (Davies et al., 1996). Finally, the *sac1* mutants exhibit normal responses when limited for P or N.

The inability of the *sac1* strain to properly respond to S limitation is also reflected in a rapid decline in the viability of mutant cells following exposure to S deprivation; the change in cell viability is sensitive to light levels and photosynthetic electron transport activity. Since the *sac1* strain cannot suppress photosynthetic electron transport activity during S deprivation, the electron transport chain would become hyper-reduced (highly reduced PQ pool), which would provoke the generation of damaging ROS. There are SAC1 homologs encoded on the genome of several other algae, and it is likely that similar regulatory mechanisms control SO_4^{2-} acquisition and assimilation in other photosynthetic organisms.

The SAC1 polypeptide has significant homology to Na^+/SO_4^{2-} type transporters (SLC13 family proteins). TMHMM2 programs predict that SAC1 has 10 transmembrane domains (TMD) with a relatively long intracellular loop between

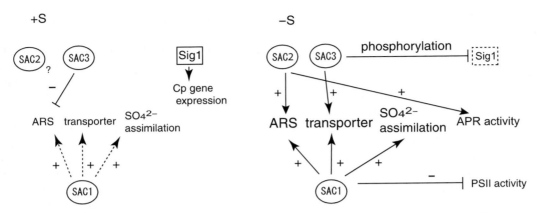

Fig. 5. Regulation of S acclimation in *C. reinhardtii*. Left: Under S-replete conditions (+S), the SAC3 protein negatively regulates *ARS* transcript accumulation. In the *sac3* mutant ARS is expressed, at a low constitutive level, even when sufficient SO_4^{2-} is present in the medium (Davies et al., 1999). Some other genes (those enoding the SO_4^{2-} transporter) may also be negatively regulated by SAC3 under S-sufficient conditions (Gonzalez-Ballester and Grossman, unpublished). SAC1 appears to positively regulate expression of ARS, genes of the SO_4^{2-} assimilation pathway, and those encoding the SO_4^{2-} transporters; in the *sac1* mutant the accumulation of transcripts for *ARS*, the SO_4^{2-} transporters and the S assimilation enzymes are lower than in the wild-type cells grown in S-replete medium (Yildiz et al., 1996,; Zhang, 2004; Pootakham and Grossman, unpublished). Furthermore, transcription of chloroplast (Cp) genes by Sig1 occurs under these (+S) conditions. It is not known if *SAC2* is expressed in + S while *SAC1* and *SAC3* appear to be expressed in + S from their transcript accumulation data and mutant phenotypes. Right: During S-deprivation (−S), the SAC3 protein positively regulates the activity of the transporters; in the *sac3* mutant there is little increase in SO_4^{2-} uptake capacity (although the affinity of the cells for SO_4^{2-} increases) (Davies et al., 1994; Davies et al., 1999). SAC3 also appears to elicit degradation of Sig1 through its kinase activity (probably by an indirect mechanism) and Cp gene expression is not maintained during S starvation; in the *sac3* mutant, Sig1 levels remain high and there is no inhibition of Cp gene transcription (Irihimovitch and Stern, 2006). SAC1 positively regulates the accumulation of transcripts encoding ARS, SO_4^{2-} transporters and many of the genes encoding enzymes involved in the assimilation of SO_4^{2-}; in the *sac1* mutant there is little accumulation of these transcripts (Davies et al., 1994; Davies et al., 1999; Zhang et al., 2004; Yildiz et al., 1996). However, SAC1 is not the only factor that controls the accumulation of transcripts that increase during S deprivation since some genes appear to be properly regulated in the *sac1* mutant. SAC1 is also involved in the down-regulation of photosynthetic activity. The inability of the *sac1* mutant to inactivate photosynthetic electron transport leads to a high light-sensitive phenotype of the mutant strain (Davies et al., 1996). SAC2 positively regulates the activity of APR at the post-transcriptional level and also *ARS* expression; the *sac2* mutants shows no induction of APR activity under −S conditions, although the APR1 transcript level goes up to the same extent as in wild-type cells, while induction of *ARS1* transcript accumulation is less in the *sac2* mutant than in wild type (Davies et al., 1994; Ravina et al., 2002).

the fourth and fifth TMD (Fig. 2). This loop is predicted to contain two tandem TrkA_C motifs. The function of TrkA_C is largely unknown, although it has been suggested to bind to an unidentified ligand (Anantharaman et al., 2001). This motif is also present in bacterial Na^+/H^+ antiporters and animal and plant potassium channels. The *C. reinhardtii* genome has three genes encoding SAC1 like proteins that may be involved in SO_4^{2-} transport, as mentioned in section II.A.

Based on SAC1 structure and the phenotype of *sac1* mutants, this transporter-like polypeptide may function as a sensor that responds to the level of extracellular SO_4^{2-}, analogous to the activity of Snf3 of *Saccharomyces cerevisiae* in sensing extracellular glucose concentrations. Based on this hypothesis, SAC1 is expected to be integral to cytoplasmic membranes, although there is still no experimental evidence that supports this possibility. In vascular plants, responses to S deprivation appear to be regulated by internal S conditions since the genes that respond to S deprivation become active when cells experience S limitation as a consequence of defective SO_4^{2-} transport (when cells are maintained in S-rich medium). Thus plants are likely to have an internal S sensor(s). No *SAC1*-like genes have been associated with sensing extracellular S conditions in vascular plants and, at this point, it is not clear whether or not there are analogous mechanisms in plants and algae for sensing and responding to S starvation.

B.   SAC2

The *sac2* mutants do not secrete active ARS following a shift to S-deficient conditions (Davies et al., 1994). Induction of transcript accumulation in response to S limitation is not impaired in *sac2* mutants, except for those from *ARS1* and *ARS2*. However, the *sac2* mutant does exhibit less APR1 activity as the cells experience limiting S levels, in spite of the fact that the APR1 mRNA accumulates to the same level as in wild-type cells. These results suggest that the function defined by the *sac2* lesion may involve translational or post-translational events that control protein biogenesis/activities. Aberrant protein biogenesis may in some cases feed back to regulate transcript abundance (e.g. possibly the case for *ARS1* and *ARS2*).

C.   SAC3

SAC3 is a putative type 2 SNF1-related serine-threonine kinase (SnRK2) (Davies et al., 1999). The *sac3* mutants accumulate low levels of *ARS1* and *ARS2* transcripts under S-replete conditions (Ravina et al., 2002); the levels of these transcripts still increase in the mutant in response to S limitation. These findings suggest that SAC3 is not involved in controlling the induced gene expression, but may be responsible for repression of the *ARS* genes under S-replete conditions. Interestingly, SAC3 has also been found to be involved in repression of chloroplast transcription through destabilizing of the plastidic transcription factor Sig1 when cells are deprived of SO_4^{2-} (Irihimovitch and Stern, 2006). The phenotype of *sac3* mutants with respect to chloroplast RNA and Sig1 protein accumulation upon S starvation can be mimicked by general kinase inhibitors, suggesting that the influence of SAC3 on plastidic transcription is exerted by its kinase activity. There may be unidentified, linking regulatory elements between SAC3 phosphorylation and plastidic sigma factor regulation. One of the candidates is in a kinase gene similar to *SAC3* which was identified in a screen for mutants that do not synthesize extracellular ARS activity in response to S deprivation (Pollock et al., 2005).

D.   Other Mutants

Recent work has also demonstrated that the acclimation of *C. reinhardtii* to S deprivation involves another serine-threonine kinase (in addition to SAC3), designated ARS11, and putative components that function in secretory processes (Pollock et al., 2005). The ARS11 kinase is required for expression of all genes that have been tested (~10), associated with the response of cells to S starvation (Gonzalez-Ballester, Pollock and Grossman, unpublished). A number of other mutants defective in their response to S deprivation have also been identified, but more work is required to define their precise role in acclimation.

VII.   Interactions between Nutrient Pools

The cell integrates the use of nutrient resources, and changes in the status of one nutrient may influence the way in which the cell allocates other nutrients. For example, when *C. reinhardtii* cells are starved for P, there is degradation of phospholipids while sulpholipid synthesis increases; this shift in lipid biosynthesis probably reflects conservation of cellular P and compensatory replacement of phospholipids by sulfolipids. Furthermore, P deprivation leads to a 25-fold increase in the concentration of cysteine relative to that of unstarved cells (Bolling and Fiehn, 2005). This increase does not appear to result from enhanced SO_4^{2-} assimilation since the transcripts encoding proteins associated with SO_4^{2-} assimilation do not increase (Moseley et al., 2006); it may simply reflect reduced cell growth and maintenance of an elevated cysteine pool which could be marshaled for the synthesis of antioxidants such as the cysteine-containing tripeptide GSH. Interestingly, a failure to acclimate to P starvation as a consequence of a lesion in the *PSR1* gene (Wykoff et al., 1999) results in aberrant expression of genes of the SO_4^{2-} assimilation pathway (Moseley et al., 2006). Unlike wild-type cells, the *psr1* mutant cannot activate expression of genes encoding phosphate (Pi) scavenging enzymes (alkaline phosphatase, Pi transporters) or 'electron valve' proteins during P deprivation (Moseley et al., 2006). These results suggest that when normal acclimation to P deprivation cannot occur, the cells exhibit signs of S deprivation. This may occur because the *psr1* mutant cannot manage the P budget in the cell and ATP levels may dramatically decline leading to an inability of the cells to take up an adequate amount of SO_4^{2-}, causing the cell to initiate S-deprivation responses. If this is the case, it would suggest that S-deprivation can be sensed from inside the cell. Alternatively, an inability to acclimate to P deprivation may lead to elevated oxidative stress and a demand for reduced S in order to synthesize GSH. Whether or not this process is controlled by SAC1 is not known at this point. On the other hand, S starvation appears to result in a relatively high intracellular P abundance. S starved *D. salina* exhibits elevated accumulation of Pi as well as nitrate and ammonium (Giordano et al., 2000). S-starved *C. reinhardtii* accumulates a number of phosphorylated metabolite intermediates (Bolling and Fiehn, 2005), and decreased sulpholipid synthesis is compensated for by increasing the levels of phopholipids (see section III.A). It has been long known that S-starvation causes deposition of condensed Pi granules (polyphosphates) in cyanobacteria (Grillo and Gibson, 1979; Lawry and Jensen, 1979; Lawry and Simon, 1982), heterotrophic bacteria (Harold and Sylvan, 1963; Smith et al., 1954), and the red algae *Rhodella maculate* (Callow and Evans, 1979). *C. reinhardtii* also synthesizes polyphosphate granules (Ruiz et al., 2001).

VIII. Concluding Remarks

C. reinhardtii has been developed as a model organism that can be analyzed using sophisticated tools and technologies that strengthen genetic and molecular studies, including transformation and RNA interference technology, the full sequences for the chloroplast and mitochondrial genomes, targeted gene disruptions of chloroplast genes, random insertion of marker genes into the nuclear genome to generate tagged mutant libraries, a well developed genetic map of the 17 linkage groups (chromosomes) that has been integrated with the physical map, a high representation of cDNA sequences and a nearly complete and significantly annotated full nuclear genome sequence (Grossman, 2005; Grossman et al., 2003; Gutman and Niyogi, 2004; Kathir et al., 2003; Lefebvre and Silflow, 1999; Shrager et al., 2003). High density DNA microarrays (Eberhard et al., 2006; Ledford et al., 2004; Moseley et al., 2006; Zhang et al., 2004) and both proteomic (Stauber and Hippler, 2004; Wagner et al., 2006) and metabolomic (Bolling and Fiehn, 2005) procedures are adding to the ways that we can explore biological process of *C. reinhardtii*. Furthermore, the genomes of a number of other algae, including *Ostreococcus tauri* (Derelle et al., 2006) (http://www.iscb.org/ismb2004/posters/stromATpsb.ugent.be_844.html), *Cyanidioschyzon merolae* (Matsuzaki et al., 2004) (http://merolae.biol.s.u-tokyo.ac.jp/), *Thallasiossira pseudonana* (http://genome.jgi-psf.org/thaps1/thaps1.home.html) and *Phaeodactylum tricornutum* (http://spider.jgi-psf.org/sequencing/DOEprojseqplans.html), have all also been sequenced. This new direction in biology has added an enormous body of information that is helping develop new and better ways to perform map-based cloning of genes, rapid identification of the disrupted sequences in mutant strains and genome-wide analysis of gene expression following changes in environmental conditions and in specific mutants. The exploitation of these new tools is having a profound effect on the quality and extent of information that we have gathered with respect to acclimation responses in photosynthetic organisms and is also providing insights into interactions among the various acclimation responses, the mechanisms involved in these responses, and the differences and similarities by which the various organisms adjust to their dynamic environment. These new technological advances combined with the ever increasing information gleaned from sequences of algal genomes and high coverage cDNA libraries has added depth and excitement to studies focused on dissecting the ways in which photosynthetic organisms sense and respond to their nutrient environment.

Acknowledgements

The authors would like to thank NSF for supporting ARG (award MCB 0235878) to develop the genomics of *C. reinhardtii* and both NS and ARG would like to thank all the students and postdoctorals of the Grossman laboratory who have participated in discussions that have helped improve this article; these include Wirulda Pootakham, David Gonzalez-Ballester, Jeffrey Moseley and Chung-Soon Im.

References

Aguilera, A. and Amils, R. (2005) Tolerance to cadmium in *Chlamydomonas* sp. (Chlorophyta) strains isolated from an extreme acidic environment, the Tinto River (SW, Spain). *Aquat Toxicol*, **75**, 316–329.

Anantharaman, V., Koonin, E.V. and Aravind, L. (2001) Regulatory potential, phyletic distribution and evolution of ancient, intracellular small-molecule-binding domains. *J Mol Biol*, **307**, 1271–1292.

Andreae, M.O. and Crutzen, P.J. (1997) Atmospheric aerosols: biogeochemical sources and role in atmospheric chemistry. *Science*, **276**, 1052–1058.

Andreae, M.O. and Raemdonch, H., 1983. (1983) Dimethyl sulfide in the surface ocean and the marine atmosphere – A global view. *Science*, **221**, 744.

Antal, T.K. and Lindblad, P. (2005) Production of H_2 by sulphur-deprived cells of the unicellular cyanobacteria *Gloeocapsa alpicola* and *Synechocystis* sp. PCC 6803 during dark incubation with methane or at various extracellular pH. *J Appl Microbiol*, **98**, 114–120.

Arz, H.E., Gisselmann, G., Schiffmann, S. and Schwenn, J.D. (1994) A cDNA for adenylyl sulphate (APS)-kinase from *Arabidopsis thaliana*. *Biochim Biophys Acta*, **1218**, 447–452.

Ashida, H., Sawa, Y. and Shibata, H. (2005) Cloning, biochemical and phylogenetic characterizations of gamma-glutamylcysteine synthetase from *Anabaena* sp. PCC 7120. *Plant Cell Physiol*, **46**, 557–562.

Aviles, C., Torres-Marquez, M.E., Mendoza-Cozatl, D. and Moreno-Sanchez, R. (2005) Time-course development of Cd^{2+} hyper-accumulating phenotype in *Euglena gracilis*. *Arch Microbiol*, **184**, 83–92.

Bacic, M.K. and Yoch, D.C. (1998) In vivo characterization of dimethylsulfoniopropionate lyase in the fungus *Fusarium lateritium*. *Appl Environ Microbiol*, **64**, 106–111.

Ball, S.G., Dirick, L., Decq, A., Martiat, J.C. and Matagne, R.F. (1990) Physiology of starch storage in the monocellular alga *Chlamydomonas reinhardtii*. *Plant Sci*, **66**, 1–9.

Balmer, Y., Koller, A., del Val, G., Manieri, W., Schurmann, P. and Buchanan, B.B. (2003) Proteomics gives insight into the regulatory function of chloroplast thioredoxins. *Proc Natl Acad Sci U S A*, **100**, 370–375.

Barnes, I., Becker, K.H. and Patroescu, I. (1994) The tropospheric oxidation of dimethyl sulfide: A new source of carbonyl sulfide. *Geophys Res Lett*, **21**, 2389.

Bates, T.S., Cline, J.D., Gammon, R.H. and Kelly-Hansen, S. (1987) Regional and seasonal variations in the flux of oceanic dimethylsulfide to the atmosphere. *J Geophys Res*, **92**, 2930–2938.

Baudouin-Cornu, P., Surdin-Kerjan, Y., Marliere, P. and Thomas, D. (2001) Molecular evolution of protein atomic composition. *Science*, **293**, 297–300.

Beeor-Tzahar, T., Ben-Hayyim, G., Holland, D. and Faltin, Z. (1995) A stress-associated citrus protein is a distinct plant phospholipid hydroperoxide glutathione peroxidase. *FEBS Lett*, **366**, 151–155.

Benning, C. (1998) Biosynthesis and function of the sulfolipid sulfoquinovosyl diacylglycerol. *Annu Rev Plant Physiol Plant Mol Biol*, **49**, 53–75.

Bick, J.A., Dennis, J.J., Zylstra, G.J., Nowack, J. and Leustek, T. (2000) Identification of a new class of 5′-adenylylsulfate (APS) reductases from sulfate-assimilating bacteria. *J Bacteriol*, **182**, 135–142.

Bolling, C. and Fiehn, O. (2005) Metabolite profiling of *Chlamydomonas reinhardtii* under nutrient deprivation. *Plant Physiol*, **139**, 1995–2005.

Bouarab, K., Potin, P., Correa, J. and Kloareg, B. (1999) Sulfated oligosaccharides mediate the interaction between a marine red alga and its green algal pathogenic endophyte. *Plant Cell*, **11**, 1635–1650.

Buchanan, B.B. and Balmer, Y. (2005) Redox regulation: a broadening horizon. *Annu Rev Plant Biol*, **56**, 187–220.

Buchner, P., Takahashi, H. and Hawkesford, M.J. (2004) Plant sulphate transporters: co-ordination of uptake, intracellular and long-distance transport. *J Exp Bot*, **55**, 1765–1773.

Buetow, D.E. and Buchanan, P.J. (1964) Isolation of mitochondria from *Euglena gracilis*. *Exp Cell Res*, **36**, 204–207.

Callow, M.E. and Evans, L.V. (1979) Polyphosphate accumulation in sulphur-starved cells of *Rhodella maculata*. *Br Phycol J*, **14**, 327–337.

Cantoni, G.L. and Anderson, D.G. (1956) Enzymatic cleavage of dimethylpropiothetin by *Polysiphonia lanosa*. *J Biol Chem*, **222**, 171–177.

Casagrande, S., Bonetto, V., Fratelli, M., Gianazza, E., Eberini, I., Massignan, T., Salmona, M., Chang, G., Holmgren, A. and Ghezzi, P. (2002) Glutathionylation of human thioredoxin: a possible crosstalk between the glutathione and thioredoxin systems. *Proc Natl Acad Sci U S A*, **99**, 9745–9749.

Challenger, F. and Simpson, M.I. (1948) Studies on biological methylation. Part XII. A precursor of dimethyl sulfide evolved by *Polysiphonia fastigiata*. Dimethyl-2-carboxyethyl sulphonium hydroxide and its salts. *J Chem Soc*, **1948**, 1591–1597.

Charlson, R.J., Lovelock, J.E., Andreae, M.O. and Warren, S.G. (1987) Oceanic phytoplankton, atmospheric sulphur, cloud albedo and climate. *Nature*, **326**, 655–661.

Chen, H.C. and Melis, A. (2004) Localization and function of SulP, a nuclear-encoded chloroplast sulfate permease in *Chlamydomonas reinhardtii*. *Planta*, **220**, 198–210.

Chen, H.C., Newton, A.J. and Melis, A. (2005) Role of SulP, a nuclear-encoded chloroplast sulfate permease, in sulfate transport and H2 evolution in *Chlamydomonas reinhardtii*. *Photosynth Res*, **84**, 289–296.

Chen, H.C., Yokthongwattana, K., Newton, A.J. and Melis, A. (2003) SulP, a nuclear gene encoding a putative chloroplast-targeted sulfate permease in *Chlamydomonas reinhardtii*. *Planta*, **218**, 98–106.

Clemens, S. (2006) Evolution and function of phytochelatin synthases. *J Plant Physiol*, **163**, 319–332.

Coates, J.B., Medeiros, J.S., Thanh, V.H. and Nielsen, N.C. (1985) Characterization of the subunits of beta-conglycinin. *Arch Biochem Biophys*, **243**, 184–194.

Cohen, I., Knopf, J.A., Irihimovitch, V. and Shapira, M. (2005) A proposed mechanism for the inhibitory effects of oxidative stress on Rubisco assembly and its subunit expression. *Plant Physiol*, **137**, 738–746.

Cohen, I., Sapir, Y. and Shapira, M. (2006) A conserved mechanism controls translation of Rubisco large subunit in different photosynthetic organisms. *Plant Physiol*, **141**, 1089–1097.

Collier, J.L. and Grossman, A.R. (1992) Chlorosis induced by nutrient deprivation in *Synechococcus* sp. strain PCC 7942: not all bleaching is the same. *J Bacteriol*, **174**, 4718–4726.

Collier, J.L. and Grossman, A.R. (1994) A small polypeptide triggers complete degradation of light-harvesting phycobiliproteins in nutrient-deprived cyanobacteria. *Embo J*, **13**, 1039–1047.

Copley, S.D. and Dhillon, J.K. (2002) Lateral gene transfer and parallel evolution in the history of glutathione biosynthesis genes. *Genome Biol*, **3**, research0025.

Craigie, J. (1990) Cells walls. In Cole, K. and Sheath, R. (eds.), *Biology of the Red Algae*. Cambridge University Press, Cambridge.

Croft, M.T., Lawrence, A.D., Raux-Deery, E., Warren, M.J. and Smith, A.G. (2005) Algae acquire vitamin B12 through a symbiotic relationship with bacteria. *Nature*, **438**, 90–93.

Dacey, J.W.H., Howse, F.A., Michaels, A.F. and Wakeham, S.G. (1998) Temporal variability of dimethylsulfide and dimethylsulfoniopropionate in the Sargasso Sea. *Deep-Sea Research*, **45**, 2085–2104.

Danon, A. and Mayfield, S.P. (1991) Light regulated translational activators: identification of chloroplast gene specific mRNA binding proteins. *Embo J*, **10**, 3993–4001.

Danon, A. and Mayfield, S.P. (1994a) ADP-dependent phosphorylation regulates RNA-binding in vitro: implications in light-modulated translation. *Embo J*, **13**, 2227–2235.

Danon, A. and Mayfield, S.P. (1994b) Light-regulated translation of chloroplast messenger RNAs through redox potential. *Science*, **266**, 1717–1719.

Dashwood, R.H. (1998) Indole-3-carbinol: anticarcinogen or tumor promoter in brassica vegetables? *Chem Biol Interact* **110**, 1–5.

Davies, J.P., Weeks, D.P. and Grossman, A.R. (1992) Expression of the arylsulfatase gene from the ß₂-tubulin promoter in *Chlamydomonas reinhardtii*. *Nucleic Acids Res*, **20**, 2959–2965.

Davies, J.P., Yildiz, F. and Grossman, A.R. (1994) Mutants of chlamydomonas with aberrant responses to sulfur deprivation. *Plant Cell*, **6**, 53–63.

Davies, J.P., Yildiz, F.H. and Grossman, A. (1996) Sac1, a putative regulator that is critical for survival of *Chlamydomonas reinhardtii* during sulfur deprivation. *Embo J*, **15**, 2150–2159.

Davies, J.P., Yildiz, F.H. and Grossman, A.R. (1999) Sac3, an Snf1-like serine/threonine kinase that positively and negatively regulates the responses of Chlamydomonas to sulfur limitation. *Plant Cell*, **11**, 1179–1190.

de Hostos, E.L., Schilling, J. and Grossman, A.R. (1989) Structure and expression of the gene encoding the periplasmic arylsulfatase of *Chlamydomonas reinhardtii*. *Mol Gen Genet*, **218**, 229–239.

de Hostos, E.L., Togasaki, R.K. and Grossman, A.R. (1988) Purification and biosynthesis of a derepressible periplasmic arylsulfatase from *Chlamydomonas reinhardtii*. *J Cell Biol*, **106**, 29–37.

de Souza, M.P. and Yoch, D.C. (1995a) Comparative physiology of dimethyl sulfide production by dimethylsulfoniopropionate lyase in *Pseudomonas doudoroffii* and *Alcaligenes* sp. strain M3A. *Appl Environ Microbiol*, **61**, 3986–3991.

de Souza, M.P. and Yoch, D.C. (1995b) Purification and characterization of dimethylsulfoniopropionate lyase from an *Alcaligenes*-like dimethyl sulfide-producing marine isolate. *Appl Environ Microbiol*, **61**, 21–26.

de Souza, M.P. and Yoch, D.C. (1996) Differential metabolism of dimethylsulphoniopropionate and acrylate in saline and brackish intertidal sediments. *Microbial Ecol*, **31**, 319–330.

Derelle, E., Ferraz, C., Rombauts, S., Rouze, P., Worden, A.Z., Robbens, S., Partensky, F., Degroeve, S., Echeynie, S., Cooke, R., Saeys, Y., Wuyts, J., Jabbari, K., Bowler, C., Panaud, O., Piegu, B., Ball, S.G., Ral, J.P., Bouget, F.Y., Piganeau, G., De Baets, B., Picard, A., Delseny, M., Demaille, J., Van de Peer, Y. and Moreau, H. (2006) Genome analysis of the smallest free-living eukaryote *Ostreococcus tauri* unveils many unique features. *Proc Natl Acad Sci U S A*, **103**, 11647–11652.

Dickson, D.M.J. and Kirst, G.O. (1987) Osmotic adjustment in marine eukaryotic algae: The role of inorganic ions, quaternary ammonium, tertiary sulphonium and carbohydrate solutes. II. Prasinophytes and haptophytes. *New Phytologist*, **106**, 657–666.

Domínguez, M.J., Gutierrez, F., Leon, R., Vilchez, C., Vega, J.M. and Vigara, J. (2003) Cadmium increases the activity levels of glutamate dehydrogenase and cysteine synthase in *Chlamydomonas reinhardtii*. *Plant Physiol Biochem*, **41**, 828–832.

Eberhard, S., Jain, M., Im, C.S., Pollock, S., Shrager, J., Lin, Y., Peek, A.S. and Grossman, A.R. (2006) Generation of an oligonucleotide array for analysis of gene expression in *Chlamydomonas reinhardtii*. *Curr Genet*, **49**, 106–124.

El-Naggar, A.H. and El-Sheekh, M.M. (1998) Abolishing cadmium toxicity in *Chlorella vulgaris* by ascorbic acid, calcium, glucose and reduced glutathione. *Environ Pollution*, **101**, 169–174.

Fabregas, J., Garca, D., Fernandez-Alonso, M., Rocha, A.I., Gomez-Puertas, P., Escribano, J.M., Otero, A. and Coll, J.M. (1999) In vitro inhibition of the replication of haemorrhagic septicaemia virus (VHSV) and African swine fever virus (ASFV) by extracts from marine microalgae. *Antiviral Research*, **44**, 67–73.

Fahey, R.C., Buschbacher, R.M. and Newton, G.L. (1987) The evolution of glutathione metabolism in phototrophic microorganisms. *J Mol Evol*, **25**, 81–88.

Faller, P., Kienzler, K. and Krieger-Liszkay, A. (2005) Mechanism of Cd^{2+} toxicity: Cd^{2+} inhibits photoactivation of Photosystem II by competitive binding to the essential Ca^{2+} site. *Biochim Biophys Acta*, **1706**, 158–164.

Ferrari, D.M. and Soling, H.D. (1999) The protein disulphide-isomerase family: unravelling a string of folds. *Biochem J*, **339 (Pt 1)**, 1–10.

Florencio, F.J., Yee, B.C., Johnson, T.C. and Buchanan, B.B. (1988) An NADP/thioredoxin system in leaves: purification and characterization of NADP-thioredoxin reductase and thioredoxin h from spinach. *Arch Biochem Biophys*, **266**, 496–507.

Fong, C.L., Lentz, A. and Mayfield, S.P. (2000) Disulfide bond formation between RNA binding domains is used to regulate mRNA binding activity of the chloroplast poly(A)-binding protein. *J Biol Chem*, **275**, 8275–8278.

Fouchard, S., Hemschemeier, A., Caruana, A., Pruvost, J., Legrand, J., Happe, T., Peltier, G. and Cournac, L. (2005) Autotrophic and mixotrophic hydrogen photoproduction in sulfur-deprived chlamydomonas cells. *Appl Environ Microbiol*, **71**, 6199–6205.

Fu, L.H., Wang, X.F., Eyal, Y., She, Y.M., Donald, L.J., Standing, K.G. and Ben-Hayyim, G. (2002) A selenoprotein in the plant kingdom. Mass spectrometry confirms that an opal codon (UGA) encodes selenocysteine in *Chlamydomonas reinhardtii* gluththione peroxidase. *J Biol Chem*, **277**, 25983–25991.

Gaber, A., Tamoi, M., Takeda, T., Nakano, Y. and Shigeoka, S. (2001) NADPH-dependent glutathione peroxidase-like proteins (Gpx-1, Gpx-2) reduce unsaturated fatty acid hydroperoxides in *Synechocystis* PCC 6803. *FEBS Lett*, **499**, 32–36.

Gaber, A., Yoshimura, K., Tamoi, M., Takeda, T., Nakano, Y. and Shigeoka, S. (2004) Induction and functional analysis of two reduced nicotinamide adenine dinucleotide phosphate-dependent glutathione peroxidase-like proteins in *Synechocystis* PCC 6803 during the progression of oxidative stress. *Plant Physiol*, **136**, 2855–2861.

Gaffron, H. and Rubin, J. (1942) Fermentation and photochemical production of hydrogen in algae. *J Gen Physiol*, **26**, 219–241.

Gage, D.A., Rhodes, D., Nolte, K.D., Hicks, W.A., Leustek, T., Cooper, A.J. and Hanson, A.D. (1997) A new route for synthesis of dimethylsulphoniopropionate in marine algae. *Nature*, **387**, 891–894.

Gao, Y., Schofield, O.M. and Leustek, T. (2000) Characterization of sulfate assimilation in marine algae focusing on the enzyme 5′-adenylylsulfate reductase. *Plant Physiol*, **123**, 1087–1096.

Gayler, K.R. and Sykes, G.E. (1985) Effects of nutritional stress on the storage proteins of soybeans. *Plant Physiol*, **78**, 582–585.

Geigenberger, P., Kolbe, A. and Tiessen, A. (2005) Redox regulation of carbon storage and partitioning in response to light and sugars. *J Exp Bot*, **56**, 1469–1479.

Gekeler, W., Grill, E., Winnacker, E.-L. and Zenk, M.H. (1988) Algae sequester heavy metals via synthesis of phytochelatin complexes. *Arch Microbiol*, **150**, 197–202.

Ghirardi, M.L., Zhang, L., Lee, J.W., Flynn, T., Seibert, M., Greenbaum, E. and Melis, A. (2000) Microalgae: a green source of renewable H$_2$. *Trends Biotechnol*, **18**, 506–511.

Giordano, M., Pezzoni, V. and Hell, R. (2000) Strategies for the allocation of resources under sulfur limitation in the green alga *Dunaliella salina*. *Plant Physiol*, **124**, 857–864.

Gladyshev, V.N. and Kryukov, G.V. (2001) Evolution of selenocysteine-containing proteins: significance of identification and functional characterization of selenoproteins. *Biofactors*, **14**, 87–92.

Gleason, F.K., Whittaker, M.M., Holmgren, A. and Jornvall, H. (1985) The primary structure of thioredoxin from the filamentous cyanobacterium *Anabaena* sp. 7119. *J Biol Chem*, **260**, 9567–9573.

Goodman, N.S. and Schiff, J.A. (1964) Studies of sulfate utilization by algae. 3. Products formed from sulfate by euglena. *J Protozool*, **11**, 120–127.

Gounaris, K., Whitford, D. and Barber, J. (1985) Isolation and characterization of a photosystem II reaction center lipoprotein complex. *FEBS Lett*, **188**, 68–72.

Green, L.S. and Grossman, A.R. (1988) Changes in sulfate transport characteristics and protein composition of *Anacystis nidulans* R2 during sulfur deprivation. *J Bacteriol*, **170**, 583–587.

Green, L.S., Laudenbach, D.E. and Grossman, A.R. (1989) A region of a cyanobacterial genome required for sulfate transport. *Proc Natl Acad Sci U S A*, **86**, 1949–1953.

Grillo, J.F. and Gibson, J. (1979) Regulation of phosphate accumulation in the unicellular cyanobacterium *Synechococcus*. *J Bacteriol*, **140**, 508–517.

Groene, T. (1995) Biogenic production and consumption of dimethylsulfide (DMS) and dimethylsulfoniopropionate (DMSP) in the marine epipelagic zone: a review. *J Marine Systems*, **6**, 191–209.

Gromer, S., Johansson, L., Bauer, H., Arscott, L.D., Rauch, S., Ballou, D.P., Williams, C.H., Jr., Schirmer, R.H. and Arner, E.S. (2003) Active sites of thioredoxin reductases: why selenoproteins? *Proc Natl Acad Sci U S A*, **100**, 12618–12623.

Gröne, T. and Kirst,G.O (1992) The effect of nitrogen deficiency, methionine and inhibitors of methionine metabolism on the DMSP contents of *Tetraselmis subcordiformis* (Stein). .*Mar Biol*, **112**, 497–503.

Grossman, A. and Takahashi, H. (2001) Macronutrient utilization by photosynthetic eukaryotes and the fabric of interactions. *Annu Rev Plant Physiol Plant Mol Biol*, **52**, 163–210.

Grossman, A.R. (2005) Paths toward algal genomics. *Plant Physiol*, **137**, 410–427.

Grossman, A.R., Harris, E.E., Hauser, C., Lefebvre, P.A., Martinez, D., Rokhsar, D., Shrager, J., Silflow, C.D., Stern, D., Vallon, O. and Zhang, Z. (2003) *Chlamydomonas reinhardtii* at the crossroads of genomics. *Eukaryot Cell*, **2**, 1137–1150.

Guler, S., Seeliger, A., Hartel, H., Renger, G. and Benning, C. (1996) A null mutant of Synechococcus sp. PCC7942 deficient in the sulfolipid sulfoquinovosyl diacylglycerol. *J Biol Chem*, **271**, 7501–7507.

Gutman, B.L. and Niyogi, K.K. (2004) Chlamydomonas and Arabidopsis. A dynamic duo. *Plant Physiol*, **135**, 607–610.

Habicht, K.S., Gade, M., Thamdrup, B., Berg, P. and Canfield, D.E. (2002) Calibration of sulfate levels in the archean ocean. *Science*, **298**, 2372–2374.

Hamilton, B.S., Nakamura, K. and Roncari, D.A. (1992) Accumulation of starch in *Chlamydomonas reinhardtii* flagellar mutants. *Biochem Cell Biol*, **70**, 255–258.

Harada, E., von Roepenack-Lahaye, E. and Clemens, S. (2004) A cyanobacterial protein with similarity to phytochelatin synthases catalyzes the conversion of glutathione to gamma-glutamylcysteine and lacks phytochelatin synthase activity. *Phytochemistry*, **65**, 3179–3185.

Harold, F.M. and Sylvan, S. (1963) Accumulation of inorganic polyphosphate in *Aerobacter aerogenes* II. Environmental control and the role of sulfur compounds. *J Bacteriol*, **86**, 222–231.

Harrison, A., Sakato, M., Tedford, H.W., Benashski, S.E., Patel-King, R.S. and King, S.M. (2002) Redox-based control of the gamma heavy chain ATPase from Chlamydomonas outer arm dynein. *Cell Motil Cytoskeleton*, **52**, 131–143.

Harwood, J.L. and Nicholls, R.G. (1979) The plant sulpholipid– a major component of the sulphur cycle. *Biochem Soc Trans*, **7**, 440–447.

Hetland, G. (2003) Anti-infective action of immuno-modulating polysaccharides (-glucan and *Plantago Major* L. Pectin) against intracellular (*Mycobacteria* sp.) and extracellular (*Streptococcus Pneumoniae* sp.) respiratory pathogens. *Current Med Chem – Anti-Infective Agents*, **2**, 135–146.

Hirata, K., Tsujimoto, Y., Namba, T., Ohta, T., Hirayanagi, N., Miyasaka, H., Zenk, M.H. and Miyamoto, K. (2001) Strong induction of phytochelatin synthesis by zinc in marine green alga, *Dunaliella tertiolecta*. *J Biosci Bioeng*, **92**, 24–29.

Hodson, R.C. and Schiff, J.A. (1971a) Studies of sulfate utilization by algae: 8. The ubiquity of sulfate reduction to thiosulfate. *Plant Physiol*, **47**, 296–299.

Hodson, R.C. and Schiff, J.A. (1971b) Studies of sulfate utilization by algae: 9. Fractionation of a cell-free system from *Chlorella* into two activities necessary for the reduction of adenosine 3′-phosphate 5′-phosphosulfate to acid-volatile radioactivity. *Plant Physiol*, **47**, 300–305.

Hodson, R.C., Schiff, J.A. and Mather, J.P. (1971) Studies of sulfate utilization by algae: 10. Nutritional and enzymatic characterization of *Chlorella* mutants impaired for sulfate utilization. *Plant Physiol*, **47**, 306–311.

Howe, G. and Merchant, S. (1992) Heavy metal-activated synthesis of peptides in *Chlamydomonas reinhardtii*. *Plant Physiol*, **98**, 127–136.

Hu, S., Lau, K.W.K. and Wu, M. (2001) Cadmium sequestration in *Chlamydomonas reinhardtii*. *Plant Sci*, **161**, 987–996.

Huleihel, M., Ishanu, V., Tal, J. and Arad, S. (2002) Activity of *Porphyridium* sp. against herpes simplex viruses in vitro and in vivo. *J Biochem Biophys Methods*, **50**, 189–200.

Huppe, H.C., de Lamotte-Guery, F., Jacquot, J.P. and Buchanan, B.B. (1990) The ferredoxin-thioredoxin system of a green alga, Chlamydomonas reinhardtii: identification and characterization of thioredoxins and ferredoxin-thioredoxin reductase components. *Planta*, **180**, 341–351.

Huppe, H.C., Picaud, A., Buchanan, B.B. and Miginiac-Maslow, M. (1991) Identification of an NADP/thioredoxin system in *Chlamydomonas reinhardtii*. *Planta*, **186**, 115–121.

Irihimovitch, V. and Shapira, M. (2000) Glutathione redox potential modulated by reactive oxygen species regulates translation of Rubisco large subunit in the chloroplast. *J Biol Chem*, **275**, 16289–16295.

Irihimovitch, V. and Stern, D.B. (2006) The sulfur acclimation SAC3 kinase is required for chloroplast transcriptional repression under sulfur limitation in *Chlamydomonas reinhardtii*. *Proc Natl Acad Sci U S A*, **103**, 7911–7916.

Ito, H., Iwabuchi, M. and Ogawa, K. (2003) The sugar-metabolic enzymes aldolase and triose-phosphate isomerase are targets of glutathionylation in *Arabidopsis thaliana*: detection using biotinylated glutathione. *Plant Cell Physiol*, **44**, 655–660.

James, F., Paquet, L., Sparace, S.A., Gage, D.A. and Hanson, A.D. (1995) Evidence implicating dimethylsulfoniopropionaldehyde as an intermediate in dimethylsulfoniopropionate biosynthesis. *Plant Physiol*, **108**, 1439–1448.

Jeanjean, R. and Broda, E. (1977) Dependence of sulphate uptake by *Anacystis nidulans* on energy, on osmotic shock and on sulphate stravation. *Arch Microbiol*, **114**, 19–23.

Jensen, T.E. and Rachlin, J.W. (1984) Effect of varying sulfur deficiency on structural components of a cyanobacterium *Synechococcus leopoliiensis*: a morphometric study. *Cytobios*, **41**, 35–46.

Jez, J.M., Cahoon, R.E. and Chen, S. (2004) *Arabidopsis thaliana* glutamate-cysteine ligase: functional properties, kinetic mechanism, and regulation of activity. *J Biol Chem*, **279**, 33463–33470.

Jo, J.H., Lee, D.S. and Park, J.M. (2006) Modeling and optimization of photosynthetic hydrogen gas production by green alga Chlamydomonas reinhardtii in sulfur-deprived circumstance. *Biotechnol Prog*, **22**, 431–437.

Jordan, P.A. and Gibbins, J.M. (2006) Extracellular disulfide exchange and the regulation of cellular function. *Antioxid Redox Signal*, **8**, 312–324.

Kanno, N., Nagahisa, E., Sato, M. and Sato, Y. (1996) Adenosine 5-prime-phosphosulfate sulfotransferase from the marine macroalga *Porphyra yezoensis* Ueda (Rhodophyta): stabilization, purification, and properties. *Planta*, **198**, 440–446.

Kaplan, D., Heimer, Y.M., Abeliovich, A. and Goldbrough, P.B. (1995) Cadmium toxicity and resistance in *Chlorella* sp. *Plant Sci*, **109**, 129–137.

Karsten, U., Wiencke, C. and G.O., K. (1992) Dimethylsulphoniopropionate (DMSP) accumulation in green macroalgae from polar to temperate regions: interactive effects of light versus salinity and light versus temperature. *Pol Biol*, **12**, 603–607.

Karsten, U., Wiencke, C. and Kirst, G.O. (1990) The effect of light intensity and daylength on the β-dimethylsulphoniopropionate (DMSP) content of marine green macroalgae from Antarctica. *Plant Cell Environ*, **12**, 989–993.

Kathir, P., LaVoie, M., Brazelton, W.J., Haas, N.A., Lefebvre, P.A. and Silflow, C.D. (2003) Molecular map of the Chlamydomonas reinhardtii nuclear genome. *Eukaryot Cell*, **2**, 362–379.

Kawaguchi, H., Hashimoto, K., Hirata, K. and Miyamoto, K. (2001) H2 production from algal biomass by a mixed culture of *Rhodobium marinum* A-501 and *Lactobacillus amylovorus*. *J Biosci Bioeng*, **91**, 277–282.

Keller, M.D., Kiene, R.P., Matrai, P.A. and Bellows, W.K. (1999) Production of glycine betaine and dimethylsulfoniopropionate in marine phytoplankton. II. N-limited chemostat cultures. *Mar Biol*, **135**, 249–257.

Kieber, D.J., Jiao, J., Kiene, R.P. and Bates, T.S. (1996) Impact of dimethylsulfide photochemistry on methyl sulfur cycling in the equatorial Pacific ocean. *J Geophys Res*, **101**, 3715–3722.

Kim, J. and Mayfield, S.P. (1997) Protein disulfide isomerase as a regulator of chloroplast translational activation. *Science*, **278**, 1954–1957.

Kim, J. and Mayfield, S.P. (2002) The active site of the thioredoxin-like domain of chloroplast protein disulfide isomerase, RB60, catalyzes the redox-regulated binding of chloroplast poly(A)-binding protein, RB47, to the 5′ untranslated region of psbA mRNA. *Plant Cell Physiol*, **43**, 1238–1243.

Kirst, G., Thiel, C., Wolff, H., Nothnagel, J., Wanzek, M. and Ulmke, R. (1991) Dimethylsulfoniopropionate (DMSP) in ice-algae and its possible biological role. *Mar Chem*, **35**, 381–388.

Klein, U. (1987) Intracellular carbon partitioning in *Chlamydomonas reinhardtii*. *Plant Physiol*, **85**, 892–897.

Kloareg, B. and Quatrano, R.S. (1988) Structure of the cell walls of marine algae and ecophysiological functions of the matrix polysaccharides. *Oceanogr Mar Biol Annu Res*, **26**, 259–315.

Kobayashi, I., Fujiwara, S., Saegusa, H., Inouhe, M., Matsumoto, H. and Tsuzuki, M. (2006) Relief of arsenate toxicity by Cd-stimulated phytochelatin synthesis in the green alga Chlamydomonas reinhardtii. *Mar Biotechnol (NY)*, **8**, 94–101.

Kocsis, M.G., Nolte, K.D., Rhodes, D., Shen, T.L., Gage, D.A. and Hanson, A.D. (1998) Dimethylsulfoniopropionate biosynthesis in *Spartina alterniflora*1. Evidence that S-methylmethionine and dimethylsulfoniopropylamine are intermediates. *Plant Physiol*, **117**, 273–281.

Kopriva, S., Buchert, T., Fritz, G., Suter, M., Benda, R., Schunemann, V., Koprivova, A., Schurmann, P., Trautwein, A.X., Kroneck, P.M. and Brunold, C. (2002) The presence of an iron-sulfur cluster in adenosine 5′-phosphosulfate reductase separates organisms utilizing adenosine 5′-phosphosulfate and phosphoadenosine 5′-phosphosulfate for sulfate assimilation. *J Biol Chem*, **277**, 21786–21791.

Kopriva, S., Buchert, T., Fritz, G., Suter, M., Weber, M., Benda, R., Schaller, J., Feller, U., Schurmann, P., Schunemann, V., Trautwein, A.X., Kroneck, P.M. and Brunold, C. (2001) Plant adenosine 5′-phosphosulfate reductase is a novel iron-sulfur protein. *J Biol Chem*, **276**, 42881–42886.

Kopriva, S. and Koprivova, A. (2004) Plant adenosine 5′-phosphosulphate reductase: the past, the present, and the future. *J Exp Bot*, **55**, 1775–1783.

Kruse, O., Hankamer, B., Konczak, C., Gerle, C., Morris, E., Radunz, A., Schmid, G.H. and Barber, J. (2000) Phosphatidylglycerol is involved in the dimerization of photosystem II. *J Biol Chem*, **275**, 6509–6514.

Kugita, M., Kaneko, A., Yamamoto, Y., Takeya, Y., Matsumoto, T. and Yoshinaga, K. (2003) The complete nucleotide sequence of the hornwort (*Anthoceros formosae*) chloroplast genome: insight into the earliest land plants. *Nucleic Acids Res*, **31**, 716–721.

Kurvari, V., Qian, F. and Snell, W.J. (1995) Increased transcript levels of a methionine synthase during adhesion-induced activation of *Chlamydomonas reinhardtii* gametes. *Plant Mol Biol*, **29**, 1235–1252.

Lauchli, A. (1993) Selenium in plants: Uptake, functions, and environmental toxicity. *Bot Acta*, **106**, 455–468.

Laudenbach, D.E. and Grossman, A.R. (1991) Characterization and mutagenesis of sulfur-regulated genes in a cyanobacterium: evidence for function in sulfate transport. *J Bacteriol*, **173**, 2739–2750.

Lawry, N.H. and Jensen, T.E. (1979) Deposition of condensed phosphate as an effect of varying sulfur deficiency in the cyanobacterium *Synechococcus* sp. (*Anacystis nidulans*). *Arch Microbiol*, **120**, 1–7.

Lawry, N.H. and Simon, R.D. (1982) The normal and induced occurrence of cyanophycin inclusion bodies in several blue-green algae. *J Phycol*, **18**, 391–399.

Le Faucheur, S., Behra, R. and Sigg, L. (2005) Phytochelatin induction, cadmium accumulation, and algal sensitivity to free cadmium ion in *Scenedesmus vacuolatus*. *Environ Toxicol Chem*, **24**, 1731–1737.

Ledford, H.K., Baroli, I., Shin, J.W., Fischer, B.B., Eggen, R.I. and Niyogi, K.K. (2004) Comparative profiling of lipid-soluble antioxidants and transcripts reveals two phases of photo-oxidative stress in a xanthophyll-deficient mutant of *Chlamydomonas reinhardtii*. *Mol Genet Genomics*, **272**, 470–479.

LeDuc, D.L., Tarun, A.S., Montes-Bayon, M., Meija, J., Malit, M.F., Wu, C.P., AbdelSamie, M., Chiang, C.Y., Tagmount, A., deSouza, M., Neuhierl, B., Bock, A., Caruso, J. and Terry, N. (2004) Overexpression of seleno-cysteine methyltransferase in Arabidopsis and Indian mustard increases selenium tolerance and accumulation. *Plant Physiol*, **135**, 377–383.

Lee, S. and Leustek, T. (1998) APS kinase from Arabidopsis thaliana: genomic organization, expression, and kinetic analysis of the recombinant enzyme. *Biochem Biophys Res Commun*, **247**, 171–175.

Lefebvre, P.A. and Silflow, C.D. (1999) Chlamydomonas: the cell and its genomes. *Genetics*, **151**, 9–14.

Leisinger, U., Rufenacht, K., Fischer, B., Pesaro, M., Spengler, A., Zehnder, A.J. and Eggen, R.I. (2001) The glutathione peroxidase homologous gene from *Chlamydomonas reinhardtii* is transcriptionally up-regulated by singlet oxygen. *Plant Mol Biol*, **46**, 395–408.

Lemaire, S., Keryer, E., Stein, M., Schepens, I.I., Issakidis-Bourguet, E., Gerard Hirni, C., Miginiac-Maslow, M. and Jacquot, J.P. (1999a) Heavy-metal regulation of thiore-doxin gene expression in *Chlamydomonas reinhardtii*. *Plant Physiol*, **120**, 773–778.

Lemaire, S.D. (2004) The glutaredoxin family in oxy-genic photosynthetic organisms. *Photosynth Res*, **79**, 305–318.

Lemaire, S.D., Guillon, B., Le Marechal, P., Keryer, E., Miginiac-Maslow, M. and Decottignies, P. (2004) New thioredoxin targets in the unicellular photosynthetic eukaryote *Chlamydomonas reinhardtii*. *Proc Natl Acad Sci U S A*, **101**, 7475–7480.

Lemaire, S.D. and Miginiac-Maslow, M. (2004) The thioredoxin superfamily in *Chlamydomonas reinhardtii*. *Photosynth Res*, **82**, 203–220.

Lemaire, S.D., Quesada, A., Merchan, F., Corral, J.M., Igeno, M.I., Keryer, E., Issakidis-Bourguet, E., Hiras-awa, M., Knaff, D.B. and Miginiac-Maslow, M. (2005) NADP-malate dehydrogenase from unicellular green alga *Chlamydomonas reinhardtii*. A first step toward redox regulation? *Plant Physiol*, **137**, 514–521.

Lemaire, S.D., Stein, M., Issakidis-Bourguet, E., Keryer, E., Benoit, V.V., Pineau, B., Gerard-Hirne, C., Miginiac-Maslow, M. and Jacquot, J.P. (1999b) The complex regulation of ferredoxin/thioredoxin-related genes by light and the circadian clock. *Planta*, **209**, 221–229.

Lemieux, C., Otis, C. and Turmel, M. (2000) Ancestral chloroplast genome in *Mesostigma viride* reveals an early branch of green plant evolution. *Nature*, **403**, 649–652.

Levitan, A., Trebitsh, T., Kiss, V., Pereg, Y., Dangoor, I. and Danon, A. (2005) Dual targeting of the protein disulfide isomerase RB60 to the chloroplast and the endoplasmic reticulum. *Proc Natl Acad Sci U S A*, **102**, 6225–6230.

Li, J.J., Saidha, T. and Schiff, J.A. (1991) Purification and properties of two forms of ATP sulfurylase from *Euglena*. *Biochim Biophys Acta*, **1078**, 68–76.

Li, J.Y. and Schiff, J.A. (1991) Purification and properties of adenosine 5′-phosphosulphate sulphotransferase from *Euglena*. *Biochem J*, **274** (Pt 2), 355–360.

Libessart, N., Maddelein, M.L., Koornhuyse, N., Decq, A., Delrue, B., Mouille, G., D'Hulst, C. and Ball, S. (1995) Storage, photosynthesis, and growth: The conditional nature of mutations affecting starch synthesis and struc-ture in Chlamydomonas. *Plant Cell*, **7**, 1117–1127.

Lien, T. and Schreiner, Ø. (1975) Purification of a derepress-ible arylsulfatase from *Chlamydomonas reinhardtii*. Pro-perties of the enzyme in intact cells and in purified state. *Biochim Biophys Acta*, **384**, 168–179.

Lindahl, M. and Florencio, F.J. (2003) Thioredoxin-linked processes in cyanobacteria are as numerous as in chloro-plasts, but targets are different. *Proc Natl Acad Sci U S A*, **100**, 16107–16112.

Loewus, F., Wagner, G., Schiff, J.A. and Weistrop, J. (1971) The incorporation of ^{35}S-labeled sulfate into carrageenan in *Chondrus crispus*. *Plant Physiol*, **48**, 373–375.

Loll, B., Kern, J., Saenger, W., Zouni, A. and Biesiadka, J. (2005) Towards complete cofactor arrangement in the 3.0 A resolu-tion structure of photosystem II. *Nature*, **438**, 1040–1044.

Lomans, B.P., van der Drift, C., Pol, A. and Op den Camp, H.J. (2002) Microbial cycling of volatile organic sulfur compounds. *Cell Mol Life Sci*, **59**, 575–588.

Lovelock, J.E. (1972) Gaia as seen through the atmosphere. *Atmos Environ*, **6**, 579–580.

Lyi, S.M., Heller, L.I., Rutzke, M., Welch, R.M., Kochian, L.V. and Li, L. (2005) Molecular and biochemical charac-terization of the selenocysteine Se-methyltransferase gene and Se-methylselenocysteine synthesis in broccoli. *Plant Physiol*, **138**, 409–420.

Malea, P., Rijstenbil, J.W. and Haritonidis, S. (2006) Effects of cadmium, zinc and nitrogen status on nonprotein thiols in the macroalgae *Enteromorpha* spp. from the Scheldt Estuary (SW Netherlands, Belgium) and Thermaikos Gulf (Greece, N Aegean Sea). *Marine Environ Res*, **62**, 45–60.

Malin, G. and Kirst, G.O. (1997) Algal production of dimethyl sulfide and its atmospheric role. *J Phycol*, **33**, 889–896.

Marrs, K.A. (1996) The functions and regulation of glutath-ione S-transferase in plants. *Annu Rev Plant Physiol Plant Mol Biol*, **47**, 127–158.

Maruyama-Nakashita, A., Inoue, E., Watanabe-Takahashi, A., Yamaya, T. and Takahashi, H. (2003) Transcriptome pro-filing of sulfur-responsive genes in Arabidopsis reveals global effects of sulfur nutrition on multiple metabolic pathways. *Plant Physiol*, **132**, 597–605.

Maruyama-Nakashita, A., Nakamura, Y., Yamaya, T. and Takahashi, H. (2004) Regulation of high-affinity sulphate transporters in plants: towards systematic analysis of sulphur signalling and regulation. *J Exp Bot*, **55**, 1843–1849.

Matagne, R.F., Loppes, R. and Deltour, R. (1976) Phosphatase of *Chlamydomonas reinhardi*: biochemical and cytochemical approach with specific mutants. *J Bacteriol*, **126**, 937–950.

Matrai, P.A. and Keller, M.D. (1994) Total organic sulfur and dimethylsulfoniopropionate (DMSP) in marine phytoplankton: Intracellular variation. *Mar Biol*, **119**, 61–68.

Matsuzaki, M., Misumi, O., Shin, I.T., Maruyama, S., Takahara, M., Miyagishima, S.Y., Mori, T., Nishida, K., Yagisawa, F., Nishida, K., Yoshida, Y., Nishimura, Y., Nakao, S., Kobayashi, T., Momoyama, Y., Higashiyama, T., Minoda, A., Sano, M., Nomoto, H., Oishi, K., Hayashi, H., Ohta, F., Nishizaka, S., Haga, S., Miura, S., Morishita, T., Kabeya, Y., Terasawa, K., Suzuki, Y., Ishii, Y., Asakawa, S., Takano, H., Ohta, N., Kuroiwa, H., Tanaka, K., Shimizu, N., Sugano, S., Sato, N., Nozaki, H., Ogasawara, N., Kohara, Y. and Kuroiwa, T. (2004) Genome sequence of the ultrasmall unicellular red alga *Cyanidioschyzon merolae* 10D. *Nature*, **428**, 653–657.

Melis, A. and Chen, H.C. (2005) Chloroplast sulfate transport in green algae – genes, proteins and effects. *Photosynth Res*, **86**, 299–307.

Melis, A. and Happe, T. (2001) Hydrogen production. Green algae as a source of energy. *Plant Physiol*, **127**, 740–748.

Melis, A. and Happe, T. (2004) Trails of green alga hydrogen research – from hans gaffron to new frontiers. *Photosynth Res*, **80**, 401–409.

Melis, A., Seibert, M. and Happe, T. (2004) Genomics of green algal hydrogen research. *Photosynth Res*, **82**, 277–288.

Meyer, A.J. and Hell, R. (2005) Glutathione homeostasis and redox regulation of sulfhydryl groups. *Photosyn Res*, **86**, 435–457.

Michelet, L., Zaffagnini, M., Marchand, C., Collin, V., Decottignies, P., Tsan, P., Lancelin, J.M., Trost, P., Miginiac-Maslow, M., Noctor, G. and Lemaire, S.D. (2005) Glutathionylation of chloroplast thioredoxin f is a redox signaling mechanism in plants. *Proc Natl Acad Sci U S A*, **102**, 16478–16483.

Morelli, E., Cruz, B.H., Somovigo, S. and Scarano, G. (2002) Speciation of cadmium-glutamyl peptides complexes in cells of the marine microagla *Phaeodactylum tricornutum*. *Plant Sci*, **163**, 807–813.

Morelli, E. and Scarano, G. (2001) Synthesis and stability of phytochelatins induced by cadmium and lead in the marine diatom *Phaeodactylum tricornutum*. *Marine Environ Res*, **52**, 383–395.

Moseley, J.L., Chang, C.W. and Grossman, A.R. (2006) Genome-based approaches to understanding phosphorus deprivation responses and PSR1 control in *Chlamydomonas reinhardtii*. *Eukaryot Cell*, **5**, 26–44.

Munavalli, S., Parker, D.V. and Hamilton, F.D. (1975) Identification of NADPH-thioredoxin reductase system in *Euglena gracilis*. *Proc Natl Acad Sci U S A*, **72**, 4233–4237.

Murata, N., Higashi, S.I. and Fujimura, Y. (1990) Glycerolipids in. various preparations of photosystem II from spinach chloroplasts. *Biochim Biophys Acta*, **1019**, 261–268.

Murata, N. and Sato, N. (1983) Analysis of lipids in *Prochloron*. sp.: occurrence of monoglucosyl diacylglycerol. *Plant Cell Physiol*, **24**, 133–138.

Nagel, K., Adelmeier, U. and Voigt, J. (1996) Subcellular distribution of. cadmium in the unicellular green alga *Chlamydomonas reinhardtii*. *J Plant Physiol*, **149**, 86–90.

Nagel, K. and Voigt, J. (1995) Impaired photosynthesis in a cadmium-tolerant Chlamydomonas mutant strain. *Microbiol Res*, **150**, 105–110.

Navarro, F. and Florencio, F.J. (1996) The cyanobacterial thioredoxin gene is required for both photoautotrophic and heterotrophic growth. *Plant Physiol*, **111**, 1067–1075.

Nicholson, M.L., Gaasenbeek, M. and Laudenbach, D.E. (1995) Two enzymes together capable of cysteine biosynthesis are encoded on a cyanobacterial plasmid. *Mol Gen Genet*, **247**, 623–632.

Nicholson, M.L. and Laudenbach, D.E. (1995) Genes encoded on a cyanobacterial plasmid are transcriptionally regulated by sulfur availability and CysR. *J Bacteriol*, **177**, 2143–2150.

Nishiguchi, M.K. and Somero, G.N. (1992) Temperature- and concentration-dependence of compatibility of the organic osmolyte beta-dimethylsulfoniopropionate. *Cryobiology*, **29**, 118–124.

Nishikawa, K., Onodera, A. and Tominaga, N. (2006) Phytochelatins do not correlate with the level of Cd accumulation in *Chlamydomonas* spp. *Chemosphere*, **63**, 1553–1559.

Noctor, G. and Foyer, C.H. (1998) Ascorbate and Glutathione: Keeping active oxygen under control. *Annu Rev Plant Physiol Plant Mol Biol*, **49**, 249–279.

Novoselov, S.V., Rao, M., Onoshko, N.V., Zhi, H., Kryukov, G.V., Xiang, Y., Weeks, D.P., Hatfield, D.L. and Gladyshev, V.N. (2002) Selenoproteins and selenocysteine insertion system in the model plant cell system, *Chlamydomonas reinhardtii*. *Embo J*, **21**, 3681–3693.

Ohta, N., Matsuzaki, M., Misumi, O., Miyagishima, S.Y., Nozaki, H., Tanaka, K., Shin, I.T., Kohara, Y. and Kuroiwa, T. (2003) Complete sequence and analysis of the plastid genome of the unicellular red alga *Cyanidioschyzon merolae*. *DNA Res*, **10**, 67–77.

Ohyama, K., Fukuzawa, H., Kohchi, T., Sano, T., Sano, S., Shirai, H., Umesono, K., Shiki, Y., Takeuchi, M., Chang, Z. and et al. (1988) Structure and organization of *Marchantia polymorpha* chloroplast genome. I. Cloning and gene identification. *J Mol Biol*, **203**, 281–298.

Omata, T. and Murata, N. (1983) Isolation and characterization of the cytoplasmic membranes from the blue-green. alga (cyanobacterium) *Anacystis nidulans*. *Plant Cell Physiol*, **24**, 1101–1112.

Perez-Rama, M., Torres Vaamonde, E. and Abalde Alonso, J. (2006) Composition and production of thiol constituents induced by cadmium in the marine microalga *Tetraselmis suecica*. *Environ Toxicol Chem*, **25**, 128–136.

Plumley, F.G. and Schmidt, G.W. (1989) Nitrogen-dependent regulation of photosynthetic gene expression. *Proc Natl Acad Sci U S A*, **86**, 2678–2682.

Pollock, S.V., Pootakham, W., Shibagaki, N., Moseley, J.L. and Grossman, A.R. (2005) Insights into the acclimation of *Chlamydomonas reinhardtii* to sulfur deprivation. *Photosynth Res*, **86**, 475–489.

Posewitz, M.C., King, P.W., Smolinski, S.L., Zhang, L., Seibert, M. and Ghirardi, M.L. (2004a) Discovery of two novel radical S-adenosylmethionine proteins required for the assembly of an active [Fe] hydrogenase. *J Biol Chem*, **279**, 25711–25720.

Posewitz, M.C., Smolinski, S.L., Kanakagiri, S., Melis, A., Seibert, M. and Ghirardi, M.L. (2004b) Hydrogen photoproduction is attenuated by disruption of an isoamylase gene in *Chlamydomonas reinhardtii*. *Plant Cell*, **16**, 2151–2163.

Quisel, J.D., Wykoff, D.D. and Grossman, A.R. (1996) Biochemical characterization of the extracellular phosphatases produced by phosphorus-deprived *Chlamydomonas reinhardtii*. *Plant Physiol*, **111**, 839–848.

Ramus, J. (1974) In vivo molybdate inhibition of sulfate transfer to porphyridium capsular polysaccharide. *Plant Physiol*, **54**, 945–949.

Ramus, J. and Groves, S.T. (1972) Incorporation of sulfate into the capsular polysaccharide of the red alga Porphyridium. *J Cell Biol*, **54**, 399–407.

Ravina, C.G., Barroso, C., Vega, J.M. and Gotor, C. (1999) Cysteine biosynthesis in *Chlamydomonas reinhardtii*. Molecular cloning and regulation of O-acetylserine (thiol)lyase. *Eur J Biochem*, **264**, 848–853.

Ravina, C.G., Chang, C.I., Tsakraklides, G.P., McDermott, J.P., Vega, J.M., Leustek, T., Gotor, C. and Davies, J.P. (2002) The *sac* mutants of *Chlamydomonas reinhardtii* reveal transcriptional and posttranscriptional control of cysteine biosynthesis. *Plant Physiol*, **130**, 2076–2084.

Rea, P.A., Vatamaniuk, O.K. and Rigden, D.J. (2004) Weeds, worms, and more. Papain's long-lost cousin, phytochelatin synthase. *Plant Physiol*, **136**, 2463–2474.

Reed, R. (1983) Measurement and osmotic significance of β-dimethylsulfoniopropionate in marine microalgae. *Mar Biol Lett*, **34**, 173–181.

Riekhof, W.R., Ruckle, M.E., Lydic, T.A., Sears, B.B. and Benning, C. (2003) The sulfolipids 2'-O-acyl-sulfoquinovosyldiacylglycerol and sulfoquinovosyldiacylglycerol are absent from a *Chlamydomonas reinhardtii* mutant deleted in SQD1. *Plant Physiol*, **133**, 864–874.

Rijstenbil, J.W., Sandee, A., Van Drie, J. and Wijnholds, J.A. (1994) Interaction of toxic trace metals and mechanisms of detoxification in the planktonic diatoms *Ditylum brightwellii* and *Thalassiosira pseudonana*. *FEMS Microbiol Rev*, **14**, 387–396.

Romanyuk, N.D., Rigden, D.J., Vatamaniuk, O.K., Lang, A., Cahoon, R.E., Jez, J.M. and Rea, P.A. (2006) Mutagenic definition of a papain-like catalytic triad, sufficiency of the N-terminal domain for single-site core catalytic enzyme acylation, and C-terminal domain for augmentative metal activation of a eukaryotic phytochelatin synthase. *Plant Physiol*, **141**, 858–869.

Ruiz, F.A., Marchesini, N., Seufferheld, M., Govindjee and Docampo, R. (2001) The polyphosphate bodies of *Chlamydomonas reinhardtii* possess a proton-pumping pyrophosphatase and are similar to acidocalcisomes. *J Biol Chem*, **276**, 46196–46203.

Ryan, J., McKillen, M. and Mason, J. (1987) Sulphate/molybdate interactions: in vivo and in vitro studies on the group VI oxyanion transport system in ovine renal tubule epithelial cells. *Ann Rech Vet*, **18**, 47–55.

Saito, K. (2000) Regulation of sulfate transport and synthesis of sulfur-containing amino acids. *Curr Opin Plant Biol*, **3**, 188–195.

Sato, N. (2004) Roles of the acidic lipids sulfoquinovosyl diacylglycerol and phosphatidylglycerol in photosynthesis: their specificity and evolution. *J Plant Res*, **117**, 495–505.

Sato, N., Sonoike, K., Tsuzuki, M. and Kawaguchi, A. (1995) Impaired photosystem II in a mutant of *Chlamydomonas reinhardtii* defective in sulfoquinovosyl diacylglycerol. *Eur J Biochem*, **234**, 16–23.

Sato, N., Sugimoto, K., Meguro, A. and Tsuzuki, M. (2003) Identification of a gene for UDP-sulfoquinovose synthase of a green alga, *Chlamydomonas reinhardtii*, and its phylogeny. *DNA Res*, **10**, 229–237.

Schiff, J.A. (1959) Studies on sulfate utilization by *Chlorella pyrenoidosa* using sulfate-S; the occurrence of S-adenosyl methionine. *Plant Physiol*, **34**, 73–80.

Schiff, J.A. (1979) Pathways of assimilatory sulphate reduction in plants and microorganisms. *Ciba Found Symp*, 49–69.

Schmidt, A. (1972) An APS-sulfotransferase from *Chlorella*. *Arch Mikrobiol*, **84**, 77–86.

Schmidt, A. (1973) Sulfate reduction in a cell-free system of *Chlorella*. The ferredoxin-dependent reduction of a protein-bound intermediate by a thiosulfonate reductase. *Arch Microbiol*, **93**, 29–52.

Schmidt, A. and Trüper, H. (1977) Reduction of adenylylsulfate and 3'-phosphoadenylylsulfate in phototrophic bacteria. *Experientia*, **33**, 1008–1009.

Schoeler, B., Delorme, N., Doench, I., Sukhorukov, G.B., Fery, A. and Glinel, K. (2006) Polyelectrolyte films based on polysaccharides of different conformations: effects on multilayer structure and mechanical properties. *Biomacromolecules*, **7**, 2065–2071.

Schreiner, Ø., Lien, T. and Knutsen, G. (1975) The capacity for arylsulfatase synthesis in synchronous and synchronized cultures of *Chlamydomonas reinhardtii*. *Biochim Biophys Acta*, **384**, 180–193.

Schultze, M., Quiclet-Sire, B., Kondorosi, E., Virelizer, H., Glushka, J.N., Endre, G., Gero, S.D. and Kondorosi, A. (1992) *Rhizobium meliloti* produces a family of sulfated lipooligosaccharides exhibiting different degrees of plant host specificity. *Proc Natl Acad Sci U S A*, **89**, 192–196.

Selstam, E. and Campbell, D. (1996) Membrane lipid composition of the unusual cyanobacterium *Gloeobacter violaceus* sp. PCC 7421, which lacks sulfoquinovosyl. *Arch Microbiol*, **166**, 132–135.

Shapira, M., Lers, A., Heifetz, P.B., Irihimovitz, V., Osmond, C.B., Gillham, N.W. and Boynton, J.E. (1997) Differential regulation of chloroplast gene expression in Chlamydomonas reinhardtii during photoacclimation: light stress transiently suppresses synthesis of the Rubisco LSU protein while enhancing synthesis of the PS II D1 protein. *Plant Mol Biol*, **33**, 1001–1011.

Shrager, J., Hauser, C., Chang, C.W., Harris, E.H., Davies, J., McDermott, J., Tamse, R., Zhang, Z. and Grossman, A.R. (2003) *Chlamydomonas reinhardtii* genome project. A guide to the generation and use of the cDNA information. *Plant Physiol*, **131**, 401–408.

Sieburth, J.M. (1960) Acrylic acid, an "antibiotic" principle in *Phaeocystis* blooms in antarctic waters. *Science*, **132**, 676–677.

Sigrist, M., Zwilliengerg, C., Giroud, C.H., Eichenberger, W. and Boschetti, A. (1988) Sulfolipid associated with the light-harvesting complex associated with photosystem II apoproteins of *Chlamydomonas reinhardtii*. *Plant Sci*, **58**, 15–23.

Simo, R. and Pedros-Alio, C. (1999) Role of vertical mixing in controlling the oceanic production of dimethyl sulphide. *Nature*, **402**, 396–399.

Siripornadulsil, S., Traina, S., Verma, D.P. and Sayre, R.T. (2002) Molecular mechanisms of proline-mediated tolerance to toxic heavy metals in transgenic microalgae. *Plant Cell*, **14**, 2837–2847.

Smith, I.W., Wilkinson, J.F. and Duguid, J.P. (1954) Volutin production in *Aerobacter aerogenes* due to nutrient imbalance. *J Bacteriol*, **68**, 450–463.

Stauber, E.J. and Hippler, M. (2004) *Chlamydomonas reinhardtii* proteomics. *Plant Physiol Biochem*, **42**, 989–1001.

Stefels, J. (2000) Physiological aspects of the production and conversion of DMSP in marine algae and higher plants. *J Sea Res*, **43**, 183–197.

Stefels, J. and van Boekel, W.H.M. (1993) Production of DMS from dissolved DMSP in axenic cultures of the marine phytoplankton species *Phaeocystis* sp. *Mar Ecol Prog Ser*, **97**, 11–18.

Sugimoto, K., Sato, N., Watanabe, A. and Tsuzuki, M. (2005) Effects of sulfur deprivation on sulfolipid metabolism in *Chlamydomonas reinhardtii*. In Saito, K., De Kok, L.J.,

Stulen, I., Hawksford, M.J., Schnug, E. and Rennenberg, H. (eds.), *Sulfur Transport and Assimilation in Plants in the Postgenomic Era*. Backhuys Publishers, Leiden.

Summers, P.S., Nolte, K.D., Cooper, A.J.L., Borgeas, H., Leustek, T., Rhodes, D. and Hanson, A.D. (1998) Identification and stereospecificity of the first three enzymes of 3-dimethylsulfoniopropionate biosynthesis in a chlorophyte alga. *Plant Physiol*, **116**, 369–378.

Sunda, W., Kieber, D.J., Kiene, R.P. and Huntsman, S. (2002) An antioxidant function for DMSP and DMS in marine algae. *Nature*, **418**, 317–320.

Tabe, L. and Higgins, T.J.V. (1998) Engineering plant protein composition for improved nutrition. *Trends in Plant Science*, **3**, 282–286.

Takahashi, H., Braby, C.E. and Grossman, A.R. (2001) Sulfur economy and cell wall biosynthesis during sulfur limitation of *Chlamydomonas reinhardtii*. *Plant Physiol*, **127**, 665–673.

Takahashi, H., Noji, M., Hirai, M.Y. and Saito, K. (2003) [Molecular regulation of assimilatory sulfur metabolism in plants]. *Tanpakushitsu Kakusan Koso*, **48**, 2121–2129.

Takeda, T., Miyao, K., Tamoi, M., Kanaboshi, H., Miyasaka, H. and Shigeoka, S. (2003) Molecular characterization of glutathione peroxidase-like protein in halotolerant *Chlamydomonas* sp. W80. *Physiol Plant*, **117**, 467–475.

Torricelli, E., Gorbi, G., Pawlik-Skowronska, B., Sanità di Toppi, L. and Grazia Corradi, M. (2004) Cadmium tolerance, cysteine and thiol peptide levels in wild type and chromium-tolerant strains of *Scenedesmus acutus* (Chlorophyceae). *Aquatic Toxicol*, **68**, 315–323.

Trebitsh, T., Levitan, A., Sofer, A. and Danon, A. (2000) Translation of chloroplast psbA mRNA is modulated in the light by counteracting oxidizing and reducing activities. *Mol Cell Biol*, **20**, 1116–1123.

Trebitsh, T., Meiri, E., Ostersetzer, O., Adam, Z. and Danon, A. (2001) The protein disulfide isomerase-like RB60 is partitioned between stroma and thylakoids in *Chlamydomonas reinhardtii* chloroplasts. *J Biol Chem*, **276**, 4564–4569.

Trémoliéres, A., Dainese, P. and Bassi, R. (1994) Heterogeneous lipid distribution among chloroplast-binding proteins of photosystem II in maize mesophyll chloroplasts. *Eur J Biochem*, **221**, 721–730.

Trossat, C., Rathinasabapathi, B., Weretilnyk, E.A., Shen, T.L., Huang, Z.H., Gage, D.A. and Hanson, A.D. (1998) Salinity promotes accumulation of 3-dimethylsulfoniopropionate and its precursor S-methylmethionine in chloroplasts. *Plant Physiol*, **116**, 165–171.

Tsang, M.L. (1981) Thioredoxin/glutaredoxin system of Chlorella: Chlorella adenosine 5′-phosphosulfate sulfotransferase cannot use thioredoxin or glutaredoxin as cofactors. *Plant Physiol*, **68**, 1098–1104.

Tsuji, N., Hirayanagi, N., Iwabe, O., Namba, T., Tagawa, M., Miyamoto, S., Miyasaka, H., Takagi, M., Hirata, K. and Miyamoto, K. (2003) Regulation of phytochelatin synthesis by zinc and cadmium in marine green alga, *Dunaliella tertiolecta*. *Phytochemistry*, **62**, 453–459.

Tsuji, N., Nishikori, S., Iwabe, O., Matsumoto, S., Shiraki, K., Miyasaka, H., Takagi, M., Miyamoto, K. and Hirata, K. (2005) Comparative analysis of the two-step reaction catalyzed by prokaryotic and eukaryotic phytochelatin synthase by an ion-pair liquid chromatography assay. *Planta*, **222**, 181–191.

Tsuji, N., Nishikori, S., Iwabe, O., Shiraki, K., Miyasaka, H., Takagi, M., Hirata, K. and Miyamoto, K. (2004) Characterization of phytochelatin synthase-like protein encoded by alr0975 from a prokaryote, *Nostoc* sp. PCC 7120. *Biochem Biophys Res Commun*, **315**, 751–755.

Turmel, M., Otis, C. and Lemieux, C. (1999) The complete chloroplast DNA sequence of the green alga *Nephroselmis olivacea:* insights into the architecture of ancestral chloroplast genomes. *Proc Natl Acad Sci U S A*, **96**, 10248–10253.

Turmel, M., Otis, C. and Lemieux, C. (2005) The complete chloroplast DNA sequences of the charophycean green algae Staurastrum and Zygnema reveal that the chloroplast genome underwent extensive changes during the evolution of the Zygnematales. *BMC Biol*, **3**, 22.

Utkilen, H.C. (1976) Thiosulphate as electron donor in the blue-green alga *Anacystis nidulans*. *J Gen Microbiol*, **95**, 177–180.

van Waasbergen, L.G., Dolganov, N. and Grossman, A.R. (2002) *nblS*, a gene involved in controlling photosynthesis-related gene expression during high light and nutrient stress in *Synechococcus elongatus* PCC 7942. *J Bacteriol*, **184**, 2481–2490.

Visviki, I. and Rachlin, J.W. (1994) Acute and chronic exposure of *Dunaliella salina* and *Chlamydomonas bullosa* to copper and cadmium: effects of ultrastructure. *Arch Environ Contam Toxicol*, **26**, 154–162.

Vivares, D., Arnoux, P. and Pignol, D. (2005) A papain-like enzyme at work: native and acyl-enzyme intermediate structures in phytochelatin synthesis. *Proc Natl Acad Sci U S A*, **102**, 18848–18853.

Vlamis-Gardikas, A. and Holmgren, A. (2002) Thioredoxin and glutaredoxin isoforms. *Methods Enzymol*, **347**, 286–296.

Wagner, V., Gessner, G., Heiland, I., Kaminski, M., Hawat, S., Scheffler, K. and Mittag, M. (2006) Analysis of the phosphoproteome of *Chlamydomonas reinhardtii* provides new insights into various cellular pathways. *Eukaryot Cell*, **5**, 457–468.

Wagner, W. and Follmann, H. (1977) A thioredoxin from green algae. *Biochem Biophys Res Commun*, **77**, 1044–1051.

Wakasugi, T., Nagai, T., Kapoor, M., Sugita, M., Ito, M., Ito, S., Tsudzuki, J., Nakashima, K., Tsudzuki, T., Suzuki, Y., Hamada, A., Ohta, T., Inamura, A., Yoshinaga, K. and Sugiura, M. (1997) Complete nucleotide sequence of the chloroplast genome from the green

alga *Chlorella vulgaris*: the existence of genes possibly involved in chloroplast division. *Proc Natl Acad Sci U S A*, **94**, 5967–5972.

Watanabe, M. and Suzuki, T. (2004) Cadmium induced synthesis of HSP70 and a role of glutathione in *Euglena gracilis*. *Redox Report*, **9**, 349–353.

Weber, M., Suter, M., Brunold, C. and Kopriva, S. (2000) Sulfate assimilation in higher plants: characterization of a stable intermediate in the adenosine 5'-phosphosulfate reductase reaction. *European Journal of Biochemistry*, **267**, 3647–3653.

Whittaker, M.M. and Gleason, F.K. (1984) Isolation and characterization of thioredoxin f from the filamentous cyanobacterium, *Anabaena* sp. 7119. *J Biol Chem*, **259**, 14088–14093.

Wilkinson, B. and Gilbert, H.F. (2004) Protein disulfide isomerase. *Biochim Biophys Acta*, **1699**, 35–44.

Wolfe, G.V., Steinke, M. and Kirst, G.O. (1997) Grazing-activated chemical defence in a unicellular marine alga. *Nature*, **387**, 894–897.

Wykoff, D.D., Davies, J.P., Melis, A. and Grossman, A.R. (1998) The regulation of photosynthetic electron transport during nutrient deprivation in *Chlamydomonas reinhardtii*. *Plant Physiol*, **117**, 129–139.

Wykoff, D.D., Grossman, A.R., Weeks, D.P., Usuda, H. and Shimogawara, K. (1999) Psr1, a nuclear localized protein that regulates phosphorus metabolism in Chlamydomonas. *Proc Natl Acad Sci U S A*, **96**, 15336–15341.

Yamanaka, G., Glazer, A.N. and Williams, R.C. (1980) Molecular architecture of a light-harvesting antenna. Comparison of wild type and mutant *Synechococcus* 6301 phycobilisomes. *J Biol Chem*, **255**, 11104–11110.

Yildiz, F.H., Davies, J.P. and Grossman, A. (1996) Sulfur availability and the *SAC1* gene control adenosine triphosphate sulfurylase gene expression in *Chlamydomonas reinhardtii*. *Plant Physiol*, **112**, 669–675.

Yildiz, F.H., Davies, J.P. and Grossman, A.R. (1994) Characterization of sulfate transport in *Chlamydomonas reinhardtii* during sulfur-Limited and sulfur-sufficient growth. *Plant Physiol*, **104**, 981–987.

Yoch, D.C., Ansede, J.H. and Rabinowitz, K.S. (1997) Evidence for intracellular and extracellular dimethylsulfoniopropionate (DMSP) lyases and DMSP uptake sites in two species of marine bacteria. *Appl Environ Microbiol*, **63**, 3182–3188.

Yokota, A., Shigeoka, S., Onishi, T. and Kitaoka, S. (1988) Selenium as inducer of glutathione peroxidase in low-CO_2-grown *Chlamydomonas reinhardtii*. *Plant Physiol*, **86**, 649–651.

Yosef, I., Irihimovitch, V., Knopf, J.A., Cohen, I., Orr-Dahan, I., Nahum, E., Keasar, C. and Shapira, M. (2004) RNA binding activity of the ribulose-1,5-bisphosphate carboxylase/oxygenase large subunit from *Chlamydomonas reinhardtii*. *J Biol Chem*, **279**, 10148–10156.

Yoshimura, K., Miyao, K., Gaber, A., Takeda, T., Kanaboshi, H., Miyasaka, H. and Shigeoka, S. (2004) Enhancement of stress tolerance in transgenic tobacco plants overexpressing Chlamydomonas glutathione peroxidase in chloroplasts or cytosol. *Plant J*, **37**, 21–33.

Zhang, L., Happe, T. and Melis, A. (2002) Biochemical and morphological characterization of sulfur-deprived and H$_2$-producing Chlamydomonas reinhardtii (green alga). *Planta*, **214**, 552–561.

Zhang, L. and Melis, A. (2002) Probing green algal hydrogen production. *Philos Trans R Soc Lond B Biol Sci*, **357**, 1499–1507; discussion 1507–1411.

Zhang, Z., Shrager, J., Jain, M., Chang, C.W., Vallon, O. and Grossman, A.R. (2004) Insights into the survival of Chlamydomonas reinhardtii during sulfur starvation based on microarray analysis of gene expression. *Eukaryot Cell*, **3**, 1331–1348.

Chapter 14

Systematics of Anoxygenic Phototrophic Bacteria

Johannes F. Imhoff
*Institut für Meereswissenschaften IFM-GEOMAR,
Düsternbrooker Weg 20, 24105 Kiel, Germany*

Summary

Many of the anoxygenic phototrophic bacteria, in particular green sulfur bacteria and purple sulfur bacteria are actively involved in the dissimilatory sulfur cycle by oxidizing reduced sulfur compounds. An introduction to the current state of the systematics of anoxygenic phototrophic bacteria is given here. With the introduction of 16S rDNA sequences, the consideration of genetic relatedness of these bacteria and a great deal of chemotaxonomic properties, the systematic treatment of many of these bacteria has changed over the past decades. Many species and strains have been reclassified and new higher taxa were established that harbour the phototrophic genera.

Four major phylogenetic groups that have significant phenotypic characteristics can be distinguished: (1) the Heliobacteria (*Heliobacteriaceae*), which are Gram-positive bacteria, (2) the filamentous and gliding green bacteria (*Chloroflexaceae*), and (3) the green sulfur bacteria (*Chlorobiaceae*), which each form separate phylogenetic lines within the eubacteria, and (4) the purple sulfur and nonsulfur bacteria (various taxa among the Alpha-, Beta- and Gammaproteobacteria).

This chapter concentrates on the following groups of which major properties and representative genera and species are treated: The **purple sulfur bacteria** are Gammaproteobacteria and treated as *Chromatiales* (*Ectothiorhodospiraceae* and *Chromatiaceae* families). The **purple nonsulfur bacteria** are Betaproteobacteria (*Comamonadaceae* of the Burkholdriales and *Rhodocyclaceae* of Rhodocyeclales) and Alphaproteobacteria (*Rhodospirillaceae, Acetobacteraceae, Rhodobacteraceae, Bradyrhizobiaceae,*

Corresponding Author, Fax: X431/600 4452, E-mail: jimhoff@ifm-geomar.de

Rüdiger Hell et al. (eds.), Sulfur Metabolism in Phototrophic Organisms, 269–287.
© 2008 *Springer.*

Hyphomicrobiaceae, Rhodobiaceae) and are closely related to non-phototrophic purely chemotrophic relatives and to so-called aerobic bacteriochlorophyll-containg bacteria. The **green sulfur bacteria** form a closely related cluster of genera united in a single family the *Chlorobiaceae*, which is according to its isolated position within the phylogenetic tree of bacteria recognised as a separate phylum, the Chlorobi. Because of the difficulties related to taxonomic treatment and phylogenetic groupings, a list of strains is given indicating old and new taxonomic names of species and genus.

I. Introduction

Phototrophic bacteria are found in major eubacterial branches. The most important common property of these bacteria is the possession of photosynthetic pigments, which is visible in their light absorption spectra, and a photosynthetic apparatus, which enables the performance of light-dependent energy transfer processes. A characteristic unifying property of all of them is the performance of chlorophyll-mediated energy transformation.

On the basis of fundamental physiological differences we distinguish between (i) oxygenic phototrophic bacteria that use water as photosynthetic electron donor and produce molecular oxygen (cyanobacteria), (ii) anoxygenic phototrophic bacteria that use reduced substrates such as sulfide, hydrogen, ferrous iron and a number of simple organic substrates as photosynthetic electron donors but do not produce

oxygen during photosynthesis, and (iii) aerobic bacteriochlorophyll-containing bacteria ("ABC-bacteria"), which represent chemoheterotrophic bacteria with the potential to produce photosynthetic pigment-protein complexes and to perform light-mediated photosynthetic electron transport. The cyanobacteria represent a separate major phylogenetic line. The anoxygenic phototrophic bacteria are found in four different major phylogenetic branches, including the filamentous green and gliding bacteria with *Chloroflexus* and relatives, the green sulfur bacteria with *Chlorobium*, the *Heliobacteriaceae* (classified among the *Clostridiales*) and the phototrophic purple bacteria belonging to the Alpha-, Beta- and Gammaproteobacteria. The "ABC-bacteria" are also purple bacteria, closely related to the anoxygenic phototrophic bacteria of the Alpha- and Betaproteobacteria. Characteristic differences in the structure and function of the photosynthetic

Table 1. Diagnostic properties of major groups of phototrophic prokaryotes.

	Chlorobiaceae	*Chloroflexaceae*	Purple bacteria	Heliobacteria	Cyanobacteria (including prochlorophytes)
Photosynthesis	anoxygenic	anoxygenic	anoxygenic	anoxygenic	oxygenic
Type of (bacterio) chlorophyll	bchl c, d, e (+ bchl a)	bchl c, d (+ bchl a)	bchl a, b	bchl g	chl a, chl b*
Phycobilins present	−	−	−	−	+/−
Type of reaction center	I	II	II	I	I + II
Reduction of NAD^+ by primary photosynthetic electron acceptor	+	−	−	+	+
Location of antenna	chlorosomes	chlorosomes	ICM	CM	ICM, phycobilisomes**
Pathway of autotrophic CO_2-fixation	reductive TCA cycle	Hydroxypropionate-pathway	Calvin cycle	−	Calvin cycle
Preferred electron donor	H_2S, H_2	organic compounds (H_2S)	H_2S, H_2 organic compounds	organic compounds	H_2O
Chemotrophic growth	−	−	+/−	−	−/(+)

Abbreviations: − characteristic absent; +/− characteristic present or absent; −/(+) characteristic absent or weak activity present;

* present in prochlorophytes; ** absent in prochlorophytes; CM cytoplasmic membrane; ICM internal membranes bchl bacteriochlorophyll, chl chlorophyll

apparatus, in the pigment content, in important physiological properties as well as 16S rDNA sequence similarities distinguish these major branches of phototrophic bacteria (see Table 1). Quite remarkable is the high variation of the light harvesting structures and of different CO_2-fixation pathways among these groups.

This chapter is concerned with the anaerobic phototrophic bacteria that perform anoxygenic photosynthesis. In these bacteria photosynthesis depends on anoxic or oxygen-deficient conditions, because synthesis of the photosynthetic pigments and the formation of the photosynthetic apparatus are repressed by oxygen. These bacteria are unable to use water as an electron donor, but need more reduced compounds. Most characteristically, sulfide and other reduced sulfur compounds, but also hydrogen and a number of small organic molecules are used as photosynthetic electron donors. Also, the growth with reduced iron as electron donor has been demonstrated in some phototrophic purple bacteria (Widdel et al., 1993; Ehrenreich and Widdel, 1994; Straub et al., 1999).

The anoxygenic phototrophic bacteria are represented by predominantly aquatic bacteria that are able to grow under anoxic conditions by photosynthesis without oxygen production. The various photosynthetic pigments, which function in the transformation of light into chemical energy, give the cell cultures a distinct coloration depending on the pigment content from green, yellowish-green, brownish-green, brown, brownish-red, red, pink, purple, and purple-violet to even blue (carotinoidless mutants of certain purple bacteria containing bacteriochlorophyll a). In particular green sulfur bacteria and purple sulfur bacteria are key players in the biological sulfur cycle and form massive blooms under appropriate conditions, when both reduced sulfur compounds and light are available but oxygen is deficient or lacking.

A.   A Short Overview on the Groups of Anoxygenic Phototrophic Bacteria

The major groups of phototrophic bacteria (Table 1) are well distinguished on the basis of fundamental structural, molecular and physiological properties and characterized by sequence comparison of the 16S rDNA molecule. A fairly complete database of 16S rDNA sequences is available of type strains

and additional other strains of the *Chlorobiaceae* (Overmann and Tuschack, 1997; Alexander et al., 2002), of purple sulfur bacteria (Imhoff and Süling, 1996; Guyoneaud et al., 1998; Imhoff et al., 1998b) of filamentous green and gliding bacteria (Keppen et al., 2000), of Heliobacteria (Bryantseva et al., 1999, 2000; Madigan, 2001) and of purple non-sulfur bacteria (Hiraishi and Ueda, 1994; Hiraishi et al., 1995; Imhoff et al., 1998a; Kawasaki et al., 1993). These data have revealed the phylogenetic relationships of the anoxygenic phototrophic bacteria based on the 16S rDNA sequences and are used to determine the phylogenetic position of new isolates.

The ***Heliobacteriaceae*** are anoxygenic phototrophic bacteria that contain bacteriochlorophyll g and carotenoids. They are highly sensitive to oxygen and some species form heat resistant endospores. They are phylogenetically related to Gram-positive bacteria and now are classified among the Clostridiales. They grow photoheterotrophically. Growth with reduced sulfur sources has not been observed, although often sulfide is oxidized to sulfur if added to growing cultures (Madigan, 2001).

The **filamentous green and gliding bacteria** (*Chloroflexaceae*) are anoxygenic phototrophic bacteria that contain bacteriochlorophyll c or d in light-harvesting complexes located in special light-harvesting organelles, the chlorosomes. They move by gliding, are tolerant to oxygen, and grow preferably as photoheterotrophs. Representative species of *Chloroflexus* and *Roseiflexus* are adapted to hot freshwater environments. Some representatives have the capability to oxidize reduced sulfur compounds.

The **green sulfur bacteria** (*Chlorobiaceae*) are anoxygenic phototrophic bacteria that contain bacteriochlorophyll c, d, or e in light-harvesting complexes located in special light-harvesting organelles, the chlorosomes. They are obligately phototrophic, require strictly anoxic growth conditions, and have a low capacity to assimilate organic compounds. Depending on the pigment content, a number of green-colored and corresponding brown-colored strains are known from several species. Reduced sulfur compounds, in particular sulfide and sulfur, are common photosynthetic electron donors in this group of bacteria. In addition, thiosulfate is used by a number of representatives. Species of this group have been found in marine and hypersaline environments, where under appropriate conditions intensively colored, visible mass developments are formed.

The **phototrophic purple bacteria comprise the purple sulfur bacteria** (*Chromatiaceae* and *Ectothiorhodospiraceae*) and the purple nonsulfur bacteria. Quite characteristic is their capability to grow photoautotrophically and/or photoheterotrophically. The major pigments are bacteriochlorophyll a or b and various carotenoids of the spirilloxanthin, rhodopinal, spheroidene, and okenone series (Schmidt, 1978). The photosynthetic pigments and the structures of the photosynthetic apparatus are located within a more or less extended system of internal membranes that is considered as originating from and being continuous with the cytoplasmic membrane. These intracellular membranes consist of small fingerlike intrusions, vesicles, tubules or lamellae parallel to or at an angle to the cytoplasmic membrane. They carry the photosynthetic apparatus, the reaction centers and light-harvesting pigment-protein complexes surrounding the reaction center in the plane of the membrane (Drews and Imhoff, 1991). Bacteria of this group are the most prominent and abundant anoxygenic phototrophic bacteria in aquatic environments.

– The *Chromatiaceae* are Gammaproteobacteria and grow well under photoautotrophic conditions and use sulfide as photosynthetic electron donor, which is oxidized to sulfate via intermediate accumulation of elemental sulfur, microscopically visible as globules inside the cells. Photoheterotrophic, chemoautotrophic, and chemoheterotrophic growth is possible by several species.

– The *Ectothiorhodospiraceae* are Gammaproteobacteria and can be distinguished from the *Chromatiaceae* by deposition of elemental sulfur outside the cells or in the peripheral periplasmic part of the cells, and their preference for alkaline and saline growth conditions. Some species may also grow chemotrophically under aerobic dark conditions in the dark.

– The purple nonsulfur bacteria are Alpha- and Betaproteobacteria and preferentially grow under photoheterotrophic conditions, though most of them have the ability to grow photoautotrophically with hydrogen and several also with reduced sulfur compounds as electron donors. Only few species are able to completely oxidize sulfide to sulfate.

II. Phototrophic Purple Sulfur Bacteria – Chromatiales

The families *Chromatiaceae* and *Ectothiorhodospiraceae* are classified with the *Chromatiales* (Imhoff, 2005a, b, c). The *Ectothiorhodospiraceae* are purple sulfur bacteria that form sulfur globules outside the cells, while the *Chromatiaceae* exclusively comprise those phototrophic sulfur bacteria able to deposit elemental sulfur inside their cells (Imhoff, 1984a), which is in agreement with Molisch's (1907) definition of the Thiorhodaceae. Quite interestingly, already Pelsh (1937) had differentiated the "Ectothiorhodaceae" from the "Endothiorhodaceae" on a family level. However, because of their illegitimacy, these family names had no standing in nomenclature. Historical aspects of the taxonomy of anoxygenic phototrophic bacteria have been discussed in more detail elsewhere (Imhoff, 1992, 1995, 1999, 2001c).

The two families can also clearly be distinguished by a number of chemotaxonomic properties. Significant differences between *Chromatiaceae* and *Ectothiorhodospiraceae* occur in quinone, lipid and fatty acid composition (see Imhoff and Bias-Imhoff, 1995). Characteristic glucolipids are present in *Chromatiaceae* species, but absent from *Ectothiorhodospiraceae* (Imhoff et al., 1982). While C-16 fatty acids (in particular C-16:1) are the major components in *Chromatiaceae*, C-18 fatty acids (in particular C-18:1) are dominant in *Ectothiorhodo-spiraceae*, and C-16:1 is only a minor component in this latter group. In addition, the lipopolysaccharides are significantly different in members of the two families (Weckesser et al., 1979, 1995). The lipid A of investigated *Chromatiaceae* (*Allochromatium vinosum*, *Thermochromatium tepidum*, *Thiocystis violacea*, *Thiocapsa roseopersicina* and *Thiococcus pfennigii*) is characterized by a phosphate-free backbone with D-glucosamine as the only amino sugar, which has terminally attached D-mannose and amide-bound 3-OH-C-14:0. In the lipid A of all tested *Ectothiorhodospiraceae* (*Ectothiorhodospira vacuolata*, *Ect. shaposhnikovii*, *Ect. haloalkaliphila* and *Halorhodospira halophila*), phosphate is present, 2,3-diamino-2,3-dideoxy-D-glucose is the major amino sugar (D-glucosamine is also present), D-mannose is lacking (D-galacturonic acid and D-glucuronic acid are present instead), and quite remarkably, 3-OH-C-10:0 is present as an amide-bound fatty acid (Zahr et al., 1992; Weckesser et al., 1995). These distinctive properties of the lipid A appear to be characteristic features of the two families.

A.   Ectothiorhodospiraceae

Ectothiorhodospiraceae (Imhoff, 1984a) represent a group of haloalkaliphilic purple sulfur bacteria that form a separate line of phylogenetic descent related to the *Chromatiaceae*. *Ectothiorhodospiraceae* are clearly separated from the *Chromatiaceae* by sequence similarity and signature sequences of their 16S rDNA (Imhoff and Süling, 1996; Imhoff et al., 1998b). In a phylogenetic tree based on 16S rDNA data both families form separate but related groups within the Gammaproteobacteria (Fowler et al., 1984; Stackebrandt et al., 1984; Woese et al., 1985; Imhoff and Süling, 1996).

Ectothiorhodospiraceae have been distinguished from the *Chromatiaceae* on the basis of both phenotypic and molecular information (Imhoff, 1984a). During oxidation of sulfide, the *Ectothiorhodospiraceae* deposit elemental sulfur outside their cells. Furthermore, they are distinguished from the *Chromatiaceae* by lamellar intracellular membrane structures, by significant differences of the polar lipid composition (Imhoff et al., 1982; Imhoff and Bias-Imhoff, 1995), and by the dependence on saline and alkaline growth conditions (Imhoff, 1989). *Halorhodospira halophila* is the most halophilic eubacterium known and even grows in saturated salt solutions.

On the basis of sequence similarities and by a number of characteristic signature sequences, two major phylogenetic groups were recognized and classified as separate genera. The extremely halophilic species were reassigned to the genus *Halorhodospira*, including the species *Halorhodospira halophila*, *Halorhodospira halochloris* and *Halorhodospira abdelmalekii*. A new species of this genus, *Halorhodospira neutriphila*, which grows at neutral pH has been described recently (Hirschler-Rea et al., 2003). Among the slightly halophilic species, the classification of strains belonging to *Ectothiorhodospira mobilis* and *Ectothiorhodospira shaposhnikovii* was improved and a close relationship between

Ect. shaposhnikovii and *Ect. vacuolata* was demonstrated. Based on genetic results, *Ectothiorhodospira marismortui* has been confirmed as a distinct species closely related to *Ect. mobilis*. Several strains which previously had been tentatively identified as *Ectothiorhodospira mobilis* formed a separate cluster on the basis of their 16S rDNA sequences and were recognized as two new species: *Ectothiorhodospira haloalkaliphila*, which includes the most alkaliphilic strains originating from strongly alkaline soda lakes and *Ectothiorhodospira marina*, describing isolates from the marine environment (Imhoff and Süling, 1996).

New alkaliphilic isolates from Siberian and Mongolian soda lakes were found to be distinct from described species of the genera *Ectothiorhodospira* and *Halorhodospira* but more closely related to *Ectothiorhodospira*. Both are regarded as species of the new genera *Thiorhodospira* (Bryantseva et al., 1999) and *Ectothiorhodosinus* (Gorlenko et al., 2004). (In contradiction to these authors, the genus gender should be masculine and therefore the correct species designation is *Ectothiorhodosinus mongolicus*.) In *Thiorhodospira sibirica* sulfur globules remain attached to the cells and according to microscopic observations are located in the cell periphery or the periplasmic space of the cells (Bryantseva et al., 1999).

In addition to the phototrophic genera (*Ectothiorhodospira*, *Thiorhodospira*, *Halorhodospira*, *Ectothiorhodosinus*), the *Ectothiorhodospiraceae* include genera of purely chemotrophic bacteria unable to perform anoxygenic photosynthesis (*Arhodomonas*, *Nitrococcus*, *Alkalispirillum*).

B.   Chromatiaceae

The *Chromatiaceae* (Bavendamm, 1924) (emended description Imhoff, 1984a) comprise those phototrophic purple sulfur bacteria that, under the proper growth conditions, deposit globules of elemental sulfur inside their cells (Imhoff, 1984a). The family represents a quite coherent group of species, based on physiological properties, on the similarity of 16S rDNA sequences (Fowler et al., 1984; Guyoneaud et al., 1998; Imhoff et al., 1998b) and on chemotaxonomic markers such as fatty acid and quinone composition (Imhoff and Bias-Imhoff, 1995) and lipopolysaccharide structures

(Meißner et al., 1988; Weckesser et al., 1995). In addition to the photoautotrophic mode of growth with reduced sulfur compounds as most important photosynthetic electron donors, several species also are able to grow under photoheterotrophic conditions, some even as chemoautotrophs, and a few species also can grow chemoheterotrophically (Gorlenko, 1974; Kondratieva et al., 1976; Kämpf and Pfennig, 1980). All of them oxidize sulfide and elemental sulfur, and some also oxidize thiosulfate and sulfite (Trüper, 1981). Sulfide is oxidized to sulfate as the final oxidation product. During growth of *Chromatiaceae* on sulfide and thiosulfate, sulfur appears in the form of globules inside the bacterial cells. During oxidation of thiosulfate, the sulfur of these globules is entirely derived from the sulfane group of thiosulfate (Smith, 1965; Trüper and Pfennig, 1966). The sulfur in the globules exists in a metastable state and is not true elemental sulfur. It mainly consists of long sulfur chains very probably terminated by organic residues (mono-/bis-organyl polysulfanes) in purple and also in green sulfur bacteria. Most probably, the organic residue at the end of the sulfur chains in the sulfur globules is glutathione or very similar to glutathione (Prange et al., 2002). For a detailed discussion of this topic see the chapters Dahl (chapter 15) and Prange et al. (chapter 23). The sulfur globules are surrounded by a protein monolayer consisting of three different proteins in *Allochromatium vinosum* and two proteins in *Thiocapsa roseopersicina* (Brune, 1995). Evidence is presented, that these sulfur globule proteins contain amino-terminal signal peptides pointing to an extracytoplasmic localization of the sulfur globules (Pattaragulwanit et al., 1998).

During aerobic dark growth, elemental sulfur may support respiration and serve as electron donor for chemolithotrophic growth (Breuker, 1964; Kämpf and Pfennig, 1986). During anaerobic dark, fermentative metabolism, intracellular sulfur serves as an electron sink during oxidation of stored carbohydrates and is reduced to sulfide (Van Gemerden, 1968a, 1968b, 1974). Though growth under these conditions is very poor in *Chromatiaceae*, several species have the capability of a fermentative metabolism that at least allows survival in the absence of light and oxygen (Van Gemerden, 1968a, 1968b; Krasilnikova et al., 1975, 1983; Krasilnikova, 1976).

Approaches to the phylogeny of the *Chromatiaceae* were made using full length 16S rDNA sequences. The first complete 16S rDNA sequences were obtained for *Allochromatium vinosum* (DeWeerd et al., 1990) and *Thermochromatium tepidum* (Madigan, 1986). With the description of the new species and genera *Rhabdochromatium marinum* (Dilling et al., 1995), *Halochromatium glycolicum* (Caumette et al., 1997) and *Thiorhodococcus minus* (Guyoneaud et al., 1997), more 16S rDNA sequences became available. With the analysis of complete 16S rDNA sequences from most *Chromatiaceae* species (Guyoneaud et al., 1998; Imhoff et al., 1998b) the phylogenetic relationship of these bacteria was analysed and the existence of major groups of species was established. As a consequence, the reclassification of a number of these bacteria, based on their genetic relationship and supported by diagnostic phenotypic properties was proposed (Guyoneaud et al., 1998; Imhoff et al., 1998b). However, morphological and a number of physiological properties used so far in the classification of these bacteria, have little relevance in a genetically oriented classification system. Apparently, ecological aspects and adaptation of bacteria to specific factors of their habitat, like salinity, are of importance in a phylogenetically oriented taxonomy (see below).

The genetic relatedness determined on the basis of 16S rDNA nucleotide sequences revealed that major phylogenetic branches of the *Chromatiaceae* contain (1) truly marine and halophilic species, (2) species that are motile by polar flagella, do not contain gas vesicles, and are primarily freshwater species, and (3) species with ovoid to spherical cells, the majority of which are non-motile freshwater species containing gas vesicles.

The marine branch includes the genera *Marichromatium*, *Halochromatium*, *Rhabdochromatium*, *Thiorhodococcus*, *Thiococcus*, *Thioflavicoccus*, *Thioalkalicoccus*, *Thiorhodovibrio*, *Thiohalocapsa* and *Isochromatium*. Three genetically related species of this group, which are adapted to the lower range of salt concentrations of brackish and marine habitats (*Thiococcus pfennigii*, *Thioalkalicoccus limnaeus* and *Thioflavicoccus mobilis*), are clearly distinct from all others by containing bacteriochlorophyll *b* and by the presence of tubular internal membranes (Bryantseva et al., 2000; Imhoff and Pfennig, 2001).

The second major branch includes the genera *Chromatium*, *Allochromatium*, *Thermochromatium* and *Thiocystis*, which are motile forms without gas vesicles.

The third branch includes the genera *Thiocapsa*, *Thiolamprovum*, *Thiobaca* and *Lamprocystis* and others as reclassified by Guyoneaud et al. (1998), namely *Thiocapsa pendens* (formerly *Amoebobacter pendens*), *Thiocapsa rosea*, (formerly *Amoebobacter roseus*), *Thiolamprovum pedioformis* (formerly *Amoebobacter pedioformis*) and *Lamprocystis purpurea* (formerly *Amoebobacter purpureus*, Imhoff, 2001b). Unfortunately, *Amoebobacter purpureus* was reclassified on the basis of purely nomenclatural aspects but without supporting data as *Pfennigia purpurea* (see Bergey's Manual of Systematic Bacteriology, Vol. 2B). Because available data were in disagreement with this classification, it had to be reclassified as a species of *Lamprocystis*, *Lamprocystis purpurea* (Imhoff, 2001b). (Attention also has to be given to the different strains assigned to this bacterium (named *Amoebobacter purpureus* or *Lamprocystis purpurea*), because one of the strains according to its 16S rDNA sequence available in databases is misclassified and belongs to a different species.) *Lamprocystis roseopersicina* is one of the rare cases where gas vesicles are formed and the cells are in addition motile by flagella. *Thiobaca trueperi* is a motile rod without gas vesicles (Rees et al., 2002). *Thiocapsa roseopersicina*, one of the best known species of this group, does not form gas vesicles. So far, unpublished sequences of *Thiodictyon* species indicate their association to the group around *Thiocapsa* species.

Because 16S rDNA sequences from *Thiospirillum jenense*, *Lamprobacter modestohalophilus* and *Thiopedia rosea* are presently not available, the phylogenetic assignment of these bacteria is still uncertain.

It was suggested that the salt response is one important taxonomic criterion in such a taxonomic system of the *Chromatiaceae* (Imhoff et al., 1998b), because both the genetic relationship and the salt responses distinguish major phylogenetic branches of the *Chromatiaceae* and single genera. Both the genetic relationship and the salt responses enable to distinguish between, e.g., the halophilic *Halochromatium salexigens* and *Halochromatium glycolicum*, and the marine *Marichromatium gracile* and *Marichromatium purpuratum* from each other and from freshwater species such as *Thiocapsa roseopersicina* and *Allochromatium vinosum* and their relatives. This implies a separate phylogenetic development in the marine and in the freshwater environment and points to the general importance of salt responses and possibly other ecological parameters defining ecological niches for species formation and evolution (Imhoff, 2001a).

III. Phototrophic Purple Nonsulfur Bacteria

Purple nonsulfur bacteria are affiliated with the Alphaproteobacteria and the Betaproteobacteria. The analysis of 16S rDNA sequences revealed a close relationship of these phototrophic bacteria to purely chemotrophic bacteria in numerous cases (e.g. Gibson et al., 1979; Woese et al., 1984a, b; Woese, 1987; Kawasaki et al., 1993; Hiraishi and Ueda, 1994). In addition, a great number of so called "aerobic bacteriochlorophyll-containing bacteria or ABC-bacteria" (not treated here) is associated with these groups.

Bacteria of the phototrophic purple nonsulfur bacteria are able to perform anoxygenic photosynthesis with bacteriochlorophylls and carotenoids as photosynthetic pigments. None of the described species contains gas vesicles. Internal photosynthetic membranes are continuous with the cytoplasmic membrane and consist of vesicles, lamellae, or membrane stacks. Color of cell suspensions is green, beige, brown, brown-red, red or pink.

This is in strict contrast to the "ABC-bacteria". The "aerobic bacteriochlorophyll-containing Alphaproteobacteria" such as *Erythrobacter longus* and others exhibit physiological properties and occupy ecological niches clearly distinct from the phototrophic purple nonsulfur bacteria, because oxygen does not repress synthesis of photosynthetic pigments in these bacteria. Furthermore, these bacteria are strictly aerobic bacteria.

The purple nonsulfur bacteria (*Rhodospirillaceae*, Pfennig and Trüper, 1971) represent by far the most diverse group of the phototrophic purple bacteria (Imhoff and Trüper, 1989). The high diversity of these bacteria is reflected in the organization of the internal membrane systems,

16S rDNA sequence similarities, carotenoid composition, utilization of carbon sources and electron donors. Furthermore, this high diversity is well documented by a number of chemotaxonomic observations, such as cytochrome c_2 amino acid sequences, lipid, quinone and fatty acid composition, as well as lipid A structures (Ambler et al., 1979; Weckesser et al., 1979, 1995; Dickerson, 1980; Hiraishi et al., 1984; Imhoff, 1984b, 1991, 1995; Imhoff et al., 1984; Imhoff and Bias-Imhoff, 1995). As a consequence, it is not appropriate to assign new species to the genera only on the basis of physiological and morphological properties. Chemotaxonomic characteristics and sequence information also have to be taken into consideration. In addition, environmental aspects and ecological distribution should be considered. An outline on recommendations on the description of new species of anoxygenic phototrophic bacteria is given by Imhoff and Caumette (2004).

It was the recognition of the close genetic relationship between phototrophic purple bacteria and chemotrophic bacteria on the basis of 16S rRNA oligonucleotide catalogues and 16S rDNA sequences, respectively, which led C.R. Woese to call the Proteobacteria the Purple Bacteria and their relatives and to discuss the role of phototrophic purple nonsulfur bacteria as ancestors of numerous chemotrophic representatives of these Proteobacteria groups (Woese et al., 1984a, b; Woese et al., 1985; Woese, 1987). With the recognition of their genetic relationships and with the support from chemotaxonomic data and ecophysiological properties, purple nonsulfur bacteria of the Alphaproteobacteria and Betaproteobacteria were taxonomically separated and rearranged according to the proposed phylogeny. Despite the fact that many of the phototrophic purple nonsulfur bacteria are closely related to strictly chemotrophic relatives, the phototrophic capability and the content of photosynthetic pigments is included in the genus definitions of these bacteria.

Members of this group are widely distributed in nature and have been found in freshwater, marine and hypersaline environments that are exposed to the light. They live preferably in aquatic habitats with significant amounts of soluble organic matter, low oxygen tension and moderate temperatures, but also in thermal springs and alkaline soda lakes. They rarely form colored blooms, which are characteristically formed by representatives of purple sulfur bacteria and phototrophic green sulfur bacteria. The preferred mode of growth is photoheterotrophically under anoxic conditions in the light, but photoautotrophic growth with molecular hydrogen and sulfide may be possible, and most species are capable of chemotrophic growth under microoxic to oxic conditions in the dark. While some species are very sensitive to oxygen, others grow equally well aerobically in the dark.

A. Anaerobic Phototrophic Alphaproteobacteria

1. Phototrophic alpha-1 Proteobacteria (Rhodospirillales)

All genera of this group are classified with the *Rhodospirillales*, most of them with the *Rhodospirillaceae* family. *Rhodopila* is classified with the *Acetobacteraceae* (Table 2). Based on 16S rDNA sequence the phototrophic alpha-1 Proteobacteria are phylogenetically distinct from other groups of phototrophic Alphaproteobacteria, though they are closely related to several purely chemotrophic representatives of this group.

Most of the species of the phototrophic alpha-1 Proteobacteria have been previously known as *Rhodospirillum* species and are of spiral shape. They belong to the genera *Rhodospirillum*, *Phaeospirillum*, *Roseospira*, *Rhodocista*, *Rhodovibrio*, *Rhodospira* and *Roseospirillum* (Imhoff et al., 1998a). The only non-spiral representative of this group is *Rhodopila globiformis*. This acidophilic phototrophic bacterium is phylogenetically closely related to acidophilic chemotrophic bacteria of the genera *Acetobacter* and *Acidiphilium* (Sievers et al., 1994). Also other phototrophic alpha-1-Proteobacteria are closely related to different chemotrophic representatives. *Phaeospirillum* species, e.g. demonstrate close sequence similarity to *Magne-tospirillum magnetotacticum* (Burgess et al., 1993) and *Rhodocista centenaria* reveals strong relations to *Azospirillum* species (Xia et al., 1994; Fani et al., 1995). *Rhodovibrio* and *Rhodothalassium* are distantly related to the other genera and their assignment to higher taxa is currently not without problems (Table 2). *Rhodothalassium* (Imhoff 2005 m) is certainly misclassified within the *Rhodobacteraceae*.

Table 2. Genera of anoxygenic phototrophic bacteria and their classification in higher taxa[a].

Class	Order	Family	Genera[b]
Chloroflexi	*Chloroflexales*	*Chloroflexaceae*	*Chloroflexus, Oscillochloris, Heliothrix, Chloronema, Roseiflexus*
Chlorobi	*Chlorobiales*	*Chlorobiaceae*	*Chlorobium, Chlorobaculum, Prosthecochloris, Chloroherpeton*
Clostridia	*Clostridiales*	*Heliobacteriaceae*	*Heliobacterium, Heliobacillus, Heliophilum, Heliorestis*
Alphaproteobacteria	*Rhodospirillales*	*Rhodospirillaceae*	*Rhodospirillum*
			Phaeospirillum, Rhodocista, Roseospira, Roseospirillum, Rhodospira
			Rhodovibrio[c], Rhodothalassium[c]
		Acetobacteraceae	*Rhodopila*
	Rhodobacterales	*Rhodobacteraceae*	*Rhodobacter*
			Rhodobaca, Rhodovulum
	Rhizobiales	*Bradyrhizobiaceae*	*Rhodopseudomonas, Rhodoblastus*
			Blastochloris[c], Rhodoplanes[c]
		Hyphomicrobiaceae	*Rhodomicrobium*
		Rhodobiaceae	*Rhodobium*
Betaproteobacteria	*Burkholderiales*	*Comamonadaceae*	*Rhodoferax*
			Rubrivivax[c]
	Rhodocyclales	*Rhodocyclaceae*	*Rhodocyclus*
Gammaproteobacteria	*Chromatiales*	*Chromatiaceae*	*Chromatium, Thermochromatium, Allochromatium, Thiocystis*
			Thiocapsa, Thiolamprovum, Thiobaca, Lamprocystis
			Marichromatium, Halichromatium, Rhabdochromatium, Thiococcus
			Thiorhodococcus, Thioflavicoccus, Thioalkalicoccus, Thiorhodovibrio
			Thiohalocapsa, Isochromatium
			(Thiospirillum, Lamprobacter, Thiodictyon, Thiopedia)[d]
		Ectothiorhodospiraceae	*Ectothiorhodospira*
			Halorhodospira,
			Ectothiorhodosinus, Thiorhodospira

[a] According to Bergey's Manual on Systematic Bacteriology, 2nd edn. (2001 and 2005).

[b] Only anoxygenic phototrophic genera of these higher taxa are listed here.

[c] In the 2nd edn. of Bergey's Manual of Systematic Bacteriology, unfortunately these genera were assigned as incertae sedis or even misplaced.

[d] These genera have no clear phylogenetic standing, because 16S rDNA sequences (and pure cultures) are not available.

Several chemotaxonomic properties distinguish the phototrophic alpha-1 Proteobacteria from other phototrophic Alphaproteobacteria. Ubiquinones, menaquinones, and rhodoquinones may be present, and the length of their side chain may vary from 7 to 10 isoprene units. They have characteristic phospholipid and fatty acid composition with C-18:1 as the dominant fatty acid and either C-16:1 and C-16:0, C-16:0 and C-18:0, or just C-16:0 as additional major components.

On the basis of distinct phenotypic properties and 16S rDNA sequence similarities of all recognized spiral-shaped purple nonsulfur Alphaproteobacteria, a reclassification of these bacteria had been proposed (Imhoff et al., 1998a). Phylogenetic relations on the basis of 16S rDNA sequences of these bacteria are in good correlation with differences in major quinone and fatty acid composition and also with their growth requirement for NaCl or sea salt. This is in accordance with different phylogenetic lines forming freshwater and salt water representatives. Therefore, these properties were considered of importance in defining and differentiating these genera. Four of these genera are defined as salt-dependent and three as fresh water bacteria. Only *Rsp. rubrum* and *Rsp. photometricum* were maintained as species of the genus *Rhodospirillum*.

2. Phototrophic Alpha-2 Proteobacteria (Rhizobiales)

Most of the species of the phototrophic alpha-2 Proteobacteria have been previously known as *Rhodopseudomonas* species and have rod-shaped motile cells. They belong to the genera *Rhodopseudomonas, Rhodobium, Rhodoplanes, Rhodoblastus, Blastochloris* and *Rhodomicrobium*. Based on analysis of 16S rDNA sequences, the phototrophic Alphaproteobacteria of the order *Rhizobiales* are well separated from other groups of phototrophic Alphaproteobacteria; however, they are closely related to several purely chemotrophic Alphaproteobacteria of the order *Rhizobiales*. *Rhodopseudomonas palustris*, for example, is most closely related to *Nitrobacter* species.

A number of morphological and chemotaxonomic properties distinguishes the phototrophic alpha-2 Proteobacteria from other purple nonsulfur bacteria. Most characteristic is their budding mode of growth and cell division which comes along with lamellar internal membranes that are lying parallel to the cytoplasmic membrane. There is variation in the presence of either ubiquinone alone, ubiquinone together with either rhodoquinone or menaquinone, or ubiquinone with both menaquinone and rhodoquinone as major components. Most species have 10 isoprenoid units in their side chains (except *Blastochloris* species). As far as known, either small or large "mitochondrial type" cytochrome c_2 is present. Characteristic phospholipids are present and among the fatty acids C-18:1 is the dominant fatty acid and either C-16:1 and C-16:0, C-16:0 and C-18:0 or just C-16:0 are additional major components (see Imhoff and Bias-Imhoff, 1995).

Outstanding properties which distinguish *Rhodomicrobium* from other alpha-2 Proteobacteria are the filament formation and the characteristic growth cycle. Other distinguishing characteristics are the composition of the lipid A, of polar lipids and fatty acids. Among the closest phylogenetic relatives of *Rhodomicrobium* based on 16S rDNA

sequence analysis is *Hyphomicrobium vulgare* (Kawasaki et al., 1993).

In the second edition of Bergey's Manual of Systematic Bacteriology, *Rhodobium* (Imhoff and Hiraishi, 2005) is assigned to the *Rhodobiaceae*, *Rhodomicrobium* (Imhoff, 2005j) to the *Hyphomicrobiaceae* and *Rhodoblastus* and *Rhodopseudomonas* (Imhoff, 2005g, h) to the *Bradyrhizobiaceae*. Unfortunately and in contrast to the author's opinion, *Blastochloris* and *Rhodoplanes* (Imhoff, 2005i; Hiraishi and Imhoff, 2005b) have been misplaced in this edition. Both should be included into the *Bradyrhizobiaceae* together with the genera *Rhodoblastus* and *Rhodopseudomonas*.

3. Phototrophic Alpha-3 Proteobacteria (Rhodobacterales)

Phototrophic alpha-3 Proteobacteria have a number of characteristic chemotaxonomic properties that enable their diagnosis. All investigated species have a large type cytochrome c_2 (Ambler et al., 1979; Dickerson, 1980) and as sole quinone component Q-10 (Imhoff, 1984b; Hiraishi et. al., 1984). Those species that are able to assimilate sulfate use the pathway via 3′-phosphoadenosine-5′-phosphosulfate (PAPS, Imhoff, 1982). C-18 and C-16 saturated and monounsaturated fatty acids are major fatty acids, C-18:1 the predominant component (Imhoff, 1991). The lipopolysaccharides of investigated species contain in their lipid A moieties glucosamine as sole amino sugar, have phosphate, amide-linked 3-OH-14:0 and/or 3-oxo-14:0 and ester-linked 3-OH-10:0 (Weckesser et al., 1995). A differentiation of the genera and species of *Rhodobacter*, *Rhodovulum* and *Rhodobaca* is possible on the basis of 16S rDNA sequences and by DNA–DNA hybridization.

The phototrophic alpha-3 Proteobacteria are phylogenetically well separated from other groups of phototrophic Alphaproteobacteria, though they are closely related to purely chemotrophic alpha-3 Proteobacteria. The majority belongs to the genera *Rhodobacter* and *Rhodovulum* (Pfennig and Trüper, 1974; Imhoff et al., 1984; Hiraishi and Ueda, 1994). The former are freshwater bacteria and the latter true marine bacteria and species of both genera have distinct 16S rDNA sequences (Hiraishi and Ueda, 1994, 1995; Hiraishi et al.,

1996; Straub et al., 1999). The recently described new bacterium *Rhodobaca bogoriensis* is an alkaliphilic slightly halophilic bacterium from African soda lakes and is phylogenetically associated to *Rhodobacter* (Milford et al., 2000). All three genera (Imhoff, 2005d, e, f) are classified with the Rhodobacteraceae of the *Rhodobacterales*.

Characteristic properties of *Rhodobacter*, *Rhodobaca* and *Rhodovulum* species are the ovoid to rod-shaped cell morphology, the presence of vesicular internal membranes (except *Rba. blasticus*) and the content of carotenoids of the spheroidene series. Most *Rhodobacter* species are distinct from *Rhodovulum* species by the lack of a substantial NaCl requirement for optimal growth, i.e. they show a typical response of freshwater bacteria. The salt requirement for optimal growth of *Rhodovulum* species on the other hand does not preclude that some of these bacteria also may grow in the absence of salt.

Species of *Rhodobacter* and *Rhodovulum* not only are well characterized by phenotypic properties, but also are established on the basis of 16S rDNA sequences and DNA–DNA hybridization studies. Comprehensive DNA/DNA hybridization studies have been performed both with *Rhodobacter* and *Rhodovulum* species. A first detailed study including 21 strains of the species known at that time gave support for the recognition of strains of *Rba. veldkampii* as a new species and in addition revealed the diversity of marine isolates of this group (DeBont et al., 1981). This study also demonstrated the identity on the species level of a denitrifying isolate of *Rba. sphaeroides* and other non-denitrifying strains of this species (Satoh et al., 1976; DeBont et al., 1981). Similarly, several marine and halophilic isolates were shown to be related to *Rhv. euryhalinum* by DNA–DNA hybridization but significantly distinct from *Rhv. sulfidophilum*, *Rba. sphaeroides* and *Rba. capsulatus* (Ivanova et al., 1988). DNA–DNA hybridization also allowed the genetic distinction of 4 strains of the denitrifying *Rba. azotoformans* from *Rba. sphaeroides* and other *Rhodobacter* species (Hiraishi et al., 1996). Several strains of *Rhodovulum strictum*, which according to 16S rDNA sequence is most similar to *Rhv. euryhalinum* (96.8%), were shown to have low DNA–DNA homology (less than 30%) to type strains of all other *Rhodovulum* species, including *Rhv. euryhalinum* (Hiraishi and Ueda, 1995).

B. Anaerobic Phototrophic Betaproteobacteria (Rhodocyclales & Burkholderiales)

The phototrophic Betaproteobacteria comprise species of three genera, *Rhodocyclus*, *Rubrivivax* and *Rhodoferax*. Based on 16S rDNA sequences, the phototrophic Betaproteobacteria represent different phylogenetic lines within the Betaproteobacteria (Hiraishi, 1994). *Rhodocyclus* (Imhoff, 2005l) is classified in the family *Rhodocyclaceae* of the order *Rhodocyclales*, *Rhodoferax* (Hiraishi and Imhoff, 2005a) is classified with the *Comamonadaceae* of the order *Burkholderiales*, and *Rubrivivax* (Imhoff, 2005k) is presently classified as genus incertae sedis.

Sequences of 16S rDNA clearly classify these bacteria as belonging to the Betaproteobacteria (Hiraishi, 1994; Maidak et al., 1994). They are freshwater bacteria common in stagnant waters that are exposed to the light and have an increased load of organic compounds and nutrients and are deficient in oxygen. Internal photosynthetic membranes are much less developed than in other phototrophic purple bacteria appearing as small fingerlike intrusions and are not always evident. Growth preferably occurs under photoheterotrophic conditions, anaerobically in the light. All species known so far do not use reduced sulfur compounds as photosynthetic electron donor and sulfide is growth inhibitory already at low concentrations. Sulfate can be assimilated as sole sulfur source and is reduced with adenosine-5′-phosphosulfate (APS) as an intermediate (Imhoff, 1982). NADH is used as a cosubstrate in the GS/GOGAT reactions and HIPIP is present (Ambler et al., 1979).

Prior to the establishment of the phylogenetic relationship among the phototrophic Betaproteobacteria, these species had been included in the *Rhodospirillaceae* together with the phototrophic Alphaproteobacteria (Pfennig and Trüper, 1974). Three species were known as *Rhodopseudomonas gelatinosa*, *Rhodospirillum tenue* (Pfennig and Trüper, 1974) and *Rhodocyclus purpureus* (Pfennig, 1978). In addition to a clear phylogenetic separation (Hiraishi, 1994), both of these groups show significant differences in a number of chemotaxonomic properties. As a consequence, *Rhodospirillum tenue* (Pfennig, 1969) was transferred to *Rhodocyclus tenuis* (Imhoff et al., 1984). *Rhodopseudomonas gelatinosa* was transferred to *Rhodocyclus gelatinosus* (Imhoff et al., 1984) and later assigned to a new genus as *Rubrivivax gelatinosus* (Willems et al., 1991). Additional new bacteria have been isolated since then that are also members of the Betaproteobacteria and have been described as the new species and genus *Rhodoferax fermentans* (Hiraishi and Kitamura, 1984; Hiraishi et al., 1991) and as additional species of this genus, *Rhodoferax antarcticus* (Madigan et al., 2000) and *Rhodoferax ferrireducens* (Finneran et al., 2003).

The phototrophic Betaproteobacteria have ubiquinone and menaquinone (or rhodoquinone) derivatives with eight isoprenoid units in the side chain (Q-8, RQ-8 and MK-8); they have a "small type" cytochrome c_{551} (in contrast to the phototrophic Alphaproteobacteria; Ambler et al., 1979; Dickerson, 1980); they have characteristic phospholipid and fatty acid compositions with the highest proportions of C-16 fatty acids (16:0 and 16:1) among all phototrophic purple bacteria and correspondingly very low ones of 18:1 (Hiraishi et al., 1991; Imhoff, 1984b; Imhoff and Bias-Imhoff, 1995; Imhoff and Trüper, 1989). Lipopolysaccharides of phototrophic Betaproteobacteria characteristically contain significant amounts of phosphate and amide-linked 3-OH-capric acid (3-OH-C-10) in their lipid A moiety (Weckesser et al., 1995). In *Rfx. fermentans* 3-OH-C-8:0 was found instead (Hiraishi et al., 1991).

IV. Phototrophic Green Sulfur Bacteria – *Chlorobiales*

The green sulfur bacteria, represented by the family *Chlorobiaceae*, form a branch of bacteria that is phylogenetically distinct from other main phylogenetic lines, and is therefore treated as a separate phylum (the Chlorobi) in *Bergey's Manual of Systematic Bacteriology* (Overmann, 2001). Traditionally, the taxonomic classification of these bacteria was based on morphological and easily recognizable phenotypic properties (Pfennig, 1989; Pfennig and Overmann, 2001a, b). Such properties include cell morphology, pigment composition and absorption spectra, and metabolic properties. However, some of these properties appear to be problematic or even misleading in a phylogenetically oriented systematic system.

In particular, (i) the formation of gas vesicles has been used to distinguish between genera; (ii) brown-colored forms have been distinguished as species from their green-colored counterparts, and are distinct in their bacteriochlorophyll and carotenoid composition; and (iii) subspecies were recognized on the basis of utilization of thiosulfate as a photosynthetic electron donor. Although these properties are easily recognizable and have allowed a phenotypic differentiation, they are not in accord with the phylogenetic relationship of these bacteria (Figueras et al., 1997; Overmann and Tuschak, 1997). Therefore, in a systematic taxonomy of green sulfur bacteria based on phylogenetic relationships, they can not be used for the differentiation of species.

Phylogenetic relationships of green sulfur bacteria were established by using 16S rRNA and *fmo* (Fenna–Matthews–Olson protein, FMO protein) gene sequences, including important signatures of amino acid and nucleotide sequences (Alexander et al., 2002). Both 16S rRNA and *fmo* gene sequence information is available for most of the type strains. In addition, a larger number of 16S rDNA sequences and *fmo* gene sequences of non-type strains are known (Figueras et al., 1997; Overmann and Tuschak, 1997; Alexander et al., 2002). The congruent phylogenetic relationships found with two independent gene sequences provide a solid basis for the phylogeny of these bacteria, and for a phylogeny-based taxonomy (Imhoff, 2003).

The phylogenetic studies revealed almost-identical grouping in trees constructed from 16S rDNA and *fmoA* sequences. This suggests a largely congruent evolution of FMO and 16S rDNA (Alexander et al., 2002) and gives strong support to the 16S rDNA-based phylogeny of these bacteria. The assignment of strains into phylogenetic groups is further supported by characteristic signatures in the amino acid sequence of the FMO protein (Alexander et al., 2002). The phylogenetic grouping of the green sulfur bacteria is not in accord with their traditional classification. Therefore, the reclassification of these bacteria was necessary and included a complete reassignment of strains and species (Imhoff, 2003) (Table 3). Three examples demonstrate this: (i) species and strains formerly assigned to the genus *Chlorobium* were found in all phylogenetic groups, those of the genus

Table 3. Species names and properties of strains of *Chlorobium*, *Prosthecochloris* and *Chlorobaculum* species[a].

Genus and species name	old name	Strain number	Thio sulfate used	Cell size [μm]	Salt required	Vitamins	Major Bchl	G + C values mol%	Gas vesicles	carotenoid
Chlorobium										
Chl. limicola	Chl. limicola	DSM 245^T	–	0.7–1.1	no	–	c	51.0	–	clb
Chl. limicola	Chl. limicola	DSM 246	–		no	B$_{12}$	c	52.0	–	clb
Chl. limicola	Chl. limicola f. thios.	1630	+		no	B$_{12}$	c	52.5	–	
Chl. limicola	Chl. limicola f. thios.	9330	+		no	–	c	52.0	–	iso
Chl. limicola	Chl. phaeobacteroides	DSM 1855	–		no	B$_{12}$	e	0	–	
Chl. limicola	Chl. limicola	DSM 247	–		no	B$_{12}$	c	51.5	–	
Chl. limicola	Chl. limicola	DSM 248	–		no	B$_{12}$	c	51.5	–	
Chl. limicola	Chl. limicola f. thios.	DSM 257	+		no	–	c	52.5	–	
Chl. phaeobacteroides	Chl. phaeobacteroides	DSM 266^T	–	0.6–0.8	no	B$_{12}$	e	49.0	–	iso
Chl. phaeobacteroides	Chl. phaeobacteroides	DSM 267	–		no	B$_{12}$	e	50.0	–	iso
Chl. clathratiforme	Pld. phaeoclathrathiforme	DSM 5477^T	+	0.8–1.1	no	B$_{12}$	e	47.9	+	iso
Chl. clathratiforme	Pld. clathratiforme	PG	–	0.7–1.2	no	0	e	(48.5)	+	clb
Chl. ferrooxidans	Chl. ferrooxidans	DSM 13031^T	0	0.5	no	0	c	0	–	clb
Chl. luteolum	Pld. luteolum	DSM 273^T	–	0.6–0.9	>1%	B$_{12}$	c	58.1	+	clb
Chl. luteolum	Chl. vibrioforme	DSM 262	–	0.5–0.7	>1%	B$_{12}$	d	57.1	–	clb
Chl. phaeovibrioides	Chl. phaeovibrioides	DSM 269^T	–	0.3–0.4	>1%	B$_{12}$	e	53.0	–	clb
Chl. phaeovibrioides	Chl. vibriof. f.thios.	DSM 265	–	0.5–0.7	>1%	B$_{12}$	d+c	53.5	–	iso
Chl. phaeovibrioides	Chl. vibrioforme	DSM 261	–		>1%	B$_{12}$	d	52.0	–	
Chl. phaeovibrioides	Chl. phaeovibrioides	DSM 270	–		>1%	B$_{12}$	e	52.0	–	
Prosthecochloris										
Ptc. aestuarii	Ptc. aestuarii	DSM 271^T	–	0.5–0.7	2–5%	B$_{12}$	c	52–56.1	–	clb
Ptc. spec.	Ptc. aestuarii	2K	–		+	0			–	
Ptc. vibrioformis	Chl. vibrioforme	DSM 260^T	–	0.5–0.7	>1%	B$_{12}$	d+c	53.5	–	clb
Ptc. vibrioformis	Chl. phaeovibrioides	DSM 1678	–		>1%	0	e	0	–	
Ptc. vibrioformis	Chl. vibrioforme	CHP 3402	0		>1%	0	e		–	
Chlorobaculum										
Chlorobaculum tepidum	Chl. tepidum	ATCC49652^T	+	0.6–0.8	no		c	56.5	–	clb
Cba. limnaeum	Chl. phaeobacteroides	DSM 1677^T	–		no	B$_{12}$	e	0	–	
Cba. limnaeum	Chl. phaeobac.	1549	–		no				–	

(continued)

Table 3. (continued)

Genus and species name	old name	Strain number	Thio sulfate used	Cell size [µm]	Salt required	Vitamins	Major Bchl	G + C values mol%	Gas vesicles	carotenoid
O	*Chl. limic.*	UdG 6040	o	0.8–1.0			c		–	clb
O	*Chl. limic.*	UdG 6042	o	0.8–1.0			c		–	clb
O	*Chl. limic.*	UdG 6045	o	0.7–1.0			c		–	clb
O	*Chl. limic.*	UdG 6038	o	0.7–1.0			c		–	clb
Cba. thiosulfatiphilum	*Chl. limic. f.thios.*	DSM 249[T]	+	0.7–1.0	no	–	c	58.1	–	clb
Cba. thiosulfatophilum	*Chl. limic. f.thios.*	1430	+		no	–	c	58.1	–	
Cba. parvum	*Chl. vibriof. f.thios.*	DSM 263[T]	+	0.7–1.1	>1%	o	d	56.6	–	clb
Cba. parvum	*Chl. vibriof. f.thios.*	NCIB 8346	+		>1%	o	d	56.1	–	
Cba. chlorovibrioides	*Chl. chlorovibrioides*	UdG 6026	–	0.3–0.4*	2–3%	B_{12}	c	54.0	–	clb
Cba. chlorovibrioides	*Chl. vibrioforme*	UdG 6043	o	0.7–0.8	5%	o	c		–	clb

[a] According to Imhoff, 2003.

Pelodictyon in two of them (Table 3); (ii) prior to their reassignment, strains of several of the old species (including *Chlorobium limicola, Chlorobium phaeobacteroides, Chlorobium vibrioforme* and *Chlorobium phaeovibrioides*) appeared in at least two of the phylogenetic groups (Table 3); (iii) *Chlorobium limicola* subsp. *thiosulfatophilum* and *Chlorobium vibrioforme* subsp. *thiosulfatophilum*, as represented by their defined type strains (again prior to reassignment), were different from their reference species at the genus level (Table 3).

The major groups of species of green sulfur bacteria recognized by Alexander et al. (2002) were used as a basis for the definition of genera. *Chloroherpeton thalassium* forms a clearly separate phylogenetic line from all other species and genera of green sulfur bacteria and therefore was not involved in the rearrangement of strains and species. The type strain of the recognized type species *Prosthecochloris aestuarii* of group 1 (Alexander et al., 2002) formed the basis of assigning other species clustering with this species to the genus *Prosthecochloris*. The groups 2 and 3 of Alexander et al. (2002) were not significantly separated from each other and treated as a single genus. The recognized type strain of *Chlorobium limicola* (included in group 3) was the basis to maintain *Chlorobium* as the genus name for this group with *Chlorobium limicola* as type species. Group 4 was represented by a number of strains and species formerly assigned to the genus *Chlorobium*, but phylogenetically distant from the type strain and species *Chlorobium limicola* and the group of bacteria that clusters with this species. Consequently, the bacteria of group 4 were assigned to a novel genus, for which the name *Chlorobaculum* was proposed (Imhoff, 2003) (Table 3). The type species of this genus is *Chlorobaculum tepidum*, the former *Chlorobium tepidum* (Wahlund et al., 1991). On the basis of 16S rDNA and fmoA sequences, the phylogenetic groups and their representatives can be recognized and distinguished in natural samples and their specific distribution in nature has been studied (Alexander and Imhoff, 2006).

References

Alexander B and Imhoff JF (2006) Communities of green sulfur bacteria in marine and saline habitats analyzed by gene sequences of 16S rRNA and Fenna–Matthews–Olson protein. Int Microbiol 9: 259–266

Alexander B, Andersen JH, Cox RP and Imhoff JF (2002) Phylogeny of green sulfur bacteria on the basis of gene sequences of 16S rRNA and of the Fenna-Matthews-Olson protein. Arch Microbiol 178: 131–140

Ambler RP, Daniel M, Hermoso J, Meyer TE, Bartsch RG and Kamen MD (1979) Cytochrome c$_2$ sequence variation among the recognized species of purple nonsulfur bacteria. Nature 278: 659–660

Breuker E (1964) Die Verwertung von intrazellulärem Schwefel durch *Chromatium vinosum* im aeroben und anaeroben Licht- und Dunkelstoffwechsel. Zentralbl Bakteriol Parasitenkd Hyg Abt 2 118: 561–568

Brune DC (1995) Isolation and characterization of sulfur globule proteins from *Chromatium vinosum* and *Thiocapsa roseopersicina*. Arch Microbiol 163: 391–399

Bryantseva IA, Gorlenko VM, Kompantseva EI, Imhoff JF, Süling J and Mityushina L (1999) *Thiorhodospira sibirica* gen. nov., sp. nov., a new alkaliphilic purple sulfur bacterium from a Siberian soda lake. Int J Syst Bacteriol 49: 697–703

Bryantseva IA, Gorlenko VM, Kompantseva EI and Imhoff JF (2000) *Thioalkalicoccus limnaeus* gen. nov., sp. nov., a new alkaliphilic purple sulfur bacterium with bacteriochlorophyll b. Int J Syst Bacteriol 50: 2157–2163

Burgess JG, Kawaguchi R, Sakaguchi T, Thornhill RH and Matsunaga T (1993) Evolutionary relationships among *Magnetospirillum* strains inferred from phylogenetic analysis of 16S rDNA sequences. J Bacteriol 175: 6689–6694

Caumette P, Imhoff JF, Süling J and Matheron R (1997) *Chromatium glycolicum* sp. nov., a moderately halophilic purple sulfur bacterium that uses glycolate as substrate. Arch Microbiol 167: 11–18

DeBont JAM, Scholten A and Hansen TA (1981) DNA–DNA hybridization of *Rhodopseudomonas capsulata, Rhodopseudomonas sphaeroides,* and *Rhodopseudomonas sulfidophila* strains. Arch Microbiol 128: 271–274

DeWeerd KA, Mandelco L, Tanner RS, Woese CR and Suflita JM (1990) *Desulfomonile tiedjei* gen. nov. and sp. nov., a novel anaerobic, dehalogenating, sulfate-reducing bacterium. Arch Microbiol 154: 23–30

Dickerson RE (1980) Evolution and gene transfer in purple photosynthetic bacteria. Nature 283: 210–212

Dilling W, Liesack W and Pfennig N (1995) *Rhabdochromatium marinum* gen. nom. rev., sp. nov., a purple sulfur bacterium from a salt marsh microbial mat. Arch Microbiol 164: 125–131

Drews G and Imhoff JF (1991) Phototrophic purple bacteria. In: Shively JM and Barton LL (eds) Variations in Autotrophic life, pp 51–97. Academic, London

Ehrenreich A and Widdel F (1994) Anaerobic oxidation of ferrous iron by purple bacteria, a new type of phototrophic metabolism. Appl Environ Microbiol 60: 4517–4526

Fani R, Bandi C, Bazzicalupo M, Ceccherini MT, Fancelli S, Gallori E, Gerace L, Grifoni A, Miclaus N and Damiani G (1995) Phylogeny of the genus *Azospirillum* based on 16S rDNA sequence. FEMS Microbiol Lett 129: 195–200

Figueras JB, Garcia-Gil LJ and Abella CA (1997) Phylogeny of the genus *Chlorobium* based on 16S rDNA sequence. FEMS Microbiol Lett 152: 31–36

Finneran KT, Johnson CV and Loveley DR (2003) *Rhodoferax ferrireducens* sp. nov., a psychrototerant, facultatively anaerobic bacterium that oxidizes acetate with the reduction of Fe(III). Int J Syst Bacteriol 53: 669–673

Fowler VJ, Pfennig N, Schubert W and Stackebrandt E (1984) Towards a phylogeny of phototrophic purple sulfur bacteria – 16S rRNA oligonucleotide cataloguing of 11 species of Chromatiaceae. Arch Microbiol 139: 382–387

Gibson J, Stackebrandt E, Zablen LB, Gupta L and Woese CR (1979) A phylogenetic analysis of the purple photosynthetic bacteria. Curr Microbiol 3: 59–64

Gorlenko VM (1974) Oxidation of thiosulfate by *Amoebobacter roseus* in the darkness under microaerobic conditions. Microbiologiya 43: 729–731

Gorlenko VM, Bryantseva IA, Panteleeva EE, Tourova TP, Kolganova TV, Makhneva ZK and Moskalenko AA (2004) *Ectothiorhodosinus mongolicum* gen. nov., sp. nov., a new purple bacterium from a soda lake in Mongolia. Microbiology 73: 66–73 (translated from Mikrobiologiya)

Guyoneaud R, Matheron R, Liesack W, Imhoff JF and Caumette P (1997) *Thiorhodococcus minus*, gen. nov., sp. nov. a new purple sulfur bacterium isolated from coastal lagoon sediments. Arch Microbiol 168: 16–23

Guyoneaud R, Süling J, Petri R, Matheron R, Caumette P, Pfennig N and Imhoff JF (1998) Taxonomic rearrangements of the genera *Thiocapsa* and *Amoebobacter* on the basis of 16S rDNA sequence analyses and description of *Thiolamprovum* gen. nov. Int J Syst Bacteriol 48: 957–964.

Hiraishi A (1994) Phylogenetic affiliations of *Rhodoferax fermentans* and related species of phototrophic bacteria as determined by automated 16S rDNA sequencing. Cur Microbiol 28: 25–29

Hiraishi A and Imhoff JF (2005a) Genus *Rhodoferax*. In: Brenner DJ, Krieg NR and Staley JT (eds) Bergey's Manual of Systematic Bacteriology, 2nd edn., Vol. 2, Part C, pp 727–732. Springer, New York

Hiraishi A and Imhoff JF (2005b) Genus *Rhodoplanes*. In: Brenner DJ, Krieg NR and Staley JT (eds) Bergey's Manual of Systematic Bacteriology, 2nd edn., Vol. 2, Part C, pp 545–549. Springer, New York

Hiraishi A and Kitamura H (1984) Distribution of phototrophic purple nonsulfur bacteria in activated sludge systems and other aquatic environments. Bull Jpn Soc Sci Fish 50: 1929–1937

Hiraishi A and Ueda Y (1994) Intrageneric structure of the genus *Rhodobacter*: transfer of *Rhodobacter sulfidophilus* and related marine species to the genus *Rhodovulvum* gen. nov. Int J Syst Bacteriol 44: 15–23.

Hiraishi A and Ueda Y (1995) Isolation and characterization of *Rhodovulum strictum* sp. nov. and some other members of purple nonsulfur bacteria from colored blooms in tidal and seawater pools. Int J System Bacteriol 45: 319–326

Hiraishi A, Hoshino Y and Kitamura H (1984) Isoprenoid quinone composition in the classification of Rhodospirillaceae. J Gen Appl Microbiol 30: 197–210

Hiraishi A, Hoshino Y and Satoh T (1991) *Rhodoferax fermentans* gen. nov., sp. nov., a phototrophic purple nonsulfur bacterium previously referred to as the "*Rhodocyclus gelatinosus*- like" group. Arch Microbiol 155: 330–336

Hiraishi A, Urata K and Satoh T (1995) A new genus of marine budding phototrophic bacteria, *Rhodobium* gen. nov., which includes *Rhodobium orientis* sp. nov. and *Rhodobium marinum* comb. nov. Int J Syst Bacteriol 45: 226–234

Hiraishi A, Muramatsu K and Ueda Y (1996) Molecular genetic analyses of *Rhodobacter azotoformans* sp. nov. and related species of phototrophic bacteria. System Appl Microbiol 19: 168–177

Hirschler-Rea A, Matheron R, Riffaud C, Moune S, Eatock C, Herbert A, Willison JC and Caumette P (2003) Isolation and characterization of spirilloid purple phototrophic bacteria forming red layers in microbial mats of the Mediterranean salterns: description of *Halorhodospira neutriphila* sp. nov. and emendation of the genus *Halorhodospira*. Int J Syst Bacteriol 53: 153–163

Imhoff JF (1982) Occurrence and evolutionary significance of two sulfate assimilation pathways in Rhodospirillaceae. Arch Microbiol 132: 197–203

Imhoff JF (1984a) Reassignment of the genus *Ectothiorhodospira* Pelsh 1936 to a new family, *Ectothiorhodospiraceae* fem. nov., and emended description of the Chromatiaceae Bavendamm 1924. Int J Syst Bacteriol 134: 338–339

Imhoff JF (1984b) Quinones of phototrophic purple bacteria. FEMS Microbio Lett 25: 85–89

Imhoff JF (1989) The genus *Ectothiorhodospira*. In: Staley JT, Bryant MP, Pfennig N and Holt JG (eds) Bergey's Manual of Systematic Bacteriology, 1st edn., Vol. 3, pp 1654–1658. Williams and Wilkins, Baltimore.

Imhoff JF (1991) Polar lipids and fatty acids in the genus *Rhodobacter*. Syst Appl Microbiol 14: 228–234

Imhoff JF (1992) Taxonomy, phylogeny and general ecology of anoxygenic phototrophic bacteria. In: Carr NG and Mann NH (eds) Biotechnology Handbook Photosynthetic Prokaryotes, pp 53–92, Plenum, London, New York

Imhoff JF (1995) Taxonomy and physiology of phototrophic purple bacteria and green sulfur bacteria. In: Blankenship RE, Madigan MT and Bauer CE (eds) Anoxygenic Photosynthetic Bacteria, pp 1–15. Kluwer Academic, The Netherlands

Imhoff JF (1999) A phylogenetically oriented taxonomy of anoxygenic phototrophic bacteria. In: Peschek GA, Löffelhardt W and Schmetterer G (eds) The Phototrophic

Prokaryotes, pp 763–774. Kluwer Academic/Plenum, New York

Imhoff JF (2001a) True marine and halophilic anoxygenic phototrophic bacteria. Arch Microbiol 176: 243–254

Imhoff JF (2001b) Transfer of *Pfennigia purpurea* Tindall 1999 (*Amoebobacter purpureus* Eichler and Pfennig 1988) to the genus *Lamprocystis* as *Lamprocystis purpurea*. Int J Syst Evol Microbiol 51: 1699–1701

Imhoff JF (2001c) The anoxygenic phototrophic purple bacteria. In: Boone DR and Castenholz RW (eds) Bergey's Manual of Systematic Bacteriology, 2nd edn., Vol. 1, pp 621–627. Springer, New York

Imhoff JF (2003) A phylogenetic taxonomy of the family Chlorobiaceae on the basis of 16S rRNA and fmo (Fenna–Matthews–Olson protein) gene sequences. Int J Syst Evol Microbiol 53: 941–951

Imhoff JF (2005a) Order *Chromatiales*. In: Brenner DJ, Krieg NR and Staley JR (eds) Bergey's Manual of Systematic Bacteriology, 2nd edn., Vol. 2, Part B, pp 1–3. Springer, New York

Imhoff JF (2005b) Family *Chromatiaceae*. In: Brenner DJ, Krieg NR and Staley JR (eds) Bergey's Manual of Systematic Bacteriology, 2nd edn., Vol. 2, Part B, pp 3–9. Springer, New York

Imhoff JF (2005c) Family *Ectothiorhodospiraceae*. In Brenner DJ, Krieg NR and Staley JR (eds) Bergey's Manual of Systematic Bacteriology, 2nd edn., Vol. 2, Part B, pp 41–43. Springer, New York

Imhoff JF (2005d) Genus *Rhodobacter*. In: Brenner DJ, Krieg NR and Staley JT (eds) Bergey's Manual of Systematic Bacteriology, 2nd edn., Vol. 2, Part C, pp 161–167. Springer, New York

Imhoff JF (2005e) Genus *Rhodobaca*. In: Brenner DJ, Krieg NR and Staley JT (eds) Bergey's Manual of Systematic Bacteriology, 2nd edn., Vol. 2, Part C, pp 204–205. Springer, New York

Imhoff JF (2005f) Genus *Rhodovulum*. In: Brenner DJ, Krieg NR and Staley JT (eds) Bergey's Manual of Systematic Bacteriology, 2nd edn., Vol. 2, Part C, pp 205–209. Springer, New York

Imhoff JF (2005g) Genus *Rhodoblastus*. In: Brenner DJ, Krieg NR and Staley JT (eds) Bergey's Manual of Systematic Bacteriology, 2nd edn., Vol. 2, Part C, pp 471–473. Springer, New York

Imhoff JF (2005h) Genus *Rhodopseudomonas*. In: Brenner DJ, Krieg NR and Staley JT (eds) Bergey's Manual of Systematic Bacteriology, 2nd edn., Vol. 2, Part C, pp 473–476. Springer, New York

Imhoff JF (2005i) Genus *Blastochloris*. In: Brenner DJ, Krieg NR and Staley JT (eds) Bergey's Manual of Systematic Bacteriology, 2nd edn., Vol. 2, Part C, pp 506–507. Springer, New York

Imhoff JF (2005j) Genus *Rhodomicrobium*. In: Brenner DJ, Krieg NR and Staley JT (eds) Bergey's Manual of Systematic Bacteriology, 2nd edn., Vol. 2, Part C, pp 543–545. Springer, New York

Imhoff JF (2005k) Genus *Rubrivivax*. In: Brenner DJ, Krieg NR and Staley JT (eds) Bergey's Manual of Systematic Bacteriology, 2nd edn., Vol. 2, Part C, pp 749–750. Springer, New York

Imhoff JF (2005l) Genus *Rhodocyclus*. In: Brenner DJ, Krieg NR and Staley JT (eds) Bergey's Manual of Systematic Bacteriology, 2nd edn., Vol. 2, Part C, pp 887–890. Springer, New York

Imhoff JF and Bias-Imhoff U (1995) Lipids, quinones and fatty acids of anoxygenic phototrophic bacteria. In: Blankenship RE, Madigan MT and Bauer CE (eds) Anoxygenic Photosynthetic Bacteria, pp 179–205. Kluwer Academic, Netherlands

Imhoff JF and Caumette P (2004) Recommended standards for the description of new species of anoxygenic phototrophic bacteria. Int J Syst Evol Microbiol 54: 1415–1421

Imhoff JF and Hiraishi A (2005) Genus *Rhodobium*. In: Brenner DJ, Krieg NR and Staley JT (eds), Bergey's Manual of Systematic Bacteriology, 2nd edn., Vol. 2, Part C, pp 671–574. Springer, New York

Imhoff JF and Pfennig N (2001) *Thioflavicoccus mobilis* gen. nov., sp. nov., a novel purple sulfur bacterium with bacteriochlorophyll b. Int J Syst Evol Microbiol 51: 105–110

Imhoff JF and Süling J (1996) The phylogenetic relationship among *Ectothiorhodospiraceae*. A reevaluation of their taxonomy on the basis of rDNA analyses. Arch Microbiol 165: 106–113.

Imhoff JF and Trüper HG (1989) The purple nonsulfur bacteria. In: Staley JT, Bryant MP, Pfennig N and Holt JG (eds), Bergey's Manual of Systematic Bacteriology, Vol. 3, pp 1658–1661. Williams and Wilkins, Baltimore

Imhoff JF, Kushner DJ, Kushawa SC and Kates M (1982) Polar lipids in phototrophic bacteria of the Rhodospirillaceae and Chromatiaceae families. J Bacteriol 150: 1192–1201

Imhoff JF, Trüper HG and Pfennig N (1984) Rearrangement of the species and genera of the phototrophic "purple nonsulfur bacteria." Int J Syst Bacteriol 34: 340–343

Imhoff JF, Petri R and Süling J (1998a) Reclassification of species of the spiral-shaped phototrophic purple nonsulfur bacteria of the alpha-Proteobacteria: description of the new genera *Phaeospirillum* gen. nov., *Rhodovibrio* gen. nov., *Rhodothalassium* gen. nov. and *Roseospira* gen. nov. as well as transfer of *Rhodospirillum fulvum* to *Phaeospirillum fulvum* comb. nov., of *Rhodospirillum molischianum* to *Phaeospirillum molischianum* comb. nov., of *Rhodospirillum salinarum* to *Rhodovibrio salinarum* comb. nov., of *Rhodospirillum sodomense* to *Rhodovibrio sodomensis* comb. nov., of *Rhodospirillum salexigens* to *Rhodothalassium salexigens* comb. nov., and of *Rhodospirillum mediosalinum* to *Roseospira mediosalina* comb. nov. Int J Syst Bacteriol 48: 793–798

Imhoff JF, Süling J and Petri R (1998b) Phylogenetic relationships among the *Chromatiaceae*, their taxonomic reclassi-

fication and description of the new genera *Allochromatium,*
Halochromatium, Isochromatium, Marichromatium, Thio-
coccus, Thiohalocapsa, and *Thermochromatium.* Int J Syst
Bacteriol 48: 1129–1143

Ivanova TL, Turova TP and Antonov AS (1988) DNA–DNA
hybridization studies on some purple nonsulfur bacteria.
System Appl Microbiol 10: 259–263

Kämpf C and Pfennig N (1980) Capacity of Chromatiaceae
for chemotrophic growth. Specific respiration rates of
Thiocystis violacea and *Chromatium vinosum.* Arch
Microbiol 127: 125–135

Kämpf C and Pfennig N (1986) Isolation and characterization
of some chemoautotrophic Chromatiaceae. J Basic Microbiol 26: 507–515

Kawasaki H, Hoshino Y and Yamasato K (1993) Phylogenetic diversity of phototrophic purple non-sulfur bacteria
in the Proteobacteria alpha group. FEMS Microbiol Lett
112: 61–66

Keppen OI, Tourova TP, Kuznetsov BB, Ivanovsky RN
and Gorlenko VM (2000) Proposal of Oscillochlori-
daceae fam.nov. on the basis of a phylogenetic analysis
of the filamentous anoxygenic phototrophic bacteria, and
emended description of *Oscillochloris* and *Oscillochloris
trichoides* in comparison with further new isolates. Int J
Syst Evol Microbiol 50: 1529–1537

Kondratieva EN, Zhukov VG, Ivanowsky RN, Petruskova
YP and Monosov EZ (1976) The capacity of the phototrophic sulfur bacterium *Thiocapsa roseopersicina* for
chemosynthesis. Arch Microbiol 108: 287–292

Krasilnikova EN (1976) Anaerobic metabolism of *Thiocapsa
roseopersicina* (in Russian, with English summary).
Mikrobiologiya 45: 372–376

Krasilnikova EN, Petushkova YP and Kondratieva EN (1975)
Growth of purple sulfur bacterium *Thiocapsa roseopersicina*
under anaerobic conditions in the darkness (in Russian, with
English summary). Mikrobiologiya 44: 700–703

Krasilnikova EN, Ivanovskii RN and Kondratieva EN (1983)
Growth of purple bacteria utilizing acetate under anaerobic conditions in darkness. Mikrobiologiya (English
translation edition) 52: 189–194

Madigan MT (1986) *Chromatium tepidum* sp. nov., a thermophilic photosynthetic bacterium of the family Chromatiaceae. Int J Syst Bacteriol 36: 222–227

Madigan MT (2001) Family VI. Heliobacteriaceae. In:
Boone DR and Castenholz RW (eds) Bergey's Manual of
Systematic Bacteriology, 2nd edn., Vol. 1, pp 627–630.
Springer, New York

Madigan MT, Jung DO, Woes, CR and Achenbach A (2000)
Rhodoferax antarcticus sp. nov., a moderately psy-
chrophilic purple nonsulfur bacterium isolated from an
Antarctic microbial mat. Arch Microbiol 173: 269–277

Maidak BL, Larsen N, McCaughey J, Overbeck R, Olsen GJ,
Fogel K, Blandy J and Woese CR (1994) The ribosomal
database project. Nucleic Acid Res 22: 3483–3487

Meißner J, Pfennig N, Krauss JH, Mayer H and Weckesser J
(1988) Lipopolysaccharides of the Chromatiaceae species

Thiocystis violacea, Thiocapsa pfennigii and *Chromatium
tepidum.* J Bacteriol 170: 3267–3272

Milford AD, Achenbach LA, Jung DO, Madigan MT (2000)
Rhodobaca bogoriensis gen. nov. and sp. nov., an alkaliphilic purple nonsulfur bacterium from African Rift Val-
ley soda lakes. Arch Microbiol 174:18–27

Molisch H (1907) Die Purpurbakterien nach neueren Untersuchungen. G. Fischer, Jena, pp 1–95

Overmann J (2001) Green sulfur bacteria. In: Boone DR and
Castenholz RW (eds) Bergey's Manual of Systematic Bacteriology, 2nd edn., Vol. 1, pp 601–605. Springer, New York

Overmann J and Tuschak C (1997) Phylogeny and molecular
fingerprinting of green sulfur bacteria. Arch Microbiol
167: 302–309

Pattaragulwanit K, Brune DC, Trüper HG and Dahl C (1998)
Molecular evidence for extracytoplasmic localization of
sulfur globules in *Chromatium vinosum.* Arch Microbiol
169: 434–444

Pelsh AD (1937) Photosynthetic sulfur bacteria of the eastern
reservoir of Lake Sakskoe. Mikrobiologiya 6: 1090–1100

Pfennig N (1969) *Rhodospirillum tenue* sp. n., a new species
of the purple nonsulfur bacteria. J Bacteriol 99: 619–620

Pfennig N (1978) *Rhodocyclus purpureus* gen. nov. and sp.
nov., a ring-shaped vitamin B_{12}-requiring member of the
family Rhodospirillaceae. Int J Syst Bacteriol 28: 283–288

Pfennig N (1989) Green sulfur bacteria. In: Staley JT, Bryant
MP, Pfennig N and Holt JG (eds) Bergeys Manual of
Systematic Bacteriology, 1st edn., Vol. 3, pp 1682–1697.
Williams & Wilkins, Baltimore

Pfennig N and Overmann J (2001a) Genus I. *Chlorobium.* In:
Boone DR and Castenholz RW (eds), Bergey's Manual of
Systematic Bacteriology, 2nd edn., Vol. 1, pp 605–610.
Springer, New York

Pfennig N and Overmann J (2001b) Genus IV. *Pelodictyon.*
In: Boone DR and Castenholz RW (eds), Bergey's Manual
of Systematic Bacteriology, 2nd edn., Vol. 1, pp 614–617.
Springer, New York

Pfennig N and Trüper HG (1971) Higher taxa of the phototrophic bacteria. Int J Syst Bacteriol 21: 17–18

Pfennig N and Trüper HG (1974) The phototrophic bacteria.
In: Buchanan RE and Gibbons NE (eds) Bergey's Manual
of Determinative Bacteriology, 8th edn., pp 24–75. The
Williams & Wilkins Co., Baltimore

Prange A, Chauvistre R, Modrow H, Hormes J, Trüper HG
and Dahl C (2002) Quantitative speciation of sulfur in
bacterial sulfur globules: x-ray absorption spectroscopy
reveals at least three different speciations of sulfur.
Microbiology 148: 267–276

Rees GN, Harfoot CG, Janssen PH, Schoenborn L, Kuever J
and Lünsdorf H (2002) *Thiobaca trueperi* gen. nov., sp.
nov., a phototrophic bacterium isolated from freshwater
lake sediment. Int J Syst Evol Microbiol 52: 671–678

Satoh T, Hoshino Y and Kitamura H (1976) *Rhodopseudomonas sphaeroides f sp. denitrificans,* a denitrifying
strain as a subspecies of *Rhodopseudomonas sphaeroides.*
Arch Microbiol 108: 265–269

Schmidt K (1978) Biosynthesis of carotenoids. In: Clayton RK and Sistrom WR (eds) The Photosynthetic Bacteria, pp 729–750. Plenum, New York

Sievers M, Ludwig W and Teuber M (1994) Phylogenetic positioning of *Acetobacter, Gluconobacter, Rhodopila* and *Acidiphilium* species as a branch of acidophilic bacteria in the alpha-subclass of proteobacteria based on 16S ribosomal DNA sequences. Syst Appl Microbiol 17: 189–196

Smith AJ (1965) The discriminative oxidation of the sulphur atoms of thiosulphate by a photosynthetic sulphur bacterium – *Chromatium* strain D. Biochem J 94: 27

Stackebrandt E, Fowler VJ, Schubert W and Imhoff JF (1984) Towards a phylogeny of phototrophic purple sulfur bacteria – the genus *Ectothiorhodospira*. Arch Microbiol 137: 366–370

Straub KL, Rainey FA and Widdel F (1999) *Rhodovulum iodosum* sp. nov. and *Rhodovulum robiginosum* sp. nov., two new marine phototrophic ferrous-iron-oxidizing purple bacteria. Int J Syst Bacteriol 49: 729–735

Trüper HG (1981) Photolithotrophic sulphur oxidation, In: Bothe H and Trebst A (eds) Biology of Inorganic Nitrogen and Sulfur, pp 199–211. Springer, Berlin

Trüper HG and Pfennig N (1966) Sulphur metabolism in Thiorhodaceae. III. Storage and turnover of thiosulphate sulphur in *Thiocapsa floridana* and *Chromatium* species. Antonie van Leeuwenhoek. J Microbiol Serol 32: 261–276

Van Gemerden H (1968a) Utilization of reducing power in growing cultures of *Chromatium*. Arch Microbiol 65: 111–117

Van Gemerden H (1968b) On the ATP generation by *Chromatium* in darkness. Arch Mikrobiol 64: 118–124

Van Gemerden H (1974) Coexistence of organisms competing for the same substrate: An example among the purple sulfur bacteria. Microb Ecol 1: 19–23

Wahlund TM, Woese CR, Castenholz RW and Madigan MT (1991) A thermophilic green sulfur bacterium from New zealand hot springs, *Chlorobium tepidum* sp. nov. Arch Microbiol 156: 81–90

Weckesser J, Drews G and Mayer H (1979) Lipopolysaccharides of photosynthetic prokaryotes. Annu Rev Microbiol 33: 215–239

Weckesser J, Mayer H and Schulz G (1995) Anoxygenic phototrophic bacteria: Model organisms for studies on cell wall macromolecules. In: Blankenship RE, Madigan MT and Bauer CE (eds) Anoxygenic Photosynthetic Bacteria, pp 207–230. Kluwer Academic, The Netherlands

Widdel F, Schnell S, Heising S, Ehrenreich A, Assmus B and Schink B (1993) Ferrous iron oxidation by anoxygenic phototrophic bacteria. Nature 362: 834–836

Willems A, Gillis M and de Ley J (1991) Transfer of *Rhodocyclus gelatinosus* to *Rubrivivax gelatinosus* gen. nov., comb. nov., and phylogenetic relationships with *Leptothrix, Sphaerotilus natans, Pseudomonas saccharophila,* and *Alcaligenes latus.* Int J Syst Bacteriol 41: 65–73

Woese CR (1987) Bacterial evolution. Microbiol Rev 51: 221–271

Woese CR, Stackebrandt E, Weisburg WG, Paster BJ, Madigan MT, Fowler VJ, Hahn CM, Blanz P, Gupta R, Nealson KH and Fox GE (1984a) The phylogeny of purple bacteria: the alpha subdivision. Syst Appl Microbiol 5: 315–326

Woese CR, Weisburg WG, Paster BJ, Hahn CM, Tanner RS, Krieg NR, Koops HP, Harms H and Stackebrandt E (1984b) The phylogeny of purple bacteria: The beta subdivision. System Appl Microbiol 5: 327–336

Woese CR, Weisburg WG, Hahn CM, Paster BJ, Zablen LB, Lewis BJ, Macke TJ, Ludwig W and Stackebrandt E (1985) The phylogeny of purple bacteria: The gamma subdivision. Syst Appl Microbiol 6: 25–33

Xia Y, Embley TM and O'Donnell AG (1994) Phylogenetic analysis of *Azospirillum* by direct sequencing of PCR amplified 16S rDNA. Syst Appl Microbiol 17: 197–201

Zahr M, Fobel B, Mayer H, Imhoff JF, Campos V and Weckesser J (1992) Chemical composition of the lipopolysaccharides of *Ectothiorhodospira shaposhnikovii, Ectothiorhodospira mobilis,* and *Ectothiorhodospira halophila.* Arch Microbiol 157: 499–504

Chapter 15

Inorganic Sulfur Compounds as Electron Donors in Purple Sulfur Bacteria

Christiane Dahl
*Institut für Mikrobiologie & Biotechnologie, Rheinische Friedrich-Wilhelms-Universität Bonn,
Meckenheimer Allee 168, D-53115 Bonn, Germany*

Summary

Most anoxygenic phototrophic bacteria can use inorganic sulfur compounds (e.g. sulfide, elemental sulfur, polysulfides, thiosulfate, or sulfide) as electron donors for reductive carbon dioxide fixation during photolithoautotrophic growth. In these organisms, light energy is used to transfer electrons from sulfur compounds to the level of the more highly reducing electron carriers $NAD(P)^+$ and ferredoxin. In this chapter the sulfur oxidizing capabilities of the different groups of anoxygenic phototrophic bacteria are briefly

Corresponding Author, Phone: +49 228 732119, Fax: +49 228 737576, E-mail: chdahl@uni-bonn.de summarized. This chapter then focuses on the pathways of sulfur compound oxidation in purple sulfur bacteria of the families *Chromatiaceae* and *Ectothiorhodospiraceae*. A variety of enzymes catalyzing sulfur oxidation reactions have been isolated from members of this group and *Allochromatium vinosum*, a representative of the *Chromatiaceae*, has been especially well characterized also on a molecular genetic level. In this organism intracellular sulfur globules are an obligate intermediate during the oxidation of thiosulfate and sulfide to sulfate. Thiosulfate oxidation is strictly dependent on the presence of three periplasmic Sox proteins encoded by the *soxBXA* and *soxYZ* genes. Sulfide oxidation does not appear to require the presence of Sox proteins. Flavocytochrome *c* is also not essential leaving sulfide:quinone oxidoreductase as the probably most important sulfide-oxidizing enzyme. Polysulfides are intermediates en route of sulfide to stored sulfur. Sulfur is deposited in the periplasm and present as long chains probably terminated by organic residues at one or both ends. The oxidation of stored sulfur is completely dependent on the proteins encoded in the *dsr* operon. These include siroamide-containing sulfite reductase (DsrAB), a transmembrane electron-transporting complex (DsrMKJOP) and a iron–sulfur flavoprotein with NADH:acceptor oxidoreductase activity (DsrL). The last step of reduced sulfur compound oxidation in purple sulfur bacteria is the oxidation of sulfite. This can occur either via the enzymes adenosine 5′-phosphosulfate (APS) reductase and ATP sulfurylase which are non-essential in *Alc. vinosum* or via direct oxidation to sulfate. The nature of the enzyme catalyzing the latter step is still unresolved in purple sulfur bacteria.

I. Introduction

Anoxygenic phototrophic bacteria are generally not able to use water as an electron-donating substrate for photosynthetic CO_2 reduction. The common property of these bacteria is the ability to carry out light-dependent, (bacterio)chlorophyll-mediated processes, a property shared with cyanobacteria, prochlorophytes, algae and green plants. In contrast to the latter, reduced sulfur compounds, molecular hydrogen, reduced iron or simple organic molecules typically serve as photosynthetic electrons in anoxygenic phototrophic bacteria. Bacteriochlorophylls are present not only in facultatively and obligately anaerobic anoxygenic phototrophic bacteria but also in large numbers of bacterial species that are strictly dependent on energy generation by respiratory electron transport processes. These organisms are called aerobic phototrophic bacteria (Yurkov, 2006) or "aerobic bacteriochlorophyll-containing (ABC) bacteria (Imhoff and Hiraishi, 2005).

The utilization of reduced sulfur compounds as photosynthetic electron donors is – though to a different extent – common to almost all groups of phototrophic prokaryotes. Classical in this respect are the purple (families *Chromatiaceae* and *Ectothiorhodospiraceae*) and green sulfur bacteria (family *Chlorobiaceae*) all of which utilize reduced sulfur compounds as electron donors. A number of classical purple "nonsulfur" bacteria, some members of the filamentous anoxygenic phototrophs (also termed green gliding bacteria or green non-sulfur bacteria) of the family *Chloroflexaceae*, and a few representatives of the strictly anaerobic gram-positive Heliobacteria are also able to oxidize reduced sulfur compounds during photosynthesis. Even certain species of the cyanobacteria can perform anoxygenic photosynthesis at the expense of sulfide as electron donor (rf Chapter Hauska/Shahak). Photoautotrophic growth with sulfur compounds has so far not been described for any of the ABC bacteria.

One purpose of this chapter is to briefly introduce researchers not specializing in bacterial sulfur metabolism to the sulfur-oxidizing capabilities of the various groups of anoxygenic phototrophic bacteria. Ecology and taxonomy of anoxygenic phototrophic bacteria are described in detail in the chapters by Imhoff and Overmann. It should be emphasized that some older reviews still serve as a valuable source of information especially regarding sulfur oxidation patterns by whole cells of anoxygenic phototrophic bacteria (Brune, 1989; Brune, 1995b).

Abbreviations: *Acd.* – Acidiphilium; *Alc.* – Allochromatium; APS – adenosine 5′-phosphosulfate; *Ect.* – Ectothiorhodospira; EC – extracellular; FAPs – filamentous anoxygenic phototrophs; HiPIP – high potential iron-sulfur protein; *Hlr.* – Halorhodospira; IC – intracellular; *Mch.* – Marichromatium; nd – not determined; *Pcs.* – Paracoccus; SQR-sulfide: quinone oxidoreductase *Rba.* – Rhodobacter; *Tca.* – Thiocapsa; *Tcs.* – Thiocystis

II. Sulfur Oxidation Capabilities of Anoxygenic Phototrophic Bacteria

In the following section the sulfur oxidation capabilities of the various groups of anoxygenic phototrophic bacteria will be briefly described. Sulfur oxidation capabilities in the aerobic bacteriochlorophyll-containing bacteria, the Heliobacteria and the anoxygenic filamentous phototrophs are rather limited. Information about the enzymes involved is in most cases not available. The sulfur oxidation pathways in the other groups are far more complex. Therefore, separate chapters concentrate on the biochemistry, molecular genetics, genomics and proteomics of sulfur oxidation in the green sulfur bacteria (Hanson, Frigaard). The sulfur metabolism in purple nonsulfur bacteria is reviewed in a forthcoming volume of this series (Sander and Dahl, 2008) This chapter focuses on sulfur compound oxidation in the purple sulfur bacteria of the families *Chromatiaceae* and *Ectothiorhodospiraceae*.

A. Aerobic Anoxygenic Bacteriochlorophyll-Containing Bacteria

ABC bacteria are probably very important as destructors of organic compounds in a broad range of habitats (Yurkov, 2006). Although this increasingly large group of bacteria is very heterogeneous phylogenetically, morphologically and physiologically, all share the inability to use bacteriochlorophyll for anaerobic photosynthetic growth and the presence of photochemical reactions in cells only under aerobic conditions (Hiraishi and Shimada, 2001). Furthermore they share aerobic chemoorganotrophy as the preferred mode of growth, low levels of bacteriochlorophylls and strong inhibition by light of bacteriochlorophyll synthesis under normal growth conditions. While fully active reaction center and LH1 complexes with bacteriochlorophyll are present in all species studied so far, peripheral antenna (LH2) are absent in most species (Hiraishi and Shimada, 2001).

All species of the ABC bacteria, except the β-Proteobacterium *Roseotales depolymerans*, belong to the α-Proteobacteria (class *Alphaproteobacteria*) where they do not form a homogeneous cluster but are closely interspersed with phototrophic and non-phototrophic species (Imhoff and Hiraishi, 2005). Differentiation and taxonomy of ABC bacteria is difficult to understand even for experts in the field as several species not containing bacteriochlorophyll have been placed in genera of the aerobic anoxygenic bacteria.

None of the ABC bacteria are able to grow photolithoautotrophically with sulfur compounds as electron donors. However, the ability to oxidize inorganic sulfur compounds has been described for several representatives of this group. Examples are *Roseinatronobacter thiooxidans*, a strictly aerobic obligately heterotrophic alkaliphile that can oxidize sulfide, thiosulfate, sulfite and elemental sulfur to sulfate in the presence of organic compounds (Sorokin et al., 2000). In another study, Yurkov et al. (1994) showed thiosulfate-oxidizing activity in *Erythromicrobium hydrolyticum*, strain E4(1) and *Rosoecoccus thiosulfatophilus*, strain RB-7. The most pronounced oxidative sulfur metabolism is present in species of the genus *Acidiphilium*. A number of studies have demonstrated sulfur-dependent chemolithotrophy of and sulfur oxidation by *Acd. acidophilum* (formerly *Thiobacillus acidophilus*) (Pronk et al., 1990; Meulenberg et al., 1992b; Hiraishi et al., 1998). In *Acd. acidophilum* the utilization of thiosulfate is initiated by the oxidative condensation of two molecules of thiosulfate yielding tetrathionate. This step is catalyzed by the periplasmic enzyme thiosulfate:cytochrome c oxidoreductase (Meulenberg et al., 1993). The details of the further oxidation of tetrathionate to sulfate are largely unclear. Meulenberg et al. (1993) obtained indications that tetrathionate oxidation takes place in the periplasm in *Acd. acidophilum*. Furthermore, a tetrathionate hydrolase (de Jong et al., 1997), a trithionate hydrolase (Meulenberg et al., 1992a) and a sulfite:cyctochrome c oxidoreductase (de Jong et al., 2000) have been characterized from the organism.

B. Heliobacteria

Heliobacteria are anoxygenic phototrophic bacteria that contain bacteriochlorophyll g as the sole chlorophyll pigment. This unique Bchl, found only in the heliobacteria, distinguishes them from all other anoxygenic phototrophic bacteria (Madigan, 2001b). They lack differentiated photosynthetic internal membranes, such as the membrane vesicles or lamellae of purple bacteria or the chlorosomes of green bacteria. Representatives of the heliobacteria mainly occur in soils and are phylogenetically related with gram-positive bacteria, specifically the *Bacillus/Clostridium* lineage. As far as is currently known, heliobacteria are obligate anaerobes. However, they can grow both photo- and chemotrophically. Photoheterotrophic growth occurs on a restricted number of organic compounds as carbon sources. Chemotrophic growth in the dark occurs by fermentation of pyruvate or lactate. Photoautotrophic growth has not been demonstrated with any species of heliobacteria. If sulfide is added to the culture media, it is frequently oxidized to elemental sulfur that appears in the medium (Bryantseva et al., 2000; Madigan, 2001a). *Heliobacterium sulfidophilum* and *Heliobacterium undosum* are especially tolerant to sulfide (up to 2 mM at pH 7.5). Many but not all members of the *Heliobacteriaceae* can assimilate sulfate as the sole source of sulfur (Madigan, 2001a).

C. Filamentous Anoxygenic Phototrophs (Chloroflexaceae)

Chloroflexus, Chloronema, Oscillochloris, Roseiflexus and *Heliothrix* are well described genera of the filamentous anoxygenic phototrophs (FAPs) (Hanada and Pierson, 2002). Filamentous morphology and gliding motility are typical features of these anoxygenic phototrophic organisms. Three of the genera, *Chloroflexus, Chloronema* and *Oscillochloris* contain chlorosomes, structural elements that are attached to the cellular membranes and contain the light-harvesting bacteriochlorophylls *c* and *d*. All five genera contain bacteriochlorophyll *a*. The filamentous anoxygenic bacteria are not closely related to the green sulfur bacteria and belong into one of the deepest bacterial phyla (*Chloroflexi*) of the Bacteria. This phylum also harbours non-phototrophic filamentous gliding bacteria. Most but not all anoxygenic filamentous bacteria are facultatively aerobic and preferentially utilize organic substrates in their phototrophic or chemotrophic metabolism.

The biochemically best characterized member of the *Chloroflexaceae* is *Chloroflexus aurantiacus*, a thermophilic organism that prefers photoheterotrophic growth. Slow photoautrophic with hydrogen or sulfide as electron donors has been observed for some strains of the species (Madigan and Brock, 1977). Photoautotrophic growth on sulfide has also been described for *Oscillochloris trichoides* and appears to be present also in marine

and hypersaline filamentous anoxygenic bacteria (Keppen et al., 1993; Hanada and Pierson, 2002). Sulfur appearing in the medium (often affixed to the cells) is the end product of sulfide oxidation. *Chloroflexus aurantiacus* is able to cover its need for sulfur for biosynthetic purposes by the assimilation of sulfate or thiosulfate, as evidenced by the presence in the genome of a gene cluster encoding proteins involved in assimilatory sulfate reduction (e.g. ATP sulfurylase CaurDraft_0193, APS kinase CaurDraft_0191, PAPS reductase CaurDraft_0192, sulfite reductase CaurDraft_0197).

D. Green Sulfur Bacteria

All green sulfur bacteria fall into a coherent taxonomic group that forms a separate bacterial phylum, the *Chlorobi* (Garrity and Holt, 2001). Besides bacteriochlorophyll *a* in the reaction center bacteriochlorophyll *c*, *d*, or *e* and various carotenoids of the chlorobactene and isorenrieratene series are used as photosynthetic pigments. Intracytoplasmic membranes are not formed, the light harvesting complexes reside on chlorosomes. All green sulfur bacteria have similar metabolic properties. They are strictly anaerobic and obligately phototrophic and can grow with CO_2 as only carbon source. In contrast to the purple bacteria CO_2 is fixed via the reductive tricarbonic acid cycle. Sulfide is used as electron donor by almost all of these species (the iron-oxidizing *Chlorobium ferrooxidans* is the only known exception) and oxidized to sulfate with intermediary accumulation of extracellular sulfur. The chemical speciation of the deposited sulfur is discussed in section VI.E. Many species are able to grow with elemental sulfur and some species also use thiosulfate (Frigaard and Bryant, 2008). *Chlorobaculum parvum* (formerly *Chlorobium vibrioforme* subsp. *thiosulfatophilum* (Imhoff, 2003)) NCIB 8346 and a strain described by Helge Larsen as *Chlorobium thiosulfatophilum* can use tetrathionate as electron donor (Larsen, 1952; Khanna and Nicholas, 1982). Sulfite utilization has not yet been described for any green sulfur bacterium.

E. Purple Nonsulfur Bacteria

The purple "nonsulfur" bacteria are an extremely heterogeneous group of bacteria. Representatives are found within the *Alpha-* and the *Betaproteobacteria* (Imhoff et al., 2005). The species in this group vary not only with respect to their cell morphology, the structure of intracytoplasmic membrane systems, the carotenoid composition and the carbon sources used but also with respect to the electron donors that can be used for photosynthesis. All species prefer photoheterotrophic growth under anaerobic conditions. In addition, many species can grow photoautotrophically with hydrogen or sulfide as electron donor, many of which do not oxidize sulfide completely to sulfate but form sulfur as the end product instead. However, in many other species, among them the species of the genus *Rhodovulum*, *Rhodopseudomonas palustris* or *Blastochloris sulfoviridis*, sulfate is the end product of sulfide oxidation (reviewed in Brune, 1995b; Imhoff et al., 2005). Thiosulfate is also used by many species, and oxidized either to tetrathionate (*Rhodopila globiformis* (Then and Trüper, 1981)) or completely to sulfate (e.g. *Rhodovulum* species (Brune, 1995b; Appia-Ayme et al., 2001; Imhoff et al., 2005)). Under microoxic to oxic conditions in the dark most representatives of the purple "nonsulfur" bacteria can grow chemoorganotrophically, some are also capable of chemolithoautotrophic growth. In addition, some species are able to metabolize sugars in the dark in the absence of oxygen by using nitrate, dimethyl sulfoxide or trimethylamine-N-oxide as electron acceptors.

F. Purple Sulfur Bacteria

The purple sulfur bacteria belong to the *Gammaproteobacteria* and fall in two families, the *Chromatiaceae* and the *Ectothiorhodospiraceae*. Both form coherent groups on the basis of their 16S rRNA sequences. During phototrophic growth in batch cultures with sulfide as electron donor, the oxidation of sulfide and sulfur follow each other. The most important and easily recognized distinguishing feature between the members of these two families is the site of sulfur deposition during growth on sulfide. In *Chromatiaceae* sulfur globules appear inside the cells while they are formed outside the cells in *Ectothiorhodospiraceae*. A notable exception among the *Ectothiorhodospiraceae* is *Thiorhodospira sibirica*. This organism deposits sulfur not only outside of the cell in the medium but also attached to the cells or in the periplasm (Bryantseva et al., 1999). The sulfur-metabolizing capabilities of the purple sulfur bacteria are summarized in Table 1.

1. Chromatiaceae

Generally, two physiological groups can be differentiated within the *Chromatiaceae*: The large-celled species (eg. *Chromatium okenii*, *Allochromatium warmingii* and *Isochromatium buderi*) are strictly anaerobic, obligately phototrophic and require sulfide or elemental sulfur as photosynthetic electron donors and as sources of sulfur for biosynthesis. The other group includes most of the small-celled species (eg. *Alc. vinosum*, *Allochromatium minutissimum*) which are metabolically much more versatile. In addition to sulfide and elemental sulfur these organisms use thiosulfate and some also use sulfite as electron donors (Imhoff, 2005a). Some organic sulfur compounds can also serve as electron donors for photosynthetic growth of *Chromatiaceae*: *Thiocapsa roseopersicina* splits mercaptomalate and mercaptopropionate to fumarate and H_2S and acrylate and H_2S, respectively and then uses the liberated H_2S as electron donor (Visscher and Taylor, 1993). This organism furthermore oxidizes dimethyl sulfide to dimethyl sulfoxide (Visscher and van Gemerden, 1991). Most of the small-celled representatives of the *Chromatiaceae* are able to assimilate sulfate for biosynthetic purposes, can grow photoorganoheterotrophically in the absence of reduced sulfur compounds and are able to grow as chemolithotrophs on reduced sulfur compounds. Some species can even grow as chemoorganotrophs in which case the addition of sulfide or thiosulfate as a sulfur source is required because the assimilation of sulfate is repressed under aerobic conditions (Kondratieva et al., 1981). During fermentative dark metabolism of *Chromatiaceae* sulfur compounds (elemental sulfur) can serve as acceptors of electrons liberated by the oxidation of stored carbon compounds (polyhydroxyalcanoic acid).

2. Ectothiorhodospiraceae

Almost all members of the *Ectothiorhodospiraceae* are halophilic and alkaliphilic bacteria. The family comprises four phototrophic genera (*Ectothiorhodospira*, *Halorhodospira*, *Thiorhodospira*, *Ectothiorhodosinus*). Formally, the genus *Ectothiorhodosinus* (Gorlenko et al., 2004) has no standing in nomenclature.

Table 1. Sulfur metabolizing capabilities of purple sulfur bacterial genera.

Genus	Sulfur substrates	Intermediates	End product	Sulfate assimilation	Chemoautotrophic growth
Chromatiaceae					
Allochromatium	Sulfide, sulfur, thiosulfate, sulfite, (latter two not in *Alc. warmingii*)	Sulfur, IC	Sulfate	+ (not in *Alc. warmingii*)	Some species
Chromatium	Sulfide, sulfur	Sulfur, IC	Sulfate	−	−
Halochromatium	Sulfide, thiosulfate, sulfur, sulfite	Sulfur, IC	Sulfate	−	+ (sulfide, thiosulfate)
Isochromatium	Sulfide, sulfur	Sulfur, IC	Sulfate	−	−
Lamprobacter	Sulfide, thiosulfate, sulfur	Sulfur, IC	S^0 and sulfate	−	+
Lamprocystis	Sulfide, thiosulfate, sulfur	Sulfur, IC	Sulfate	−/nd	+/−
Marichromatium	Sulfide, thiosulfate, sulfur, sulfite (only *Mch. gracile*)	Sulfur, IC	Sulfate	+/−	+/−
Lamprobacter	Sulfide, thiosulfate, sulfur	Sulfur, IC	S^0 and sulfate	−	+
Lamprocystis	Sulfide, thiosulfate, sulfur	Sulfur, IC	Sulfate	−/nd	+/−
Rhabdochromatium	Sulfide, thiosulfate, sulfur	Sulfur, IC	Sulfate	nd	−
Thermochromatium	Sulfide, sulfur	Sulfur, IC	Sulfate	nd	−
Thioalkalicoccus	Sulfide, sulfur	Sulfur, IC	Sulfate	nd	−
Thiobaca	Sulfide	Sulfur, IC	Sulfate	nd	nd
Thiocapsa	Sulfide, thiosulfate, sulfur, sulfite (only *Tca. litoralis* and *Tca. pendens*)	Sulfur, IC	Sulfate	+/−	+/−
Thiococcus	Sulfide, sulfur	Sulfur, IC	Sulfate	−	−
Thiocystis	Sulfide, thiosulfate (not in *Tcs. gelatinosa*), sulfur, sulfite in some strains	Sulfur, IC	Sulfate	+in some strains	+
Thiodictyon	Sulfide, sulfur	Sulfur, IC	Sulfate	nd	−
Thioflaviococcus	Sulfide, sulfur	Sulfur, IC	Sulfate	nd	−
Thiohalocapsa	Sulfide, thiosulfate, sulfur, sulfite	Sulfur, IC	Sulfate	−	+
Thiolamprovum	Sulfide, thiosulfate, sulfur	Sulfur, IC	Sulfate	−	+
Thiopedia	Sulfide, sulfur	Sulfur, IC	Sulfate	−	−
Thiorhodococcus	Sulfide, thiosulfate, sulfur	Sulfur, IC	Sulfate	−	+/−
Thiorhodovibrio	Sulfide, sulfur	Sulfur, IC	Sulfate	nd	+
Thiospirillum	Sulfide, sulfur	Sulfur, IC	Sulfate	nd	−
Ectothiorhodospiraceae					
Ectothiorhodospira	Sulfide, thiosulfate (not in *Ect. marismortui*), sulfur, sulfite (nd for some species)	Polysulfide, sulfur, EC	Sulfate	+in some species	+in some species
Halorhodospira	Sulfide, thiosulfate only in *Hlr. halophila*	Sulfur, EC	Sulfur or sulfate	+in *Hlr. halochloris*	−

(continued)

Table 1. (continued)

Genus	Sulfur substrates	Intermediates	End product	Sulfate assimilation	Chemoautotrophic growth
Thiorhodospira	Sulfide, sulfur	Sulfur, EC & IC	Sulfate	nd	−
Ectothiorhodosinus[*]	Sulfide, thiosulfate	Sulfur, EC	Sulfate	nd	nd

The tabulated data were mostly taken from Imhoff (2005a; 2005b). Additional information was taken from Rees et al. (2002); Zaar et al. (2003); Gorlenko et al. (2004); Arunasri et al. (2005).

IC, intracellular; EC, extracellular; nd, not determined.

[*] The genus *Ectothiorhodosinus* has no standing in nomenclature.

All species of the genus *Ectothiorhdodospira* grow well under anoxic conditions in the light with reduced sulfur compounds as photosynthetic electron donors and in the presence of organic carbon sources and inorganic carbonate. Under the alkaline growth conditions which are optimal for *Ectothiorhodospira* species, polysulfides are stable intermediates during sulfide oxidation. As a result, polysulfides have been described as the first measurable oxidation products almost 25 years ago (Then and Trüper, 1983; Then and Trüper, 1984). When grown on elemental sulfur *Ect. halochloris* does not oxidize this compound to sulfate, but reduces it to sulfide and polysulfide (Then and Trüper, 1984). Several species of the genus *Ectothiorhodospira* are also able to grow chemolithotrophically on sulfur compounds (Table 1). Members of the genus *Halorhodospira* oxidize sulfide to sulfur which is further oxidized to sulfate by some species. Thiosulfate is only used by *Hlr. halophila* (Raymond and Sistrom, 1969) and poorly by *Halorhodospira neutriphila* (Hirschler-Rea et al., 2003). Sulfur can also be used by some species (Imhoff, 2005b).

III.   Electron Transport in Purple Sulfur Bacteria

During photoautotrophic growth of purple sulfur bacteria reduced sulfur compounds yield electrons for the reduction of CO_2. The electrons from the sulfur compounds are transferred to CO_2 via the photosynthetic electron transport chain and NAD^+. Photosynthetic electron transport and CO_2 fixation are therefore intimately intertwined with the oxidation of reduced sulfur compounds and will be briefly presented.

Light-driven electron flow in purple sulfur bacteria is essentially cyclic and involves two membrane-embedded complexes, the reaction center and the cytochrome $bc1$ complex (Fig. 1). In most purple bacteria the reaction center is intimately associated with a tetraheme cytochrome binding two heme c with a relatively low redox potential (10 mV) and two heme c with high redox potential (330 and 360 mV, respectively) (Nitschke et al., 1993). The reaction center uses light energy to transfer electrons from a mobile periplasmic or membrane-associated donor protein with a positive redox potential to quinone in the membrane. The reduction of the quinone occurs with incorporation of two protons from the cytoplasm close to the cytoplasmic membrane surface. The cycle is complete when the electrons are transferred back to the mobile electron-carrying protein via the cytochrome bc_1 complex.

The periplasmic electron carrier protein is cytochrome c_2 in several of the well studied purple nonsulfur bacteria, e.g. *Rhodobacter sphaeroides*, *Rhodobacter capsulatus*, and *Blastochloris viridis* (see, for example Drepper and Mathis (1997)). Surveys of photosynthetic electron transfer among other proteobacterial species, however, showed that the participation of HiPIP (high potential iron–sulfur protein), a ferredoxin-like [4Fe–4S] protein with a redox potential of +350 mV, instead of soluble cytochrome c is the rule rather than the exception (Menin et al., 1998). The idea has been put forward, that HiPIP is the electron carrier of choice in the purple sulfur bacteria in the families *Chromatiaceae* and *Ectothiorhodospiraceae*, but that the majority of purple nonsulfur bacteria are likely to utilize cytochrome c_2 (van Driessche et al., 2003). Other soluble cytochromes, such

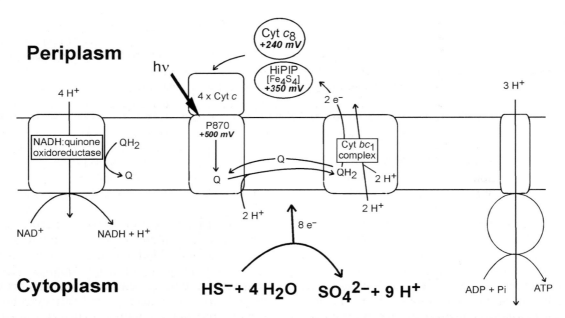

Fig. 1. Schematic representation of photosynthetic electron flow in the purple sulfur bacterium *Allochromatium vinosum*. Oxidation of sulfur compounds as the electron source for NAD$^+$ reduction is shown for sulfide in a simplified fashion. It should be noted that only part of the several steps involved in sulfur compound oxidation take indeed place in the cytoplasm (see Section IV).

as cytochrome c_8 or the membrane-associated cytochrome c_y can also mediate electron flow from the reaction center to the cytochrome $bc1$ complex in some species (Jenney et al., 1994; Samyn et al., 1996; Kerfeld et al., 1996). In *Allochromatium vinosum*, our model organism for the investigation sulfur oxidation pathways, both, HiPIP and cytochrome c_8 can serve as reductants of the high potential reaction center heme. In this purple sulfur bacterium the growth conditions influence the identity of the electron donor that is preferentially used. Cells grown autotrophically in the presence of sulfide and thiosulfate appear to use almost exclusively HiPIP while cytochrome c_8 is used in cells grown with organic compounds (Vermeglio et al., 2002).

The oxidation of quinol at the bc_1 complex results in the release of two protons into the periplasm. The resulting proton gradient drives ATP synthesis and the reduction of NAD$^+$ to NADH with quinol as the reductant. NADH and CO$_2$ are required for the reduction of CO$_2$ to carbohydrates via the Calvin cycle. Electrons drained from photosynthetic electron flow for the reduction of CO$_2$, are replaced by electrons released from oxidizable substrates. Taking into consideration the redox potential of the sulfur compounds

used as photosynthetic electron donors by purple sulfur bacteria, the respective electrons could in principle be transferred to periplasmic c-type cytochromes or directly into the quinone pool. Periplasmic cytochromes, such as flavocytochrome c and cytochrome c_{551} (SoxA), may feed electrons from sulfide or thiosulfate (see below) into the photosynthetic pathways via the same soluble carriers as are part of the cyclic system. Sulfide:quinone oxidoreductase would directly reduce quinone with electrons from sulfide. Electrons resulting from cytoplasmic oxidation of sulfite via APS reductase may also be directly transferred to quinone. It should be noted that the dissimilatory APS reductase is an iron–sulfur flavoprotein that bares no resemblance to the APS/PAPS reductases of the assimilatory sulfate reduction pathway.

IV. Biochemistry of Sulfur Oxidation Pathways in Purple Sulfur Bacteria

The last comprehensive reviews about the biochemistry of sulfur oxidation pathways in purple sulfur bacteria were published by Daniel C Brune (Brune, 1989; Brune, 1995b). In these excellent articles a wealth of information on sulfur oxidation patterns by whole cells of anoxygenic phototrophic bacteria was presented. In addition, the available data on enzymes potentially involved in sulfur transformations in these organisms was summarized. At that time, it was very difficult to unify the available information into a valid scheme, mostly due to the facts that a whole array of different organisms had been investigated and that molecular genetic information was essentially not present. In order to obtain a better picture of sulfur oxidation in purple sulfur bacteria we concentrated on one model bacterium, *Allochromatium vinosum* DSMZ 180^T and developed reverse genetics for this organism (Pattaragulwanit and Dahl, 1995). The following discussion on the oxidation of different reduced sulfur compounds, the properties of sulfur globules and their degradation therefore focuses on this organism. Results obtained with other purple sulfur bacteria are discussed but a complete survey of all available information is not attempted. Further relevant information is available through the genome sequence of *Halorhodospira halophila* SL1 (NZ_AAOQO1000001.1), a member of the *Ectothiorhodospiraceae*. In this organism many genes encoding proteins potentially involved in sulfur oxidation are clustered (genes Hhal1932 through Hhal1967). In Fig. 2 a comparison between the arrangement of these genes and those currently known for *Alc. vinosum* is presented. The genome sequence of *Alc. vinosum* has not yet been determined.

A. Oxidation of Thiosulfate

Thiosulfate ($S_2O_3^{2-}$) is a rather stable and environmentally abundant sulfur compound of intermediate oxidation state. It fulfils an important role in the natural sulfur cycle and is used by many phototrophic and chemotrophic sulfur oxidizers (Jørgensen, 1990; Sorokin et al., 1999). Two completely different pathways of thiosulfate oxidation appear to exist in purple sulfur bacteria. In one form tetrathionate is produced by oxidation of two thiosulfate anions via thiosulfate dehydrogenase (thiosulfate:acceptor oxidoreductase, EC 1.8.2.2). In the second form thiosulfate is completely oxidized to sulfate via several different mechanisms.

1. Thiosulfate Dehydrogenase

The formation of tetrathionate from thiosulfate has been mainly studied in chemoorganotrophic bacteria that use thiosulfate as a supplemental but not as the sole energy source (Jørgensen, 1990; Sorokin et al., 1999; Podgorsek and Imhoff, 1999). The pathway occurs only in a few purple sulfur bacteria including *Alc. vinosum* (Smith and Lascelles, 1966; Hensen et al., 2006).

In *Alc. vinosum* the ratio between tetrathionate and sulfate formed from thiosulfate is strongly pH-dependent with more tetrathionate as the product under slightly acidic conditions (Smith, 1966). In *Alc. vinosum* thiosulfate dehydrogenase is a periplasmic 30-kDa monomer with an isoelectric point of 4.2. The enzyme contains heme *c* and is reduced by thiosulfate at pH 5.0 but not at pH 7.0. In accordance, the pH optimum of the enzyme was determined to be 4.25 (Hensen et al., 2006). An examination of the kinetic properties of *Alc. vinosum* thiosulfate dehydrogenase with ferricyanide as artificial electron acceptor was initiated but interpretation of experimental results is complicated by the fact that enzymes that use two molecules of the same substrate do not follow regular Michaelis–Menten kinetics. However, some important constants could be estimated: the limiting V_{max} is about 34,000 units (mg protein)$^{-1}$ (corresponding to a k_{cat} of 1.7×10^4 s^{-1}) and the $[S]_{0.5}$ for ferricyanide is about 0.5. $[S]_{0.5}$ is the substrate concentration that yields half maximal velocity. It is important to note that it is not identical to K_m as a K_m cannot be given for reactions not following Michaelis–Menten kinetics (Segel, 1993). While thiosulfate did not display strong substrate inhibition at any of the experimental ferricyanide levels, ferricyanide did show substrate inhibition on *Alc. vinosum* thiosulfate dehydrogenase (Hensen et al., 2006). Furthermore, the enzyme was significantly inhibited by sulfite (50% inhibition at 80 μM sulfite). Under optimized assay conditions cytochrome *c* from yeast is used as electron acceptor instead of ferricyanide by the enzyme, whereas horse heart cytochrome *c* is not accepted. The properties of *Alc. vinosum* thiosulfate dehydrogenase described by Hensen et al. (2006) are compatible with older data presented by Smith (1966) and Fukumori and Yamanka (1979). In both reports a tetrathionate-forming activity with a pH optimum

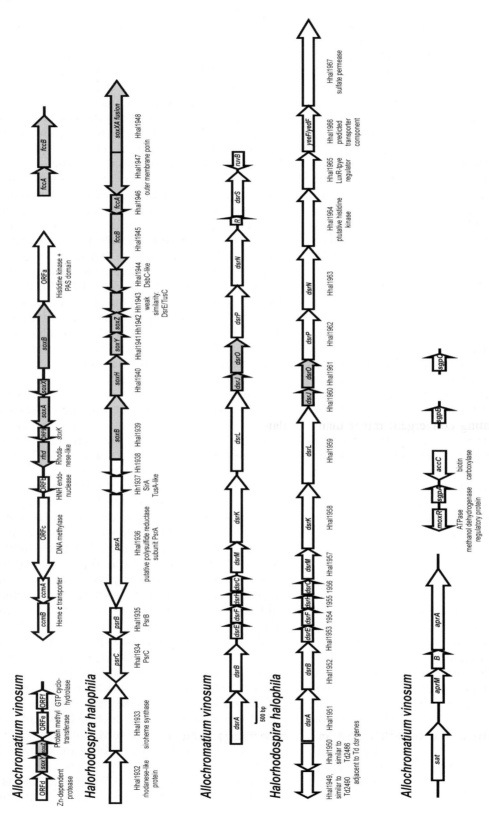

Fig. 2. Comparison of the established sulfur oxidation genes in *Allochromatium vinosum* DSMZ 180[T] and the genes of the sulfur oxidation cluster in the genome of *Halorhodospira halophila* SL1. Related genes/gene clusters are immediately compared where possible. It should be noted that the *Hlr. halophila* genes are all connected, while in *Alc. vinosum sox* genes are found in two independent loci and the *dsr*, *fcc* and *sat-apr* genes each form independent entities. The three *Alc. vinosum sgp* genes also represent independent transcriptional units. The *Allochromatium sat-apr* and *sgp* genes do not have equivalents in the *Halorhodospira* genome. Genes encoding proteins predicted to be localized in the periplasm are highlighted in grey.

in the acidic range was described. Fukumori and Yamanaka (1979) found that *Alc. vinosum* thiosulfate dehydrogenase used HiPIP isolated from the same organism as an efficient electron acceptor. This is in complete agreement with the fact that both, thiosulfate dehydrogenase and HiPIP are located in the periplasm of *Alc. vinosum* (Brüser et al., 1997) where HiPIP is photooxidized by the reaction center (van Driessche et al., 2003).

With our current analysis we cannot confirm the presence of any tetrathionate-forming enzyme operating at pH 8.0 in *Alc. vinosum* as has been claimed earlier (Schmitt et al., 1981; Knobloch et al., 1981). Of the tetrathionate-forming enzymes characterized so far, thiosulfate dehydrogenase from *Acidithiobacillus thiooxidans* (Nakamura et al., 2001) most closely resembles the enzyme from *Alc. vinosum*. Both species belong to the *Gammaproteobacteria*. The protein from *Acidithiobacillus* has been described as a monomeric 27.9-kDa *c*-type cytochrome with a pH optimum at 3.5. Thiosulfate dehydrogenases from other sources show remarkable heterogeneity with respect to structural properties and catalytic characteristics (Kusai and Yamanaka, 1973; Then and Trüper, 1981; Visser et al., 1996) which has been interpreted as indicating convergent rather than divergent evolution (Visser et al., 1996). A gene sequence encoding a heme-containing thiosulfate dehydrogenase has not yet been reported. A Blast search with the amino-terminal sequence of the enzyme from *Alc. vinosum* yielded only one significantly related sequence, a hypothetical *c*-type cytochrome from *Cupriavidus (Ralstonia, Wautersia) metallidurans* (Hensen et al., 2006).

2. Oxidation of Thiosulfate to Sulfate

Many purple sulfur bacteria can oxidize thiosulfate completely to sulfate (Table 1). In batch cultures of purple sulfur bacteria growing on thiosulfate the formation of sulfur globules is sometimes – but not always – observed. It is therefore very important to note, that the formation of sulfur globules is known to be an obligatory step during the oxidation of thiosulfate to sulfate in *Alc. vinosum* and probably also in other purple sulfur bacteria. Two independent lines of evidence prove that sulfur formation is an essential step: (1) An *Alc. vinosum* mutant unable to form sulfur globules due to the lack of

sulfur globule proteins cannot grow on thiosulfate (Prange et al., 2004) and (2) *Alc. vinosum* mutants blocked in sulfur oxidation form intracellular sulfur globules from thiosulfate as a dead end product (Pott and Dahl, 1998). In addition, studies with radioactively labelled thiosulfate demonstrated very clearly that the more reduced sulfane and the more oxidized sulfone sulfur atoms are processed differently in purple sulfur bacteria (Smith and Lascelles, 1966; Trüper and Pfennig, 1966). Only the sulfane sulfur accumulates as stored sulfur [S^0] before further oxidation, whereas the sulfone sulfur is rapidly converted into sulfate and excreted. The formation of sulfur as an intermediate in purple sulfur bacteria is different from the thiosulfate-oxidizing pathway (Sox pathway) that occurs in a wide range of facultatively chemo- or photolithotrophic bacteria like *Paracoccus pantotrophus* or *Rhodovulum sulfidophilum* (Appia-Ayme et al., 2001; Friedrich et al., 2001). In the latter, both sulfur atoms of thiosulfate are oxidized to sulfate without the appearance of sulfur deposits as intermediates.

In spite of this fundamental difference similar proteins appear to be essential for thiosulfate oxidation to sulfate in organisms forming sulfur as an intermediate and those not producing sulfur. Gene inactivation and complementation studies clearly showed that the *soxBXA* and *soxYZ* genes, located in two independent gene regions (Fig. 2), are essential for thiosulfate oxidation in *Alc. vinosum* (Hensen et al., 2006). Three periplasmic Sox proteins were purified from *Alc. vinosum*: the heterodimeric *c*-type cytochrome SoxXA (SoxX 11 kDa, SoxA 29 kDa; one covalently bound heme is present in each subunit), the heterodimeric SoxYZ (SoxY 12.7 kDa, SoxZ 11.2 kDa) and the monomeric SoxB (62 kDa, predicted to bind two manganese atoms) (Hensen et al., 2006).

In *Alc. vinosum* the genes *soxB* and *soxXA* are transcribed divergently. Upstream of *soxB* a gene encoding a potential regulator protein is located and immediately downstream of *soxA* two further interesting genes are found: The first (ORF9) encodes a hypothetical 12.2-kDa (9.2 kDa after processing) protein with a signal peptide. A homologous gene (*orf1020* or *soxK*) is present in all currently known *sox* gene clusters of thiosulfate-oxidizing green sulfur bacteria (Frigaard and Bryant, 2008), however, a homolog is not present in the *Hlr. halophila* sulfur gene cluster (Fig. 2). The second (*rhd*) encodes a putative periplasmic protein (22.2 kDa after processing) containing a conserved domain typical for rhodaneses. A homologous gene is neither found close to *sox* genes of green sulfur bacteria nor in *Hlr. halophila*. *In vitro*, rhodaneses (thiosulfate:sulfur transferases) can catalyze the transfer of the sulfane sulfur atom of thiosulfate to cyanide yielding thiocyanate (rhodanide, SCN^-) and sulfite. This is, however, not the physiological role in most cases. In the past, the detection of rhodanese and thiosulfate reductase activity in phototrophic sulfur bacteria led to the assumption that thiosulfate would be cleaved into sulfite and sulfide in the presence of suitable reduced thiol acceptors like glutathione and dihydrolipoic acid, and that the H_2S formed during the proposed reaction would be immediately oxidized to sulfur stored in sulfur globules (Brune, 1989; Brune, 1995b; Dahl, 1999). However, gene inactivation showed that the *Alc. vinosum rhd* product does not play such a vital role and is dispensable for thiosulfate oxidation (Hensen et al., 2006). The physiological role of the *rhd*-encoded protein remains to be elucidated. The deduced properties of other genes encoded in immediate vicinity of the *Alc. vinosum sox* genes were described in detail by Hensen et al. (2006). A function in oxidative sulfur metabolism of these hypothetical proteins is not obvious.

In *Hlr. halophila* putative *sox* genes are clustered but not organized in a single operon (Fig. 2). The gene *soxH* which is apparently not present close to the sequenced *sox* genes in *Alc. vinosum* is not required for lithotrophic growth on thiosulfate in *Pcs. pantotrophus* (Rother et al., 2001). In *Hlr. halophila soxBHYZ* appear to be co-transcribed. They are separated from a gene encoding a fusion of SoxXA by a divergently oriented cluster of four genes, among them *fccAB* possibly encoding a flavocytochrome *c* (sulfide dehydrogenase). The derived FccB polypeptide also shows similarity to SoxF, an important though not essential component of the *Pcs. pantotrophus* Sox system (Bardischewsky et al., 2006). However, the similarity is significantly lower than that to the flavoprotein subunit FccB of *Alc. vinosum* flavocytochrome *c* (Dolata et al., 1993; Reinartz et al., 1998). The gene immediately upstream of *fccB* in *Hlr. halophila* is clearly related to *fccA*

encoding the cytochrome *c* subunit of *Alc. vinosum* flavocytochrome *c* while similarity to *soxE* from *Pcs. pantotrophus* is below detection limits in searches using the BLAST algorithm (Altschul et al., 1990).

Occurrence and arrangement of *sox* genes in both purple sulfur bacteria is different from that in *Pcs. pantotrophus* in which the *sox* gene cluster comprises 15 genes organized into three transcriptional units, *soxRS*, *soxVW* and *sox XYZABCDEFGH*. In this organism the periplasmic proteins SoxXA, SoxYZ, SoxB and Sox(CD)$_2$ are essential for thiosulfate oxidation *in vivo* and *in vitro*. Currently, a model has been proposed that SoxXA initiates oxidation and covalent attachment of thiosulfate to a conserved cysteine (residue 138) in SoxY of the SoxYZ complex. SoxB would then hydrolytically release sulfate leaving a cysteine-138-persulfide in SoxYZ, which is proposed to be oxidized by the hemomolybdoenzyme Sox(CD)$_2$ yielding a cysteine-S-sulfonate. In the final step SoxB would again release sulfate and thereby recycle SoxYZ. In green sulfur bacteria, a *orf1015(soxJ)-soxXYZA-orf1020(soxK)-soxBW* genomic arrangement is generally found (Frigaard and Bryant, 2008). We were not able to detect the genes *soxCD* in *Alc. vinosum* and they are also not present in the genome of *Hlr. halophila*. Furthermore, these genes are absent in the magnetotactic *Magnetococcus* sp. MC1 and *Thiobacillus denitrificans*. *Alc. vinosum* and the latter two organisms have in common that they form sulfur as intermediate during thiosulfate oxidation, either as globules or as finely dispersed membrane-associated sulfur (Schedel and Trüper, 1980; Williams et al., 2006). The genomes of thiosulfate-oxidizing green sulfur bacteria (Frigaard and Bryant, 2008) also do not contain *soxCD*. Sulfur formation during thiosulfate oxidation has been described for one of these species, *Chlorobaculum parvum* DSM 263 (Steinmetz and Fischer, 1982). Sulfur may be an intermediate also in the other green sulfur bacteria, though may not be detectable due to a high turnover rate. Polysulfides have also been suggested as intermediates occurring in the periplasm of green sulfur bacteria during thiosulfate oxidation (Frigaard and Bryant, 2008). Summarizing the observation that the lack of *soxCD* appears to correlate with the formation of sulfur or possibly polysulfides as metabolic intermediates we suggested the following model (Fig. 3) (Hensen et al., 2006): The initial oxidation and covalent binding of thiosulfate to SoxYZ would be brought about by SoxXA and sulfate would then be hydrolytically released by SoxB just as proposed for *Pcs. pantotrophus* (Friedrich et al., 2001). However, in organisms like *Alc. vinosum* that lack "sulfur dehydrogenase" the sulfane sulfur atom linked to SoxY cannot be directly further oxidized. We suggest that the sulfur is instead transferred to growing sulfur globules (or polysulfide). Such a suggestion is feasible because the sulfur globules in *Alc. vinosum* and in many if not all other organ-

isms forming intracellular sulfur deposits reside in the bacterial periplasm (Pattaragulwanit et al., 1998; Dahl and Prange, 2006) (see also below) and therefore in the same cellular compartment as the Sox proteins (Hensen et al., 2006). Such a mechanism would require the transfer of SoxY-bound sulfur to the sulfur globules, a process that is currently unclear. The sulfur transferase encoded by the *rhd* gene has the capacity to play such a role however its inactivation did not lead to a detectable phenotype. Possibly, other sulfur transferases present in the cells function as a back up system.

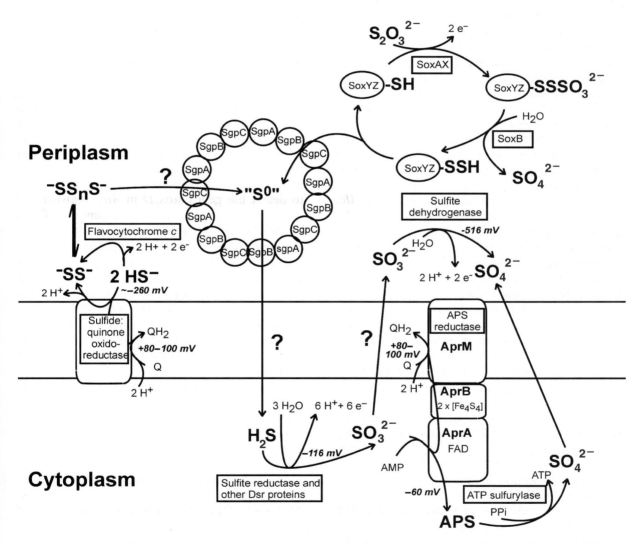

Fig. 3. Model of the sulfur oxidation pathway in *Allochromatium vinosum*. Direct oxidation of sulfite to sulfate is hypothesized to occur periplasmically by a classical sulfite dehydrogenase. However, as elaborated in the text, the possibility of direct cytoplasmic oxidation of sulfite cannot be excluded. *APS*, adenosine 5′-phosphosulfate.

B. Oxidation of Sulfide to "Elemental" Sulfur

In purple sulfur bacteria, the main enzymes that have been discussed as catalyzing the oxidation of sulfide are the periplasmic FAD-containing flavocytochrome c and the membrane-bound sulfide: quinone oxidoreductase (SQR) (Brune, 1995b; Reinartz et al., 1998) (Fig. 3).

The distribution of flavocytochrome c among the anoxygenic phototrophic bacteria and its chemical and catalytic properties have been discussed in detail elsewhere (Brune, 1989; Brune, 1995b; Frigaard and Bryant, 2008). The protein is located in the periplasm, consists of a FAD-binding (FccB, 46–47 kDa) and a smaller heme c-binding subunit (FccA, 21 kDa, two heme c in *Alc. vinosum* (van Beeumen et al., 1991)). The genome of *Hlr. halophila* contains three copies of potential *fccAB* genes (Hhal1945 and 1946, Hhal1162 and 1163, Hhal1330 and 1331). In vitro, flavocytochromes can efficiently catalyze electron transfer from sulfide to a variety of small c-type cytochromes (e.g. cytochrome c_{550} from *Alc. vinosum* (Davidson et al., 1985)) that may then donate electrons to the photosynthetic reaction center. However, the *in vivo* role of flavocytochrome c is unclear. It occurs in many purple and green sulfur bacteria but there are also many species that lack this protein. Moreover, an *Alc. vinosum* mutant deficient in flavocytochrome c exhibits sulfide oxidation rates similar to those of the wild type (Reinartz et al., 1998).

As an alternative to sulfide oxidation via flavocytochrome c, the transfer of electrons from sulfide primarily into the quinone pool was proposed, based on energetic considerations as well as on the inhibitory effect of rotenone, CCCP, and antimycin A on NAD photoreduction by sulfide (Brune and Trüper, 1986; Brune, 1989). Sulfide: quinone oxidoreductase (SQR) activity has in the meantime been described for many phototrophic organisms including the cyanobacterium *Oscillatoria limnetica* (Arieli et al., 1994), the purple nonsulfur bacterium *Rhodobacter capsulatus* (Schütz et al., 1997), green sulfur bacteria (Shahak et al., 1992) and also *Alc. vinosum* (Reinartz et al., 1998). The properties of this enzyme from diverse sources are described in detail in the chapter by Hauska and Shahak. Although *Alc. vinosum* membranes exhibit SQR activity, my laboratory has so far neither been able to detect a *sqr*-related gene via Southern hybridization with heterologous probes or heterologous PCR nor could we detect the protein with antibodies directed against the *Rba. capsulatus* protein (M. Reinartz and C. Dahl, unpublished). We therefore hypothesize that the enzyme from *Alc. vinosum* and possibly other purple sulfur bacteria has properties distinct from those of characterized SQRs. In accordance, the *Hlr. halophila* genome contains one only distantly related homolog (Hhal1665) of the biochemically well characterized SQR from *Rhodobacter capsulatus* (Schütz et al., 1999; Griesbeck et al., 2002).

In *Rba. capsulatus*, SQR is a peripherally membrane-bound flavoprotein with its active site located in the periplasm (Schütz et al., 1999). The primary product of the SQR reaction is soluble polysulfide whereas elemental sulfur does not appear to be formed *in vitro* (Griesbeck et al., 2002). Very probably, disulfide (or possibly a longer chain polysulfide) is the initial product of sulfide oxidation, which is released from the enzyme. Polysulfide anions of different chain lengths are in equilibrium with each other and longer-chain polysulfides can be formed by disproportionation reactions from the initial disulfide (Steudel, 1996). When whole cells of *Rba. capsulatus* grow with sulfide, elemental sulfur is formed as the final product. In principle, elemental sulfur can form spontaneously from polysulfides (Steudel, 1996).) In experiments using isolated spheroplasts from *Chlorobium vibrioforme* and *Allochromatium minutissimum*, soluble polysulfides have been detected as the product of sulfide oxidation (Blöthe and Fischer, 2000). Polysulfides were also detected as primary products of sulfide oxidation by whole cells of *Alc. vinosum* (Prange et al., 2004) and have been reported as intermediates of the oxidation of sulfide to extracellular sulfur by species of the purple sulfur bacterial family *Ectothiorhodospiraceae* (Trüper, 1978; Then and Trüper, 1983). While transient formation of polysulfide by the latter organism species has originally been attributed to chemical reaction between H_2S and elemental sulfur promoted by the alkaline culture medium (Trüper, 1978), it now appears more likely that they present biochemically generated intermediates.

It remains difficult to assign any function to flavocytochrome c, a protein that is constitutive in *Alc. vinosum* (Bartsch, 1978). Based on the large difference of redox potential between flavocytochrome c and the photosynthetic reaction center, Brune (1995b) suggested that flavocytochrome c may represent a high affinity system for sulfide oxidation that might be of advantage for the cells especially at very low sulfide concentrations. At present such a function cannot be excluded and flavocytochrome c could indeed supplement the energetically more efficient system involving electron transfer from sulfide to quinone via SQR.

In *Alc. vinosum*, sulfite reductase operating in reverse, i.e. in the direction of sulfite formation, has also been discussed to be involved in sulfide oxidation (Schedel et al., 1979). However, we have shown that this protein is not essential for sulfide oxidation but rather absolutely required for oxidation of intracellularly stored sulfur (Pott and Dahl, 1998). In the purple nonsulfur bacterium *Rhodovulum sulfidophilum* the Sox enzyme system that catalyzes the oxidation of thiosulfate to sulfate (see above), is also indispensable for the oxidation of sulfide in vivo (Appia-Ayme et al., 2001). However, in *Alc. vinosum* mutants deficient of either flavocytochrome c (Reinartz et al., 1998), *sox* genes or both (D. Hensen, B. Franz and C. Dahl, unpublished) sulfide oxidation proceeds with wild-type rates indicating that that SQR plays the main role in sulfide oxidation in this organism.

It should be noted that cytochromes without flavin groups have also been proposed to mediate electron transfer from sulfide to the reaction center in some purple sulfur bacteria (Fischer, 1984; Brune, 1989; Leguijt, 1993).

C.   Oxidation of Polysulfides

As outlined above, polysulfides appear to be the primary product of the oxidation of sulfide in purple sulfur bacteria. It is therefore not astonishing that those members of the *Chromatiaceae* that have been studied with respect to the utilization of externally added polysulfides with an average chain length of 3–4 sulfur atoms (*Alc. vinosum* and *Tca. roseopersicina*) readily used these compounds as photosynthetic electron donors (van Gemerden, 1987; Steudel et al., 1990; Visscher et al.,

1990). It is currently unknown how polysulfides are converted into sulfur globules. Theoretically this could be a purely chemical, spontaneous process as longer polysulfides are in equilibrium with elemental sulfur (Steudel et al., 1990). However, we have shown that *Alc. vinosum* sulfur globules do not contain major amounts of sulfur rings but probably consist of long-chains of sulfur with organic residues at one or both ends (Prange et al., 1999; Prange et al., 2002a). Such organylsulfanes must eventually be formed by an unknown (enzymatic) mechanism.

D.   Uptake of External Sulfur

Very many purple sulfur bacteria including *Alc. vinosum* are able to oxidize externally supplied solid, virtually insoluble elemental sulfur (Table 1). This step – although very important in the global sulfur cycle – is hardly understood.

The formal valence of elemental sulfur is zero. Elemental sulfur tends to catenate and to form chains with various lengths (polymeric sulfur) or ring sizes (Steudel and Eckert, 2003). All sulfur and allotropes are hydrophobic, not wetted by water and hardly dissolvable in water (Steudel, 1989). The most stable form of elemental sulfur at ambient pressure and temperature is cyclic, orthorhombic α-sulfur (α-S_8) (Steudel, 2000). Polymeric sulfur consists mainly of chain-like macromolecules but the presence of large S_n rings with $n>50$ is likely (Steudel and Eckert, 2003). Commercially available elemental sulfur sublimed at ambient temperature ("flowers of sulfur") consists of S_8 rings, traces of S_7 rings which are responsible for the yellow colour and varying amounts of polymeric sulfur. The bonding energy between S–S bonds in polymeric sulfur is $2.4\,kJ\,mol^{-1}$ weaker than in *cyclo*-octasulfur (Steudel and Eckert, 2003) and it might therefore be more accessible for sulfur-oxidizing bacteria (Franz et al., 2006).

Enzymes catalyzing the uptake and oxidation of externally added elemental sulfur have not yet been isolated from any species of phototrophic sulfur bacteria. The process must include binding and/or activation of the sulfur as well as transport inside of the cells. In principle, two different strategies would be possible: physical contact of the cells to their insoluble substrate and direct electron transfer from the cell envelope to the substrate via outer membrane proteins (Myers and Myers, 2001) or excretion of reducing substances, e.g. low molecular weight thiols that can act on substrate distant from the cells. Both possibilities are discussed in detail in the chapter by Hanson. Generally, little information is available about adhesion to and attack of extracellular sulfur. Leaching sulfur-oxidizing bacteria like *Acidithiobacillus ferrooxidans* appear to follow the first pathway and attach to sulfur by extracellular polymeric substances, specifically, lipopolysaccharides (Gehrke et al., 1998). Structures, attached to the cell wall (the so-called "spinae") have been postulated to mediate adhesion of a green sulfur bacterium to extracellularly deposited sulfur (Pibernat and Abella, 1996). In all cases so far, a reaction activating elemental sulfur prior to its oxidation is postulated, due to the stability and low water solubility of the substrate. In case of *cyclo*-octasulfur this activation reaction could be an opening of the S_8 ring by nucleophilic reagents, resulting in the formation of linear inorganic or organic polysulfanes. In addition, the reduction of elemental sulfur to water-soluble sulfide is discussed. Both reactions could be carried out by thiol groups of cysteine residues. Along this line, it was proposed for *Acidithibacillus* and *Acidiphilium* that extracellular elemental sulfur is mobilized by thiol groups of special outer membrane proteins and transported into the periplasmic space as persulfide sulfur (Rohwerder and Sand, 2003). Experimetal evidence for the existence of an outer membrane protein involved in cell-sulfur adhesion in this organism was obtained by Ramírez et al. (2004). In this respect it might be interesting to note that a gene encoding a potential outer membrane porin is found in the sulfur gene cluster of *Hlr. halophila* where it is situated immediately upstream of genes encoding a potential flavocytochrome *c* (Fig. 2). For *Alc. vinosum* we recently obtained first experimental evidence that an intimate physical cell-sulfur contact is indeed a prerequisite for uptake of elemental sulfur (Franz et al., 2006).

In our model organism *Alc. vinosum* the first step during oxidation of externally supplied sulfur is the accumulation of sulfur in intracellular sulfur globules which are then further oxidized to sulfate. XANES measurements provided evidence that *Alc. vinosum* uses only or at least strongly prefers the polymeric sulfur (sulfur chains) fraction of commercially available elemental sulfur and is probably unable to take up and form sulfur globules from *cyclo*-octasulfur (Franz et al., 2006). We did not find evidence for the formation of intermediates like sulfide or polysulfides during uptake of elemental sulfur. One might speculate that "sulfur chains" rather than the more stable "sulfur rings" are the microbiologically preferred form of elemental sulfur also for other sulfur-oxidizing bacteria.

E. Sulfur Globules and Their Properties

In anoxygenic phototrophic sulfur bacteria, sulfur appears to be generally deposited outside of the cytoplasm. Green sulfur bacteria and purple sulfur bacteria of the family *Ectothiorhodospiraceae* form extracellular sulfur globules while the globules are located in the periplasmic space in members of the family *Chromatiaceae* (Pattaragulwanit et al., 1998).

Despite the different site of deposition (outside or inside the confines of the cell) the sulfur appears to be of a similar speciation in the different groups of phototrophic sulfur bacteria: The exact chemical nature of the "elemental sulfur" in bacterial sulfur globules has been a matter of debate for many years (for a detailed historical account consult Dahl and Prange (2006). In most investigations, methods were used that required extraction of the sulfur globules from the cells prior to analysis (e.g. X-ray diffraction, (Hageage et al., 1970)) which causes changes in the chemical structure of the sulfur (Prange et al., 2002a). Only recently, X-ray absorption near-edge structure (XANES) spectroscopy at the sulfur K-edge using synchrotron radiation was introduced as an in situ approach to investigate the sulfur speciation in intact bacterial cells (Prange et al., 1999; Pickering et al., 2001; Prange et al., 2002a). A detailed description of these methods is given in the chapter by Prange et al. XANES spectroscopy yielded the following results for phototrophic sulfur bacteria: irrespective of whether the sulfur is accumulated in globules inside or outside the cells, it mainly consists of long sulfur chains very probably terminated by organic residues (mono-/bis-organyl polysulfanes) in purple and also in green sulfur bacteria. Most probably, the organic residue at the end of the sulfur chains present in the sulfur globules is glutathione or very similar to glutathione (Prange et al., 2002a). This hydrophilic residue could be responsible for maintaining the sulfur in a "liquid" state at ambient pressure and temperature. Earlier speculations and proposals that reduced glutathione (probably in its amidated form) could act as a carrier molecule of sulfur to and from the globules (Bartsch et al., 1996; Pott and Dahl, 1998) are supported by the XANES spectroscopy results (Prange et al., 2002a). Furthermore, XANES spectroscopy yielded evidence that the sulfur chains in globules of *Alc.vinosum* are gradually shortened during oxidation of intracellularly stored sulfur to sulfate (Prange et al., 2002b). It should be mentioned that some controversy has arisen about the interpretation of data acquired by XANES spectroscopy: Investigations of phototrophic sulfur bacteria by two different groups (Pickering et al., 2001; Prange et al., 2002a) yielded partly comparable experimental data but were interpreted in quite a different way. Pickering et al. (2001) concluded on the basis of theoretical considerations that the sulfur is "simply solid S_8". The discrepancies are mainly based on the measurement mode (George et al., 2002; Prange et al., 2002c). The model for the sulfur globules of *Alc.vinosum* that corresponds best with the available experimental data consists of long sulfur chains terminated by organic groups as was suggested by Prange et al. (Kleinjan et al., 2003). Sulfur of sulfur globules isolated in the presence of oxygen from anaerobically grown *Alc.vinosum* was found as S_8 rings (Prange et al., 2002a), indicating the influence of oxygen and the necessity of in situ methods like XANES spectroscopy that can be applied to avoid destruction of the original sulfur environment.

While sulfur globules appear to be more or less evenly distributed in many species of the *Chromatiaceae*, they can have very special and conspicuous localizations in other species. In *Allochromatium warmigii* for example, globules are predominantly located at the two poles of the cell. Dividing cells form additional sulfur globules near the central division plane. In *Lamprobacter modestohalophilus* the sulfur globules appear in the center of cells, while they are found in the peripheral part of the cells that is free of gas vesicles in species of the genera *Lamprocystis* and *Thiodictyon*. Sulfur globules are also found in the cell periphery in *Thiopedia rosea* (Imhoff, 2005a). For *Thiorhodovibrio winogradskyi* a formation of up to ten small sulfur globules in

a row along the long cell axis has been reported (Overmann et al., 1992). The specialized arrangement of sulfur inclusions suggests an important structure function relationship.

The sulfur globules in the *Chromatiaceae* are enclosed by a protein envelope, a feature shared by most if not all of the chemotrophic sulfur-oxidizing bacteria that form intracellular sulfur globules (Brune, 1995a; Dahl, 1999; Dahl and Prange, 2006). In *Alc. vinosum* this envelope is a monolayer of 2–5 nm consisting of three different hydrophobic "sulfur globule proteins" (Sgps) of 10.5 kDa, 10.6 kDa (SgpA and SgpB) and 8.5 kDa (SgpC), while that of the related *Thiocapsa roseopersicina* contains only two proteins of 10.7 and 8.7 kDa (Brune, 1995a; Pattaragulwanit et al., 1998). In *Alc. vinosum* the sulfur globule proteins are synthesized with cleavable amino-terminal signal sequences implying Sec-dependent transport across the cytoplasmic membrane and finally a periplasmic localization of the proteins and therefore the whole sulfur globules. The targeting process was experimentally verified with *phoA* fusions in *E. coli* (Pattaragulwanit et al., 1998) and also in *Alc. vinosum* (Prange et al., 2004). Electron micrographs of two other species of the family *Chromatiaceae* (*Thiocystis violaceae* and *Tca. roseopersicina*) provided further support for an extracytoplasmic localization of the sulfur globules (Pattaragulwanit et al., 1998).

The two larger sulfur globule proteins (SgpA and SgpB) of *Alc.vinosum* are homologous to each other and to the larger protein of *Tca. roseopersicina*. The smaller sulfur globule proteins (SgpC) in *Alc. vinosum* and *Tca. roseopersicina* are also homologous, indicating that these proteins are highly conserved between different species of the family *Chromatiaceae*. Interestingly, all three sulfur globule proteins are rich in glycine and aromatic amino acids, particularly tyrosine. The amino acid sequences contain tandem repeats typically found in cytoskeletal keratin or plant cell wall proteins suggesting that they are structural proteins rather than enzymes involved in sulfur metabolism (Brune, 1995a). A direct/covalent attachment of chains of stored sulfur to the proteins enclosing the globules is unlikely as none of the Sgp proteins sequenced so far contains cysteine residues.

Little is known about the function of the sulfur globule proteins. Proteinaceous envelopes have never been reported for extracellular sulfur globules. Consistent with this observation, neither the complete genome sequences of several green sulfur bacteria (Frigaard and Bryant, 2008) nor the *Hlr. halophila* genome contain homologues of *Alc. vinosum sgp* genes. As outlined above, the sulfur speciation in sulfur globules of anoxygenic phototrophic bacteria is nearly identical irrespective whether it is accumulated in globules inside or outside the cells. It therefore appears that the Sgp proteins themselves are not responsible for keeping the sulfur in a certain chemical structure. Ideas have been promoted, that the protein envelope serves as a barrier to separate the sulfur from other cellular constituents (Shively et al., 1989) and/or that it provides binding sites for sulfur-metabolizing enzymes (Schmidt et al., 1971). In *Alc. vinosum* mutants SgpA and SgpB can replace each other in the presence of SgpC (Pattaragulwanit et al., 1998; Prange et al., 2004). A mutant possessing SgpA and SgpB but lacking SgpC can grow on sulfide and thiosulfate. This mutant forms significantly smaller sulfur globules. SgpC therefore probably plays an important role in sulfur globule expansion. SgpA and SgpB are not fully competent to replace each other as sulfur globule formation is not possible in mutants possessing solely SgpA or SgpB. Experiments with a *sgpBC⁻* double mutant clearly showed that an envelope is indispensable for the formation and deposition of intracellular sulfur. Neither sulfide nor thiosulfate is oxidized by this mutant (Prange et al., 2004). In *Alc. vinosum* cell survival is absolutely dependent on the presence of at least SgpA even under conditions that do not allow sulfur globule formation (Prange et al., 2004). All three *sgp* genes of *Alc.vinosum* form separate transcriptional units (Pattaragulwanit et al., 1998). All are constitutively expressed, however, the expression of *sgpB* and *sgpC* is significantly enhanced under photolithoautotrophic compared to photoorganoheterotrophic conditions. The *sgpB* gene is expressed ten times less than *sgpA* and *sgpC* implying that SgpA and SgpC are the "main proteins" of the sulfur globule envelope (Prange et al., 2004).

Sulfur globules can also serve as an electron acceptor reserve that allows a rudimentary anaerobic respiration with sulfur. Under anoxic conditions in the absence of light purple sulfur bacteria like *Alc. vinosum* can reduce stored sulfur back to sulfide (van Gemerden, 1968; Trüper, 1978). Nothing is known about the enzymatic mechanisms underlying these processes.

F. Oxidation of Stored Sulfur to Sulfite

The oxidative degradation of these sulfur deposits is one of the most poorly understood areas of sulfur metabolism. In the case of extracellularly deposited sulfur, this process does not only involve oxidation of the sulfur but must include binding, activation and transport into the cells (see above).

The only gene region known so far to be essential for oxidation of stored sulfur was localized by interposon mutagenesis in *Alc. vinosum* (Pott and Dahl, 1998; Dahl et al., 2005). Fifteen open reading frames, designated *dsrABEFHCMKLJOPNRS*, were identified (Figs. 2 and 3). A very similar gene cluster is found in *Hlr. halophila* (Fig. 2), which contains in addition, genes encoding putative regulatory proteins and proteins possibly involved in sulfate transport downstream of *dsrN*. In *Alc. vinosum*, the *dsrAB* products form the cytoplasmic $\alpha_2\beta_2$-structured sulfite reductase. This protein is closely related to the dissimilatory sulfite reductases from sulfate-reducing bacteria and archaea (Hipp et al., 1997). The prosthetic group of DsrAB is siroamide-$[Fe_4S_4]$ with siroamide being an amidated form of the classical siroheme (Lübbe et al., 2006). The *dsrN*-encoded protein resembles cobyrinic acid *a, c* diamide synthases and catalyzes the glutamine-dependent amidation of siroheme. A $\Delta dsrN$ mutant showed a reduced sulfur oxidation rate. *Alc. vinosum* is apparently able to incorporate siroheme instead of siroamide into sulfite reductase, thereby retaining some function of the enzyme (Lübbe et al., 2006). Adjacent to *dsrAB* the *dsrEFH* genes are located. The products of these three genes show significant similarity to each other. DsrEFH were purified from the soluble fraction and constitute a soluble $\alpha_2\beta_2\gamma_2$-structured 75-kDa holoprotein (Dahl et al., 2005). DsrC is a small soluble cytoplasmatic protein with a highly conserved C-terminus including two conserved cysteine residues. Proteins closely related to DsrEFH and DsrC have recently been shown to act as parts of a sulfur relay system involved in thiouridine biosynthesis at tRNA wobble positions in *E. coli* (Numata et al., 2006; Ikeuchi et al., 2006). The *dsrM*-encoded protein is predicted to be a membrane-bound *b*-type cytochrome and shows similarities to a subunit of heterodisulfide reductases from methanogenic archaea. The cytoplasmic iron–sulfur protein DsrK exhibits relevant similarity to the catalytic subunit of heterodisulfide reductases. DsrK is predicted to reside in the cytoplasm. DsrP is another integral membrane protein. The periplasmic proteins DsrJ and DsrO are a triheme *c*-type cytochrome and an iron–sulfur protein, respectively. DsrKJO were co-purified from membranes pointing at the presence of a transmembrane electron-transporting complex consisting of DsrMKJOP (Dahl et al., 2005). Individual in frame deletions of the *dsrMKJOP* genes lead to the complete inability of the mutants to oxidize stored sulfur (Sander et al., 2006). In accordance with the suggestion that related complexes from dissimilatory sulfate reducers transfer electrons to sulfite reductase (Pires et al., 2006), the *Alc. vinosum* Dsr complex is co-purified with sulfite reductase, DsrEFH and DsrC (Dahl et al., 2005). DsrL is a cytoplasmic iron–sulfur flavoprotein with NADH: acceptor oxidoreductase activity (Y. Lübbe and C. Dahl, unpublished). *In frame* deletion of *dsrL* completely inhibited the oxidation of stored sulfur (Lübbe et al., 2006). DsrR and DsrS are soluble cytoplasmic proteins of unknown function. The *dsr* genes, with the exception of the constitutively expressed *dsrC*, are expressed and the encoded proteins are formed at a low basic level even in the absence of sulfur compounds. An increased production of all Dsr proteins is induced by sulfide and/or stored sulfur (Dahl et al., 2005).

The mechanism by which the periplasmically stored sulfur is made available to the cytoplasmic sulfite reductase is unclear. In sulfate-reducing bacteria dissimilatory sulfite reductase catalyzes the six electron reduction of sulfite to sulfide. It has therefore been proposed that the sulfur is reductively activated, transported to and further oxidized in the cytoplasm by sulfite reductase operating in reverse. Different models have been suggested to explain the roles of the *dsr*-encoded proteins in such a scenario (Dahl et al., 2005; Pott and Dahl 1998). A modified model is shown in Fig. 4. Here, the NADH: acceptor oxidoreductase activity of DsrL is taken into account. Interestingly,

the protein carries a thioredoxin motif CysXX-Cys immediately preceding the carboxy-terminal iron–sulfur cluster binding sites. This indicates a potential disulfide reductase activity. Therefore, the possibility exists that DsrL uses NADH as electron donor for reduction of a di- or persulfidic compound. DsrL could be involved in the reductive release of sulfide from a carrier molecule – probably an organic perthiol – that may transport sulfur from the periplasmic sulfur globules to the cytoplasm where it is further metabolized by Dsr proteins (Dahl et al., 2005). Glutathione amide is a likely candidate for carrying sulfur from the periplasm to the cytoplasm. Glutathione amide bears an amide group at the glycyl moiety of glutathione and is especially resistant to autoxidation. The compound was found to be largely in the persulfidic state when *Alc. vinosum* was cultured photoautotrophically on sulfide (Bartsch et al., 1996). Recently, transporters have been characterized in *E. coli* mediating export (Pittman et al., 2005) and import (Suzuki et al., 2005) of glutathione. Shuttling of glutathione amide between cytoplasm and periplasm in purple sulfur bacteria like *Alc. vinosum*, therefore also appears feasible. DsrL, being an essential protein for sulfur oxidation, is co-purified with the sulfite reductase (Y. Lübbe and C. Dahl, unpublished). Sulfide released from the perthiol could therefore be directly passed to *dsrAB*-encoded sulfite reductase thereby reducing losses caused by evaporation of gaseous H_2S. Obviously, *Alc. vinosum* sulfite reductase specifically interacts with the soluble protein DsrL on one hand and with membrane-bound Dsr proteins and DsrE-FHC on the other hand. Electrons released from the oxidation of sulfide by sulfite reductase may be fed into photosynthetic electron transport via DsrC and DsrMKJOP, which would be analogous to the pathway postulated for sulfate reducers, operating in the reverse direction. DsrM could operate as a quinone reductase, DsrP as a quinol oxidase and finally the *c*-type cytochrome DsrJ would be reduced (Dahl et al., 2005). From here, electrons could be transferred to HiPIP, the primary electron donor to the photosynthetic reaction center (Vermeglio et al., 2002). The function of DsrEFH remains unclear, but as it occurs exclusively in sulfur oxidizers and shows some interaction with DsrC, it may be important for the pathway to operate in the sulfide oxidizing direction. On the

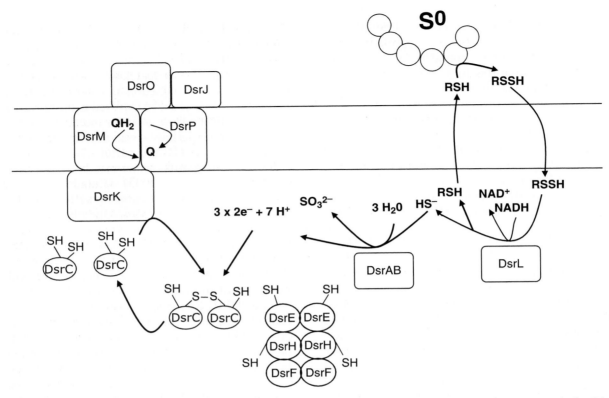

Fig. 4. Schematic presentation of Dsr proteins from *Allochromatium vinosum*. The scheme is based on sequence analysis of the encoding genes and on biochemical information where available. The products of the *dsrS* and *dsrR* genes are not shown for clarity because biochemical information is not available and possible functions cannot be predicted on the basis of sequence homologies. Both proteins are predicted to be soluble and to reside in the cytoplasm. DsrN is also not shown as it does not participate in redox or sulfur transfer reactions but is involved in biosynthesis of siroamide. Siroamide-[4Fe–4S] is a prosthetic group of sulfite reductase.

other hand, sulfur transfer reactions as performed by the related TusBCD and TusE proteins in *E. coli* (Ikeuchi et al., 2006) could be important for the Dsr-catalyzed sulfite formation pathway. As there is no experimental evidence available in this direction so far, this possibility is not taken into account in the model presented in Fig. 4.

G. Oxidation of Sulfite to Sulfate

In the final step of sulfur compound oxidation in purple sulfur bacteria, sulfite is oxidized to sulfate. Some purple sulfur bacteria can also grow on externally supplied sulfite (Table 1). As evident from Fig. 4 sulfite arising from the oxidation of more reduced sulfur compounds is generated in the bacterial cytoplasm. Two fundamentally different pathways for sulfite

oxidation have been rather well characterized in a number of chemotrophic and phototrophic sulfur oxidizers (Kappler and Dahl, 2001): (a) direct oxidation by a, probably molybdenum-containing, sulfite dehydrogenase (EC 1.8.2.1); and (b) indirect, AMP-dependent oxidation via the intermediate adenylylsulfate (adenosine 5′-phosphosulfate, APS).

The simultaneous presence of both enzymatic activities has been established for a number of chemo- and photolithotrophic sulfur oxidizers belonging to the β- and γ-Proteobacteria (e.g. *Thiobacillus denitrificans*, *Thiobacillus thioparus*, *Allochromatium vinosum*, strains of *Thiocapsa roseopersicina*) and green sulfur bacteria (Trüper and Fischer, 1982; Brune, 1995b; Kappler and Dahl, 2001). So far, there is no evidence for an occurrence of the sulfite-oxidizing form of the

APS reductase pathway in *Alphaproteobacteria* or in *Ectothiorhodospiraceae*. In accordance, potential APS reductase genes (*aprBA*, see below) are not found in the genome of *Hlr. halophila*. It has to be kept in mind that in some cases (*Beggiatoa*, *Chromatiaceae*, green sulfur bacteria) the occurrence of one or both sulfite oxidation pathways can vary between different strains of the same genus or between genera of the same family (Kappler and Dahl, 2001; Frigaard and Bryant, 2008).

1.   Indirect Pathway via Adenylylsulfate (APS)

During indirect sulfite oxidation, APS is formed from sulfite and AMP by APS reductase (EC 1.8.99.2). In a second step the AMP moiety of APS is transferred either to pyrophosphate by ATP sulfurylase (ATP:sulfate adenylyltransferase, EC 2.7.7.4), or to phosphate by adenylylsulfate:phosphate adenylyltransferase (APAT, formerly ADP sulfurylase (Brüser et al., 2000)), resulting in the formation of ATP or ADP, respectively. Since ADP can be converted to ATP and AMP by adenylate kinase, both sulfate-liberating enzymes catalyze substrate phosphorylations, which are of energetic importance, especially in chemolithoautotrophic bacteria (Peck, 1968). The APS pathway can also function in sulfate reduction, serving assimilatory and dissimilatory purposes. While the APS reductases from dissimilatory sulfate reducers resemble the enzymes found in sulfur oxidizers (Hipp et al., 1997), the APS reductases functioning in assimilatory sulfate reduction studied so far are completely different enzymes related to 3'-phosphoadenosine-5'-phosphosulfate (PAPS) reductases (Bick et al., 2000; Kopriva et al. 2001). Concerning this topic, consult also the chapter by Kopriva et al. in this book.

Indirect AMP-dependent oxidation of sulfite to sulfate via APS (Fig. 3) occurs in the bacterial cytoplasm with APS reductase being membrane-bound (e.g. in many *Chromatiaceae*) or soluble, and ATP sulfurylase and APAT being soluble enzymes (Brune, 1995a; Brüser et al., 2000). In *Alc. vinosum* the genes for ATP sulfurylase (*sat*) and APS reductase (*aprMBA*, with *aprM* encoding a putative membrane anchor) form an operon ((Hipp et al., 1997), A. Wynen, H. G. Trüper, C. Dahl, unpublished, GenBank No. U84759, Fig. 2).

In the genomes of four green sulfur bacteria the genes for ATP sulfurylase and APS reductase are located directly adjacent to each other (Frigaard and Bryant, 2008). Genes related with *aprM* are not present. Instead, the green sulfur bacterial APS reductase and ATP sulfurylase genes are always clustered with genes encoding a Qmo complex (*qmoABC*). A closely related complex was biochemically characterized from the sulfate reducer *Desulfovibrio vulgaris*, shown to have quinol-oxidizing activity and proposed to deliver electrons form membrane-bound quinols to APS reductase (Pires et al., 2003). In phototrophic sulfur oxidizers containing *qmo* genes, the situation could just be opposite and the Qmo complex could accept electrons from APS reductase operating in the sulfite-oxidizing direction. We propose that the membrane protein AprM serves an analogous function in *Alc. vinosum*.

APS reductase activity is usually measured as AMP-dependent sulfite oxidation with ferricyanide or *c*-type cytochromes. Substrate inhibition by AMP is characteristic for APS reductases (Taylor, 1994; Hagen and Nelson, 1997). All investigated dissimilatory APS reductases irrespective of metabolic type have been characterized as heterodimers with one α-subunit of 70–75-kDa (1 FAD) and one β-subunit of 18–23 kDa (2 [4Fe-4S] centers) (Fritz et al., 2000). Additional subunits mediating membrane association may be present (Hipp et al., 1997). The heme groups originally reported for the enzyme from the purple sulfur bacterium *Thiocapsa roseopersicina* were due to a contaminating protein (Brune, 1995b). A catalytic mechanism has been proposed in which sulfite initially forms a complex with the flavin (Brune (1995b) and references therein). This then reacts with AMP to yield APS, releasing two electrons that are transferred via the flavin to the iron–sulfur centers.

The best characterized ATP sulfurylase (Sat) from a sulfur-oxidizing bacterium is the enzyme from the endosymbiont of the hydrothermal vent worm *Riftia pachyptila* (Renosto et al., 1991; Beynon et al., 2001). Like all other ATP sulfurylases the enzyme is strictly Mg^{2+}-dependent. The V_{max} of ATP synthesis is seven times higher than that of molybdolysis, the assay used for measuring the APS-producing reaction. The *Riftia* symbiont enzyme also has a higher k_{cat} for the ATP synthesis direction ($257 s^{-1}$ compared to $64 s^{-1}$

for the assimilatory enzyme from *Penicillium chrysogenum* that works in the sulfate activating direction (Renosto et al., 1991)). The native enzyme appears to be a dimer (MW 90 kDa) composed of identical size subunits (396 residues). The ATP sulfurylase from *Alc. vinosum* is isolated as a monomer with an apparent molecular mass of 45 kDa (A. Wynen, C., Dahl, H. G. Trüper, unpublished). More information is available for ATP sulfurylases from sulfate-assimilating or sulfate-reducing organisms in which the activation of the chemically extremely inert sulfate by adenylylation is the relevant reaction. Two completely different, unrelated types of ATP sulfurylase can be distinguished: The heterodimeric CysDN type occurs exclusively in sulfate-assimilating prokaryotes, e.g., *E. coli* (Leyh, 1993). The other ATP sulfurylases characterized in sufficient detail are monomers or homo-oligomers of 41–69 kDa (Sperling et al., 1998; Gavel et al., 1998; Yu et al., 2007). Size variations are due to APS kinase or PAPS-binding allosteric domains residing on the same polypeptide in some cases. Five highly conserved regions are present, two of which are rich in basic amino acids, suggesting that they may participate in binding of $MgATP^{2-}$ and SO_4^{2-}.

The existence of APAT as an independent entity has been questioned for a long time. In 2000 the enzyme was finally purified from *Thiobacillus denitrificans* (Brüser et al., 2000): The enzyme is a homodimer of 41.4-kDa subunits. The K_M values for APS and phosphate are 300 μM and 12 mM, respectively. The pH optimum is 8.5–9.0. Catalysis is strictly unidirectional and occurs by a Ping-Pong mechanism with a covalently bound AMP as intermediate. Histidine modification suggested a histidine as the nucleotide binding residue. APAT is related to galactose-1-phosphate uridylyltransferase and diadenosine 5′, 5‴-P^1, P^4-tetraphosphate (Ap_4A) phosphorylase. Ap_4′A phosphorylase from yeast also has APAT activity while APAT from *Thiobacillus denitrificans* does not exhibit Ap_4A phosphorylase activity. The *in vivo* function of the latter enzyme may therefore indeed be the formation of ADP and sulfate from phosphate and APS. However, genetic evidence for this assumption is currently missing. The in vivo role of APAT is especially difficult to assign because all organisms with significant APAT activity (> 100 mU mg^{-1} in crude extracts) also contain ATP sulfurylase. It has been hypothesized that APAT may serve to ensure a high turnover of APS under pyrophosphate limiting conditions as this enzyme is independent of the energy-rich pyrophosphate molecule (Brüser et al., 2000). In *Alc. vinosum* APAT does not appear to be present while significant activity was found in strains of *Tca. roseopersicina* (Dahl and Trüper, 1989).

2. Direct Pathway

Two types of enzymes catalyzing direct oxidation of sulfite to sulfate are well characterized, the sulfite oxidases that can transfer electrons to oxygen, ferricyanide and sometime cytochrome *c* and the sulfite dehydrogenases that can use one or both of the latter electron acceptors but not oxygen (Kappler and Dahl, 2001; Kappler, 2007). The oxygen-dependent enzymes are not relevant in anoxygenic phototrophic bacteria.

All sulfite dehydrogenases characterized to date belong to the sulfite oxidase family of molybdoenzymes comprising established sulfite-oxidizing enzymes and proteins related to these as well as assimilatory nitrate reductases from plants (Hille, 1996). The active site is formed by a single molydopterin cofactor. Additional redox active centers may be present. The best characterized sulfite-oxidizing enzymes from the sulfite oxidase family are those from avian and mammalian sources (Kisker et al., 1997) that are homodimers containing heme *b* and molybdenum coordinated via an MPT-type molybdenum pterin cofactor, a conserved cysteine residue from the enzyme and two oxo groups. The SorAB protein from *Starkeya novella* (formerly *Thiobacillus novellus*) was the first true bacterial sulfite-oxidizing enzyme to be characterized in detail (Kappler et al., 2000; Feng et al., 2003; Kappler and Bailey, 2005; Raitsimring et al., 2005; Doonan et al., 2006). It is a periplasmic heterodimer of a large MoCo-dimer domain (40.2 kDa) and a small cytochrome *c* subunit (8.8 kDa). Its molybdenum pterin cofactor is of the MPT-type with a 1:1 ratio between Mo and MPT. During catalysis, electrons are sequentially transferred to a single heme c_{552} ($E_{m8.0} = +280$ mV) located on the smaller subunit and passed on from there to a cytochrome c_{550} from the same organism, thought to be the enzyme's natural electron acceptor. The enzyme exhibits the Ping–Pong mechanism that is also found in eukaryotic sulfite oxidases and is non-competitively inhibited by sulfate. It is encoded by the *sorAB* genes, which appear to form an operon by themselves. Characterized related proteins also appear to be localized in the periplasm and to contain a heme *c*-binding subunit (Myers and Kelly, 2005).

Biochemical studies and most importantly the sequencing of a large number of bacterial genomes in the past few years revealed that many bacterial genes exist that encode proteins belonging into the sulfite oxidase family (Kappler, 2007). While the well-characterized bacterial sulfite dehydrogenases are soluble proteins, membrane-bound bacterial sulfite-oxidizing enzymes have also been reported in the literature (reviewed in Kappler and Dahl, 2001; Kappler, 2007). Most of the established or predicted soluble members of the sulfite oxidase family are periplasmic enzymes, however, some of the proteins belonging to this group (however without a biochemically characterized function) are predicted to reside in the bacterial cytoplasm (Kappler, 2007). Direct oxidation of sulfite to sulfate in the bacterial cytoplasm can, therefore, not generally be excluded.

3. Sulfite Oxidation in Purple Sulfur Bacteria: an Unresolved Question

Although enzymes participating in the indirect sulfite oxidation pathway in purple sulfur bacteria have been studied for more than 30 years (Trüper and Rogers, 1971) their in vivo role is still questionable. In *Alc. vinosum* APS reductase is clearly dispensable (Dahl, 1996): The growth rates of the wild type and an APS-reductase-deficient mutant show little differences under light-limiting conditions. A difference is observed only at saturating irradiances. Under these conditions, the wild type grows considerably faster, indicating that the presence of a second pathway of sulfite oxidation allows a higher rate of supply of reducing power (Sanchez et al., 2001).

Experiments with cultures grown in the presence of the molybdate antagonist tungstate indicated that APS reductase-independent sulfite oxidation in *Alc. vinosum* is catalyzed by a molybdenum-containing enzyme. Sulfite oxidation was severely inhibited by tungstate in an APS-reductase deficient mutant, suggesting the involvement of a classical molydopterin-containing enzyme of the sulfite oxidase family (Dahl, 1996). However, it should be noted that genes related to *sorAB*

cannot be detected in *Alc. vinosum* nor have the proteins been detected using antibodies (e.g. against SorAB from *Starkeya novella*, U. Kappler and C. Dahl, unpublished). This finding appears even more interesting when we realize that a gene homologous to those encoding proteins of the sulfite oxidase family is neither present in the genome of *Hlr. halophila* nor in any of the green sulfur bacterial genome sequences. As *Hlr. halophila* and some of the green sulfur bacteria do not possess genes encoding for the APS pathway, they must have a different means for sulfite oxidation. Frigaard and Bryant (2008) present the very attractive speculation that a potential protein encoded by three genes resembling those for polysulfide reductase from *Wolinella succinogenes* (Krafft et al., 1992) could play this role. In this regard, it appears rather conspicuous that three related genes (Hhal1934, 1935 and 1936) are also found in the sulfur gene cluster of *Hlr. halophila* (Fig. 2). Similar to the situation in green sulfur bacteria, the molydopterin-binding putative active site-bearing subunit (PsrA) would be localized in the cytoplasm. On the other hand, we have some indications that a *soxY*-deficient mutant of *Alc. vinosum* is severely impaired in the oxidation of sulfite (D. Hensen, B. Franz and C. Dahl, unpublished). Clearly, the question of sulfite oxidation in phototrophic sulfur bacteria will require special attention in the future.

Acknowledgements

Support by the Deutsche Forschungsgemeinschaft to CD is gratefully acknowledged. I thank Johannes Sander for help with analyzing genomic data.

References

Altschul SF, Gish W, Miller W, Myers EW and Lipman DJ (1990) Basic local alignment search tool. J Mol Biol 215: 403–410

Appia-Ayme C, Little PJ, Matsumoto Y, Leech AP and Berks BC (2001) Cytochrome complex essential for photosynthetic oxidation of both thiosulfate and sulfide in *Rhodovulum sulfidophilum*. J Bacteriol 183: 6107–6118

Arieli B, Shahak Y, Taglicht D, Hauska G and Padan E (1994) Purification and characterization of sulfide-quinone reductase, a novel enzyme driving anoxygenic photosynthesis in *Oscillatoria limnetica*. J Biol Chem 269: 5705–5711

Arunasri K, Sasikala C, Ramana CV, Süling J and Imhoff JF (2005) *Marichromatium indicum* sp. nov., a novel purple sulfur gammaproteobacterium from mangrove soil of Goa, India. Int J Syst Evol Microbiol 55: 673–679

Bardischewsky F, Quentmeier A and Friedrich CG (2006) The flavoprotein SoxF functions in chemotrophic thiosulfate oxidation of *Paracoccus pantotrophus in vivo* and *in vitro*. FEMS Microbiol Lett 258: 121–126

Bartsch RG (1978) Cytochromes. In: Clayton RK and Sistrom WR (eds) The Photosynthetic Bacteria, pp 249–279. Plenum, New York

Bartsch RG, Newton GL, Sherrill C and Fahey RC (1996) Glutathione amide and its perthiol in anaerobic sulfur bacteria. J Bacteriol 178: 4742–4746

Beynon JD, MacRae IJ, Huston SL, Nelson DC, Segel IH and Fisher AJ (2001) Crystal structure of ATP sulfurylase from the bacterial symbiont of the hydrothermal vent tubeworm *Riftia pachyptila*. Biochemistry 40: 14509–14517

Bick JA, Dennis JJ, Zylstra GJ, Nowack J and Leustek T (2000) Identification of a new class of 5′-adenylylsulfate (APS) reductases from sulfate-assimilating bacteria. J Bacteriol 182: 135–142

Blöthe, M and Fischer, U (2000) New insights in sulfur metabolism of purple and green phototrophic sulfur bacteria and their spheroplasts. BIOspektrum, Special edition 1st Joint Congress of DGHM, ÖGHMP and VAAM: "Microbiology 2000", Munidi, p.62

Brune DC (1989) Sulfur oxidation by phototrophic bacteria. Biochim Biophys Acta 975: 189–221

Brune DC (1995a) Isolation and characterization of sulfur globule proteins from *Chromatium vinosum* and *Thiocapsa roseopersicina*. Arch Microbiol 163: 391–399

Brune DC (1995b) Sulfur compounds as photosynthetic electron donors. In: Blankenship RE, Madigan MT and Bauer CE (eds) Anoxygenic Photosynthetic Bacteria, pp 847–870, Vol 2 of Advances in Photosnythesis (Govindjee ed.). Kluwer Academic Publishers (now Springer), Dordrecht

Brune DC and Trüper HG (1986) Noncyclic electron transport in chromatophores from photolithoautotrophically grown *Rhodobacter sulfidophilus*. Arch Microbiol 145: 295–301

Brüser T, Selmer T and Dahl C (2000) "ADP sulfurylase" from *Thiobacillus denitrificans* is an adenylylsulfate:phosphate adenylyltransferase and belongs to a new family of nucleotidyltransferases. J Biol Chem 275: 1691–1698

Brüser T, Trüper HG and Dahl C (1997) Cloning and sequencing of the gene encoding the high potential iron–sulfur protein (HiPIP) from the purple sulfur bacterium *Chromatium vinosum*. Biochim Biophys Acta 1352: 18–22

Bryantseva IA, Gorlenko VM, Kompantseva EI, Imhoff JF, Süling J and Mityushina L (1999) *Thiorhodospira sibirica* gen. nov., sp. nov., a new alkaliphilic purple sulfur bacterium from a Siberian Soda lake. Int J Syst Bacteriol 49: 697–703

Bryantseva IA, Gorlenko VM, Kompantseva EI, Tourova TP, Kuznetsov B and Osipov GA (2000) Alkaliphilic heliobacterium *Heliorestis baculata* sp. nov. and emended description of the genus *Heliorestis*. Arch Microbiol 174: 283–291

Dahl C (1996) Insertional gene inactivation in a phototrophic sulphur bacterium: APS-reductase-deficient mutants of *Chromatium vinosum*. Microbiology 142: 3363–3372

Dahl C (1999) Deposition and oxidation of polymeric sulfur in prokaryotes. In: Steinbüchel, A (ed) Biochemical Principles and Mechanisms of Biosynthesis and Biodegradation of Polymers, pp 27–34. Wiley, Weinheim

Dahl C and Prange A (2006) Bacterial sulfur globules: occurrence, structure and metabolism. In: Shively JM (ed) Inclusions in Prokaryotes, pp 21–51. Springer, Heidelberg

Dahl C and Trüper HG (1989) Comparative enzymology of sulfite oxidation in *Thiocapsa roseopersicina* strains 6311, M1 and BBS under chemotrophic and phototrophic conditions. Z Naturforsch 44c: 617–622

Dahl C, Engels S, Pott-Sperling AS, Schulte A, Sander J, Lübbe Y, Deuster O and Brune DC (2005) Novel genes of the *dsr* gene cluster and evidence for close interaction of Dsr proteins during sulfur oxidation in the phototrophic sulfur bacterium *Allochromatium vinosum*. J Bacteriol 187: 1392–1404

Davidson MW, Gray GO and Knaff DB (1985) Interaction of *Chromatium vinosum* flavocytochrome *c*-552 with cytochromes *c* studied by affinity chromatography. FEBS Lett 187: 155–159

de Jong GAH, Hazeu W, Bos P and Kuenen JG (1997) Isolation of the tetrathionate hydrolase from *Thiobacillus acidophilus*. Eur J Biochem 243: 678–683

de Jong GAH, Tang JA, Bos P, de Vries S and Kuenen GJ (2000) Purification and characterization of a sulfite:cytochrome *c* oxidoreductase from *Thiobacillus acidophilus*. J Mol Catal B 8: 61–67

Dolata MM, van Beeumen JJ, Ambler RP, Meyer TE and Cusanovich MA (1993) Nucleotide sequence of the heme subunit of flavocytochrome c from the purple phototrophic bacterium, *Chromatium vinosum*. A 2.6-kilobase pair DNA fragment contains two multiheme cytochromes, a flavoprotein, and a homolog of human ankyrin. J Biol Chem 268: 14426–14431

Doonan CJ, Kappler U and George GN (2006) Structure of the active site of sulfite dehydrogenase from *Starkeya novella*. Inorg Chem 45: 7488–7492

Drepper F and Mathis P (1997) Structure and function of cytochrome c(2) in electron transfer complexes with the photosynthetic reaction center of *Rhodobacter sphaeroides*: Optical linear dichroism and EPR. Biochemistry 36: 1428–1440

Feng C, Kappler U, Tollin G and Enemark JH (2003) Intramolecular electron transfer in a bacterial sulfite dehydrogenase. J Am Chem Soc 125: 14696–14697

Fischer U (1984) Cytochromes and iron sulfur proteins in sulfur metabolism of phototrophic sulfur bacteria. In:

Müller A and Krebs B (eds) Sulfur, Its Significance for Chemistry, for the Geo-, Bio- and Cosmosphere and Technology, pp 383–407. Elsevier Science, Amsterdam

Franz B, Lichtenberg H, Hormes J, Modrow H, Dahl C and Prange A (2006) Utilization of solid "elemental" sulfur by the phototrophic purple sulfur bacterium *Allochromatium vinosum*: a sulfur K-edge XANES spectroscopy study. Microbiology 153: 1268–1274

Friedrich CG, Rother D, Bardischewsky F, Quentmeier A and Fischer J (2001) Oxidation of reduced inorganic sulfur compounds by bacteria: emergence of a common mechanism? Appl Environ Microbiol 67: 2873–2882

Frigaard NU and Bryant DA (2008) Genomic insights into the sulfur metabolism of phototrophic green sulfur bacteria. In: Hell R, Dahl C, Knaff DB and Leustek T (eds), Sulfur Metabolism in Phototrophic Organisms, (Advances in Photosynthesis and Respiration, Vol 27), pp 337–355. Springer, New York

Fritz G, Buchert T, Huber H, Stetter KO and Kroneck PMH (2000) Adenylylsulfate reductases from archaea and bacteria are 1: 1 alpha beta-heterodimeric iron–sulfur flavoenzymes – high similarity of molecular properties emphasizes their central role in sulfur metabolism. FEBS Lett 473: 63–66

Fukumori Y and Yamanaka T (1979) A high-potential non-heme iron protein (HiPIP)-linked, thiosulfate-oxidizing enzyme derived from *Chromatium vinosum*. Curr Microbiol 3: 117–120

Garrity GM and Holt G (2001) *Chlorobi* phy. nov. In: Boone, DR, Castenholz, RW and Garrity GM (eds) Bergey's Manual of Systematic Bacteriology, Vol 1, pp 601–623. Springer, New York

Gavel OY, Bursakov SA, Calvete JJ, George GN, Moura JJG and Moura I (1998) ATP sulfurylases from sulfate-reducing bacteria of the genus *Desulfovibrio*. A novel metalloprotein containing cobalt and zinc. Biochemistry 37: 16225–16232

Gehrke T, Telegdi J, Thierry D and Sand W (1998) Importance of extracellular polymeric substances from *Thiobacillus ferrooxidans* for bioleaching. Appl Environ Microbiol 64: 2743–2747

George, GN, Pickering, IJ, Yu, EY and Prince, RC (2002) X-ray absorption spectroscopy of bacterial sulfur globules. Microbiology 148: 2267–2268

Gorlenko VM, Bryantseva IA, Panteleeva EE, Tourova TP, Kolganova TV, Makhneva ZK and Moskalenko AA (2004) *Ectothiorhodosinus mongolicum* gen. nov., sp. nov., a new purple bacterium from a soda lake in Mongolia. Microbiology 73: 66–73

Griesbeck C, Schütz M, Schödl T, Bathe S, Nausch L, Mederer N, Vielreicher M and Hauska G (2002) Mechanism of sulfide-quinone oxidoreductase investigated using site-directed mutagenesis and sulfur analysis. Biochemistry 41: 11552–11565

Hageage GJ, Jr., Eanes ED and Gherna RL (1970) X-ray diffraction studies of the sulfur globules accumulated by *Chromatium* species. J Bacteriol 101: 464–469

Hagen KD and Nelson DC (1997) Use of reduced sulfur compounds by *Beggiatoa* spp.: Enzymology and physiology of marine and freshwater strains in homogeneous and gradient cultures. Appl Environ Microbiol 63: 3957–3964

Hanada S and Pierson BK (2002) The family Chloroflexaceae. In: Dworkin M (ed) The Prokaryotes: an Evolving Electronic Resource for the Microbiological Community, http://link.springer-ny.com/link/service/books/10125/. Springer, Berlin, Heidelberg, New York

Hensen D, Sperling D, Trüper HG, Brune DC and Dahl C (2006) Thiosulfate oxidation in the phototrophic sulfur bacterium *Allochromatium vinosum*. Mol Microbiol 62: 794–810

Hille R (1996) The mononuclear molybdenum enzymes. Chem Rev 96: 2757–2816

Hipp WM, Pott AS, Thum-Schmitz N, Faath I, Dahl C and Trüper HG (1997) Towards the phylogeny of APS reductases and sirohaem sulfite reductases in sulfate-reducing and sulfur-oxidizing prokaryotes. Microbiology 143: 2891–2902

Hiraishi A and Shimada K (2001) Aerobic anoxygenic photosynthetic bacteria with zinc-bacteriochlorophyll. J Gen Appl Microbiol 47: 161–180

Hiraishi A, Nagashima KVP, Matsuura K, Shimada K, Takaichi S, Wakao N and Katayama Y (1998) Phylogeny and photosynthetic features of *Thiobacillus acidophilus* and related acidophilic bacteria: its transfer to the genus *Acidiphilium* as *Acidiphilium acidophilum* comb. nov. Int J Syst Bacteriol 48: 1389–1398

Hirschler-Rea A, Matheron R, Riffaud C, Moune S, Eatock C, Herbert RA, Willison JC and Caumette P (2003) Isolation and characterization of spirilloid purple phototrophic bacteria forming red layers in microbial mats of Mediterranean salterns: description of *Halorhodospira neutriphila* sp. nov. and emendation of the genus *Halorhodospira*. Int J Syst Evol Microbiol 53: 153–163

Ikeuchi Y, Shigi N, Kato J, Nishimura A and Suzuki T (2006) Mechanistic insights into sulfur relay by multiple sulfur mediators involved in thiouridine biosynthesis at tRNA wobble positions. Mol Cell 21: 97–108

Imhoff JF (2003) Phylogenetic taxonomy of the family *Chlorobiaceae* on the basis of 16S rRNA and *fmo* (Fenna–Matthews–Olson protein) gene sequences. Int J Syst Evol Microbiol 53: 941–951

Imhoff JF (2005a) Family I. Chromatiaceae Bavendamm 1924, 125[AL] emend. Imhoff 1984b, 339. In: Brenner, DJ, Krieg NR, Staley JT and Garrity GM (eds) Bergey's Manual of Systematic Bacteriology, Vol 2, part B, pp 3–40. Springer, New York

Imhoff JF (2005b) Family II. Ectothiorhodospiraceae Imhoff 1984b, 339[VP]. In: Brenner DJ, Krieg NR, Staley JT and Garrity GM (eds) Bergey's Manual of Systematic Bacteriology, Vol 2, Part B, pp 41–57. Springer, New York

Imhoff JF and Hiraishi A (2005) Aerobic bacteria containing bacteriochlorophyll and belonging to the *Alphaproteobacteria*. In: Brenner DJ, Krieg NR, Staley JT and Garrity GM (eds) Bergey's Manual of Systematic Bacteriology, Vol 2, part A, pp 135. Springer, New York

Imhoff JF, Hiraishi A and Süling J (2005) Anoxygenic phototrophic purple bacteria. In: Brenner DJ, Krieg NR, Staley JT and Garrity GM (eds) Bergey's Manual of Systematic Bacteriology, Vol 2, part A, pp 119–132. Springer, New York

Jenney FE, Prince RC and Daldal F (1994) Roles of soluble cytochrome c(2) and membrane-associated cytochrome c(y) of *Rhodobacter capsulatus* in photosynthetic electron transfer. Biochemistry 33: 2496–2502

Jørgensen BB (1990) The sulfur cycle of freshwater sediments: Role of thiosulfate. Limnol Oceanogr 35: 1329–1342

Kappler U (2007) Bacterial sulfite-oxidizing enzymes – enzymes for chemolithotrophy only? In: Dahl C and Friedrich CG (eds) Microbial Sulfur Metabolism, pp 151–169. Springer, Heidelberg

Kappler, U and Bailey, S (2005) Molecular basis of intramolecular electron transfer in sulfite-oxidizing enzymes is revealed by high resolution structure of a heterodimeric complex of the catalytic molybdopterin subunit and a c-type cytochrome subunit. J Biol Chem 280: 24999–245007

Kappler U and Dahl C (2001) Enzymology and molecular biology of prokaryotic sulfite oxidation (minireview). FEMS Microbiol Lett 203: 1–9

Kappler U, Bennett B, Rethmeier J, Schwarz G, Deutzmann R, McEwan AG and Dahl C (2000) Sulfite: cytochrome c oxidoreductase from *Thiobacillus novellus* – Purification, characterization, and molecular biology of a heterodimeric member of the sulfite oxidase family. J Biol Chem 275: 13202–13212

Keppen OI, Baulina OI, Lysenko AM and Kondrateva EN (1993) A new green bacterium belonging to the Chloroflexaceae family. Microbiology 62: 179–185

Kerfeld CA, Chan C, Hirasawa M, Kleis-SanFrancisco S, Yeates TO and Knaff DB (1996) Isolation and characterization of soluble electron transfer proteins from *Chromatium purpuratum*. Biochemistry 35: 7812–7818

Khanna S and Nicholas DJD (1982) Utilization of tetrathionate and ^{35}S-labelled thiosulphate by washed cells of *Chlorobium vibrioforme* f. sp. *thiosulfatophilum*. J Gen Microbiol 128: 1027–1034

Kisker C, Schindelin H and Rees DC (1997) Molybdenum-cofactor-containing enzymes: Structure and mechanism. Ann Rev Biochem 66: 233–267

Kleinjan WE, de Keizer A and Janssen AJH (2003) Biologically produced sulfur. In: Steudel R (ed) Elemental Sulfur and Sulfur-Rich Compounds I., pp 167–187. Springer, Berlin

Knobloch K, Schmitt W, Schleifer G, Appelt N and Müller H (1981) On the enzymatic system thiosulfate-cytochrome c-oxidoreductase. In: Bothe H and Trebst A (eds) Biology of Inorganic Nitrogen and Sulfur, pp 359–365. Springer, Berlin

Kondratieva EN, Zhukov VG, Ivanovskii RN, Petushkova YP and Monosov EZ (1981) Light and dark metabolism in purple sulfur bacteria. Sov Sci Rev 2: 325–364

Kopriva S, Büchert T, Fritz G, Suter M, Weber M, Benda R, Schaller J, Feller U, Schürmann P, Schünemann V, Trautwein AX, Kroneck PMH and Brunold C (2001) Plant adenosine 5'-phosphosulfate reductase is a novel iron–sulfur protein. J Biol Chem 276: 42881–42886

Krafft T, Bokranz M, Klimmek O, Schröder I, Fahrenholz F, Kojro E and Kröger A (1992) Cloning and nucleotide sequence of the *psrA* gene of *Wolinella succinogenes* polysulphide reductase. Eur J Biochem 206: 503–510

Kusai K and Yamanaka T (1973) The oxidation mechanisms of thiosulphate and sulphide in *Chlorobium thiosulphatophilum*: roles of cytochrome c-551 and cytochrome c-553. Biochim Biophys Acta 325: 304–314

Larsen H (1952) On the culture and general physiology of the green sulfur bacteria. J Bacteriol 64: 187–196

Leguijt T (1993) Photosynthetic electron transfer in *Ectothiorhodospira*. PhD Dissertation, University of Amsterdam

Leyh TS (1993) The physical biochemistry and molecular genetics of sulfate activation. Crit Rev Biochem Mol Biol 28: 515–542

Lübbe YJ, Youn H-S, Timkovich R and Dahl C (2006) Siro(haem)amide in *Allochromatium vinosum* and relevance of DsrL and DsrN, a homolog of cobyrinic acid *a,c* diamide synthase for sulfur oxidation. FEMS Microbiol Lett 261: 194–202

Madigan MT (2001a) Family VI. "Heliobacteriaceae" Beer-Romero and Gest 1987, 113. In: Garrity G (ed) Bergey's Manual of Systematic Bacteriology, Vol 1, pp 625–630. Springer, New York

Madigan MT (2001b) The Family Heliobacteriaceae. In: Dworkin M (ed) The Prokaryotes: An Evolving Electronic Resource for the Microbiological Community, http://link.springer-ny.com/link/service/books/10125/. Springer, New York

Madigan MT and Brock TD (1977) CO_2 fixation in photosynthetically-grown *Chloroflexus auranticus*. FEMS Microbiol Lett 1: 301–304

Menin L, Gaillard J, Parot P, Schoepp B, Nitschke W and Verméglio A (1998) Role of HiPIP as electron donor to the RC-bound cytochrome in photosynthetic purple bacteria. Photosyn Res 55: 343–348

Meulenberg R, Pronk JT, Frank J, Hazeu W, Bos P and Kuenen JG (1992a) Purification and partial characterization of a thermostable trithionate hydrolase from the acidophilic sulfur oxidizer *Thiobacillus acidophilus*. Eur J Biochem 209: 367–374

Meulenberg R, Pronk JT, Hazeu W, van Dijken JP, Frank J, Bos P and Kuenen JG (1993) Purification and partial characterization of thiosulphate dehydrogenase from *Thiobacillus acidophilus*. J Gen Microbiol 139: 2033–2039

Meulenberg R, Pronk JT, Hazeu W, Bos P and Kuenen JG (1992b) Oxidation of reduced sulphur compounds by intact cells of *Thiobacillus acidophilus*. Arch Microbiol 157: 161–168

Myers JD and Kelly DJ (2005) A sulphite respiration system in the chemoheterotrophic human pathogen *Campylobacter jejuni*. Microbiology151: 233–242

Myers JM and Myers CR (2001) Role for outer membrane cytochromes OmcA and OmcB of *Shewanella putrefaciens* MR-1 in reduction of manganese dioxide. Appl Environ Microbiol 67: 260–269

Nakamura K, Nakamura M, Yoshikawa H and Amano Y (2001) Purification and properties of thiosulfate dehydrogenase from *Acidithiobacillus thiooxidans* JCM7814. Biosci Biotechnol Biochem 65: 102–108

Nitschke W, Jubault-Bregler M and Rutherford AW (1993) The reaction center associated tetraheme cytochrome subunit from *Chromatium vinosum* revisited: a reexamination of its EPR properties. Biochemistry 32: 8871–8879

Numata T, Fukai S, Ikeuchi Y, Suzuki T and Nureki O (2006) Structural basis for sulfur relay to RNA mediated by heterohexameric TusBCD complex. Structure 14: 357–366

Overmann J, Fischer U and Pfennig N (1992) A new purple sulfur bacterium from saline littoral sediments, *Thiorhodovibrio winogradskyi* gen. nov. and sp. nov. Arch Microbiol 157: 329–335

Pattaragulwanit K and Dahl C (1995) Development of a genetic system for a purple sulfur bacterium: conjugative plasmid transfer in *Chromatium vinosum*. Arch Microbiol 164: 217–222

Pattaragulwanit K, Brune DC, Trüper HG and Dahl C (1998) Molecular genetic evidence for extracytoplasmic localization of sulfur globules in *Chromatium vinosum*. Arch Microbiol 169: 434–444

Peck HD, Jr. (1968) Energy-coupling mechanisms in chemolithotrophic bacteria. Annu Rev Microbiol 22: 489–518

Pibernat IV and Abella CA (1996) Sulfide pulsing as the controlling factor of spinae production in *Chlorobium limicola* strain UdG 6038. Arch Microbiol 165: 272–278

Pickering IJ, George GN, Yu EY, Brune DC, Tuschak C, Overmann J, Beatty JT and Prince RC (2001) Analysis of sulfur biochemistry of sulfur bacteria using x-ray absorption spectroscopy. Biochemistry 40: 8138–8145

Pires RH, Lourenco AI, Morais F, Teixeira M, Xavier AV, Saraiva LM and Pereira IAC (2003) A novel membrane-bound respiratory complex from *Desulfovibrio desulfuricans* ATCC 27774. Biochim Biophys Acta 1605: 67–82

Pires, RH, Venceslau, SS, Morais, F, Teixeira, M, Xavier, AV and Pereira, IAC (2006) Characterization of the *Desulfovibrio desulfuricans* ATCC 27774 DsrMKJOP complex – a membrane-bound redox complex involved in the sulfate respiratory pathway. Biochemistry 45: 249–262

Pittman MS, Robinson HC and Poole RK (2005) A bacterial glutathione transporter (*Escherichia coli* CydDC) exports reductant to the periplasm. J Biol Chem 280: 32254–32261

Podgorsek L and Imhoff JF (1999) Tetrathionate production by sulfur oxidizing bacteria and the role of tetrathionate in the sulfur cycle of Baltic Sea sediments. Aquat Microb Ecol 17: 255–265

Pott AS and Dahl C (1998) Sirohaem-sulfite reductase and other proteins encoded by the *dsr* locus of *Chromatium vinosum* are involved in the oxidation of intracellular sulfur. Microbiology 144: 1881–1894

Prange A, Arzberger I, Engemann C, Modrow H, Schumann O, Trüper HG, Steudel R, Dahl C and Hormes J (1999) *In situ* analysis of sulfur in the sulfur globules of phototrophic sulfur bacteria by X-ray absorption near edge spectroscopy. Biochim Biophys Acta 1428: 446–454

Prange A, Chauvistre R, Modrow H, Hormes J, Trüper HG and Dahl C (2002a) Quantitative speciation of sulfur in bacterial sulfur globules: X-ray absorption spectroscopy reveals at least three different speciations of sulfur. Microbiology 148: 267–276

Prange A, Dahl C, Trüper HG, Behnke M, Hahn J, Modrow H and Hormes J (2002b) Investigation of S–H bonds in biologically important compounds by sulfur K-edge X-ray absorption spectroscopy. Eur Phys J D 20: 589–596

Prange A, Dahl C, Trüper HG, Chauvistre R, Modrow H and Hormes J (2002c) X-ray absorption spectroscopy of bacterial sulfur globules: a detailed reply. Microbiology 148: 2268–2270

Prange A, Engelhardt H, Trüper HG and Dahl C (2004) The role of the sulfur globule proteins of *Allochromatium vinosum*: mutagenesis of the sulfur globule protein genes and expression studies by real-time RT PCR. Arch Microbiol 182: 165–174

Pronk JT, Meulenberg R, Hazeu W, Bos P and Kuenen JG (1990) Oxidation of reduced inorganic sulphur compounds by acidophilic thiobacilli. FEMS Microbiol Rev 75: 293–306

Raitsimring AM, Kappler U, Feng CJ, Astashkin AV and Enemark JH (2005) Pulsed EPR studies of a bacterial sulfite-oxidizing enzyme with pH-invariant hyperfine interactions from exchangeable protons. Inorg Chem 44: 7283–7285

Ramírez, P, Guiliani, N, Valenzuela, L, Beard, S and Jerez, CA (2004) Differential protein expression during growth of *Acidithiobacillus ferrooxidans* on ferrous iron, sulfur compounds, or metal sulfides. Appl Enivron Microbiol 70: 4491–4498

Raymond JC and Sistrom WR (1969) *Ectothiorhodospira halophila* – a new species of the genus *Ectothiorhodospira*. Arch Mikrobiol 69: 121–126

Rees GN, Harfoot CG, Janssen PH, Schoenborn L, Kuever J and Lunsdorf H (2002) *Thiobaca trueperi* gen. nov., sp. nov., a phototrophic purple sulfur bacterium isolated from freshwater lake sediment. Int J Syst Evol Microbiol 52: 671–678

Reinartz M, Tschäpe J, Brüser T, Trüper HG and Dahl C (1998) Sulfide oxidation in the phototrophic sulfur bacterium *Chromatium vinosum*. Arch Microbiol 170: 59–68

Renosto F, Martin RL, Borrell JL, Nelson DC and Segel IH (1991) ATP sulfurylase from trophosome tissue of *Riftia pachyptila* (hydrothermal vent tube worm). Arch Biochem Biophys 290: 66–78

Rohwerder T and Sand W (2003) The sulfane sulfur of persulfides is the actual substrate of the sulfur-oxidizing enzymes from *Acidithiobacillus* and *Acidiphilium* spp. Microbiology 149: 1699–1709

Rother D, Heinrich HJ, Quentmeier A, Bardischewsky F and Friedrich CG (2001) Novel genes of the *sox* gene cluster, mutagenesis of the flavoprotein SoxF, and evidence for a general sulfur-oxidizing system in *Paracoccus pantotrophus* GB17. J Bacteriol 183: 4499–4508

Samyn B, DeSmet L, van Driessche G, Meyer TE, Bartsch RG, Cusanovich MA and van Beeumen JJ (1996) A high-potential soluble cytochrome *c*-551 from the purple phototrophic bacterium *Chromatium vinosum* is homologous to cytochrome c(8) from denitrifying pseudomonads. Eur J Biochem 236: 689–696

Sanchez O, Ferrera I, Dahl C and Mas J (2001) *In vivo* role of APS reductase in the purple sulfur bacterium *Allochromatium vinosum*. Arch Microbiol 176: 301–305

Sander J and Dahl C (2008) Metabolism of inorganic sulfur compounds in purple bacteria. In: Hunter CN, Daldal, F, Thurnauer, MC and Beatty, JT (eds) Purple Bacteria (Advances in Photosynthesis and Respiration), in press. Springer, New York

Sander J, Engels-Schwarzlose S and Dahl C (2006) Importance of the DsrMKJOP complex for sulfur oxidation in *Allochromatium vinosum* and phylogenetic analysis of related complexes in other prokaryotes. Arch Microbiol 186: 357–366

Schedel M and Trüper HG (1980) Anaerobic oxidation of thiosulfate and elemental sulfur in *Thiobacillus denitrificans*. Arch Microbiol 124: 205–210

Schedel M, Vanselow M and Trüper HG (1979) Siroheme sulfite reductase from *Chromatium vinosum*. Purification and investigation of some of its molecular and catalytic properties. Arch Microbiol 121: 29–36

Schmidt GL, Nicolson GL and Kamen MD (1971) Composition of the sulfur particle of *Chromatium vinosum*. J Bacteriol 105: 1137–1141

Schmitt W, Schleifer G and Knobloch K (1981) The enzymatic system thiosulfate: cytochrome *c* oxidoreductase from photolithoautotrophically grown *Chromatium vinosum*. Arch Microbiol 130: 334–338

Schütz M, Maldener I, Griesbeck C and Hauska G (1999) Sulfide-quinone reductase from *Rhodobacter capsulatus*: requirement for growth, periplasmic localization, and extension of gene sequence analysis. J Bacteriol 181: 6516–6523

Schütz M, Shahak Y, Padan E and Hauska G (1997) Sulfide-quinone reductase from *Rhodobacter capsulatus*. J Biol Chem 272: 9890–9894

Segel IH (1993) Enzyme kinetics: behaviour and analysis of rapid equilibrium and steady-state enzyme systems. Wiley-Interscience, New York

Shahak Y, Arieli B, Padan E and Hauska G (1992) Sulfide quinone reductase (SQR) activity in *Chlorobium*. FEBS Lett 299: 127–130

Shively JM, Bryant DA, Fuller RC, Konopka AE, Stevens SE and Strohl WR (1989) Functional inclusions in prokaryotic cells. Int Rev Cytol 113: 35–100

Smith AJ (1966) The role of tetrathionate in the oxidation of thiosulphate by *Chromatium* sp. strain D. J Gen Microbiol 42: 371–380

Smith AJ and Lascelles J (1966) Thiosulphate metabolism and rhodanese in *Chromatium* sp. strain D. J Gen Microbiol 42: 357–370

Sorokin DY, Teske A, Robertson LA and Kuenen JG (1999) Anaerobic oxidation of thiosulfate to tetrathionate by obligately heterotrophic bacteria, belonging to the *Pseudomonas stutzeri* group. FEMS Microbiol Ecol 30: 113–123

Sorokin DY, Tourova TP, Kuznetsov BB, Bryantseva IA and Gorlenko VM (2000) *Roseinatronobacter thiooxidans* gen. nov., sp. nov., a new alkaliphilic aerobic bacteriochlorophyll a – containing bacterium isolated from a soda lake. Microbiology 69: 75–82

Sperling D, Kappler U, Wynen A, Dahl C and Trüper HG (1998) Dissimilatory ATP sulfurylase from the hyperthermophilic sulfate reducer *Archaeoglobus fulgidus* belongs to the group of homo-oligomeric ATP sulfurylases. FEMS Microbiol Lett 162: 257–264

Steinmetz MA and Fischer U (1982) Cytochromes of the green sulfur bacterium *Chlorobium vibrioforme* f. *thiosulfatophilum*. Purification, characterization and sulfur metabolism. Arch Microbiol 19: 19–26

Steudel R (1989) On the nature of the "elemental sulfur" (S^0) produced by sulfur-oxidizing bacteria- a model for S^0 globules. In: Schlegel HG and Bowien B (eds) Autotrophic Bacteria, pp 289–303. Science Tech Publishers, Madison, WI

Steudel R (1996) Mechanism for the formation of elemental sulfur from aqueous sulfide in chemical and microbiological desulfurization processes. Ind Eng Chem Res 35: 1417–1423

Steudel R (2000) The chemical sulfur cycle. In: Lens P and Hulshoff Pol W (eds) Environmental Technologies to Treat Sulfur Pollution, pp 1–31. IWA Publishing, London

Steudel R and Eckert B (2003) Solid sulfur allotropes. In: Steudel R (ed) Elemental Sulfur and Sulfur-Rich Compounds, pp 1–79. Springer, Berlin

Steudel R, Holdt G, Visscher PT and van Gemerden H (1990) Search for polythionates in cultures of *Chromatium vinosum* after sulfide incubation. Arch Microbiol 155: 432–437

Suzuki H, Koyanagi T, Izuka S, Onishi A and Kumagai H (2005) The *yliA*, -*B*, -*C*, and -*D* genes of *Escherichia coli* K-12 encode a novel glutathione importer with an ATP-binding cassette. J Bacteriol 187: 5861–5867

Taylor BF (1994) Adenylylsulfate reductases from thiobacilli. Meth Enzymol 243: 393–400

Then J and Trüper HG (1981) The role of thiosulfate in sulfur metabolism of *Rhodopseudomonas globiformis*. Arch Microbiol 130: 143–146

Then J and Trüper HG (1983) Sulfide oxidation in *Ectothiorhodospira abdelmalekii*. Evidence for the catalytic role of cytochrome *c*-551. Arch Microbiol 135: 254–258

Then J and Trüper HG (1984) Utilization of sulfide and elemental sulfur by *Ectothiorhodospira halochloris*. Arch Microbiol 139: 295–298

Trüper HG (1978) Sulfur metabolism. In: Clayton RK and Sistrom WR (eds) The Photosynthetic Bacteria, pp 677–690. Plenum, New York

Trüper HG and Fischer U (1982) Anaerobic oxidation of sulphur compounds as electron donors for bacterial photosynthesis. Phil Trans R Soc Lond B 298: 529–542

Trüper HG and Pfennig N (1966) Sulphur metabolism in Thiorhodaceae. III. Storage and turnover of thiosulphate sulphur in *Thiocapsa floridana* and *Chromatium* species. Antonie van Leeuwenhoek Int J Gen Mol Microbiol 32: 261–276

Trüper HG and Rogers LA (1971) Purification and properties of adenylyl sulfate reductase from the phototrophic sulfur bacterium, *Thiocapsa roseopersicina*. J Bacteriol 108: 1112–1121

van Beeumen JJ, Demol H, Samyn B, Bartsch RG, Meyer TE, Dolata MM and Cusanovich MA (1991) Covalent structure of the diheme cytochrome subunit and amino-terminal sequence of the flavoprotein subunit of flavocytochrome c from *Chromatium vinosum*. J Biol Chem 266: 12921–12931

van Driessche G, Vandenberghe I, Devreese B, Samyn B, Meyer TE, Leigh R, Cusanovich MA, Bartsch RG, Fischer U and van Beeumen JJ (2003) Amino acid sequences and distribution of high-potential iron–sulfur proteins that donate electrons to the photosynthetic reaction center in phototropic proteobacteria. J Mol Evol 57: 181–199

van Gemerden H (1968) On the ATP generation by *Chromatium* in the dark. Arch Mikrobiol 64: 118–124

van Gemerden H (1987) Competition between purple sulfur bacteria and green sulfur bacteria: role of sulfide, sulfur and polysulfides. In: Lindholm T (ed) Ecology of Photosynthetic Prokaryotes with Special Reference to Meromictic Lakes and Coastal Lagoons, pp 13–27. Abo Academy, Abo

Vermeglio A, Li J, Schoepp-Cothenet B, Pratt N and Knaff DB (2002) The role of high-potential iron protein and cytochrome c(8) as alternative electron donors to the reaction center of *Chromatium vinosum*. Biochemistry 41: 8868–8875

Visscher PT, Nijburg JW and van Gemerden H (1990) Polysulfide utilization by *Thiocapsa roseopersicina*. Arch Microbiol 155: 75–81

Visscher PT and Taylor BF (1993) Organic thiols as organolithotrophic substrates for growth of phototrophic bacteria. Appl Environ Microbiol 59: 93–96

Visscher PT and van Gemerden H (1991) Photoautotrophic growth of *Thiocapsa roseopersicina* on dimethyl sulfide. FEMS Microbiol Lett 81: 247–250

Visser JM, de Jong GAH, Robertson LA and Kuenen JG (1996) Purification and characterization of a periplasmic thiosulfate dehydrogenase from the obligately autotrophic *Thiobacillus* sp. W5. Arch Microbiol 166: 372–378

Williams TJ, Zhang CL, Scott JH and Bazylinski DA (2006) Evidence for autotrophy via the reverse tricarboxylic acid cycle in the marine magnetotactic coccus strain MC-1. Appl Environ Microbiol 72: 1322–1329

Yu Z, Lansdon EB, Segel IH and Fisher AJ (2007) Crystal structure of the bifunctional ATP sulfurylase – APS kinase from the chemolithotrophic thermophile *Aquifex aeolicus*. J Mol Biol 365: 732–743

Yurkov VV (2006) Aerobic phototrophic proteobacteria. In: Dworkin M, Falkow, S, Rosenberg, E, Schleifer, K-H and Stackebrandt, E (eds) The Prokaryotes, Vol 5, pp 562–584. Springer, New York

Yurkov VV, Krasil'nikova EN and Gorlenko VM (1994) Thiosulfate metabolism in the aerobic bacteriochlorophyll-a-containing bacteria *Erythromicrobium hydrolyticum* and *Roseococcus thiosulfatophilus*. Microbiology 63: 91–94

Zaar A, Fuchs G, Golecki JR and Overmann J (2003) A new purple sulfur bacterium isolated from a littoral microbial mat, *Thiorhodococcus drewsii* sp. nov. Arch Microbiol 179: 174–183

Chapter 16

Sulfide Oxidation from Cyanobacteria to Humans: Sulfide–Quinone Oxidoreductase (SQR)

Yosepha Shahak
Institute of Plant Sciences, The Volcani Center, Bet-Dagan 50250, Israel

Günter Hauska*
*Lehrstuhl für Botanik, Fakultät für Biologie und Vorklinische Medizin,
Universität Regensburg, Germany*

---

* Corresponding author, E-mail: guenther.hauska@biologie.uni-regensburg.de, Fax: +49 941 943 3352

Rüdiger Hell et al. (eds.), Sulfur Metabolism in Phototrophic Organisms, 319–335.
© 2008 *Springer.*

Summary

After the discovery that anoxygenic, sulfidotrophic photosynthesis can be induced in cyanobacteria, sulfide–quinone reductase (SQR) was identified and characterized in *Oscillatoria limnetica*. This was closely followed by the study of SQR in the purple bacterium *Rhodobacter capsulatus*. Subsequently the genes of the purple bacterium and of two cyanobacteria, as well as of the hyperthermophilic hydrogen bacterium *Aquifex aeolicus* were cloned, sequenced and expressed in *Escherichia coli*, and the enzymes were characterized.

Sequence analysis showed that SQR belongs to the disulfide oxidoreductase flavoprotein family, together with flavocytochrome *c* (FCC), another sulfide oxidizing enzyme. All the members of this family are characterized by two redox active cysteines which cooperate with the flavin in the redox cycle. A redox mechanism for SQR is proposed on the basis of site directed mutations of the cysteins and of other amino acid residues. Furthermore, a 3d-structural model is derived from the crystal structure of FCC.

The search into the genomes accessible in the internet documents a widely spread occurrence of SQR-genes in bacteria. From the 19 completed canobacterial genomes, five contain the gene. Phylogenetic analysis classifies these genes into at least two clades – SQR-type I and SQR-type II. However, SQR-like enzymes are not confined to prokaryotes. They occur in the mitochondria of some fungi, as well as of all animals for which the genomes have been sequenced. From these eukaryotic SQR-like proteins (SQRDL) only the one of fission yeast was isolated and was enzymatically characterized. It is involved in heavy metal tolerance, and has therefore been denoted HMT2. Since sulfide has been indentified as a gaseous transmitter substance in animals, a possible role for SQRDL signalling is considered.

Finally, phylogenetic scenarios for the descent of SQR from a common ancestor are discussed. Two observations are of special interest: (i) The mitochondrial SQRDL is of type II, although the endosymbiontic ancestor of mitochondria is considered to be a proteobacterium, which should have had a type I-SQR. (ii) The two essential cysteines among the flavoprotein family must have changed positions in the primary structure during evolution, thus constituting an example of functional plasticity within phylogenies.

I. Introduction

In the dawn of biological evolution, before cyanobacteria developed their photosynthetic water oxidation, the world was anaerobic and sulfidic (Shen et al., 2001; Anbar and Knoll, 2002). Photosynthetic bacteria largely used sulfide as the hydrogen donor for the assimilation of CO_2 and nitrogen into organic matter. Sulfidotrophy is still wide spread among extant bacteria and archaea (Blankenship et al., 1995; Brune, 1995; Friedrich, 1998; see also the chapter 15 by C. Dahl in this volume) and also some cyanobacteria that retained the ability to shift back from water to sulfide oxidation in photosynthesis under stress conditions (Garlick et al., 1977).

The first enzymatic step of sulfidotrophy was initially attributed to flavocytochrome *c*, which oxidizes sulfide with cytochrome *c* (Kusai and Yamanaka, 1973). However, further evidence (Brune and Trüper, 1986; Reinartz et al., 1999; Schütz et al., 1999) is in favour of sulfide–quinone oxidoreductase (SQR[1]), a flavoenzyme belonging to the large disulfide oxidoreductase family, like glutathione reductase or lipoamide dehydrogenase (Shahak et al., 1999; Griesbeck et al., 2000). SQR feeds electrons from sulfide into the quinone pools of the electron transport chains for energy conservation, either in photosynthesis or respiration.

Meanwhile, genes for SQR have been identified in numerous genomes, including all domains of organisms, except plants. Surprisingly, it is

---

Abbreviations: FCC – flavocytochrome *c* sulfide dehydrogenase; HMT2 – heavy metal tolerance factor 2; IEU – international enzyme unit (μmoles substrate reacted per mg protein and min); SQR – sulfide–quinone reductase; SQRDL – sulfide–quinone reductase like protein

---

[1] Unfortunately the acronym "SQR" is also in use for succinate–quinone oxidoreductase of respiratory chains.

even found in the human genome. The recent discovery that hydrogen sulfide, like NO and CO, functions as a "gasotransmitter" substance in smooth muscle relaxation and neuronal signalling (Wang, 2002; Boehning and Snyder, 2003), tempts into speculating that SQR may be involved in adjusting an appropriate sulfide level.

This review updates our earlier accounts on the structure and function of cyanobacterial and other prokaryotic SQRs (Shahak et al., 1999; Bronstein et al., 2000; Griesbeck et al., 2000) in complementation to the part that deals with sulfide oxidation in purple sulfur bacteria in chapter 15 by C. Dahl (in this volume). In addition we give a brief account on SQR-like enzymes in eukaryotes, from fungi to humans. Thus we intend to widen the view of an ancient enzyme, which is gaining new momentum by its possible involvement in cell signalling.

II.   Discovery and Development of Studies

The ability of cyanobacteria to shift from oxygenic to anoxygenic photosynthesis using sulfide as hydrogen donor in place of water was an exciting discovery. It was regarded as a return to an earlier, more primitive form of photosynthesis (Padan, 1979, 1989), since it is found widely distributed among other extant phototrophic bacteria. Van Niel's ingenious generalization of photosynthesis experienced another impressive manifestation. The shift occurs under light and oxygen stress and is induced by sulfide in various cyanobacteria (Garlick et al., 1977). Further studies concentrated on the filamentous cyanobacterium *Oscillatoria limnetica* which inhabits the salty ponds of the Negev desert. The sulfide-induced cells catalyzed sulfide-dependent CO_2-fixation, as well as nitrogen fixation (Belkin et al., 1982) and hydrogen evolution (Belkin and Padan, 1978), under appropriate growth conditions. Tracking the inducible factor enabling sulfide-dependent photosynthesis led to the discovery and characterization of SQR (Shahak et al., 1987, 1992a, 1993; Belkin et al., 1988; Arieli et al., 1991, 1994), an enzyme (E.C.1.8.5.-) that feeds electrons into the quinone pool (reviewed by Shahak et al., 1999; see also Bronstein et al., 2000). Subsequently the SQR-genes from *O. limnetica* as well as from the unicellular

cyanobacterium *Aphanothece halophytica* have been cloned, sequenced and were expressed in *E. coli* (Bronstein et al., 2000), in collaboration with an equivalent study on *Rhodobacter capsulatus* (Schütz et al., 1997, 1999).

SQR activity has been detected in the membranes of several addional photosynthetic (*Chlorobium limicola* – Shahak et al., 1992b; *Allochromatium vinosum* – Reinartz et al., 1998) as well as chemosynthetic bacteria (*Paracoccus denitrificans* – Schütz et al., 1998; *Aquifex aeolicus* – Nübel et al., 2000). Most remarkably, however, SQR activity was also found in the mitochondria of invertebrates (Grieshaber and Völkel, 1998; Parrino et al., 2000) and vertebrates (Furne et al., 2001; Yong and Searcy, 2001). Moreover, a SQR-like enzyme from the mitochondria of the fission yeast *Schizosaccharomycis pombe* has been cloned and sequenced, and was expressed in *E. coli* as a His-tag protein, that after purification by Ni-chelate chromatography has been characterized in detail. It is known as HMT2, since its gene was detected by compensating a mutant defect in heavy metal tolerance (Vande Weghe and Ow, 1999, 2001).

A search through current databases reveals that the genes for SQR and SQR-like proteins (SQRDL) are present in the genomes throughout all domains of life, by far exceeding the phemomenon of sulfidotrophy. A comprehensive phylogenetic analysis of SQR-genes by Theissen et al. (2003) will be discussed below.

III.   Characterization

A.   Occurrence and Comparison of SQR-Genes

In the early study of Garlick et al. (1977) 11 out of 21 cyanobacterial strains where found to be capable of facultative anoxygenic photosynthesis with sulfide as the electron donor (reviewed by Padan, 1979). The phenomenon included strains of rather different habitats (aerobic, anaerobic, marine and fresh water) and of filamentous as well as unicellular types. At present, an inspection of the NCBI-website with 405 completed plus 644 unfinished microbial genome projects yields 19 completed cyanobacterial genomes and 25 in progress. Among these 34 cyanobacteria 10 strains contain a total of 12 SQR-genes. They are

j

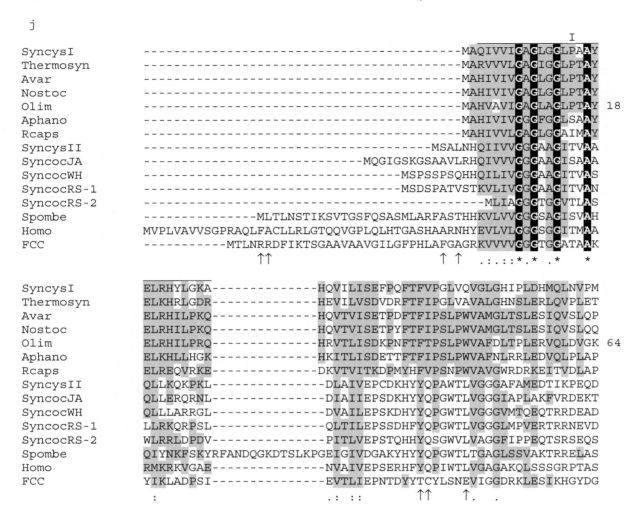

Fig. 1. (continued)

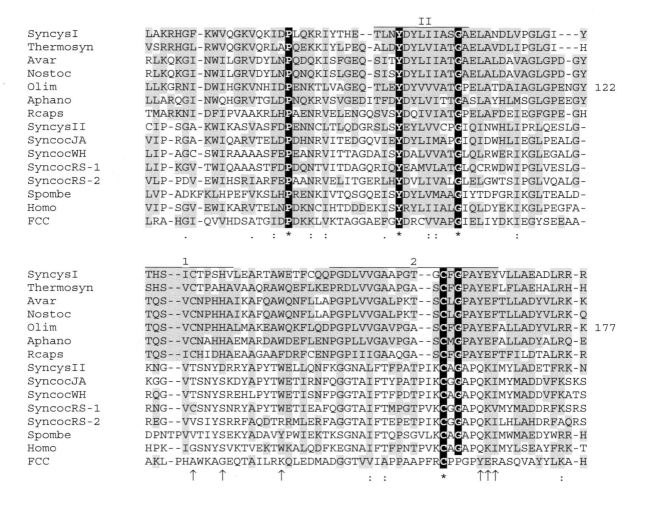

Fig. 1. (continued)

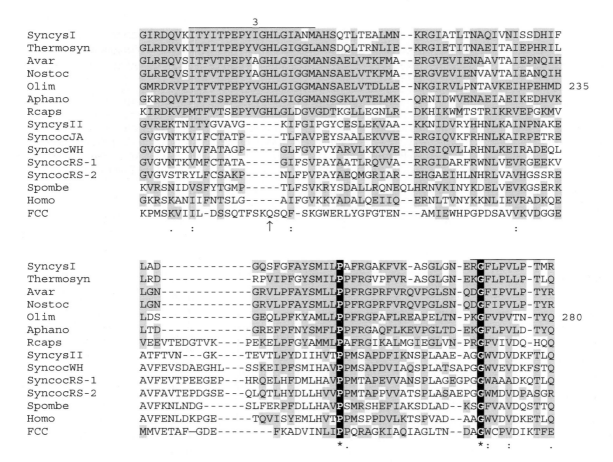

Fig. 1. (continued) Multiple alignment of SQR-sequences and comparison to FCC. The alignment shows the Clustal format obtained via the program T-Coffee (Notredame et al., 2000); the amino acid residues are given in single letter code; residues conserved in all types of SQR are underlayed in black, those which are addionally shared by FCC are indicated by a star at the bottom of the alignment; similar residues are indicated in two ways: at the bottom by single or double dots for low and high similarity in all sequences according to T-Coffee, and by shading the residues of high similarity which are typical for the cyanobacteria, but partially also extend into the other sequences; homologous peptide regions are lined and numbered on the top, I-III indicating FAD-binding in the disulfide oxidoreductase family, 1–6 indicating homology within SQR-type I (Griesbeck et al., 2000); arrows at the bottom indicate positions addressed in the text; abbreviations on the left stand for: SyncysI – *Synechocystis* PCC 6803 SQR-type I (accession number NP_942192), Thermosyn – *Thermosynechococcus elongates* (acc.no. NP_681079), Avar – *Anabaena variabilis* (acc.no. ABA22985), Nostoc – *Nostoc* PCC 7120 (acc.no. NP_488552), Olim – *Oscillatoria limnetica* (acc.no. AAF72962), Aphano – *Aphanotece halophytica* (acc.no. AAF72963), Rcaps – *Rhodobacter capsulatus* (acc.no. CAA66112), SyncysII – *Synechocystis* PCC 6803 SQR-type II (accession number NP_440916), SyncocJA – *Synechococcus* strain JA-3-3Ab (acc. no. ABD00861), SyncocWH – *Synechococcus* strain WH 5701 (acc.no. EAQ74835), SyncocRS – *Synechococcus* strain RS 9917 (acc.no. EAQ69368), Spombe – *Schizosaccharomyces pombe* (acc.no. CAA21882), Homo – *homo sapiens* (acc.no. AAH16836.1), FCC – flavocytocrome *c* from *Allochromatium vinosum* preprotein (acc.no. AAB86576); numbers at the right edge of the alignment blocks refer to the position in the sequence of *O. limnetica*, the count starting at the N-terminal M.

aligned in Fig. 1, together with the sequences of the SQR from *R. capsulatus*, of the mitochondrial SQRDL proteins from fission yeast and man, and of FCC from *Allochromatium vinosum*. In previous alignments of fewer sequences we had defined certain peptide regions that are significant for

SQR-proteins (Bronstein et al., 2000; Griesbeck et al., 2000; Griesbeck et al., 2002). These are pointed out again by horizontal lines above of the alignment blocks in Fig. 1. Three FAD-binding regions which are characteristic for all members in the disulfide oxidoreductase family are numbered with I to III, six additional regions specific for SQRs are numbered 1–6. Obviously, while the FAD-binding peptides are similar throughout the alignment of the 15 sequences, the other peptides are not. These peptides and over-all similarities divide the SQR-sequences into two groups, which are represented by the upper and the lower half of the blocks in Fig. 1. The upper part contains six cyanobacterial SQRs plus the sequence of the purple bacterium *R. capsulatus*, while the lower part has closer resemblance to FCC and joins five cyanobacterial sequences with the mitochondrial SQRDL of fission yeast and humans. It is remarkable that both types are found among cyanobacteria. *O. limnetica*, *Thermosynchococcus elongatus*, *Aphanotece halophytica*, *Anabaena variabilis* and *Nostoc* PCC 7120 contain type I, while all the strains of *Synechococcus* contain type II. Two cyanobacteria contain two SQR-genes. *Synechococcus* strain RS 9917 has two versions of type II, and interestingly *Synechocystis* PCC 6803 contains one gene for type I, as well as one for type II.

The majority of the 19 completed cyanobacterial genomes lacks any SQR gene, however. Only five contain it – *Synechocystis* PCC 6803 (type I + type II), *Anabaena variabilis* (type I), *Thermosynechocooccus elongatus* (type I), *Nostoc* PCC 7120 (type I), and *Synechococcus* strain JA-3-3AB (type II). Seven other genomes of the genus *Synechococcus* lack the gene. It has also not been detected in the genomes of *Gloeobacter*, *Trichodesmium*, or in any *Prochlorophyta* so far. A gene for SQR is also absent from green plants, and from bakers yeast *Saccharomyces cerevisiae*. Nevertheless, it is present as a preprotein gene targeting for mitochondria in the genomes of fission yeast and other fungi, and in all animal genomes available so far in the NCBI-website (see also Vande Weghe and Ow, 1999). Noteworthy, the SQR gene is also found in the genome for the diatom *Thalassiosira pseudonana*.

Fifteen amino acid residues are fully conserved among both SQR-types in the alignment of Fig. 1, twelve of them are also found in FCC. They are underlayed in black in Fig. 1, and are shown in the 3d-structural model of Fig. 3 as well. Most significant are the two conserved cyteines, at positions 159 and 346 in the sequence of *O. limnetica*, which form a redox active disulfide bridge close to the isoalloxazine ring of FAD (see

Fig. 2). A few further interesting observations are highlighted by arrows on the bottom of the alignment blocks (Fig. 1) and are listed below according to the position numbers in the sequence of *O. limnetica*:

- At position −3 lies the putative start of the mature SQRDL protein for *S.pombe* (Vande Weghe and Ow, 1999) and possibly also for *homo*, after cleavage of the mitochondrial targeting N-terminal extensions (Von Heijne et al., 1989). At position −1 lies the aminoterminus of mature FCC, the preprotein carrying the double-arginine signal (at −26 and −27), for export into the periplasmic space (Berks, 1996; see below).

- At positions 41, 47, 139 and 166 aromatic/hydrophobic residues are found in all SQRs, unlike the charged/hydrophilic residues that are present in the FCC-sequence. Thus they may be involved in binding the quinone which is not the substrate for FCC.

- At position 42 an additional cysteine is present in FCC which covalently binds the FCC via the 8-methyl group of the isoalloxazine ring (Chen et al., 1994).

- The presence of a cysteine at position 127 is characteristic for type I-SQR with its high substrate affinities (see Fig. 1 and text above). However, a Cys is also found in the first sequence of *Synechococcus* RS 9917, which clearly belongs to type II, as judged by overall homology and the presence of an aspartate instead of valine at position 291. It would be interesting to test the sulfide affinity of this SQR.

- An aspartate instead of valine at position 291 is not characteristic for type II-SQR only. It is also found in FCC, and even in all other members of the disulfide oxidoreductase family (see Griesbeck et al., 2002). In the crystal structures available position 291 is close to the ribityl moiety of FAD (see V291 in Fig. 3).

- Y164E165 might represent the active site base in type I (Fig. 2). The two residues are replaced by QK in type II, yet are found in FCC. This is intriguing in view of the low affinity of type II for sulfide, while high affinity of both, type I SQR and FCC (12.5 μM; Cusanovich et al., 1991).

- The histidines at positions 131 and 196 have been considered to contribute to quinone binding, as suggested from inspection of quinone binding sites in 3d-structures of membrane proteins (Rich and Fisher, 1999). However, the histidines are not conserved in type II. Furthermore, site-directed mutation did not affect the affinity for quinone. Rather the affinity for sulfide was decreased (Griesbeck et al., 2002).

B. Properties of the Protein

1. Isolation

SQR was solubilized from thylakoids of sulfide-induced cells of *O. limnetica* and *A. halophytica* by mild detergent treatment and was purified by ammonium sulfate fractionation and HPLC (Arieli et al., 1994; Bronstein et al., 2000). A similar procedure was developed for the solubilization and purification of the enzyme from chromatophores of induced cells of the purple bacterium *R. capsulatus* (Schütz et al., 1997). Later the SQR enzymes were isolated as expression proteins from *E. coli* membranes, with or without His-tags (Vande Weghe and Ow, 1999; Bronstein et al., 2000; Griesbeck et al., 2002). A particularly convenient isolation protocol was developed for the isolation of the expression protein of the hyperthermophilic hydrogen bacterium *Aquifex aeolicus* from *E. coli* membranes, since this SQR could be purified to homogeneity in a single heat step after solubilization (Schödl, 2003).

2. FAD – the Prosthetic Group

One FAD is bound non-covalently to each protomer, as suggested by the FAD-binding motifs found in the primary structure (see Fig. 2), especially by the N-terminal $\beta\alpha\beta$-fold (Wieringa et al., 1986). Accordingly, partial reconstitution of activity of the denatured enzyme was achieved with FAD only (Griesbeck, 2001). The redox potential E_0' for FAD/FADH$_2$ is -60 mV (n = 2), which is about 160 mV more positive than for free flav in (Thauer et al., 1977). This has only been determined for SQR from *Aquifex aeolicus* so far (Schödl, 2003), but is expected to be similar for the SQRs from *O. limnetica, A. halophytica* and *R. capsulatus* which belong to type I (see below).

The optical spectra of the purified SQRs are characteristic for a flavoprotein, with absorption/excitation peaks at 280, 375 and 460 nm and an emission maximum at 520 nm. The fluorescence of SQR is quenched by sulfide at micromolar concentrations, and is recovered by the addition of quinone (Schütz et al., 1997; Griesbeck et al., 2002). On this basis a sensitive microsensor

for sulfide may be developed. However, our efforts to do so with the thermostable SQR from *Aq. aeolicus* have failed so far (Schödl, 2003).

3. Enzymatic Activity and Inhibitors

The activity of SQR-type I is characterized by its high affinities for both substrates, with K_m-values in the µM range. Membranes, isolated enzymes and expression proteins have been studied extensively for the cyanobacterium *O. limnetica* (Arieli et al., 1991, 1994), the purple bacterium *R. capsulatus* (Schütz et al, 1997; Griesbeck, 2001) and the hydrogen-oxidizing bacterium *Aq. aeolicus* (Nübel et al., 2000; Schödl, 2003). Data are also available for membranes of the α-proteobacterium *Paracoccus denitrificans* (Schütz et al., 1998), the green sulfur bacterium *Chlorobium limicola* (Shahak et al., 1992b) and the purple sulfur bacterium *Allochromatium vinosum* (Reinartz et al., 1998), as well as for the isolated enzyme from the cyanobacterium *A. halophytica* (Bronstein et al., 2000). The specific activities in the membranes of these organisms, as compiled by Griesbeck et al. (2000), range from 0.02 IEU for *O. limnetica* to 3.5 IEU for *Aq. aeolicus*, reflecting different amounts and possibly also different turnover numbers for the various SQR-type I enzymes. Highly effective as inhibitors are quinone analogs, like stigmatellin and the akyl-hydroxyquinoline-N-oxides, which are known to block also the quinone interaction sites of the cytochrome complexes in photosynthesis and respiration. In this context it is remarkable that antimycin A and myxothiazole are ineffective on SQR-activity of the cyanobacterium *O. limnetica*, which operates with plastoquinone, together with the cytochrome *b6f*-complex. Both are efficient inhibitors, in the µM range, of the SQRs operating with ubiquinone and of cytochrome *bc*1-complexes. Antimycin as well as myxothiazole are well known to inhibit quinol/quinone-interaction with the *bc*1-complexes, but are ineffective with the *b6f*-complex. Thus there must be something common with respect to the structure of the quinone interaction sites of SQR and the cytochrome complexes.

The final product of sulfide oxidation in sulfidotrophic organisms is either sulfate or sulfur (see Blankenship et al., 1995; and the chapter by C. Dahl in this volume), while it is thiosulfate in mitochondria

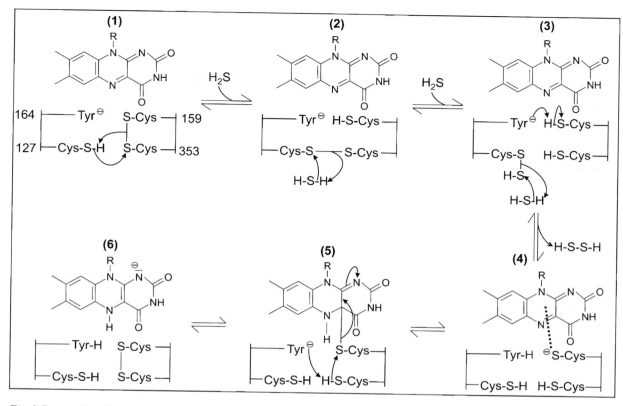

Fig. 2. Proposed mechanism of the reductive half-reaction of SQR. The proposal is a modified version of scheme 1 in Griesbeck et al.(2002), and is based on site-directed Cys/Ser-mutants; see text for explanation; the numbers correspond to the positions in the sequence of *R. capsulatus* (Fig. 1).

(Grieshaber and Völkel, 1998; Furne et al., 2001). For the SQR-reaction proper, the product is elemental sulfur. However, since this is an insoluble compound and is deposited outside the cell, the immediate products are soluble polysulfides (see contribution by C. Dahl), as documented for *R. capsulatus* (Griesbeck et al., 2002).

The only SQR of type II that was studied as an isolated expression protein is HMT2 from fission yeast (Vande Weghe and Ow, 1999). In view of its low affinity to sulfide, with K_m-values in the mM range we were initially reluctant to call it SQR (Griesbeck et al., 2000). However, since type II-SQR lacks the third conserved Cys (Fig. 1), it is possible that the enzyme cooperates with another SH-carrying cofactor *in vivo*. In this respect conserved cysteines in neighbouring ORFs of several SQR-genes have been considered (Theissen et al., 2003). In some cases these ORFs have been fused with the SQR-gene. Furthermore, work by H. Shibata and S. Kobayashi, Meji University/Japan, in collaboration with David W. Ow/USDA – unfortunately still unpublished – showed that the affinity of HMT2 as well as of SQR-type II from the cyanobacterium *Synechocystis* PCC 6803 (see below) towards the substrates could be substantially increased by adding mercaptoethanol or cyanide to the assay mixture (personal communication).

A word of caution, on the other hand, regarding the activity measurement is appropriate in this context: Quinones are directly reduced by sulfide, and this background rate increases with pH, the concentrations of the reaction partners, and – often not considered – with the presence of certain detergents. Thus, at substrate concentrations in the mM range, if detergent has been used to solubilize the SQR and is introduced together with SQR it may stimulate the reaction, thus mimicking

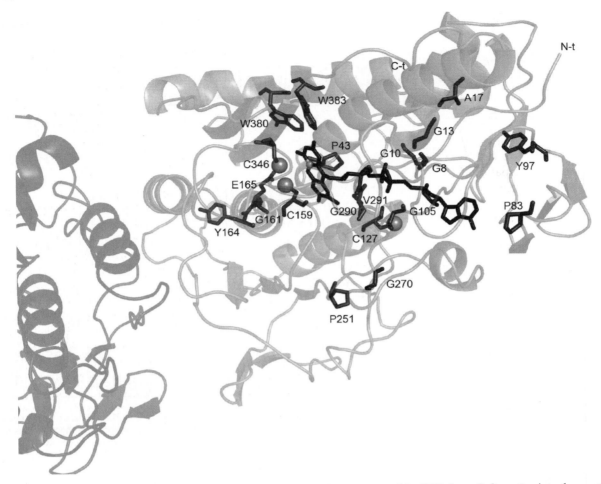

Fig. 3. A structural model. The model was obtained by threading the sequence of the SQR from *O. limnetica* into the crystal structure of FCC from *Allochromatium vinosum* (Chen et al., 1994; NCBI protein data base PDB: 1FCD), using the LOOPP server which is available online at http://cbsuapps.tc.cornell.edu/loopp.aspx (Teodorescu et al., 2004); locations of residues conserved and/or discussed for the alignment of Fig. 1 are indicated using the numbering of the sequence from *O. limnetica*; N-t and C-t stand for N- and C-terminus, respectively; the sulphur atoms of the three conserved cysteines are shown as spheres, the C4a-atom of the isoalloazine ring is indicated by a small sphere; on the left, part of the second protomer folding which forms the dimeric structure of SQR is shown in darker grey.

the enzyme. In contrast to the enzyme this artefact is heat stable (G. Hauska, unpublished).

4. Site-Directed Mutants

The three cysteines C127, C159 and C346, which are strictly conserved in SQRs of type I, and several other conserved residues have been changed in site-directed mutants of *R. capsulatus*, and the effect on activity has been tested in expression proteins (Griesbeck et al., 2002). All three cysteines are essential for the SQR activity, and for the quenching of FAD-fluorescence by sulfide.

For the other members of the disulfide oxidoreductase flavoprotein family, only two cysteines are essential for activity (Williams, 1992), as documented by mutations of the gene for lipoamide dehydrogenase (Hopkins and Williams, 1995). Perhaps the third cysteine in SQR-type I is making up for the second sulfur atom in the substrates of disulfide oxidoreductases, allowing an efficient oxidation–reduction reaction.

Based on the results of site-directed mutagenesis with *R. capsulatus* a mechanism has been suggested (Griesbeck et al., 2002), which very likely also applies to the enzyme from cyanobacteria.

5.   A Proposed Mechanism

The redox cycle of the disulfide oxidoreductases like glutathione reductase or lipoamide dehydrogenase involves a thiolate-FAD charge transfer complex and a covalent C4a-thioadduct of the flavin moiety (Williams, 1992). In analogy Fig. 2 shows our proposal for the reaction mechanism of the reductive interaction of sulfide and flavin in SQR-type I, with the participation of the three essential cysteines. The numbering of the essential residues in Fig. 2 corresponds to the SQR from *R. capsulatus*. In addition to the coordinated action of the cysteines an active site base is involved, which subtracts a proton from the SH-group in the cysteine vicinal to the flavin ring. Originally we considered E165 for this role. However, mutation of Y164 to leucine led to a substantially higher decrease in activity and affinity (C. Griesbeck and G. Hauska, unpublished).

The oxidized state cotains a disulfide bridge between C159 and C353 (1), or between C127 and C353. The reductive half reaction is considered to start with the breaking of the disulfide bridge between C127 and C353 by sulfide (2). The formed persulfide is split by a second sulfide molecule to form free persulfide, which undergoes chemical disproportionation to H_2S and polysulfides (not shown). At the same time a proton is taken from the vicinal C153 (3). The resulting thiolate forms a charge transfer complex to the flavin (4), followed by covalent attachment to the C4a-position which is accompanied by proton transfer from the active site base to N5 of the flavin ring (5). The C4a-adduct is in equilibrium with the anion of the reduced flavin (6), which is reoxidized by quinone (not shown).

A more extensive discussion and the description of further mutations can be found in Griesbeck et al. (2002). However, one more mutant should be considered. As pointed out above, SQR-type I differs from type II, FCC and the other members of the disulfide oxidoreductase family not only by its third conserved Cys, but also in the third FAD-binding region. At position 291/300 for the sequences of *O. limnetica/R. capsulatus*, respectively, a valine takes this place in SQR-types I instead of the aspartate found in all other cases. This may be another reason for the low substrate affinities of HMT2 from *S.pombe* (see above).

Indeed a V300D-mutant of the SQR-gene from *R. capsulatus* showed drastically increased K_m-values for sulfide as well as for ubiquinone (Griesbeck et al., 2002). It is of note, however, that FCC which has an aspartate at this site (Fig. 1), has a high affinity for sulfide with a K_m-value of 12.5 µM (Cusanovich et al., 1991), in spite of having the aspartate in this position.

6.   Structural Aspects

The purified and functional SQRs consist of a single protein which migrates as a band of about 55 kDa in SDS-PAGE. From gel filtration the size was estimated to range from 67 to 80 kDa, which was taken to suggest that the active enzyme is composed as a single polypeptide, surrounded by 20–46 detergent molecules (Arieli et al., 1994). Alternatively, however, since the actual M is only about 47 kDa when calculated from the primary structures (Fig. 1), the functional state of SQR may well be dimeric, as it is the case for the other members of the disulfide oxidoreductase flavoprotein family (Williams, 1992). Indeed, ultrafiltration experiments of SQR from *R. capsulatus* are in favour of a functional state that is larger than the monomer (Schütz et al., 1997).

In Fig. 3 a structural model for SQR from *O. limnetica*, which was obtained by threading the sequence into the crystal structure of FCC (Chen et al., 1994) is depicted. Noteworthy, a similar model for the human SQRDL can be found in the website http.//www.expasy.org/, via the link ModBase, where structures for all the proteins in the human genomes are suggested. In our model for SQR the location of three conserved cysteines and of the other residues pointed out for the alignment in Fig. 1 are shown with respect to the FAD in extended structure. The distance from the sulfur atom of the vicinal cysteine (C159 in SQR and C161 in FCC) to the C4a-atom in the isoalloxazine ring is 3.71 Å. By analogy to the blue light receptor phototropin, a flavoprotein with FMN (Fedorov et al., 2003), this distance should shorten to about 1.8 Å upon formation of the covalent thio-adduct, pulling the C4a-atom out of the ring plane.

C.   Cellular Location

The SQR-enzymes are usually isolated by detergent treatment and are therefore considered to be membrane bound. For chromatophores of R. capsulatus also the chaotropic agent NaBr sufficed to detach the enzyme (Schütz et al., 1997). It was concluded that in this case the enzyme is bound peripherally to the membrane, although by hydrophobic interaction. The enzymes of O. limnetica (Arieli et al., 1994), Allochromatioum vinosum (Reinartz et al., 1998) and Aquifex aeolicus (Nübel et al., 2000) are bound more tightly, and an integral insertion into the membrane has been suggested (Shahak et al., 1999). In view of the other similarities within SQR-type I, and since solubilization with NaBr has not been tried out systematically, further investigation is required to establish such a profound difference firmly.

Another problematic aspect with respect to function in membrane bound form is the dimeric state of functional isolates (see above), in particular since the dimer is central symmetric (Fig. 3). Such a structure on a membrane surface implies that only one of the two active centers is functional at a given time, with a possible switching between the two centers.

It was demonstrated by PhoA-fusion constructs that SQR of R. capsulatus faces the periplasmic space (Schütz et al., 1999), in line with the deposition of elemental sulfur outside the cells (Hansen and Van Gemerden, 1972), as it is known for many of the sulfide-oxidizing photosynthetic bacteria (Brune, 1995). However, no N-terminal extension as export signal into the periplasmic space is present in the bacterial SQR-squences, neither for the Sec-dependent nor for the Sec-independent pathway (Berks, 1996). The latter is characterized by a double RR-motiv, which is found-for FCC only (Fig. 1). Translocation of SQR depends on the C-terminal region and obviously uses a third pathway (Schütz et al., 1999), as discussed more extensively by Griesbeck et al. (2000).

In spite of the notion that cyanobacteria deposite elemental sulfur outside the cell (Garlick et al., 1977), we initially concluded that SQR in O. limnetica is inserted into the thylakoid membranes with its sulfide oxidation site facing the cytoplasm (Arieli et al., 1991, Shahak et al., 1999). This would require transport of elemental sulfur or at least of polysulfides through the cytoplasm to the outside. Alternatively, if in accordance with the location of SQR in purple bacteria the sulfide oxidation by the cyanobacterial SQR faces the intrathylakoid space, contact sites between the

cytoplasmic and the thylakoid membranes would be required (Gantt, 1994). Further investigations should clarify this point.

The eukaryotic typeII-SQRDLs are directed into mitochondria by N-terminal signal sequences, as can be seen for fission yeast and homo in Fig. 1. Whether sulfide oxidation by analogy to the bacteria faces the intramembrane space remains to be established.

IV. Physiological Considerations

A. Bacteria

1. Sulfidotrophy and Sulfide Tolerance

Cyanobacteria thrive in two H_2S-rich ecosystems – in the "thiobios" of marine benthic mats (Fenchel and Riedl, 1970), and in sulfidic hot springs (Castenholtz, 1977), as reviewed extensively by Padan (1979, 1989). Other strains are not sulfidotrophic but rather sulfide tolerant – they can use sulfide for photosynthetic CO_2-fixation but do not grow on sulfide as the sole hydrogen donor (Garlick et al., 1977). Significantly the various strains of cyanobacteria which can shift from oxygenic to anoxygenic photosynthesis can fix CO_2 optimally with sulfide of a rather wide concentration range. At super optimal concentrations sulfide becomes inhibitory. The optima were found at 0.1, 0.7 and 3.5 mM for Lyngbya 7140, Aphanotece halophytica and O. limnetica, respectively. Since sulfide is rather toxic, the major susceptible target being the respiratory cytochrome oxidase (Grieshaber and Völkel, 1998), SQR in addition to energy conservation obviously serves for sulfide detoxification.

B. Fungi and Animals

1. Detoxification

The role of SQR in sulfide detoxification certainly holds also for mitochondrial SQRDL, although ATP formation with sulfide as the electron donor has been observed in mitochondria (Grieshaber and Völkel, 1998; Parrino et al., 2000; Yong and Searcy, 2001). Noteworthy, the colon mucosa of vertebrates is specialized for the mitochondrial oxidation of the sulfide produced by the anaerobic metabolism of enterobacteria (Furne et al., 2001), possibly via SQRDL. Sulfide has been implicated in the etiology of ulcerative colitis, and recently it has been identified as the "first inorganic substrate" for mitochondria of human colon cells (Goubern et al., 2007).

A more general role of detoxification by SQRDL could be linked to the turnover of FeS-centers. The biosynthesis of FeS-centers in eukaryotes is a mitochondrial process, which was inherited from the parent bacterial endosymbiont (Mühlenhoff and Lill, 2000). Nothing is known though about the fate of sulfide which is liberated during degradation of FeS-centers. Mitochondria containing a large number of FeS-centers, possibly liberating toxic amounts of sulfide during FeS-turnover, should benefit from the detoxifying function of SQRDL. However, since SQRDL is confined to animals and some fungi, this protective mechanism cannot be universal.

2. Heavy Metal Tolerance

The SQRDL of the fission yeast *Saccharomyces pombe* is denoted HMT2 for its subtle role in heavy metal tolerance (Vande Weghe and Ow, 1999, 2001). Resistance requires the complexation of heavy metals in the cytoplasm by phytochelatins and sulfide in a defined stoichiometry, suitable for efficient uptake into the vacuole by the ABC-transporter HMT1. If sulfide concentrations become too high heavy metal sulfide precipitates in the cytoplasm, and becomes lethal. The major role of HMT2 in this system is adjusting the optimal level of sulfide by oxidation with ubiquinone in the mitochondria.

3. A Role in Sulfide Signalling

It has been put forward recently that besides its toxicity, at low concentrations sulfide plays a role in animal signallig. This double role is shared with NO and CO. Sulfide is therefore recognized as the third "gasotransmitter" (Wang, 2002; Boehning and Snyder, 2003). Besides its local formation by the intestinal microflora sulfide originates from pyridoxal phosphate dependent cysteine metabolism in our body. It causes swelling of blood vessels by smooth muscle relaxation, much like and in cooperation with NO. Sulfide may be involved in long term potentiation of the central nervous system as well. Efficient signalling requires well controlled adjustment of the signal transmitters. Thus it is conceivable that sulfide oxidation by the SQRDL enzymes in mitochondria plays a role in the fine tuning of the sulfide level, reminiscent of the function of HMT2 in fungal heavy metal tolerance. Such a central role would explain why SQRDL proteins are ubiquitous in animals, but are absent from plants.

V.   Phylogenetic Aspects

A.   Cladograms

An extensive phylogenetic analysis of SQR-genes was provided by Theissen et al., in 2003. From 37 SQR-genes the peptide regions containing the two essential cysteines C159 and C346 (peptide 2 and 5 in Fig. 2), plus the FAD-binding peptide III, with the distinctive GV or GD-pair (see above) from 37 SQR-genes were taken for the calculation of phylogenetic distances. In the resulting cladograms three major groups of SQR-genes could be discerned. Types I and II were well separated and correspond to the two types specified in Fig. 1. Type III, which is prevailing in archaea and green sulfur bacteria, was not as well defined. Type I spreaded over the proteo- and cyanobacteria, but also included the phylogenetically distant bacterium *Aquifex aeolicus*. Type II comprised *Chloroflexus*, Bacilli and the eukaryotic, mitochondrial SQRDL proteins, and surprisingly also one SQR from a cyanobacterium (SyncysII in Fig. 1). The relationship between the mitochondrial and this special cyanobacterial protein had previously been recognized by Vande Weghe and Ow (1999). It was put forward that SQR represents an ancient enzyme originating from a common ancestor which enabled photo- and/or chemosulfidotrophy in the archaic, sulfidic environment. Furthermore, SQR should probably have entered the eukaryotic world by the monophyletic event of endosymbiosis creating the mitochondria. The alternative event of a later horizontal gene transfer could not be excluded, however.

The occurrence in cyanobacteria has been extended since, as documented in the NCBI-website and outlined above already. Both SQR-types occur among cyanobacteria, either type I or type

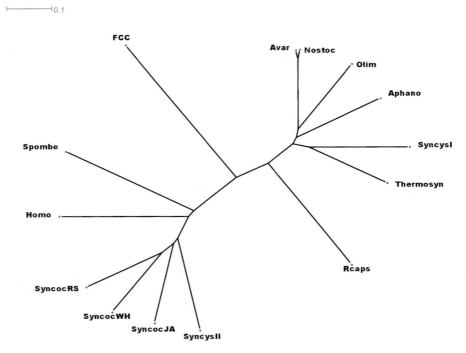

Fig. 4. Cladogram of the SQR-sequences presented in Fig. 1. The cladogram was obtained by using the SplitsTree4 implementation of the NJ-algorithm based on UncorrectedP distances. Different algorithms resulted in similar trees (Huson and Bryant 2006).

II, while both types occur in *Synechocystis* PCC 6803 (Syncys in Fig. 1). The distance cladogram of Fig. 4 shows the relatedness of the sequences presented in Fig. 1.

The split into the two groups with types I on the right and types II on the left, and FCC branching off in between the two, is obvious. Thus we support the analysis of Theissen et al. (2003), and further specify the conclusions in two aspects: (1) The closer relation of bacterial SQR-type II with the mitochondrial SQRDL enzymes and with the notion that the mitochondrial ancestor came from proteobacteria that have SQR-type I, is suggesting the acquirement of SQR by horizontal gene transfer from a Bacillus or cyanobacterium. (2) The absence of SQR from the plant kingdom, and its obligatory occurrence in all animals, may reflect the attainment of new functions during evolution. One such possible function is an involvement in animal type signalling, which may have evolved somehow from fungal heavy metal tolerance. Plants and green algae may have lost their SQR-typeII gene, but the diatoms, and

possibly other orders of algae, interestingly, have retained this gene, for an unknown reason. A more comprehensive phylogenetic treatment of SQR, in particular with regard to the closer relationship between type II and FCC will be given elsewhere (R. Merkl and G. Hauska, in preparation).

B. Convergence within a Phylogeny – the Position of the Conserved Cysteines

The SQR proteins and FCC represent an additional interesting aspect within the phylogeny of the disulfide oxidoreductase family, previously noticed for FCC (Todd et al., 2002). In the crystal structures of glutathione reductase and lipoamide dehydrogenase, as well as of FCC (Mattevi et al., 1992; Chen et al., 1994) two cysteins are located as a redox active disulfide bridge, close to the isoalloxazine ring of the FAD. We concluded above, that this is also the case for SQR (Figs. 2 and 3). However, the positions are rather different in FCC and the disulfide oxidoreductases proper. While in the disulfide oxidoreductases the

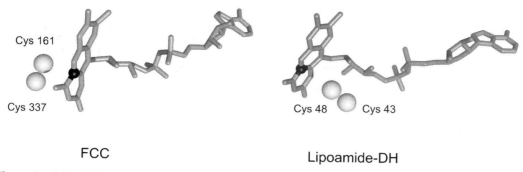

Cys 161

Cys 337

FCC

Cys 48   Cys 43

Lipoamide-DH

Fig. 5. Change in the spatial position of the two essential cysteines within the disulfide oxidoreductase flavoprotein family. The figure was derived from the crystal structures of FCC from *Allochromatium vinosum* (Chen et al., 1994; PDB: 1FCD) and lipoamide dehydrogenase from *Pseudomonas putida* (Mattevi et al., 1992; PDB: 1LVL). The sulfur atoms of the cysteines are shown by spheres, the C4a-position of the isoalloxazine ring in FAD is shown as a small black sphere.

two cysteines are located close together in the primary sequence, spaced by four amino acid residues, in the N-terminal region, in both FCC and SQRs one cysteine comes from the middle and one from the C-terminal region. As shown in Fig. 5 these are C48 and C53 for a bacterial lipoamide dehydrogenase, while C161 and C337 for FCC. Moreover, the spatial location of the Cys-pairs differ. In FCC the pair approaches the isoalloxazine ring from the opposite side of the ribityl-ADP moiety, while from the same side in lipoamide dehydrogenase (Fig. 5).

Such plasticity of catalytically essential positions within phylogenies is puzzling at first sight, yet known for other enzyme families as well (Todd et al., 2002). How could this be reconciled with the origin from a common ancestral enzyme? A reasonable assumption is that the ancestor contained both pairs of cysteines. Perhaps C72 in the N-terminal region of FCC (Fig. 1), a third cysteine, which is conserved in bacterial FCCs for covalent binding of FAD (Chen et al., 1994), represents a relic of the ancestral enzyme.

VI.   Concluding Remark

Finally we would like to express our own surprise about the remarkable extension of the biological relevance for the SQR enzymes, hoping that our review encourages future investigations, with all the precautions for the still speculative parts of it. Own efforts focus on the human SQRDL which has been expressed in *E. coli* after engineering for codon usage (T. Schödl, unpublished). Studies on activity, as well as on expression in rat and human tissues are in progress (M. Ackermann, K. Maier, A.-L. Piña and G.Hauska, in preparation).

Acknowledgements

We are primarily grateful to our collaborators, in first place to Etana Padan from the Hebrew University in Jerusalem, who had introduced us to the subject, and walked along the SQR trail with us for many years. Boaz Arieli, Danny Taglicht and Michal Bronstein on the Israeli side, and Michael Schütz, Christoph Griesbeck and Thomas Schödl on the German side, have all substantially contributed to our current knowledge. The bioinformatic assistance by Rainer Merkl, Matthias Zwick and Daniel Schneider from the Institute of Biophysics and Physical Biochemistry, University of Regensburg is gratefully acknowledged. Part of this work was carried out by using the resources of the Computational Biology Service Unit from Cornell University which is partially funded by Microsoft Corporation.

References

Anbar AD and Knoll AH (2002) Proterozoic ocean chemistry and evolution: A bioinorganic bridge. Science 297: 1137–1142

Arieli B, Padan E and Shahak Y (1991) Sulfide induced sulfide–quinone reductase activity in thylakoids of *Oscillatoria limnetica*. J Biol Chem 266: 104–111

Arieli B, Shahak Y, Taglicht D, Hauska G and Padan E (1994) Purification and characterization of sulfide–quinone reductase (SQR), a novel enzyme driving anoxygenic photosynthesis in *Oscillatoria limnetica*. J Biol Chem 269: 5705–5711

Belkin S and Padan E (1978) Sulfide dependent hydrogen evolution in the cyanobacterium *Oscillatoria limnetica*. FEBS Lett 94: 291–294

Belkin S, Arieli B and Padan E (1982) Sulfide dependent electron transport in *Oscillatoria limnetica*. Isr J Botany 31: 199–200

Belkin S, Shahak Y and Padan E (1988) Anoxygenic photosynthetic electron transport. Meth Enzymol 167: 380–386

Berks BC (1996) A common export pathway for proteins baring complex redox factors. Mol Microbiol 22: 393–404

Blankenship RE, Madigan MT and Bauer CE (1995) Anoxygenic Photosynthetic Bacteria, Vol 2 of Advances in Photosynthesis (Govindjee ed) Kluwer Academic (now Springer), Dordrecht, Boston, London

Boehning D and Snyder SH (2003) Novel neural modulators. Annu Rev Neurosci 26: 105–131

Bronstein M, Schütz M, Hauska G, Padan E and Shahak Y (2000) Cyanobacterial sulfide–quinone reductase: Cloning and heterologous expression. J Bacteriol 182: 3336–3344

Brune DC (1995) Sulfur compounds as photosynthetic electron donors. In: Blankenship R, Madigan MT and Bauer CE (eds) Anoxygenic Photosynthetic Bacteria, Vol II, pp 847–870. Advances in Photosynthesis (Govindjee ed). Kluwer Academic (now Springer), Dordrecht, Boston, London

Brune DC and Trüper HG (1986) Noncyclic electron transport in chromatophores of photolithotrophically grown *Rhodobacter sulfidophilus*. Arch Microbiol 145: 295–301

Castenholtz RW (1977) The effect of sulfide on the blue-green algae of hot springs II. Yellowstone National Park. Microb Ecol 3: 79–105

Chen ZW, Koh M, Van Driesche G, Van Beeumen JJ, Bartsch RG, Meyer TE, Cusanovich MA and Mathews FS (1994) The structure of flavocytochrome c from a purple phototrophic bacterium. Science 266: 430–432

Cusanovich MA, Meyer TE and Bartsch RG (1991) Flavocytochrome c. In: Müller F (ed) Chemistry and Biochemistry of Flavoenzymes, Vol II, pp377–393. CRC, Boca Raton, FL

Fedorov R, Schlichting I, Hartmann E, Domaratcheva T, Fuhrmann M and Hegemann P (2003) Crystal structures and molecular mechanism of a light-induced signalling switch: The Phot-LOV1 domain from *Chlamydomonas reinhardtii*. Biophysics J 84: 501–508

Fenchel TM and Riedl RJ (1970) The sulfide system: a new biotic community underneath the oxidized layer of marine sandy bottoms. Mar Biol 7: 255–268

Friedrich CG (1998) Physiology and genetics of bacterial sulphur oxidation. In: Poole RK (ed) Advances in Microbial Physiology, pp 236–289. Academic, London

Furne J, Springfield J, Koenig T, DeMaster E and Levitt MD (2001) Oxidation of hydrogen sulfide and methanethiol to thiosulfate by rat tissues: a specialized function of the colonic mucosa. Biochem Pharmacol 62: 255–259

Gantt E (1994) Supramolecular membrane organization. In: Bryant DA (ed) The Molecular Biology of Cyanobacteria, Vol 1, pp 119–138. Advances in Photosynthesis (Govindjee ed.). Kluwer Academic (now Springer), Dordrecht, Boston, London

Garlick S, Oren A and Padan E (1977) Occurrence of facultative anoxygenic photosynthesis among filamentous unicellular cyanobacteria. J Bacteriol 129: 623–629

Goubern M, Andriamihaja M, Nübel T, Blachier F and Bouillaud F (2007) Sulfide, the first inorganic substrate for human cells. FASEB J 21: 1699–1706

Griesbeck C (2001) Sulfid-Chinon Reduktase (SQR) aus *Rhodobacter capsulatus*: Physikochemische Charakterisierung und Studien zum katalytischen Mechanismus. Thesis, Universität Regensburg, Germany

Griesbeck C, Hauska G and Schütz M (2000) Biological sulfide oxidation: Sulfide–quinone reductase (SQR), the primary reaction. In: Pandalai SG (ed) Recent Research Development in Microbiology, Vol 4, pp 179–203. Research Signpost, Trivandrum, India

Griesbeck C, Schütz M, Schödl T, Bathe S, Nausch L, Mederer N, Vielreicher M and Hauska G (2002) Mechanism of sulfide–quinone reductase investigated using site-directed mutagenesis and sulphur analysis. Biochemistry 41: 11552–11565

Grieshaber MK and Völkel S (1998) Animal adaptations for tolerance and exploitation of poisonous sulfide. Annu Rev Physiol 60: 33–53

Hansen TA and Van Gemerden H (1972) Sulfide utilization by purple non-sulfur bacteria. Arch Microbiol 86: 49–56

Hopkins N and Williams CH (1995) Characterization of lipoamide dehydrogenase from *Escherichia coli* lacking the redox active disulfide: C44S and C49S. Biochemistry 34: 11757–11765

Huson DH and Bryant D (2006) Application of phylogenetic networks in evolutionary studies. Mol Biol Evol 23: 254–267

Kusai A and Yamanaka T (1973) The oxidation mechanism of thiosulphate and sulphide in *Chlorobium thiosulfatophilum*: Roles of cytochromes c-551 and c-553. Biochim Biophys Acta 325: 304–314

Mattevi A, Obmolova G, Sokatch JR, Betzel C and Hol WG (1992) The refined crystal structure of *Pseudomonas putida* lipoamide dehydrogenase complexed with NAD+ at 2.45 Å resolution. Proteins 13: 336–351

Mühlenhoff U and Lill R (2000) Biogenesis of iron sulfur proteins in eukaryotes: A novel task of mitochondria that is inherited from bacteria. Biochim Biophys Acta 1459: 370–382

Notredame C, Higgins D and Heringa J (2000) T-Coffee: a novel method for multiple sequence alignments. J Mol Biol 302: 205–217

Nübel T, Klughammer C, Huber R, Hauska G and Schütz M (2000) Sulfide–quinone oxidoreductase in membranes of the hyperthermophilic bacterium *Aquifex aeolicus* (VF5). Arch Microbiol 173: 233–244

Padan E (1979) Facultative anoxygenic photosynthesis in cyanobacteria. Annu Rev Plant Physiol 30: 27–40

Padan E (1989) Combined molecular and physiological approach to anoxygenic photosynthesis in cyanobacteria. In: Cohen Y and Rosenberg E (eds.) Microbial Mats: Physiological Ecology and Benthic Microbial Communities, pp 277–282. American Society of Microbiology, Washington, DC

Parrino V, Kraus DW and Doeller JE (2000) ATP production from the oxidation of sulfide in gill mitochondria of the ribbed mussel *Geukensia demissa*. J Exp Biol 203: 2209–2218

Reinartz M, Tschäpe T, Brüser T, Trüper HG and Dahl C (1998) Sulfide oxidation in the phototrophic bacterium *Chromatium vinosum*. Arch Microbiol 170: 59–68

Rich P and Fisher N (1999) Quinone-binding sites in membrane proteins: Structure, function and applied aspects. Biochem Soc Trans 27: 561–565

Schödl T (2003) Sulfid-Chinon Reduktase (SQR) aus *Aquifex aeolicus*: Gensynthese, Expression, Reinigung und biochemische Charakterisierung. Thesis, Universität Regensburg, Germany

Schütz M, Shahak Y, Padan E and Hauska G (1997) Sulfide–quinone reductase from *Rhodobacter capsulatus* – purification, cloning and expression. J Biol Chem 272: 9890–9894

Schütz M, Klughammer C, Griesbeck C, Quentmeier A, Friedrich CG and Hauska G (1998) Sulfide–quinone reductase activity in membranes of the chemotrophic bacterium *Paracoccus denitrificans*. Arch Microbiol 170: 353–360

Schütz M, Maldener I, Griesbeck C and Hauska G (1999) Sulfide–quinone reductase from *Rhodobacter capsulatus*: requirement for growth, periplasmic localization and extension of sequence analysis. J Bacteriol 181: 6516–6523

Shahak Y, Arieli B, Binder B and Padan E (1987) Sulfide-dependent photosynthetic electron flow coupled to proton translocation in thylakoids of the cyanobacterium *Oscillatoria limnetica*. Arch Biochem Biophys 259: 605–615

Shahak Y, Hauska G, Herrmann I, Arieli B, Taglicht D and Padan E (1992a) Sulfide–quinone reductase (SQR) drives anoxygenic photosynthesis in prokaryotes. In: Murata N (ed) Research in Photosynthesis, Vol II, pp 483–486. Kluwer Academic, The Netherlands

Shahak Y, Arieli B, Padan E and Hauska G (1992b) Sulfide quinone reductase (SQR) activity in Chlorobium. FEBS Lett 299: 127–130

Shahak Y, Arieli B, Hauska G, Herrmann I and Padan E (1993) Isolation of sulfide–quinone reductase (SQR) from prokaryotes. Phyton 32: 133–137

Shahak Y, Schütz M, Bronstein M, Griesbeck C, Hauska G and Padan E (1999) Sulfide-dependent anoxygenic photosynthesis in prokaryotes – sulfide–quinone reductase (SQR), the initial step. In: Peschek GA, Löffelhardt WL and Schmetterer G (eds) The Phototrophic Prokaryotes, pp 217–228. Kluwer Academic/Plenum, New York

Shen Y, Buick R and Canfield DE (2001) Isotopic evidence for microbial sulphate reduction in the early Archaean era. Nature 410: 77–81

Teodorescu O, Galor T, Pillardy J and Elber R (2004) Enriching the sequence substitution matrix by structural information. Protein Structure Function Genet 54: 41–48

Thauer RK, Jungermann K and Decker K (1977) Energy conservation in chemotrophic anaerobic bacteria. Bacteriol Rev 41: 100–180

Theissen U, Hoffmeister M, Grieshaber M and Martin W (2003) Single eubacterial origin of eukaryotic sulfide:quinone oxidoreductase, a mitochondrial enzyme conserved from the early evolution of eukaryotes during anoxic and sulfidic times. Mol Biol Evol 20: 1564–1574

Todd AE, Orengo CA and Thornton JM (2002) Plasticity of enzyme active sites. Trends Biochem Sci 27: 419–426

Vande Weghe JG and Ow DW (1999) A fission yeast gene for mitochondrial sulfide oxidation. J Biol Chem 274: 13250–13257

Vande Weghe JG and Ow DW (2001) Accumulation of metal-binding peptides in fission yeast requires hmt2+. Mol Microbiol 42: 29–36

Von Heijne G, Steppuhn J and Herrman R (1989) Domain structure of mitochondrial and chloroplast targeting peptides. Eur J Biochem 180: 535–545

Wang R (2002) Two's a company, three's a crowd: can H_2S be the third endogenous gasotransmitter? FASEB J 16: 1792–1798

Wieringa RK, Terpstra P and Hol WGJ (1986) Prediction of the occurrence of the ADP-binding βαβ-fold in proteins, using an amino acid finger print. J Mol Biol 187: 101–107

Williams CH (1992) Lipoamide dehydrogenase, glutathione reductase, thioredoxin reductase and mercuric ion reductase: a familiy of flavoenzyme transhydrogenases. In: Müller F (ed) Chemistry and Biochemistry of Flavoenzymes, Vol III, pp 121–211. CRC, Boca Raton, FL

Yong R and Searcy DG (2001) Sulfide oxidation coupled to ATP synthesis in chicken liver mitochondria. Comp Biochem Physiol Part B: Biochem Mol Biol 129: 129–137

Genomic Insights into the Sulfur Metabolism of Phototrophic Green Sulfur Bacteria

Niels-Ulrik Frigaard*

Department of Molecular Biology, University of Copenhagen, Ole Maaløes Vej 5, 2200 Copenhagen N, Denmark

Donald A. Bryant

Department of Biochemistry and Molecular Biology, The Pennsylvania State University, University Park, Pennsylvania, 16802, USA

*Corresponding author, Phone: +45 35 32 20 31 Fax: +45 35 32 21 28, E-mail: nuf@mermaid.molbio.ku.dk

Rüdiger Hell et al. (eds.), Sulfur Metabolism in Phototrophic Organisms, 337–355.
© 2008 *Springer.*

Summary

Green sulfur bacteria (GSB) utilize various combinations of sulfide, elemental sulfur, thiosulfate, ferrous iron, and hydrogen for anaerobic photoautotrophic growth. Genome sequence data is currently available for 12 strains of GSB. We present here a genome-based survey of the distribution and phylogenies of genes involved in oxidation of sulfur compounds in these strains. Sulfide:quinone reductase, encoded by *sqr*, is the only known sulfur-oxidizing enzyme found in all strains. All sulfide-utilizing strains contain the dissimilatory sulfite reductase *dsrABCEFHLNMKJOPT* genes, which appear to be involved in elemental sulfur utilization. All thiosulfate-utilizing strains have an identical *sox* gene cluster (*soxJX-YZAKBW*). The *soxCD* genes found in certain other thiosulfate-utilizing organisms like *Paracoccus pantotrophus* are absent from GSB. Genes encoding flavocytochrome *c* (*fccAB*), adenosine-5'-phospho-sulfate reductase (*aprAB*), ATP-sulfurylase (*sat*), a homolog of heterodisulfide reductase (*qmoABC*), and other enzymes related to sulfur utilization are found in some, but not all sulfide-utilizing strains. Other than *sqr*, *Chlorobium ferrooxidans*, a Fe^{2+}-oxidizing organism that cannot grow on sulfide, has no genes obviously involved in oxidation of sulfur compounds. Instead, *Chl. ferrooxidans* possesses genes involved in assimilatory sulfate reduction (*cysIHDNCG*), a trait that is not found in most other GSB. Given the irregular distribution of certain enzymes (such as FccAB, AprAB, Sat, QmoABC) among GSB strains, it appears that different enzymes may produce the same sulfur oxidation phenotype in different strains. Finally, even though the GSB are closely related, sequence analyses show that the sulfur metabolism gene content in these bacteria is substantially influenced by gene duplication and elimination and by lateral gene transfer both within the GSB phylum and with prokaryotes from other phyla.

I. Introduction

Inorganic sulfur metabolism in prokaryotic organisms is a complicated topic due to the complex chemistry of sulfur and the multitude of enzymes that have evolved to catalyze this chemistry. Nevertheless, the ability to use inorganic sulfur compounds as electron sources for growth is widespread among very different prokaryotes of both archaeal and bacterial affiliation. Phototrophic sulfur bacteria characteristically oxidize reduced inorganic sulfur compounds for photoautotrophic growth under anaerobic conditions. These bacteria are traditionally divided into the green sulfur bacteria (GSB) and the purple sulfur bacteria (PSB). The ecology of GSB and PSB is to some extent similar (van Gemerden and Mas, 1995; Over-

mann, 2007) and their oxidative sulfur metabolism probably shares many characteristics as well (Brune, 1989, 1995). However, other aspects of their physiology, evolution, and taxonomy are rather different. These differences are reflected in current taxonomic assignments of these two groups of *Bacteria*: The GSB comprise the phylum *Chlorobi*, whereas all PSB are members of the physiologically highly diverse phylum *Proteobacteria* (Boone and Castenholz, 2001).

GSB are obligately anaerobic and obligately photoautotrophic, and they form a phylogenetically and physiologically distinct group (Overmann, 2000; Garrity and Holt, 2001; Imhoff, 2007). They are commonly found in anoxic and sulfide-rich freshwater and estuarine environments, either in the water column, in sediments, or within microbial mats. They have also recently been found in the anoxic zone 100 m below the surface of the Black Sea (Overmann et al., 1992; Manske et al., 2005), on deep-sea hydrothermal vents in the Pacific Ocean (Beatty et al., 2005), and in the microbial mats of Octopus and Mushroom Springs in Yellowstone National Park (Ward et al., 1998). All GSB characterized to date have unique light-harvesting organelles known as chlorosomes, which allow highly efficient

Abbreviations: Alc. – Allochromatium; APS – adenosine-5'-phosphosulfate (also called adenylylsulfate); *Cba. – Chlorobaculum*; *Chl. – Chlorobium*; GSB – green sulfur bacteria; PAPS – 3'-phosphoadenosine-5'-phosphosulfate (also called 3'-phosphoadenylylsulfate); *Pld. – Pelodictyon*; PSB – purple sulfur bacteria; PSR – polysulfide reductase; PSRLC – polysulfide reductase-like complex; *Ptc. – Prosthecochloris*; SQR – sulfide:quinone reductase; SQRLP – sulfide:quinone reductase-like protein

Table 1. Strains of green sulfur bacteria selected for genome sequencing.

Strain designations[a]	Names used[b]	Phenotype[c]	Genome sequence status[d]
ATCC 35110[T]	*Chloroherpeton thalassium*	BChl *c*	3.25 Mbp (1 chromosome; 3 gaps)
TLS, ATCC 49652[T]	*Chlorobaculum tepidum*, (*Chlorobium tepidum*)	BChl *c*, Tio[+]	2,154,946 bp (1 chromosome)
BS1	*Prosthecochloris* sp., (*Chlorobium phaeobacteroides*)	BChl *e*	2,736,402 bp (1 chromosome)
CaD3	*Chlorobium chlorochromatii*, ("*Chlorochromatium aggregatum*" epibiont)	BChl *c*, Tio[-]	2,572,079 bp (1 chromosome)
DSMZ 245[T], 6330	*Chlorobium limicola*	BChl *c*, Tio[-]	2,763,183 bp (1 chromosome)
DSMZ 263[T], NCIMB 8327	*Chlorobaculum parvum*, (*Chlorobium vibrioforme* subsp. *thiosulfatophilum*, *Chlorobium thiosulfatophilum*)	BChl *d*, Tio[+]	2,289,236 bp (1 chromosome)
DSMZ 265, 1930	*Chlorobium phaeovibrioides*, (*Chlorobium vibrioforme* subsp. *thiosulfatophilum*)	BChl *c* + *d*, Tio[+]	1,966,858 bp (1 chromosome)
DSMZ 266[T], 2430	*Chlorobium phaeobacteroides*	BChl *e*, Tio[-]	3,133,902 bp (1 chromosome)
DSMZ 271[T], SK-413	*Prosthecochloris aestuarii*	BChl *c*, Tio[-]	2,512,923 bp (1 chromosome)
DSMZ 273[T], 2530	*Chlorobium luteolum*, (*Pelodictyon luteolum*)	BChl *c*, Tio[-]	2,364,842 bp (1 chromosome)
DSMZ 5477[T], BU-1	*Chlorobium clathratiforme*, (*Pelodictyon phaeoclathratiforme*)	BChl *e*, Tio[+]	2,018,240 bp (1 chromosome)
DSMZ 13031[T]	*Chlorobium ferrooxidans*	BChl *c*, Tio[-]	2.6 Mbp (draft available; 5 gaps)

[a] Superscript "T" denotes type strain.
[b] Names are according to Imhoff (2003) except those in parenthesis, which represents alternative names.
[c] Tio[+] indicates that thiosulfate is utilized for growth.
[d] Status as of November 2007.

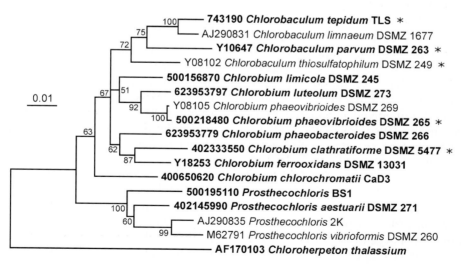

Fig. 1. Unrooted neighbor-joining phylogenetic tree of the 16S rRNA gene of selected GSB strains. Strains that have been selected for genome sequencing are shown in bold. Strains that have been shown to grow on thiosulfate are indicated with an asterisk. Sequence accession numbers are either from the JGI data base (digits only) or from GenBank. The tree is based on 1119 nucleotide positions and was made with MEGA version 3.1 (Kumar et al., 2004). Bootstrap values in percentage are shown for 1000 replications. Except for the position of *Chl. chlorochromatii* CaD3, whose position is not resolved, minimum evolution and maximum parsimony analyses also support the topology of this tree.

capture of light energy; therefore, these organisms can grow at remarkably low light intensities. All characterized strains use the reductive (also called reverse) tricarboxylic acid cycle for CO_2 fixation. Most strains use electrons derived from oxidation of sulfide, thiosulfate, elemental sulfur, and H_2, but a few characterized strains can also oxidize Fe^{2+}.

In an effort to learn more about the physiology and evolution of this unique group, and especially about their photosynthesis and carbon and sulfur metabolism, 12 strains of GSB have been selected for genome sequencing (Table 1). Figure 1 shows a 16S rRNA phylogenetic tree of these and other GSB strains. The genome of one of the best characterized strains, *Chlorobaculum tepidum* TLS (previously known as *Chlorobium tepidum* TLS), was sequenced and annotated in 2002 by The Institute for Genomic Research (TIGR; Eisen et al., 2002). Eleven other strains are currently at various stages of genome sequencing and annotation at the Joint Genome Institute (United States Department of Energy) and in the laboratories of Donald A. Bryant (The Pennsylvania State University, United States) and Jörg Overmann (Ludwig-Maximilians-Universität, Germany) (Table 1). The genome sequences can be accessed and analyzed on the websites of the Integrated Microbial Genomes resource (http://img.jgi.doe.gov) and the National Center for Biotechnology Information (http://www.ncbi.nlm.nih.gov). Recent reviews on information derived from genome sequence data of GSB are available (Frigaard et al., 2003, 2006; Frigaard and Bryant, 2004). Here, we present an overview of the content and phylogeny of known or putative enzymes involved in inorganic sulfur metabolism in the 10 strains of the GSB for which sufficient genome sequence information is currently available (all strains listed in Table 1, except *Chlorobaculum parvum* DSMZ 263 and *Chloroherpeton thalassium* ATCC 35110). Sulfur metabolism in GSB is also discussed in the chapter by Hanson (2008). Current knowledge on the oxidation of inorganic sulfur compounds in PSB is reviewed in the chapter by Dahl (2008). At present, genome sequence information is publicly available only for one PSB: the halophilic *Halorhodospira halophila* SL1 (DSMZ 244) of the *Ectothiorhodospiraceae* family (http://www.ncbi.nlm.nih.gov).

II. Sulfur Compounds Oxidized for Growth

Almost all GSB are capable of oxidizing sulfide and elemental sulfur to sulfate. GSB have a high affinity for sulfide, and sulfide is usually the preferred substrate even if other sulfur substrates are available (Brune, 1989, 1995). Initially, sulfide is typically incompletely oxidized to elemental sulfur, which is deposited extracellularly as sulfur globules. These sulfur globules are oxidized completely to sulfate when the sulfide has been consumed. Some of the sulfur compounds utilized by the strains for which genome sequence data are available are shown in Table 2.

Several strains of GSB are also capable of growing on thiosulfate ($S_2O_3^{2-}$) (see overview in Imhoff, 2003). Some of these strains can grow on thiosulfate in the absence of any other reduced sulfur compound (N.-U. Frigaard, unpublished data). Two thiosulfate-utilizing strains of GSB have been reported to utilize tetrathionate ($S_4O_6^{2-}$) (Brune et al., 1989). No GSB has been reported to utilize sulfite (SO_3^{2-}).

III. Sulfur Compound Oxidation Enzymes

Many enzymes potentially involved in sulfur metabolism can readily be identified in the genome sequences by sequence homology with known enzymes. Table 2 shows a detailed survey of genes known to be involved or potentially involved in the oxidative sulfur metabolism in the genome-sequenced strains of GSB. Specific enzymes are discussed below. An overview of the metabolic network formed by these enzymes is shown in Fig. 2.

A. Sulfide:Quinone Reductase

Sulfide:quinone reductase (SQR; EC 1.8.5.-) catalyzes oxidation of sulfide with an isoprenoid quinone as the electron acceptor. This enzyme occurs in both chemotrophic and phototrophic prokaryotes as well as in some mitochondria (Griesbeck et al., 2000; Theissen et al., 2003). Membrane-bound SQR activity has been biochemically demonstrated in GSB and presumably feeds electrons into the photosynthetic electron transfer chain

Table 2. Phenotypes and genotypes of green sulfur bacteria for which genome sequence data is available.

Strain	Electron donor[a]			Genotype[b]													
	S²⁻	S⁰	S₂O₃²⁻	sqr	dsr	fcc	soyYZ	sox	apr	sat	qmo	hdr-hyd	PSRLC1	PSRLC2	PSRLC3	hup	cys
Chlorobaculum tepidum TLS	+	+	+	+	+	+	−	+	+	+	+	+	+	−	−	−	−
Chlorobium chlorochromatii CaD3	+	−	−	+	+	+	−	+	+	+	+	+	+	+	+	−	−
Chlorobium clathratiforme DSMZ 5477	+	+	+	+	+	+	+	+	+	+	+	+	+	+	−	+	−
Chlorobium ferrooxidans DSMZ 13031	−	−	−	+	−	−	+	−	−	−	−	+	−	−	−	+	+
Chlorobium limicola DSMZ 245	+	+	−	+	+	+	+	−	−	−	−	+	−	+	+	+	−
Chlorobium luteolum DSMZ 273	+	+	−	+	+	−	+	−	−	−	−	+	+	−	+	+	+
Chlorobium phaeobacteroides DSMZ 266	+	+	−	+	+	+	+	−	−	−	−	+	−	+	+	+	−
Chlorobium phaeovibrioides DSMZ 265	+	+	+	+	+	+	−	+	−	−	−	+	+	−	+	−	−
Prosthecochloris sp. BS1	+	−	−	+	+	+	−	−	+	+	+	−	+	+	−	+	−
Prosthecochloris aestuarii DSMZ 271	+	+	−	+	+	+	−	−	−	+	+	−	−	+	+	+	−

[a] Garrity and Holt 2001, Heising et al. 1999, Vogl et al. 2006.
[b] The following abbreviations designates more than one gene: apr: *aprBA*, cys: *cysHDNCG-cysPTWA*, dsr: *dsrABCEFHLNMKJOPT*, fcc: *fccAB*, hdr-hyd: *hdrDA-orf1247-orf1248-hydB2G2*, hup: *hupSLCD*, sox: *soxJXYZAKBW*, qmo: *qmoABC*.

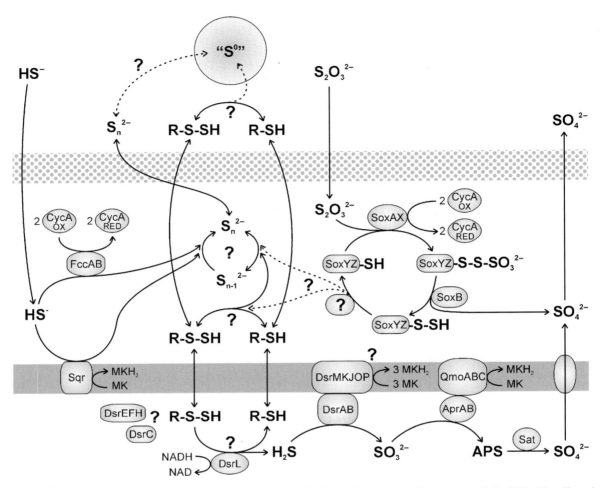

Fig. 2. Overview of known or hypothesized pathways in the oxidation of inorganic sulfur compounds in GSB. Not all strains have all pathways shown here. This figure is derived from information in Eisen et al., 2002 and Dahl, 2008. (See text for details.)

via a quinol-oxidizing Rieske iron-sulfur protein/ cytochrome *b* complex (Shahak et al., 1992). The genome sequences of all GSB strains encode one (CT0117 in *Cba. tepidum* TLS) or two homologs of the biochemically characterized SQRs from *Rhodobacter capsulatus* (CAA66112) and *Oscillatoria limnetica* (AAF72962) (Fig. 3). This includes *Chl. ferrooxidans* (ZP_01385816), which cannot grow on sulfide as the sole electron donor (Heising et al., 1999). This organism may benefit from SQR activity as a supplement to its energy metabolism. It could also use SQR as a protective mechanism to remove sulfide, which prevents growth when present in high concentrations. The SQR homologs of GSB are flavoproteins with predicted masses of about 53 kDa, and each con-

tains all three conserved cysteine residues that are essential for sulfide oxidation in *Rhodobacter capsulatus* SQR (Griesbeck et al., 2002).

Some GSB additionally contain distantly related homologs of SQR with no assigned or obvious function (here denoted SQRLP1 and SQRLP2 for SQR-like proteins type 1 and 2) (Fig. 3). SQRLP1 is present in *Cba. tepidum* TLS (CT1087) and in three other GSB. Among the GSB, SQRLP2 is only present in *Cba. tepidum* TLS (CT0876). Interestingly, SQRLP2 clusters in phylogenetic analyses with proteins of unknown function from various archaea. In addition, SQRLP2 from *Cba. tepidum* TLS shares 54% amino acid sequence identity with a protein from *Sulfitobacter* strain NAS14.1, which is a marine

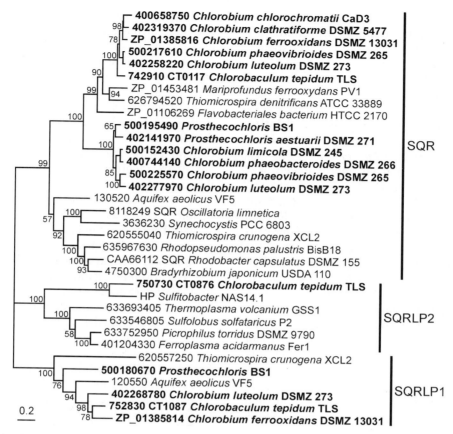

Fig. 3. Unrooted neighbor-joining phylogenetic tree of sulfide: quinone reductase (SQR) and two types of SQR-like proteins (SQRLP1 and SQRLP2) in selected organisms. Strains of GSB are shown in bold. The tree was made with MEGA version 3.1 (Kumar et al., 2004) and shows bootstrap values for 100 replications.

Roseobacter-like strain that grows aerobically on dimethylsulfoniopropionate (https://research. venterinstitute.org/moore/). Otherwise, SQRLP2 has few homologs in the databases.

B.   Flavocytochrome c

Flavocytochrome *c* is usually a soluble, periplasmic enzyme consisting of a large sulfide-binding FccB flavoprotein subunit and a small FccA cytochrome *c* subunit (Brune, 1995). Although this protein efficiently oxidizes sulfide and reduces cytochrome *c* in vitro, the exact function and significance of this protein in vivo is still not clear. Although many sulfide-utilizing organisms produce flavocytochrome *c*, some sulfide-utilizing GSB and PSB do not, which clearly demonstrates that flavocytochrome *c* is not essential for sulfide

oxidation (Brune, 1995). For example, the pattern of sulfide oxidation and the concomitant formation of elemental sulfur that is subsequently oxidized to sulfate upon sulfide depletion, is similar in *Chl. luteolum* DSMZ 273, which does not contain flavocytochrome *c*, and in *Chl. limicola* DSMZ 245, which does contain flavocytochrome *c* (Steinmetz and Fischer, 1982). In addition, a mutant of the PSB *Allochromatium vinosum* DSMZ 180, in which flavocytochrome *c* has been eliminated genetically, exhibits sulfide and thiosulfate oxidation rates similar to the wild type (Reinartz et al., 1998). If indeed the FccAB flavocytochrome *c* oxidizes sulfide in vivo, both GSB and PSB apparently have alternative sulfide-oxidizing enzyme systems, possibly sulfide:quinone reductase (section III.A) and the Dsr system (section III. C), that may be quantitatively more important.

However, it is also possible that flavocytochrome *c* is advantageous under certain growth conditions and that such conditions have not yet been identified in these bacteria.

An *fccAB*-encoded flavocytochrome *c* is found in all GSB strains for which genome sequence data is available, except *Chl. ferrooxidans* DSMZ 13031 and *Chl. luteolum* DSMZ 273. The GSB flavocytochrome *c* consists of a 10-kDa FccA cytochrome c_{553} subunit (CT2080 in *Cba. tepidum* TLS) that binds a single heme and an approximately 47-kDa sulfide-binding FccB flavoprotein subunit (CT2081 in *Cba. tepidum* TLS). The FccB subunit has high sequence similarity (approximately 50% amino acid sequence identity) to a flavoprotein (SoxJ) encoded in the *sox* gene cluster (section III. G.1). FccAB is constitutively expressed in *Cba. thiosulfatiphilum* (formerly *Chl.limicola* subsp. *thiosulfatophilum*) DSMZ 249 (Verté et al., 2002). In this strain, FccAB was reported to be membrane-bound, possibly due to an unusual signal peptide in the FccA subunit that is not cleaved but supposedly anchors the protein in the cytoplasmic membrane. The predicted signal peptide in FccA from GSB is followed by a highly variable, 15- to 25-residue sequence, which is rich in alanine and proline and which is suggested to act as a flexible arm (Verté et al., 2002).

C. Dissimilatory Sulfite Reductase

The well-studied PSB, *Alc. vinosum,* contains a gene cluster with high sequence similarity to the dissimilatory sulfite reductase *dsr* gene cluster of sulfate-reducing bacteria (Dahl et al., 2005). The *dsr* gene cluster in *Alc. vinosum, dsrABEFHCMKLJOPNRS*, is essential for the oxidation of intracellular sulfur globules, and thus it is assumed that the Dsr enzyme system in this organism functions in the oxidative direction to produce sulfite (Pott and Dahl, 1998; Dahl et al., 2005; Sander et al., 2006). All GSB, except *Chl. ferrooxidans*, contain the *dsrABCE-FHLNMKJOPT* genes. Thus, despite the absence of *dsrRS* and the presence of *dsrT* in the Dsr system in GSB, this system most likely functions very similarly to the Dsr enzyme system in PSB. The absence of *dsr* genes in *Chl. ferrooxidans* is consistent with the observation that this bacterium is incapable of growth on elemental sulfur and sulfide.

In *Cba. tepidum* TLS the *dsr* genes are split into two clusters, and three functional *dsr* genes are duplicated (*dsrA*, *dsrC*, and *dsrL*). This may be due to a frameshift mutation in the *dsrB* gene in a recent ancestor of the TLS strain that rendered the gene nonfunctional. This could have selected for a duplication, rearrangement and subsequent frameshift mutation of a small segment of the genome, which restored a functional *dsrB* gene but also resulted in a duplication of the *dsrCABL* gene cluster. The two regions that contain a *dsrCABL* cluster in *Cba. tepidum* TLS are 99.4% identical at the nucleotide level. From the currently available data it appears that the *dsr* genes only occur as a single cluster in all other genome-sequenced GSB.

Based upon phylogenetic analyses, the cytoplasmic DsrAB sulfite reductase and other cytoplasmic Dsr proteins in GSB are related to the Dsr proteins from other sulfide-oxidizing prokaryotes (Sander et al., 2006). However, the subunits of the membrane-bound DsrMKJOP complex are related to the DsrMKJOP proteins from sulfate-reducing prokaryotes. In addition, the DsrT protein (unknown function) is only found in GSB and sulfate-reducing prokaryotes and not in other sulfide-oxidizers. This suggests an intriguing chimeric nature of the Dsr system in GSB, possibly generated by lateral gene transfer of *dsrTMKJOP* from a sulfate-reducing prokaryote to a common ancestor of GSB. An interesting possibility is that it might have been the acquisition of the ability to oxidize sulfur to sulfite that led to the relatively recent, explosive radiation of the GSB.

D. Sulfite Oxidation

Although GSB cannot grow on sulfite as sole sulfur source and electron donor, sulfite appears to be the product of the Dsr enzyme system (section III.C, Fig. 2). Sulfite can be oxidized by adenosine-5′-phosphosulfate (APS) reductase (also called adenylylsulfate reductase, EC 1.8.99.2) in a reaction that consumes sulfite and AMP and generates APS and reducing equivalents. Two non-homologous types of such enzymes are known: the AprAB type that functions in dissimilatory sulfur metabolism and the CysH type that functions in assimilatory sulfur metabolism. An Apr-type APS reductase is found in the genomes of four GSB (Table 2). In *Cba. tepidum* TLS the genes are *aprA/CT0865* and *aprB/CT0864*. A CysH-type APS reductase is found in *Chl. luteolum* DSMZ 273 and *Chl. ferrooxidans* DSMZ 13031, in which this enzyme probably functions in assimilatory sulfate reduction (see section VI). No APS reductase has been identified in the genomes of other GSB. Despite the absence of a recognizable APS reductase in its genome sequence, an APS reductase activity has been purified from *Chl. limicola* DSMZ 245 and biochemically characterized although not sequenced (Kirchhoff and Trüper, 1974). This APS reductase from *Chl. limicola* DSMZ 245 was reported to have a molecular mass of about 200 kDa and to contain one flavin per molecule and non-heme iron, but it contained no heme. Homologs of the molybdopterin-binding SorAB sulfite:cytochrome *c* oxidoreductase (EC 1.8.2.1) from *Starkeya novella* (formerly *Thiobacillus novellus*) (Kappler et al., 2000) are not found in any GSB genome sequence. An alternative putative molybdopterin-binding enzyme that may function in sulfite oxidation in GSB is described in section IV.B.

E.   Release of Sulfate from APS

AMP-dependent oxidation of sulfite by APS reductase produces APS (section III.D). APS can be hydrolyzed to AMP and sulfate by adenylylsulfatase (EC 3.6.2.1) but there is no evidence for this enzyme in GSB. Alternatively, the energy of the phosphosulfate anhydride bond in APS can be conserved by the action of sulfate adenylyltransferase (also called ATP-sulfurylase; EC 2.7.7.4) encoded by the *sat* gene. ATP-sulfurylase generates ATP and sulfate from APS and pyrophosphate. ADP-sulfurylase (EC 2.7.7.5) generates ADP and sulfate from APS and phosphate (this enzyme is also called adenylylsulfate:phosphate adenylyltransferase, APAT). Some GSB strains have been reported biochemically to contain either ATP-sulfurylase activity (strains DSMZ 249 and DSMZ 257) or ADP-sulfurylase activity (strains DSMZ 263 and NCIMB 8346) but not both activities (Khanna and Nicholas, 1983; Bias and Trüper, 1987); however, the genome of none of these strains has been sequenced. Four GSB genomes encode highly similar, *sat*-encoded ATP-sulfurylases (Table 2; CT0862 in *Cba. tepidum* TLS).

Another type of ATP-sulfurylase, which is not homologous with the Sat enzyme, is the heterodimeric CysDN known from assimilatory sulfate reduction in *Alc. vinosum*, *Escherichia coli*, and other prokaryotes (Kredich, 1996; Neumann et al., 2000). Genes encoding a CysDN-like complex are found in *Chl. ferrooxidans* and *Chl. luteolum* DSMZ 273 as part of an assimilatory sulfate reduction gene cluster (section VI). *Chl. phaeovibrioides* DSMZ 265 and *Ptc. aestuarii* DSMZ 271 also contain *cysC* and *cysN* homologs. But in these two latter organisms the genes occur in a cluster that appears to be involved in another aspect of sulfur metabolism. This cluster also contains a *cysQ* homolog that may encode a phosphatase acting on APS or PAPS (Neuwald et al., 1992) and a homolog of the ArsB/NhaD superfamily of permeases that translocates Na^+ and various anions such as sulfate across the cytoplasmic membrane (500231320 and 500231330, respectively, in *Chl. phaeovibrioides* DSMZ 265). It is therefore possible that *Chl. phaeovibrioides* DSMZ 265 and *Ptc. aestuarii* DSMZ 271 possess a system that processes APS or sulfite (or both) differently than in other GSB and that actively excretes sulfate. No homologs of *sat*, *cysN*, or *cysD* have been identified in *Chl. limicola* DSMZ 245 or *Chl. phaeobacteroides* DSMZ 266 and it is not clear how these strains convert APS to sulfate, if they do at all.

F.   Qmo Complex

Cba. tepidum TLS and three other GSB strains (Table 2) contain a membrane-bound electron-transfer complex (QmoA/CT0866-QmoB/CT0867-QmoC/CT0868) that shares homology with subunits of heterodisulfide reductases. The Qmo complex from GSB has the same subunit structure and the same putative cofactor-binding sites as the Qmo complex in the sulfate-reducing organisms, *Desulfovibrio desulfuricans* and *Archaeoglobus fulgidus* (Pires et al., 2003). It appears that in all these organisms QmoA and QmoB are cytoplasmic, nucleotide- and iron–sulfur-cluster-binding subunits and QmoC is a membrane-bound, heme-*b*-binding subunit that exchanges electrons with the isoprenoid quinone pool. The Qmo complex from *Desulfovibrio desulfuricans* was biochemically characterized and shown to have quinol-oxidizing activity (Pires et al., 2003). In all four cases in which the genes encoding a Qmo complex (*qmoABC*), an ATP-

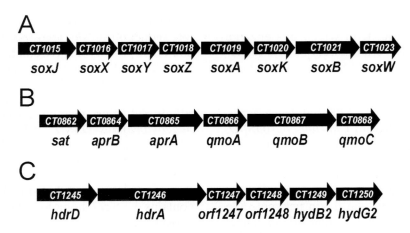

Fig. 4. Gene clusters in *Cba. tepidum* TLS of (A) *sox* genes, (B) *sat-apr-qmo* genes, and (C) *hdr-hyd* genes. These clusters are conserved in all GSB strains in which the genes occur.

sulfurylase (*sat*), and an APS reductase (*aprAB*) occur in GSB, these genes are clustered in one apparent *sat-aprBA-qmoABC* operon (Table 2 and Fig. 4B). A similar *aprBA-qmoABC* operon with high sequence similarity occurs in *Desulfovibrio desulfuricans* and other *Desulfovibrio*-like strains. It therefore seems likely that the four GSB that have this operon may have obtained it from a *Desulfovibrio*-like organism by lateral gene transfer. The enzymes encoded by the *sat-aprBA-qmoABC* operon in principle allow oxidation of sulfite to sulfate via an APS intermediate with concomitant reduction of membrane-bound quinones (Fig. 2). How this reaction occurs in GSB that do not have the *sat-aprBA-qmoABC* genes is not clear (see Section IV.B for a possible alternative sulfite oxidation system).

G. Thiosulfate Oxidation

1. Sox System

In the chemolithoautotrophic α-proteobacterium *Paracoccus pantotrophus* the *sox* gene cluster comprises 15 genes that encode proteins involved in the oxidation of thiosulfate and possibly other sulfur compounds (Friedrich et al., 2001, 2005). The Sox proteins are transported to the periplasm, either by encoding a signal peptide recognized by the Sec system or the Tat system, or by forming a complex with another Sox protein that encodes a signal peptide. The products

of seven *sox* genes, *soxXYZABCD*, are induced by thiosulfate and are sufficient to reconstitute a thiosulfate-oxidizing enzyme system in vitro. Friedrich et al. (2001) proposed the following model for thiosulfate oxidation: The sulfane atom of thiosulfate is bound to the SoxYZ complex (SoxYZ-SH) by an oxidation reaction, catalyzed by the SoxAX *c*-type cytochrome, which results in a thiocysteine-*S*-sulfate residue (SoxYZ-S-S-SO_3^{2-}). A conserved cysteine residue in the SoxY subunit constitutes the substrate-carrying site. Sulfate is liberated by hydrolysis catalyzed by SoxB to yield a persulfide intermediate (SoxYZ-S-S$^-$), which subsequently is oxidized by another *c*-type cytochrome, SoxCD, to form a cysteine-*S*-sulfate residue (SoxYZ-S-SO_3^{2-}). Hydrolysis by SoxB releases sulfate and regenerates the SoxYZ complex (SoxYZ-SH).

The cluster of *sox* genes in *Cba. tepidum* TLS, *CT1015-soxXYZA-CT1020-soxBW* (Fig. 4A), is conserved in the genomes of three other strains (Table 2). Because of the organizational conservation and the congruent phylogeny of the eight genes in this cluster (see section VII), the genes *CT1015* and *CT1020* are likely involved in the Sox system. Thus, these two genes are now denoted as *soxJ* and *soxK*, respectively. The *soxJXYZAKBW* cluster is present in all three thiosulfate-utilizing strains (TLS, DSMZ 265, DSMZ 5477) and one strain (CaD3) that has not been reported to grow on thiosulfate. All SoxY proteins in GSB have the conserved C-terminal

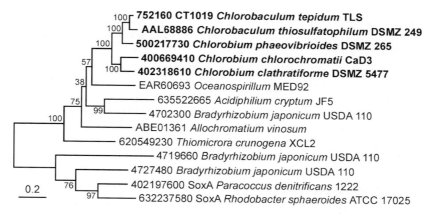

Fig. 5. Unrooted neighbor-joining phylogenetic tree of SoxA proteins. Strains of GSB are shown in bold. The tree was made with MEGA version 3.1 (Kumar et al., 2004) and shows bootstrap values for 100 replications.

motif GGCGG-COOH with the substrate-binding cysteine residue. (The genes encoding the SoxYZ complex have been duplicated in some GSB; see section III.H.) The gene *soxJ* encodes a putative FAD-containing dehydrogenase related to the sulfide-binding flavoprotein subunit of flavocytochrome *c*. The gene *soxK* encodes a hypothetical 11-kDa protein with a signal peptide and with a homolog encoded in the *sox* cluster of the purple sulfur bacterium *Alc. vinosum* (ABE01362), but has no identified homologs in other organisms. SoxA from bacteria such as *P. pantotrophus* and *Rhodovulum sulfidophilum* binds heme groups in two conserved CXXCH motifs (Appia-Ayme et al., 2001; Bamford et al., 2002). The C-terminal heme presumably participates in oxidation of the thiosulfate-SoxYZ complex; the N-terminal heme is presumably too far from the C-terminal heme to allow electron transfer between the two heme groups, and its function is not known. SoxA in GSB and some other bacteria such as *Alc. vinosum* and *Starkeya novella* only has the C-terminal heme-binding motif. This difference is reflected in a phylogenetic sequence analysis, in which SoxA sequences from GSB and *Alc. vinosum* group separately from the SoxA sequences of *P. denitrificans* and *Rhodobacter sphaeroides* that have two heme-binding motifs (Fig. 5).

The *soxCD* genes, which are essential components of the Sox system in *P. pantotrophus*, do not occur in the genome sequences of GSB. The *soxCD* genes have also not been found in the purple sulfur bacterium *Alc. vinosum* (Hensen

et al., 2006). This observation suggests (1) that the persulfide-form of the SoxYZ carrier protein (SoxYZ-S-S⁻) is transformed back to the unmodified form (SoxYZ-SH) differently in GSB and in *P. pantotrophus,* and (2) that the sulfane moiety from the thiosulfate molecules that become attached to the SoxYZ carrier protein are not completely oxidized to sulfate in the GSB Sox system as they are in the *P. pantotrophus* Sox system. This is consistent with experimental evidence in *Alc. vinosum,* which shows that the sulfane moiety from thiosulfate is found as elemental sulfur when sulfur globules are formed by oxidation of thiosulfate (Smith and Lascelles, 1966; Trüper and Pfennig, 1966). In this process, a net electron gain in the GSB Sox system is only accomplished by the SoxAX-dependent oxidation (Fig. 2). The SoxCD-independent reaction in GSB that regenerates the SoxYZ complex may involve the SoxJ and SoxK proteins due to the conservation of their genes in the GSB *sox* gene cluster (Fig. 4A).

2. Rhodaneses

Rhodaneses (thiosulfate sulfurtransferases, EC 2.8.1.1) are enzymes that catalyze the transfer of the sulfane sulfur atom of thiosulfate to cyanide (CN⁻) to generate thiocyanate (SCN⁻) and SO_3^{2-}. However, this is often not the physiological role of the enzymes. Rhodaneses are common in many organisms, including phototrophic sulfur bacteria, but their roles in lithotrophic sulfur metabolism are not clear (Brune, 1995). Two rhodaneses with unknown function have been purified from the thiosulfate-utilizing *Cba. parvum* DSMZ 263 of which the most abundant and active rhodanese was a 39-kDa basic protein with an isoelectric point of 9.2 (Steinmetz and Fischer, 1985).

All GSB, other than *Chl. ferrooxidans*, contain a 17-kDa rhodanese (CT0843 in *Cba. tepidum* TLS) as part of the *dsr* gene cluster (section III.C). Other putative rhodaneses are found scattered among the genome sequences of GSB in a manner that does not obviously correspond with their ability to use thiosulfate. A putative periplasmic 22-kDa rhodanese with an isoelectric point of 8.8 is found in the thiosulfate-utilizing *Chl. clathratiforme* DSMZ 5477 (ZP_00590525). A homolog of this protein is found in *Chl. chlorochromatii* CaD3 (ABB28218), which has not been reported to use thiosulfate, but homologs are not found in the genome sequences of other GSB. *Chl. limicola* DSMZ 245 and *Prosthecochloris* sp. BS1 contain an acidic, putatively cytoplasmic, ~ 50-kDa rhodanese (ZP_00512484 and ZP_00530387, respectively), which is homologous with a putative rhodanese found in *Salinibacter ruber* (53% sequence identity) and several strains of *Escherichia coli* (37% sequence identity).

3. Plasmid-Encoded Thiosulfate Oxidation?

A 15-kb plasmid, named pCL1, has been isolated from the thiosulfate-utilizing *Cba. thiosulfatiphilum* DSMZ 249 (Méndez-Alvarez et al., 1994). When this plasmid was transferred to *Chl. limicola* DSMZ 245, which cannot grow on thiosulfate, the resulting transformants were reported to utilize thiosulfate as the sole electron donor for growth. The plasmid has been sequenced (NC_002095), but surprisingly it does not encode genes known to be involved in thiosulfate utilization (C. Jakobs et al., unpublished data). No putative enzymes involved in sulfur chemistry are encoded by the plasmid, but it is possible that the plasmid somehow allows cellular transport of certain sulfur compounds. The plasmid contains a cluster of seven genes (similar to *cpaBCEF-tadBC-pilD*), which are homologs of genes involved in a type II secretion system that functions in pilus formation in many bacteria. The other genes on the pCL1 plasmid are apparently involved in plasmid maintenance. Three of the seven genes in the *cpaBCEF-tadBC-pilD*-like

cluster on the plasmid contain frame-shift mutations that probably cause non-functional proteins. (Alternatively, these mutations could be due to sequencing errors.) A highly similar cluster, but with the reading frame of all genes intact, is found in the genome sequences of all three GSB that can grow on thiosulfate (strains TLS, DSMZ 265, DSMZ 5477), as well as in one strain that cannot grow on thiosulfate (DSMZ 273), but not in other strains. Since *Chl. chlorochromatii* CaD3 is the only strain that has the *Chlorobium*-type *sox* cluster (Fig. 4A) but cannot grow on thiosulfate (Table 2), one can speculate that the inability of this strain to grow on thiosulfate is due to the absence of the *cpaBCEF-tadBC-pilD*-like genes. The GSB-type *sox* cluster is also present in *Cba. thiosulfatiphilum* DSMZ 249 (AY074395) from which plasmid pCL1 was isolated, and presumably is located on the chromosome (Verté et al., 2002). It is not clear from the genome sequence how *Chl. limicola* DSMZ 245 can grow on thiosulfate after receiving the pCL1 plasmid (Méndez-Alvarez et al., 1994). Strain DSMZ 245 does not have *sox* genes and has no other obvious candidate for a thiosulfate-metabolizing enzyme, other than a putative cytoplasmic rhodanese (see above) and the putative novel enzyme system FccAB-SoyYZ (see section III.H). Identification of the oxidation product of thiosulfate in the pCL1 transformants of *Chl. limicola* DSMZ 245 could help clarify the biochemical mechanism of its thiosulfate utilization.

H. A Potential Novel Complex: SoyYZ

The heterodimeric SoxYZ complex carries sulfur substrates on a conserved cysteine residue in the SoxY subunit (section III.G; Quentmeier and Friedrich, 2001). The *soxYZ* gene cluster has been duplicated in four GSB and is here denoted *soyYZ* (Table 2). (In *Chl. limicola* DSMZ 245, SoyY and SoyZ have the accession numbers EAM43192 and EAM43152, respectively.) A signal sequence at the amino termini of the SoyY sequences suggests that, like SoxYZ, SoyYZ is a periplasmic complex. Neither SoxZ nor SoyZ has a signal sequence, and both are probably transferred across the cytoplasmic membrane as part of complexes with SoxY or SoyY, respectively. The presence of *soyYZ* does not correlate with thiosulfate utilization (Table 2). In all GSB

752140 CT1017 *Cba. tepidum*	YRATREVKVTIGGCGG	SoxY
500217750 *Chl. phaeovibrioides*	YRAVREVKVTIGGCGG	
402131650 *Ptc. aestuarii*	YRATVDVAVTAQAC--	SoyY
402332320 *Chl. clathratiforme*	YVTSVQVEVVAQAC--	
500168720 *Chl. limicola*	YAASVHVDVLAQAC--	
400755570 *Chl. phaeobacteroides*	YRASVHINVLAQAC--	
626264690 *Alk. ehrlichei*	YRGASFVRIGAAGC--	

Fig. 6. Alignment of the C-terminal region of two SoxY proteins and all currently known SoyY proteins.

that have *soyYZ*, these genes are located immediately upstream of the *fccAB* genes in an apparent operon. Therefore, it is attractive to propose that SoyY and SoyZ form a complex in the periplasm that carries a sulfur substrate and that this complex reacts with the periplasmic FccAB flavocytochrome *c*. However, not all GSB that encode *fccAB* also encode *soyYZ*.

In most organisms of all taxonomic affiliations that have SoxY, the sulfur-substrate-carrying cysteine residue of SoxY is located at the C-terminus within the motif GGC(G_{1-2})–COOH. SoyY differs from all known examples of SoxY by having a C-terminus in which the putative sulfur-substrate-binding cysteine residue is the terminal residue (Fig. 6). The proximity of the C-terminal carboxyl group and the thiol group of the substrate-carrying cysteine residue in SoyY is likely to affect the chemistry at this site in a manner that does not occur in SoxY. If this is the case, this might explain the evolution of this particular motif in SoyY. In GSB the conserved motif in SoyY is VXAQAC-COOH. The *soyY* gene has only been found in one other organism other than GSB: the anaerobic, sulfide-oxidizing, chemoautotrophic *Alkalilimnicola ehrlichei* MLHE-1, which based on ribosomal RNA phylogeny is closely related to PSB of the *Ectothiorhodospiraceae* family. In *A. ehrlichei*, *soyY* (EAP35245) is located upstream of a *soxZ* homolog in a cluster of genes related to dimethyl sulfoxide utilization and cytochrome *c* biogenesis.

IV. Other Enzymes Related to Sulfur Compound Oxidation

A. RuBisCO-Like Protein

The enzyme RuBisCO (ribulose 1,5-bisphosphate carboxylase/oxygenase) catalyzes the key step in the Calvin–Benson–Bassham CO_2 fixation pathway in many phototrophic and chemotrophic organisms (Tabita, 1999). However, genome sequencing has revealed that some bacteria and archaea contain homologs of RuBisCO, called RuBisCO-like proteins (RLPs), which do not have the same enzymatic activity as *bona fide* RuBisCO. For example, in *Bacillus subtilis* RLP functions as a 2, 3-diketomethythiopentyl-1-phosphate enolase in the methionine salvage pathway of this organism (Ashida et al., 2003). All GSB genomes sequenced to date contain an RLP, but they do not have other recognizable genes for a methionine salvage pathway. A mutant of *Cba. tepidum* TLS lacking RLP (CT1772) has a pleiotropic phenotype with increased levels of oxidative stress proteins and defects in photopigment content, photoautotrophic growth rate, carbon fixation rates, and the ability to oxidize thiosulfate and elemental sulfur (Hanson and Tabita, 2001, 2003). Notably, sulfide oxidation is not affected in the *rlp* mutant of *Cba. tepidum* TLS. Hanson and Tabita subsequently suggested that RLP is involved in the biosynthesis of a low-molecular-weight thiol, which is essential for oxidation of thiosulfate and elemental sulfur. The possible role of such a hypothetical thiol as a carrier of sulfane sulfur is illustrated in Fig. 2. However, the function of GSB RLP is probably not limited to oxidation of inorganic sulfur compounds because *Chl. ferrooxidans* contains an RLP very similar to the RLP in other GSB, even though this organism cannot grow on inorganic sulfur compounds and does not contain genes thought to be involved in oxidation of thiosulfate (*sox* genes) and elemental sulfur (*dsr* genes) (Table 2).

B. Polysulfide-Reductase-Like Complexes

Three types of complexes, here denoted polysulfide-reductase-like complex 1, 2, and 3 (PSRLC1, PSRLC2, and PSRLC3), with sequence similarity to the characterized polysulfide reductase (PSR) in *Wolinella succinogenes* (Krafft et al., 1992) are found in the genome sequences of GSB (Table 2). The *W. succinogenes* PSR is encoded by the *psrABC* genes and consists of two periplasmic subunits, a molybdopterin-containing PsrA subunit and a [4Fe–4S]-cluster-binding PsrB subunit, and a membrane-anchoring PsrC subunit that binds an isoprenoid quinone and exchanges electrons with PsrB. In *W. succinogenes*, PSR and a hydrogenase allow respiration on polysulfide using H_2 as electron donor. However, homologs of PSR are also involved in metabolizing thiosulfate, tetrathionate, and other inorganic and organic compounds. Thus, the function of the PSR-like complexes in GSB cannot easily be established from sequence analysis alone.

Similar to the case of *W. succinogenes* PSR, polysulfide reductase-like complex 1 (PSRLC1, comprising CT0494, CT0495, and CT0496 in *Cba. tepidum* TLS) and PSRLC2 (comprising 400748340, 400748350, and 400748360 in *Chl. phaeobacteroides* DSM 266) are encoded by three genes. For both PSRLC1 and PSRLC2, the PsrA-like subunits with the catalytic site, have a Tat signal sequence and thus should be translocated into the periplasm. Homologs of PSRLC1 and PSRLC2 are found in many other organisms; for example, *Carboxydothermus hydrogenoformans* has a PSRLC1 homolog that has an overall amino acid sequence identity of approximate 50% with the PSRLC1 of *Cba. tepidum* TLS.

Two genes encode PSRLC3 in GSB. One is homologous with *psrA* (400751650 in *Chl. phaeobacteroides* DSMZ 266), and the other is homologous to a fusion of *psrB* and *psrC* (400751660 in *Chl. phaeobacteroides* DSMZ 266). Sequence analysis of the PsrBC-like subunit of PSRLC3 suggests that the PsrC-like domain has an orientation in the cytoplasmic membrane that is opposite that of the PsrC subunit of *W. succinogenes* PSR such that the PsrB-like domain of PSRLC3 is in the cytoplasm. In addition, the PsrA-like subunit of PSRLC3 does not have any obvious signal sequence. Thus, the catalytic PsrA-like catalytic subunit and the PsrB-like domain of PSRLC3 are probably located in the cytoplasm. Interestingly, the genes in GSB encoding PSRLC3 are immediately upstream of the *dsr* gene cluster. It is therefore an attractive possibility that PSRLC3 is involved in cytoplasmic oxidation of the sulfite produced by the Dsr system (section III.C). If so, PSRLC3 could provide all of the Dsr-containing GSB strains that lack the putative Sat-Apr-Qmo sulfite oxidation system (sections III.D–F, Fig. 2, Table 2) with a means to oxidize sulfite. Many known and puta-

tive prokaryotic sulfite oxidases are thought to be distantly related molybdopterin-containing enzymes that oxidize sulfite directly to sulfate (Kappler and Dahl, 2001). However, PSRLC3 is not widespread among other organisms, but a homologous complex with an overall amino acid sequence identity of approximate 50% is found in *Chloroflexus aurantiacus*, *Roseiflexus* sp. RS-1, and a few members of the high-GC *Firmicutes*.

C. Sulfhydrogenase-Like and Heterodisulfide-Reductase-Like Complexes

A putative cytoplasmic $\alpha\beta\gamma\delta$-heterotetrameric, bi-directional hydrogenase, which resembles *Pyrococcus furiosus* hydrogenase II that catalyzes H_2 production, H_2 oxidation, as well as the reduction of elemental sulfur and polysulfide to sulfide (Ma et al., 2000), is present in all sequenced GSB genomes, except those of *Chl. phaeovibrioides* DSM 265 and *Chl. luteolum* DSM 273. The genes encoding this putative sulfhydrogenase form a conserved *hyd1* cluster, *hydB1G1DA* (*CT1891–CT1894* in *Cba. tepidum* TLS), except in *Chl. chlorochromatii* CaD3 in which the genes are split into two clusters, *hydB1G1* and *hydDA*. Since *Cba. tepidum* TLS is unable to grow on H_2, these genes apparently do not confer the ability to oxidize large amounts of H_2. Likewise, the presence of this enzyme in *Chl. ferrooxidans* DSMZ 13031 and its absence from *Chl. phaeovibrioides* DSM 265 and *Chl. luteolum* DSM 273 suggests that its primary role is not related to elemental sulfur or polysulfide metabolism.

There are two types of complexes with sequence homology to heterodisulfide reductases encoded in the genomes of the sequenced GSB. One is the Qmo complex, which is probably involved in intracellular sulfite oxidation as discussed above (section III.F). The other is encoded by genes that form a conserved cluster with genes encoding a putative hydrogenase. This *hdr-hyd2* gene cluster, *hdrD-hdrA-orf1247-orf1248-hydB2-hydG2*, is conserved in seven of the sequenced strains (Fig. 4C). The *hdrD* gene in GSB is a fusion of the *hdrC* and *hdrB* genes found in other organisms. As with the Hyd1 complex mentioned above, the presence of Hdr-Hyd2 in *Chl. ferrooxidans* DSMZ 13031 and its absence from two other GSB strains, suggests that its presence is not essential in elemental sulfur or polysulfide metabolism.

V. Non-Sulfurous Compounds Oxidized for Growth

A. Hydrogen

Many strains of GSB can grow on H_2 as electron donor (Overmann, 2000; Garrity and Holt, 2001). In general, cultures growing on H_2 need to be supplemented with a small amount of a reduced sulfur compound (such as sulfide or thiosulfate), probably to fulfill biosynthetic needs. This is not the case for strains capable of assimilatory sulfate reduction (section VI). The genomes of seven sequenced GSB strains contain a *hupSLCD* gene cluster that encodes a HupSL-type Ni–Fe uptake hydrogenase, a membrane-bound HupC cytochrome *b* subunit and the HupD maturation protein (Table 2). *Cba. tepidum* TLS contains a similar gene cluster, but in this organism *hupS* and a part of *hupL* have been deleted (Frigaard and Bryant, 2003). This deletion probably explains why *Cba. tepidum* TLS cannot grow on H_2 (T.E. Hanson and F.R. Tabita, personal communication).

Warthmann et al. (1992) found that *Chl. phaeovibrioides* DSMZ 265 produces H_2 and elemental sulfur from sulfide or thiosulfate under diazotrophic conditions in the light. When this strain was grown syntrophically with the sulfur-reducing bacterium *Desulfuromonas acteoxidans*, 3.1 mol (78% of the theoretical maximum) of H_2 were produced in a nitrogenase-dependent manner per mole of acetate consumed. This high efficiency in comparison to other GSB strains is possibly due to the absence of the *hupSLCD* genes from the genome of the DSMZ 265 strain (Table 2).

No GSB genome encodes genes homologous to *hoxEFUYH*, which together encode the subunits of a putative bidirectional NAD-reducing hydrogenase found in some cyanobacteria and the purple sulfur bacterium *Thiocapsa roseopersicina* (Tamagnini et al., 2002; Rakhely et al., 2004).

B. Ferrous Iron

Chlorobium ferrooxidans DSMZ 13031 uses Fe^{2+} as the sole electron donor for growth (Heising et al.,

1999). This strain appears to have lost the ability to oxidize sulfur compounds because it does not grow on sulfide, elemental sulfur, or thiosulfate. This phenotype is largely confirmed by the absence of many genes related to oxidation of sulfur compounds in its genome (Table 2). Interestingly, *Chl. ferrooxidans* and *Chl. luteolum* DSMZ 273 are the only sequenced GSB whose genomes encode a bacterioferritin homolog (EAT59385 in *Chl. ferrooxidans*). Bacterioferritin binds heme *b* and non-heme iron and may be involved in the intracellular redox chemistry and storing of iron (Carrondo, 2003). *Chl. ferrooxidans* and *Chl. luteolum* DSMZ 273 also have a dicistronic operon encoding two *c*-type cytochromes that are not found in other GSB genomes. Although paralogs of the smaller cytochrome, a putative membrane-bound cytochrome of the c_5/c_{555} family, are observed in other GSB genomes, the larger, cytochrome is uniquely found in these two strains (EAT58010 in *Chl. ferrooxidans*). The N-terminal region of this protein bears a single *c*-type, heme-binding sequence (CAACH), and this domain has weak sequence similarity to several other *c*-type cytochromes. Since *Chl. ferrooxidans* and *Chl. luteolum* DSMZ 273 differ from other sequenced GSB by having bacterioferritin and an identical cluster of assimilatory sulfate reduction and sulfate permease genes, it is possible that both strains can grow with Fe^{2+} as the sole electron donor and by assimilatory sulfate reduction. Because the cytochromes mentioned above are also present only in *Chl. ferrooxidans* and *Chl. luteolum* DSMZ 273, they may be part of the Fe^{2+}-oxidizing enzyme system.

C. Arsenite

To our best knowledge, no GSB has been demonstrated to grow on arsenite. However, two GSB strains (*Chl. limicola* DSMZ 245 and *Prosthecochloris* sp. BS1) contain an enzyme consisting of a large molybdopterin-binding subunit and a small Rieske-type [2Fe–2S]-cluster-binding subunit (EAM42933 and EAM42934 in strain DSMZ 245, respectively) not found in any other genome-sequenced GSB strain. The small subunit contains a Tat signal peptide that presumably translocates the enzyme to the periplasm. The enzyme has high sequence similarity (an overall amino acid sequence identity of approximately 40%) with the well-characterized arsenite oxidase from *Alcaligenes faecalis* (Anderson et al., 1992). This enzyme oxidizes arsenite (AsO_3^{3-}) to arsenate (AsO_4^{3-}) and donates the electrons to soluble, periplasmic cytochrome *c*. The presumed primary function of the enzyme in many organisms is to detoxify arsenite. Although the concentration of arsenite in most natural environments is low, the reduction of periplasmic cytochrome *c* in GSB would contribute to the photosynthetic electron transport and thus the growth of the organism. It is also possible that the enzyme in GSB oxidizes a different substrate (e.g. nitrite to nitrate).

VI. Assimilatory Sulfur Metabolism

It is generally thought that GSB can not perform assimilatory sulfate reduction (Lippert and Pfennig, 1969). Nevertheless, *Chl. ferrooxidans* DSMZ 13031 grows with sulfate as the sole sulfur source and cannot utilize sulfide, thiosulfate, or elemental sulfur (Heising et al., 1999). Thus, *Chl. ferrooxidans* must be capable of assimilatory sulfate reduction. In agreement with this observation, the *Chl. ferrooxidans* genome encodes a single gene cluster that includes the assimilatory sulfate reduction genes *cysIHDNCG* and the sulfate permease genes *cysPTWA*, which are transcribed in opposite directions. These assimilatory sulfate reduction genes share a high degree of sequence similarity with those of the clostridia *Clostridium thermocellum* and *Desulfitobacterium hafniense*. However, sequence analyses show that the APS reductase encoded by *cysH* in *Chl. ferrooxidans* is the plant-type enzyme that uses APS and not PAPS as substrate. An identical *cys* gene cluster is observed in *Chl. luteolum* DSMZ 273, but these genes are not found in any other GSB genome. This raises the possibility that *Chl. luteolum* DSMZ 273 also is capable of assimilatory sulfate reduction and growth in the absence of reduced sulfur compounds using electron donors such as H_2 and Fe^{2+}.

VII. Evolution of Sulfur Metabolism

Thiosulfate utilization by the Sox system in GSB is an interesting case study of lateral transfer of an ability that presumably confers a strong competitive advantage in certain natural environments. The Sox proteins are only present in some GSB strains and have phylogenies that are incongruent with that for ribosomal RNAs (Figs. 1 and 5). This indicates that the *sox* genes in GSB were not inherited vertically from a common ancestor. However, the Sox proteins from GSB form a monophyletic cluster in phylogenetic analyses (Fig. 5), which strongly implies that the currently known *sox* genes in GSB have only been laterally exchanged within the GSB. In addition, all eight genes in the *sox* gene cluster of GSB (Fig. 4A) have congruent phylogenies (data not shown). This suggests that all eight *sox* genes were transferred simultaneously as one conserved cluster to each recipient strain. How might such a transfer have occurred? Two GSB strains, strains DSMZ 273 and DSMZ 265, which are very closely related in terms of ribosomal RNA phylogeny and genome organization, differ in one important respect: the latter strain contains the *sox* gene cluster whereas the former strain does not (Table 2). Analysis of the genome sequences reveals that the *sox* cluster in strain DSMZ 265 resides on an island that contains four additional genes (Fig. 7). This island appears to have been inserted into a region of the genome in a recent ancestor that was not involved in sulfur metabolism. This ancestor was likely similar to strain DSMZ 273 and unable to use thiosulfate. The genes on the island include a transposase (EAO15044), an integrase (EAO15046), and an RNA-directed DNA polymerase (EAO15045),

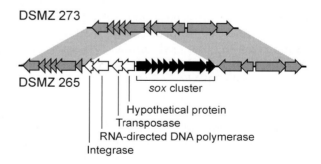

Fig. 7. Alignment of genomic regions from *Chl. luteolum* DSMZ 273, which cannot utilize thiosulfate, and *Chl. phaeovibrioides* DSMZ 265, which can utilize thiosulfate. Syntenic regions (i.e., regions having identical gene arrangements) are connected with gray trapezoids. Genes found in both organisms are shown in gray, *sox* genes are shown in black, and other genes are shown in white.

all of which are indicative of a mobile element. The RNA-directed DNA polymerase and the integrase are indicative of an RNA virus, and thus one possible scenario is that the island is a remnant structure derived from an RNA viral genome. The *sox* cluster could have been transferred into the viral genome by the transpose in a previous host and then integrated laterally into the genome of strain DSMZ 265. Highly similar RNA-directed DNA polymerases and integrases are found in several other GSB genomes in different genetic clusters, which suggests the existence of a GSB-specific RNA virus. To our knowledge, no virus of any kind that infects GSB has yet been isolated but there is no reason to believe such viruses do not exist.

Even though the GSB are closely related and exhibit limited apparent variation in physiology, it is obvious from genome sequence analyses that their gene contents are highly dynamic. For example, *Chl. clathratiforme* DSMZ 5477 and *Chl. ferrooxidans* DSMZ 13031 are closely related based on ribosomal RNA phylogeny (Fig. 1). However, whereas the former has the highest number of known and putative genes involved in oxidation of sulfur compounds among the GSB strains investigated, the latter has lost nearly all of these genes and in their place acquired the ability to reduce sulfate for assimilatory purposes (Table 2). On a similar note, *Chl. phaeobacteroides* DSMZ 266 has a single 3.1-Mbp chromosome and encodes roughly 1000 genes more than *Chl. phaeovibrioides* DSMZ 265 that only has a 2.0-Mbp chromosome (Table 1). These observations illustrate the dynamic structures of prokaryotic genomes and in addition demonstrate that organisms that are very closely related on the basis of their cellular core machinery nevertheless can have unexpected differences in physiology and life style.

Acknowledgements

N.-U.F. gratefully acknowledges support from the Danish Natural Science Research Council (grant 21–04–0463). D.A.B. gratefully acknowledges support for genomics studies from the United States Department of Energy (grant DE-FG02–94ER20137) and the National Science Foundation (grant MCB-0523100).

References

Anderson GL, Williams J and Hille R (1992) The purification and characterization of arsenite oxidase from *Alcaligenes faecalis*, a molybdenum-containing hydroxylase. J Biol Chem 267:23674–23682

Appia-Ayme C, Little PJ, Matsumoto Y, Leech AP and Berks BC (2001) Cytochrome complex essential for photosynthetic oxidation of both thiosulfate and sulfide in *Rhodovulum sulfidophilum*. J Bacteriol 183:6107–6118

Ashida H, Saito Y, Kojima C, Kobayashi K, Ogasawara N and Yokota A (2003) A functional link between RuBisCO-like protein of *Bacillus* and photosynthetic RuBisCO. Science 302:286–290

Bamford VA, Bruno S, Rasmussen T, Appia-Ayme C, Cheesman MR, Berks BC and Hemmings AM (2002) Structural basis for the oxidation of thiosulfate by a sulfur cycle enzyme. EMBO J 21:5599–5610

Beatty JT, Overmann J, Lince MT, Manske AK, Lang AS, Blankenship RE, Van Dover CL, Martinson TA and Plumley FG (2005) An obligately photosynthetic bacterial anaerobe from a deep-sea hydrothermal vent. Proc Natl Acad Sci USA 102:9306–9310

Bias U and Trüper HG (1987) Species specific release of sulfate from adenylyl sulfate by ATP sulfurylase or ADP sulfurylase in the green sulfur bacteria *Chlorobium limicola* and *Chlorobium vibrioforme*. Arch Microbiol 147:406–410

Boone DR and Castenholz RW (2001) Bergey's Manual of Systematic Bacteriology, 2nd edn., Vol 1, Springer, Berlin

Brune DC (1989) Sulfur oxidation by phototrophic bacteria. Biochim Biophys Acta 975:189–221

Brune DC (1995) Sulfur compounds as photosynthetic electron donors. In: Blankenship RE, Madigan MT, and Bauer CE (eds) Anoxygenic Photosynthetic Bacteria, pp 847–870, Vol 2,of Advances in Photosynthesis (Govindjee ed.), Kluwer Academic (now Springer), Dordrecht

Carrondo MA (2003) Ferritins, iron uptake and storage from the bacterioferritin viewpoint. EMBO J 22: 1959–1968

Dahl C, Engels S, Pott-Sperling AS, Schulte A, Sander J, Lübbe Y, Deuster O and Brune DC (2005) Novel genes of the dsr gene cluster and evidence for close interaction of Dsr proteins during sulfur oxidation in the phototrophic sulfur bacterium *Allochromatium vinosum*. J Bacteriol 187:1392–1404

Dahl C (2008) Inorganic sulfur compounds as electron donors in purple sulfur bacteria. In: Hell R, Dahl C, Knaff DB, and Leustek T (eds) Sulfur Metabolism in Phototrophic Organisms, in press, Vol xxvii of Advances in Photosynthesis and Respiration (Govindjee ed.), Springer, New York

Eisen JA, Nelson KE, Paulsen IT, Heidelberg JF, Wu M, Dodson RJ, Deboy R, Gwinn ML, Nelson WC, Haft DH,

Hickey EK, Peterson JD, Durkin AS, Kolonay JL, Yang F, Holt I, Umayam LA, Mason T, Brenner M, Shea TP, Parksey D, Nierman WC, Feldblyum TV, Hansen CL, Craven MB, Radune D, Vamathevan J, Khouri H, White O, Gruber TM, Ketchum KA, Venter JC, Tettelin H, Bryant DA and Fraser CM (2002) The complete genome sequence of *Chlorobium tepidum* TLS, a photosynthetic, anaerobic, green-sulfur bacterium. Proc Natl Acad Sci USA 99:9509–9514

Friedrich CG, Rother D, Bardischewsky F, Quentmeier A and Fischer J (2001) Oxidation of reduced inorganic sulfur compounds by bacteria: Emergence of a common mechanism? Appl Environm Microbiol 67:2873–2882

Friedrich CG, Bardischewsky F, Rother D, Quentmeier A and Fischer J (2005) Prokaryotic sulfur oxidation. Curr Opin Microbiol 8:253–259

Frigaard N-U and Bryant DA (2004) Seeing green bacteria in a new light: genomics-enabled studies of the photosynthetic apparatus in green sulfur bacteria and filamentous anoxygenic phototrophic bacteria. Arch Microbiol 182: 265–276

Frigaard N-U, Gomez Maqueo Chew A, Li H, Maresca JA and Bryant DA (2003) *Chlorobium tepidum*: Insights into the structure, physiology, and metabolism of a green sulfur bacterium derived from the complete genome sequence. Photosynth Res 78: 93–117

Frigaard N-U, Gomez Maqueo Chew A, Maresca JA and Bryant DA (2006) Bacteriochlorophyll biosynthesis in green bacteria. In: Grimm B, Porra R, Rüdiger W, and Scheer H (eds) Advances in Photosynthesis and Respiration, pp 201–221, Vol 25, Springer, Dordrecht

Garrity GM and Holt JG (2001) Phylum BXI. *Chlorobi* phy. nov. In: Boone DR and Castenholz RW (eds) Bergey's Manual of Systematic Bacteriology, 2nd edn., pp 601–623, Vol 1, Springer, New York

Griesbeck C, Hauska G and Schütz M (2000) Biological sulfide oxidation: Sulfide-quinone reductase (SQR), the primary reaction. In: Pandalai SG (ed) Recent Research Developments in Microbiology, pp 179–203, Vol 4, Research Signpost, Trivandrum, India

Griesbeck C, Schütz M, Schödl T, Bathe S, Nausch L, Mederer N, Vielreicher M and Hauska G (2002) Mechanism of sulfide-quinone reductase investigated using site-directed mutagenesis and sulfur analysis. Biochemistry 41:11552–11565

Hanson TE (2008) Proteome analysis of phototrophic sulfur bacteria with emphasis on sulfur metabolism. In: Hell R, Dahl C, Knaff DB and Leustek T (eds) Sulfur Metabolism in Phototrophic Organisms, in press, Vol xxvii of Advances in Photosynthesis and Respiration (Govindjee ed.), Springer, New York

Hanson TE and Tabita FR (2001) A ribulose-1, 5-bisphosphate carboxylase/oxygenase (RubisCO)-like protein from *Chlorobium tepidum* that is involved with sulfur metabolism and the response to oxidative stress. Proc Natl Acad Sci USA 98:4397–4402

Hanson TE and Tabita FR (2003) Insights into the stress response and sulfur metabolism revealed by proteome analysis of a *Chlorobium tepidum* mutant lacking the Rubisco-like protein. Photosynth Res 78:231–248

Heising S, Richter L, Ludwig W and Schink B (1999) *Chlorobium ferrooxidans* sp. nov., a phototrophic green sulfur bacterium that oxidizes ferrous iron in coculture with a "*Geospirillum*" sp. strain. Arch Microbiol 172:116–124

Hensen D, Sperling D, Trüper HG, Brune DC and Dahl C (2006) Thiosulphate oxidation in the phototrophic sulphur bacterium *Allochromatium vinosum*. Molec Microbiol, 62: 794–810

Imhoff JF (2003) Phylogenetic taxonomy of the family *Chlorobiaceae* on the basis of 16S rRNA and *fmo* (Fenna–Matthews–Olson protein) gene sequences. Intl J Syst Evol Microbiol 53:941–951

Imhoff JF (2008) Systematics of anoxygenic phototrophic bacteria. In: Hell R, Dahl C, Knaff DB and Leustek T (eds) Sulfur Metabolism in Phototrophic Organisms, in press, Vol xxvii of Advances in Photosynthesis and Respiration (Govindjee ed.), Springer, New York

Kappler U and Dahl C (2001) Enzymology and molecular biology of prokaryotic sulfite oxidation. FEMS Microbiol Lett 203:1–9

Kappler U, Bennett B, Rethmeier J, Schwarz G, Deutzmann R, McEwan AG and Dahl C (2000) Sulfite:cytochrome *c* oxidoreductase from *Thiobacillus novellus* – Purification, characterization, and molecular biology of a heterodimeric member of the sulfite oxidase family. J Biol Chem 275:13202–13212

Khanna S and Nicholas DJD (1983) Substrate phosphorylation in *Chlorobium vibrioforme* f. sp. *thiosulfatophilum*. J Gen Microbiol 129:1365–1370

Kirchhoff J and Trüper HG (1974) Adenylyl sulfate reductase of *Chlorobium limicola*. Arch Microbiol 100:115–120

Krafft T, Bokranz M, Klimmek O, Schroder I, Fahrenholz F, Kojro E and Kröger A (1992) Cloning and nucleotide-sequence of the *psrA* gene of *Wolinella succinogenes* polysulfide reductase. Eur J Biochem 206:503–510

Kredich NM (1996) Biosynthesis of cysteine. In: Neidhardt FC (eds) *Escherichia coli* and *Salmonella*, 2nd edn., Vol 1, ASM

Kumar S, Tamura K and Nei M (2004) MEGA3: Integrated software for molecular evolutionary genetics analysis and sequence alignment. Briefings Bioinformatics 5:150–163

Lippert KD and Pfennig N (1969) Die Verwertung von molekularem Wasserstoff durch *Chlorobium thiosulfatophilum*. Arch Microbiol 65:29–47

Ma KS, Weiss R and Adams MWW (2000) Characterization of hydrogenase II from the hyperthermophilic archaeon *Pyrococcus furiosus* and assessment of its role in sulfur reduction. J Bacteriol 182:1864–1871

Manske AK, Glaeser J, Kuypers MAM and Overmann J (2005) Physiology and phylogeny of green sulfur bacteria forming a monospecific phototrophic assemblage at a depth of 100 meters in the Black Sea. Appl Environ Microbiol 71:8049–8060

Méndez-Alvarez S, Pavón V, Esteve I, Guerrero R and Gaju N (1994) Transformation of *Chlorobium limicola* by a plasmid that confers the ability to utilize thiosulfate. J Bacteriol 176:7395–7397

Neumann S, Wynen A, Trüper HG and Dahl C (2000) Characterization of the *cys* gene locus from *Allochromatium vinosum* indicates an unusual sulfate assimilation pathway. Molec Biol Rep 27:27–33

Neuwald AF, Krishnan BR, Brikun I, Kulakauskas S, Suziedelis K, Tomcsanyi T, Leyh TS and Berg DE (1992) *cysQ*, a gene needed for cysteine synthesis in *Escherichia coli* K-12 only during aerobic growth. J Bacteriol 174:415–425

Overmann J (2000) The family Chlorobiaceae. In: The Prokaryotes: an Evolving Electronic Resource for the Microbiological Community, 3rd edn., release 3.1, Springer, New York, http://link.springer-ny.com/link/service/books/10125/

Overmann J (2008) Ecology of phototrophic sulfur bacteria. In: Hell R, Dahl C, Knaff DB, and Leustek T (eds) Advances in Photosynthesis and Respiration, Vol xxvii, Sulfur Metabolism in Phototrophic Organisms, in press, Springer, New York

Overmann, J, Cypionka H and Pfennig N (1992) An extremely low-light-adapted phototrophic sulfur bacterium from the Black Sea. Limnol Oceanogr 37:150–155

Pires RH, Lourenco AI, Morais F, Teixeira M, Xavier AV, Saraiva LM and Pereira IAC (2003) A novel membrane-bound respiratory complex from *Desulfovibrio desulfuricans* ATCC 27774. Biochim Biophys Acta 1605:67–82

Pott AS and Dahl C (1998) Siroheam sulfite reductase and other proteins encoded by genes at the dsr locus of *Chromatium vinosum* are involved in the oxidation of intracellular sulfur. Microbiology 144:1881–1894

Quentmeier A and Friedrich CG (2001) The cysteine residue of the SoxY protein as the active site of protein-bound sulfur oxidation of *Paracoccus pantotrophus* GB17. FEBS Lett 503:168–172

Rákhely G, Kovács AT, Maróti G, Fodor BD, Csanádi G, Latinovics D and Kovács KL (2004) Cyanobacterial-type, heteropentameric, NAD^+ -reducing NiFe hydrogenase in the purple sulfur photosynthetic bacterium *Thiocapsa roseopersicina*. Appl Environ Microbiol 70: 722–728

Reinartz M, Tschäpe J, Brüser T, Trüper HG and Dahl C (1998) Sulfide oxidation in the phototrophic sulfur bacterium *Chromatium vinosum*. Arch Microbiol 170:59–68

Sander J, Engels-Schwarzlose S and Dahl C (2006) Importance of the DsrMKJOP complex for sulfur oxidation in *Allochromatium vinosum* and phylogenetic analysis of related complexes in other prokaryotes. Arch Microbiol, 186: 357–366

Shahak Y, Arieli B, Padan E and Hauska G (1992) Sulfide quinone reductase (SQR) activity in *Chlorobium*. FEBS Lett 299:127–130

Smith AJ and Lascelles J (1966) Thiosulphate metabolism and rhodanese in *Chromatium* sp. strain D. J Gen Microbiol 42:357–370

Steinmetz MA and Fischer U (1982) Cytochromes, rubredoxin, and sulfur metabolism of the non-thiosulfate-utilizing green sulfur bacterium *Pelodictyon luteolum*. Arch Microbiol 132:204–210

Steinmetz MA and Fischer U (1985) Thiosulfate sulfur transferases (rhodaneses) of *Chlorobium vibrioforme f. thiosulfatophilum*. Arch Microbiol 142:253–258

Tabita F (1999) Microbial ribulose-1, 5-bisphosphate carboxylase/oxygenase: a different perspective. Photosynth Res 60:1–28

Tamagnini P, Axelsson R, Lindberg, P, Oxelfelt, F, Wünschiers R and Lindblad P (2002) Hydrogenases and hydrogen metabolism of cyanobacteria. Microbiol Mol Biol Rev 66:1–20

Theissen U, Hoffmeister M, Grieshaber M and Martin W (2003) Single eubacterial origin of eukaryotic sulfide:quinone oxidoreductase, a mitochondrial enzyme conserved from the early evolution of eukaryotes during anoxic and sulfidic times. Molec Biol Evol 20:1564–1574

Trüper HG and Pfennig N (1966) Sulphur metabolism in Thiorhodaceae. III. Storage and turnover of thiosulphate sulphur in *Thiocapsa floridana* and *Chromatium* species. Antonie van Leeuwenhoek 32:261–276

Van Gemerden H and Mas J (1995) Ecology of phototrophic sulfur bacteria. In: Blankenship RE, Madigan MT, Bauer CE (eds) Anoxygenic photosynthetic bacteria, pp 49–85, Vol 2 of Advances in Photosynthesis (Govindjee, ed.), Kluwer Academic (now Springer), Dordrecht

Verté F, Kostanjevecki V, De Smet L, Meyer TE, Cusanovich MA and Van Beeumen JJ (2002) Identification of a thiosulfate utilization gene cluster from the green phototrophic bacterium *Chlorobium limicola*. Biochemistry 41:2932–2945

Vogl K, Glaeser J, Pfannes KR, Wanner G and Overmann J (2006) *Chlorobium chlorochromatii* sp. nov., a symbiotic green sulfur bacterium isolated from the phototrophic consortium "*Chlorochromatium aggregatum*". Arch Microbiol 185: 363–372

Ward DM, Ferris MJ, Nold SC and Bateson MM (1998) A natural view of microbial biodiversity within hot spring cyanobacterial mat communities. Microbiol Molec Biol Rev 62:1353–1370

Warthmann R, Cypionka H and Pfennig N (1992) Photoproduction of H_2 from acetate by syntrophic cocultures of green sulfur bacteria and sulfur-reducing bacteria. Arch Microbiol 157:343–348

Genetic and Proteomic Studies of Sulfur Oxidation in *Chlorobium tepidum* (syn. *Chlorobaculum tepidum*)

Leong-Keat Chan, Rachael Morgan-Kiss, and Thomas E. Hanson*
*College of Marine and Earth Studies and Delaware Biotechnology Institute,
University of Delaware, 15 Innovation Way, Newark, DE 19711 USA*

Summary

The oxidation of reduced sulfur compounds is perhaps the most poorly understood physiological process carried out by the green sulfur bacteria (the *Chlorobiaceae*). My laboratory is testing models of sulfur oxidation pathways in the model system *Chlorobium tepidum* (ATCC 49652 syn. *Chlorobaculum tepidum* (Imhoff, 2003)) by the creation and analysis of mutant strains lacking specific gene products. The availability of a complete, annotated genome sequence for *C. tepidum* enables this approach, which will specify targets for biochemical analysis by indicating which genes are important in an organismal context. This is particularly important when several potentially redundant enzymes are encoded by

*Corresponding author, Phone: +(302) 831 3404, Fax: +(302) 831 3447, E-mail: tehanson@udel.edu

the genome for a particular reaction, such as sulfide oxidation. Additionally, we are using proteomics approaches to define the subcellular locations of proteins involved in sulfur oxidation pathways. The results produced by this research will refine models of anaerobic sulfur oxidation pathways and their integration into the global physiology of the *Chlorobiaceae*.

I. Introduction

The *Chlorobiaceae* are anoxygenic phototrophs characterized by the oxidation of reduced sulfur compounds as electron donors for photosynthetic electron transport and anabolic metabolism (Overmann, 2000). The *Chlorobiaceae* participate in global elemental cycles through carbon dioxide (CO_2) assimilation, dinitrogen (N_2) fixation, and the oxidation of reduced sulfur compounds (sulfide, thiosulfate and elemental sulfur) in anaerobic environments. The *Chlorobiaceae* thus provide a valuable ecosystem service: the transformation of compounds, like sulfide, that can be potently toxic to a wide variety of aerobic organisms, including humans, to relatively innocuous forms.

Features that make the *Chlorobiaceae* of general biological interest include metabolic pathways like the reductive tricarboxylic acid (rTCA) cycle of autotrophic CO_2 fixation (Buchanan and Arnon, 1990) and the formation of specific symbiotic associations with other bacteria (Overmann and van Gemerden, 2000). The existence of the rTCA cycle, now considered one of the most ancient metabolic cycles, was first proposed based on experiments with *Chlorobium limicola* f sp. *thiosulfatophilum* (DSM 249 syn. *Chlorobaculum thiosulfatophilum* (Imhoff, 2003)) (Evans et al., 1966). The *Chlorobiaceae* also possess unique quinones, carotenoids (Powls et al., 1968; Jensen et al., 1991; Takaichi et al., 1997; Cho et al., 1998), and a complex light harvesting apparatus called the chlorosome (Frigaard and Bryant, 2004) that is shared with the *Chloroflexaceae*, the green gliding bacteria. The chlorosome is thought to allow the *Chlorobiaceae* to grow at extremely low light intensities, which they appear to be uniquely suited for among phototrophs. Strains of *Chlorobiaceae* have been isolated from extremely light limited environments including deep sea hydrothermal vents (Beatty et al., 2005) and anoxic basins in the Black Sea (Overmann et al., 1992; Manske et al., 2005), supporting this hypothesis.

The wide distribution and gross physiological attributes of the *Chlorobiaceae* indicate that they are significant players in the global sulfur cycle, but few details are known about their direct contributions to that cycle or the enzymes involved. The *Chlorobiaceae* are found at high densities in diverse anoxic environments where dynamic sulfur cycling occurs including geothermal hot springs (Castenholz et al., 1990; Wahlund et al., 1991), freshwater hypolimnia (Butow and Bergsteinbendan, 1992; Baneras et al., 1999; Tuschak et al., 1999; Jung et al., 2000; Vila et al., 2002), and estuaries (Imhoff, 2001). Phylogenetic signatures similar to the *Chlorobiaceae* are found in the open ocean (Gordon and Giovannoni, 1996), another active sulfur transformation environment (Gonzalez et al., 1999), albeit an aerobic one. Clearly, understanding sulfur oxidation in the *Chlorobiaceae* has general implications for global sulfur cycling in anaerobic environments and biotechnological applications in wastewater and industrial gas stream treatment schemes (Kim and Chang, 1991; Basu et al., 1994; An and Kim, 2000; Henshaw and Zhu, 2001).

Prior studies of sulfur oxidation in the *Chlorobiaceae* have produced more contradictions than generalities (Paschinger et al., 1974; Shahak et al., 1992; Prange et al., 1999; Blöthe and Fischer, 2001; Prange et al., 2002; Verté et al., 2002). Some of these discrepancies may be explained by valid strain differences, but a coherent and detailed picture of the enzymes involved in sulfur oxidation in the *Chlorobiaceae* has yet to emerge. A summary of the state of knowledge of sulfur oxidation in *C. tepidum*, with appropriate comparisons to other systems follows. The conservation of sulfur oxidation

---

Abbreviations: IVTM – in vitro transposition mutagenesis; SQR – sulfide:quinone oxidoreductase; Dsr – dissimilatory sulfite reductase; Gm – gentamycin; PCR – polymerase chain reaction; MS – mass spectrometry; LMW – thiol low molecular weight organic thiol; CoA – coenzyme A

genes in multiple *Chlorobiaceae* strains as well as characterized strain preferences can be found in the chapter by Frigaard and Bryant elsewhere in this volume.

A.   Organization of Putative Sulfur Oxidation Genes in C. tepidum

Many of the predicted *C. tepidum* sulfur oxidation genes are physically associated on the genome as parts of gene and operon clusters, which we term Sulfur Islands, ranging from 3 to 26 genes in size (Fig. 1). Clustering may indicate a common evolutionary origin and functional significance for clustered genes. Potential transposase genes are associated with the Sox operon and Sulfur Island I, supporting the hypothesis that some of these genes were inherited as complete clusters by horizontal transfer (Eisen et al., 2002).

Two related models for *C. tepidum* sulfur oxidation pathways have been proposed based on the presence of between 52 and 94 genes (Fig. 1) encoding proteins that are recognizably similar to sulfur oxidation systems in other organisms (Eisen et al., 2002; Hanson and Tabita, 2003). These genes account for 2–4% of the 2,288 predicted genes encoded by *C. tepidum* (Eisen et al., 2002), a significant genetic investment for the organism. The lower bound is the number of open reading frames with recognizable similarity to sulfur oxidation genes in other microbes (shaded ORF's only in Fig. 1). The upper bound is the total number of genes that are associated with these recognized genes in sulfur islands (shaded and white ORF's in Fig. 1). The uncertainty in the number of genes for these pathways reflect the difficulty in correctly assigning gene or protein function in the absence of experimental verification. These numbers are also likely underestimates as they do not include genes of assimilatory sulfur metabolism or low molecular weight (LMW) thiol biosynthesis, which may play important roles in facilitating or regulating sulfur metabolism in *C. tepidum*. The genes for the former are fairly obvious, while the latter are completely unknown probably owing to the fact that the *Chlorobiaceae* contain structurally novel LMW thiols (Fahey et al., 1987; Fahey, 2001). This suggests that model systems like *C. tepidum* have much to tell us about microbial

processes controlling biogeochemical cycles (Friedrich et al., 2001).

When it was isolated (Wahlund et al., 1991), *C. tepidum* was found to utilize sulfide and thiosulfate as electron donors for photosynthesis. A typical profile of these compounds as well as elemental sulfur through a batch culture growth curve clearly shows that *C. tepidum* consumes sulfide first, followed by elemental sulfur and finally thiosulfate (Fig. 2). Sulfate accumulation only begins once the oxidation of elemental sulfur has started, indicating that it is an obligate intermediate in the conversion of sulfide to sulfate in *C. tepidum*. Brief discussions of the *C. tepidum* genes potentially responsible for the oxidation of sulfide, elemental sulfur and thiosulfate follow.

1.   Sulfide Oxidation

C. tepidum oxidizes sulfide (HS^-) to elemental sulfur (S^0) that is accumulated extracellularly (Brune, 1989; Brune, 1995). Two enzymes are likely candidates for sulfide oxidation, sulfide:quinone oxidoreductase (SQR) and a flavocytochrome *c* sulfide dehydrogenase. The *C. tepidum* genome contains genes for both activities and functionality for both has been implicated by different studies in different strains of *Chlorobiaceae* (Shahak et al., 1992; Verté et al., 2002).

Studies of other phototrophs have failed to provide evidence for flavocytochrome *c* activity in sulfide oxidation. A mutant strain of the purple sulfur bacterium *Allochromatium vinosum* lacking flavocytochrome *c* was competent for sulfide oxidation via SQR (Reinartz et al., 1998). Furthermore, *Rhodobacter capsulatus* contains SQR, but no flavocytochrome *c*, and efficiently oxidizes sulfide (Schütz et al., 1997; Schütz et al., 1999). Conversely, *Rhodopseudomonas palustris*, which contains flavocytochrome *c* and lacks SQR (Larimer et al., 2004), is inhibited by very low levels of sulfide (>0.25 mM) (Hansen and van Gemerden, 1972). These observations suggest that SQR is likely the most important route for sulfide oxidation in *C. tepidum* (Fig. 1). *C. tepidum* has been reported to require sulfide for early stages of growth both in batch (Wahlund et al., 1991) and reactor scale (Mukhopadhyay et al., 1999) cultures unlike other *Chlorobiaceae* that can utilize hydrogen gas or ferric iron (Fe^{2+}) as electron donors (Heising et al., 1999).

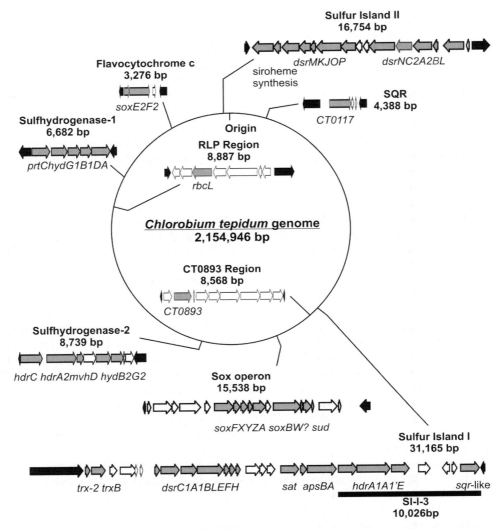

Fig. 1. Sulfur oxidation genes and their locations on the *C. tepidum* genome. The *C. tepidum* genome is depicted as a circle with the origin of replication (nucleotide 1 of the genomic sequence) at the top of the circle. The locations of sulfur oxidation gene clusters are indicated as are the structure of each cluster. Open reading frames are represented by arrows. Grey arrows are ORF's with discernible homology to sulfur oxidizing gene products in other microbes, black arrows indicate the borders of sulfur oxidation gene clusters, and white arrows indicate ORF's with unknown functions in sulfur oxidation. Shaded genes inside the genome have been experimentally implicated in sulfur oxidation in *C. tepidum*, while those outside the genome have been identified strictly by homology.

2. *Elemental Sulfur Oxidation*

When sulfide levels are low, the *Chlorobiaceae* oxidize extracellular elemental sulfur, the product of sulfide oxidation, to sulfate. The identity of enzymes mediating this process in the *Chlorobiaceae* is not clear (Brune, 1989; Brune, 1995; Pickering et al., 2001; Prange et al., 2002), though speculation exists. The *C. tepidum* genome anno-

tation (Eisen et al., 2002) predicts that a sulfhydrogenase/polysulfide reductase activity will oxidize H_2 and reduce elemental sulfur to sulfide for subsequent oxidation via SQR (Ma et al., 1993). However, *C. tepidum* is normally grown in the absence of H_2 under which conditions it is proficient at elemental sulfur oxidation (Wahlund et al., 1991; Hanson and Tabita, 2001). Furthermore, studies on *C. limicola* suggest that elemental sulfur

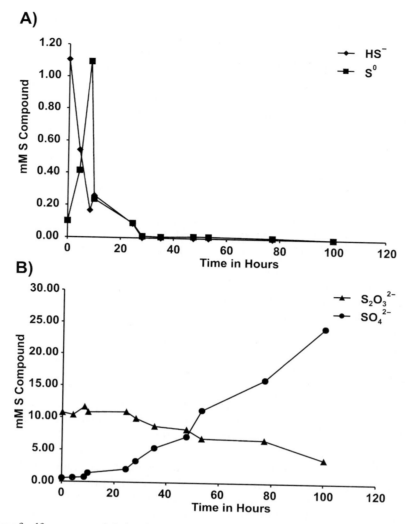

Fig. 2. Typical profiles of sulfur compounds in batch cultures of *C. tepidum*. The concentrations of each of four compounds are plotted as a function of time in a photomixotrophic culture of *C. tepidum* WT2321. Data points are the average of three independent cultures. (A) Sulfide (HS^-) and elemental sulfur (S^0). (B) Thiosulfate ($S_2O_3^{2-}$) and sulfate (SO_4^{2-}). Note that sulfate is much higher than other compounds as two moles of sulfate are produced for each mole of thiosulfate oxidized.

is disproportionated to sulfide and sulfate in a light dependent reaction that is independent of H_2 as a source of reductant (Paschinger et al., 1974).

Other possible routes for phototrophic oxidation of elemental sulfur exist in anoxygenic phototrophic bacteria. Gene products of the dissimilatory sulfite reductase (Dsr) system have been genetically implicated in elemental sulfur oxidation in the purple sulfur bacterium *Allochromatium vinosum* (Pott and Dahl, 1998; Dahl et al., 2005). *C. tepidum* encodes homologs of this system including a duplication of the *dsrCABL*

genes (Eisen et al., 2002). One copy of the *dsr* operon along with genes encoding ATP sulfurylase, APS reductase, heterodisulfide reductase, thioredoxin reductase and thioredoxin homologs constitute Sulfur Island I, one of a number of gene clusters encoding potential sulfur oxidizing activities (Fig. 1). A more detailed discussion on the involvement of the *dsr* genes in sulfur oxidation by the *Chromatiaceae* is given by Dahl elsewhere in this volume.

Low molecular weight thiols (LMW thiols) have been postulated to be involved in elemental sulfur oxidation in phototrophic bacteria that contain Dsr complexes (Brune, 1989; Brune, 1995). Specifically, glutathione amide appears to cycle between the thiol (R-SH) and perthiol (R-S-SH) forms when *A. vinosum* oxidizes stored elemental sulfur globules (Bartsch et al., 1996). A similar process is proposed for *C. tepidum*, but the structure of the thiols and details of their involvement is still unclear (see section B.3 of this chapter).

Chlorobium limicola and *Chlorobium phaeobacteroides* have also been reported to use the metal sulfides MnS and FeS as growth substrates (Borrego and Garcia-Gil, 1995). Metal sulfides are analogous to elemental sulfur in that they are largely insoluble. This capability would presumably enable survival in sulfide poor environments. It is not clear whether *C. tepidum* has this activity or not, nor what genes/proteins would be required to carry out this function.

3. Thiosulfate Oxidation

Some of the *Chlorobiaceae*, including *C. tepidum*, also oxidize thiosulfate ($S_2O_3^{2-}$) to sulfate (Overmann, 2000). The *C. tepidum* genome contains a thirteen gene cluster encoding a system similar to the well characterized Sox sulfur oxidation system of *Paracoccus pantotrophus* strain GB17 (Fig. 1) (Friedrich et al., 2001). However, *C. tepidum* lacks genes encoding the SoxCD protein complex. In *P. pantotrophus*, SoxCD acts as a sulfur dehydrogenase on a SoxY-bound sulfur atom liberating six electrons for use in anabolism or energy conservation (Quentmeier and Friedrich, 2001). As six electrons represents 75% of the available reducing power from thiosulfate, it seems likely that *C. tepidum* possesses an alternative mechanism for the complete oxidation of SoxY-bound sulfur atoms.

B. Gaps and Redundancies in Proposed Sulfur Oxidation Pathways

1. Extracellular Elemental Sulfur Oxidation

Extracellular elemental sulfur is oxidized by *C. tepidum* during growth leading to the production of sulfate. This is a challenging biochemical problem in that elemental sulfur is extremely hydrophobic and insoluble. Clearly, a mechanism

must exist that allows *C. tepidum* to access this relatively rich source of reductant, but there are no obvious candidate proteins for either the activation or transport of extracellular elemental sulfur encoded by the *C. tepidum* genome (Eisen et al., 2002). This is a particularly difficult problem if elemental sulfur oxidation proceeds via a periplasmic sulfhydrogenase, as was proposed in the genome annotation.

A similar problem conceptually is the reduction of insoluble electron acceptors such as elemental sulfur, metallics and/or humics by anaerobic and facultatively anaerobic microbes. At least two general strategies have evolved to account for this conundrum (Lloyd, 2003). The first is direct electron transfer from the cell envelope to the acceptor as has been proposed for various metal reducing bacteria. In *Shewanella oneidensis* MR-1, this process is thought to involve cytochromes that span the outer membrane of the organism and provide a direct path for electrons from cytoplasmic or periplasmic reductants to the acceptor (Myers and Myers, 2001). A prediction of this mechanism is that cells should be in intimate physical contact with the substrate to facilitate direct electron transfer. In the case of *C. tepidum*, physical association with elemental sulfur would also facilitate capture of the reduced product, sulfide or polysulfide. However, our observations do not support this proposed strategy of tight associations between the cell and elemental sulfur, since *C. tepidum* does not appear to frequently or conspicuously associate with sulfur globules in cultures that are actively oxidizing elemental sulfur (Chan and Hanson, unpublished).

The second strategy for acquisition of these insoluble sources of sulfur and reducing power is the use of electron shuttling compounds to mediate extracellular electron transfer at sites distant from the cell. One particular example among several is the apparent use of phenazine antibiotics by some soil microbes to facilitate electron transfer to metals or other soil components (Hernandez et al., 2004). A more recent study reported that *S. oneidensis* MR-1 also has the ability to reduce iron at a distance as well as by direct contact (Lies et al., 2005). This concept could be translated to elemental sulfur oxidation via a low molecular weight thiol that could react with exposed –SH or –S-S- groups on the surface of elemental sulfur globules to generate a linear polysulfide that is tagged with an organic molecule. The organic could then be used as a specific recognition tag for cell surface receptors, which could in effect reserve the elemental resource for use only by organisms that have the specific receptor.

Neither the direct contact nor action at a distance model has been conclusively tested in the *Chlorobiaceae*, but the subcellular fractionation approach discussed in Section III below will test the direct electron transfer model by examining the outer membrane proteome of *C. tepidum* for potential electron transfer proteins.

2. Missing and Duplicated Genes

Duplicated genes include subunits of sulfhydro-genase (*hydBG*, CT1249-50 and CT1891-92), heterodisulfide reductase (*hdrA*, CT0866 and CT1246), sulfide:quinone reductase (CT0117, CT0876, CT1087), polysulfide reductase C and B subunits (CT0495-96 and CT2241-40) and a long duplication of Dsr complex subunits stretching >5 kb at > 99.5% DNA sequence identity (*dsrCABL*, CT0851-54 and CT2250-46). However, one of these copies (CT2250-46) contains an authentic frame shift in the *dsrB* gene (Eisen et al., 2002). Whether both copies of *dsrCABL* are expressed differentially or are indeed functional is currently unknown. Other duplicated or trip-licated genes appear to encode viable gene products, though their functions have not been experimentally verified. In particular, the SQR homologs appear to have diverged significantly from one another, raising the question of whether all three can be involved in sulfide oxidation (Chan and Hanson, unpublished data).

Some genes are unexpectedly missing includ-ing the previously discussed *soxCD* and *shxV* (Friedrich et al., 2001). The *shxV* gene in *P. pan-totrophus* encodes a product similar to CcdA proteins that are involved in the maturation of periplasmic cytochromes. A *P. pantotrophus* *shxV* mutant is unable to oxidize either thiosul-fate or molecular hydrogen, but is not generally defective for periplasmic cytochrome biogenesis (Bardischewsky and Friedrich, 2001). A substi-tute function for ShxV in *C. tepidum* could be encoded by CT1559. The CT1559 gene product is similar to the plastid encoded *ccsA/ycf5* gene product of *Chlamydomonas reinhardtii*, which was

genetically shown to be involved in cytochrome maturation (Xie and Merchant, 1996). CT1559 is not associated with any other predicted sulfur oxidation genes on the *C. tepidum* genome (Eisen et al., 2002).

3. Other Aspects

The model of sulfur oxidation put forward with the *C. tepidum* genome annotation does not propose any mechanism for the transport and assimilation of reduced sulfur or thiosulfate into methionine and cysteine (Eisen et al., 2002). As *C. tepidum* does not require these amino acids for growth, this process must occur. Cysteine synthase activity has been directly assayed in *C. tepidum* (Hanson and Tabita, 2001) and sufficient levels of activity were found to account for the ability to grow in medium lacking cysteine supplementation. Genes for both methionine and cysteine biosynthesis from sulfide and precursor metabolites have been identified (Eisen et al., 2002). Presumably the sulfide for biosynthesis must be generated in the cytoplasm, but reductive sulfate assimilation has never been observed in the *Chlorobiaceae* (Over-mann, 2000) suggesting transport of sulfide by an unknown mechanism as a route for providing this nutrient. It is unclear whether or not cysteine or methionine biosynthetic activities are regulated by exogenous substrate.

Models of sulfur oxidation in *C. tepidum* predict that sulfate will be generated in the cytoplasm. Sulfate accumulates extracellularly (Fig. 2) in *C. tepidum* cultures, implying a transport mechanism as the divalent anion is not likely to cross the cytoplasmic membrane by diffusion. However, the *C. tepidum* genome lacks a clearly identifiable sulfate transporter (Eisen et al., 2002). The lack of a recognizable sulfate transporter also supports the notion that sulfide for sulfur amino acid biosynthesis may be transported from outside the cell when it is available.

As noted briefly above, the *Chlorobiaceae* also lack clearly identifiable genes for the biosynthesis of typical redox buffering LMW thiol compounds, with the exception of CoA, which is used by some Gram positive bacteria. Genes have been identified for the biosynthesis of glutathione, mycothiol, and ergothioneine in various microbes (Fahey, 2001), and these have been used to search the *C. tepidum* genome for homologs without any success (Hanson, unpublished data). *C. limicola* has been shown to contain a LMW thiol named U11, for an unknown compound with a retention time of 11 minutes in an HPLC separation, that does not correspond to the retention time of common thiol compounds (Fahey et al., 1987). *C. tepidum* appears to contain U11 and as many as four other novel LMW thiols (Hanson, unpublished data). Thus, LMW thiol biosynthetic genes should be part of a *Chlorobiaceae* specific suite of genes shared amongst these genomes, but not by other microbes. It will be interesting to see what roles these compounds play in the function and regulation of sulfur oxidation pathways in the *Chlorobiaceae*.

II.　Genetic Studies

A.　*Genetics in the* Chlorobiaceae

The ability to genetically manipulate members of the *Chlorobiaceae* has been developed to a quite useful level, but it is not by any means mature. It will become increasingly important to develop additional techniques for subtle genetic manipulations and ectopic gene expression as more genomic information is rapidly developed for this group of organisms (see the chapter of Frigaard and Bryant in this volume).

The first instance of genetic manipulation in the *Chlorobiaceae* was reported by John Ormerod in 1987 at the EMBO Workshop on Green and Heliobacteria in Nyborg, Denmark (Ormerod, 1988). This involved the transfer of a spontaneously arising streptomycin resistance allele between two strains of *Chlorobium limicola* strain Tassajara (DSM 249 syn. *Chlorobaculum thiosulfatiphilum* (Imhoff, 2003)) and 8327 (DSM 263 syn. *Chlorobaculum parvum* (Imhoff, 2003)) by natural transformation (the ability of certain microbes to assimilate exogenously provided DNA into their genome by homologous recombination).

Natural transformation is still the preferred method of genetic manipulation the *Chlorobiaceae* and has been used to inactivate genes encoding chlorosome proteins in *Chlorobium vibrioforme* as well (Chung et al., 1998). The ability to inactivate specific genes in *C. tepidum* was reported in 2001 in two separate reports describing the insertional

inactivation of genes encoding a nitrogenase subunit (Frigaard and Bryant, 2001) and a RubisCO-like protein (Hanson and Tabita, 2001). Subsequently, gene inactivation has been used with great effect to delineate pathways of bacteriochlorophyll and carotenoid biosynthesis in *C. tepidum* (Vassilieva et al., 2002a; Frigaard et al., 2004a; Frigaard et al., 2004b; Maresca et al., 2004). While gene by gene inactivation is certainly useful, additional genetic approaches to increase the rate of analysis of the *C. tepidum* genome need to be developed and one of these is briefly described below.

B.　*In Vitro Transposition Mutagenesis of* C. tepidum *Sulfur Islands*

To take advantage of the large gene clusters encoded by the Sulfur Islands evident in the *C. tepidum* genome (Fig. 1), we have employed in vitro transposition mutagenesis (IVTM). IVTM is a molecular genetic technique that constructs a library of DNA fragments each containing a unique transposon insertion. By using specific fragments of the *C. tepidum* genome produced by PCR, a transposon mutant library can be constructed that is tightly focused on a specific genomic region. Each transposon insertion that occurs in a gene creates a knock-out mutation of that gene, rendering it incapable of producing the protein it normally encodes. Transfer of these mutated fragments back into *C. tepidum* creates mutant strains lacking specific proteins, which can then be analyzed for defects in sulfur oxidation. IVTM was originally developed for the genetic analysis of various enteric and mycobacteria (Rubin et al., 1999; Wong and Mekalanos, 2000) and has been utilized in the genetic analysis of various other microbes as reviewed by Hayes (Hayes, 2003). IVTM has now been commercialized and transposases are available from several companies. IVTM should be adaptable to insert nearly any fragment of DNA that can be PCR amplified as the only requirement is that the transposase recognition sites flank the section of DNA to be inserted. The transposase recognition sites are easily added by including them at the ends of the PCR primers used to amplify the fragment of interest.

C. tepidum Sulfur Island I (Fig. 1) is a 33 kb section of the genome that contains sixteen genes encoding for potential sulfur oxidation enzymes: thioredoxin and thioredoxin reductase, *trx2* (*CT0841*) and *trxB* (*CT0842*), dissimilatory sulfite reductase and functionally associated proteins, *dsrC1A1BLEFH* (*CT0851-57*), sulfate adenylyltransferase, *sat* (*CT0862*), adenylylsulfate reductase, *apsBA* (*CT0864-65*, known as *aprAB* in many other organisms), heterodisulfide reductase, *hdrA-1A-1'E* (*CT0866-68*), and a sulfide:quinone oxidoreductase (*sqr*)-like (*CT0876*) homolog. Sulfur Island I contains twenty one ORF's encoding hypothetical and conserved hypothetical proteins (*CT0843-50*, *CT0858-61*, *CT0863*, *CT0869-75*, and *CT0877*), well more than half of the thirty seven ORF's in this region. In addition, a predicted transposase is encoded by *CT0878* suggesting that this region of the *C. tepidum* genome has been modified by lateral transfer.

An ~11kb subsection of Sulfur Island I (SI-I-3, Fig. 3) containing genes *CT0866-CT0876* was chosen to provide a proof of concept that IVTM could be successfully applied to *C. tepidum*. This fragment was amplified by PCR and cloned into *E. coli* followed by IVTM using a synthetic transposon, TnOGm, produced from pTnMod-OGm (Dennis and Zylstra, 1998) by PCR. This transposon was selected because it carries the *aacC1* gentamycin resistance marker that had previously been used to disrupt the *C. tepidum nifD* gene (Frigaard and Bryant, 2001). TnOGm also carries a conditional origin of replication that will allow self cloning of TnOGm insertions in future experiments and is one of a large series of such transposons with varying resistance markers and structures available from Gerben Zylstra at Rutgers University (Dennis and Zylstra, 1998).

Following cloning of the SI-I-3 fragment, IVTM was performed with TnOGm and a library of mutated plasmids was recovered in *E. coli* by virtue of the Gm resistance marker of TnOGm. DNA was purified from this library and then used to transform *C. tepidum*. Randomly selected gentamycin resistant colonies of *C. tepidum* were screened by PCR of the SI-I-3 region to verify that TnOGm insertion had occurred and broadly localize the site of transposon insertion. The specific site of TnOGm insertion in mutant strains displaying growth defects was analyzed by sequencing out from TnOGm in the SI-I-3 PCR product produced from each strain (Fig. 3A). Strains AK and AJ contained transposon

insertions in gene *CT0868* encoding an HdrE homolog and *CT0867* encoding a fusion of HdrA and HdrD proteins called HdrA-1', respectively. These genes will certainly be renamed in light of recent work on homologs in the sulfate reducer *Desulfovibrio desulfuricans* described in the next section. Genes from 3' of *hdrA-1'* to the 5' end of the *sqr*-like ortholog (*CT0867-76*) were deleted and replaced with TnOGm in strain C5. Details of the IVTM procedures and conditions will be described fully elsewhere (Chan et al., manuscript in preparation).

C.   Physiological Characterization of Mutant Strains

Initial characterization of IVTM generated SI-I-3 mutant strains indicated that the Gm resistance marker in TnOGm appears to be temperature sensitive (Fig. 3B). Both single insertion mutant strains are incapable of growth at 48°C in the presence of Gm, while the insertion deletion strain C5 is clearly defective for growth at 48°C in the presence or absence of Gm. All mutant strains grew well with or without Gm at 42°C indicating that the Gm marker was functional. The wild type was incapable of growth in the presence of Gm at either 42 or 48°C. The temperature sensitivity of a selectable marker can complicate some physiological analyses and will need to be addressed for each marker that will be used in *C. tepidum*. Because of this, physiological measurements are routinely made on cells grown at 48°C, the reported optimum growth temperature for *C. tepidum*, in the absence of Gm selection. Despite the absence of antibiotic selection, the genotype of insertion mutants including the TnOGm mutants is stable under these conditions (Chan et al., manuscript in preparation). Starter cultures for physiological analysis and those for the preparation of frozen permanents are always grown at 42°C under appropriate selection.

Strain C5, the insertion-deletion strain, displayed a markedly temperature sensitive phenotype exhibiting comparable growth in comparison to the wild type at 42°C, but exhibiting a significant growth deficiency at a growth temperature of 48°C (Fig. 3B). The single mutant strains both had more subtle phenotypes, but also displayed reproducible growth yield defects compared to the wild type. All mutant strains

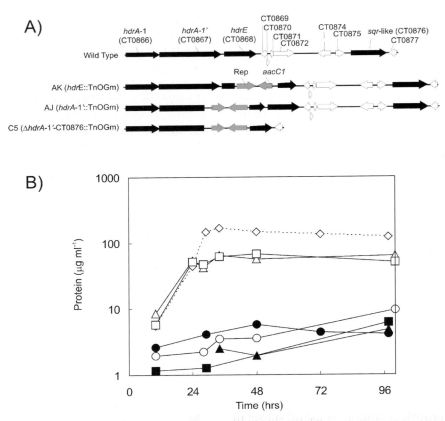

Fig. 3. Single mutations in SI–I–3 result in growth yield defects, while an insertion–deletion results in drastic reductions in growth rate and growth yield. A. Physical map of SI–I–3 (see Fig. 1 for context) and genotypes of relevant mutants. ORF's are depicted as arrows and gene names provided. The genes *hdrA-1*, *hdrA-1'* and *hdrE* encode proteins similar to subunits of the QmoABC complex of *D. desulfuricans* as discussed in the text. White arrows depict ORF's encoding hypothetical and conserved hypothetical proteins, while grey arrows depict regions of the transposon, TnOGm. B. Growth of wild type and SI-I-3 mutant strains. Strains were grown in acetate containing Pf-7 medium as previously described (Hanson and Tabita, 2001) in the presence (filled symbols) or absence (open symbols) of 4 µm Gm ml⁻¹. Symbols: wild type ◊, strain AJ □, strain AK Δ, insertion–deletion strain C5 ○.

exhibited defects in sulfur oxidation, particularly in thiosulfate oxidation. The genes affected include homologs of a membrane protein complex, the quinone interacting membrane bound oxidoreductase (Qmo) as named in *Desulfovibrio desulfuricans* (Pires et al., 2003). In *C. tepidum,* the Qmo homologs (*CT0866-0868*, referred to as *hdrA-1*, *hdrA-1'*, and *hdrE* above and in Fig. 3A) are part of Sulfur Island I and the results indicate that these genes are important for optimal growth of *C. tepidum*. The insertion-deletion strain C5 has also lost a number of hypothetical open reading frames and the first part of a gene encoding one of three SQR homologs in *C. tepidum*. These mutant strains provide the first experimental evidence for specific sulfur oxidation genes being

important to the overall function of the *Chlorobiaceae*. The detailed characterization of these strains and their phenotypes is being prepared for publication (Chan et al., manuscript in preparation). While the precise function of the Qmo complex is still not known, it appears to be present in sulfate reducing bacteria and archaea and it has been suggested that it may function to mediate electron transfer between the membrane quinone pool and APS reductase and thereby contribute to energy conservation.

The results taken together strongly suggest that IVTM is a viable route for making insertion mutations in tightly focused regions of the *C. tepidum* genome providing another tool for the genetic dissection of metabolism in the *Chloro-*

biaceae to complement gene by gene inactivation. The development of additional tools, including complementation of mutations, heterologus gene expression, reporter gene development, and others are actively being pursued and should provide yet more tools to understand the biology of these organisms. *C. tepidum* will apparently maintain certain broad host range plasmids (Wahlund and Madigan, 1995) indicating that mutant complementation by ectopic gene expression should be feasible, although this has yet to be demonstrated.

III. Proteomic Studies

A. *Proteomics in the* Chlorobiaceae

Proteomics can be broadly defined as the determination of the entire suite of proteins synthesized by a population of cells in response to prevailing environmental conditions. In practice, the entire suite of proteins is rarely resolved and usually a subset is examined to determine the identity of specific proteins that are differentially synthesized in cells faced with a particular challenge or growth condition. A useful subset of proteomics is subcellular proteomics, or defining the protein complement of specific subcellular components.

The primary tools of proteomics since the coining of the term in 1995 (Patterson and Aebersold, 2003) have been two dimensional polyacrylamide gel electrophoresis (2D PAGE), gel image analysis, and mass spectrometric (MS) methods of protein identification. It should be noted that 2D-PAGE in particular has a long and proven track record in defining protein complements before the term proteomics was coined, for example, see the work of Neidhardt and co-workers in the 1970s and 1980s (Bloch et al., 1980; Phillips et al., 1980; Neidhardt et al., 1989). More recently, 2D-PAGE is being supplanted by liquid chromatographic separations of proteolytically digested cellular extracts followed by direct MS identification of specific peptides (Patterson and Aebersold, 2003). Isotopic and dye based peptide and protein labeling systems have also been developed to facilitate quantitative comparisons of proteomic samples (Patton, 2002).

2D-PAGE has only recently been applied to *C. tepidum* to analyze the proteome under standard growth conditions. Tsiotis and co-workers have examined both the membrane proteome (Aivaliotis et al., 2004a, 2004b, 2006a) as well as cytoplasmic protein complexes (Aivaliotis et al., 2006b). Subcellular proteomics has been applied in a focused manner to examine the protein complement of the chlorosome in *C. tepidum* (Chung et al., 1994; Bryant et al., 2002; Vassilieva et al., 2002b). Both 2D-PAGE and subcellular proteomics have been utilized to study widespread changes in the *C. tepidum* proteome resulting from the loss of the RLP (Hanson and Tabita, 2003).

B. *Subcellular Fractionation of* C. tepidum *to Identify Secreted and Outer Membrane Proteins*

Starting from existing protocols for subcellular fractionation of *C. tepidum* (Vassilieva et al., 2002b; Hanson and Tabita, 2003), my laboratory is now attempting to specifically isolate both the secreted and outer membrane proteome of *C. tepidum* to identify proteins involved in elemental sulfur oxidation. As outlined above, the outer membrane may contain electron transfer proteins if direct cellular contact is the primary mode of elemental sulfur oxidation. Secreted proteins may be involved in either the direct contact or action at a distance model of sulfur oxidation. Genes encoding proteins identified by this approach will be analyzed for their expression patterns followed by mutagenesis as described above.

Prior membrane proteome studies of *C. tepidum* have provided a limited number of outer membrane protein identifications, but ~70% of proteins previously identified in membrane preparations are known or predicted to be cytoplasmic (Aivaliotis et al., 2006a). Acidic glycine extraction to selectively enrich outer membrane proteins was somewhat more successful with ~37% of proteins identified as being either periplasmic or outer membrane proteins.

A variety of methods have been used over the last 30 years to separate inner and outer membrane protein fractions from Gram negative bacteria (Mizuno and Kageyama, 1978; Shen et al., 1989; Buonfiglio et al., 1993; Link et al., 1997; Sagulenko et al., 2001; Yarzabal et al., 2002; Baik

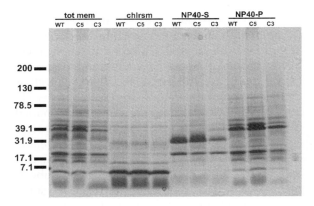

Fig. 4. Subcellular fractionation of C. tepidum strains. Chlorosome depleted membranes were prepared via fractionation on sucrose gradients from the wild type (WT) and two IVTM generated mutant strains (C3 and C5) of C. tepidum and then fractionated by solubilization with the detergent Nonidet P40 into detergent soluble (NP40-S) and insoluble (NP40-P) fractions. Proteins in each fraction were separated by 1D SDS-PAGE and stained with colloidal Coomassie blue. The NP40–S fraction is enriched in inner membrane proteins while the NP40–P fraction is enriched in outer membrane proteins.

et al., 2004; Huang et al., 2004). A number of these methods were used to fractionate chlorosome depleted membranes of C. tepidum (Vassilieva et al., 2002b). Solubilization with the detergent Nonidet P-40, used to extract outer membrane proteins from Hyphomonas jannaschiana (Shen et al., 1989), was found to reliably produce distinct banding patterns when NP-40 soluble and insoluble fractions were compared by 1D SDS-PAGE (compare the NP40-S and NP40-P fractions in Fig. 4). Protein identification by mass spectrometry is proceeding from these fractions and preliminary data indicates that NP-40 soluble proteins are predominantly inner membrane proteins including reaction center and ATPase subunits, while the NP-40 insoluble proteins are outer membrane proteins (Snellinger-O'Brien, Morgan-Kiss, Johnston and Hanson, unpublished).

C. Proteomics of Sulfur Oxidation Mutants

The improved fractionation protocol outlined above is currently being applied to C. tepidum in our laboratory along with additional fractionation and detection techniques. Heme staining via peroxidase activity (Thomas et al., 1976) has

been applied to subcellular fractions from C. tepidum WT2321 (Fig. 5) and mutant strains including C5 and an additional mutant strain, C3. The mutation in strain C3 is localized to Sulfur Island II (Fig. 1) and the strain appears to be defective for elemental sulfur oxidation. Current data (Fig. 5) indicate C. tepidum synthesizes at least a dozen hemoproteins with distinct subcellular localizations and that strain C5 overexpresses a ~25 kDa outer membrane heme protein relative to WT2321. Two heme stained proteins are apparent in secreted protein fractions at ~15 kDa and 5 kDa, which are distinct from hemoproteins observed in other fractions. The 15 kDa protein appears to be overexpressed in strain C3 while the 5 kDa protein appears to be overexpressed in strain C5. As the identities of these hemoproteins are resolved, they should provide great insight into mechanisms of sulfur oxidation in C. tepidum. In addition, there appear to be clear changes in protein expression profiles in the inner and outer membranes in both strains C5 and C3 when total proteins are visualized by Coomassie staining (compare lanes within each fraction across strains in Fig. 4). Identification of the hemoproteins and differentially expressed proteins along with detailed studies of growth and sulfur compound utilization by these particular mutant strains are underway.

IV. Conclusions

A combination of approaches will be required to fully understand anaerobic sulfur oxidation pathways in the Chlorobiaceae. The combination of genetic and proteomic approaches outlined above will serve as a starting point to identify the proteins critical for this process in the model system C. tepidum. Parallel approaches including comparative genomics as described by Frigaard and Bryant elsewhere in this volume will determine the distribution of these proteins across not only the Chlorobiaceae, but in other sulfur oxidizing microbes as well and thus translate to an overall improved understanding of biogeochemical cycling. Biochemical, crystallographic and more refined genetic approaches applied to genes/proteins identified in C. tepidum, including gene fusions, site directed mutagenesis and expression of exogenous genes in C. tepidum for mutant

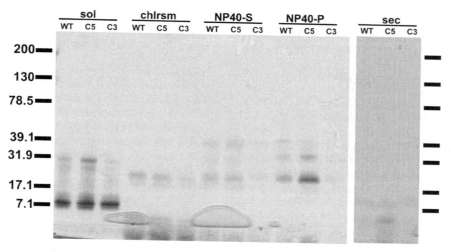

Fig. 5. Cellular distribution of hemoproteins in *C. tepidum* wild type and mutant strains. Cytosolic (sol), chlorosome (chlrsm), inner membrane (NP40–S), outer membrane (NP–40P), and secreted (sec) protein fractions were prepared from cultures of wild type (WT) and two IVTM generated mutant strains (C3 and C5). Proteins were separated by 1D SDS-PAGE without prior sample heating or addition of reducing agent and stained for heme peroxidase activity.

complementation will also bring about a more detailed understanding of anaerobic sulfur oxidation in general and further establish *C. tepidum* as one of the most useful and relevant model systems for the genetics of strictly anaerobic processes.

Acknowledgements

The authors would like to thank Joy Lawani, Jessica Martin, Egle Burbaite, Tim Weber and Michele Madorma for excellent technical assistance over the course of the project. This project was supported by a grant from the National Science Foundation (MCB-0447649 to T.E.H.) and utilized common instrumentation facilities provided in part by the National Institutes of Health (P20-RR116472-04 from the IDeA Networks of Biomedical Research Excellence program of the National Center for Research Resources).

References

Aivaliotis M, Corvey C, Tsirogianni I, Karas M and Tsiotis G (2004a) Membrane proteome analysis of the green-sulfur bacterium *Chlorobium tepidum*. Electrophoresis 25: 3468–3474

Aivaliotis M, Neofotistou E, Remigy HW, Tsimpinos G, Lustig A, Lottspeich F and Tsiotis G (2004b) Isolation and characterization of an outer membrane protein of *Chlorobium tepidum*. Photosynth Res 79: 161–166

Aivaliotis M, Haase W, Karas M and Tsiotis G (2006a) Proteomic analysis of chlorosome-depleted membranes of the green sulfur bacterium *Chlorobium tepidum*. Proteomics 6: 217–232

Aivaliotis M, Karas M and Tsiotis G (2006b) High throughput two-dimensional blue-native electrophoresis: A tool for functional proteomics of cytoplasmatic protein complexes from *Chlorobium tepidum*. Photosynth Res 88: 143–157

An JY and Kim BW (2000) Biological desulfurization in an optical-fiber photobioreactor using an automatic sunlight collection system. J Biotechnol 80: 35–44

Baik SC, Kim KM, Song SM, Kim DS, Jun JS, Lee SG, Song JY, Park JU, Kang HL, Lee WK, Cho MJ, Youn HS, Ko GH and Rhee KH (2004) Proteomic analysis of the sarcosine-insoluble outer membrane fraction of *Helicobacter pylori* strain 26695. J Bacteriol 186: 949–955

Baneras L, Rodriguez-Gonzalez J and Garcia-Gil LJ (1999) Contribution of photosynthetic sulfur bacteria to the alkaline phosphatase activity in anoxic aquatic ecosystems. Aquat Microb Ecol 18: 15–22

Bardischewsky F and Friedrich CG (2001) The *shxVW* locus is essential for oxidation of inorganic sulfur and molecular hydrogen by *Paracoccus pantotrophus GB17*: A novel function for lithotrophy. FEMS Microbiol Lett 202: 215–220

Bartsch R, Newton G, Sherril C and Fahey R (1996) Glutathione amide and its perthiol in anaerobic sulfur bacteria. J Bacteriol 178: 4742–4746

Basu R, Klasson KT, Clausen EC and Gaddy JL (1994) Removal of carbonyl sulfide and hydrogen sulfide from synthesis gas by *Chlorobium thiosulfatophilum*. Appl Biochem Biotechnol 45: 787–797

Beatty JT, Overmann J, Lince MT, Manske AK, Lang AS, Blankenship RE, Van Dover CL, Martinson TA and Plumley FG (2005) An obligately photosynthetic bacterial anaerobe from a deep-sea hydrothermal vent. Proc Natl Acad Sci USA 102: 9306–9310

Bloch PL, Phillips TA and Neidhardt FC (1980) Protein identifications of O'Farrell two-dimensional gels: Locations of 81 escherichia coli proteins. J Bacteriol 141: 1409–1420

Blöthe M and Fischer U (2001) Occurrence of polysulfides during anaerobic sulfide oxidation by whole cells and spheroplasts of *Chlorobium vibrioforme* and *Allochromatium minutissimum* and photooxidation of sulfite. 101st General Meeting of the American Society for Microbiology 101: 465

Borrego C and Garcia-Gil J (1995) Photosynthetic oxidation of MnS and FeS by *Chlorobium* spp. Microbiologia 11: 351–358

Brune DC (1989) Sulfur oxidation by phototrophic bacteria. Biochim Biophys Acta 975: 189–221

Brune DC (1995) Sulfur compounds as photosynthetic electron donors. In: Blankenship RE, Madigan MT, Bauer CE (eds) Anoxygenic Photosynthetic Bacteria, 847–870, Vol 2 of Advances in Photosynthesis and Respiration (Govindjee ed.) Kluwer Academic (now Springer), Dordrecht

Bryant DA, Vassilieva EV, Frigaard NU and Li H (2002) Selective protein extraction from *Chlorobium tepidum* chlorosomes using detergents. Evidence that CsmA forms multimers and binds bacteriochlorophyll a. Biochemistry 41: 14403–14411

Buchanan BB and Arnon DI (1990) A reverse Krebs cycle in photosynthesis: Consensus at last. Photosynth Res 24: 47–53

Buonfiglio V, Polidoro M, Flora L, Citro G, Valenti P and Orsi N (1993) Identification of two outer membrane proteins involved in the oxidation of sulphur compounds in *Thiobacillus ferrooxidans*. International Symposium on Advances on Biohydrometallurgy: Microbiology and Applications 11: 43–50

Butow B and Bergsteinbendan T (1992) Occurrence of *Rhodopseudomonas palustris* and *Chlorobium phaeobacteroides* blooms in Lake Kinneret. Hydrobiologia 232: 193–200

Castenholz RW, Bauld J and Jorgenson BB (1990) Anoxygenic microbial mats of hot springs – thermophilic *Chlorobium* sp. FEMS Microbiol Ecol 74: 325–336

Chan LK, Weber TS, Morgan-kiss RM, and Hanson TE (2007) A genomic region required for phototrophic thiosulfate oxidation in the green sulfur bacterium chlorobium tepidum (syn. *Chlorobaculum tepidum*). Microbiology In press

Cho SH, Na JU, Youn H, Hwang CS, Lee CH and Kang SO (1998) Tepidopterin, 1-o-(l-threo-biopterin-2′-yl)-beta-n-acetylglucosamine from *Chlorobium tepidum*. Biochim Biophys Acta 1379: 53–60

Chung S, Frank G, Zuber H and Bryant DA (1994) Genes encoding two chlorosome components from the green sulfur bacteria *Chlorobium vibrioforme* strain 83271d and *Chlorobium tepidum*. Photosynth Res 41: 261–275

Chung S, Shen G, Ormerod J and Bryant DA (1998) Insertional inactivation studies of the *csmA* and *csmC* genes of the green sulfur bacterium *Chlorobium vibrioforme* 8327: The chlorosome protein CsmA is required for viability but CsmC is dispensable. FEMS Microbiol Lett 164: 353–361

Dahl C, Engels S, Pott-Sperling AS, Schulte A, Sander J, Lübbe Y, Deuster O and Brune DC (2005) Novel genes of the *dsr* gene cluster and evidence for close interaction of the Dsr proteins during sulfur oxidation in the phototrophic sulfur bacterium *Allochromatium vinosum*. J Bacteriol 187: 1392–1404

Dennis JJ and Zylstra GJ (1998) Plasposons: Modular self-cloning minitransposon derivatives for rapid genetic analysis of gram-negative bacterial genomes. Appl Environ Microbiol 64: 2710–2715

Eisen JA, Nelson KE, Paulsen IT, Heidelberg JF, Wu M, Dodson RJ, Deboy R, Gwinn ML, Nelson WC, Haft DH, Hickey EK, Peterson JD, Durkin AS, Kolonay JL, Yang F, Holt I, Umayam LA, Mason T, Brenner M, Shea TP, Parksey D, Nierman WC, Feldblyum TV, Hansen CL, Craven MB, Radune D, Vamathevan J, Khouri H, White O, Gruber TM, Ketchum KA, Venter JC, Tettelin H, Bryant DA and Fraser CM (2002) The complete genome sequence of *Chlorobium tepidum* TLS, a photosynthetic, anaerobic, green-sulfur bacterium. Proc Nat Acad Sci USA 99: 9509–9514

Evans MCW, Buchanan BB and Arnon DI (1966) A new ferredoxin-dependent carbon reduction cycle in a photosynthetic bacterium. Proc Nat Acad Sci USA 55: 928–934

Fahey RC (2001) Novel thiols of prokaryotes. Annu Rev Microbiol 55: 333–356

Fahey RC, Buschbacher RM and Newton GL (1987) The evolution of glutathione metabolism in phototrophic microorganisms. J Mol Evol 25: 81–88

Friedrich CG, Rother D, Bardischewsky F, Quentmeier A and Fischer J (2001) Oxidation of reduced inorganic sulfur compounds by bacteria: Emergence of a common mechanism? Appl Environ Microbiol 67: 2873–2882

Frigaard NU and Bryant DA (2001) Chromosomal gene inactivation in the green sulfur bacterium *Chlorobium tepidum* by natural transformation. Appl Environ Microbiol 67: 2538–2544

Frigaard NU and Bryant DA (2004) Seeing green bacteria in a new light: Genomics-enabled studies of the photosynthetic apparatus in green sulfur bacteria and filamentous anoxygenic phototrophic bacteria. Arch Microbiol 182: 265–276

Frigaard NU, Li H, Milks KJ and Bryant DA (2004a) Nine mutants of *Chlorobium tepidum* each unable to synthesize a different chlorosome protein still assemble functional chlorosomes. J Bacteriol 186: 646–653

Frigaard NU, Maresca JA, Yunker CE, Jones AD and Bryant DA (2004b) Genetic manipulation of carotenoid biosynthesis in the green sulfur bacterium *Chlorobium tepidum*. J Bacteriol 186: 5210–5220

Gonzalez JM, Kiene RP and Moran MA (1999) Transformation of sulfur compounds by an abundant lineage of marine bacteria in the alpha-subclass of the class *Proteobacteria*. Appl Environ Microbiol 65: 3810–3819

Gordon DA and Giovannoni SJ (1996) Detection of stratified microbial populations related to *Chlorobium* and *Fibrobacter* species in the Atlantic and Pacific oceans. Appl Environ Microbiol 62: 1171–1177

Hansen TA and van Gemerden H (1972) Sulfide utilization by purple nonsulfur bacteria. Arch Mikrobiol 86: 49–56

Hanson TE and Tabita FR (2001) A ribulose-1, 5-bisphosphate carboxylase/oxygenase (rubisco)-like protein from *Chlorobium tepidum* that is involved with sulfur metabolism and the response to oxidative stress. Proc Natl Acad Sci USA 98: 4397–4402

Hanson TE and Tabita FR (2003) Insights into the stress response and sulfur metabolism revealed by proteome analysis of a *Chlorobium tepidum* mutant lacking the Rubisco-like protein. Photosynth Res 78: 231–248

Hayes F (2003) Transposon-based strategies for microbial functional genomics and proteomics. Annu Rev Genet 37: 3–29

Heising S, Richter L, Ludwig W and Schink B (1999) *Chlorobium ferrooxidans* sp. nov., a phototrophic green sulfur bacterium that oxidizes ferrous iron in coculture with a "*Geospirillum*" sp. Strain. Arch Microbiol 172: 116–124

Henshaw PF and Zhu W (2001) Biological conversion of hydrogen sulphide to elemental sulphur in a fixed-film continuous flow photo-reactor. Water Res 35: 3605–3610

Hernandez ME, Kappler A and Newman DK (2004) Phenazines and other redox-active antibiotics promote microbial mineral reduction. Appl Environ Microbiol 70: 921–928

Huang F, Hedman E, Funk C, Kieselbach T, Schroder WP and Norling B (2004) Isolation of outer membrane of *Synechocystis* sp. Pcc 6803 and its proteomic characterization. Mol Cell Proteomics 3: 586–595

Imhoff JF (2001) True marine and halophilic anoxygenic phototrophic bacteria. Arch Microbiol 176: 243–254

Imhoff JF (2003) Phylogenetic taxonomy of the family Chlorobiaceae on the basis of 16S rRNA and *fmo* (Fenna–Matthews–Olson protein) gene sequences. Int J Syst Evol Microbiol 53: 941–951

Jensen MT, Knudsen J and Olson JM (1991) A novel aminoglycosphingolipid found in *Chlorobium limicola f thiosulfatophilum*-6230. Arch Microbiol 156: 248–254

Jung DO, Carey JR, Achenbach LA and Madigan MT (2000) Phototrophic green sulfur bacteria from permanently frozen antarctic lakes. 100th General Meeting of the American Society for Microbiology 100: 388

Kim BW and Chang HN (1991) Removal of hydrogen sulfide by *Chlorobium thiosulfatophilum* in immobilized-cell and sulfur-settling free-cell recycle reactors. Biotechnol Prog 7: 495–500

Larimer FW, Chain P, Hauser L, Lamerdin J, Malfatti S, Do L, Land ML, Pelletier DA, Beatty JT, Lang AS, Tabita FR, Gibson JL, Hanson TE, Bobst C, Torres JL, Peres C, Harrison FH, Gibson J and Harwood CS (2004) Complete genome sequence of the metabolically versatile photosynthetic bacterium *Rhodopseudomonas palustris*. Nat Biotechnol 22: 55–61

Lies DP, Hernandez ME, Kappler A, Mielke RE, Gralnick JA and Newman DK (2005) *Shewanella oneidensis* MR-1 uses overlapping pathways for iron reduction at a distance and by direct contact under conditions relevant for biofilms. Appl Environ Microbiol 71: 4414–4426

Link AJ, Hays LG, Carmack EB and Yates JR, 3rd (1997) Identifying the major proteome components of *Haemophilus influenzae* type-strain NCTC 8143. Electrophoresis 18: 1314–1334

Lloyd JR (2003) Microbial reduction of metals and radionuclides. FEMS Microbiol Rev 27: 411–425

Ma K, Schicho RN, Kelly RM and Adams MW (1993) Hydrogenase of the hyperthermophile *Pyrococcus furiosus* is an elemental sulfur reductase or sulfhydrogenase: Evidence for a sulfur-reducing hydrogenase ancestor. Proc Natl Acad Sci USA 90: 5341–5344

Manske AK, Glaeser J, Kuypers MMM and Overmann J (2005) Physiology and phylogeny of green sulfur bacteria forming a monospecific phototrophic assemblage at a depth of 100 m in the Black Sea. Appl Environ Microbiol 71: 8049–8060

Maresca JA, Gomez Maqueo Chew A, Ponsati MR, Frigaard NU, Ormerod JG and Bryant DA (2004) The *bchU* gene of *Chlorobium tepidum* encodes the C-20 methyltransferase in bacteriochlorophyll *c* biosynthesis. J Bacteriol 186: 2558–2566

Mizuno T and Kageyama M (1978) Separation and characterization of the outer membrane of *Pseudomonas aeruginosa*. J Biochem (Tokyo) 84: 179–191

Mukhopadhyay B, Johnson EF and Ascano MJ (1999) Conditions for vigorous growth on sulfide and reactor-scale cultivation protocols for the thermophilic green sulfur bacterium *Chlorobium tepidum*. Appl Environ Microbiol 65: 301–306

Myers JM and Myers CR (2001) Role for outer membrane cytochromes OmcA and OmcB of *Shewanella putrefaciens* MR-1 in reduction of manganese dioxide. Appl Environ Microbiol 67: 260–269

Neidhardt FC, Appleby DB, Sankar P, Hutton ME and Phillips TA (1989) Genomically linked cellular protein databases derived from two-dimensional polyacrylamide gel electrophoresis. Electrophoresis 10: 116–122

Ormerod J (1988) Natural genetic transformation in *Chlorobium*. In: Olson JM, Ormerod J, Amesz J, Stackebrandt E, Trüper HG (eds) Green Photosynthetic Bacteria, 315–319. Plenum, New York

Overmann J (2000) The family *Chlorobiaceae*. In: Dworkin M (ed) The Prokaryotes: An Evolving Electronic Resource for the Microbiological Community. Springer, New York

Overmann J and van Gemerden H (2000) Microbial interactions involving sulfur bacteria: Implications for the ecology and evolution of bacterial communities. FEMS Microbiol Rev 24: 591–599

Overmann J, Cypionka H and Pfennig N (1992) An extremely low-light-adapted phototrophic sulfur bacterium from the black sea. Limnol Oceanogr 37: 150–155

Paschinger H, Paschinger J and Gaffron H (1974) Photochemical disproportionation of sulfur into sulfide and sulfate by *Chlorobium limicola forma thiosulfatophilum*. Arch Microbiol 96: 341–351

Patterson SD and Aebersold RH (2003) Proteomics: The first decade and beyond. Nat Genet 33 Suppl: 311–323

Patton WF (2002) Detection technologies in proteome analysis. J Chromatogr B Analyt Technol Biomed Life Sci 771: 3–31

Phillips TA, Bloch PL and Neidhardt FC (1980) Protein identifications on O'Farrell two-dimensional gels: Locations of 55 additional *Escherichia coli* proteins. J Bacteriol 144: 1024–1033

Pickering IJ, George GN, Yu EY, Brune DC, Tuschak C, Overmann J, Beatty JT and Prince RC (2001) Analysis of sulfur biochemistry of sulfur bacteria using x-ray absorption spectroscopy. Biochemistry 40: 8138–8145

Pires RH, Lourenco AI, Morais F, Teixeira M, Xavier AV, Saraiva LM and Pereira IA (2003) A novel membrane-bound respiratory complex from *Desulfovibrio desulfuricans* ATCC 27774. Biochim Biophys Acta 1605: 67–82

Pott AS and Dahl C (1998) Sirohaem sulfite reductase and other proteins encoded by genes at the d*sr* locus of *Chromatium vinosum* are involved in the oxidation of intracellular sulfur. Microbiology 144: 1881–1894

Powls R, Redfearn E and Trippett S (1968) The structure of chlorobiumquinone. Biochem Biophys Res Commun 33: 408–411

Prange A, Arzberger I, Engemann C, Modrow H, Schumann O, Trüper HG, Steudel R, Dahl C and Hormes J (1999) In situ analysis of sulfur in the sulfur globules of phototrophic sulfur bacteria by x-ray absorption near edge spectroscopy. Biochim Biophys Acta 1428: 446–454

Prange A, Chauvistre R, Modrow H, Hormes J, Trüper HG and Dahl C (2002) Quantitative speciation of sulfur in bacterial sulfur globules: x-ray absorption spectroscopy reveals at least three different species of sulfur. Microbiology 148: 267–276

Quentmeier A and Friedrich CG (2001) The cysteine residue of the SoxY protein as the active site of protein-bound

sulfur oxidation of *Paracoccus pantotrophus* GB17. FEBS Lett 503: 168–172

Reinartz M, Tschape J, Brüser T, Trüper HG and Dahl C (1998) Sulfide oxidation in the phototrophic sulfur bacterium chromatium vinosum. Arch Microbiol 170: 59–68

Rubin EJ, Akerley BJ, Novik VN, Lampe DJ, Husson RN and Mekalanos JJ (1999) In vivo transposition of mariner-based elements in enteric bacteria and mycobacteria. Proc Natl Acad Sci USA 96: 1645–1650

Sagulenko V, Sagulenko E, Jakubowski S, Spudich E and Christie PJ (2001) Virb7 lipoprotein is exocellular and associates with the *Agrobacterium tumefaciens* T pilus. J Bacteriol 183: 3642–3651

Schütz M, Shahak Y, Padan E and Hauska G (1997) Sulfide-quinone reductase from *Rhodobacter capsulatus*. Purification, cloning and expression. J Biol Chem 272: 9890–9894

Schütz M, Maldener I, Griesbeck C and Hauska G (1999) Sulfide-quinone reductase from *Rhodobacter capsulatus*: Requirement for growth, periplasmic localization and extension of gene sequence analysis. J Bacteriol 181: 6516–6523

Shahak Y, Arieli B, Padan E and Hauska G (1992) Sulfide quinone reductase (sqr) activity in *Chlorobium*. FEBS Lett 299: 127–130

Shen N, Dagasan L, Sledjeski D and Weiner RM (1989) Major outer membrane proteins unique to reproductive cells of *Hyphomonas jannaschiana*. J Bacteriol 171: 2226–2228

Takaichi S, Wang ZY, Umetsu M, Nozawa T, Shimada K and Madigan MT (1997) New carotenoids from the thermophilic green sulfur bacterium *Chlorobium tepidum*: 1′,2′-dihydro-gamma-carotene, 1′,2′-dihydrochlorobactene, and OH-chlorobactene glucoside ester, and the carotenoid composition of different strains. Arch Microbiol 168: 270–276

Thomas PE, Ryan D and Levin W (1976) An improved staining procedure for the detection of the peroxidase activity of cytochrome P-450 on sodium dodecyl sulfate polyacrylamide gels. Anal Biochem 75: 168–176

Tuschak C, Glaeser J and Overmann J (1999) Specific detection of green sulfur bacteria by in situ hybridization with a fluorescently labeled oligonucleotide probe. Arch Microbiol 171: 265–272

Vassilieva EV, Ormerod JG and Bryant DA (2002a) Biosynthesis of chlorosome proteins is not inhibited in acetylene-treated cultures of *Chlorobium vibrioforme*. Photosynth Res 71: 69–81

Vassilieva EV, Stirewalt VL, Jakobs CU, Frigaard NU, Inoue-Sakamoto K, Baker MA, Sotak A and Bryant DA (2002b) Subcellular localization of chlorosome proteins in *Chlorobium tepidum* and characterization of three new chlorosome proteins: CsmF, CsmH, and CsmX. Biochemistry 41: 4358–4370

Verté F, Kostanjevecki V, De Smet L, Meyer TE, Cusanovich MA and Van Beeumen JJ (2002) Identification of a

thiosulfate utilization gene cluster from the green phototrophic bacterium *Chlorobium limicola*. Biochemistry
41: 2932–2945

Vila X, Guyoneaud R, Cristina XP, Figueras JB and Abella
CA (2002) Green sulfur bacteria from hypersaline
Chiprana Lake (Monegros, Spain): Habitat description
and phylogenetic relationship of isolated strains. Photosynth Res 71: 165–172

Wahlund TM and Madigan MT (1995) Genetic transfer by
conjugation in the thermophilic green sulfur bacterium
Chlorobium tepidum. J Bacteriol 177: 2583–2588.

Wahlund TM, Woese CR, Castenholz RW and Madigan MT
(1991) A thermophilic green sulfur bacterium from New

Zealand hot springs, *Chlorobium tepidum* sp. nov. Arch
Microbiol 156: 81–90

Wong SM and Mekalanos JJ (2000) Genetic footprinting with
mariner-based transposition in *Pseudomonas aeruginosa*.
Proc Natl Acad Sci USA 97: 10191–10196

Xie Z and Merchant S (1996) The plastid-encoded *ccsA*
gene is required for heme attachment to chloroplast *c*-type
cytochromes. J Biol Chem 271: 4632–4639

Yarzabal A, Brasseur G, Ratouchniak J, Lund K, Lemesle-
Meunier D, DeMoss JA and Bonnefoy V (2002) The high-
molecular-weight cytochrome *c* Cyc2 of *Acidithiobacillus
ferrooxidans* is an outer membrane protein. J Bacteriol
184: 313–317

Chapter 19

Ecology of Phototrophic Sulfur Bacteria

Jörg Overmann

Bereich Mikrobiologie, Department Biologie I, Ludwig-Maximilians-Universität München, Maria-Ward-Str. 1a, D-80638 München, Germany

Summary

Anoxygenic phototrophic sulfur bacteria florish where light reaches sulfidic water layers or sediments. Their often dense communities have continuously attracted the attention of microbiologists. Although the major fraction of the existing diversity of phototrophic sulfur bacteria remains to be explored, ecophysiological studies have revealed a number of selective factors which govern the growth and the survival of phototrophic sulfur bacteria in the environment. Some novel aspects of the ecology of phototrophic sulfur bacteria have become apparent recently. Representing the most extremely low-light adapted photosynthetic organisms known to date, a brown-colored *Chlorobium* strain colonizes the chemocline of the Black Sea and is capable of maintaining a stable population at 0.0007% of surface light intensity. Besides the light intensity, the spectral composition of ambient light is a selective factor for the composition of anoxygenic phototrophic communities. A strong competition for infrared light occurs in laminated microbial benthic mats where phototrophic sulfur bacteria occupy their niches according to their long wavelength absorption properties. During evolution this apparently has led to the formation of a novel type of pigment-protein complex which was recently detected in a benthic *Chromatiaceae* species. Thirdly, the capability to establish a highly specialized symbiosis with motile Proteobacteria enabled some species of green sulfur bacteria to acquire motility. In these phototrophic consortia, a rapid

Corresponding Author, Phone: +49 (0)89 2180 6123, Fax: +49 (0)89 2180 6125, E-mail: j.overmann@LRZ.uni-muenchen.de

Rüdiger Hell et al. (eds.), Sulfur Metabolism in Phototrophic Organisms, 375–396.
© 2008 *Springer.*

signal transfer exists between the two partners and permits a scotophobic response toward light required by the immotile green sulfur bacterial epibiont. The isolation and characterization of dominant species of phototrophic sulfur bacteria and an improved understanding of their particular niche has also implications for the interpretation of molecular fossils of these bacteria which have been detected in sedimentary rocks of all geological eras and interpreted as evidence for the existence of extended oceanic anoxia in the past.

I. Introduction

Anoxygenic phototrophic sulfur bacteria occur where light reaches anoxic layers in the water column or aquatic sediments. Since the antiquity, colored waters or sediments occurring in various natural environments were described by natural scientists. First observations of blood-red lakes and swamps were reported from the Nile area. Red coloration of a crater lake near Rome was described by Pliny in 208 BC and reddish waters were observed at the seashore near Venice in the year 586 AD (Kondratieva 1965).

The first to describe unicellular motile phototrophic sulfur bacteria was Ch. G. Ehrenberg (1883), who discovered dense accumulations of purple sulfur bacteria, then named *Monas okenii* (now *Chromatium okenii*) at the sediment surface of a small polluted pond near Jena in Eastern Germany. Since then, dense communities of purple and green sulfur bacteria (Fig. 1) have continuously attracted the attention of microbiologists due to their conspicuous reddish, green or brown coloration and have stimulated numerous investigations of their environments, their morphology, physiology as well as repeated cultivation attempts (e.g., Winogradsky, 1887; Engelmann, 1988; Bavendamm, 1924; van Niel, 1931).

Earlier investigations of the community composition and physiology of these bacteria included measurements of relevant environmental parameters and physiological rates in situ (e.g., Sorokin, 1970). Elaborate cultivation techniques were developed based upon the insights into their ecological niches (Pfennig, 1993). More recently, a suite of culture-independent molecular methods have been established und permitted novel insights into the ecophysiology and population biology of phototrophic sulfur bacteria.

The current chapter will focus on the ecology of purple sulfur bacteria (members of the *Chromatiaceae* and *Ectothiorhodospiraceae*) and the green sulfur bacteria (*Chlorobiaceae*). Besides providing a condensed view on the ecology of these groups, novel aspects are addressed including extreme low-light adaptation, low maintenance energy requirements, the formation of symbioses in phototrophic consortia and, finally, the analysis of fossil phototrophic communities.

II. Habitats and Natural Populations of Phototrophic Sulfur Bacteria

A. Ecological Niches

1. Light Quantity and Quality

Typically, accumulations of phototrophic sulfur bacteria have been observed between 2 and 20 m, rarely down to 30 m depth in pelagic environments (Montesinos et al., 1983; Guerrero et al., 1987b; Gorlenko, 1988; van Gemerden and Mas, 1995; Herbert et al., 2005). In such environments, values for the light transmission to populations of phototrophic sulfur bacteria range from 0.015 to 10% (Parkin and Brock, 1980a; van Gemerden and Mas, 1995). *Chromatiaceae* so far have been found in chemocline environments down to depths of ≤ 20 m. The tight correlation between anoxygenic photosynthesis and the available irradiance suggests that light is the main environmental variable controlling the activity of phototrophic sulfur bacteria.

Since the accumulation of phototrophic sulfur bacterial cells results in an increased self-shading, they can only extend over a limited vertical distance, which is reciprocally related to the amount of biomass present. Accordingly, the densest pelagic communities of

---

Abbreviations: *Bchl* bacteriochlorophyll

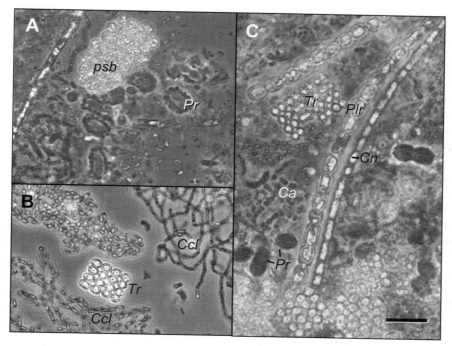

Fig. 1. Typical composition of a pelagic community of phototrophic and chemotrophic bacteria. Samples from the chemocline of dimictic Lake Dagow (North Brandenburg, Eastern Germany; obtained in September 2006) were left overnight after sampling for sedimentation of microbial cells. A. Purple sulfur bacteria (*psb*) containing numerous yellowish sulfur droplets. *Pr*, brown-colored phototrophic consortium "*Pelochromatium roseum*" in a partically disaggregated state. B. Brown- and green colored forms of *Chlorobium clathratiforme* (*Ccl*) and platelet-like microcolony of the purple sulfur bacterium *Thiopedia rosea* (*Tr*). C. Intact "*Pelochromatium roseum*" (*Pr*), one disintegrated phototrophic consortium "*Chlorochromatium aggrega-tum*" (*Ca*). Green-colored epibionts surround the central chemotrophic bacterium. *Tr*, *Thiopedia rosea*; *Cn*, filamentous green bacterium *Chloronema* sp. (*Chloroflexaceae*); *Plr*, filamentous cyanobacterium *Planktothrix rubescens*. Highly refractile and irregular intracellular regions in the latter three bacteria are gas vacuoles. Bar, 10 μm (*See Color Plates*).

phototrophic sulfur bacteria (up to 28 mg bacteriochlorophyll·l^{-1}; Overmann et al., 1994) extend over a depth range of 10 cm (Overmann et al., 1991a) whereas the least dense population in the Black Sea (0.068–0.94 μg BChl$e$·l^{-1}) is spread out over a depth interval of 30 m (Repeta et al., 1989; Manske et al., 2005). Communities of phototrophic sulfur bacteria in littoral sediments of sandy beaches, salt marshes or intertidal mudflats live in a significantly steeper light gradient and growth is limited to the uppermost 1.5–5 mm (van Gemerden and Mas, 1995). At the same time, biomass densities of 900 mg bacteriochlorophyll·dm^{-3} can be attained in these latter systems (van Gemerden et al., 1989).

In purple bacteria, the size of the photosynthetic antenna is in the range of 20–200 bacteriochlorophyll a per reaction center (Zuber and Cogdell, 1995). The photosynthetic antenna of green sulfur bacteria consist of specialized intracellular structures (so-called chlorosomes) and are significantly larger than those of other anoxygenic phototrophs with about 5000–8000 bacteriochlorophyll molecules connected to one reaction center (Frigaard et al., 2003). In addition, the theoretical quantum requirement for the CO_2-fixation of purple sulfur bacteria is 8–10.5 mol quanta·(mol CO_2)$^{-1}$, but only 3.5–4.5 mol quanta·(mol CO_2)$^{-1}$ for green sulfur bacteria (Brune, 1989). While the values for purple sulfur bacteria have been verified experimentally, much higher values than theoretically expected were reported so far for green sulfur bacteria. This discrepancy warrants further investigations. In addition to their larger antenna, green sulfur bacteria exhibit lower maintenance energy requirements and higher sulfide tolerance than other phototrophic sulfur bacteria (Overmann and Garcia-Pichel, 2000) and hence are especially well adapted to low-light habitats.

In the chemocline of the Black Sea, brown-colored green sulfur bacteria form an extremely dilute, but detectable population. The uppermost sulfidic water layers were detected at 80–120 m depth, while light attenuation in the overlying water layers is comparable to other stratified aquatic systems. As a consequence, the Black Sea chemocline is characterized by a very extreme low-light situation. According to recent measurements conducted with an integrating quantum meter, maximum in situ light intensities reach only 0.0022–0.00075 μmol Quanta × m^{-2} s^{-1} during winter, corresponding to 0.0007% of surface light intensity (Manske et al., 2005), while other environmental factors correspond to those prevailing in other oxic/anoxic habitats of phototrophic sulfur bacteria (Overmann and Manske, 2006). Combining the data on available light intensities and on concentrations of photosynthetic pigments, it can be calculated that each bacteriochlorophyll e molecule of the green sulfur bacteria on average absorbs one photon every 8 hours (Overmann and Manske, 2006).

Culture-independent 16S rRNA gene sequence analyses of the bacterial community present in the Black Sea chemocline revealed that one single and novel phylotype (BS-1) of green sulfur bacteria persisted over more than 13 years (Manske et al., 2005). These bacteria thus form a single population under the extreme conditions in the Black Sea chemocline, while other types of anoxygenic phototrophic sulfur bacteria could not be detected. Its continuous presence suggests a specific adaptation of phylotype BS-1 to the specific environmental conditions in the Black Sea chemocline and a high competitive advantage in situ. Subsequent $H^{14}CO_3^-$ - incorporation studies indicated that phylotype BS-1 is in fact capable of exploiting the minute light quantum flux available in situ.

From water samples obtained during the US-Turkish expedition of the RV *Knorr*, the first successful enrichment of the chemocline bacterium could be established (Overmann et al., 1992). This bacterium was isolated again from chemocline water samples recovered in 2001 from a depth of 95 m in the central western basin (Manske et al., 2005) and permitted first insights into its specific mechanisms of adaptation to extreme low-light conditions. In comparison to all other green sulfur bacteria tested, the Black Sea isolate incorporated $H^{14}CO_3^-$, oxidized sulfide and grew significantly faster at light intensities ≤1 μmol Quanta × m^{-2} s^{-1}.

Acclimation to very low light intensities in most phototrophic organisms involves an increase in the size of the photosynthetic unit (Göbel, 1978; Drews and Golecki, 1995; Sanchez et al., 1998). In the Black Sea isolate, the intracellular concentration of light-harvesting pigments is twice as high than in other green sulfur bacteria (Overmann et al., 1992) and chlorosomes are twofold larger than in another strain investigated (Fuhrmann et al., 1993). A conspicuous feature of the low-light-adaptation of the green sulfur bacterium from the Black Sea is the presence of geranyl homologs of BChl$e$, which had never been described for any other isolate of green sulfur bacteria. The structure of the esterifying alcohols in the bacteriochlorophylls of green sulfur bacteria may influence the function of the light-harvesting chlorosomes (Steensgard et al., 2000). Interestingly, geranyl ester isobutyl/ethyl [I, E]-Bchle_G an unusual Bchl e- homologue, and minor amounts of ethyl/methyl, ethyl/ethyl and propyl/ethyl-Bchle_G were detected in the chemocline as well as in cultures of strain BS-1. Upon low-light-adaptation of the culture of BS-1, the composition of these homologues changes towards a strong dominance of the higher alkylated [I, E]-Bchle_G (Manske et al., 2005). The alkyl side chains of the BChl tetrapyrrol system are directly involved in the aggregation of BChl molecules (van Rossum et al., 2001). Accordingly, a higher degree of alkylation leads to a red shift of the Q_y absorption maximum by 7–11 nm, which has been hypothesized to increase the energy transfer efficiency of the chlorosomes (Borrego and Garcia-Gil, 1995). The extremely efficient low light utilization of the Black Sea isolate comes at a price, however, since its specific physiological rates at saturating light intensities are much lower than in other phototrophic sulfur bacteria. The low specific metabolic rates reached under light-saturation may be caused by a reduction in the intracellular enzyme levels which may represent a way to decrease maintenance energy demand of the BS-1 cells (see section II.B).

Besides the available light intensity, the composition of the underwater light spectrum is a selective factor for the composition of anoxygenic phototrophic communities and differs considerably between pelagic and benthic habitats. In many lacustrine habitats, light absorption by phytoplankton exceeds that of humic substances or water itself and light of the bluegreen to green wavelength range reaches layers of phototrophic bacteria. In contrast, infrared light is an important source of energy in benthic microbial mats.

In general light absorption by anoxygenic phototrophs in the free water column is mediated by carotenoids and the short wavelength (Soret) bands of bacteriochlorophylls. In coastal and most lacustrine waters, the in vivo-absorption spectrum of *Chromatiaceae* which contain the carotenoid okenone matches the available light of the green wavelength range. Accordingly, okenone-bearing *Chromatiaceae* dominate in 63% of all natural communities investigated (van Gemerden and Mas, 1995) which has been explained by their higher efficiency of light absorption compared to species containing other types of carotenoids (Guerrero et al., 1987b; Overmann et al., 1991a). Dominant *Chromatiaceae* are obligately photolithotrophic, lack assimilatory sulfate-reduction, cannot reduce nitrate, and assimilate only few organic carbon sources. Obviously, metabolically versatile species of the *Chromatiaceae* have no selective advantage in most pelagic habitats.

Humic substances in lakes are of terrestrial origin and absorb light of the ultraviolet and blue portion of the spectrum. As a consequence, light of the red wavelength range prevails in lakes containing humic substance as the major light-absorbing constituents. Under these conditions, green-colored species of green sulfur bacteria have a selective advantage over their brown-colored counterparts, or over purple sulfur bacteria (Parkin and Brock, 1980b).

In benthic microbial mats, radiation of the visible wavelength range is strongly attenuated by mineral and biogenic particles. In quartz sand, light attenuation occurs preferentially in the wavelength range of blue light due to the reflection by sand grains (Kühl et al., 1994), while the absorption of infrared light by the sediment particles is low and absorption by water is negligible due to the short optical pathlength. As a consequence, the red and infrared portion of the spectrum penetrate the deepest in benthic environments; the irradiance reaching phototrophic sulfur bacteria may be reduced to <1% of the surface value for light in the visible region, while >10% of the near infrared light is still available (Kühl and Jørgensen, 1992). Under these conditions, the long wavelength (Q_y) absorption bands are significant for light-harvesting in anoxic sediment layers, and variations in the type of bacteriochlorophyll (Bchl$a$ or Bchl b) and in the fine structure of the pigment-protein complexes thus are the means of ecological niche separation.

Populations of phototrophic microorganisms impose strong absorption signatures on the spectrum of the scalar irradiances, creating different niches in a vertical sequence by the successive absorption of different wavelength bands in the red and infrared portion of electromagnetic radiation (Pierson et al., 1987). This may lead to the formation of up to five distinctly colored layers which (from top to bottom) comprise diatoms and cyanobacteria, cyanobacteria alone, *Chromatiaceae* containing Bchl$a$, *Chromatiaceae* containing Bchl$b$, and *Chlorobiaceae* (Nicholson et al., 1987). Only these benthic habitats are known to harbor distinct blooms of Bchl$b$-containing *Chromatiaceae*. The absorption spectra of whole cells of phototrophic bacteria seem to have evolved in such a way that almost the entire electromagnetic spectrum suitable for electrochemical reactions can be exploited.

Yet, no phototrophic sulfur bacterium was known which could absorb light of the wavelengths between 900 and 1020 nm until recently. Because of the strong competition for infrared light in sediment ecosystems, an effective absorption in this wavelength range by other types of photosynthetic antenna complexes would be expected to be of selective advantage in microbial mats. The isolation of purple nonsulfur α-Proteobacteria with long wavelength absorption maxima at 911 nm (Glaeser and Overmann, 1999) and 986 nm (Pfennig et al., 1997) indicates that the diversity of the pigment-protein complexes in photosynthetic Proteobacteria is greater than previously assumed. More recently, a purple photosynthetic sulfur bacterium with an absorption maximum at 970 nm was isolated from a littoral microbial mat (Permentier et al., 2001). Since this bacterium contain bacteriochlorophyll a as the photosynthetic pigment like most other members of the *Chromatiaceae*, the different in vivo absorption spectrum must be the result of differences in the non-covalent binding of Bchl$a$ to the light-harvesting proteins.

Deep sea hydrothermal vents represent a novel potential habitat of phototrophic sulfur bacteria identified recently (Beatty et al., 2005). Black smokers are thought to emit geothermal radiation at wavelengths commensurate with the absorption spectrum of phototrophic organisms (Van Dover et al., 1996). Indeed, a novel phylotype of obligately photolithoautotrophic green sulfur bacterium could be isolated from water samples originating from the TY black smoker located at 2391 m depths on the East Pacific Rise (Beatty et al., 2005). It remains to be elucidated whether the isolated bacterium is a typical and long-term resident of hydrothermal vents or whether it also thrives in habitats known for other phototrophic sulfur bacteria.

2. Reduced Sulfur Compounds and Redox Potential

A combination of two photosystems as in oxygenic phototrophs is required for the thermodynamically unfavorable utilization of water as an electron donor for photosynthesis. Due to the simpler architecture of their photosystems, all anoxygenic phototrophic bacteria depend on electron donors which exhibit standard redox potentials more negative than water (e.g., H_2S, H_2, acetate). This molecular feature thus is one major reason for the narrow ecological niche of anoxygenic phototrophic bacteria in extant ecosystems. Most phototrophic sulfur bacteria grow preferentially by photolithoautotrophic oxidation of reduced sulfur compounds. Other inorganic electron donors utilized include H_2, polysulfides, elemental sulfur, thiosulfate, sulfite and iron. In the green sulfur bacteria, polysulfide utilization is inhibited by sulfide. In addition to reduced sulfur compounds, molecular hydrogen serves as electron donor in the majority of green sulfur bacteria, and in the metabolically more versatile species of purple sulfur bacteria like *Allochromatium vinosum* and *Thiocapsa roseopersicina*.

In green sulfur bacteria which lack assimilatory sulfate reduction, a reduced sulfur source is required during the growth with molecular hydrogen as electron donor. In microbial mats, polysulfides and organic sulfur compounds may be significant as photosynthetic electron donor. Polysulfide oxidation has been reported for *Chlorobium limicola*, *Ach. vinosum* and *Tca. roseopersicina* while dimethylsulfide is utilized and oxidized to dimethylsulfoxide by the purple sulfur bacteria *Thiocystis* sp. and *Tca. roseopersicina* (van Gemerden and Mas, 1995).

Sulfide frequently becomes the growth-limiting factor at the top of the phototrophic sulfur bacterial layers where light intensities are highest, but sulfide has to diffuse through the remainder of the community. The affinity for sulfide during photolithoautotrophic growth varies between the different groups of anoxygenic phototrophs and has been shown to be of selective value during competition experiments. *Chlorobiaceae* and *Ectothiorhodospiraceae* exhibit five to seven times higher affinities for sulfide than *Chromatiaceae* (van Gemerden and Mas, 1995). On the contrary, affinities for polysulfides are similar for *Chlorobiaceae* and *Chromatiaceae*.

Because light and sulfide occur in opposing gradients, growth of phototrophic sulfur bacteria is confined to a narrow zone of overlap and only possible if the chemical gradient of sulfide is sufficiently stabilized against vertical mixing. In open water, like lakes or lagoons, stratification of oxic and anoxic water layers is maintained by density differences. Stratification can be transient if caused by temperature differences, or permanent (as in so-called meromictic lakes) if caused by higher salt concentrations of the bottom water layers. Benthic environments of phototrophic sulfur bacteria are characterized by a lower frequency of turbulent mixing and by diffusion as the dominant means of mass transport. As a result, and because of the higher rates of sulfate reduction, gradients of sulfide are much steeper in these environments.

In some habitats of phototrophic sulfur bacteria, redox conditions change rapidly within hours. This is particularly true for intertidal sediments. Certain small-celled species of the *Chromatiaceae* (*Allochromatium vinosum*, *Marichromatium gracile*, *Thiocapsa roseopersicina*, *Tca. rosea*, *Thiocystis minor*, *Tcs. violascens*, *Tcs. violacea*, *Thiorhodovibrio winogradskyi*) which are typical inhabitants of these fluctuating environments, as well as most of the *Ectothiorhodospiraceae* have adapted to these conditions and can switch to an aerobic chemolithotrophic growth mode and oxidize sulfide or thiosulfate with molecular oxygen. Under oxic conditions, the synthesis of pigments and of pigment-binding proteins of the photosynthetic apparatus ceases and the cells become colorless. Concomitantly, the activities of the respiratory enzymes NADH dehydrogenase and cytochrome oxidase and respiratory activity are increased. However, growth affinities of chemolithoautotrophically growing cells of *Tca. roseopersicina* are lower than for the directly competing colorless sulfur bacteria which may explain why no natural populations of purple sulfur bacteria are known which grow permanently by chemotrophy. All *Chlorobiaceae* are obligate anaerobes.

3. Temperature and Salinity

Although green and purple sulfur bacteria typically form conspicuous blooms in non-thermal aquatic ecosystems, moderately thermophilic members have been described from hot spring mats (Castenholz et al., 1990). *Chlorobaculum tepidum* (formerly *Chlorobium tepidum*) occurs in only a few New Zealand hot springs at pH values of 4.3 and 6.2 and at temperatures up to 56°C. *Thermochromatium tepidum* (formerly *Chromatium tepidum*) was found in several hot springs of western North America at temperatures up to 58°C and might represent the most thermophilic proteobacterium (Castenholz and Pierson, 1995).

Of the purple sulfur bacteria, most members of the *Chromatiaceae* are typically found in freshwater and marine environments, whereas the *Ectothiorhodospiraceae* inhabit hypersaline waters. About 10 species of *Chromatiaceae* are halophilic (Imhoff, 2005a). Members of the marine subgroup I of the green sulfur bacteria forming extraordinarily dense blooms could be isolated from a hypersaline (30–70&ppercnt; salinity) athalassohaline lake in the semi-arid Ebro region (Spain) (Vila et al., 2002).

4. Mixotrophy and Organotrophy

Organic carbon as it is present in microbial biomass is considerably more reduced than CO_2.

Given the high energy demand of CO_2-fixation, the capability for assimilation of organic carbon compounds would be expected to be of selective advantage in natural populations of phototrophic sulfur bacteria if limited by light or low sulfide concentrations. Acetate represents one of the most important intermediates during the degradation of organic matter and almost all phototrophic sulfur bacteria are capable of assimilating this compound. At limiting concentrations of sulfide, the cell yield of green sulfur bacteria is increased three times in the presence of acetate, i.e. under mixotrophic growth conditions.

Green sulfur bacteria are the least versatile of all phototrophic sulfur bacteria with all species growing obligately photolithoautotrophic. Only acetate, propionate and pyruvate are assimilated as carbon compounds during mixotrophic growth and a few strains are capable of using fructose or glutamate in addition. A number of *Chromatiaceae*, like *Allochromatium vinosum* and other small-celled members of the family, as well as the *Ectothiorhodospiraceae* are capable of using organic carbon compounds not only as carbon source but also as the only electron donating substrate (Imhoff, 2005a, b). These latter versatile *Chromatiaceae* utilize a wide range of organic carbon compounds and usually are capable of assimilatory sulfate reduction. The affinity for acetate is 30 times higher than that of green sulfur bacteria. Still, the metabolic flexible *Chromatiaceae* rarely form dense blooms under natural conditions, from that organotrophy confers only a limited selective advantage to the cells.

5. Motility and Taxis

Sedimentation represents a significant loss process for natural populations of phototrophic sulfur bacteria in pelagic habitats. The minimum buoyant density which has been determined for phototrophic cells devoid of gas vesicles was $1010 \, kg \cdot m^{-3}$ (Overmann et al., 1991b). Actively growing cells which contain storage carbohydrate and elemental sulfur can easily attain much higher buoyant densities of up to $1046 \, kg \cdot m^{-3}$ (Overmann and Pfennig, 1992) whereas freshwater has a considerably lower density (in the order of $996 \, kg \cdot m^{-3}$). In the stably stratified pelagic habitats of phototrophic sulfur bacteria, the difference in buoyant density of the cells to that of the surrounding water would result a sedimentation of bacteria out of the photic zone towards the lake bottom. Many species of phototrophic sulfur bacteria use vertical migration, mediated by tactic responses and/or the formation of gas vesicles to change their vertical position in the light and sulfide gradients of their environment.

In its pelagic habitat, *Chromatium okenii* may display diurnal migrations with a vertical amplitude of about 2 m (Sorokin, 1970). Vertical migrations of *Thiocystis minor* extended over a vertical distance of 30–35 cm (Pedrós-Alió and Sala, 1990). Planktonic anoxygenic phototrophs, unlike some planktonic cyanobacteria, do not seem to perform vertical migrations mediated by changes in gas vesicle content, but rather employ these cell organelles to maintain their vertical position within the chemocline (Overmann et al., 1991b; Overmann et al., 1994).

About two thirds of the *Chromatiaceae* species and all known species of the *Ectothiorhodopiraceae* swim by means of flagella, whereas only one benthic species of the *Chlorobiaceae*, *Chloroherpeton thalassium*, moves by gliding. True phototaxis is the ability to move towards or away from the direction of light, but is not found in phototrophic sulfur bacteria. Instead, these bacteria employ the scotophobic response to accumulate in regions of higher light intensity by changing the direction of movement in reaction to abrupt changes in light intensity (Armitage, 1997). As a result, cells accumulate in the light and at wavelengths corresponding to the absorption maxima of photosynthetic pigments. The formation of flagella in *Chromatium* and *Allochromatium* species is induced by low sulfide concentrations and low light intensities.

In laboratory cultures of *Chromatium* sp. and *Marichromatium gracile*, a combined effect of chemotaxis and photoresponses can be observed under the microscope: the cells accumulate around air bubbles in the absence of light but move away if illuminated (Armitage, 1997; Thar and Kühl, 2001). Motile *Chromatiaceae* are found in many microbial mats and exhibit diurnal vertical migrations in response to the recurrent changes in environmental conditions. Vertical migrations of *Chromatium* spp. and of *Thermochromatium tepidum* have been documented for populations in

ponds, and intertidal or hot spring microbial mats (Castenholz and Pierson, 1995). In these environments, cells migrate upwards to the surface of the mat and enter the overlaying water as a result of a positive aerotaxis during the night. It is assumed that this migration into the microoxic layers enables the cells to grow chemoautotrophically by oxidation of sulfide or intracellular sulfur with molecular oxygen. In contrast, microbial mats of intertidal sediments are typically colonized by the immotile purple sulfur bacterium *Thiocapsa roseopersicina*. Cells form aggregates together with sand grains, apparently as an adaptation to the hydrodynamic instability of the habitat (van den Ende et al., 1996).

Although all known pelagic species of green sulfur bacteria are nonflagellated, some of them have acquired motility by forming highly specific symbioses with a chemoheterotrophic motile Betaproteobacterium (Overmann and Schubert, 2002). These associations, termed phototrophic consortia (see section II.A.6), exhibit a scotophobic response and accumulate in a spot of white light. The action spectrum of this response corresponds to the absorption spectrum of the green sulfur bacterial epibionts, indicating that the scotophobic behavior is based on a rapid signal transfer between green sulfur bacterial epibionts and the colorless motile bacterium (Fröstl and Overmann, 1998).

One third of the species of *Chromatiaceae* (including *Lamprobacter*, *Lamprocystis*, *Thiocapsa*, *Thiodictyon*, *Thiopedia* and *Thiolamprovum* spp.) (Fig. 1), some green sulfur bacteria (Fig. 1B) but only one species of *Ectothiorhodopiraceae* (*Ets. vacuolata*) habor gas vesicles. This pattern reflects the distribution of these bacteria in nature, with gas vesicle-bearing *Chromatiaceae* typically colonizing low-light stratified environments, and *Ectothiorhodospiraceae* usually inhabiting more shallow saline ponds and sediments. Gas vesicles are cylindrical structures with conical ends and species-specific lengths and widths and are filled with a gas mixture which corresponds to that in the surrounding medium. Gas vesicle formation in *Chlorobium clathratiforme* is detected exclusively at light intensities $<5\,\mu$mol Quanta$\cdot$m$^{-2}\cdot$s^{-1} (Overmann et al., 1991b) which appears to be the reason for the rare observation of gas vesicles in pure cultures that are routinely incubated at higher light intensities. Similarly, a transfer of the purple sulfur bacterium *Lamprocystis purpurea* to the dark was found to result in an increase in the specific gas vesicle content by a factor of 9 (Overmann and Pfennig, 1992). *Ectothiorhodospira vacuolata* forms gas vesicles during the stationary phase.

By comparison, the vertical migration based on flagellar movement and gas vesicle formation have different advantages under natural conditions. Whereas the movement by flagella requires a continuous supply of metabolic energy (the proton motive force), gas vesicle formation requires an initially higher, but one-time investment for the phototrophic cell. Gas vesicles, once formed, help to keep the bacterial cell at the appropriate vertical position without any further demand for energy. In accordance with this view, species like *Lamprobacter modestohalophilus* or *Ectothiorhodospira vacuolata* which are capable of both, gas vesicle synthesis as well as flagellar movement, use flagella during exponential growth but become immotile and form gas vesicles upon entry in the stationary phase. Gas vesicle formation therefore may represent an adaptation to conditions of starvation in these species. It has been estimated that flagellar movement is sustained at underwater irradiances of $0.2\,\mu mol$ Quanta$\cdot m^{-2}\cdot s^{-1}$ (Overmann and Garcia-Pichel, 2000). Indeed, a dominance of gas-vacuolated forms over motile *Chromatiaceae* is usually observed in lakes where irradiances are below $1\,\mu mol$ Quanta$\cdot m^{-2}\times s^{-1}$ (Fig. 1).

However, a lower limit appears to exist, below which gas vesicle formation does not represent a selective advantage for phototrophic sulfur bacteria due to its metabolic burden. The extremely low-light adapted *Chlorobium* BS-1 from the Black Sea chemocline exhibits an extremely low maintenance energy requirement but is not capable of gas vesicle synthesis. Apparently, synthesis of the proteinaceous gas vesicle sheaths becomes too energy-demanding under the severe light limitation in the Black Sea chemocline.

6. Syntrophy and Symbioses

In the laboratory, stable associations between green sulfur bacteria and sulfur- or sulfate-reducing bacteria can be established readily (Warthmann et al., 1992). These associations are based upon a cycling of sulfur compounds but not carbon. Simultaneous growth of the two partner bacteria is fueled by the oxidation of organic carbon substrates and light. In a similar manner, cocultures of *Chromatiaceae* with sulfate-reducing bacteria have been established in the laboratory. Interestingly, cellular aggregates consisting of the sulfate-reducing Proteobacterium *Desulfocapsa thiozymogenes* and small-celled *Chromatiaceae* were observed in the chemocline of a meromictic alpine lake (Tonolla et al., 2000).

A commensalistic relationship may exist between coccoid epibiotic bacteria and the purple sulfur bacterium *Chromatium weissei* (Clark et al., 1993). The unidentified epibionts attach to healthy *Chromatium* cells but do lyse the host cells like the morphologically similar parasite *Vampirococcus* (Guerrero et al., 1987a). Possibly, the epibiont grows chemotrophically on carbon compounds excreted by the purple sulfur bacterium.

The most spectacular type of association involving phototrophic bacteria is represented by the so-called phototrophic consortia. Phototrophic consortia (Fig. 2) consist of epibionts arranged in a regular fashion around a central chemotrophic bacterium and are regarded as the most highly developed interactions between different species of prokaryotes (Overmann and Schubert, 2002). Eight different types of motile phototrophic consortia can be distinguished based on the overall morphology of the association and the color of the epibionts (Glaeser and Overmann, 2004). In addition, two immotile forms ("*Chloroplana vacuolata*" and "*Cylindrogloea bacterifera*") are recognized (Fig. 2). Fluorescence in situ hybridization identified the epibionts as green sulfur bacteria (Tuschak et al., 1999) and the central rod-shaped colorless and motile bacterium as a member of the Betaproteobacteria (Fröstl and Overmann, 2000). Based on their distinct morphology, intact phototrophic consortia can be specifically collected from natural communities by micromanipulation and the 16S rRNA gene sequences of the green sulfur bacterial epibionts can be determined. Employing this technique, an unexpected diversity of 19 different types of epibionts have recently been identified in a culture-independent manner. All epibionts represent distinct and novel phylotypes that are often only distantly related to known species of green sulfur

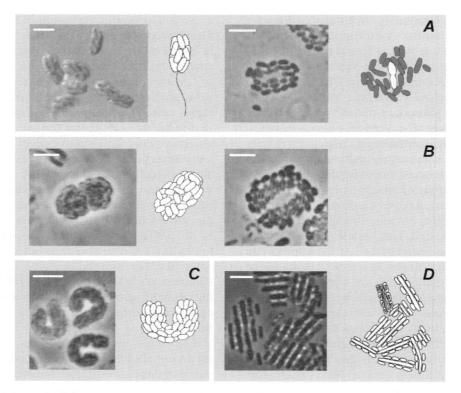

Fig. 2. Light microscopic and schematic views of five different types of phototrophic consortia. A. Differential interference contrast image and schematic view of "*Chlorochromatium aggregatum*", phase contrast photomicrograph of the disaggregated state and schematic view of "*Pelochromatium roseum*" (from left to right). B. Phase contrast photomicrograph in the intact state, schematic view, and the disaggregated state of "*Pelochromatium latum*". C. Phase contrast photomicrograph and schematic view of "*Chlorochromatium glebulum*". D. Phase contrast photomicrograph and schematic view of "*Chloroplana vacuolata*". Bars, 5 μm. Taken from Overmann (2006).

bacteria (Glaeser and Overmann, 2004). None of the epibiont 16S rRNA sequences have so far been detected in free-living green sulfur bacteria, suggesting that the interaction in phototrophic consortia is an obligate one. Based on the comparative phylogenetic analysis, the epibiont sequences are not monophyletic. Thus, the ability to form symbiotic associations either arose independently from different ancestors or was present in a common ancestor prior to the radiation of green sulfur bacteria and the transition to the free-living state in independent lineages. With regard to the phylogenetic affiliation of the central bacterium, a recent molecular analysis of the phototrophic consortium "*Chlorochromatium aggregatum*" revealed that this bacterium represents an isolated phylogenetic lineage distantly related to *Rhodoferax* spp., *Polaromonas vacu-*

olata and *Variovorax paradoxus* (Kanzler et al., 2005).

Maximum rates of light-dependent $H^{14}CO_3^-$ fixation were observed in a natural population of phototrophic consortia, suggesting that the green sulfur bacterial epibionts grow autotrophically like their free-living relatives. This conclusion was substantiated by the stable carbon isotope ratios ($\delta^{13}C$) of farnesol, tetradecanol, hexadecanol and hexadecenol which are esterifying alcohols of BChl*e* and biomarkers of the epibionts (Glaeser and Overmann, 2003a). Intact phototrophic consortia exhibit a scotophobic response in which the bacteriochlorophylls of the epibionts function as light sensors, whereas the central bacterium confers motility. Hence, a rapid signal transfer exists between the two partners and permits phototrophic consortia to accumulate at preferred light intensities and wavelengths (Fröstl and Overmann, 1998). Phototrophic consortia are attracted by sulfide and 2-oxoglutarate, which indicates a potential role of these compounds in the metabolism of the consortia. Microautoradiography of consortia in natural water samples revealed that 2-oxoglutarate is incorporated only in the presence of both light and sulfide (Glaeser and Overmann, 2003b). Because the green sulfur bacterial epibionts grow autotrophically, 2-oxoglutarate most likely is taken up and utilized by the central bacterium while sulfide is the electron-donating substrate of the epibionts. These results indicate that incorporation of 2-oxoglutarate by the central bacterium is regulated by the metabolic state of the epibiont cells.

In phototrophic consortia, the immotile green sulfur bacteria not only functionally attain motility like their purple sulfur bacterial competitors, but are obviously capable of controlling the chemotactic and physiological response of the chemotrophic partner bacterium. The high numbers of phototrophic consortia found in many lakes, the fact that in some environments all cells of green sulfur bacteria occur in the associated state (Glaeser and Overmann, 2003a), and the repeated advent of epibionts during the green sulfur bacterial radiation indicate that this strategy must be of high competitive value under certain environmental conditions. Since phototrophic consortia have recently become available in enrichment cultures, they can now serve as suitable model systems for the investigation of the molecular mechanisms of cell–cell recognition and signal exchange, and for studies of the coevolution of nonrelated prokaryotes.

B.   Physiology in Situ as Opposed to Growth in the Test Tube: Growth Rates, Low Maintenance Energy Requirements and Survival

Doubling times of phototrophic sulfur bacteria under natural conditions are significantly lower than in laboratory cultures. As an example, values for *Chromatiaceae* in lakes have been estimated to range from 1.5 to 238 days (Garcia-Cantizano et al., 2005). These data suggest that adaptation to low growth rates and survival may have played a significant role in the evolution of phototrophic sulfur bacteria.

In a careful study, the maintenance energy requirement of the purple nonsulfur Alphaproteobacteria *Rhodobacter capsulatus* and *Rba. acidophilus* was determined to amount to 0.012 mol quanta (g dry weightċh)$^{-1}$ (Göbel, 1978). Based on indirect estimates, the maintenance energy requirements of green sulfur bacteria are significantly lower than that of purple sulfur bacteria (van Gemerden and Mas, 1995). This may be explained by the fact that biosynthesis of proteins requires a major fraction of the energy expenditure of the bacterial cell and that the protein content of green sulfur bacterial antenna is much lower than that in purple sulfur bacteria: the paracrystalline rod-like structure of bacteriochlorophyll aggregates in the chlorosomes of green sulfur bacteria features a significantly lower protein:pigment mass ratio (0.5–2.2; Overmann and Garcia-Pichel, 2000) than the light-harvesting complexes of other anoxygenic phototrophs (3.9–6.7) or cyanobacteria (22.4).

Based on the value for the maintenance energy requirement of *Rba. capsulatus* and *Rba. acidophilus* given above it has been calculated that the minimum irradiance which would be required for survival of photosynthetic bacteria is 2 μmol Quanta·m^{-2}·s^{-1} (Overmann and Garcia-Pichel, 2000). By comparison, irradiances of this order or lower prevail in the natural habitats of phototrophic sulfur bacteria (Overmann and Manske, 2006). Yet, growth of natural populations can be observed under these conditions, indicating that phototrophic sulfur bacteria under natural conditions must exhibit significantly lower maintenance energy requirements than those of the laboratory cultures studied to date.

In particular, this holds true for the green sulfur bacterium from the Black Sea chemocline which forms the slowest growing population of anoxygenic phototrophic bacteria known to date (Overmann and Manske, 2006). As extrapolated from the laboratory growth rates or carbon fixation rates, green sulfur bacteria in the Black Sea attain doubling times of 3 years in summer and 26 years in winter (Overmann et al., 1992; Manske et al., 2005; Overmann and Manske, 2006). The green sulfur bacterium from the Black Sea chemocline thus represents an excellent model system for the study of low-light adaptation and, from a more fundamental perspective, a model system for studies of mechanisms of adaptation towards extreme energy-limited environments.

Commensurate with the extremely long doubling times, the maintenance energy requirement of the Black Sea *Chlorobium* is significantly lowered compared to other green sulfur bacteria (Overmann et al., 1992). At the same time, the Black Sea strain exhibits significantly decreased specific rates of photosynthetic CO_2-fixation, sulfide oxidation and growth at saturating light intensities (section II.A.1). These observations suggest that the extraordinarily low maintenance energy requirement of this bacterium at least in part is accomplished by lowering the intracellular concentrations of metabolic enzymes.

The threshold light intensity supporting the photosynthetic growth of green sulfur bacteria is decreased in the presence of suitable organic carbon substrates like acetate (Bergstein et al., 1981). Although the environmental concentrations and the turnover rates of potential carbon substrates of phototrophic sulfur bacteria have rarely been determined (Bergstein et al., 1979), these compounds may support the survival of the cells.

C. Diversity of Phototrophic Sulfur Bacteria

In almost all freshwater and marine photic anoxic environments, green and/or purple sulfur bacteria (*Chlorobiaceae* and *Chromatiaceae*) represent the dominant anoxygenic phototrophs. *Ectothiorhodospiraceae* dominate in saline habitats. Only very few, and atypical, ecosystems have been described in which phototrophic Alphaproteobacteria (purple nonsulfur bacteria) outnumber the phototrophic sulfur bacteria. These latter habitats are aquatic systems heavily polluted with organic wastewaters in which low-molecular organic compounds occur at millimolar concentrations (Okubo et al., 2006).

Based on the current status of taxonomy (Overmann, 2001; Imhoff, 2003; Imhoff, 2005a, b), 41 different species of *Chromatiaceae*, 12 species of *Ectothiorhodospiraceae* and 17 species of *Chlorobiaceae* are currently recognized. In contrast, the ribosomal database project (rdp) database lists 429, 460 and 339 16S rRNA gene sequences for the above three families of phototrophic sulfur bacteria (Cole et al., 2006).

Culture independent analyses of natural communities based on 16S rRNA gene sequences routinely recover novel sequence types of green or purple sulfur bacteria from natural bacterial communities (Coolen and Overmann 1998; Overmann et al., 1999a; Elshahed et al., 2003; Glaeser and Overmann, 2003a; Koizumi et al., 2004; Martínez-Alonso et al., 2005; Tonolla et al., 2005). In addition, the dominant phylotypes determined by the culture-independent approach often do not match the phylotypes cultured from the same environment. It has to be concluded that the current number of species described do not reflect the full phylogenetic breadth of phototrophic sulfur bacteria and that the dominant species may differ considerably from known types with respect to physiology and ecology. The latter assumption has been substantiated by the discovery of several novel isolates of green (Vogl et al., 2006) and purple sulfur bacteria (Permentier et al., 2001) which exhibit conspicuously different physiological properties in comparison to the previously described species.

At low biomass densities, 16S rRNA gene sequences of phototrophic sulfur bacteria often cannot be detected PCR amplification methods using universal primer pairs (Coolen and Overmann, 1998; Vetriani et al., 2003), despite the presence of their specific pigment biomarkers. Therefore, specific molecular detection methods have been developed, which employ group-specific primers targeting 16S rRNA gene sequences of green or purple sulfur bacteria (Coolen and Overmann, 1998; Overmann et al., 1999a). In the case of green sulfur bacteria, the highly specific PCR method permits the detection of as little as 100 cells (Glaeser and Overmann, 2004).

III. Biogeochemical Significance of Phototrophic Sulfur Bacteria

In lakes haboring phototrophic sulfur bacteria, an average of 28.7% of the primary production is anoxygenic and a maximum fraction of 83% has been determined (Overmann, 1997). Anoxygenic photosynthesis depends on reduced inorganic sulfur compounds which originate from the anaerobic degradation of organic carbon und the concomitant sulfide production by sulfate- and sulfur-reducing bacteria. During anaerobic degradation, a large fraction of reducing equivalents become trapped in sulfide due to the low growth yield of fermenting and sulfate-reducing bacteria. Since these reducing equivalents originate from carbon already fixed by oxygenic photosynthesis, the CO_2-fixation of anoxygenic phototrophic bacteria does not lead to a net increase in organic carbon of the entire oxic/anoxic stratified ecosystem. In a sense, then, capture of light energy by anoxygenic photosynthesis merely compensates for the degradation of organic carbon in the anaerobic food chain. The CO_2-assimilation by anoxgenic phototrophs has therefore been termed "secondary primary production" (Pfennig, 1978). Geothermal sulfur springs are the only exception since their sulfide is of abiotic origin (Elshahed et al., 2003). Yet, sulfur springs are rather scarce, and anoxygenic photosynthetic carbon fixation of these ecosystems thus appears to be of minor significance on a global scale.

When reoxidizing the sulfide, anoxygenic phototrophs do not need to divert part of the electron donor towards ATP generation and almost completely transfer them to CO_2. In contrast to the chemolithoautotrophic sulfide-oxidizing bacteria, green and purple sulfur bacteria therefore efficiently recycle the reducing equivalents present in sulfide where light reaches the sulfide-containing water or sediment layers at sufficient intensities (Overmann, 1997). As a result, phototrophic sulfur bacteria often form dense microbial biomass accumulations even at moderate supply of sulfide and can attain biomass concentrations up to 28 mg bacteriochlorophyll $a \cdot l^{-1}$ in the case of *Chromatiaceae* in a saline meromictic lake (Overmann et al., 1994) and 16.7 mg bacteriochlorophyll $d \cdot l^{-1}$ in the case of green sulfur bacteria thriving in a athalassohaline hypersaline lake (Vila et al., 2002). The dense accumulations of phototrophic sulfur bacteria in turn may feed organic carbon (which would otherwise be lost) into the carbon cycle of the overlaying oxic water or sediment layers (Overmann et al., 1996, 1999b). However, several instances have been reported in which predation of phototrophic sulfur bacteria is of minor importance due to the toxicity of hydrogen sulfide for grazing organisms (van Gemerden and Mas, 1995). More recent data indicate a significant transfer of phototrophic bacterial biomass into the aerobic grazing food chain via

rotifers and calanoid copepods (Overmann et al., 1999b, c).

Thus, anoxygenic primary production does only represent a net input of organic carbon to an ecosystem if (1) the anaerobic food chain within the system is fueled by additional allochthonous carbon from outside or by geothermal sulfide, and (2) aerobic grazers have access to the biomass of phototrophic sulfur bacteria. Based on experimental evidence, these conditions are met at least in some stratified aquatic environments (Overmann, 1997) where phototrophic sulfur bacteria can substantially alter the carbon and sulfur cycles.

As the closest analogue to past sulfidic oceans, the Black Sea has repeatedly been chosen as a model system for the study of the carbon and sulfur cycles in a large stratified marine water body and of the microorganisms relevant in these environments. The biomass of green sulfur bacteria in the chemocline of the Black Sea amounts to ≤0.8 mg BChl$e$ m^{-2} (Manske et al., 2005) and hence is orders of magnitude lower than in any other environment studied so far (25–2000 mg BChl$e$ m^{-2}; van Gemerden and Mas, 1995). The green sulfur bacteria in the chemocline of the Black Sea therefore represent the most dilute population of anoxygenic phototrophs known to date which can be attributed to the highly (i.e. four orders of magnitude more) sensitive method employed for the detection of their bacteriochlorophylls. Attempts to quantify photosynthetic activity in natural samples from the Black Sea chemocline have failed (Jørgensen et al., 1991) and stimulation of sulfide oxidation by light could not be detected in natural water samples (Repeta et al., 1989). Based on an extrapolation of light-limited rates of $H^{14}CO_3^-$ fixation, integrated anoxygenic photosynthesis contributes well below 1% to total photosynthetic carbon fixation in the Black Sea (Manske et al., 2005). Similarly, anoxygenic phototrophic sulfide oxidation accounts for ≤0.1% of total sulfide oxidation in the Black Sea chemocline. These estimates support the results of the modeling of sulfide fluxes which revealed that direct and indirect oxidation by molecular oxygen accounts for most, if not all of the sulfide removal in and beneath the chemocline of the Black Sea (Konovalov et al., 2001, 2003). Due to their low population density in the chemocline of the Black Sea, the green sulfur bacteria most likely are not significant in the carbon and sulfur cycles of this environment.

IV. Phototrophic Sulfur Bacteria in the Past: Interpretation of Molecular Fossils

Today, habitats of phototrophic sulfur bacteria are restricted to a limited number of lacustrine environments, coastal lagoons and intertidal sandflats. Of these, the Black Sea currently represents the largest anoxic water body on Earth and covers 0.083% of the area of the planet. Its stratified water column comprises a ~60-m thick oxic top layer, a ~40-m-thick suboxic intermediate zone devoid of sulfide and oxygen, and a ~2000-m-deep sulfidic bottom zone (Murray et al., 1989). As a consequence, between 87 and 92% of the Black Sea water body remain permanently anoxic (Codispoti et al., 1991; Konovalov et al., 2001; Sorokin, 2002).

Due to the usually small size, the limited number, and the pronounced light limitation (see section II.A.1) of their contemporary habitats, the contribution of phototrophic sulfur bacteria to global photosynthetic CO_2-fixation is very small and has been estimated to amount to less than 1% (Overmann and Garcia-Pichel, 2000). In contrast, the entire Proterozoic ocean may have consisted of sulfidic deep water covered by a possibly 100 m-thick oxygenated surface layer (Anbar and Knoll, 2002), and its sulfidic pelagial may have persisted over 1000 million years. Furthermore, extended water column anoxia may also have occurred during the Phanerozoic, starting with the Ordovician, and including the Upper Devonian, Permian, Mid-Triassic, early Jurassic and Miocene (Messinian). The most distinct, and probably global, Mesozoic anoxic oceanic events occurred during the Toarcian (~187 Myr ago), Early Aptian (~132 Myr ago) and latest Cenomanian (~94 Myr ago). These events have been explained by an increase in atmospheric CO_2 due to an increased volcanism, with the associated greenhouse effect leading to increased continental weathering, and the resulting increased nutrient supply stimulating the productivity in Mesozoic Oceans together with an increased supply of biolimiting metals by submarine hydrothermal activity (Erba, 2004). Since an upper mixed layer of less than 20 m is relatively common in the present warm coastal and even open ocean (Kara et al., 2003), the existence of such large scale photic zone anoxia would appear to be a reasonable assumption.

All known green sulfur bacteria and about half of the species of purple sulfur bacteria are obligate anaerobic photolithoautotrophs (Overmann and Garcia-Pichel, 2000; Overmann, 2001). These bacteria only grow in an environment providing both, light and reduced sulfur compounds, and therefore represent suitable indicator organisms also for past photic zone anoxia of aquatic ecosystems. Three specific carotenoids (isorenieratene, β-isorenieratene, chlorobactene) occur in the different green sulfur bacteria (Overmann, 2001) and represent highly specific biomarkers which occur almost exclusively in this group (however, β-isorenieratane can also be formed by aromatization from β-carotene, which is widely distributed among different photosynthetic organisms; Koopmans et al., 1996a). Okenone so far has only been found in nine species of *Chromatiaceae* (Imhoff, 2005a). These biomarkers of phototrophic sulfur bacteria may survive over extended geological time periods and even some of their degradation products can be identified with sufficient reliability. Due to their specificity and chemical stability, pigment biomarkers thus offer the opportunity to reconstruct past ecosystems and numerous studies have taken the presence of isorenieratene and its geochemical derivatives like sulfurized isorenieratane (Repeta, 1993; Sinninghe-Damsté et al., 1993; Wakeham et al., 1995; Passier et al., 1999; Menzel et al., 2002), or degradation products of bacteriochlorophylls (Grice et al., 1996) as evidence for extended water column anoxia of ancient oceans, so called Oceanic Anoxic Events (Fig. 3).

Numerous marine deep-sea sediments and sedimentary rocks (some of them presently even located on land; Kohnen et al., 1991) were shown to contain such green sulfur bacterial biomarkers (Fig. 3). Since it may represent a modern analogue of past water column anoxia, the development of anoxic conditions in the Black Sea has been studied extensively. Bottom water anoxia was initiated 7000 to 8000 years ago by the intrusion of saltwater from the Mediterranean via the Bosporus strait (Ross and Degens, 1974). Within the subsequent 3000 years, the O_2–H_2S interface rose from

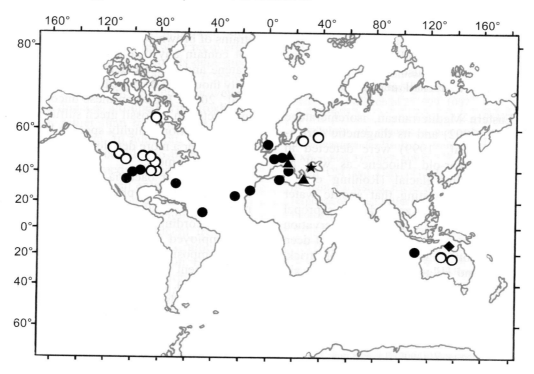

Fig. 3. Contemporary geographical location of fossil sediments and sedimentary rocks containing biomarkers of phototrophic sulfur bacterial as evidence for Oceanic Anoxic Events. Locations are distinguished according to their geological age.

★ Extant population of green sulfur bacteria in the chemocline (Manske et al., 2005) and isorenieratene in the Black Sea sediments (Sinninghe-Damsté et al., 1995).

▲ Cenozoic environments:, Pliocene/Pleistocene (isorenieratene; Rohling et al., 2006) and Pleistocene/Holocene (isorenieratene and 16S rRNA gene sequences; Coolen and Overmann, 2007) deposits in Eastern Mediterranean sapropels; C_{11}–C_{19} trimethyl substituted aryl isoprenoids in late Eocene/Early Oligocene shales of the Austrian Molasse Basin (Schulz et al., 2002), Northern Italian Marl sediments of the Messinian (Miocene) (Kohnen et al., 1991; Sinninghe-Damsté et al., 1995).

● Mesozoic environments: black shales from the eastern equatorial Atlantic of the Coniacian-Santonian (Deep Ivorian basin, Lake Cretaceous; Wagner et al., 2005), from the southern part of the proto-North Atlantic Ocean of the late Cenomanian (Cretaceous) (Kuypers et al., 2002; Kuypers et al., 2004); late Cenomanian marls of the Western Interior Basin (USA) from Kansas to New Mexico (Simons and Kenig, 2001); late Cenomanian Italian, NE Tunesian and Indian Ocean black shales, early Aptian black shales from the Italian Marche-Umbria basin, Toarcian (Jurassic) black shales of the same region, of the Belluno through and the Paris basin, as well as Middle Lias German Basin Posidonia Shale (Pancost et al., 2004); early Jurassic organic rich bituminous facies from Yorkshire (UK; Bowden et al., 2006).

○ Paleozoic environments: isorenieratene derivatives in Upper Devonian Williston and Western Canada Sedimentary Basins of western Canada (Requejo et al., 1992; Hartgers et al., 1994; Koopmans et al., 1996b), in Middle and Upper Devonian black shales of the Illinois and Michigan basins (Brown and Kenig, 2004), of the Belarussian Pripyat River Basin (Clifford et al., 1998), of the Holy Cross Mountains in Poland (Joachimski et al., 2001) and of the Western Australia Canning Basin (Barber et al., 2003) and Late Ordovician Boas Oil Shale of Southampton Island (Koopmans et al., 1996b). Aryl and diaryl isoprenoids in middle Ordovician grey-green shales of the central US (Pancost et al., 1998). Isorenieratane was also detected in mid-Cambrian limestone of the Georgina Basin (Brocks et al., 2005).

◆ Proterozoic environments: 1.64-Gyr-old mid-Proterozoic records of *Chromatiaceae* (okenane) and *Chlorobiaceae* (chlorobactane and isorenieratane) (Brocks et al., 2005).

the bottom at 2200 m depth toward the surface (Degens and Stoffers, 1976). The presence of isorenieratene and its degradation products in subfossil Black Sea sediments suggests that

photic zone anoxia occurred already more than 6000 years ago (Repeta, 1993; Sinninghe-Damsté et al., 1993). Recently, fossil 16S rRNA gene sequences of the low-light adapted green sulfur

public data. Nucleic Acids Research; doi: 10.1093/nar/gkl889

Coolen MJL and Overmann J (1998) Analysis of subfossil molecular remains of purple sulfur bacteria in a lake sediment. Appl Environ Microbiol 64: 4513–4521

Coolen MJL and Overmann J (2007) 217,000-year-old DNA sequences of green sulfur bacteria in Mediterranean sapropels and their implications for the reconstruction of the paleoenvironment. Environ Microbiol 9: 238–249

Crecchio C and Stotzky G (1998) Binding of DNA on humic acids: effect on transformation of *Bacillus subtilis* and resistance to Dnase. Soil Biol Biochem 30: 1061–1067

Degens ET and Stoffers P (1976) Stratified waters as a key to the past. Nature 263: 22–27

Drews G and Golecki JR (1995) Structure, molecular organization, and biosynthesis of membranes of purple sulfur bacteria. In: Blankenship RE, Madigan MT and Bauer CE (eds) Anoxygenic photosynthetic bacteria, Advances in Photosynthesis, Vol II, pp 231–257. Kluwer Academic, Dordrecht, The Netherlands

Ehrenberg CG (1838) Die Infusionstierchen als vollkommene Organismen. Leipzig, 2 vol.

Elshahed MS, Senko JM, Najar FZ, Kenton SM, Roe BA, Dewers TA, Spear JR and Krumholz LE (2003) Bacterial diversity and sulfur cycling in a mesophilic sulfide-rich spring. Appl Environ Microbiol 69: 5609–5621

Engelmann TW (1888) Die Purpurbakterien und ihre Beziehung zum Licht. Bot Z 46: 661–669

Erba E (2004) Calcareous nannofossils and Mesozoic oceanic anoxic events. Mar Micropaleontol 52: 85–106

Frigaard N-U, Chew AGM, Li H, Maresca JA and Bryant DA (2003) *Chlorobium tepidum*: insights into the structure, physiology, and metabolism of a green sulfur bacterium derived from the complete genome sequence. Photosynth Res 78: 93–117

Fröstl JM and Overmann J (1998) Physiology and tactic response of "*Chlorochromatium aggregatum*". Arch Microbiol 169: 129–135

Fröstl JM and Overmann J (2000) Phylogenetic affiliation of the bacteria that constitute phototrophic consortia. Arch Microbiol 174: 50–58

Fuhrmann S, Overmann J, Pfennig N and Fischer U (1993) Influence of vitamin B_{12} and light on the formation of chlorosomes in green- and brown-colored *Chlorobium* species. Arch Microbiol 16: 193–198

Garcia-Cantizano J, Casamayor EO, Gasol JM, Guerrero R and Pedros-Alio (2005) Partitioning of CO_2 incorporation among planktonic microbial guilds and estimation of in situ specific growth rates. Microb Ecol 50: 230–241

Glaeser J and Overmann J (1999) Selective enrichment and characterization of *Roseospirillum parvum*, gen. nov and sp nov., a new purple nonsulfur bacterium with unusual light absorption properties. Arch Microbiol 171: 405–416

Glaeser J and Overmann J (2003a) Characterization and in situ carbon metabolism of phototrophic consortia. Appl Environ Microbiol 69: 3739–3750

Glaeser J and Overmann J (2003b) The significance of organic carbon compounds for in situ metabolism and chemotaxis of phototrophic consortia. Environ Microbiol 5: 1053–1063

Glaeser J and Overmann J (2004) Biogeography, evolution and diversity of the epibionts in phototrophic consortia. Appl Environ Microbiol 70: 4821–4830

Glaeser J, Baneras L, Rütters H and Overmann J (2002) Novel bacteriochlorophyll e structures and species-specific variability of pigment composition in green sulfur bacteria. Arch Microbiol 177: 475–485

Göbel F (1978) Quantum efficiencies of growth. In: Clayton RK and Sistrom WR (eds) The Photosynthetic Bacteria, pp 907–925. Plenum, New York

Gorlenko VM (1988) Ecological niches of green sulfur bacteria. In: Olson JM, Ormerod JG, Amesz J, Stackebrandt E and Trüper HG (eds) Green Photosynthetic Bacteria, pp 257–267. Plenum, New York, London

Grice K, Gibbison R, Atkinson JE, Schwark L, Eckhardt CB and Maxwell JR (1996) Maleimides (1*H*-pyrrole-2,5-diones) as molecular indicators of anoxygenic photosynthesis in ancient water columns. Geochim Cosmochim Acta 60: 3913–3924

Guerrero R, Esteve I, Pedros-Alio C and Gaju N (1987a) Predatory bacteria in prokaryotic communities. The earliest trophic relationship. Ann NY Acad Sci 503: 238–250

Guerrero R, Pedrós-Alió C, Esteve I and Mas J (1987b) Communities of phototrophic bacteria in lakes of the Spanish Mediterranean region. Acta Acad Aboensis 47: 125–151

Hartgers WA, Sinninghe Damsté JS, Requejo AG, Allan J, Hayes JM and de Leeuw JW (1994) Evidence for only minor contributions from bacteria to sedimentary organic carbon. Nature 369: 224–227

Hebting Y, Schaeffer P, Behrens A, Adam P, Schmitt G, Schneckenburger P, Bernasconi SM and Albrecht P (2006) Biomarker evidence for a major preservation pathway of sedimentary organic carbon. Science 312: 1627–1631

Herbert RA, Ranchou-Peyruse A, Duran R, Guyoneaud R and Schwabe S (2005) Characterization of purple sulfur bacteria from the South Andros Black Hole cave system: highlights taxonomic problems for ecological studies among the genera *Allochromatium* and *Thiocapsa*. Environ Microbiol 7: 1260–1268

Imhoff JF (2003) Phylogenetic taxonomy of the family *Chlorobiaceae* on the basis of 16S rRNA and *fmo* (Fenna–Matthews–Olson protein) gene sequences. Int J Syst Evol Microbiol 53: 941–951

Imhoff JF (2005a) Family I. Chromatiaceae. In: Garrity JM (ed) Bergey's Manual of Systematic Bacteriology, Vol 2B, pp 3–40. Williams and Wilkins, Baltimore

Imhoff JF (2005b) Family II. Ectothiorhodospiraceae. In: Garrity JM (ed) Bergey's Manual of Systematic Bacteriology, Vol 2B, pp 41–48. Williams and Wilkins, Baltimore

Joachimski MM, Ostertag-Henning C, Pancost RD, Strauss H, Freeman KH, Littke R, Sinninghe Damsté JS and Racki G (2001) Water column anoxia, enhanced productivity and concomitant changes in $\delta^{13}C$ and $\delta^{34}S$ across the Frasnian–Famennian boundary (Kowala Holy Cross Mountains/Poland). Chem Geol 175: 109–131

Jørgensen BB, Fossing H, Wirsen CO and Jannasch HW (1991) Sulfide oxidation in the Black Sea chemocline. Deep-Sea Res 38 (Suppl): S1083–S1103

Kanzler BEM, Pfannes KR, Vogl K and Overmann J (2005) Molecular characterization of the nonphotosynthetic partner bacterium in the consortium "Chlorochromatium aggregatum". Appl Environ Microbiol 71: 7434–7441

Kara AN, Rochford PA and Hurlburt HE (2003) Mixed layer depth variability over the global ocean. J Geophys Res 108: 3079 (doi:10.1029/2000JC000736)

Kohnen MEL, Sinninghe Damsté JS and DeLeeuw JW (1991) Biases from natural sulphurization in palaeoenvironmental reconstruction based on hydrocarbon biomarker distribution. Nature 349: 775–778

Koizumi Y, Kojima H, Oguri K, Kitazato H and Fukui M (2004) Vertical and temporal shifts in microbial communities in the water column and sediment of saline meromictic Lake Kaiike (Japan), as determined by a 16S rDNA-based anmalysis, and related to physicochemical gradients. Environ Microbiol 6: 622–637

Kondratieva EN (1965) Photosynthetic Bacteria. Akademiya Nauk SSSR Institut Mikrobiologii. S. Monson, Jerusalem

Konovalov SK, Ivanov LI and Samodurov AS (2001) Fluxes and budgets of sulphide and ammonia in the Black Sea anoxic layer. J Mar Syst 31: 203–216

Konovalov SK, Luther GW, Friederich GE, Nuzzio DB, Tebo BM, Murray JW, Oguz T, Glazer B, Trouwborst RE, Clement B, Murray KJ and Romanov AS (2003) Lateral injection of oxygen with the Bosporus plume – fingers of oxidizing potential in the Black Sea. Limnol Oceanogr 48: 2369–2376

Koopmans MP, Schouten S, Kohnen MEL and Sinninghe Damsté JS (1996a) Restricted utility of aryl isoprenoids as indicators for photic zone anoxia. Geochim Cosmochim Acta 60: 4873–4876

Koopmans MP, Köster J, van Kaam-Peters HME, Kenig F, Schouten S, Hartgers WA, de Leeuw JW and Sinninghe Damsté JS (1996b) Diagenetic and catagenic products of isorenieratene: Molecular indicators for photic zone anoxia. Geochim Cosmochim Acta 60: 4467–4496

Krubasik P and Sandmann G (2000) A carotenogenic gene cluster from Brevibacterium linens with novel lycopene cyclase genes involved in the synthesis of aromatic carotenoids. Mol Gen Genet 263: 423–432

Krügel H, Krubasik P, Weber K, Saluz HP and Sandmann G (1999) Functional analysis of genes from Streptomyces griseus involved in the synthesis of isorenieratene, a carotenoid with aromatic end groups, revealed a novel type of carotenoid desaturase. Biochim Biophys Acta 1439: 57–64

Kühl M and Jørgensen BB (1992) Spectral light measurements in microbenthic phototrophic communities with a fibre-optic microprobe coupled to a sensitive diode array detector. Limnol Oceanogr 37: 1813–1823

Kühl M, Lassen C and Jørgensen BB (1994) Light penetration and light intensity in sandy marine sediments measured with irradiance and scalar irradiance fiber-optic microprobes. Mar Ecol Prog Ser 105, 139–148

Kuypers MMM, Pancost RD, Nijenhuis IA and Sinninghe Damsté JS (2002) Enhanced productivity led to increased organic carbon burial in the euxinic North Atlantic basin during the late Cenomanian oceanic anoxic event. Paleoceanography 17: 1051 (doi:10.1029/2000PA000569)

Kuypers MMM, Lourens LJ, Rijpstra IC, Pancost RD, Nijenhuis IA and Sinninghe Damsté JS (2004) Orbital forcing of organic carbon burial in the proto-North Atlantic durign oceanic anoxic event 2. Earth Planet Sci Lett 228: 465–482

Lindahl T (1993) Instability and decay of the primary structure of DNA. Nature 362: 709–715

Manske AK, Glaeser J, Kuypers MMM and Overmann J (2005) Physiology and phylogeny of green sulfur bacteria forming a monospecific phototrophic assemblage at 100 m depth in the Black Sea. Appl Environ Microbiol 71: 8049–8060

Manske AK, Henßge U, Glaeser J, Overmann J (2008) Subfossil 16S rRNA gene sequences of green sulfur bacteria as biomarkers for photic zone anoxia in the ancient Black Sea. Appl Environ Microbiol (in press)

Martínez-Alonso M, Van Bleijswijk J, Gaju N and Muyzer G (2005) Diversity of anoxygenic phototrophic sulfur bacteria in the microbial mats of Ebro Delta: a combined morphological and molecular approach. FEMS Microbiol Ecol 52: 339–350

Menzel D, Hopmans EC, van Bergen PF, de Leeuw JW and Sinninghe Damsté JS (2002) Development of photic zone euxinia in the eastern Mediterranean basin during deposition of pliocene sapropels. Mar Geol 189: 215–226

Montesinos E, Guerrero R, Abella C and Esteve I (1983) Ecology and physiology of the competition for light between Chlorobium limicola and Chlorobium phaeobacteroides in natural habitats. Appl Environ Microbiol 46: 1007–1016

Murray JW, Jannasch HW, Honjo S, Anderson RF, Reeburgh WS, Top Z, Friederich GE, Codispoti LA and Izdar E (1989) Unexpected changes in the oxic/anoxic interface in the Black Sea. Nature 338: 411–413

Nicholson JAM, Stolz JF and Pierson BK (1987) Structure of a microbial mat at Great Sippewissett Marsh, Cape Cod, Massachusetts. FEMS Microbiol Ecol 45: 343–364

Okubo Y, Futamata H and Hiraishi A (2006) Characterization of phototrophic purple nonsulfur bacteria forming colored microbial mats in a swine wastewater ditch. Appl Environ Microbiol 72: 6225–6233

Overmann J (1997) Mahoney Lake: a case study of the ecological significance of phototrophic sulfur bacteria. Adv Microbial Ecol 15: 251–288

Overmann J (2001) Green sulfur bacteria. In Garrity JM (ed) Bergey's Manual of Systematic Bacteriology, Vol 1, pp 601–623. Williams and Wilkins, Baltimore

Overmann J (2006) The symbiosis between nonrelated bacteria in phototrophic consortia. Chapter II. In: Overmann J (ed) Molecular Basis of Symbiosis. Progress in Molecular Subcellular Biology, pp 21–37. Springer, Berlin, Heidelberg

Overmann J and Garcia-Pichel F (2000) The phototrophic way of life. In: Dworkin M, Falkow S, Rosenberg E, Schleifer K-H, Stackebrandt E (eds) The Prokaryotes: An Evolving Electronic Resource for the Microbiological Community, 3rd edition (latest update release 3.11, September 2002), Springer, New York, 2000 (http://ep.springer-ny.com:6336/contents/)

Overmann J and Manske A (2006) Anoxygenic phototrophic bacteria in the Black Sea chemocline. In: Neretin L (ed) Past and Present Water Column Anoxia. NATO ASI series, pp 543–561. Springer, Berlin

Overmann J and Pfennig N (1992) Buoyancy regulation and aggregate formation in Amoebobacter purpureus from Mahoney Lake. FEMS Microbiol Ecol 101: 67–79

Overmann J and Schubert K (2002) Phototrophic consortia: model systems for symbiotic interrelations between prokaryotes. Arch Microbiol 177: 201–208

Overmann J and Tuschak C (1997) Phylogeny and molecular fingerprinting of green sulfur bacteria. Arch Microbiol 167: 302–309

Overmann J, Beatty JT, Hall KJ, Pfennig N and Northcote TG (1991a) Characterization of a dense, purple sulfur bacterial layer in a meromictic salt lake. Limnol Oceanogr 36: 846–859

Overmann J, Lehmann S and Pfennig N (1991b) Gas vesicle formation and buoyancy regulation in Pelodictyon phaeoclathratiforme (Green sulfur bacteria). Arch Microbiol 157: 29–37

Overmann J, Cypionka H and Pfennig N (1992) An extremely low-light adapted phototrophic sulfur bacterium from the Black Sea. Limnol Oceanogr 37: 150–155

Overmann J, Beatty JT and Hall KJ (1994) Photosynthetic activity and population dynamics of Amoebobacter purpureus in a meromictic saline lake. FEMS Microbiol Ecol 15: 309–320

Overmann J, Beatty JT and Hall KJ (1996) Purple sulfur bacteria control the growth of aerobic heterotrophic bacterioplankton in a meromictic salt lake. Appl Environ Microbiol 62: 3251–3258

Overmann J, Coolen MJL and Tuschak C (1999a) Specific detection of different phylogenetic groups of chemocline bacteria based on PCR and denaturing gradient gel electrophoresis of 16S rRNA gene fragments. Arch Microbiol 172: 83–94

Overmann J, Hall KJ, Northcote TG and Beatty JT (1999b) Grazing of the copepod Diaptomus connexus on purple sulfur bacteria in a meromictic salt lake. Environ Microbiol 1: 213–222

Overmann J, Hall KJ, Northcote TG, Ebenhöh W, Chapman MA and Beatty JT (1999c) Structure of the aerobic food chain in a meromictic lake dominated by purple sulfur bacteria. Arch Hydrobiol 144: 127–156

Pancost RD, Freeman KH, Patzkowsky E, Wavrek DA and Collister JW (1998) Molecular indicators of redox and marine photoautotroph composition in the late Middle Ordovivian of Iowa, U.S.A. Org Geochem 29: 1649–1662

Pancost RD, Crawford N, Magness S, Turner A, Jenkyns HC and Maxwell JR (2004) Further evidence for the development of photic-zone euxinic conditions during Mesozoic oceanic anoxic events. J Geol Soc London 161: 353–364

Parkin TB and Brock TD (1980a) Photosynthetic bacterial production in lakes: the effects of light intensity. Limnol Oceanogr 25: 711–718

Parkin TB and Brock TD (1980b) The effects of light quality on the growth of phototrophic bacteria in lakes. Arch Microbiol 125: 19–27

Passier HF, Bosch H-J, Nijenhuis LJL, Böttcher ME, Leenders A, Sinninghe Damsté JS, de Lange GJ and de Leeuw JW (1999) Sulphidic Mediterranean surface waters during Pliocene sapropel formation. Nature 397: 146–149

Pedrós-Alió C and Sala MM (1990) Microdistribution and diel vertical migration of flagellated vs. gas-vacuolate purple sulfur bacteria in a stratified water body. Limnol Oceanogr 35: 1637–1644

Permentier HP, Neerken S, Overmann J and Amesz J (2001) A bacteriochlorophyll a antenna complex from purple bacteria absorbing at 963 nm. Biochemistry 40: 5573–5578

Pfennig N (1978) General physiology and ecology of photosynthetic bacteria. In: Clayton RK and Sistrom WR (eds) The Photosynthetic Bacteria, pp 3–18. Plenum, New York

Pfennig N (1993) Reflections of a microbiologist, or how to learn from the microbes. Annu Rev Microbiol 47: 1–29

Pfennig N, Lünsdorf H, Süling J and Imhoff JF (1997) Rhodospira trueperi gen. nov., spec. nov., a new phototrophic Proteobacterium of the alpha group. Arch Microbiol 168: 39–45

Pierson B, Oesterle A and Murphy GL (1987) Pigments, light penetration, and photosynthetic activity in the multilayered microbial mats of Great Sippewissett Salt Marsh, Massachusetts. FEMS Microbiol Ecol 45: 365–376

Poinar HN, Höss M, Bada JL and Pääbo S (1996) Amino acid racemization and the preservation of ancient DNA. Science 272: 864–866

Repeta DJ (1993) A high resolution historical record of Holocene anoxygenic primary production in the Black Sea. Geochim Cosmochim Acta 57: 4337–4342

Repeta DJ, Simpson DJ, Jørgensen BB and Jannasch HW (1989) Evidence for anoxygenic photosynthesis from the distribution of bacterichlorophylls in the Black Sea. Nature 342: 69–72

Requejo AG, Allan J, Creaney S, Gray NR and Cole KS (1992) Aryl isoprenoids and aromatic carotenoids in Paleozoic rocks and oils from the Western Canada and Williston Basins. Org Geochem 19: 245–264

Rohling EJ and Hilgen FJ (1991) The eastern Mediterranean climate at times of sapropel formation: a review. Geol Mijnbouw 70: 253–264

Rohling EJ, Hopmans EC and Sinninghe Damsté JS (2006) Water column dynamics during the last interglacial anoxic event in the Mediterranean (sapropel S5). Paleoceanography 21: PA2018, doi:10.1029/2005PA001237

Romanowski G, Lorenz MG and Wackernagel W (1991) Adsorption of plasmid DNA to mineral surfaces and protection against Dnase I. Appl Environ Microbiol 57: 1057–1061

Ross DA and Degens ET (1974) In: Degens E and Ross D (eds), The Black Sea – Geology, Chemistry and Biology, pp 183–199 (Memoir No. 20. Am Ass. Petrol. Geol., Tulsa, Oklahoma)

Rossignol-Strick M (1985) Mediterranean Quaternary sapropels, an inmediate response of the African Monsoon to variation of insolation. Paleogeogr Paleoclimatol Paleoecol 49: 237–263

Sanchez O, van Gemerden and Mas J (1998) Acclimation of the photosynthetic response of *Chromatium vinosum* to light-limiting conditions. Arch Microbiol 170: 405–410

Schulz H-M, Sachsenhofer RF, Bechtel A, Polesny H and Wagner L (2002) The origin and hydrocarbon source rocks in the Austrian Molasse Basin (Eocene–Oligocene transition). Mar Petrol Geol 19: 683–709

Simons D-JH and Kenig F (2001) Molecular fossil constraints on the water column structure of the Cenomanian–Turonian Western Interior Seaway, USA. Palaeogr Palaeoclimatol Palaeoecol 169: 129–152

Sinninghe Damsté JS, Wakeham SG, Kohnen MEL, Hayes JM and de Leeuw JW (1993) A 6,000-year sedimentary molecular record of chemocline excursion in the Black Sea. Nature 362: 827–829

Sinninghe Damsté JS, Frewin NL, Kenig F and de Leeuw JW (1995) Molecular indictaors for palaeoenvironmental change in a Messinian evaporitic sequence (Vena del Gesso, Italy). I: Variations in extractable organic matter of ten cyclically deposited marl beds. Org Geochem 23: 471–483

Sorokin YuI (1970) Interrelations between sulphur and carbon turnover in meromictic lakes. Arch Hydrobiol 66: 391–446

Sorokin YuI (2002) The Black Sea. Ecology and oceanography. Backhuys, Leiden, The Netherlands

Steensgaard DB, Wackerbarth H, Hildebrandt P and Holzwarth AR (2000) Diastereoselective control of bac-teriochlorophyll *e* aggregation. 3^1-S-BChl *e* is essential for the formation of chlorosome-like aggregates. J Phys Chem B 104: 10379–10386

Thar R and Kühl M (2001) Motility of *Marichromatium gracile* in response to light, oxygen, and sulfide. Appl Environ Microbiol 67: 5410–5419

Tonolla M, Demarta A, Peduzzi S, Hahn D and Peduzzi R (2000) In situ analysis of sulfate-reducing bacteria related to *Desulfocapsa thiozymogenes* in the chemocline of meromictic Lake Cadagno (Switzerland). Appl Environ Microbiol 66: 820–824

Tonolla M, Peduzzi R and Hahn D (2005) Long-term population dynamics of phototrophic sulfur bacteria in the chemocline of Lake Cadagno, Switzerland. Appl Environ Microbiol 71: 3544–3550

Tuschak C, Glaeser J and Overmann J (1999) Specific detection of green sulfur bacteria by in situ-hybridization with a fluorescently labeled oligonucleotide probe. Arch Microbiol 171: 265–272

van den Ende FP, Laverman AM and van Gemerden H (1996) Coexistence of aerobic chemotrophic and anaerobiv phototrophic sulfur bacteria under oxygen limitation. FEMS Microbiol Ecol 19: 141–151

Van Dover CL, Reynolds GT, Chave AD and Ryson JA (1996) Light at deep-sea hydrothermal vents. Geophys Res Lett 23: 2049–2052

van Gemerden H and Mas J (1995) Ecology of phototrophic sulfur bacteria. In: Blankenship RE, Madigan MT and Bauer CE (eds) Anoxygenic Photosynthetic Bacteria, Advances in Photosynthesis, Vol II, pp 49–85. Kluwer Academic, Dordrecht, The Netherlands

van Gemerden H, Tughan CS, De Wit R and Herbert RA (1989) Laminated microbial ecosystems on sheltered beaches in Scapa Flow, Orkney Islands. FEMS Microbiol Ecol 62: 87–102

van Niel (1931) On the morphology and physiology of the purple and green sulphur bacteria. Arch Microbiol 3: 1–112

van Rossum B-J, Steensgaard DB, Mulder FM, Boender GJ, Schaffner K, Holzwarth AR and de Groot HJM (2001) A refinded model of the chlorosomal antennae of the green sulfur bacterium *Chlorobium tepidum* from proton chemical shift constraints obtained with high-field 2-D and 3-D MAS NMR dipolar correlation spectroscopy. Biochemistry 40: 1587–1595

Vetriani C, Tran HV and Kerkhof LJ (2003) Fingerprinting microbial assemblages from the oxic/anoxic chemocline of the Black Sea. Appl Environ Microbiol 69: 6481–6488

Vila X, Guyoneaud R, Cristina XP, Figueras JB and Abella CA (2002) Green sulfur bacteria from hypersaline Chiprana Lake (Monegros, Spain): habitat description and phylogenetic relationship of isolated strains. Photosynth Res 71: 165–172

Vogl K, Glaeser J, Pfannes KR, Wanner G and Overmann J (2006) *Chlorobium chlorochromatii* sp. nov., a symbiotic green sulfur bacterium isolated from the phototrophic consortium "*Chlorochromatium aggregatum*". Arch Microbiol 185: 363–372

Wagner T, Sinninghe Damsté JS, Hofmann P and Beckmann B (2005) Euxinia and primary production in Late Cretaceous eastern equatorial Atlantic surface waters fostered orbitally driven formation of marine black shales. Palaeoceanography 19:PA3009

Wakeham SG, Sinninghe Damsté JS, Kohnen MEL and DeLeeuw JW (1995) Organic sulfur compounds formed during early diagenesis in Black Sea sediments. Geochim Cosmochim Acta 59: 521–533

Warthmann R, Cypionka H and Pfennig N (1992) Photoproduction of H_2 from acetate by syntrophic cocultures of green sulfur bacteria and sulfur-reducing bacteria. Arch Microbiol 157: 343–348

Willerslev E, Hansen AJ, and Poinar HN (2004) Isolation of nucleic acids and cultures from fossil ice and permafrost. TRENDS Ecol Evol 19: 141–147

Winogradsky S (1887) Über Schwefelbakterien. Bot Ztg 45: 489–508

Zuber H and Cogdell RJ (1995) Structure and organization of purple bacterial antenna complexes. In: Blankenship RE, Madigan MT and Bauer CE (eds) Anoxygenic Photosynthetic Bacteria, pp 315–348. Kluwer Academic, Dordrecht, The Netherlands

Chapter 20

Role of Sulfur for Algae: Acquisition, Metabolism, Ecology and Evolution

Mario Giordano*, Alessandra Norici, and Simona Ratti
Laboratorio di Fisiologia delle Alghe, Dipartimento di Scienze del Mare, Università Politecnica delle Marche, Via Brecce Bianche, 60131 Ancona, Italy

John A. Raven
University of Dundee at SCRI, Scottish Crop Research Institute, Invergowrie, Dundee DD2 5DA, UK

Summary

Algae, like most photolithotrophs, acquire sulfur as sulfate. In the oceans sulfate concentration is never limiting and is consistently very high (29 mM). Freshwaters, however, are characterized by daily and seasonal variations and by concentrations that cover a very broad range and can, in some cases, be very low. The decrease in anthropogenic sulfur emission may reduce further these concentrations and in the future sulfur may becomes limiting in some lakes. The strategies that algae adopt in acquiring and assimilating sulfur in different environments often reflect these differences. The availability of sulfate also has repercussion on the overall metabolism of algal cells, because of its many pivotal roles in cell physiology. The maintenance of homeostasis and the responses to sulfate deprivation have a strong impact on photosynthesis, and carbon and nitrogen acquisition and metabolism. In a number of algae, most sulfur is allocated into dimethylsulfonioproprionate, whose cleavage into acrylate and dimethylsulfide is of great relevance for the ecology of extant phytoplankton and may have been important in the radiation of some algal groups. Dimethylsulfide, furthermore, is the main source of biogenic atmospheric sulfur and it is believed to have a major role in the control of global climate. These and other related matters are discussed in this review.

---

*Corresponding author, Phone: +39 071 220 4652, Fax: +39 071 220 4650, E-mail: m.giordano@univpm.it

Rüdiger Hell et al. (eds.), Sulfur Metabolism in Phototrophic Organisms, 397–415.
© 2008 *Springer.*

I. Introduction

The yearly sulfur assimilation by phytoplankton is in the order of 1.3 Pg and requires 11.3×10^{15} to 15.9×10^{15} kJ yr^{-1}, depending on the assimilation pathway (Norici et al., 2005) and not taking into consideration the energy used to make sulfate esters in the cell (Bates et al., 1994). The net primary production of aquatic photolithotrophs is about 47.5 Pg of C per year (Field et al., 1998) and the amount of energy necessary to support this C fixation is approximately 854×10^{15} kJ yr^{-1}, if all C is fixed via the Calvin cycle. From this derives that sulfur assimilation by phytoplankton, on a global scale, uses 1.3 to 1.9% of the energy used for CO_2 assimilation. In spite of this non-trivial amount of energy that sulfur assimilation by phytoplankton requires, and the fact that some of the earliest and most detailed evidence on the biochemistry (references in Schmidt and Jäger, 1992 and Schiff et al., 1993) and regulation (Davies and Grossman, 1998; Lilly et al., 2002; Zhang et al., 2004) of sulfur metabolism were obtained from algae, the literature on sulfur in algae is rather scant (Giordano et al., 2005b; Norici et al., 2005). The data available for algae, moreover, mostly concern organisms of little ecological significance and very limited value as a model for all algal systems. The release of biogenic reduced sulfur into the atmosphere as dimethylsulfide (DMS) is the only portion of the sulfur cycle for which the role of phytoplankton has been investigated in detail. The interest in this topic is probably due to the proposed influence that DMS may exert on global climate (Lovelock et al., 1972, 1974; Charlson et al., 1987). This review provides an overview of the ecological and physiological aspects of the interaction between sulfur and phytoplankton.

Abbreviations: APS – 5′-adenylsulfate, DMS – dimethylsulfide, DMSHB – 4-dimethylsulfonio-2-hydroxybutyrate, DMSP – dimethylsulfonioproprionate, GSH – glutathione, MTHB – 4-methylthio-2-hydroxybutyrate, MTOB – 4-methylthio-2-oxobutyrate, OAS-TLO – -acetylserine (thiol) lyase, PAPS – phosphoadenosine-5′-phosphosulfate, PAPR – PAPS – reductase, psu – practical salinity units, SAT – serine acetyltransferase

II. Sulfur Availability in Aquatic Ecosytems

In algae, as in most photosynthetic organisms, sulfur is usually taken up and assimilated as sulfate. The availability of sulfate may vary substantially in different aquatic environments, determining remarkable differences in the strategies algae use in its acquisition and utilization.

Freshwater Lakes – In lakes, sulfur primary provenience are the weathering of rocks in the catchment and the oxidation of organic sulfur from terrestrial sources. As a consequence of the industrial revolution, anthropogenic atmospheric SO_2 emission rose, causing a large increase in lacustrine sulfate concentrations centred on industrialized areas (Brimblecombe et al., 1989; Zhao et al., 1998; Fig. 1). The extent of this increase depended on the level of industrialization of the surrounding areas, on the atmospheric circulation, and on water chemistry; consequently, the concentration of sulfate in lakes is highly variable, with values ranging from 0.01 to 1 mM (Tipping et al., 1998; Holmer and Storkholm, 2001). The increase in sulfur deposition, in many cases, caused acidification that was often associated with a reduction in biodiversity and organism abundance (Brown, 1983; Driscoll and Schecher, 1990; Charles, 1991; Driscoll et al., 2001 and references therein). The increase of sulfate concentrations in lakes may also have affected the availability of other nutrients. Increased sulfate in the anoxic parts of the hypolimnion stimulates sulfate reduction, and insoluble iron sulfide tends to precipitate in sediments. As a consequence of this, iron availability in the water column declines, potentially causing a reduction in productivity (Nürnberg, 1996; Kleeberg, 1997). However, the formation of iron sulfide competes with the binding of phosphate to iron oxides (Murray, 1995; Nürnberg, 1996; Kleeberg, 1997), making phosphate, often the limiting nutrient in lakes (Holmer and Storkholm, 2001 and references therein; Maberly et al., 2003; Petaloti et al., 2004), more available in the water column.

In recent years, the enforcement of legislation limiting SO_2 emissions (Cape et al., 2003) caused the trends of sulfate concentrations in freshwaters to reverse, in some areas of Europe and North America (Tipping et al., 1998; Stoddard et al., 1999). This is starting to have an impact

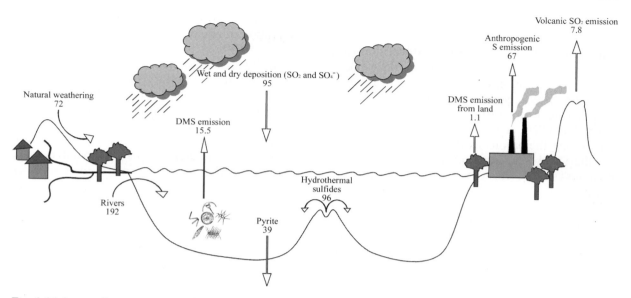

Fig. 1. Main contributors to the sulfur budget for aquatic and terrestrial environments. All values are in 10^{12} g S yr^{-1}. The values are derived from Schlesinger, 1997 and Giordano et al. 2005 b and references therein. The figure is modified from Schlesinger, 1997.

on phytoplankton communities: the addition of sulfate relieved growth limitation in natural phytoplankton assemblages from an oligotrophic lake in northern England, once phosphorus was replenished (M. Giordano, A. Norici, V. Pezzoni; S.C. Maberly, unpublished). If sulfur deposition continues to decline, sulfur may thus become limiting for phytoplankton growth, in some lakes, either directly or by reducing the effect of iron sulfide formation on phosphorus availability (Holmer and Storkholm, 2001 and references there in).

Sulfur availability in water basins is also influenced by the presence of pollutants such as selenate and molybdate; these anions, in fact, compete with sulfate for uptake via common transporters and may therefore reduce the ability of organisms to acquire sulfate (Ramaiah and Shanmugasundaram, 1962; Tweedie and Segel, 1970; Wheeler et al., 1982; Vandermeulen and Foda, 1988; Jonnalagadda and Prasada Rao, 1993; Ogle and Knight, 1996; Riedel et al., 1996; Riedel and Sanders, 1996; Bagchi and Verma, 1997; Baines and Fisher, 2001; Neumann et al., 2003; Obata et al., 2004). Cadmium, when present in high concentrations, may elicit a physiological syndrome resembling the effect of sulfur deficiency (Mosulén et al., 2003). To cope with

heavy metals, algae produce phytochelatins. Phytochelatins derive from glutathione (GSH), which is constituted for a substantial quota by cysteine. In response to high cadmium levels, thus, GSH production can be dramatically enhanced, using up large amount of cysteine (Cobbett, 2000; Saito, 2000; Dominguez-Solis et al., 2004; Mendoza-Cozatl et al., 2005) and energy (two moles of ATP per mole of GSH; Noctor et al., 2002; Mullineaux and Rausch, 2005). Under these conditions, especially in oligotrophic lakes, algae may experience sulfur limitation, which may be exacerbated by the energy shortage deriving from the high energetic requirements of GSH synthesis.

Sulfur limitation has left an evolutionary imprint on the proteome of organisms. Mazel and Marlière (1989) found that sulfur-starved cyanobacteria express versions of light-harvesting phycobilins with a lower content of cysteine and methionine than the versions expressed in sulfur-replete cells. Subsequent work has shown proteomic responses of this type in organisms of various trophic modes from habitats which could be subject to sulfur deficiency (Baudouin-Cornu et al., 2001, 2004; Raven et al., 2005; Bragg et al., 2006).

Inland Seas and Saline Lakes – Inland seas and saline lakes are water basins with a salinity in excess of 3 psu (practical salinity units: seawater averages about 35 psu). These saline inland waters have a global volume of 0.104 million km³, compared to 0.125 million km³ in freshwaters, and 1,322 million km³ in the ocean (Horne and Goldman, 1994). Some very saline lakes (about 10 times seawater salinity), e.g. the Great Salt Lake, UT, USA, have the same fraction of sulfate among the anions as does seawater. Others, e.g. the Dead Sea, Israel/Jordan, have relatively much less sulfate than the sea, although sulfate is still present in abundance for the few photosynthetic organisms (mainly the chlorophyte flagellate *Dunaliella*). While salt lakes like the two mentioned above, similarly to the sea, have chloride as the dominant anion, others are dominated by carbonate or sulfate (Last and Ginn, 2005). Macroalgae in sulfate-dominated lakes can have sulfate as a dominant vacuolar anion (Hoffmann and Bisson, 1986). Other algal habitats which are not typically saline, but where sulfate is naturally the dominant anion, are the acidic volcanic habitats typically inhabited by red microalgae such as *Cyanidium*, *Cyanidioschyzon* and *Galdieria* (Graham and Wilcox, 2000).

High-sulfate habitats may have impacts on the availability of essential elements acquired as oxyanions which are structurally similar to sulfate, and with which sulfate may compete for uptake sites. Examples of such possible competitive interactions are selenate, providing an element used in a selenium-requiring glutathione peroxidase which occurs in some algae (Raven et al., 1999), and molybdate, a component of such enzymes as nitrate reductase in cyanobacteria and eukaryotic algae and the most common form of nitrogenase in cyanobacteria (Marino et al., 2003). While sulfate limits the stimulation of nitrogen fixation by marine cyanoplankton by addition of molybdate to mesocosms, the effect of sulfate is not entirely competitive with molybdate (Marino et al., 2003). Such effects are the inverse of the inhibition of sulfate uptake by selenate and molybdate in polluted freshwaters discussed above.

Oceans – In oceans, the amount of sulfur increased substantially during the course of Earth's history (e.g.: Canfield et al., 2000; Habicht et al., 2002; Shen et al., 2003). Today, the oceans are one of the main reservoirs of dissolved sulfur (Fig. 1) (Strauss, 1997), with a sulfate concentration of about 29 mmol per l (Strauss, 1997; Pilson, 1998), 2–3 orders of magnitude more than in freshwaters (Holmer and Storkholm, 2001). The difference in sulfur concentration between oceans and freshwaters may be what permits the cell walls of unrelated marine algae and marine angiosperms (seagrasses, but not, as far as the literature known to us is concerned, emergent halophytes such as mangroves and salt marsh plants), and extracellular structures of marine invertebrates to have high concentrations of sulfated polysaccharides absent in freshwater organisms and land plants. In marine angiosperms, sulfated galactan is in the plant cell walls, mostly in rhizomes and roots. This localization may reflect a relationship with nutrient absorption and a possible structural function. The occurrence of sulfated galactan in marine organisms may results from a convergent adaptation due to common pressures in the marine environment, e.g. maintaining the structure and function of anionic polysaccharides in an environment of high ionic strength (Aquino et al., 2005; cf. sulfated polysaccharides in mammalian connective tissue, also in a high ionic strength environment: Fitzgerald, 1976). It is equally possible, however, that freshwater organisms lost their ability to synthesize sulfated galactans in response to a sulfur-poor environment There are some intriguing observations on the metabolism of these sulfated polysaccharides, such as the apparent rapid (three exchanges per hour) and energy-dependent turnover of the peripheral, but not the central sulfate moieties of fucoidan in the brown macroalga *Pelvetia canaliculata* (Léstang-Bremond and Quillèt, 1976), which deserve more investigation.

Marine macroalgae use sulfate as a major vacuolar osmoticum to a greater extent than most freshwater algae. In some cases magnesium serves as the major counter-ion; in a larger number of organisms, however, protons are the major counter-ions, giving vacuolar pH values as low as 1.0 (Peters et al., 1997; Sasaki et al., 1999, 2005a, 2005b). The high sulfur content of marine organisms may be related to its proposed role in the deterrence of grazers (Pelletreau and Muller-Parker, 2002). Sulfate is the densest of the common inorganic anionic constituents of algal vacuoles, while protons are among the least dense cations (Raven, 1997; Boyd and Gradmann,

2002). This has implications for the regulation of overall cell density and hence of buoyancy in phytoplankton with a large fraction of the cells occupied by vacuoles (e.g. many large-celled diatoms), as well as for the posture of benthic macroalgae (Raven, 1997).

Unfortunately, most published data on the modulation of sulfur acquisition and metabolism in algae concern sulfur limitation. While this is not a condition that marine phytoplankton often faces, it may concern algae inhabiting some freshwater environments. In the oceans, the abundance of sulfur may allow a more liberal use of S, which can be used in metabolites of various kinds (osmolytes, structural components, compounds contributing to cell density and hence buoyancy, etc.). Since algae maintain a relatively constant cell composition (Sterner and Elser, 2002), the acquisition and assimilation of sulfur is controlled by the availability of other nutrients and, in turns, affects their use. The relative abundance of S, C, N, for instance, can have major repercussion on cell metabolism (Giordano et al., 2000). Therefore, in spite of the high and relatively invariant sulfur availability, marine algae must be ready to modulate their sulfur acquisition and metabolism to respond to disturbances in the interactions between sulfur-related pathways and the rest of cell activities.

III. Sulfur Acquisition by Algae

Sulfate transporters of green algae and cyanobacteria are integral membrane proteins that, in freshwater organisms, promote sulfate/H^+ or sulfate/Na^+ co-transport (Weiss et al., 2001; Palenik et al., 2003). Several genes encoding sulfate transporters have been cloned from plants. Heterologous expression in yeast reveals high ($K_m^{sulfate}$ 1–10 μM) and low affinity ($K_m^{sulfate}$ 0.1–1 mM) subtypes (Hawkesford and Wray, 2000; Saito, 2000). Algal sulfate transporters have been characterized in *Chlamydomonas reinhardtii* (Yildiz et al., 1994), in which both the V_{max} (10-fold) and the affinity (7-fold) for sulfate transport were greatly increased in this organism under sulfur limited conditions (Davies et al., 1994). Similar results were also obtained for *Dunaliella salina* (M. Giordano and V. Pezzoni, unpublished data). Despite the findings from

D. salina which normally lives in high-S habitats, it is currently unclear whether these findings can be generalized to marine algae and other algal taxa; apparently, the high sulfate content of the oceans makes inducible uptake systems unnecessary. It should be noted that inducible sulfate transporters do not seem to be present in prokaryotic algae such as freshwater *Synechococcus* spp., where a single transport system for sulfate has been found, which is probably directly powered by ATP rather than by coupling to potassium, proton, sodium or chloride transport (Green and Grossman, 1988; Laudenbach and Grossman, 1991; Ritchie, 1996). Ritchie (1996) carefully defined the chemical concentration and electrical potential difference driving sulfate across the plasmalemma in *Synechococcus*. The energy released in sulfate efflux from the cells defines the minimum energy input (kJ) needed to transport a mole of sulfate into the cells. For a marine alga, with 29 mM sulfate in the medium, an assumed 1 mM free sulfate in the cytosol, the free energy difference favoring sulfate entry would be about 8.4 kJ per mole. The driving force tending to move a divalent anion like sulfate out of the cell, with a 'typical' electrical potential difference across the plasmalemma of 60 mV, inside negative, would be approximately 11.6 kJ per mole. Combining the chemical and electrical components gives a driving force tending to move sulfate out of the cell of 3.2 kJ per mole: this is the minimum energy needed to move one mole of sulfate into the cells.

It has been shown that some eukaryotic organisms can also cleave ester bonds between sulfate and organic substances via the action of arylsulfatases (*ARS*). The *ARS* genes are mostly induced upon sulfate deficiency; the enzymes are targeted to the outside of the cell, making sulfate from organic sources available for import. Among algae, ARS enzymes have been found in the freshwater chlorophytes *Chlamydomonas reinhardtii*, whose protein complement contains two periplasmic arylsulfatases with similar regulatory and kinetic properties (Davies and Grossman, 1998; Ravina et al., 2002), and *Volvox carteri*, whose genome includes only one ARS gene (Hallmann and Sumper, 1994). Further investigations are needed to reveal whether the presence of ARS occur in other taxa and environments.

IV. Assimilation and Reduction of Sulfate by Algae

In the green alga *Chlamydomonas*, the metabolic pathways for sulfur assimilation appears to be very similar to that of higher plants (Hawkesford and Wray, 2000; Leustek et al., 2000; Saito, 2000), and the sequence similarity between the genes encoding the key enzymes in this alga and their orthologues from higher plants is very high (Ravina et al., 2002). Phylogenetically, the algal enzymes seem to have diverged from the higher plant genes prior to the gene amplification that generated the gene families of higher plants (Ravina et al., 2002). Sulfate, after uptake into the cytoplasm, is transported into the plastids, or, if present in excess, stored in vacuoles.

The chemically stable sulfate anion must be activated by ATP to 5'-adenylsulfate (APS) before it can be reduced. This reaction is catalyzed by ATP sulfurylase (ATP-S). Such activation is also required, apparently involving the cytidine analogue of APS, when sulfur is incorporated, at the sulfate redox level, into sulfated polysaccharides (Léstang and Quillet, 1973). Sulfate activation is also needed for incorporation, with partial reduction, of sulfate into the sulfonyl residue of sulfoquinovose, the polar group of the thylakoid membrane sulfolipid sulfoquinovosyl diacylglyceride found in all O_2-evolving organisms examined, apart from the cyanobacterium *Gloeobacter violaceous*, that lacks thylakoids and performs the 'thylakoid reactions' in the cell membrane (Güler et al., 1996; Selstam and Campbell, 1996; Benning, 1998; Khotimchenko, 2002; Sato, 2004). Van Mooy et al. (2006) have shown that *Prochlorococcus*, a marine cyanobacterium with a very low cell phosphorus content, has a very high ratio of sulfolipid to phospholipid, suggesting that sulfolipid substitutes for phospholipid in the plasmalemma as well as being, as usual, the dominant polar lipid in the thylakoid membranes. These data are from laboratory cultures with sulfate to phosphate concentration ratios in the culture medium that are three orders of magnitude higher than in some *Prochlorococcus* habitats, and Van Mooy et al. (2006) speculate, following Benning (1998), that even higher sulfolipid to phospholipid ratios may occur in some natural populations of *Prochlorococcus*.

In the mainstream pathway, APS is then reduced to sulfite by APS reductase, receiving two electrons from GSH (Bick and Leustek, 1998). In both plants and algae, APS reductase is a primary regulation site for the sulfate assimilation pathway (Gao et al., 2000; Kopriva et al., 2002). Remarkably, in some algae APS reductase activity is 400-fold higher than typically found in plants, it correlates with growth rates and depends on N availability (Gao et al., 2000). Sulfite reduction to sulfide is catalyzed by sulfite reductase, an enzyme structurally and functionally similar to nitrite reductase (Bork et al., 1998). The resulting free sulfide is rapidly incorporated into cysteine, the first stable organic sulfur compound in the pathway.

Early studies of sulfate assimilation using *Chlorella pyrenoidosa* assumed the existence of an APS-sulfotransferase that used APS to transfer the sulfate group to a carrier-bound SH-group ($RS:SO_3H$), possibly GSH (Schmidt, 1972). The presence of this activity was later also demonstrated for freshwater *Euglena* and the marine macroalga *Porphyra* (Schiff et al., 1993, and references therein; Kanno et al., 1996), suggesting that bound-sulfite was as an intermediate of sulfate reduction. In the freshwater flowering plant *Lemna,* however, the APS-sulfotransferase activity was identical to the enzyme APS reductase (Suter et al., 2000). A thioredoxin-dependent reduction of phosphoadenosine-5'-phosphosulfate (PAPS) by a PAPS reductase (PAPR) and further reduction of free sulfite by sulfite reductase was also proposed (Schwenn, 1989). Reduction via PAPS requires the activation of APS to PAPS by an APS kinase, whose enzymatic activity and the corresponding genes are well characterized in terrestrial plants (Schwenn, 1994). The resulting reaction sequence would closely resemble that found in cyanobacteria and enterobacteria. Interestingly, APS reductase-like proteins were also found in eubacteria such as *Pseudomonas* (Bick et al., 2000). Recently, a targeted knockout of the single APS-reductase gene (*APR*) in the bryophyte *Physcomitrella patens* revealed the co-existence of a functional *PAPR* gene (Koprivova et al., 2000), provided unambiguous molecular-genetic evidence of the existence of such enzymatic activity in eukaryotic photosynthetic organism. The genome databases of higher plants have so far not revealed sequences with significant homology to either bacterial or *Physcomitrella* PAPR genes. It is therefore likely that APR and PAPR genes were present in the common ancestor of mosses and higher plants and that PAPR genes were subsequently lost during the evolution of vascular plants (Koprivova et al., 2002). The evolutionary trajectories of this assimilatory pathways among the algae remains to be elucidated (Gao et al., 2000).

In photoautotrophic organisms, essentially all reduced sulfur is incorporated into cysteine, that, directly or indirectly, serves as the precursor for all compounds containing reduced sulfur. Cysteine synthesis is afforded by the consecutive action of serine acetyltransferase (SAT) and O-acetylserine (thiol) lyase (OAS-TL). In higher plants these two enzymes occur in the cysteine synthase complex, which is a pivotal position in the regulatory system that mediates between sulfate acquisition at the plasmalemma and cysteine formation (Wirtz et al., 2000; Hell and Hillebrand, 2001; Berkowitz et al., 2002; Hell et al., 2002). In *Chlamydomonas reinhardtii*, all the messengers involved in cysteine synthesis (at least those isolated so far) appear to code for proteins with chloroplast transit peptides; this suggests that, in this alga, differently from vascular plants, cysteine synthesis takes place exclusively in the chloroplast (Ravina et al., 2002). If this holds true, it would agree with the finding that the genes of this pathways identified so far in algae are not diversified into families, in contrast to those of higher plants. The distribution of the enzyme of cysteine synthesis in different cell compartments may thus be a recent event, possibly associated with tissue specialization (Ravina et al., 2002).

Photosynthetic organisms, but not animals, use cysteine as a precursor of methionine (Ravanel et al., 1998; Hell and Hillebrand, 2001; Hesse and Hoefgen, 2003). Until recently, only cytosolic forms of methionine synthase had been isolated or cloned (Eichel et al., 1995). This would imply that methionine is only synthesized in the cytosol and must therefore be re-imported into plastids. Ravanel et al (2004), however, reported on a plastidial methionine synthase in *Arabidopsis thaliana*, which would make the chloroplast autonomous with respect to methionine synthesis. These findings still await confirmation for algae, since significant differences in the plastid proteome and metabolome exists between higher plants and at least some algae. We refer to other chapters of this book for a more thorough discussion of these aspects of sulfur metabolism.

As the result of the assimilatory steps described above, phytoplankton assimilates inorganic sulfate into organic compounds at an average global rate of $260 \mu moles\ m^{-2}\ d^{-1}$. About 40% of the assimilated sulfur is allocated into low molecular weight compounds such as GSH and dimethylsulfonioproprionate (DMSP) at the sulfide redox level ($195 \mu mol\ m^{-2}\ d^{-1}$), 35% into proteins, again at the sulfide redox level ($90 \mu moles\ m^{-2}\ d^{-1}$), 21% into ester-sulfate, and 4% into sulfolipids at the sulfonate redox level (Bates et al., 1994).

V.   Interactions Between S and C, N, P Metabolism

Very little is known about the cross-talk between sulfur metabolism and the other major mineral assimilation pathways. Since carbon, nitrogen and sulfur are essential components of proteins, the limitation of any one of them stalls protein biosynthesis in all compartments. The cell reacts by increasing the acquisition of the limiting nutrient and/or adjusting the capacity, flux rates and intermediate storage options of the pathways of the non-limiting nutrients. Under conditions of prolonged or severe deficiency of carbon, nitrogen or sulfur, these processes eventually affect the photosynthetic apparatus and consequently the photosynthetic efficiency. In organisms such as *C. reinhardtii* and in *Synechococcus* PCC 7942, sulfur limitation typically causes a rapid decline in the expression of major components of the photosynthetic apparatus (Lilly et al., 2002; Zhang et al., 2002, 2004). Also, upon S-deprivation, D1 biosynthesis slows down and the PSII repair process is stopped (Vasilikiotis and Melis, 1994; Wykoff et al., 1998; Melis, 1999). As a consequence, a gradual loss of PSII activity and of oxygen evolution occurs as photodamaged PSII centers accumulate in the chloroplast thylakoids (Wykoff et al., 1998) and occasionally microaerobic conditions occur (Collier et al., 1994; Melis et al., 2000; Zhang et al., 2002; Melis and Chen, 2005). The reduction in photosynthetic electron flow during sulfur starvation, in *C. reinhardtii*, seems to be an active mechanism necessary to cope with this condition (Davies et al., 1996;

Wykoff et al., 1998). The importance of similar events in ecologically more relevant organisms and in natural ecosystems, remains to be ascertained.

A decrease in photosynthetic capacity and affinity for inorganic carbon is frequently consequent to sulfur deprivation (Giordano et al., 2000). In *Dunaliella salina*, the reduction in photosynthetic capacity correlates with a decline of both rubisco protein and chlorophyll *a/b*-binding proteins, indicating that both light and "dark" reactions of photosynthesis are affected by sulfur deficiency (Wykoff et al., 1998; Giordano et al., 2000). Interestingly, the reduction in photosynthetic affinity for inorganic carbon is not associated with a decline in the activity of carbonic anhydrases, enzymes usually considered strictly correlated with the activity of CO_2 concentrating mechanisms (Giordano et al., 2000; Giordano et al., 2005a). Further studies will be necessary to comprehend the mechanisms by which sulfur deprivation reduces the photosynthetic affinity for inorganic carbon. The maintenance of extracellular carbonic anhydrase activity and protein abundance (Giordano et al., 2000) suggests that protein degradation under sulfur limitation is selective, with maintenance of this enzyme that is widely held to be important in algal CO_2 concentrating mechanisms. Rubisco is arguably among the proteins whose amount is more obviously reduced under sulfur limitation. The degradation of rubisco may partially depend on the fact that this enzyme represents a very large reservoir of reduced sulfur (up to 50 mM in the chloroplast, according to Ferreira and Teixeira, 1992). A reduction in the relative abundance of rubisco could thus facilitate sulfur recycling and allocation of carbon skeletons to molecules involved in specific responses to sulfur deprivation (Giordano et al., 2000).

The effect of sulfur limitation on photosynthesis could also be related to its impact on sulfolipids. Sulfolipids are key components of photosynthetic membranes and are crucial for the function of photosystems and ATP synthase (Güler et al., 1996 and reference therein; Benning, 1998; Khotimchenko, 2002; Sato, 2004; cf. Selstam and Campbell, 1996).

Sulfur limitation causes an increase in free amino acids, most notably of glutamine and arginine, that, at least in tobacco, is paralleled by a transcriptional down-regulation of nitrate reductase and plastid glutamine synthase gene expression (Migge et al., 2000). In the green alga *Dunaliella salina,* nitrate reductase inhibition may derive from an increase of intracellular NH_4^+ (Berges, 1997) determined by a dramatic reduction in anaplerosis and by a diversion of carbon towards C3 rather than C4 compounds, subsequent to sulfur deprivation (Giordano et al., 2000).

VI. Algae are the Main Source of Biogenic Reduced Sulfur in the Environment

It was only recently that, revisiting speculations by Lovelock and co-workers (1972), the elevated SO_2 concentration above the sea surface was connected to biogenic DMS production (Fig. 1) (Nguyen et al., 1983). Most of the global emissions of DMS are indeed due to events taking place in the ocean (15.5 Tg S yr^{-1}), with only a minor contribution from land (1.1 Tg S yr^{-1}; Bates et al., 1992; Roelofs et al., 1998; Fig. 1). Once in the atmosphere, DMS reacts with NO_3^- and OH' to generate SO_2, which is then oxidized by OH' to sulfuric acid; sulfuric acid is probably quickly converted to NH_4HSO_4 (Roelofs et al., 1998). Theses product of DMS oxidation act as cloud condensation nuclei and thus affect the radiation balance of the Earth (Charlson et al., 1987; Falkowski et al., 1992) and the acid-base chemistry of the atmosphere (Charlson and Rhode, 1982). DMS also impacts the local acid-base balance in water (Cantoni and Anderson, 1956; Dacey and Blough, 1987) and may determine local changes in the equilibria of dissolved inorganic carbon with some effect on photosynthesis (Raven, 1993), especially in poorly buffered waters. In spite of the fact that the ocean is the main natural source of reduced sulfur to the atmosphere (Giordano et al. 2005b and references therein) and that most of this sulfur is in the form of algal DMS (Charlson et al., 1987; Bates et al., 1992; Andreae and Crutzen, 1997), the atmosphere remains a minor sink in the overall DMS cycle in the ocean (Wolfe et al., 1991; Gabric et al., 1999, 2001). It was estimated that less than 10% of DMS in surface waters ever enters the atmosphere (Malin, 1998); the rest is utilized by methylotrophic bacteria (Kiene and Bates, 1990;

Yoch, 2002) or is photolytically oxidized to non-volatile products (Gabric et al., 2001).

DMS, together with acrylate, is the product of the cleavage of the tertiary sulfonium compound 3-dimethylsulfonioproprionate $((CH_3)_2S^+ CH_2CH_2COO^-)$, better known by its acronym DMSP (Charlson et al., 1987; Keller et al., 1989; Malin, 1996). DMSP concentration, in DMS producers, ranges between 50 and 400 mM, equivalent to 50 to almost 100% of the total organic sulfur in the cell (Matrai and Keller, 1994; Keller et al., 1999; Wolfe, 2000; Yoch, 2002). DMSP biological turnover is 3–130 nM d^{-1} in non-bloom conditions (Kiene, 1996a; Ledyard and Dacey, 1996), with higher rates during algal blooms (Van Duyl et al., 1998). At the pH and temperature of seawater, the uncatalyzed production of DMS and acrylate from DMSP is quantitatively negligible (Dacey and Blough, 1987); most DMS production in the ocean is the result of biological activity. It should be noted, however, that 70 to 95% of algal DMSP is not degraded to DMS but is instead used to produce methanethiol in a competing pathway (Taylor and Gilchrist, 1996) that may exert a control function on DMS production (Kiene, 1996b). It is believed that DMSP in the oceans is usually much more abundant than DMS, with ratios DMSP:DMS varying from 3 to 25 (Bates et al., 1994 and references therein).

The biological production of DMS from DMSP is catalyzed by the enzyme DMSP lyase (Wolfe, 2000). DMSP lyases have been identified in bacteria (de Souza and Yoch, 1995a, b; Yoch et al., 1997), in the marine fungus *Fusarium lateritium* (Basic and Yoch, 1998), and in a number of marine phytoplankters (Stefels and van Boekel, 1993; de Souza et al., 1996, Stefels and Dijkhuizen, 1996; Steinke and Kirst, 1996; Niki et al., 2000).

Apparently, substrate and enzyme only come in contact when cells break because of (sloppy) grazing, senescence, and viral and bacterial attack (Leck et al., 1990; Stefels and van Boekel, 1993; Christaki et al., 1996; Levasseur et al., 1996; Verity and Smetacek, 1996; Wolfe et al., 1997, 2002; Hill et al., 1998; Tang, 2000; Bucciarelli and Sunda, 2003), and enzyme and substrate are often produced by different organisms (Steinke and Kirst, 1996; Steinke et al., 1996; Wolfe et al., 1997; Steinke et al., 2002a, b).

Since certain groups of algae are greater DMSP producers than others (Sunda et al., 2002; Yoch, 2002; Kasamatsu et al., 2004a, b), the species composition of phytoplankton can have a strong influence on the rate of DMS production (Liss et al., 1997; Matrai and Vernet, 1997; Bates et al., 1994 and references therein). However, there is evidence of the fact that taxonomy can often be overrun by physiology in controlling DMSP production (Matrai and Vernet, 1997). Nutrient availability, for instance, is a very strong modulator of DMSP production (Turner et al., 1996); in the South Pacific, the addition of iron (as ferrous sulfate) dramatically stimulated DMS production (NIWA Science 2006). The dependence of DMSP metabolism and DMS emission on phytoplankton composition and environmental factors is reflected by its temporal variability. A number of studies showed that average concentrations of DMS in surface seawater can vary by as much as a factor of 50 between summer and winter in the mid and high latitudes (Bates et al., 1994, and references therein). DMS production (and presumably concentrations) is probably susceptible of rather large oscillations also in the long term: according to ice core data, DMS emissions may have varied by a factor of 6 between glacial and interglacial times (Legrand et al., 1991).

The synthesis of DMSP, in all organisms, starts from methionine. The pathway for the conversion of methionine to DMSP is likely to have evolved

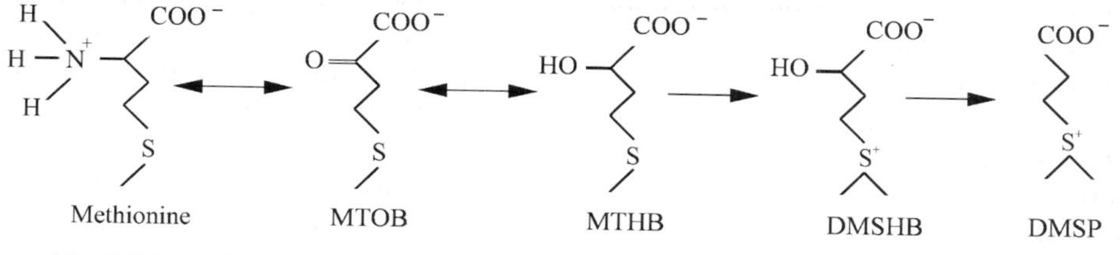

Fig. 2. Biosynthetic DMSP pathways in marine algae. Modified from Gage et al., 1997.

independently at least three times, once in algae and twice in flowering plants, as suggested by the existence of three distinct pathways (Gage et al., 1997; McNeil et al., 1999). In microalgae and macroalgae DMSP synthesis occurs via a four-step pathway (Gage et al., 1997). In these organisms, methionine is converted to the unstable 2-oxo acid, 4-methylthio-2-oxobutyrate (MTOB) (Gage et al., 1997; Fig. 2). This reaction is catalyzed by a transaminase that, differently from that of the typical pathway for methionine salvage in plants, animal and bacteria, has a higher affinity for glutamate as the amino donor than for glutamine and asparagine (Gage et al., 1997; Summers et al., 1998); also, the K_m for methionine of this enzyme is an order of magnitude lower ($30\,\mu M$) than that of most transaminases (usually in the millimolar range; Jenkins and Fonda, 1985). It is possible that this high affinity for methionine is required to compete with enzymes like methionyl-tRNA synthetase (Burbaum and Schimmel, 1992; Schröder et al., 1997), also known to have a very low K_m for methionine. It should be considered that DMSP synthesis must drain very large amounts of methionine, in consideration of the fact that, in *Enteromorpha intestinalis,* up to 90% of the reduced sulfur is in DMSP (Gage et al. 1997). The compound MTOB is then reduced to 4-methylthio-2-hydroxybutyrate (MTHB). The enzyme that catalyzes this reaction, MTOB reductase, is very active in algae that produce large quantities of DMSP, whereas its activity is usually much lower (less than 5%) in algae that do not accumulate DMSP (Summers et al., 1998). While Met aminotransferase and MTOB reductase are also present in non-DMSP producers, even if with much lower activities, the subsequent steps in the pathway of DMSP synthesis are exclusive of DMSP producers. A MTHB S-methyl transferase is responsible for the S-methylation of MTHB, with production of 4-dimethylsulfonio-2-hydroxybutyrate (DMSHB); this is possibly the committing step of this pathway (Gage et al., 1997). DMSHB has osmoprotectant properties and it has been proposed that it was the functional and evolutionary precursor of DMSP (Summers et al., 1998). The final step of the pathway consists in the oxidative decarboxylation of DMSHB to DMSP (Summers et al., 1998).

Along the lines of the "Gaia hypothesis" (Lovelock and Margulis, 1974), Charlson et al. (1987) proposed that the DMS released in the atmosphere, by contributing to sunlight-scattering and cloud condensation, is a crucial component of a feedback mechanism initiated by a stimulation of algal growth associated with increasing global temperature (Andreae and Crutzen, 1997). While a role of DMS in climate forcing cannot be excluded, the observations that (a) little DMSP is degraded to DMS by healthy algal cells, (b) only a small portion of DMSP is ever converted to DMS, (c) only a small percentage of DMS in surface seawaters ever enters the atmosphere (see above) diverted the attention of researchers towards the role that DMSP may play within or around the cells that produce it, rather than on its "gaian" contribution.

DMSP is an effective cryoprotectant (Kirst et al., 1991; Karsten et al., 1992; Nishiguchi and Somero, 1992), whose production, at least in some cases, was shown to be favored by a decrease in water temperature (Sheets and Rhodes, 1996; this is obviously in contradiction with a role of DMS in the cooling of the planet). It has also been shown that DMSP (like many other compatible solutes, but not glycine betaine: Smirnoff and Cumbes, 1989) and its breakdown products act as scavengers of hydroxyl radical and other reactive oxygen species (Sunda et al., 2002; Wolfe et al., 2002). DMSP is certainly a compatible solute, although the importance of short-term changes in DMSP concentration in response to salinity changes (i.e. an osmoregulatory role) is less clear (Malin and Kirst, 1997; Trossat et al., 1998; Stefels, 2000; Welsh, 2000). DMSP is indeed rather similar in structure and in function as a compatible solute to the N-containing osmolyte glycine betaine; this makes DMSP an excellent substitute for glycine betaine under N deficiency, especially when S is abundant, as is the case in the ocean (Andreae, 1986; Dacey et al., 1987; Turner et al., 1988; Gröne and Kirst, 1992; Liss et al., 1997; McNeil et al., 1999). Moreover, DMSP synthesis is initiated by a transamination (see above), which is favored by a depletion of cellular amino acids, hence favoring the switch to DMSP when nitrogen is limiting (Gage et al., 1997). DMSP solutions have a higher density than isosmotic glycine betaine solutions (Boyd and Gradmann, 2002), with implications for the buoyancy of phytoplankton. DMSP is also the precursor for cues for chemosensory attraction between algae and bacteria (Zimmer-Faust et al., 1996), deterrence between algae and microzooplankton (Wolfe et al., 1997; Simò, 2001 and references therein), and as a feeding attractant for top carnivore seabirds with implications for feedback effects on algal grazing (Nevitt et al., 1995). It has been hypothesized that DMSP cleavage constitutes a mechanism of grazing-activated defense (Wolfe et al., 1997; Strom et al., 2003a, b) based on the potent antimicrobial activity of acrylate (Sieburth, 1960, 1961). This idea is supported by the observation that predators preferentially feed on algae with low DMSP production or DMSP lyase activity (Wolfe et al., 1997; Strom et al., 2003a, b). It is possible that the energetic cost associated with this anti-grazing strategy is at least partially compensated by the several (e.g. cryoprotectant, antioxidant, compatible solute) functional roles of DMSP. DMSP synthesis may also represent an overflow mechanism for excess reduced sulfur and reducing power under unbalanced growth, though the excess must be excreted if osmoregulation is not to be compromised (Stefels, 2000).

The large amount of sometimes contradictory data on DMSP, DMS and their roles calls for an additional research effort. The role of DMSP as a sink for sulfur and the physiological consequences for the control of flux rates of sulfate uptake and assimilation remain to be clarified. Precise knowledge on the cellular processes affecting sulfur metabolism and DMSP synthesis will contribute to improve models for the role of phytoplankton in the emission of atmospheric sulfur at a large scale and on the biogeochemical S cycle in general.

VII.   Sulfur Availability may have Played a Role in Algal Evolution, Succession and Distribution

Algae cultured under similar nutritional regimes can have rather different C:S ratios (Ho et al., 2003). Diatoms and dinoflagellates are the groups of algae that, on average, contain the highest amount of sulfur per unit of carbon. Chlorophytes are at the other end of the range, with the highest C:S ratios. Coccolithophores are intermediates between these two extremes with respect to their C:S ratios (Keller et al., 1989; Heldal et al., 2003; Ho et al., 2003). Interestingly, the only

two species of chrysophytes tested for DMSP cell content (*Chrysamoeba* sp. and *Ochromonas* sp.; Yoch, 2002 and references therein) are among the phytoplankters with the highest cell DMSP concentration (Sunda et al., 2002). In the absence of data for a larger number of species, however, this cannot be assumed as a characteristic of this taxon. A high S use efficiency could be important in some freshwater environments, but it appears to be of minimal relevance in marine environments, where sulfate is very abundant (Pilson, 1998). It does therefore make sense that, at least in general terms, dinoflagellates, diatoms and coccolithophores (chlorophyll *a/c* algae), among the most abundant photolithotrophs in the oceans, have C:S ratios that are lower than those of chlorophytes (chlorophyll *a/b* algae), that prevail in freshwaters (e.g. Falkowski et al., 2004). High DMSP production (Keller et al., 1989 and references therein) is usually found in organisms with lower C:S ratios (Ho et al., 2003). This may be not coincidental and the species composition of phytoplankton in today's oceans could be related, to some extent, to the ability of cells to produce DMSP (and its degradation products) as a protection against grazing. What is known about Earth and ocean history also fits with the hypothesis that sulfur availability may have contributed to determine phytoplankton composition (A.H. Knoll, personal communication). Until the late Proterozoic Eon, 550 million years before present, sulfate levels in the oceans were possibly only a few percent of today's levels (Shen et al., 2003; Kah et al., 2004; Poulton et al., 2004), which is within the range for present-day freshwaters (Tipping et al., 1998; Holmer and Storkholm, 2001). During this interval, cyanobacteria and green algae (chlorophyll *a/b*) radiated and constituted most of the ocean phytoplankton. Most chlorophyll *a/c* algae probably originated before the Neoproterozoic, but did not become major components of the phytoplankton until the Triassic (Anbar and Knoll, 2002; Falkowski et al., 2004; Kah et al., 2004; Poulton et al., 2004). It is interesting that food web complexity and grazing intensity also changed during the Mesozoic (250–65 Ma). This is reflected in the changes in the structure of marine animal communities (almost all benthic, as far as the fossil record for grazers is concerned, with the exception of foraminifera) and in the repeated acquisition of skeletons by phytoplankton

(Vermeij, 1977; Bambach et al., 2002; Knoll, 2003; Falkowski et al., 2004). This increase in the grazing pressure could have exerted a selective action in favor of the phytoplankters with low C:S ratios and high capacity for DMSP productions; in an environment with abundant bio-available sulfur, such as the Mesozoic (and today's) oceans, these organisms were in a comparatively better position than their counterpart with high C:S ratios and a reduced capacity for DMSP production. The ability to produce DMSP, which had probably been acquired previously for other functions (N-saving osmolyte, cryoprotectant, antioxidant, metabolic overflow) may have become a major evolutionary discriminant in response to increased effectiveness of grazers.

Acknowledgements

Work in John Raven's laboratory on algal ecophysiology is supported by the Natural Environment Research Council (UK).

References

Anbar AD and Knoll AH (2002) Proterozoic ocean chemistry and evolution: a bioinorganic bridge? Science 297: 1137–1142

Andreae MO (1986) The ocean as a source of atmospheric sulfur compounds. In: Buat-Ménard P (ed) The Role of Air-Sea Exchange in Geochemical Cycling, pp 331–362. D. Reidel Publishing Company, Dordrecht, Holland

Andreae MO and Crutzen PJ (1997) Atmospheric aerosols: biogeochemical sources and role in atmospheric chemistry. Science 276: 1052–1058

Aquino RS, Landeira-Fernandez AM, Valente AP, Andrade LR and Mourão PAS (2005) Occurrence of sulfated galactans in marine angiosperms: evolutionary implications. Glycobiology 15: 11–20

Bagchi D and Verma D (1997) Selenate-regulation of sulfur metabolism in a cyanobacterium, *Phormidium uncinatum*. J Plant Physiol 150: 762–764

Baines SB and Fisher NS (2001) Interspecific differences in the bioconcentration of selenite by phytoplankton and their ecological implications. Mar Ecol Prog Ser 213: 1–12

Bambach RK, Knoll AH and Sepkoski JJ Jr (2002) Anatomical and ecological constraints on Phanerozoic animal diversity in the marine realm. Proc Natl Acad Sci USA 99: 6854–6859

Basic MK and Yoch DC (1998) *In vivo* characterization of dimethylsulfonioproprionate lyase in the fungus *Fusarium lateritium*. Appl Environ Microbiol 64: 106–111

Bates TS, Lamb BK, Guenther AB, Dignon J and Stoiber RE (1992) Sulfur emissions to the atmosphere from natural sources. J Atmos Chem 14: 315–337

Bates TS, Kiene RP, Wolfe GV, Matrai PA, Chavez FP, Buck KR, Blomquist BW and Cuhel RL (1994) The cycling of sulfur in surface seawater of northeast Pacific. J Geophys Res 99: 7835–7843

Baudouin-Cornu P, Surdin-Kerjan Y, Marlière P and Thomas D (2001) Molecular evolution of protein atomic composition. Science 293: 297–300

Baudouin-Cornu P, Schuerer K, Marlière P and Thomas D (2004) Intimate evolution of proteins: proteome atomic content correlates with genome base composition. J Biol Chem 279: 5421–5428

Benning C (1998) Biosynthesis and function of the sulfolipid sulfoquinovosyl diacylglycerol. Ann Rev Plant Physiol Plant Mol Biol 49: 53–75

Berges JA (1997) Miniview: algal nitrate reductase. Eur J Phycol 32: 3–8

Berkowitz O, Wirtz M, Wolf A, Kuhlmann J and Hell R (2002) Use of biomolecular interaction analysis to elucidate the regulatory mechanism of the cysteine synthase complex from *Arabidopsis thaliana*. J Biol Chem 277: 30629–30634

Bick JA and Leustek T (1998) Plant sulfur metabolism – the reduction of sulfate to sulfite. Curr Opin Plant Biol 1: 240–244

Bick JA, Dennis JJ, Zylstra GJ, Nowack J and Leustek T (2000) Identification of a new class of 5′-adenylylsulfate (APS) reductases from sulfate-assimilating bacteria. J Bacteriol 182: 135–142

Bork C, Schwenn J-D and Hell R (1998) Isolation and characterization of a gene for assimilatory sulfite reductase from *Arabidopsis thaliana*. Gene 212: 147–153

Boyd CM and Gradmann D (2002) Impact of osmolytes on buoyancy of marine phytoplankton. Mar Biol 141: 605–618

Bragg JG, Thomas D and Baudouin-Cornu P (2006) Variation among species in proteomic sulphur content is related to environmental conditions. Proc Roy Soc B 273: 1293–1300

Brimblecombe P, Hammer C, Rodhe H, Ryaboshapko A and Boutron CF (1989) Human influence on the sulfur cycle. In: Brimblecombe P and Lein AY (eds) Evolution of the Global Biogeochemical Sulfur Cycle, SCOPE 39, pp 77–121. Wiley, Chichester, UK

Brown DJA (1983) Effect of calcium and aluminium concentrations on the survival of brown trout (*Salmo trutta*) at low pH. Bull Environ Contam Toxicol 30: 582–587

Bucciarelli E and Sunda WG (2003) Influence of CO_2, nitrate, phosphate, and silicate limitation on intracellular dimethylsulfoniopropionate in batch cultures of the coastal diatom *Thalassiosira pseudonana*. Limnol Oceanogr 48: 2256–2265

Burbaum JJ and Schimmel P (1992) Amino acid binding by the class I aminoacyl-tRNA synthetase: role for a conserved proline in the signature sequence. Prot Sci 1: 575–581

Canfield DE, Habicht KS and Thamdrup B (2000) The Archean sulfur cycle and the early history of atmospheric oxygen. Science 288: 658–661

Cantoni GL and Anderson DG (1956) Enzymatic cleavage of dimethylpropriothetin by *Polysiphonia lanosa*. J Biol Chem 222: 171–177

Cape JN, Fowler D and Davison A (2003) Ecological effects of sulfur dioxide, fluorides, and minor air pollutants: recent trends and research needs. Environ Int 29: 201–211

Charles DF (1991) Acidic Deposition and Aquatic Ecosystems. Regional Case Studies. Springer, New York, USA

Charlson RJ and Rodhe H (1982) Factors controlling the acidity of rainwater. Nature 295: 683–685

Charlson RJ, Lovelock JE, Andreae MO and Warren SG (1987) Oceanic phytoplankton atmospheric sulfur, cloud albedo and climate. Nature 326: 655–661

Christaki U, Belviso S, Dolan JR and Corn M (1996) Assessment of the role of copepods and ciliates in the release to solution of particulate DMSP. Mar Ecol Prog Ser 141: 119–127

Cobbett CS (2000) Phytochelatins and their roles in heavy metal detoxification. Plant Physiol 123: 825–832

Collier JL, Herbert SK, Fork DC and Grossman AR (1994) Changes in the cyanobacterial photosynthetic apparatus in response to macronutrient deprivation. Photosynth Res 42: 173–183

Dacey JWH and Blough NV (1987) Hydroxide decomposition of DMSP to form DMS. Geophys Res Lett 14: 1246–1249

Dacey JWH, King GM and Wakeham SG (1987) Factors controlling emission of dimethylsulphide from salt marshes. Nature 330: 643–645

Davies JP, Yildiz FH and Grossman AR (1994). Mutants of *Chlamydomonas* with aberrant responses to sulfur deprivation. Plant Cell 6: 53–63

Davies JP, Yildiz FH and Grossman AR (1996) Sac1, a putative regulator that is critical for survival of *Chlamydomonas reinhardtii* during sulfur deprivation. EMBO J 15: 2150–2159

Davies JP and Grossman AR (1998) Responses to deficiencies in macronutrients. In: Rochaix J-D, Goldschmidt-Clermont M and Merchant S (eds) The Molecular Biology of Chloroplasts and Mitochondria. In: Chlamydomonas, pp 613–633. Kluwer Academic, Dordrecht, The Netherlands

de Souza MP and Yoch DC (1995a) Comparative physiology of dimethyl sulphide production by dimethylsulfoniopropionate lyase in *Pseudomonas doudoroffii* and *Alcaligenes* sp. strain M3A. Appl Environ Microbiol 61: 3986–3991

de Souza MP and Yoch DC (1995b) Purification and characterization of dimethylsulfoniopropionate lyase from an *Alcaligenes*-like dimethyl sulphide-producing marine isolate. Appl Environ Microbiol 61: 21–26

de Souza MP, Chen YP and Yoch DC (1996) Dimethylsulfoniopropionate lyase from the marine macroalga *Ulva curvata*: purification and characterization of the enzyme. Planta 199: 433–438

Dominguez-Solis JR, Lopez-Martin MC, Ager FJ, Dolores Ynsa M, Romero LC and Gotor C (2004) Increased cysteine availability is essential for cadmium tolerance and accumulation in *Arabidopsis thaliana*. Plant Biotechnol J 2: 469–476

Driscoll CT, Lawrence GB, Bulger AJ, Butler TJ, Cronan CS, Eagar C, Lambert KF, Likens GE, Stoddard JL and Weathers KC (2001) Acidic deposition in the Northeastern United States: sources and inputs, ecosystems effects, and management strategies. BioScience 51: 180–198

Driscoll CT and Schecher WD (1990) The chemistry of aluminium in the environment. Environ Geochem Hlth 12: 28–49

Eichel J, Gonzales JC, Hotze M, Matthews RG and Schröder J (1995) Vitamin-B12-independent methionine synthase from a higher plant (*Catharanthus roseus*). Molecular characterization, regulation, heterologous expression, and enzyme properties. Eur J Biochem 230: 1053–1058

Falkowski PG, Kim Y, Kolber Z, Wilson C, Wirick C and Cess R (1992) Natural versus anthropogenic factors affecting low-level cloud albedo over the North Atlantic. Science 256: 1311–1313

Falkowski PG, Katz ME, Knoll AH, Quigg A, Raven JA, Schofield O and Taylor FJR (2004) The evolution of modern eukaryotic phytoplankton. Science 305: 354–360

Ferreira RMB and Teixeira ARN (1992) Sulfur starvation in *Lemna* leads to degradation of ribulose bisphosphate carboxylase without plant death. J Biol Chem 267: 6253–6257

Field CB, Behrenfeld MJ, Randerson JT and Falkowski PG (1998) Primary production of the biosphere: integrating terrestrial and oceanic components. Science 281: 237–240

Fitzgerald JW (1976) Sulfate ester formation and hydrolysis: a potentially important yet often ignored aspect of the sulfur cycle in aerobic soils. Bact Revs 40: 698–721

Gabric AJ, Matrai P and Vernet M (1999) Modeling the production of dimethylsulfide during the vernal bloom in the Barents Sea. Tellus 51: 919–938

Gabric AJ, Gregg W, Najjar R and Erickson D (2001) Modeling the biogeochemical cycle of dimethylsulfide in the upper ocean: a review. Glob Change Sci 3: 377–392

Gage DA, Rhodes D, Nolte KD, Hicks WA, Leustek T, Cooper AJL and Hanson AD (1997) A new route for synthesis of dimethylsulphoniopropionate in marine algae. Nature 387: 891–894

Gao Y, Schofield OME and Leustek T (2000) Characterization of sulfate assimilation in marine algae focusing on the enzyme 5'-adenylylsulfate reductase. Plant Physiol 123: 1087–1096

Giordano M, Pezzoni V and Hell R (2000) Strategies for the allocation of resources under sulfur limitation in the green alga *Dunaliella salina*. Plant Physiol 124: 857–864

Giordano M, Beardall J and Raven JA (2005a) CO$_2$ concentrating mechanisms in algae: mechanisms, environmental modulation, and evolution. Annu Rev Physiol 56: 99–131

Giordano M, Norici A and Hell R (2005b) Sulfur and phytoplankton: acquisition, metabolism and impact on the environment. New Phytol 166: 371–382

Graham LE and Wilcox LW (2000) Algae. Prentice-Hall, Upper Saddle River, NJ

Green RM and Grossman AR (1988) Changes in sulfate transport characteristics and protein composition of *Anacystis nidulans* R2 during sulfur deprivation. J Bacteriol 170: 583–587

Gröne T and Kirst GO (1992) The effect of nitrogen deficiency, methionine and inhibitors of methionine metabolism on the DMSP contents of *Tetraselmis subcordiformis* (Stein). Mar Biol 112: 497–503

Güler S, Seeliger A, Härtel, Renger G and Benning C (1996) A null mutant of *Synechococcus* sp. PCC7942 deficient in the sulfolipid sulfoquinovosyl diacylglycerol. J Biol Chem 271: 7501–7507

Habicht KS, Gade M, Thamdrup B, Berg P and Canfield DE (2002) Calibration of sulfate levels in the Archean Ocean. Science 298: 2372–2374

Hallmann A and Sumper M (1994) An inducible arylsulfatase of *Volvox carteri* with properties suitable for a reporter gene system: purification, characterization and molecular cloning. Eur J Biochem 221: 143–150

Hawkesford MJ and Wray JL (2000) Molecular genetics of sulfur assimilation. Adv Bot Res 33: 159–223

Heldal M, Scanlan DJ, Norland S, Thingstad F and Mann NH (2003) Elemental composition of single cells of various strains of marine *Prochlorococcus* and *Synechococcus* using X-ray microanalysis. Limnol Oceanogr 48: 1732–1743

Hell R and Hillebrand H (2001) Plant concepts for mineral acquisition and assimilation. Curr Opin Biotechnol 12: 161–168

Hell R, Jost R, Berkowitz O and Wirtz M (2002) Molecular and biochemical analysis of the enzymes of cysteine biosynthesis in the plant *Arabidopsis thaliana*. Amino Acids 22: 245–257

Hesse H and Hoefgen R (2003) Molecular aspects of methionine biosynthesis. Trends Plant Sci 8: 259–262

Hill RW, White BA, Cottrell MT and Dacey JWH (1998) Virus-mediated total release of dimethylsulfopropionate from marine phytoplankton: a potential climate process. Aquat Microb Ecol 14: 1–6

Ho T-Y, Quigg A, Finkel ZV, Milligan AJ, Wyman K, Falkowski PG and Morel FMM (2003) The elemental composition of some marine phytoplankton. J Phycol 39: 1145–1159

Hoffmann R and Bisson MA (1986) *Chara buckellii*, a euryhaline charophyte from an unusual saline environment. 1. Osmotic relations at steady-state. Can J Bot 64: 1599–1605

Holmer M and Storkholm P (2001) Sulfate reduction and sulfur cycling in lake sediments: a review. Freshwater Biol 46: 431–451

Horne AJ and Goldman CR (1994) Limnology, 2nd Edition. McGraw-Hill Inc., New York

Jenkins WT and Fonda ML (1985) Kinetics, eqilibria, and affinity for coenzymes and substrates. In: Christen P and Metzler DE (eds) Transaminases, Vol 2, pp. 216–234. Wiley, New York, USA

Jonnalagadda SB and Prasada Rao PVV (1993) Toxicity, bioavailability and metal speciation. Comparative Biochemistry and Physiology Part C: Pharmacol Toxicol Endocrinol 106: 585–595

Kah LC, Lyons TW and Frank TD (2004) Low marine sulphate and protracted oxygenation of the Proterozoic biosphere. Nature 431: 834–838

Kanno N, Nagahisa E, Sato M and Sato Y (1996) Adenosine 5'-phosphosulfate sulfotransferase from the marine macroalga *Porphyra yezoensis* Ueda (Rhodophyta) – stabilization, purification, and properties. Planta 198: 440–446

Karsten U, Wiencke C and Kirst GO (1992) Dimethylsulfoniopropionate (DMSP) accumulation in green macroalgae from polar to temperate regions: interactive effect of light versus salinity and light versus temperature. Polar Biol 12: 603–660

Kasamatsu N, Hirano T, Kudoh S, Odate T and Fukuchi M (2004a) Dimethylsulfoniopropionate production by psychrophilic diatom isolates. J Phycol 40: 874–878

Kasamatsu N, Kawaguchi S, Watanabe S, Odate T and Fukuchi M (2004b) Possible impacts of zooplankton grazing on dimethylsulfide production in the Antarctic Ocean. Can J Fish Aquat Sci 61: 736–743

Keller MD, Bellows WK and Guillard RL (1989) Dimethyl sulfide production in marine phytoplankton. In: Saltzman ES and Cooper WJ (eds) Biogenic Sulfur in the Environment, pp 167–182. Am Chem Soc, Washington, USA

Keller MD, Kiene RP, Matrai PA and Bellows WK (1999) Production of glycine betaine and dimethylsulfoniopropionate in marine phytoplankton. I. Batch cultures. Mar Biol 135 237–248

Khotimchenko SV (2002) Distribution of glycerolipids in marine algae and grasses. Chem Nat Prods 38: 223–229

Kiene RP and Bates TS (1990) Biological removal of dimethylsulfide from seawater. Nature 345: 702–705

Kiene RP (1996a) Production of methane thiol from dimethylsulfoniopropionate in marine surface waters. Mar Chem 54: 69–83

Kiene RP (1996b) Turnover of dissolved DMSP in estuarine and shelf waters from the Northern Gulf of Mexico. In: Kiene RP, Visscher PT, Keller MD and Kirst GO (eds) Biological and Environmental Chemistry of DMSP and

Related Sulfonium Compounds, pp 337–349. Plenum, New York, USA

Kirst GO, Thiel C, Wolff H, Nothnagel J, Wanzek M and Ulmke R (1991) Dimethylsulfoniopropionate (DMSP) in ice-algae and its possible biological role. Mar Chem 35: 381–388

Kleeberg A (1997) Interactions between benthic phosphorus release and sulfur cycling in Lake Scharmützelsee (Germany). Water Air Soil Poll 99: 391–399

Knoll AH (2003) Biomineralization and evolutionary history. Rev Mineral Geochem 54: 329–356

Kopriva S, Suter M, von Ballmoos P, Hesse H, Krähenbühl U, Rennenberg H and Brunold C (2002) Interaction of sulfate assimilation with carbon and nitrogen metabolism in *Lemna minor*. Plant Physiol 130: 1406–1413

Koprivova A, Suter M, den Camp RO, Brunold C and Kopriva S (2000) Regulation of sulfate assimilation by nitrogen in *Arabidopsis*. Plant Physiol 122: 737–746

Koprivova A, Meyer AJ, Schween G, Herschbach C, Reski R and Kopriva S (2002) Konckout of the Adenosine 5'-phosphosulfate reductase gene in Physcomitrella patens revives an old route of sulfate assimilation. J Biol Chem 277: 32195–32201

Last WM and Ginn WM (2005) Saline systems of the Great Plains of western Canada: an overview of the limnology and palaeolimnology. Saline Systems 1: 10 doi:10.1186/1746-1448-1-10

Laudenbach GE and Grossman AR (1991) Characterization and mutagenesis of sulfur-regulated genes in a cyanobacterium: evidence for function in sulfate transport. J Bacteriol 173: 2739–2750

Leck C, Larsson U, Gander LE, Johansson S and Hajdu S (1990) DMS in the Baltic Sea – annual variability in relation to biological society. J Geophys Res 95: 3353–3363

Ledyard KM and Dacey JWH (1996) Kinetics of DMSP-lyase activity in coastal seawater. In: Kiene RP, Visscher PT, Keller MD and Kirst GO (eds) Biological and Environmental Chemistry of DMSP and Related Sulfonium Compounds, pp 325–335. Plenum, New York, USA

Legrand M, Feniet-Saigne C, Saltzman ES, Germain C, Barkov NI and Petrov VN (1991) Ice-core record of oceanic emissions of dimethylsulfide during the last climate cycle. Nature 350: 144–146

Léstang G de and Quillèt M (1973) Un nouveau nucleotide: le cytosine-riboside de P-sulfate, activeur de SO_4^{2-}, isolé et characterisé chez *Pelvetia canaliculata* Dcne et Thur. Compt Ren Acad Sci Paris D 277: 2165–2167

Léstang-Bremond G and Quillèt M (1976) Etude, à l'aide de $^{35}SO_4$, du turn-over des sulphates esterificant le fucoidane de *Pelvetia canaliculata* (Dcne et Thur). Physiol Veg 14: 259–269

Leustek T, Martin MN, Bick J-A and Davies JP (2000) Pathways and regulation of sulfur metabolism revealed through molecular and genetic studies. Annu Rev Plant Phys 51: 141–165

Levasseur M, Michaud S, Egge J, Cantin G, Nejstgaard JC, Sanders R, Fernadez E, Solberg PT, Heimdal B and Gosselin M (1996) Production of DMSP and DMS during a mesocosm study of an *Emiliania huxleyi* bloom: influence of bacteria and *Calanus finmarchicus* grazing. Mar Biol 126: 609–618

Lilly JW, Maul JE and Stern DB (2002) The *Chlamydomonas reinhardtii* organellar genomes respond transcriptionally and post-transcriptionally to abiotic stimuli. Plant Cell 14: 2681–2706

Liss PS, Hatton AD, Malin G, Nightingale PD and Turner SM (1997) Marine sulfur emissions. Philos T Roy Soc B 352: 159–168

Lovelock JE, Mags RJ and Rasmussen RA (1972) Atmospheric dimethylsulfide and the natural sulfur cycle. Nature 237: 452–453

Lovelock JE and Margulis L (1974) Atmospheric homeostasis by and for the biosphere: the Gaia hypothesis. Tellus 26: 2–10

Maberly SC, King L, Gibson CE, May L, Jones RI, Dent MM and Jordan C (2003) Linking nutrient limitation and water chemistry in upland lakes to catchment characteristics. Hydrobiologia 506–509: 83–91

Malin G (1996) Biological and environmental chemistry of DMSP and related sulfonium compound. In: Kiene RP, Visscher PT, Keller MD and Kirst GO (eds) The Role of DMSP in the Global Sulfur Cycle and Climate Regulation, pp 177–189. Plenum, New York, USA

Malin G and Kirst GO (1997) Algal production of dimethyl sulfide and its atmospheric role. J Phycol 33: 889–896

Malin G (1998) Sulphur, climate and the microbial maze. Nature 387: 857–859

Marino R, Howarth RW, Chan F, Cole JJ and Likens GE (2003) Sulfate inhibition of molybdenum-dependent nitrogen fixation by planktonic cyanobacteria under seawater conditions: a non-reversible effect. Hydrobiologia 500: 277–293

Matrai PA and Keller MD (1994) Total organic sulfur and dimethylsulfoniopropionate in marine phytoplankton: intracellular variation. Mar Biol 119: 61–68

Matrai PA and Vernet M (1997) Dynamics of the vernal bloom in the marginal ice zone of the Barents Sea. Dimethylsulfide and dimethylsulfonopropionate budgets. J Geophys Res 102: 22965–22979

Mazel D and Marliere P (1989) Adaptive eradication of methionine and cysteine from cyanobacterial light-harvesting proteins. Nature 341: 245–248

McNeil SD, Nuccio ML and Hanson AD (1999) Betaines and related osmoprotectants. Targets for metabolic engineering of stress resistance. Plant Physiol 120: 945–949

Melis A (1999) Photosystem-II damage and repair cycle in chloroplasts: what modulates the rate of photodamage in vivo? Trend Plant Sci 4: 130–135

Melis A, Zhang L, Forestier M, Ghirardi ML and Seibert M (2000) Sustained photobiological hydrogen gas production upon reversible inactivation of oxygen evolution in the green alga *Chlamydomonas reinhardtii*. Plant Physiol 122: 127–136

Melis A and Chen HC (2005) Chloroplast sulfate transport in green algae – genes, proteins and effects. Photosynth Res 86: 299–307

Mendoza-Cozatl D, Loza-Tavera H, Hernandez-Navarro A and Moreno-Sanchez R (2005). Sulfur assimilation and glutathione metabolism under cadmium stress in yeast, protists and plants. FEMS Microbiol Rev 29: 653–671

Migge A, Bork C, Hell R and Becker TW (2000) Negative regulation of nitrate reductase gene expression by glutamine or asparagine accumulating in leaves of sulfur-deprived tobacco. Planta 211: 587–595

Mosulén S, Dominuez MJ, Vigara J, Vilchez C, Guiraum A and Vega JM (2003) Metal toxicity in *Chlamydomonas reinhardtii*. Effect on sulfate and nitrate assimilation. Biomol Eng 20: 199–203

Mullineaux PM and Rausch T (2005) Glutathione, photosynthesis and the redox regulation of stress-responsive gene expression. Photosynth Res 86: 459–474

Murray TE (1995) The correlation between iron sulfide precipitation and hypolimnetic phosphorus accumulation during one summer in a softwater lake. Can J Fish Aquat Sci 52: 1190–1194

Neumann PM, de Souza MP, Pickerung IJ and Terry N (2003) Rapid microalgal metabolism of selenate to volatile dimethylselenide. Plant Cell Environ 26: 897–905

Nevitt GA, Veit RR and Kareiva P (1995) Dimethyl sulfide as a foraging cue for Antarctic procellariiform seabirds. Nature 376: 680–682

Nguyen BC, Bonsang B and Gaudry A (1983) The role of the ocean in the global atmospheric sulphur cycle. J Geophys Res 88: 10903–10914

Niki T, Kunugi M and Otsuki A (2000) DMSP-lyase activity in five marine phytoplankton species: its potential importance in DMS production. Mar Biol 136: 759–764

Nishiguchi MK and Somero GN (1992) Temperature- and concentration-dependence of compatibility of the organic osmolyte β-dimethylsulfoniopropionate. Cryobiology 29: 118–124

NIWA Science (2006) Ocean atmosphere sulfur project. www.niwa.co.nz/rc/prog/oai/news/dmsrec (May 20, 2006)

Noctor G, Gomez L, Vanacker H and Foyer CH (2002) Interactions between biosynthesis, compartmentation and transport in the control of glutathione homeostasis and signaling. J Exp Bot 53: 1283–1304

Norici A, Hell R and Giordano M (2005) Sulfur and primary production in aquatic environments: an ecological perspective. Photosynth Res 86: 409–417

Nürnberg GK (1996) Comment: Phosphorus budgets and stoichiometry during the open-water season in two unmanipulated lakes in the experimental lakes area, northwestern Ontario. Can J Fish Aquat Sci 53: 1469–1471

Obata T, Araic H and Shiraiwa Y (2004) Bioconcentration mechanism of selenium by coccolithophorid, *Emiliania huxleyi*. Plant Cell Physiol 45: 1434–1441

Ogle RS and Knight AW (1996) Selenium bioaccumulation in aquatic ecosystems: 1. Effects of sulfate on the uptake and toxicity of selenate in *Daphnia magna*. Arch Environ Con Tox 30: 274–279

Palenik B, Brahamsha B, Larimer FW, Land M, Hauser L, Chain P, Lamerdin J, Regala W, Allen EE, McCarren J, Paulsen I, Dufresne A, Partensky F, Webb EA and Waterbury J (2003). The genome of a motile marine *Synechococcus*. Nature 424: 1037–1042

Pelletreau KN and Muller-Parker G (2002) Sulfuric acid in the phaeophyte alga *Desmarestia munda* deters feeding by the sea urchin *Strongylocentrotus droechachiensis*. Mar Biol 141: 1–9

Petaloti C, Voutsa D, Samara C, Sofoniou M, Stratis I and Kouimtzis T (2004) Nutrient dynamics in shallow lakes of northern Greece. Environ Sci Pollut Res Int 11: 11–17

Peters AF, van Oppen MJH, Wiencke C, Stam WT and Olsen JL (1997) Phylogeny and historical ecology of the Desmarestiales (Phaeophyceae) support a southern hemisphere origin. J Phycol 33: 294–309

Pilson MEQ (1998) An introduction to the chemistry of the sea. Prentice Hall, Upper Saddle River, USA

Poulton SW, Fralick PW and Canfield DE (2004) The transition to a sulphidic ocean ~1.84 billion years ago. Nature 431: 173–177

Ramaiah A and Shanmugasundaram ERB (1962) Effect of sulphur compounds on the uptake of molybdenum by *Neurospora crassa*. Biochim Biophys Acta 60: 386–392

Ravanel S, Gakiere B, Job D and Douce R (1998) The specific features of methionine biosynthesis and metabolism in plants. P Natl Acad Sci USA 95: 7805–7812

Ravanel S, Block MA, Rippert P, Jabrin S, Curien G, Rebeille F and Douce R (2004) Methionine metabolism in plants: chloroplasts are autonomous for *de novo* methionine synthesis and can import S-adenosylmethionine from the cytosol. J Biol Chem 279: 22548–22557

Raven JA (1993) Carbon: a phycocentric view. In: Evans GT and Fasham MJR (eds) Towards a Model of Ocean Biogeochemical Processes, pp 123–132. Springer, Berlin, Germany

Raven JA (1997) The vacuole: a cost-benefit analysis. Adv Bot Res 25: 59–86

Raven JA, Evans MCW and Korb RE (1999) The role of trace metals in photosynthetic electron transport in O_2-evolving organisms. Photosynth Res 60: 111–149

Raven JA, Andrews M and Quigg A (2005) The evolution of oligotrophy: implications for the breeding of crop plants for low input agricultural systems. Ann Appl Biol 146: 261–280

Ravina CG, Chang CI, Tsakraklides GP, McDermott JP, Vega JM, Leustek T, Gotor C and Davies JP (2002) The sac mutants of *Chlamydomonas reinhardtii* reveal transcriptional and posttranscriptional control of cysteine biosynthesis. Plant Physiol 130: 2076–2084

Riedel GF and Sanders JG (1996) The influence of pH and media composition on the uptake of inorganic selenium by *Chlamydomonas reinhardtii*. Environ Toxic Chem 15: 1577–1583

Riedel GF, Sanders JG and Gilmour CC (1996) Uptake, transformation, and impact of selenium in freshwater phytoplankton and bacterioplankton communities. Aquat Microb Ecol 11: 43–51

Ritchie RJ (1996) Sulphate transport in the cyanobacterium *Synechococcus* R2 (*Anacystis nidulans*, *S. leopolensis*) PCC 7942. Plant Cell Environm 19: 1307–1316

Roelofs GJ, Lelieveld J and Ganzeveld L (1998) Simulation of global sulfate distribution and the influence on effective cloud drop radii with a coupled phytochemistry-sulfur cycle model. Tellus 50B: 224–242

Saito K (2000) Regulation of sulfate transport and synthesis of sulfur-containing amino acids. Curr Opin Plant Biol 3: 188–195

Sasaki H, Kataoka H, Kamiya M and Kawai H (1999) Accumulation of sulfuric acid in Dictyotales (Phaeophyceae): Taxonomic distribution and ion chromatography of cell extracts. J Phycol 35: 732–739

Sasaki H, Murakami A and Kawai H (2005a) Seasonal stability of sulfuric acid accumulation in the Dictyotales (Phaeophyceae). Phycol Res 53: 124–137

Sasaki H, Murakami A and Kawai H (2005b) Inorganic ion compositions in the Ulvophyceae and the Rhodophyceae, with special reference to sulfuric acid ion accumulation. Phycol Res 53: 85–92

Sato N (2004) Roles of the acidic lipids sufoquinovosyl diacylglycerol and phosphatidylglycerol in photosynthesis: their specificity and evolution. J Plant Res 117: 496–505

Schiff JA, Stern AI, Saidha T and Li J (1993) Some molecular aspects of sulfate metabolism in photosynthetic organisms. In: De Kok LJ, Stulen I, Rennenberg H, Brunold C and Rauser WE (eds) Sulfur Nutrition and Assimilation in Higher Plants, pp 21–35. SPB Academic, The Hague, The Netherlands

Schlesinger WH (1997) Biogeochemistry: an analysis of global change. Academic, San Diego, USA

Schmidt A (1972) On the mechanisms of photosynthetic sulfate reduction. An APS-sulfotransferase from *Chlorella*. Arch Microbiol 84: 77–87

Schmidt A and Jäger K (1992) Open questions about sulfur metabolism in plants. Annu Rev Plant Physiol 43: 325–349

Schröder G, Eichel J, Breinig S and Schröder J (1997) Three differentially expressed S-adenosylmethionine synthetase from *Catharanthus roseus*: molecular and functional characterization. Plant Mol Biol 33: 211–222

Schwenn JD (1989) Sulphate assimilation in higher plants: a thioredoxin PAPS reductase from spinach leaves. Z Naturforsch 44c: 504–508

Schwenn JD (1994) Photosynthetic sulphate reduction. Z Naturforsch 49c: 531–539

Selstam E and Campbell D (1996) Membrane lipid composition of the unusual cyanobacterium *Gloeobacter*

violaceous sp. PCC 7421, which lacks sulfoquinovosyl diacylglycerol. Arch Microbiol 166: 132–135

Sheets EB and Rhodes D (1996) Determination of DMSP and other onium compounds on *Tetraselmis subcordiformis* by plasma desorption mass spectrometry. In: Kiene RP, Visscher PT, Keller MD and Kirst GO (eds) Biological and Environmental Chemistry of DMSP and Related Sulfonium Compounds, pp 317–324. Plenum, New York, NY

Shen Y, Knoll AH and Walter MR (2003) Evidence for low sulfate and anoxia in a mid-Proterozoic marine basin. Nature 423: 632–635

Sieburth JM (1960) Acrylic acid, an "antibiotic" principle in *Phaeocystis* blooms in Antartic waters. Science 132: 676–677

Sieburth JM (1961) Antibiotic properties of acrylic acid, a factor in gastrointestinal antibiosis of polar marine animals. J Bact 82: 72–79

Simò R (2001) Production of atmospheric sulfur by oceanic plankton: biogeochemical, ecological and evolutionary links. Trends Ecol Evol 16: 287–294

Smirnoff N and Cumbes QJ (1989) Hydroxyl radical scavenging activity of compatible solutes. Phytochemistry 28: 1057–1060

Stefels J and van Boekel WHM (1993) Production of DMS from dissolved DMSP in axenic cultures of the marine phytoplankton species *Phaeocystis* sp. Mar Ecol Prog Ser 97: 11–18

Stefels J and Dijkhuizen L (1996) Characteristics of DMSP-lyase in *Phaeocystis* sp. (Prymnesiophyceae). Mar Ecol Prog Ser 131: 307–313

Stefels J (2000) Physiological aspects of the production and conversion of DMSP in marine algae and higher plants. J Sea Res 43: 183–197

Steinke M and Kirst GO (1996) Enzymatic cleavage of dimethylsulfonioproprionate (DMSP) in cell-free extracts of the marine macroalgae *Enteromorpha clathrata* (Roth) Grev. (Ulvales, Chlorophyta). J Exp Mar Biol Ecol 210: 73–85

Steinke M, Daniel C and Kirst GO (1996) DMSP lyase in marine macro- and microalgae. In: Kienem RP, Visscher PT, Keller MD and Kirst GO (eds) Biological and Environmental Chemistry of DMSP and Related Sulfonium Compounds, pp 317–324. Plenum, New York, USA

Steinke M, Malin G, Stuart WG and Burkill PH (2002a) Vertical and temporal variability of DMSP lyase activity in a coccolithophorid bloom in the northern North Sea. Deep Sea Res Pt II 49: 3001–3016

Steinke M, Malin G, Archer SD, Burkill PH and Liss PS (2002b) DMS production in a coccolithophorid bloom: evidence for the importance of dinoflagellate DMSP lyases. Aquat Microb Ecol 26: 259–270

Sterner RW and Elser JJ (2002) Ecological Stoichiometry: the Biology of Elements from Molecules to the Biosphere. Princeton University Press, Princeton

Stoddard JL, Jeffries DS, Lukewille A, Clair TA, Dillon PJ, Driscoll CT, Forsius M, Johannessen M, Kahl JS and

Kellog JH (1999) Regional trends in aquatic recovery from acidification in North America and Europe. Nature 401: 575–578

Strauss H (1997) The isotopic composition of sedimentary sulfur through time. Paleogeogr Paleocl Paleoecol 132: 97–118

Strom S, Wolfe G, Holmes J, Stecher H, Shimeneck C, Lambert S and Moreno E (2003a) Chemical defense in the microplankton II: feeding and growth rates of heterotrophic protists on the DMS-producing phytoplankter *Emiliania huxleyi*. Limnol Oceanogr 48: 217–229

Strom S, Wolfe G, Slajer A, Lambert S and Clough J (2003b) Chemical defense in the microplankton II: inhibition of protist feeding by β-dimethylsulfonioproprionate (DMSP). Limnol Oceanogr 48: 230–237

Summers PS, Nolte KD, Cooper AJL, Borgeas H, Leustek T, Rhodes D and Hanson AD (1998) Identification and stereospecificity of the first three enzymes of 3-dimethylsulfonioproprionate biosynthesis in a chlorophyte alga. Plant Physiol 116: 369–378

Sunda W, Kieber DJ, Kiene RP and Huntsman S (2002) An antioxidant function for DMSP and DMS in marine algae. Nature 418: 317–319

Suter M, von Ballmoos P, Kopriva S, den Camp RO, Schaller J, Kuhlemeier C, Schurmann P and Brunold C (2000) Adenosine 5′-phosphosulfate sulfotransferase and adenosine 5′-phosphosulfate reductase are identical enzymes. J Biol Chem 275: 930–936

Tang KW (2000) Dynamics of dimethylsulfoproprionate (DMSP) in a migratory grazer: a laboratory simulation study. J Exp Marine Biol Ecol 243: 283–293

Taylor BF and Gilchrist PT (1996) Metabolic pathways involved in DMSP degradation. In: Kiene RP, Vischer PT, Keller MD and Kirst GO (eds) Biological Environmental Chemistry of DMSP and Related Sulfonium Compounds, pp 265–276. Plenum, New York, USA

Tipping E, Carrick TR, Hurley MA, James JB, Lawlor AJ, Lofts S, Rigg E, Sutcliffe DW and Woof C (1998) Reversal of acidification in upland waters of the English lake district. Environ Pollut 103: 143–151

Trossat C, Rathinasabapathi B, Weretilnyk EA, Shen TL, Huang ZH, Gage DA and Hanson AD (1998) Salinity promotes accumulation of 3-dimethylsulfoniopropionate and its precursor S-methylmethionine in chloroplasts. Plant Physiol 116: 165–171

Turner SM, Malin G, Liss PS, Harbour DS and Holligan PM (1988) The seasonal variation of dimethyl sulphide and dimethylsulfoniopropionate concentrations in nearshore waters. Limnol Oceanogr 33: 364–375

Turner SM, Nightingale PD, Spokes LJ, Iddicoat MI and Liss PS (1996) Increased dimethylsulfide concentrations in sea water from in situ iron enrichment. Nature 383: 513–517

Tweedie JW and Segel IH (1970) Specificity of transport processes for sulfur, selenium, and molybdenum anions by filamentous fungi. Biochim Biophys Acta 196: 95–106

Vandermeulen JH and Foda A (1988) Cycling of selenite and selenate in marine phytoplankton. Mar Biol 98: 115–123

Van Duyl FC, Gieskes WWC, Kop AJ and Lewis WE (1998) Biological control of short-term variation in the concentration of DMSP and DMS during *Phaeocystis* spring bloom. J Sea Res 40: 221–231

Van Mooy BAS, Rocap G, Fredricks HF, Evans CT and Devol AH (2006) Sulfolipids dramatically decrease phosphorus demand by picocyanobacteria in oligotrophic marine environments. PNAS 103: 8607–8612

Vasilikiotis C and Melis A (1994) Photosystem II reaction center damage and repair cycle: chloroplast acclimation strategy to irradiance stress. Proc Natl Acad Sci USA 91: 7222–7226

Verity PG and Smetacek V (1996) Organism life cycles, predation, and the structure of marine pelagic ecosystems. Mar Ecol Prog Ser 130: 277–293

Vermeij GJ (1977) The Mesozoic marine revolution: evidence from snails, predators and grazers. Paleobiology 3: 245–258

Weiss M, Haimovich G and Pick U (2001) Phosphate and sulfate uptake in the halotolerant alga *Dunaliella* are driven by Na$^+$ -symport mechanism. J Plant Physiol 158: 1519–1525

Wheeler E, Zingaro RA, Cox ER, Irgolic KJ and Bottino NR (1982) The effect of selenate, selenite and sulfate on the growth of six unicellular marine algae. J Exp Mar Biol Ecol 57: 181–194

Welsh DT (2000) Ecological significance of compatible solute accumulation by micro-organisms: from single cells to global climate. FEMS Microbiol Rev 24: 263–290

Wirtz M, Berkowitz O, Droux M and Hell R (2000) The cysteine synthase complex from plants: mitochondrial serine acetyltransferase from *Arabidopsis thaliana* carries a bifunctional domain for catalysis and protein–protein interaction. Eur J Biochem 268: 686–693

Wolfe GV, Bates TS and Charleson RJ (1991) Climatic and environmental implications of biogas exchange at the sea surface: modeling DMS and the marine biologic cycle. In: Mantoura RFC, Martin JM, Wollast R (eds) Ocean Margin Processes in Global Change, pp 383–400. Wiley, Chichester, UK

Wolfe GV, Steinke M and Kirst GO (1997) Grazing-activated chemical defence in a unicellular marine alga. Nature 387: 894–897

Wolfe GV (2000) The chemical defence ecology of marine unicellular plankton: constraints, mechanisms, and impacts. Biol Bull 198: 225–244

Wolfe GV, Strom SL, Holmes JL, Radzio T and Olson MB (2002) Dimethylsulfoniopropionate cleavage by marine phytoplankton in response to mechanical, chemical, or dark stress. J Phycol 38: 948–960

Wykoff DD, Davies JP, Melis A and Grossman AR (1998) The regulation of photosynthetic electron transport during nutrient deprivation in *Chlamydomonas reinhardtii*. Plant Physiol 117: 129–139

Yildiz FH, Davies JP and Grossman AR (1994) Characterization of sulfate transport in *Chlamydomonas reinhardtii* during sulfur-limited and sulfur-sufficient growth. Plant Physiol 104: 981–987

Yoch DC, Ansede JH and Rabinowitz KS (1997) Evidence for intracellular and extracellular dimethylsulfoniopropionate (DMSP) lyases and DMSP uptake sites in two species of marine bacteria. Appl Environ Microb 63: 3182–3188

Yoch DC (2002) Dimethylsulfonioproprionate: its sources, role in the marine food web, and biological degradation to dimethylsulfide. Appl Environ Microb 68: 5804–5815

Zhao FJ, Spiro B, Poulton PR and McGrath SP (1998) Use of sulfur isotope ratios to determine anthropogenic sulfur signals in a grassland ecosystem. Environ Sci Technol 32: 2288–2291

Zhang L, Happe T and Melis A (2002) Biochemical and morphological characterization of sulfur-deprived and H$_2$-producing *Chlamydomonas reinhardtii* (green alga). Planta 214: 552–561

Zhang Z, Shrager J, Jain M, Chang CW, Vallon O and Grossman A (2004) Insights into the survival of *Chlamydomonas reinhardtii* during sulfur starvation based on microarray analysis of gene expression. Eukaryot Cell 3: 1331–1348

Zimmer-Faust RK, de Souza MP and Yoch DC (1996) Bacterial chemotaxis and its potential role in marine dimethylsulfide production and biogeochemical sulfur cycling. Limnol Oceanogr 41: 1330–1334

Chapter 21

Role of Sulfur for Plant Production in Agricultural and Natural Ecosystems

Fang-Jie Zhao
Agriculture and Environment Division, Rothamsted Research, Harpenden, Hertfordshire AL5 2JQ, UK

Michael Tausz
School of Forest and Ecosystem Science, The University of Melbourne, Creswick, Victoria 3363, Australia

Luit J. De Kok*
Laboratory of Plant Physiology, University of Groningen, P.O. Box 14, 9750 AA Haren, The Netherlands

Summary

Sulfur is essential for plant growth and functioning. Sulfate taken up by the roots is the primary sulfur source for growth, but additionally plants are able to utilize absorbed sulfur gases by the shoot. Prior to its assimilation sulfur needs to be reduced and cysteine is the primary precursor or sulfur donor for other plant sulfur metabolites. Sulfur is of great significance for the structure of proteins and functioning of enzymes and it plays an important role in the defense of plants against stresses and pests. Sulfur metabolites such as glutathione provide protection of plants against oxidative stress, heavy metals and xenobiotics. Secondary sulfur compounds (viz. glucosinolates, γ-glutamyl peptides and alliins), phytoalexins, sulfur-rich proteins (thionins), localized deposition of elemental sulfur and the release of volatile sulfur compounds may provide resistance against pathogens and herbivory. Plant species vary largely in sulfur requirement, and an adequate and balanced sulfur nutrition is crucial for their production, quality and health. The assimilation of sulfur and nitrogen are strongly interrelated and sulfur deficiency in plants can be diagnosed by the nitrogen to sulfur ratio of plant tissue. In agricultural ecosystems, the

*Corresponding author, Fax: +31 503632273, E-mail: l.j.de.kok@rug.nl

Rüdiger Hell et al. (eds.), Sulfur Metabolism in Phototrophic Organisms, 417–435.
© 2008 *Springer.*

occurrence of sulfur deficiency of soils can easily be corrected by the application of sulfur fertilizers, which additionally prevents negative environmental side effects such as leakage of nitrate to drainage water. Plants in natural ecosystems generally have an adequate sulfur supply, which partly originates from atmospheric sulfur inputs. Humans and animals rely on plants for their reduced sulfur, and plant sulfur nutrition has a decisive effect on food quality, e.g., availability of methionine, breadmaking and malting quality, and on health, because some secondary sulfur compounds have significance as phyto-pharmaceuticals. A balanced sulfur diet is essential in animal feeding and deficiency negatively affects sheep wool production, though excessive sulfur may induce copper or selenium deficiency in cattle.

I. Introduction

Sulfur is an essential nutrient for plants and is considered as the fourth major plant nutrient after nitrogen, phosphorous and potassium. Total sulfur content in plants tissue ranges from 0.3% to 7.6%; the latter is found in plants from gypsum soils (Tabatabai, 1986; Ernst, 1990). In general, plants rely on sulfate taken up by roots as the sulfur source for growth. In agro-ecosystems sulfur supply is not always optimal for plant growth and quality (Schnug, 1998). The abundance of sulfate in the pedosphere varies widely and may originate from weathering of rock, mineralization of organic sulfur, ground or runoff water, atmospheric deposition of sulfur gases and fertilizers. The majority of the sulfate taken up by the plant is reduced and metabolized into cysteine and methionine, both of which are highly important in proteins (De Kok et al., 2002a; Hawkesford and De Kok, 2006; Haneklaus et al., 2007b). Accordingly, the assimilation of sulfur and nitrogen are strongly interrelated and the organic molar N/S ratio may reflect the sulfur status of the plant, which usually ranges from 30 to 35 for sulfur-sufficient crop plants (Durenkamp and De Kok, 2003; Oenema and Postma, 2003). Plants contain a large variety of other organic sulfur compounds, which play an important role in plant functioning and adaptation to the environment (De Kok et al., 2002a).

In the present chapter, the role of sulfur for plant production in agro- and natural ecosystems will be evaluated. It is evident that the sulfur supply to the plant has a decisive effect on the growth, the performance and fitness, and the resistance of plants to biotic and abiotic stresses. Furthermore, sulfur strongly affects food quality of crop plants.

II. Uptake, Assimilation and Distribution of Sulfur

The uptake and reduction of sulfate in plants and its subsequent assimilation into organic sulfur compounds is highly coordinated (Hawkesford and De Kok, 2006). The uptake and distribution of sulfate in the plant is mediated by sulfate transporter proteins, which are encoded by a sulfate transporter gene family consisting of at least 14 members. The sulfate transporters have been classified in five different groups according to their cellular and subcellular expression and possible functioning (Davidian et al., 2000; Hawkesford, 2000; Hawkesford and Wray, 2000; Hawkesford et al., 2003a,b; Buchner et al., 2004; Hawkesford and De Kok, 2006; Chapter 2). There is a distinct group of sulfate transporters (Group 1), which mediate the uptake of sulfate by the roots that have a high affinity for sulfate (Km 1.5–10 μM). Another group of sulfate transporters (Group 2) are involved in the vascular loading and unloading of sulfate, however, these transporters have a lower affinity for sulfate. There are also distinct transporters (Group 4) involved in the vacuolar exchange of sulfate, whereas the functions of other transporter groups (Groups 3 and 5) are less well characterized (Buchner et al., 2004; Hawkesford and De Kok, 2006). The uptake and distribution of sulfate and the level of expression of the Groups 1, 2 and 4 sulfate transporter genes are directly controlled by plant sulfur status (Buchner et al., 2004; Hawkesford and De Kok, 2006). It needs to be further evaluated whether the local *in situ* sulfate concentration or that of a metabolic product of sulfate assimilation, such as cysteine or glutathione, is involved as signaling of the regulatory control of the different sulfate transporters.

The uptake of sulfate by the root is dependent on the plant sulfur requirement for growth, the shoot to root biomass partitioning, the different developmental stages of the plant *viz.* vegetative growth (herbaceous and crop plants), primary and secondary growth (woody plants), fruit and seed production, and varies widely between species (Hawkesford and De Kok, 2006). The plant's growth rate (1), sulfur requirement (2) and sulfate uptake (3) may be estimated as follows (derived from Durenkamp and De Kok, 2004):

$$\text{Growth rate} = (\ln W_2 - \ln W_1)/(t_2 - t_1) \quad (1)$$

Sulfur requirement (μmol g^{-1} plant day^{-1}) =

$$\text{growth rate (g g}^{-1}\text{ plant day}^{-1}) \times \text{sulfur content}$$
$$(\mu\text{mol g}^{-1}\text{ plant)} \quad (2)$$

Sulfate uptake (μmol g^{-1} root day^{-1}) =
sulfur requirement (μmol g^{-1} plant day^{-1}) ×
$$(\text{S/R ratio} + 1) \quad (3)$$

In (1) W_1 and W_2 represent the total plant weight (g) at time t_1 and t_2, respectively, and $t_2 - t_1$ the time interval (days) between harvests, and in (3) the S/R ratio represents the shoot (S) to root (R) biomass partitioning of the plant. Sulfate uptake by the roots of different crop species may for instance range from 8 to 40 μmol g^{-1} fresh weight day^{-1} (Westerman et al., 2000; Durenkamp and De Kok, 2004; Buchner et al., 2004), whereas that for tree species is presumably much lower (<5 μmol g^{-1} fresh weight day^{-1}; Kreuzwieser et al., 1996; Herschbach et al., 2000; van der Zalm et al., 2005). The sulfur requirement of a crop might be predicted by scaling up the sulfur requirement (2) in kmol sulfur ha^{-1} day^{-1} by estimating the crop biomass density ha^{-1} (Haneklaus et al., 2007a). The sulfur requirement (2) might have to be adjusted for woody species to allow for differences in growth rate and sulfur content among the roots, stems and branches, and the foliage (Johnson, 1984). The possible significance of mycorrhiza (symbiosis between roots and fungi) in the uptake of sulfate by roots of plants from natural ecosystems, e.g., forests, needs to be evaluated further (Rennenberg, 1999; Herschbach and Rennenberg, 2001; Tausz, 2007).

Generally the major proportion of the sulfate taken up is reduced and metabolized into organic compounds essential for structural growth, whereas the remaining sulfate in plant tissue is transferred into the vacuoles. The remobilization and in some species the redistribution of the vacuolar sulfate reserves may be rather slow and sulfur-deficient plants might still contain detectable levels of sulfate (Cram, 1990; Davidian et al., 2000; Hawkesford, 2000; Buchner et al., 2004).

The chloroplast is the predominant site of sulfate reduction; however, root plastids are also able to reduce sulfate, since all enzymes of sulfate assimilation are present (Heiss et al., 1999; Lappartient et al., 1999; Lee and Leustek, 1999; Yonekura-Sakakibara et al., 2000; Chapter 3). At least for most herbaceous and crop plants, as with nitrate reduction (Scheurwater et al., 2002), sulfate reduction in the root as a proportion of the whole plant sulfur assimilation is limited, since here the shoot to root ratio generally exceeds 2 to 6. Sulfate needs to be activated by ATP to adenosine 5′ phosphosulfate (APS) catalyzed by APS sulfurylase before it is reduced to sulfite by adenosine 5′ phosphosulfate reductase with glutathione as a reductant (Leustek and Saito, 1999; Kopriva and Koprivova, 2003; Fig. 1). Subsequently the sulfite is reduced to sulfide by sulfite reductase with ferredoxin as a reductant. The formed sulfide is incorporated into cysteine with *O*-acetylserine as the substrate. This reaction is catalyzed by *O*-acetylserine(thiol)lyase, which is associated as an enzyme complex with serine acetyltransferase (the *O*-acetylserine synthesizing enzyme) named cysteine synthase (Droux et al., 1998; Hell, 2003; Chapters 4 and 5) and is the primary direct coupling step between sulfur and nitrogen assimilation in plants (Brunold, 1990, 1993; Brunold et al., 2003; Fig. 1).

The *in situ* sulfate concentration in the chloroplast (plastid) may be one of the limiting/regulatory steps in the reduction of sulfate, because the affinity of ATP sulfurylase for sulfate is rather low (K_m approximately 1 mM; Stulen and De Kok, 1993). Moreover, the expression and activity of APS reductase is highly responsive to the sulfur status of plant, with metabolic products of sulfate assimilation, such as cysteine or glutathione, as the likely regulating signals (Brunold, 1990, 1993; Leustek and Saito, 1999; Kopriva and Koprivova, 2003; Saito, 2003; Chapter 5).

The reduced sulfur in the shoot may be distributed from the source to the sink as glutathione (Rennenberg et al., 1979) or in some plant species

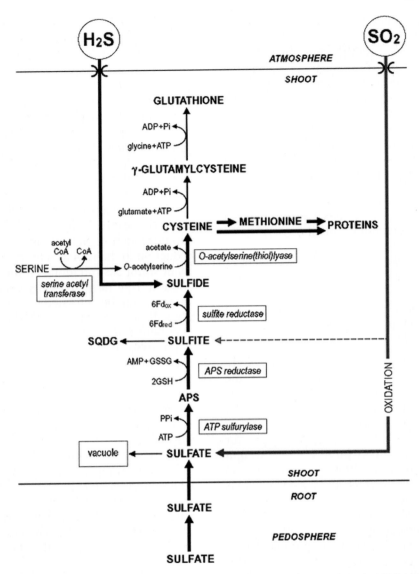

Fig. 1. Metabolism of sulfate and atmospheric sulfur gases in plants (adapted from De Kok et al., 2002a). APS, adenosine 5′-phosphosulfate; Fd_{red}, Fd_{ox}, reduced and oxidized ferredoxin; GSH, GSSG, reduced and oxidized glutathione; SQDG, sulfoquinovosyl diacylglycerol.

as *S*-methylmethionine via the phloem (Bourgis et al., 1999). The reduced sulfur formed in the roots may be transported as methionine and to a lesser extent as cysteine and glutathione to the shoot via the xylem (Pate, 1965). In contrast to annual herbaceous plants, the distribution of sulfur in perennial species and particularly trees is considerably more complex due to their specific features (Tausz, 2007). In trees there are large storage tissues in trunks, long distances between

uptake in roots and consumption/reduction in foliage, and long life spans subject to seasonal changes. Compared to fast-growing herbaceous species, sulfur distribution has only been investigated in a few tree species in detail. A comparison of beech (*Fagus sylvatica*), a deciduous broadleaf, and spruce (*Picea abies*), an evergreen conifer, revealed basic differences in sulfur distribution (Rennenberg and Herschbach, 1995; Herschbach and Rennenberg, 2001; Herschbach,

2003). *Picea abies* takes up sulfate and transports it to older needles, apparently the predominant site of sulfur reduction. Buds and young needles rely on reduced sulfur supplied by older needles as glutathione transported in xylem and phloem. Under normal conditions, spruce trees do not seem to transfer reduced sulfur from their foliage to the trunk and roots. These tissues may rely on reduced sulfur produced by root sulfur reduction. Upon exposure to high atmospheric sulfur input, however, spruce trees seem capable of transferring organic sulfur (most probably as glutathione) from the needles into the roots (Tausz et al., 2003b). In deciduous beech, both reduced organic sulfur in the form of thiols (mainly cysteine and some glutathione) and sulfate are supplied via xylem to the developing young leaves until they can meet their own sulfur reduction requirements. The cysteine seems to be supplied by the breakdown of storage proteins in the trunk. These storage proteins are synthesized during the vegetation period using sulfur imported as glutathione and sulfate (rather than cysteine) from leaves into the trunk. It should be noted that the specific modifications of sulfur distribution pathways in spruce and beech are closely related to the rhythms in flushing, shedding of foliage, and dormancy periods typical of a strictly seasonal climate, and that not much is known about the sulfur metabolism of trees in other climate zones.

In addition to sulfate taken up by the root, plant shoots are also able to absorb and metabolize sulfur gases, viz. SO_2 and H_2S, and use them as a sulfur source for growth (De Kok, 1990; De Kok et al., 1997, 1998, 2000, 2002a,b, 2007; Westerman et al., 2000, 2001; De Kok and Tausz, 2001; Tausz, 2007; Fig. 1). The foliar uptake of SO_2 is generally directly dependent on the degree of opening of the stomata, and the internal resistance to the SO_2 gas is low due to its high solubility in water. In general there is a linear relationship between the uptake of SO_2 by the plant shoot and the atmospheric concentration of SO_2. Once the SO_2 gas diffuses to the mesophyll, it dissociates in water and forms bi(sulfite). The absorbed SO_2 in the mesophyll may enter the sulfur reduction pathway as either sulfite or, after its oxidation as sulfate. Generally, SO_2 exposure results in an enhanced sulfur content of the foliage, mainly because of an accumulation of sulfate presumably in the vacuole, even at relatively low atmospheric concentrations (De Kok, 1990; De Kok and Tausz, 2001; Tausz, 2007). Plants are able to utilize atmospheric H_2S as a sulfur source. The uptake of H_2S by the shoot is largely determined by the rate of its metabolism into cysteine and exposure generally results in rapid accumulation of cysteine and glutathione in the shoot (De Kok, 1990; De Kok et al., 1998, 2002a,b; De Kok and Tausz, 2001). Exposure of plants to atmospheric sulfur gases may depress the uptake of sulfate by the root and its reduction in the shoot (De Kok and Tausz, 2001; De Kok et al., 2002a,b). It has been estimated that at atmospheric levels of $\geq 0.03\,\mu l\ l^{-1}$ SO_2 or H_2S, foliarly absorbed sulfur may contribute substantially (>10–40%) to the sulfur requirement for growth of crop plants (De Kok et al., 2007).

III. Significance of Sulfur in Plant Functioning and Adaptation to Stress and Pests

Cysteine is the sulfur donor for the synthesis of methionine, and the precursor of several other sulfur compounds, such as glutathione, coenzyme A, biotin and secondary sulfur compounds in plants (Giovanelli, 1990; Noji and Saito, 2003; Chapter 6). The predominant proportion of the organic sulfur in plant tissue is present as cysteine and methionine residues in proteins, which may account for up to 60% and 90% of the total and the organic sulfur fraction, respectively (Heinz, 1993; Stulen and De Kok, 1993). The sulfur-containing amino acids are of great significance in the structure, conformation and function of proteins and enzymes. High levels of these amino acids may also be present as storage proteins, e.g., in seeds (Tabatabai, 1986). The thiol groups of the cysteine residues are highly significant in various functional reactions. In proteins the thiol groups can form covalent bounds upon oxidation resulting in disulfide bridges with other cysteine side chains and/or linkage of polypeptides to form cystine residues. The thiol groups of cysteine residues in enzymes are also of great importance in the binding of substrates by enzymes, in metal-sulfur clusters in proteins (e.g., ferredoxins, metallothionins) and in regulatory proteins (e.g., thioredoxins (Jacquot et al, 1997; Verkleij et al., 2003).

Plants contain water-soluble non-protein thiol compounds, which account for 1–2% of the total sulfur, with concentration in plant tissue ranging from 0.1 to 3 mM. Cysteine and the tripeptide glutathione (γGlu-Cys-Gly; GSH) or its homologues, e.g., homoglutathione (γGlu-Cys-βAla) in Fabaceae, hydroxymethylglutathione (γGlu-Cys-βSer) in Poaceae, are the major water-soluble non-protein thiol compounds present in plant tissues in glutathione/cysteine ratio generally exceeding 10 (De Kok and Stulen, 1993; Rennenberg, 1997; Grill et al., 2001; Chapter 11). Glutathione and its homologues are enzymatically synthesized in two steps, both of which are ATP dependent reactions (Fig. 1). First, γ-glutamylcysteine is synthesized from cysteine and glutamate by γ-glutamylcysteine synthetase and second, glutathione is synthesized from γ-glutamylcysteine and glycine (in glutathione homologues, β-alanine or serine) catalyzed by glutathione synthetase.

Glutathione functions in sulfur metabolism in the reduction of APS (as a reductant), storage and transport of reduced sulfur and regulation of sulfate assimilation in plants (Grill et al., 2001; Chapter 11). Furthermore it functions as a reductant in the enzymatic detoxification of reactive oxygen species in the glutathione-ascorbate cycle and as thiol buffer in the protection of proteins via direct reaction with reactive oxygen species, e.g., superoxide, hydrogen peroxide and lipid hydroperoxides, or by the formation of mixed disulfides (De Kok and Stulen, 1993; Grill et al., 2001; Tausz et al., 2003a; Fig. 2). All these reactions occur via a sulfide/disulfide exchange reaction of its cysteine residue generally in combination with glutathione reductase, an NADPH-dependent enzyme. The redox state of glutathione (GSH/GSSG ratio) in plant tissue generally exceeds a value of 7 (Rennenberg, 1997; Foyer and Noctor, 2001; Tausz, 2001). Variation in glutathione levels, its redox state and the activity of glutathione reductase have been related to the adaptation of plants in agro- and natural ecosystems to stress and a

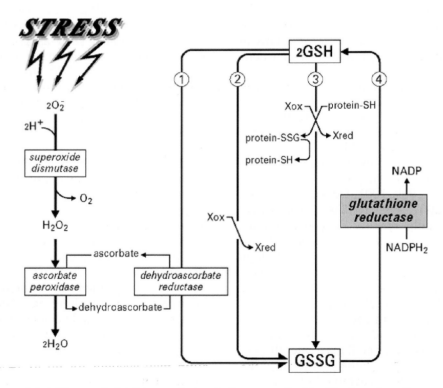

Fig. 2. Significance of glutathione and glutathione reductase in the enzymatic detoxification of reactive oxygen species in the glutathione-ascorbate cycle (1) and as thiol buffer in the protection of proteins via direct reaction with reactive oxygen species, e.g., superoxide, H_2O_2 and lipid hydroperoxides (2), or by the formation of mixed disulfides (3) (adapted from De Kok and Stulen, 1993).

changing environment, e.g., air pollution, drought, low temperature, UV-B radiation (Grill et al., 2001; Chapter 11).

Moreover, glutathione is the precursor of phytochelatins ((γGlu-Cys)$_n$Gly), which are synthesized by a inducible phytochelatin synthase. The number of γ-glutamyl-cysteine residues (γGlu-Cys)$_n$ in phytochelatins generally ranges from 2 to 5 (though sometimes up to 11). Phytochelatins play an important role in the detoxification of cadmium, and possibly also arsenic (Cobbett and Goldsbrough, 2002). Cadmium exposure was found to stimulate sulfate uptake by maize roots through enhanced expression of a high affinity sulfate transporter, as a result of the increased demand for sulfur in the biosynthesis of phytochelatins (Nocito et al., 2002, 2006). Although several other metals or metalloids (e.g., Cu, Hg, Ag, Zn and Ni) can also induce synthesis of phytochelatins, there is no direct evidence that phytochelatins are responsible for their detoxification (Cobbett and Goldsbrough, 2002). It is assumed that the cadmium-phytochelatin complex is transported into the vacuole in order to sequester the potentially toxic cadmium (Rauser, 1993, 2000, 2001). Glutathione is also involved in the detoxification of xenobiotics (Schröder, 1998, 2001; Gullner and Kömives, 2001). Different xenobiotics may induce distinct isoforms of the enzyme glutathione S-transferase, which catalyzes their conjugation with glutathione. Under natural conditions, glutathione S-transferases are assumed to have significance in the detoxification of lipid hydroperoxides, in the conjugation of endogenous metabolites, hormones and DNA degradation products, and in the transport of flavonoids, but in agro-ecosystems they may have great significance in herbicide detoxification and tolerance.

Sulfoquinovosyl diacylglycerol is the predominant sulfolipid present in plants, and in leaves it accounts for up to 3–6% of the total sulfur content (Heinz, 1993; Benning, 1998; Harwood and Okanenko, 2003). It is a constituent of plastid membranes and is likely to be involved in chloroplast (plastid) functioning. Sulfite is the likely sulfur precursor for the formation of the sulfoquinovose group of this lipid (Harwood and Okanenko, 2003). Despite quantitative and qualitative changes in sulfolipid content and its fatty acid composition upon exposure to stress and pests, its actual significance in adaptation needs further evaluation.

Some plant species contain secondary sulfur compounds, such as glucosinolates in *Brassica* (Schnug, 1990, 1993; Rosa, 1997; Graser et al., 2001; Glawisching et al., 2003) and γ-glutamyl peptides and alliins (*S*-alk(en)yl cysteine sulfoxides; Chapter 13) in *Allium* (Lancaster and Boland, 1990; Randle et al., 1993, 1995; Randle, 2000; Randle and Lancaster, 2002; Coolong and Randle, 2003a,b). Glucosinolates account for 1–6% of the total sulfur in the leaves of oilseed rape (Blake-Kalff et al., 1998). There are at least 120 glucosinolates identified in 16 families of dicotyledonous plants, which vary in the side chains and are derived from 8 different amino acids (Fahey et al., 2001; Halkier and Gershenzon, 2006). Upon tissue disruption glucosinolates are enzymatically degraded by myrosinase and yield a variety of biologically active products such as isothiocyanates, thiocyanates, nitriles and oxazolidine-2-thiones (Rosa, 1997, 1999; Kushad et al., 1999; Graser et al., 2001; Petersen et al., 2002; Reichelt et al., 2002; Wittstock and Halkier, 2002). The glucosinolate-myrosinase system is assumed to play a role in plant–herbivore and plant–pathogen interactions. Furthermore, glucosinolates are responsible for the flavor properties of Brassicaceae and recently have been shown to have significance as phytopharmaceuticals in view of their potential anti-carcinogenic properties (Zhang et al., 1992; Fahey et al., 1997, 2002; Kushad et al., 1999; Graser et al., 2001; Petersen et al., 2002; Reichelt et al., 2002).

In *Allium* the content of γ-glutamyl peptides and alliins is strongly dependent on the stage of development of the plant, temperature, water availability and the level of nitrogen and sulfur nutrition (Lancaster et al., 1986, 2000; Lancaster and Shaw, 1989, 1991; Randle et al., 1993, 1995; Randle, 2000; Randle and Lancaster, 2002; Coolong and Randle, 2003a,b; Durenkamp and De Kok, 2002, 2003, 2004). Bloem et al. (2004) observed that in onion bulbs the content of iso-alliin may account for up to 74% of the total sulfur content. γ-Glutamyl peptides may be synthesized from cysteine (via γ-glutamylcysteine or glutathione) and can be metabolized into the corresponding alliins via oxidation and subsequent hydrolyzation by γ-glutamyl transpeptidases, although other possible routes of their synthesis cannot be excluded (Granroth, 1970; Lancaster and Boland, 1990; Edwards et al., 1994; Randle and Lancaster, 2002). The alliins and their breakdown products (e.g., allicin) are the flavor precursors for the odor and taste of the *Allium* species. A wide variety of volatile and non-volatile sulfur-containing compounds are released from the tissue by alliinase, an enzyme that is released from the vacuole upon disruption of the tissue (Lancaster and Collin, 1981; Block, 1992). The physiological functions of γ-glutamyl peptides and alliins are still largely unresolved, but may have significance in chemical defense against insects and pathogens and in the storage of nitrogen and sulfur (Lancaster and Boland, 1990, 1991; Schnug, 1993; Lancaster and Shaw, 1991; Randle and Lancaster, 2002). Furthermore, these compounds may have potential value as phytopharmaceuticals (Haq and Ali, 2003).

Several other sulfur metabolites may play a role in the resistance of plants against stress and pests, e.g., phytoalexins, sulfur-rich proteins (thionins) and localized cellular deposition of elemental sulfur (Cooper and Williams, 2004; Hell and Kruse, 2007) and even the possible release of volatile sulfur compounds as H_2S (Schnug, 1997; Städler, 2000; Glawisching et al., 2003; Haneklaus et al., 2003; Haq and Ali, 2003; Hell and Kruse, 2007; Tausz, 2007). However, their significance in "sulfur-induced-resistance" is not yet fully understood and needs further to be assessed (Schnug, 1997; Haneklaus et al., 2003). For instance, plants grown under normal sulfur conditions may emit minute amounts of H_2S, which may be formed prior to or after the synthesis of cysteine, in the latter case by cysteine desulfhydrase (Schröder, 1993; Haneklaus et al., 2003; Riemenschneider et al., 2005). However, the rate H_2S release is a negligible proportion of the total sulfur flux in plants (Stulen and De Kok, 1993). The H_2S emission may be strongly enhanced when plants are previously exposed to high levels of atmospheric sulfur gases (Rennenberg, 1984; Schröder, 1993; Haneklaus et al., 2003; Tausz, 2007). However, its physiological significance under natural conditions appears be unclear (Ernst, 1990).

IV. Plant Sulfur Requirement and Nutrition in Agro- and Natural Ecosystems

Sulfur requirement varies greatly among agricultural crops. *Brassica* crops have a high demand for S (1.5–2.2 kmol ha⁻¹), followed by *Allium* crops such as leek and onion (1–1.2 kmol ha⁻¹), whereas cereals and legume crops require relatively small quantities of S (0.3–0.6 kmol ha⁻¹) (Zhao et al., 2002). The high requirements of *Brassica* and *Allium* crops are partly due to the synthesis of S-containing secondary metabolites, glucosinolates in *Brassicas* and S-alk(en)yl-L-cysteine sulfoxide in *Allium* crops. In addition, the high sulfur requirement of *Brassicas* may also be attributed to a large accumulation in the vegetative tissue of sulfate, which is remobilized slowly in response to sulfur deficiency (Blake-Kalff et al., 1998). Sulfur requirement is also dependent on crop yield; a high yielding crop requires more nutrients including S.

Sulfur deficiency occurs when sulfur supply from the environment does not match the requirement by the crop. Incidences of sulfur deficiency in agricultural crops or grassland have been reported in different regions of the world (Pasricha and Fox, 1993; Dobermann et al., 1998; Blair, 2002; Zhao et al., 2002; Edmeades et al., 2005; Malhi et al., 2005). *Brassica* crops and multiple-cut grass are generally more prone to sulfur deficiency than other crops, because of their high requirements for sulfur. In Western Europe, sulfur deficiency has become more common in recent decades mainly because of a dramatic reduction in the sulfur inputs from the atmosphere (McGrath et al., 2002). For example, in many areas in the United Kingdom, atmospheric sulfur deposition decreased from 70 kg ha⁻¹ year⁻¹ in the 1970s to less than 10 kg ha⁻¹ year⁻¹ in the early 2000s. Other contributing factors include the use of sulfur-free compound fertilizers and the increasing trend of crop yield. Sulfur deficiency usually occurs as a result of a nutrient imbalance, particularly with regard to nitrogen, and one of the common indicators that are used to diagnose sulfur deficiency is the nitrogen to sulfur ratio (Dijkshoorn and van Wijk, 1967). A nitrogen to sulfur ratio of greater than 17:1 (molar ratio 39:1) in wheat grain generally indicates that the crop had been supplied with inadequate sulfur (Randall et al., 1981). In leaf tissues, the critical value of the ratio is approximately 15:1 (molar ratio 34:1) for cereals and 6–10:1 (molar ratio 14–23:1) for oilseed rape (Spencer and Freney, 1980; McGrath and Zhao, 1996; Blake-Kalff et al., 2000; Blake-Kalff et al., 2002). Recently, Blake-Kalff et al. (2000) proposed the use of the malate to sulfate ratio in leaf tissues as a reliable diagnostic method for crops such as cereals and oilseed rape. A sulfur-deficient crop utilizes nitrogen inefficiently, which subsequently leads to increased nitrogen losses to the environment. For example, Brown et al. (2000) showed that the application of sulfur reduced nitrate leaching to drainage water by 5–72% at a sulfur-deficient grassland site. Therefore, correcting sulfur deficiency in agricultural crops not only benefits yield but also the environment.

Sulfur deficiency can be easily corrected by the application of sulfur fertilizers. Sulfur fertilizers are available in a number of chemical forms; the most common are sulfate and elemental sulfur. The sulfate form is readily available to plant uptake, but sulfate is very mobile in near neutral and alkaline soils and is thus prone to leaching losses when rainfall exceeds evatranspiration. In contrast, elemental sulfur is not available to plant uptake until it is oxidized to sulfate. Oxidation of elemental sulfur is mediated by both autotrophic chemolithotrophs, including some species of *Thiobacillus*, and a wide range of heterotrophic bacteria and fungi in soil (Lawrence and Germida, 1991; Germida and Janzen, 1993). The oxidation is affected by: (a) presence of microorganisms capable of oxidation of elemental sulfur, (b) effective surface area of the elemental sulfur particles, (c) soil temperature, (d) soil water potential and (e) soil aeration (Janzen and Bettany, 1987; McCaskill and Blair, 1987; Watkinson and Blair, 1993; Watkinson and Lee, 1994; Haneklaus et al., 2007a). For annual agricultural crops, sulfate fertilizers are generally a better option than elemental sulfur because the rate of oxidation of elemental sulfur may not match crop demand (Riley et al., 2000; Zhao et al., 2002; Malhi et al., 2005). On the other hand, elemental sulfur is suitable as a maintenance fertilizer for pasture (Blair, 2002).

The sulfur supply to plants in natural ecosystems originates from weathering of rocks, mineralization of organic sulfur and ground or runoff water (Edwards, 1998; Haneklaus et al., 2003).

Atmospheric sulfur inputs may contribute substantially to the sulfur influx in natural ecosystems, where the total of dry and wet sulfur deposition may range from 0.06 to 1 kmol ha^{-1} year^{-1}, though locally in heavily polluted areas these values can be much higher (Johnson, 1984; Cappellato et al., 1998; Edwards, 1998; Haneklaus et al., 2003). It is generally assumed that, in contrast to agroecosystems, plants in natural ecosystems have an adequate sulfur supply (Ernst, 1990, 1993, 1997; Haneklaus et al., 2003). For example, in lowland and wetland ecosystems, the sulfate concentrations in ground or runoff water and the sulfur content in the soils and sediments burials are presumed to be sufficient for natural plant growth (Haneklaus et al., 2003). The sulfur uptake necessary for tree growth ranges from 0.1 to 0.2 kmol ha^{-1} year^{-1}, whereas the total sulfur content of mineral soils in upland ecosystems ranges from 10 to 100 kmol ha^{-1} (and forest floors 0.6 to 1.8 kmol ha^{-1}) (Haneklaus et al., 2003). However, most of the soil total sulfur might not be available for plant uptake (Edwards, 1998). Soil soluble sulfate-S in forest soils ranges from 0.15 kmol ha^{-1} at a conifer site remote from pollutant inputs (*Pseudotsuga menziesii* plantation in southeastern Australia) to more than 50 kmol ha^{-1} at a deciduous forest site with high sulfur pollution input (*Quercus prinus* in TN, USA; Johnson, 1984). Because sulfur requirements of forest stands are low, sulfur deficiencies in forests have only been reported from areas remote from pollutant inputs, e.g., from southeastern Australia and northwestern United States (Johnson, 1984). Atmospheric sulfur inputs as low as 0.03 kmol ha^{-1} year^{-1} may be sufficient to sustain growth of a *Pinus radiata* stand in Australia (Johnson, 1984). Although a recent large scale survey suggested that the effect of atmospheric sulfur deposition is still measurable at many forest sites across Europe (Augustin et al., 2005), decreasing atmospheric sulfur inputs in conjunction with increasing nitrogen deposition may lead to an increased probability of sulfur limitation of forest ecosystems in the future (Johnson and Mitchell, 1998). It has to be pointed out that the cited forest studies dealt with overall ecosystem budgets and not with specific tree physiological measurements. Instantaneous sulfur requirements of forest stands may deviate significantly from annual averages, e.g., during development of new foliage or when storage proteins are laid down. Physiological studies indicated such changes and showed contributions of internal redistribution (see above), but currently do not allow quantification on whole tree or stand basis (Rennenberg and Herschbach, 1995). Adsorption/desorption properties of sulfate in forest soils lend themselves to causing temporary limitations in available sulfate during periods of increased requirements (Johnson and Mitchell, 1998). Hence, the instantaneous requirements of forest trees in relation to their growth stage warrant further interest.

In some natural ecosystems plants have to cope with excessive sulfur in oxidized (sulfate) or reduced form (sulfite, sulfide), which may be available to the plant via (i) the pedosphere from sulfur-emitting fumeroles, gypsum-rich soils and waterlogging, (ii) the hydrosphere from salinity and (iii) the atmosphere from dry and wet deposition of atmospheric sulfur. Excessive sulfur may negatively affect plant growth, for instance as the consequences of sulfate salinity. However, plants may be adapted and are able to cope with excessive sulfur by accumulating it as sulfate in the vacuole (so-called thiophores), or by its elimination (avoidance of sulfur accumulation) by secretion from the shoot via salt glands in halophytes or emission of sulfur gases as H_2S and dimethyl sulfide (DMS; Ernst, 1990, 1993, 1997; Stefels, 2000, 2007). The latter emission is highly significant in some plant species from marine ecosystems, e.g., *Spartina*, which may accumulate high levels of dimethylsulfoniopropionate (DMSP) in leaves upon exposure to excessive sulfur. DMSP may be enzymatically degraded to yield DMS emission by the shoot (Ernst, 1990, 1993, 1997; Hanson and Gage, 1996; Stefels, 2000, 2007). Specific species (e.g., *Allium* and *Brassica*) may emit a variety of other organic sulfur gases including DMS, which are likely degradation products of secondary sulfur compounds (Lanzotti, 2006).

Permanent or temporary potentially phytotoxic levels of dissolved H_2S in the rhizosphere may occur in marshes and tideland wetlands, and in poorly drained and waterlogged soils, e.g., rice paddies (Trudinger, 1986; Van Diggelen et al., 1987; Bates et al., 1992; Armstrong et al., 1996; Armstrong and Armstrong, 2005). In these anoxic soils, H_2S is produced from biological decay of organic sulfur and the activity of dissimilatory sulfate-reducing bacteria and is accumulated under anoxic conditions. Levels from 0.02 to 1.4 mM sulfide may occur in the soil solution around the root zone (Allam and Hollis, 1972; Carlson and Forrest, 1982; Van Diggelen et al., 1987). Levels as low as 0.002 mM sulfide may negatively affect root respiration and nutrient uptake (Allam and Hollis, 1972; Joshi et al., 1973, 1975). Sulfide levels higher than 0.08 mM may reduce root growth and development or result in root and bud death, lignification and blockages within the root arenchyma and vascular tissue (Ford 1973; Armstrong et al.,1996; Armstrong and Armstrong, 2005). The tolerance to high H_2S levels under anoxia is most likely determined by the *in situ* sulfide level in the roots, which is dependent on the rate of oxidation of sulfide in the rhizosphere by bacteria such as *Beggiotoa* or in the plant, and by the sulfide resistance of metabolic processes in the plant species (Joshi et al., 1973, 1975; Joshi and Hollis, 1977; Carlson and Forrest, 1982; Fry et al., 1982; Van Diggelen et al., 1987; Armstrong et al., 1996; Armstrong and Armstrong, 2005). Despite the toxicity of sulfide, there is evidence on basis of the $^{34}S/^{32}S$ ratio in plant tissue that some species from saline habitats are able to use the abundant sulfide upon anoxia directly or indirectly as sulfur source (Raven and Scrimgeour, 1997).

V. Sulfur and Food Quality

Quality requirements of agricultural products vary widely and are determined mainly by the end use of the product. The sulfur nutrition of a crop often has strong and diverse influences on the quality of the produce, because of its essential role in the synthesis of amino acids, proteins and some secondary metabolites.

Legume seeds are important sources of protein for humans and animals. However, their nutritional value is limited by the low contents of the essential amino acid methionine (Friedman, 1996). Different storage proteins of legume seeds vary considerably in their contents of the S-containing amino acids. For example, the pea storage proteins vicilin and lectin contain no cysteine and methionine, whereas legumin contains 1.7% S-containing amino acids (Spencer et al., 1990). Soybean seed has two major storage proteins, glycinin and β-conglycinin. Glycinin is rich in S-containing amino acids (1.8%), whereas β-conglycinin is poor in these amino acids (0.6%) (Shortwell and Larkins, 1989). In general, sulfur deficiency decreases the synthesis of S-rich storage proteins, but increases the synthesis of S-poor proteins concomitantly (Blagrove et al., 1976; Gayler and Sykes, 1985; Spencer et al., 1990; Naito et al., 1995). As a result, sulfur deficiency in legume crops reduces the nutritive value of the seeds (Eppendorfer, 1971; Eppendorfer and Eggum, 1995).

The disulfide and thiol groups of gluten proteins are essential for viscoelasticity of the wheat dough during breadmaking (Shewry and Tatham, 1997; Zhao et al., 1999b,c). Studies in Australia established that sulfur nutrition plays an important role in the breadmaking quality of wheat (Moss et al., 1981, 1983; Wrigley et al., 1984; MacRitchie and Gupta, 1993). These studies showed that sulfur increased synthesis of S-rich storage proteins, such as the α-, β-, and γ-gliadins and the low-molecular-weight glutenin subunits, and decreased the proportion of S-poor proteins, such as the ω-gliadins and the high-molecular-weight glutenin subunits. Furthermore, the concentration of sulfur in flour correlated positively with dough extensibility, but negatively with resistance to stretching. Similarly, several studies in Europe showed that the S status of wheat has a profound effect on the composition of gluten proteins, the rheological properties of dough and breadmaking performance (Schnug et al., 1993; Zhao et al., 1999a,b,c; Wieser et al., 2004; Flaete et al., 2005). In general, increasing S concentration in wheat grain is associated with increasing dough extensibility and increased loaf volume of bread. It is clear that maintaining an adequate sulfur status for wheat is important for breadmaking quality.

Sulfur deficiency has been found to result in a large accumulation of asparagine in cereal leaves and grain (Shewry et al., 1983; Zhao et al., 1996). A recent study has shown a startling increase in the levels of free asparagine in the grain of wheat grown under sulfur deficiency and the formation of high levels of acrylamide during heating of flour derived from sulfur-deficient grain (Muttucumaru et al., 2006). The level of acrylamide produced from the sulfur-deficient flour was approximately 5–10-fold higher than that from the sulfur-sufficient flour. The presence of

acrylamide in a range of fried and oven-cooked foods has caused worldwide concern because this compound has been classified as probably carcinogenic in humans; acrylamide also has neurological and reproductive effects (Friedman, 2003). Acrylamide found in cooked foods is formed during heat treatment of food components as a result of the Maillard reaction between amino acids, particularly asparagines, and reducing sugars (Mottram et al., 2002; Stadler et al., 2002). This explains why sulfur deficiency has such a marked effect on acrylamide formation in heat-treated wheat flour, and highlights the importance of sulfur nutrition in terms of food safety.

Sulfur deficiency can also affect the quality of sugar beet storage roots. Using a hydroponic system to grow sugar beet, Bell et al. (1995) showed that withholding sulfur supply for two months increased the concentration of α-amino-N in the beet roots more than 2-fold. The increased concentration of amino acids in roots reduces juice purity, and therefore the extraction yield of white sugar.

Malting quality of barley and brewhouse performance are assessed in terms of ease of processing and flavor characteristics (Palmer, 1989). During malting of barley grain, large molecular weight components of the endosperm cell walls, the storage proteins and starch granules are hydrolyzed (modified) enzymatically, rendering them more soluble in hot water during mashing. Variability in malting quality is due to factors such as the rate of enzyme synthesis during germination, the composition of the endosperm, and the packing of starch granules (Palmer, 1989). A recent study by Zhao et al. (2006) showed that sulfur applications significantly increased the activities of hydrolytic enzymes and improved endosperm modification during malting. As a result, the concentration of β-glucan in the wort was decreased, which is beneficial for the filtration of wort. Furthermore, sulfur applications also increased the concentration of S-methylmethionine (the precursor of the flavor compound dimethylsulfide) in kilned malt, which is expected to impact on beer flavor.

The presence of high levels of glucosinolates in rapeseed can restrict its use in animal feed. One of the predominant glucosinolates in rapeseed is 2-hydroxy-3-butenyl glucosinolate (progoitrin), which forms oxazolidine-2-thione upon hydrolysis. This hydrolytic product has goitrogenic and anti-nutritional effects in animals (Fenwick et al., 1983; Griffiths et al., 1998). The glucosinolate content in rapeseed has been much decreased by breeding of the double-low (low erucic acid and low glucosinolates) varieties. However, even in these varieties glucosinolates can account for up to 30% of the total sulfur in the seeds. Sulfur supply is one of the most important environmental factors that influence the synthesis and accumulation of glucosinolates in seeds (Schnug, 1990; Zhao et al., 1993). In general, use of sulfur fertilizer to oilseed rape at a normal recommended rate is unlikely to increase the glucosinolate concentration to an unacceptable level, but applying more S than required by the oilseed rape crop must be avoided (Zhao et al., 2002). In contrast to the undesirable effects of hydroxyalkenyl glucosinolates mentioned above, certain glucosinolates have been identified as possible cancer prevention agents in model animals and might have significance as phytopharmaceuticals. In particular, sulforaphane, the isothiocyanate produced from the hydrolysis of 4-methylsulfinylbutyl glucosinolate, which is present in broccoli, has been found to induce anticarcinogenic protective enzymes (phase II enzymes) (Zhang et al., 1992; Fahey et al., 1997; Fahey et al., 2001). Sulforaphane reduced the incidence, delayed the appearance of, and reduced the size of tumors in a rat mammary tumor model (Fahey et al., 1997), and showed potential for treating *Helicobacter pylori*-caused gastritis and stomach cancer (Fahey et al., 2002). Therefore, sulfur nutrition could play a role in enhancing the health promoting properties of *Brassica* vegetables.

The nutritional quality of sulfur deficient grass silage is poor. For animal feeding, a N:S ratio below 15:1 is considered satisfactory (Murphy and O'Donnell, 1989). Applications of sulfur fertilizer increased the proportion of total N in grass present as protein-N, and decreased the contents of nitrate and free amino N (Murphy and O'Donnell, 1989; Richards, 1990; Murphy and Quirke, 1997; Murphy et al., 2002). These effects are beneficial to animal nutrition. Sheep are more sensitive to S deficiency than cattle because of the special requirements for wool production. Early studies showed that sheep performance was improved by sulfur fertilization of forage (Rendig and Weir, 1957). Too much sulfur in grass may have an adverse effect on animal health. Sulfur may induce copper deficiency in cattle, by forming thiomolybdate compounds in the rumen which bind copper and make it unavailable to the animal (Leach and Thornton, 1987). This appears to occur when the copper status of the animal diet is marginal and the concentrations of molybdenum and sulfur are high. The critical concentrations of Mo and S in herbage that can cause an antagonism on Cu metabolism are >3 mg kg^{-1} and >0.3–0.4% on a dry matter basis, respectively. The other antagonistic effect of sulfur is to decrease selenium uptake by plants (White et al., 2007). Selenate is a chemical analogue of sulfate and is taken up by the plant root via sulfate transporters. Sulfur fertilization decreased herbage Se concentration, which subsequently resulted in a significantly lower blood Se level in the cattle grazing the S-treated pasture (Murphy and Quirke, 1997). Selenium supplementation can be practiced by addition to animal feeds or by the inclusion of selenium in fertilizers (White et al., 2007).

Acknowledgements

The authors would like to thank Mr. Dick Visser, University of Groningen, for the drawing of the figures.

References

Allam AI and Hollis JP (1972) Sulphide inhibition of oxidases in rice roots. Phytopathology 62: 634–639

Armstrong J and Armstrong W (2005) Rice: sulfide-induced barriers to root radial oxygen loss, Fe^{2+} and water uptake, and lateral root emergence. Ann Bot 96: 625–638

Armstrong J, Afreen-Zobayed F and Armstrong W (1996) *Phragmites* die-back: sulphide- and acetic acid-induced bud and root death, lignifications, and blockages with the aeration and vascular systems. New Phytol 134: 601–614

Augustin S, Bolte A, Holzhausen M and Wolff B (2005) Exceedance of critical loads of nitrogen and sulphur and its relation to forest conditions. Eur J For Res 124: 289–300

Bates TS, Lamb BK, Guenther A, Dignon J and Stoiber RE (1992) Sulfur emission to the atmosphere from natural sources. J Atmos Chem 14: 315–337

Bell C, Jones J, Franklin J, Milford G and Leigh R (1995) Sulfate supply and its effects on sap quality during growth in sugar beet storage roots. Z Pflanzenernähr Bodenk 158: 93–95

Benning C (1998) Biosynthesis and function of the sulfolipid sulfoquinovosyl diacylglycerol. Annu Rev Plant Physiol Plant Mol Biol 49: 53–75

Blagrove RJ, Gillespie JM and Randall PJ (1976) Effect of sulphur supply on the seed globulin composition of *Lupinus angustifolius*. Aust J Plant Physiol 3: 173–184

Blair GJ (2002) Sulphur fertilisers: a global perspective. Proceedings No. 498. International Fertiliser Society, York

Blake-Kalff MMA, Harrison KR, Hawkesford MJ, Zhao FJ and McGrath SP (1998) Distribution of sulfur within oilseed rape leaves in response to sulfur deficiency during vegetative growth. Plant Physiol 118: 1337–1344

Blake-Kalff MMA, Hawkesford MJ, Zhao FJ and McGrath SP (2000) Diagnosing sulfur deficiency in field-grown oilseed rape (*Brassica napus* L.) and wheat (*Triticum aestivum* L.). Plant Soil 225: 95–107

Blake-Kalff MMA, Zhao FJ and McGrath SP (2002) Sulphur deficiency diagnosis using plant tissue analysis. Proceedings No. 503. International Fertiliser Society, York

Block E (1992) The organosulfur chemistry of the genus *Allium*. Implications for the organic chemistry of sulfur. Angew Chem Int Ed Eng 31: 1135–1178

Bloem E, Haneklaus S and Schnug E (2004) Influence of nitrogen and sulfur fertilization on the alliin content of onions and garlic. J Plant Nutr 27: 1827–1839

Bourgis F, Roje S, Nuccio ML, Fisher DB, Tarczynski MC, Li CJ, Herschbach C, Rennenberg H, Pimenta MJ, Shen TL, Gage DA and Hanson AD (1999) *S*-methylmethionine plays a major role in phloem sulfur transport and is synthesized by a novel type of methyltransferase. Plant Cell 11: 1485–1497

Brown L, Scholefield D, Jewkes EC, Preedy N, Wadge K and Butler M (2000) The effect of sulphur application on the efficiency of nitrogen use in two contrasting grassland soils. J Agric Sci 135: 131–138

Brunold C (1990) Reduction of sulfate to sulfide. In: Rennenberg H, Brunold C, De Kok LJ and Stulen I (eds) Sulfur Nutrition and Sulfur Assimilation in Higher Plants: Fundamental, Environmental and Agricultural Aspects, pp 13–31, SPB Academic, The Hague

Brunold C (1993) Regulatory interactions between sulfate and nitrate assimilation. In: De Kok LJ, Stulen I, Rennenberg H, Brunold C and Rauser W (eds) Sulfur Nutrition and Sulfur Assimilation in Higher Plants: Regulatory, Agricultural and Environmental Aspects, pp 125–138, SPB Academic, The Hague

Brunold C, Von Ballmoos P, Hesse H, Fell D and Kopriva S (2003) Interactions between sulfur, nitrogen and carbon metabolism. In: Davidian J-C, Grill D, De Kok LJ, Stulen I, Hawkesford MJ, Schnug E and Rennenberg H (eds) Sulfur Transport and Assimilation in Plants: Regulation, Interaction and Signaling, pp 45–56, Backhuys Publishers, Leiden

Buchner P, Stuiver CEE, Westerman S, Wirtz M, Hell R, Hawkesford MJ and De Kok LJ (2004) Regulation of sulfate uptake and expression of sulfate transporter genes in *Brassica oleracea* L. as affected by atmospheric H_2S and pedospheric sulfate nutrition. Plant Physiol 136: 3396–3408

Cappellato R, Peters NE and Meyers TP (1998) Aboveground sulfur cycling in adjacent coniferous and deciduous forests and watershed sulfur retention in the Georgia Piedmont, U.S.A. Water Air Soil Pollut 103: 151–171

Carlson PRJr and Forrest J (1982) Uptake of dissolved sulfide by *Spartina alterniflora*: evidence from natural sulfur isotope ratios. Science 216: 633–635

Cobbett C and Goldsbrough P (2002) Phytochelatins and metallothioneins: roles in heavy metal detoxification and homeostasis. Annu Rev Plant Biol 53: 159–182

Coolong TW and Randle WM (2003a) Ammonium nitrate fertility levels influence flavor development in hydroponically grown "Granex 33" onion. J Sci Food Agric 83: 477–482

Coolong TW and Randle WM (2003b) Temperature influences flavor intensity and quality in "Granex 33" onion. J Am Soc Hort Sci 128: 176–181

Cooper RM and Williams JS (2004) Elemental sulphur as an induced antifungal substance in plant defence. J Exp Bot 55: 1947–1953

Cram WJ (1990) Uptake and transport of sulfate. In: Rennenberg H, Brunold C, De Kok LJ and Stulen I (eds) Sulfur Nutrition and Sulfur Assimilation in Higher Plants: Fundamental, Environmental and Agricultural Aspects, pp 3–11, SPB Academic, The Hague

Davidian J-C, Hatzfeld Y, Cathala N, Tagmount A and Vidmar JJ (2000) Sulfate uptake and transport in plants. In: Brunold C, Rennenberg H, De Kok LJ, Stulen I and Davidian J-C (eds) Sulfur Nutrition and Sulfur Assimilation in Higher Plants: Molecular Biochemical and Physiological Aspects, pp 19–40, Paul Haupt, Bern

De Kok LJ (1990) Sulfur metabolism in plants exposed to atmospheric sulfur. In: Rennenberg H, Brunold C, De Kok LJ and Stulen I (eds) Sulfur Nutrition and Sulfur Assimilation in Higher Plants: Fundamental, Environmental and Agricultural Aspects, pp 111–130, SPB Academic, The Hague

De Kok LJ and Stulen I (1993) Functions of glutathione in plants under oxidative stress. In: De Kok LJ, Stulen I, Rennenberg H, Brunold C and Rauser WE (eds) Sulfur Nutrition and Sulfur Assimilation in Higher Plants: Regulatory, Agricultural and Environmental Aspects, pp 125–138, SPB Academic, The Hague

De Kok LJ and Tausz M (2001) The role of glutathione in plant reaction and adaptation to air pollutants. In: Grill D, Tausz M and De Kok LJ (eds) Significance of Glutathione to Plant Adaptation to the Environment, pp 185–201, Kluwer Academic, Dordrecht

De Kok LJ, Stuiver CEE, Rubinigg M, Westerman S and Grill D (1997) Impact of atmospheric sulfur deposition on sulfur metabolism in plants: H$_2$S as sulfur source for sulfur deprived *Brassica oleracea* L. Bot Acta 110: 411–419

De Kok LJ, Stuiver CEE and Stulen I (1998) Impact of atmospheric H$_2$S on plants. In: De Kok LJ and Stulen I (eds) Responses of Plant Metabolism to Air Pollution and Global Change, pp 41–63, Backhuys Publishers, Leiden

De Kok LJ, Westerman S, Stuiver CEE and Stulen I (2000) Atmospheric H$_2$S as plant sulfur source: interaction with pedospheric sulfur nutrition – a case study with *Brassica oleracea* L. In: Brunold C, Rennenberg H, De Kok LJ, Stulen I and Davidian J-C (eds) Sulfur Nutrition and Sulfur Assimilation in Higher Plants: Molecular, Biochemical and Physiological Aspects, pp 41–56, Paul Haupt, Bern

De Kok LJ, Castro A, Durenkamp M, Stuiver CEE, Westerman S, Yang L and Stulen I (2002a) Sulphur in plant physiology. Proceedings No. 500, pp 1–26, The International Fertiliser Society, York

De Kok LJ, Stuiver CEE, Westerman S and Stulen I (2002b) Elevated levels of hydrogen sulfide in the plant environment: nutrient or toxin. In: Omasa K, Saji H, Youssefian S and Kondo N (eds) Air Pollution, Biotechnology in Plants, pp 201–213, Springer, Tokyo

De Kok LJ, Durenkamp M, Yang L and Stulen I (2007) Atmospheric sulfur. In: Hawkesford MJ and De Kok LJ (eds) Sulfur in Plants – an Ecological Perspective, pp 91–106, Springer

Dijkshoorn W and van Wijk AL (1967) The sulphur requirement of plants as evidenced by the sulphur–nitrogen ratio in the organic matter: a review of published data. Plant Soil 26: 129–157

Dobermann A, Cassman KG, Mamaril CP and Sheehy JE (1998) Management of phosphorus, potassium, and sulfur in intensive, irrigated lowland rice. Field Crops Res 56: 113–138

Droux M, Ruffet ML, Douce R and Job D (1998) Interactions between serine acetyltransferase and *O*-acetylserine (thiol) lyase in higher plants: structural, kinetic properties of the free, bound enzymes. Eur J Biochem 155: 235–245

Durenkamp M and De Kok LJ (2002) The impact of atmospheric H$_2$S on growth and sulfur metabolism of *Allium cepa* L. Phyton 42(3): 55–63

Durenkamp M and De Kok LJ (2003) Impact of atmospheric H$_2$S on sulfur and nitrogen metabolism in *Allium* species, cultivars. In: Davidian J-C, Grill D, De Kok LJ, Stulen H, Hawkesford MJ, Schnug E and Rennenberg H (eds) Sulfur Transport and Assimilation in Plants: Regulation, Interaction and Signaling, pp 197–199, Backhuys Publishers, Leiden

Durenkamp M and De Kok LJ (2004) Impact of pedospheric and atmospheric sulphur nutrition on sulphur metabolism of *Allium cepa* L. a species with a potential sink capacity for secondary sulphur compounds. J Exp Bot 55: 1821–1830

Edmeades DC, Thorrold BS and Roberts AHC (2005) The diagnosis and correction of sulfur deficiency and the management of sulfur requirements in New Zealand pastures: a review. Aust J Exp Agric 45: 1205–1223

Edwards PJ (1998) Sulfur cycling, retention, and mobility in soils: a review. USDA General Technical Report NE-250, pp 1–18, USDA Forest Services, Radnor

Edwards SJ, Britton G and Collin HA (1994) The biosynthetic pathway of the S-alk(en)yl-L-cysteine sulphoxides (flavor precursors) in species of *Allium*. Plant Cell Tissue Organ Cult 38: 181–188

Eppendorfer WH (1971) Effects of S, N and P on amino acid composition of field beans (*Vicia faba*) and responses of the biological value of the seed protein to S-amino acid content. J Sci Food Agric 22: 501–505

Eppendorfer WH and Eggum BO (1995) Sulfur amino-acid content and nutritive value of pea and cauliflower crude protein as influenced by sulfur deficiency. Z Pflanzenernähr Bodenk 158: 89–91

Ernst WHO (1990) Ecological aspects of sulfur metabolism. In: Rennenberg H, Brunold C, De Kok LJ and Stulen I (eds) Sulfur Nutrition and Sulfur Assimilation in Higher Plants: Fundamental, Environmental, and Agricultural Aspects, pp 131–144, SPB Academic, The Hague

Ernst WHO (1993) Ecological aspects of sulfur in higher plants: the impact of SO$_2$ and the evolution of the biosynthesis of organic sulfur compounds on populations, ecosystems. In: De Kok LJ, Stulen I, Rennenberg H, Brunold C and Rauser WE (eds) Sulfur Nutrition and Sulfur Assimilation in Higher Plants: Regulatory, Agricultural and Environmental Aspects, pp 125–138, SPB Academic, The Hague

Ernst WHO (1997) Life-history syndromes and the ecology of plants from high sulfur habitats. In: Cram WJ, De Kok LJ, Stulen I, Brunold C and Rennenberg H (eds) Sulfur Metabolism in Higher Plants: Molecular, Ecophysiological and Nutritional Aspects, pp 289–291, Backhuys Publishers, Leiden

Fahey JW, Zhang YS and Talalay P (1997) Broccoli sprouts: an exceptionally rich source of inducers of enzymes that protect against chemical carcinogens. Proc Nat Acad Sci USA 94: 10367–10372

Fahey JW, Zalcmann AT and Talalay P (2001) The chemical diversity and distribution of glucosinolates and isothiocyanates among plants. Phytochemistry 56: 5–51

Fahey JW, Haristoy X, Dolan PM, Kensler TW, Scholtus I, Stephenson KK, Talalay P and Lozniewski A (2002) Sulforaphane inhibits extracellular, intracellular, and antibiotic-resistant strains of *Helicobacter pylori* and prevents benzo[a]pyrene-induced stomach tumors. Proc Nat Acad Sci USA 99: 7610–7615

Fenwick GR, Heaney RK and Mullin WJ (1983) Glucosinolates and their breakdown products in food and food plants. CRC Critic Rev Food Sci Nutr 18: 123–201

Flaete NES, Hollung K, Ruud L, Sogn T, Faergestad EM, Skarpeid HJ, Magnus EM and Uhlen AK (2005) Combined nitrogen and sulphur fertilisation and its effect on wheat quality and protein composition measured by SE-FPLC and proteomics. J Cereal Sci 41: 357–369

Ford HW (1973) Levels of hydrogen sulfide toxic to citrus roots. J Am Soc Hortic Sci 98: 66–68

Foyer CH and Noctor G (2001) The molecular biology, metabolism of glutathione. In: Grill D, Tausz M and De Kok LJ (eds) Significance of Glutathione to Plant Adaptation to the Environment, pp 27–56, Kluwer Academic, Dordrecht

Friedman M (1996) Nutritional value of proteins from different food sources: a review. J Agric Food Chem 44: 6–29

Friedman M (2003) Chemistry, biochemistry, and safety of acrylamide: a review. J Agric Food Chem 51: 4504–4526

Fry B, Scalan RS, Winters JK and Parker PL (1982) Sulphur uptake by salt grasses, mangroves, and seagrasses in anaerobic sediments. Geochim Cosmochim Acta 46: 1121–1124

Gayler KR and Sykes GE (1985) Effects of nutritional stress on the storage proteins of soybeans. Plant Physiol 78: 582–585

Germida JJ and Janzen HH (1993) Factors affecting the oxidation of elemental sulfur in soils. Fertilizer Res 35: 101–114

Giovanelli J (1990) Regulatory aspects of cysteine, methionine synthesis. In: Rennenberg H, Brunold C, De Kok LJ and Stulen I (eds) Sulfur Nutrition and Sulfur Assimilation in Higher Plants: Fundamental, Environmental and Agricultural Aspects, pp 33–48, SPB Academic, The Hague

Glawisching E, Mikkelsen MD and Balkier BA (2003) Glucosinolates: biosynthesis, metabolism. In: Abrol YP and Ahmad A (eds) Sulphur in Plants, pp 145–162, Kluwer Academic, Dordrecht

Granroth B (1970) Biosynthesis and decomposition of cysteine derivatives in onion, other *Allium* species. Ann Acad Sci Fenn A2 154: 1–71

Graser G, Oldham NJ, Brown PD, Temp U and Gershenzon J (2001) The biosynthesis of benzoic acid glucosinolate esters in *Arabidopsis thaliana*. Phytochemistry 57: 23–32

Griffiths DW, Birch ANE and Hillman JR (1998) Antinutritional compounds in the Brassicaceae: analysis, biosynthesis, chemistry and dietary effects. J Hort Sci Biotechnol 73: 1–18

Grill D, Tausz M and De Kok LJ (eds) (2001) Significance of Glutathione to Plant Adaptation to the Environment. Kluwer Academic, Dordrecht

Gullner G and Kömives T (2001) The role of glutathione and glutathione-related enzymes in plant–pathogen interactions. In: Grill D, Tausz M and De Kok LJ (eds) Significance of Glutathione to Plant Adaptation to the Environment, pp 207–239, Kluwer Academic, Dordrecht

Halkier BA and Gershenzon J (2006) Biology and biochemistry of glucosinolates. Annu Rev Plant Biol 57: 303–333

Haneklaus S, Bloem E and Schnug E (2003) The global sulphur cycle and its links to plant environment. In: Abrol

YP and Ahmad A (eds) Sulphur in Plants, pp 1–28, Kluwer Academic, Dordrecht

Haneklaus S, Bloem E and Schnug E (2007a) Sulfur interactions in crop ecosystems. In: Hawkesford MJ and De Kok LJ (eds) Sulfur in Plants – an Ecological Perspective, pp 17–56, Springer

Haneklaus S, Bloem E, Schnug E, De Kok LJ and Stulen I (2007b) Sulfur. In: Barker AV and Pilbeam DJ (eds) Handbook of Plant Nutrition, pp 183–238, CRC Press, Boca Raton

Hanson AD and Gage DA (1996) 3-Dimethylsulfoniopropionate biosynthesis and the use by flowering plants. In: Kiene RP, Visscher PT, Keller MD and Kirst GO (eds) Biological and Environmental Chemistry of DMSP and Related Sulfonium Compounds, pp 75–86, Plenum, New York

Haq K and Ali M (2003) Biologically active sulphur compounds of plant origin. In: Abrol YP and Ahmad A (eds) Sulphur in Plants, pp 375–386, Kluwer Academic, Dordrecht

Harwood JL and Okanenko AA (2003) Sulphoquinovosyl diacylglycerol (SQDG) – the sulpholipid of higher plants. In: Abrol YP and Ahmad A (eds) Sulphur in Plants, pp 189–219, Kluwer Academic, Dordrecht

Hawkesford MJ (2000) Plant responses to sulfur deficiency and the genetic manipulation of sulfate transporters to improve S-utilization efficiency. J Exp Bot 51: 131–138

Hawkesford MJ and Wray JL (2000) Molecular genetics of sulphate assimilation. Adv Bot Res 33: 159–223

Hawkesford MJ and De Kok LJ (2006) Managing sulphur metabolism in plants. Plant Cell Environ 29: 382–395

Hawkesford MJ, Buchner P, Hopkins L and Howarth JR (2003a) The plant sulfate transporter family: specialized functions, integration with whole plant nutrition. In: Davidian J-C, Grill D, De Kok LJ, Stulen I, Hawkesford MJ, Schnug E and Rennenberg H (eds) Sulfur Transport and Assimilation in Plants: Regulation, Interaction and Signalling, pp 1–10, Backhuys Publishers, Leiden

Hawkesford MJ, Buchner P, Hopkins L and Howarth JR (2003b) Sulphate uptake and transport. In: Abrol YP and Ahmad A (eds) Sulphur in Plants, pp 71–86, Kluwer Academic, Dordrecht

Heinz E (1993) Recent investigations on the biosynthesis of the plant sulfolipid. In: De Kok LJ, Stulen I, Rennenberg H, Brunold C and Rauser WE (eds) Sulfur Nutrition and Sulfur Assimilation in Higher Plants: Regulatory Agricultural, Environmental Aspects, pp 163–178, SPB Academic, The Hague

Heiss S, Schäfer HJ, Haag-Kerwer A and Rausch T (1999) Cloning sulfur assimilation genes of *Brassica juncea* L.: cadmium differentially affects the expression of a putative low-affinity sulfate transporter and isoforms of ATP sulfurylase and APS reductase. Plant Mol Biol 39: 847–857

Hell R (2003) Metabolic regulation of cysteine synthesis and sulfur assimilation. The plant sulfate transporter family: specialized functions, integration with whole plant nutrition. In: Davidian J-C, Grill D, De Kok LJ, Stulen I, Hawkesford MJ, Schnug E and Rennenberg H (eds) Sulfur Transport and Assimilation in Plants: Regulation, Interaction and Signaling, pp 21–31, Backhuys Publishers, Leiden

Hell R and Kruse C (2007) Sulfur in biotic interactions of plants. In: Hawkesford MJ and De Kok LJ (eds) Sulfur in Plants – an Ecological Perspective, pp 197–224, Springer

Herschbach C (2003) Whole plant regulation of sulfur nutrition of deciduous trees – influences of the environment. Plant Biol 5: 233–244

Herschbach C and Rennenberg H (1995) Long-distance transport of ^{35}S-sulphur in 3-year-old beech trees (*Fagus sylvatica*). Physiol Plant 95: 379–386

Herschbach C and Rennenberg H (2001) Sulfur nutrition of deciduous trees. Naturwissenschaften 88: 25–36

Herschbach C, van der Zalm E, Schneider A, Jouanin L, De Kok LJ and Rennenberg H (2000) Regulation of sulfur nutrition in wild-type and transgenic polpar over-expressing γ-glutamylysteine synthase in the cytosol as affected by atmospheric H_2S. Plant Physiol 124: 461–474

Janzen HH and Bettany JR (1987) The effect of temperature and water potential on sulfur oxidation in soils. Soil Sci 144: 81–89

Jacquot J-P, Lancelin J-M and Meyer Y (1997) Tansley Review No.94. Thioredoxins: Structure and function in plant cells. New Phytol 136: 543–570

Johnson DW (1984) Sulfur cycling in forests. Biogeochemistry 1: 29–43

Johnson DW and Mitchell MJ (1998) Responses of forest ecosystems to changing sulfur inputs. In: Maynard DG (ed), Sulfur in the Environment, pp 219–262, Marcel Dekker, New York

Joshi MM and Hollis JP (1977) Interaction of *Beggiatoa* and rice plant: detoxification of hydrogen sulfide in the rice rhizosphere. Science 195: 179–180

Joshi MM, Ibrahim IKA and Hollis JP (1973) Oxygen release from rice seedlings. Physiol Plant 29: 269–271

Joshi MM, Ibrahim IKA and Hollis JP (1975) Hydrogen sulphide: effects on the physiology of rice plants and relation to straighthead disease. Phytopathology 65: 1165–1170

Kopriva S and Koprivova A (2003) Sulphate assimilation: a pathway which likes to surprise. In: Abrol YP and Ahmad A (eds) Sulphur in Plants, pp 87–112, Kluwer Academic, Dordrecht

Kreuzwieser J, Herschbach C and Rennenberg H (1996) Sulphate uptake and xylem loading of non-mycorrhizal excised roots of young *Fagus sylvatica* trees. Plant Physiol Biochem 34: 409–416

Kushad MM, Brown AF, Kurilich AC, Juvik JA, Klein BP, Wallig MA and Jeffery EH (1999) Variation of glucosinolates in vegetable crops of *Brassica oleracea*. J Agric Food Chem 47: 1541–1548

Lancaster JE and Collin HA (1981) Presence of alliinase in isolated vacuoles and of alkyl cysteine sulphoxides in the cytoplasm of bulbs of onion (*Allium cepa*). Plant Sci Lett 22: 169–176

Lancaster JE and Shaw ML (1989) γ-Glutamyl peptides in the biosynthesis of *S*-alk(en)yl-L-cysteine sulphoxides (flavour precursors) in *Allium*. Phytochemistry 28: 455–460

Lancaster JE and Boland MJ (1990) Flavor biochemistry. In: Brewster JL and Rabinowitch HD (eds) Onions, Allied Crops. Volume III: Biochemistry Food Science, Minor Crops, pp 33–72, CRC Press, Boca Raton

Lancaster JE and Shaw ML (1991) Metabolism of γ-glutamyl peptides during development, storage and sprouting of onion bulbs. Phytochemistry 30: 2857–2859

Lancaster JE, McCallion BJ and Shaw ML (1986) The dynamics of the flavour precursors the *S*-alk(en)yl-L-cysteine sulphoxides during leaf blade, scale development in the onion (*Allium cepa*). Physiol Plant 66: 293–297

Lancaster JE, Farrant JF and Shaw ML (2000) Effect of sulfur supply on alliinase, the flavour generating enzyme in onion. J Food Biochem 24: 353–361

Lanzotti V (2006) The analysis of onion and garlic. J Chromatogr A 1112: 3–22

Lappartient AG, Vidmar JJ, Leustek T, Glass AD and Touraine B (1999) Inter-organ signaling in plants: regulation of ATP sulfurylase and sulfate transporter genes expression in roots mediated by phloem-translocated compound. Plant J 18: 89–95

Lawrence JR and Germida JJ (1991) Enumeration of sulfur oxidizing populations in Saskatchewan agricultural soils. Can J Soil Sci 71: 127–136

Leach FA and Thornton I (1987) Trace elements in soils and pasture herbage on farms with bovine hypocupraemia. J Agric Sci 108: 591–597

Lee S and Leustek T (1999) The affect of cadmium on sulfate assimilation enzymes in *Brassica juncea*. Plant Sci 141: 201–207

Leustek T and Saito K (1999) Sulfate transport and assimilation in plants. Plant Physiol 120: 637–643

MacRitchie F and Gupta RB (1993) Functionality–composition relationships of wheat-flour as a result of variation in sulfur availability. Aust J Agric Res 44: 1767–1774

Malhi SS, Schoenau JJ and Grant CA (2005) A review of sulphur fertilizer management for optimum yield and quality of canola in the Canadian Great Plains. Can J Plant Sci 85: 297–307

McCaskill MR and Blair GJ (1987) Particle size and soil texture effects on elemental sulfur oxidation. Agron J 79: 1079–1083

McGrath SP and Zhao FJ (1996) Sulphur uptake, yield responses and the interactions between nitrogen and sulphur in winter oilseed rape (*Brassica napus*). J Agric Sci 126: 53–62

McGrath SP, Zhao FJ and Blake-Kalff MMA (2002) History and outlook for sulphur fertilisers in Europe. Proceedings No. 497. International Fertiliser Society, York

Moss HJ, Wrigley CW, Macritchie F and Randall PJ (1981) Sulfur and nitrogen fertilizer effects on wheat. II. Influence on grain quality. Aust J Agric Res 32: 213–226

Moss HJ, Randall PJ and Wrigley CW (1983) Alteration to grain, flour and dough quality in three wheat types with variation in soil sulfur supply. J Cereal Sci 1: 255–264

Mottram DS, Wedzicha BL and Dodson AT (2002) Acrylamide is formed in the Maillard reaction. Nature 419: 448–449

Murphy MD and O'Donnell T (1989) Sulphur deficiency in herbage in Ireland 2. Sulphur fertilisation and its effect on yield and quality of herbage. Irish J Agric Res 28: 79–90

Murphy MD and Quirke WA (1997) The effect of sulphur/nitrogen/selenium interactions on herbage yield and quality. Irish J Agric Food Res 36: 31–38

Murphy MD, Coulter BS, Noonan DG and Connolly J (2002) The effect of sulphur fertilisation on grass growth and animal performance. Irish J Agric Food Res 41: 1–15

Muttucumaru N, Halford NG, Elmore JS, Dodson AT, Parry M, Shewry PR and Mottram DS (2006) The formation of high levels of acrylamide during the processing of flour derived from sulfate-deprived wheat. J Agric Food Chem 54: 8951–8955

Naito S, Hirai MY, Inaba-Higano K, Nambara E, Fujiwara T, Hayashi H, Komeda Y and Chino M (1995) Expression of soybean seed storage protein genes in transgenic plants and their response to sulfur nutritional conditions. J Plant Physiol 145: 614–619

Noji M and Saito K (2003) Sulfur amino acids: biosynthesis of cysteine and methionine. In: Abrol YP and Ahmad A (eds) Sulphur in Plants, pp 135–144, Kluwer Academic, Dordrecht

Nocito FF, Pirovano L, Cocucci M and Sacchi GA (2002) Cadmium-induced sulfate uptake in maize roots. Plant Physiol 129: 1872–1879

Nocito FF, Lancilli C, Crema B, Fourcroy P, Davidian JC and Sacchi GA (2006) Heavy metal stress and sulfate uptake in maize roots. Plant Physiol 141: 1138–1148

Oenema O and Postma R (2003) Managing sulphur in agroecosystems. In: Abrol, YP and Ahmad A (eds) Sulphur in Plants, pp 45–70, Kluwer Academic, Dordrecht

Palmer GH (1989) Cereals in malting and brewing. In: Palmer GH (ed) Cereal Science and Technology, pp 61–242, Aberdeen University Press, Aberdeen

Pasricha NS and Fox RL (1993) Plant nutrient sulfur in the tropics and subtropics. Adv Agron 50: 209–269

Pate JS (1965) Roots as organs of assimilation of sulfate. Science 149: 547–548

Petersen BL, Chen S, Hansen CH, Olsen CE and Halkier BA (2002) Composition and content of glucosinolates in developing Arabidopsis thaliana. Planta 214: 562–571

Randall PJ, Spencer K and Freney JR (1981) Sulfur and nitrogen fertilizer effects on wheat. 1. Concentrations of sulfur and nitrogen and the nitrogen to sulfur ratio in grain, in relation to the yield response. Aust J Agric Res 32: 203–212

Randle WM (2000) Increasing nitrogen concentration in hydroponic solutions affects onion flavor, bulb quality. J Am Soc Hort Sci 125: 254–259

Randle WM and Lancaster JE (2002) Sulphur compounds in Alliums in relation to flavour quality. In: Rabinowitch HD and Currah L (eds) Allium Crop Science: Recent Advances, pp 329–356, CAB International Wallingford

Randle WM, Bussard ML and Warnock DF (1993) Ontogeny and sulfur fertility affect leaf sulfur in short-day onions. J Am Soc Hort Sci 118: 762–765

Randle WM, Lancaster JE, Shaw ML, Sutton KH, Hay RL and Bussard ML (1995) Quantifying onion flavor compounds responding to sulfur fertility. Sulfur increases levels of alk(en)yl cysteine sulfoxides, biosynthetic intermediates. J Am Soc Hort Sci 120: 1075–1081

Rauser WE (1993) Metal-binding peptides in plants. In: De Kok LJ, Stulen I, Rennenberg H, Brunold C and Rauser WE (eds) Sulfur Nutrition and Sulfur Assimilation in Higher Plants: Regulatory, Agricultural and Environmental Aspects, pp 239–251, SPB Academic, The Hague

Rauser WE (2000) The role of thiols in plants under metal stress. In: Brunold C, Rennenberg H, De Kok LJ, Stulen I and Davidian J-C (eds) Sulfur Nutrition and Sulfur Assimilation in Higher Plants: Molecular, Biochemical and Physiological Aspects, pp 169–183, Paul Haupt, Bern

Rauser WE (2001) The role of glutathione in plant reaction and adaptation to excess metals. In: Grill D, Tausz M and De Kok LJ (eds) Significance of Glutathione to Plant Adaptation to the Environment, pp 123–154, Kluwer Academic, Dordrecht

Raven JA and Scrimgeour CM (1997) The influence of anoxia on plants of saline habitats with special reference to the sulfur cycle. Ann Bot 79: 79–86

Reichelt M, Brown PD, Schneider B, Oldham NJ, Stauber E, Tokuhisa J, Kliebenstein DJ, Mitchell-Olds T and Gershenzon J (2002) Benzoic acid glucosinolate esters and other glucosinolates from Arabidopsis thaliana. Phytochemistry 59: 663–671

Rendig VV and Weir WC (1957) Evaluation of lambs feeding tests of alfalfa hay grown on low-sulphur soil. J Anim Sci 16: 451–462

Rennenberg H (1984) The fate of excess sulfur in higher plant. Annu Rev Plant Physiol 35: 121–153

Rennenberg H (1997) Molecular approaches to glutathione biosynthesis. In: Cram WJ, De Kok LJ, Brunold C and Rennenberg H (eds) Sulfur Metabolism in Higher Plants: Molecular, Ecophysiological and Nutritional Aspects, pp 59–70, Backhuys Publishers, Leiden

Rennenberg H (1999) The significance of ectomycorrhizal fungi for sulfur nutrition of trees. Plant Soil 215: 115–122

Rennenberg H and Herschbach C (1995) Sulfur nutrition of trees: a comparison of spruce (Picea abies L.) and beech (Fagus sylvatica L.). Z Pflanzenernähr Bodenk 158: 513–517

Rennenberg H, Schmitz K and Bergmann L (1979) Long-distance transport of sulfur in *Nicotiana tabacum*. Planta 147: 57–62

Riemenschneider A, Nikiforova V, Hoefgen R, De Kok LJ and Papenbrock J (2005) Impact of elevated H_2S on metabolite levels, activity of enzymes and expression of genes involved in cysteine metabolism. Plant Physiol Biochem 43: 473–483

Richards IR (1990) Sulphur as a crop nutrient in the United Kingdom. Sulphur Agric 14: 8–9

Riley NG, Zhao FJ and McGrath SP (2000) Availability of different forms of sulphur fertilisers to wheat and oilseed rape. Plant Soil 222: 139–147

Rosa E (1997) Glucosinolates from flower buds of Portuguese *Brassica* crops. Phytochemistry 44: 1415–1419

Rosa E (1999) Chemical composition. In: Gomez-Campo C (ed) Biology of *Brassica* coenospecies, pp 315–357, Elsevier Science, Amsterdam

Saito K (2003) Molecular and metabolic regulation of sulfur assimilation: initial approach by the post-genomics strategy. In: Davidian J-C, Grill D, De Kok LJ, Stulen I, Hawkesford MJ, Schnug E and Rennenberg H (eds) Sulfur Transport and Assimilation in Plants: Regulation, Interaction and Signaling, pp 11–20, Backhuys Publishers, Leiden

Scheurwater I, Koren M, Lambers H and Atkin OK (2002) The contribution of roots and shoots to whole plant nitrate reduction in fast- and slow-growing grass species. J Exp Bot 53: 1635–1642

Schnug E (1990) Glucosinolates – fundamental environmental and agricultural aspects. In: Rennenberg H, Brunold C, De Kok LJ and Stulen I (eds) Sulfur Nutrition and Sulfur Assimilation in Higher Plants: Fundamental, Environmental and Agricultural Aspects, pp 97–106, SPB Academic, The Hague

Schnug E (1993) Physiological functions and environmental relevance of sulfur-containing secondary metabolites. In: De Kok LJ, Stulen I, Rennenberg H, Brunold C and Rauser W (eds) Sulfur Nutrition and Sulfur Assimilation in Higher Plants: Regulatory, Agricultural and Environmental Aspects, pp 179–190, SPB Academic, The Hague

Schnug E (1997) Significance of sulphur for the quality of domesticated plants. In: Cram WJ, De Kok LJ, Brunold C and Rennenberg H (eds) Sulphur Metabolism in Higher Plants: Molecular, Ecophysiological and Nutritional Aspects, pp 109–130, Backhuys Publishers, Leiden

Schnug E (ed) (1998) Sulfur in Agroecosystems. Kluwer Academic, Dordrecht

Schnug E, Haneklaus S and Murphy D (1993) Impact of sulphur supply on the baking quality of wheat. Aspect Appl Biol 36: 337–345

Schröder P (1998) Halogenated air pollutants. In: De Kok LJ and Stulen I (eds) Responses of Plant Metabolism to Air pollution, Global Change, pp 131–145, Backhuys Publishers, Leiden

Schröder P (1993) Plants are sources of atmospheric sulfur. In: De Kok LJ, Stulen I, Rennenberg H, Brunold C and

Rauser W (eds) Sulfur Nutrition and Sulfur Assimilation in Higher Plants: Regulatory, Agricultural en Environmental Aspects, pp. 252–270, SPB Academic, The Hague

Schröder P (2001) The role of glutathione *S*-transferases in plant reaction and adaptation to xenobiotics. In: Grill D, Tausz M and De Kok LJ (eds) Significance of Glutathione to Plant Adaptation to the Environment, pp 155–183, Kluwer Academic, Dordrecht

Shewry PR and Tatham AS (1997) Disulphide bonds in wheat gluten proteins. J Cereal Sci 25: 207–227

Shewry PR, Franklin J, Parmar S, Smith SJ and Miflin BJ (1983) The effects of sulphur starvation on the amino acid and protein compositions of barley grain. J Cereal Sci 1: 21–31

Shortwell MA and Larkins BA (1989) The biochemistry and molecular biology of seed storage proteins. In: The Biochemistry of Plants, Vol 15, pp 297–345, Academic, New York

Spencer K and Freney JR (1980) Assessing the sulfur status of field-grown wheat by plant analysis. Agron J 72: 469–472

Spencer D, Rerie WG, Randall PJ and Higgins TJV (1990) The regulation of pea seed storage protein genes by sulfur stress. Aust J Plant Physiol 17: 355–363

Stadler RH, Blank I, Varga N, Robert F, Hau J, Guy PA, Robert MC and Riediker S (2002) Acrylamide from Maillard reaction products. Nature 419: 449–450

Städler E (2000) Secondary sulfur compounds influencing herbivorous insects. In: Brunold C, Rennenberg H, De Kok LJ, Stulen I and Davidian J-C (eds) Sulfur Nutrition and Sulfur Assimilation in Higher Plants: Molecular, Biochemical and Physiological Aspects, pp 187–202, Paul Haupt, Bern

Stefels J (2000) Physiological aspects of the production and conversion of DMSP in marine algae and higher plants. J Sea Res 43: 183–197

Stefels J (2007) Sulfur in the marine environment. In: Hawkesford MJ and De Kok LJ (eds) Sulfur in Plants – an Ecological Perspective, pp 77–90, Springer

Stulen I and De Kok LJ (1993) Whole plant regulation of sulfur metabolism. In: De Kok LJ, Stulen I, Rennenberg H, Brunold C and Rauser WE (eds) Sulfur Nutrition and Sulfur Assimilation in Higher Plants: Regulatory, Agricultural and Environmental Aspects, pp 77–91, SPB Academic, The Hague

Tabatabai MA (ed) (1986) Sulfur in Agriculture. American Society of Agronomy, Madison, Wisconsin

Tausz M (2001) The role of glutathione in plant response and adaptation to natural stress. In: Grill D, Tausz M and De Kok LJ (eds) Significance of Glutathione to Plant Adaptation to the Environment, pp 101–122, Kluwer Academic, Dordrecht

Tausz M (2007) Sulfur in forest ecosystems. In: Hawkesford MJ and De Kok LJ (eds) Sulfur in Plants – an Ecological Perspective, pp 59–75, Springer

Tausz M, Gullner G, Kömives T and Grill D (2003a) The role of thiols in plant adaptation to environmental stress.

In: Abrol YP and Ahmad A (eds) Sulphur in Plants, pp 221–244, Kluwer Academic, Dordrecht

Tausz M, Weidner W, Wonisch A, De Kok LJ and Grill D (2003b) Uptake and distribution of ^{35}S-sulfate in needles and roots of spruce seedlings as affected by exposure to SO_2 and H_2S. Environ Exp Bot 50:211–220

Trudinger PA (1986) Chemistry of the sulfur cycle. In: Tabatabai MA (ed) Sulfur in Agriculture, pp 295–323, American Society of Agronomy, Madison

Van Diggelen J, Rozema J, and Broekman R (1987) Growth and mineral relations of salt-marsh species on nutrient solutions containing various sodium sulphide concentrations. In: Huiskes AHL, Blom CWPM and Rozema J (eds) Vegetation between Land and Sea, pp 260–268, Junk Publishers, Dordrecht

van der Zalm E, Schneider A and Rennenberg H (2005) Regulation of sulfate uptake and xylem loading of poplar roots (Populus tremula x P. alba). Trees 19: 204–212

Verkleij JAC, Sneller FEC and Schat H (2003) Metallothioneins and phytochelatins: ecophysiological aspects. In: Abrol YP and Ahmad A (eds) Sulphur in Plants, pp 163–176, Kluwer Academic, Dordrecht

Watkinson JH and Blair GJ (1993) Modeling the oxidation of elemental sulfur in soils. Fert Res 35: 115–126

Watkinson JH and Lee A (1994) Kinetics of field oxidation of elemental sulfur in New Zealand pastoral soils and the effects of soil temperature and moisture. Fert Res 37: 59–68

Westerman S, De Kok LJ and Stulen I (2000) Interaction between metabolism of atmospheric H_2S in the shoot and sulfate uptake by the roots of curly kale (Brassica oleracea L.). Physiol Plant 109: 443–449

Westerman S, Stulen I, Suter M, Brunold C and De Kok LJ (2001) Atmospheric H_2S as sulfur source for Brassica oleracea: consequences for the activity of the enzymes of the assimilatory sulfate reduction pathway. Plant Physiol Biochem 39: 425–432

White PJ, Broadley MR, Bowen HC and Johnson SE (2007) Selenium and its relationship with sulfur. In: Hawkesford MJ and De Kok LJ (eds) Sulfur in Plants – an Ecological Perspective, pp 225–252, Springer

Wieser H, Gutser R and von Tucher S (2004) Influence of sulphur fertilisation on quantities and proportions of gluten protein types in wheat flour. J Cereal Sci 40: 239–244

Wittstock U and Halkier BA (2002) Glucosinolate research in the Arabidopsis era. Trends Plant Sci 7: 263–270

Wrigley CW, Ducros DL, Fullington JG and Kasarda DD (1984) Changes in polypeptide composition and grain quality due to sulfur deficiency in wheat. J Cereal Sci 2: 15–24

Yonekura-Sakakibara K, Onda Y, Ashikari T, Tanaka Y, Kusumi T and Hase T (2000) Analysis of reductant supply systems for ferredoxin-dependent sulfite reductase in photosynthetic and nonphotosynthetic organs of maize. Plant Physiol 122: 887–894

Zhang Y, Talalay P, Cho CG and Posner GH (1992) A major inducer of anticarcinogenic protective enzymes from broccoli: isolation and elucidation of structure. Proc Nat Acad Sci USA 89: 2399–2403

Zhao FJ, Evans EJ, Bilsborrow PE and Syers JK (1993) Influence of sulphur and nitrogen on seed yield and quality of low glucosinolate oilseed rape (Brassica napus L.). J Sci Food Agric 63: 29–37

Zhao FJ, Hawkesford MJ, Warrilow AGS, McGrath SP and Clarkson DT (1996) Responses of two wheat varieties to sulphur addition and diagnosis of sulphur deficiency. Plant Soil 181: 317–327

Zhao FJ, Hawkesford MJ and McGrath SP (1999a) Sulphur assimilation and effects on yield and quality of wheat. J Cereal Sci 30: 1–17

Zhao FJ, Salmon SE, Withers PJA, Evans EJ, Monaghan JM, Shewry PR and McGrath SP (1999b) Responses of breadmaking quality to sulphur in three wheat varieties. J Sci Food Agric 79: 1865–1874

Zhao FJ, Salmon SE, Withers PJA, Monaghan JM, Evans EJ, Shewry PR and McGrath SP (1999c) Variation in the breadmaking quality and rheological properties of wheat in relation to sulphur nutrition under field conditions. J Cereal Sci 30: 19–31

Zhao FJ, McGrath SP, Blake-Kalff MMA, Link A and Tucker M (2002) Crop responses to sulphur fertilisation in Europe. Proceedings No. 504. International Fertiliser Society, York

Zhao FJ, Fortune S, Barbosa VL, McGrath SP, Stobart R, Bilsborrow PE, Booth EJ, Brown A and Robson P (2006) Effects of sulphur on yield and malting quality of barley. J Cereal Sci 43: 369–377

Chapter 22

Using Anoxygenic Photosynthetic Bacteria for the Removal of Sulfide from Wastewater

Timothy J. Hurse* and Ulrike Kappler
School of Molecular and Microbial Sciences, The University of Queensland, Brisbane QLD 4072, Australia

Jürg Keller
Advanced Water Management Centre, School of Engineering, The University of Queensland, Brisbane QLD 4072, Australia

---

*Corresponding author, Fax: 61 7 3365 4620, E-mail: timhurse@yahoo.com.au

Rüdiger Hell et al. (eds.), Sulfur Metabolism in Phototrophic Organisms, 437–460.
© 2008 *Springer.*

Summary

The removal of the malodorous, corrosive and toxic compound, sulfide, from liquid waste streams is an on-going problem faced by many municipalities and industrial plants. Several systems for the treatment of such waste using phototrophic bacteria have been proposed, and the majority of these reactors rely on green sulfur bacteria as the biological agent. This chapter sets out several criteria for the selection of suitable phototrophic bacteria for use in a biological sulfide removal (BSR) process. It also discusses issues such as the supply of light to these systems and the efficiency with which different phototrophic bacteria use the supplied light for sulfide removal in the context of the different existing BSR reactors. Directions for the future development of phototrophic BSR processes are given.

I. Introduction

The element sulfur (S) is a minor component of most wastewaters, and may be present there in forms such as inorganic sulfides, various sulfoxy anions, and organic-S compounds. In domestic sewage the concentration of S lies typically between 0.3 and 1.5 mM (10–50 mg l^{-1}), which is less than the usual concentrations of C and N. Nevertheless, if sulfur is present in the form of dissolved inorganic sulfide (an equilibrium mixture of H_2S, HS^- and S^{2-}) it can be problematic even at concentrations of just 0.01 mM (0.3 mg l^{-1}), for wastewater containing dissolved inorganic sulfide tends to liberate the colorless gas H_2S, which is potentially malodorous, toxic, and corrosive. The human sense of smell detects H_2S at a very low concentration, between approximately 0.1 and 1 ppb (v/v) (Bardsley, 2002). As regards the impact of H_2S on human health, the relevant occupational exposure standards in the UK have recently been revised to 5 ppm (8 h time-weighted average) and 10 ppm (short-term exposure limit) (Costigan, 2003). Humans breathing air containing H_2S in

concentrations above approximately 500 ppm (v/v) for more than a few hours are likely to perish (Costigan, 2003). The fact that H_2S no longer triggers an olfactory response from most people if at concentrations greater than about 100 ppm makes H_2S an insidious poison and a real danger to anybody working in confined spaces through which a sulfidic wastewater stream passes – such as sewers or parts of a wastewater treatment plant. The corrosiveness of H_2S has two origins: H_2S is a weak acid, so H_2S-containing humid atmospheres are at least mildly acidic and corrosive to some metals, e.g., copper. The corrosiveness of such atmospheres is greatly exacerbated in the presence of microbial communities that oxidize the H_2S to the strong acid, H_2SO_4, which is corrosive to many metals and to cement (H_2S Technological Standing Committee, 1989). All these facets of the H_2S problem may be in evidence in those parts of an urban sewerage system where air and sulfide-containing sewage come into contact.

The sources of sulfide present in sewage are generally understood to be (1) sulfides used in certain industries (e.g., leather tanning), (2) sulfides present in the feedstocks of certain industries (petroleum refining), (3) sulfides formed during the anaerobic treatment of industrial waste before its discharge to the sewer, and (4) sulfides formed by microbially mediated sulfate reduction in anaerobic parts of the sewer (H_2S Technological Standing Committee, 1989).

The first aim of any sound strategy for addressing problems caused by H_2S in the sewerage system must be to place reasonable restrictions on the additions of sulfide, sulfate and biologically degradable organic material to the sewerage system. Wherever that measure does not sufficiently alleviate the H_2S problems, other measures are

---

Abbreviations: BChl – bacteriochlorophyll; BSR – biological sulfide removal; C – concentration of sulfide; C_{in} – concentration of sulfide in the reactor influent; C_{out} – concentration of sulfide in the reactor effluent; d – day; GSB – green sulfur bacteria; i.d. – internal diameter; PNSB – purple non-sulfur bacteria; PSB – purple sulfur bacteria; Q – volumetric flow rate of wastewater; R_V – volumetric rate of sulfide removal; R_M – amount of sulfide removed per unit time; V – volume of the reactor; VFA – volatile fatty acid; μ – cell-specific growth rate; Ω – amount of sulfide removed per unit of supplied radiant energy (or sulfide removal rate per unit of supplied radiant power)

required. One such measure is to reduce the concentration of H_2S dissolved in the sewage. Many techniques for this are applied routinely throughout the world, and commonly involve adding a chemical to the sewage so as to achieve one of the following ends: to precipitate the dissolved sulfide as insoluble non-odorous sulfide (e.g., with $FeCl_2$ or $FeCl_3$); to buffer the sewage at a mildly alkaline pH and thereby maintain most of the dissolved sulfide as the non-volatile HS^- (e.g., with $Mg(OH)_2$); to kill the sulfidogenic bacteria (e.g., with $NaOCl$ or $NaOH$); to inhibit microbial sulfate reduction by raising the redox potential of the sewage (e.g., with $Ca(NO_3)_2$); and to oxidize sulfide and to delay the onset of anaerobic conditions (e.g., with oxygen). The amount of chemical consumed being proportional to the chemical dosage and the volumetric flow rate of the sewage, chemical dosing can be a very expensive solution to sulfide problems in the larger sewers. There, the cheapest solution may be to ventilate the sewer, and to discharge this ventilation air from a tall vent, possibly after treatment to remove most of the H_2S and other odorous compounds.

From the foregoing it will be apparent that for countering sulfide-related problems in sewers, there are roles for technologies that remove sulfide or sulfate from wastewaters before discharge to the sewer, for technologies that act to reduce the levels of H_2S dissolved in the sewage flowing along the sewers, and for technologies that remove sulfide and other odorous compounds from air extracted from the sewer. Physico-chemical solutions to sulfide problems in the sewers are currently popular (see above), but are not without drawbacks. Solutions based on biological sulfide removal (BSR) have been proposed, researched and, in some cases, commercialized, particularly for removing sulfide from air. These systems rely on aerobic or microaerophilic sulfide-oxidizing chemotrophs. The potential for using anoxygenic photosynthetic bacteria to achieve the same, either in gas or wastewater, has also been recognized. This is referred to as *phototrophic BSR*. In this chapter we review the status of systems that by design use anoxygenic photosynthetic bacteria to remove sulfide from wastewater. To this end we consider the likely suitability of the various types of anoxygenic phototrophs, report on the progress made towards the development of a practical system, examine attempts to model the processes

conceptually, and present our views on ways in which research can be advanced. Systems conceived for treating gas streams to remove sulfide are referred to in passing but are not considered to be within the scope of this review.

II. Suitable Bacterial Species

For identifying the sorts of anoxygenic photosynthetic microorganisms that would form a suitable basis for a biological sulfide removal (BSR) process, the following matters are believed to be particularly pertinent: (1) an ability to carry out a dissimilatory oxidation of sulfide; (2) the influence of light availability on the rate of sulfide oxidation; (3) the quantum requirement for sulfide oxidation; (4) the products of sulfide oxidation; (5) the effect of sulfide concentration on the rate of sulfide oxidation; and (6) the influence of the presence of volatile fatty acids on the rate of sulfide oxidation.

The factors listed above shall be discussed in the following paragraphs, and by way of conclusion, the best of the known types of phototroph on which to base a sulfide removal process will be identified. In order to appreciate the practical significance of some of those six factors listed above, a familiarity with the concepts that affect the size of a reactor is advantageous. It is to these concepts that we firstly and briefly turn.

A.   Important Practical Preliminaries

The whole-of-life cost of a phototrophic BSR process is overwhelmingly determined by the volume of the reactor, the size of the radiant energy supply system, and the cost of generating the light artificially, if it is so generated. The factors that influence these should therefore be understood. In this chapter the terms *radiant energy*, *light* and *optical energy* shall be used interchangeably and should be understood as referring to electromagnetic radiation that can support photosynthesis.

1.   Reactor Size and the Volumetric Rate of Sulfide Removal

The size of a reactor for removing sulfide from wastewater depends on the following four factors (Levenspiel, 1972): the flow rate of wastewater to be treated, the concentration of sulfide in the wastewater to be treated, the proportion of sulfide that must be removed during the treatment, and a quantity called the *volumetric rate of sulfide removal*, R_V, which is defined below. Once the first three of these factors are fixed – and they are usually fixed by circumstances – the reactor size depends essentially on R_V: the larger this rate is, the smaller (and cheaper) the reactor can be. R_V is defined by Eq. 1:

$$R_V = R_M \div V \qquad (1)$$

where R_M is the amount of sulfide removed from the wastewater per unit time, and V is the volume of the reactor. R_M is given by:

$$R_M = Q \times (C_{in} - C_{out}) \qquad (2)$$

where Q is the volumetric flow rate of wastewater, C_{in} is the concentration of sulfide in the reactor influent, and C_{out} is the concentration of sulfide in the reactor effluent.

Eq. 1 and Eq. 2 can be contained to give:

$$R_v = Q \times (C_{in} - C_{out}) \div V \qquad (3)$$

The concept that is volumetric rate of sulfide removal (R_v) is evidently important, but it is not an intuitively obvious one. R_v can, however, be related to concepts familiar to most readers, and when that relation is made apparent, it will be readily appreciated how R_v can be manipulated. To this end we will consider the case of a hypothetical BSR reactor whose contents are completely mixed, and in which light availability is everywhere the same. In this restricted case all cells are exposed to the same environment, and evenly distributed throughout the reactor; and R_v can be interpreted as the product of two terms: the amount of sulfide each cell removes from the wastewater per unit time, and the number of cells per unit volume of reactor. The first of these two terms is the cell-specific rate of sulfide removal, and the second is simply the cell concentration. Hence:

$$R_v = \text{cell-specific rate of sulfide removal} \times \text{cell concentration} \qquad (4)$$

One can at once appreciate from Eq. 4 that R_v may be increased by increasing one of these factors while holding the other steady, or by increasing both the cell-specific rate of sulfide removal and the average cell concentration. As will be explained in greater detail below, anoxygenic photosynthetic bacteria remove sulfide in order to support their growth, and, consequently, the cell-specific rates of sulfide removal and cell growth are coupled, all other things being equal.

If the restriction concerning light availability were lifted so as to allow for the spatial variation of light availability that does occur in photobioreactors, the cells within the hypothetical reactor would no longer necessarily be exposed to the one and the same environment, and it may be anticipated that the cell-specific rates of growth and sulfide removal might also vary spatially within the reactor. In order to continue expressing R_v as the product of cell concentration and cell-specific rate of sulfide removal (Eq. 4), it would be more rigorous to regard the latter as a whole-reactor-average cell-specific rate of sulfide removal. If the restriction regarding the spatial distribution of cells were also abolished, so as to allow, for example, for most of the cells within the reactor to be concentrated in biofilms, then for R_v to be expressed as the product of cell concentration and cell-specific rate of sulfide removal, it would be more accurate to regard both of these terms as being akin to whole-reactor-averages.

The potential maximum cell-specific rate of sulfide removal is very much a characteristic of the species, the environmental conditions then determining how much of the potential rate is achieved on average throughout the whole reactor. The whole-of-reactor average cell concentration is not really dependent on the species, but instead is essentially dependent on the design of the reactor and on the way it is operated. We shall have more to say about these concepts when reviewing the different reactor designs.

2.   Radiant Power and the Average Radiant Power Requirement

The radiant power (i.e., the amount of radiant energy per unit time) that the phototrophic BSR reactor requires is a key factor determining the physical size of the radiant energy supply system, as well as the amount of electrical power needed to generate the radiant energy artificially. The radiant power requirement depends on many factors, and perhaps the most influential of these is the amount of sulfide that has to be removed per unit time – the mass (or molar) rate of sulfide removal (R_M). A reactor sized to remove 1 t of sulfide per day is going to have, all other things being equal, a much higher radiant power requirement than one sized to remove 1 kg of sulfide per day. We have developed a line of argument, which for brevity's sake will not be elaborated here, that suggests that the amount of sulfide removed per unit time is an appropriate basis on which to scale the radiant power requirement. Whether one expresses the scaled radiant power requirement as the sulfide removal rate per unit of supplied radiant power (which is the same thing as the amount of sulfide removed per unit of supplied radiant energy) or as the radiant power per unit rate of sulfide removal (which is the same thing as the amount of radiant energy supplied per unit of sulfide removed) is a matter of personal preference; as is the choice of units in which amounts of sulfide removed and radiant energy supplied are expressed. The quantity we have chosen to work with is the *mass of sulfide removed per unit of supplied radiant energy*, a quantity to be represented by the symbol Ω and which we choose for convenience to express in units of g W^{-1} d^{-1}. The value of Ω may be obtained from Eq. 5.

$$\Omega = \text{mass of sulfide removed by the reactor per unit time} \div \text{radiant power supplied to the reactor} \quad (5)$$

The larger the value of Ω is, the less radiant energy is consumed per unit of sulfide removed.

B.   Characteristics of Phototrophic Bacteria Relevant for Phototrophic BSR

1.   Ability to Carry Out a Dissimilatory Oxidation of Sulfide

The first of the desirable characteristics to be considered is the most obvious one, the ability to remove sulfide from the cells' surrounds. In this regard there are considerable differences amongst the types of anoxygenic phototrophs. First of all, some cannot take up sulfide from their environment. Those that can remove sulfide may do so for one or both of the following reasons: to obtain sulfur for biosynthesis (assimilatory demand) or to obtain reducing equivalents for biosynthesis (dissimilatory oxidation). Not all species can do the former. The purple sulfur bacteria (PSB), almost all green sulfur bacteria (GSB), some purple non-sulfur bacteria (PNSB), e.g., *Rho-*

dobacter capsulatus, and some of the members of the *Chloroflexaceae* are capable of the latter, as reportedly are some cyanobacterial strains in sulfidic environments (Cohen and Gurevitz, 1992; Pfennig and Trüper, 1992; Trüper and Pfennig, 1992; Imhoff, 1999; Overmann, 2000; Imhoff, 2001a, b; Hanada and Pierson, 2002, see also Chapters 14, 15, and 17 in this volume). In some species, e.g., almost all GSB, sulfide may serve in both capacities. Some strains of heliobacteria also possess an ability to oxidize sulfide by dissimilation while they grow photoheterotrophically (Bryantseva et al., 2000; Madigan, 2001).

The global relation that describes anoxygenic photosynthesis supported by dissimilatory oxidation of sulfide can be represented by the following equations in which the <CH_2O> is an approximation for synthesized cell material (van Gemerden and Mas, 1995).

$$2H_2S + CO_2 \rightarrow 2S + <CH_2O> + H_2O \quad (6)$$

$$H_2S + 2CO_2 + 2H_2O \rightarrow H_2SO_4 + 2 <CH_2O> \quad (7)$$

These equations imply that per gram of C fixed, between 1.33 and 5.33 g of sulfide-S will be removed by dissimilatory oxidation. (Marginally larger values are computed if the biomass is represented by <$C_4H_7O_3$> (Biebl and Pfennig, 1978) or by <$C_4H_8O_2N$> (Overmann et al., 1992)). If just the assimilatory demand for sulfur were relied upon, by contrast, only approximately 0.02 g of sulfide-S would be taken up per gram of C fixed. Hence, much more light would be needed to remove sulfide purely by assimilation than by dissimilatory oxidation, even after allowing for the apparent fact that the quantum requirement for photoheterotrophic growth tends to be two to three times smaller than that for photoautotrophic growth (Göbel, 1978). This is an important reason for preferring to base a phototrophic BSR system on dissimilatory sulfide oxidation. Another reason concerns the cell concentration. Even allowing for the apparent fact that the cell-specific rate of photoautotrophic growth is typically several times smaller than that of photoheterotrophic growth (Göbel, 1978), a much (10–50 times) higher cell-specific rate of sulfide removal is expected from microorganisms carrying out dissimilatory sulfide oxidation than just the assimilatory uptake of sulfide. Therefore, the cell concentration in a BSR process based purely on S-assimilation would have to be 10–50 times greater than the cell concentration in a BSR process based on dissimilatory oxidation just to allow the reactors to be the same size; and such a great concentration of cells would further complicate the task of distributing light within the reactor.

2. Influence of Light Availability on the Rate of Sulfide Removal

Since it is light that fuels the growth of these dissimilatory sulfide-oxidizing phototrophs, and which drives their oxidation of sulfide, it would be essential for the purpose of reactor design to know how light availability influences the cell-specific rate of sulfide removal, and with what efficiency absorbed light is used by a species for sulfide oxidation. These pieces of information are rarely available, however. (The report of Overmann et al. (1992) is a rare exception in this regard.) Thus, a general view of the situation must be deduced from other, available information. In view of the nexus between cell growth and sulfide oxidation, we anticipate that, for a culture growing under light limitation, the cell-specific rate of sulfide oxidation and the cell-specific rate of growth, μ, should each depend on light availability in much the same way. We are thus led to draw inferences about light availability's effects on sulfide oxidation from the known effects of light availability on light-limited cell growth, particularly under autotrophic conditions, and ideally with sulfide serving as the electron donor for photosynthesis.

Turning firstly to the relation between light availability and cell growth rate, we note at the outset that light availability is often reported as the incident irradiance. It is doubtful whether incident irradiance provides the best indication of light availability within the culture (Göbel, 1978), but except in particular circumstances, it is still believed to be indicative of that availability. The general relationship between the cell-specific growth rate, μ, of a light-limited culture and the incident irradiance on that culture is as follows (Biebl and Pfennig, 1978; Veldhuis and van Gemerden, 1986; Guyoneaud et al., 2001). There is no net growth until the incident irradiance exceeds a threshold value. As incident irradiance increases beyond that threshold value μ increases. The rate at which μ increases with respect to incident irradiance declines with increasing incident irradiance. As a result μ attains a maximum value, μ_{max}, and declines as incident irradiance climbs further. This decline is termed *photoinhibition*. Given the link between growth and sulfide oxidation, the larger the value of μ_{max}, the larger is the cell-specific rate of sulfide removal that can potentially be attained. The availability of light needed to bring about growth at μ_{max} is also a relevant consideration, but is rarely reported. Instead, it is to be inferred from a much more commonly reported parameter, the value of the incident irradiance at which μ is some sizeable and defined proportion (say 50%) of μ_{max}. The body of experimental evidence gathered so far indicates that amongst each of the PSB, PNSB and GSB, μ_{max} for photoautolithotrophic growth on sulfide varies considerably from strain to strain, but that the values of μ_{max} for the more rapidly growing strains of each group are similar (van Gemerden, 1984). Thus, species with the potential for relatively rapid sulfide removal can be found amongst each of the PSB, PNSB and GSB. As regards the irradiance necessary to realize this potential, it is the generally held view that the GSB make much better use of the available light for growth than PSB or PNSB do, especially at very low irradiances (Biebl and Pfennig, 1978; Overmann et al., 1992; van Gemerden and Mas, 1995). This would suggest that in order to attain a maximum cell-specific rate of sulfide removal, less radiant power would be needed, and consequently less cost incurred, if the BSR process were based on GSB than if based on PSB or PNSB. Nevertheless, the natural and experimental situations from which the aforementioned generally held view is derived are not necessarily going to be representative of the situation of the phototrophs in a phototrophic BSR reactor. Given the sizeable costs of supplying light at large rates, these reactors are likely to be engineered so as to enable near total absorption of light by cells. In that situation, the radiant power requirement is better judged by the quantum requirement for sulfide oxidation, and it is to this issue that we now turn.

3. The Quantum Requirement for Sulfide Oxidation

As explained above, the effectiveness with which a phototrophic organism uses absorbed light for sulfide oxidation has, for want of direct measurements, to be inferred from the effectiveness with which it uses absorbed light for a process related to sulfide oxidation. At least in the case of photoautotrophic growth CO_2 fixation is such a related process. The effectiveness with which light is used for CO_2 fixation is indicated by the number of quanta absorbed per molecule of CO_2 fixed, a quantity referred to as the *quantum requirement for CO$_2$ fixation*. Of two species capable of photoautotrophic growth with sulfide as the electron donor, it is the one having the lower minimum quantum requirement for CO_2 fixation under photoautotrophic conditions that is able to use absorbed light more effectively for sulfide oxidation – all other things being equal.

Quantum requirements for CO_2 fixation under purely autotrophic conditions have been investigated experimentally for a handful of strains only (see below). All investigations carried out so far indicate that the quantum requirement depends partly on innate characteristics of the phototroph strain and partly on cultivation conditions, notably the degree of light limitation and the wavelength of the absorbed light (Larsen, 1953; Göbel, 1978). The innate characteristics determine the minimum quantum requirement that can be attained, and the culture conditions determine how closely this minimum may be approached. The quantum requirement is at its minimum when the light is (1) sufficiently available to support light-limited growth at about half $\mu_{max,}$ and (2) supplied at a wavelength strongly absorbed by the cells' light-harvesting bacteriochlorophyll (BChl) in vivo (Larsen, 1953; Göbel, 1978).

Table 1 is a summary of measured values of the quantum requirement for CO_2 fixation by different strains under conditions of light-limitation with inorganic electron donors. As only wavelengths of light strongly absorbed by the light-harvesting BChls in vivo were used in those experiments, these values of the quantum requirement should be regarded as the minimum values attainable for the corresponding strain. Notwithstanding the report of French (1937), which has been criticized at length by Larsen (1953), the lowest minimum quantum requirement observed so far for CO_2 fixation with an inorganic electron donor is about 9–10 quanta, and this was for a GSB strain. Proceeding from biochemical theory, Brune (1989) has

predicted that the minimum quantum requirement for CO_2 fixation is 8 1/2 quanta in PSB and between 3 1/3 and 4 1/2 in GSB. Although in none of the quantum requirement measurements was sulfide used as the electron donor, Larsen (1953) observed essentially the same minimum quantum requirements with each of three different electron donors, hydrogen, thiosulfate and tetrathionate. Kim et al. (1992) have reported that sulfur globules formed by *Chlorobium* from sulfide increase light scattering more than they do light absorption. On that basis, we would anticipate that sulfur globule formation will not greatly affect the observable minimum quantum requirement.

The wavelengths strongly absorbed by the various types of light-harvesting BChls in vivo have been summarized by Overmann and Garcia-Pichel (2000). These light-harvesting BChls are responsible for at least two maxima in the spectral absorbance, between 350 nm and 1100 nm, of intact anoxygenic phototrophs. At least one of the absorbance maxima attributable to light-harvesting BChl occurs at a wavelength between 700 and 1100 nm, i.e., in the near IR region. The value of this strongly absorbed wavelength is determined by the form of light-harvesting BChl present in the cell and by the chemical environment in which the BChl is situated. To some extent the type of light-harvesting BChl is also a characteristic shared by members of a particular group of phototrophs: the light-harvesting BChl in PNSB strains and PSB strains is either BChl *a* or *b*; that in GSB strains is either BChl *c*, *d* or *e*; that in *Chloroflexaceae* is either BChl *a* or *c*; and that in heliobacteria is BChl *g*. Hence, the use of a light source emitting most of its optical energy at a near IR wavelength strongly absorbed by the light-harvesting BChl of a desired type of phototroph not only favors the effective utilization of the absorbed light by the phototroph but can also be used to select against some unwanted types of phototroph. It is opportune to mention here that wastewater is rarely sterile and may be prohibitively expensive to sterilize and keep sterile. On this account the practical BSR system will contain a microbial community of greater or lesser diversity, and the ability to favor the development of a desired type of phototroph or to hinder the growth of an undesired type is of practical benefit.

Table 1. Quantum requirement for CO_2 fixation.

Organism (name as used in reference)	Minimum quantum requirement for CO_2 fixation (quanta absorbed per molecule of CO_2 fixed)	Electron donor	Assay	Reference
Chlorobium thiosulfat-ophilum	Range: 7.8–11.8; Ave: 9.7 Range: 8.9–11.4; Ave:10.1 Range: 8.9–9.7; Ave: 9.3	Hydrogen Thiosulfate Tetrathionate	Manometry with resting cells	Larsen (1953)
Rhodopseudomonas acidophila 10050 (now *Rhodoblastus acido-philus*)	17.2 (CO_2 in cells) 11.7 (CO_2 in cells and exudates)	Hydrogen	Cells growing in continuous culture	Göbel (1978)
Chromatium strain D (now *Allochro matium vinosum*)	Range: 9.4–15.2; Ave: 12 Range: 8.5–16; Ave: 12	Thiosulfate Hydrogen	Manometry with resting cells	Wassink et al. (1942)
Streptococcus varians (now *Rhodobacter cap-sulatus*)	Range: 4.4–16.2 Values depend on pre-treatment	Hydrogen	Manometry with resting cells	French (1937)

4. The Products of Sulfide Oxidation by Phototrophic Bacteria

Sulfate is the final product of oxidation in most but by no means all cases (see summary in Brune (1989)). A zero-valent form of sulfur, recoverable as elemental sulfur, is formed extracellularly as a final product of sulfide oxidation by some species of PNSB, cyanobacteria, *Chloroflexaceae*, and heliobacteria. Zero-valent sulfur is also formed as an intermediate product by many PSB and GSB species. This intermediate zero-valent sulfur is formed extracellularly by strains belonging to the *Ectothiorhodospiraceae* and GSB strains, and intracellularly by members of the *Chromatiaceae*. The aforementioned global relations (Eq. 6 and Eq. 7) that describe anoxygenic photosynthesis supported by sulfide oxidation indicate that for each unit of <CH_2O> formed, four times more sulfide will be oxidized if the sulfide is oxidized to zero-valent sulfur than if the sulfide is oxidized to sulfate. In situations where bacterial growth is limited by the availability of light, which is, for economic reasons, likely to be the case in a large-scale system, the equations thus imply that four times less light will be used per unit mass of sulfide oxidized if the sulfide is oxidized to the zero-valent level than if the sulfide is oxidized to sulfate. Hence, the nature of the product or products of bacterial sulfide oxidation can have important ramifications for the energy efficiency and cost of using a particular strain as the basis of the phototrophic BSR process.

The nature of the products of sulfide oxidation will also determine the ease with which the products can be recovered. For the time being at least, the recovery of the elemental sulfur and sulfate is unlikely to be justified by their economic value. Recovery would, however, have to be considered if there were a risk that the release of the oxidation products would cause problems or worsen existing, sulfur-related problems downstream, e.g., due to sulfide formation by microbial sulfate reduction, or due to the blockage of drains and other orifices with elemental sulfur. The most easily recoverable product of sulfide oxidation is elemental sulfur, as this is essentially insoluble in water.

5. The Effect of Sulfide Concentration on the Rate of Sulfide Oxidation

The significant cost of providing optical energy will militate in favor of a design that makes efficient use of that energy, and, by implication, an operation in which the sulfide removal rate by the phototrophic community as a whole is strongly limited by light supply. However, even under those conditions, growth in the parts of the community very close to the light source or near the reactor exit, where the sulfide concentration is least, may be limited more strongly by sulfide availability than by light availability. The potential for process inhibition by sulfide is also a possibility to be reckoned with. The influence of sulfide availability on the rate of sulfide oxidation is therefore a relevant consideration. For want of the desired information, the broad picture about the influence of sulfide concentration on the rate of sulfide oxidation has to be inferred from the influence of sulfide concentration on a related process. From Eq. 6 and Eq. 7 we anticipate that for a culture growing autotrophically under sulfide limitation, the cell-specific rates of sulfide oxidation rate and growth (μ) will depend on sulfide concentration (C) in much the same way (except insofar as discussed in the following paragraph). (This inference finds direct support in Beeftink and van Gemerden (1979) and in Veldhuis and van Gemerden (1986)). Under sulfide limitation μ increases as C increases from zero. The rate at which μ increases with respect to C decreases with increasing C, however, with the result that μ tends to a maximum value, and declines as C climbs further. This decline is sulfide inhibition. van Gemerden (1984) has described this dependence of μ on C with a Haldane type model, $\mu = \hat{\mu} \, C/[(K_s + C)(1 + C/K_i)]$, and has provided parameter values for 17 strains (GSB, PSB and PNSB). The model parameters are: $\hat{\mu}$, the theoretical maximum value of μ that would occur if the substrate were not inhibitory; K_i, the inhibition constant, is the higher of the two values of C at which μ equals $1/2 \, \hat{\mu}$; and K_s is the lower of the two values of C at which μ equals $1/2 \, \hat{\mu}$. The lower the value of K_s, the less the cells' growth is limited by sulfide availability at a given value of C. The larger the value of K_i, the more tolerant is the organism to sulfide, and the more closely the actual maximum value of μ, $\hat{\mu}_{max}$, approaches $\hat{\mu}$. Some published values of K_s and K_i are presented in Table 2. They indicate that sulfide limitation and inhibition would be least likely to occur if the process were based on GSB or PSB (especially *Ectothiorhodospiraceae* strains). A process based on PNSB would struggle to cope with sulfide concentrations greater than 2 mM, a result confirmed generally by Hansen and van Gemerden (1972). For the sulfide-oxidizing strains of the *Chloroflexaceae* and heliobacteria the relationship between sulfide oxidation and sulfide availability has yet to be surveyed. Nevertheless, some strains can evidently oxidize sulfide at millimolar concentrations of sulfide (Keppen et al., 1993; Bryantseva et al., 2000).

In addition to affecting the cell-specific growth rate, the sulfide availability has also been shown to affect the average degree of sulfide oxidation. Working with sulfide-limited, pure cultures of representative GSB and PSB strains growing in chemostats at sulfide concentrations of the order of the K_s value, van Gemerden (1986) showed that the ratio of the two products of sulfide oxidation, sulfate and sulfur, increased as the concentration of sulfide decreased. This result suggests that it becomes difficult to stop the oxidation at the zero-valent level once the sulfide concentration in the photic zone is of the order of the K_s value. The sulfide concentration in the photic zone equals the sulfide concentration in the wastewater only in the case where the cells are suspended in the wastewater. In cases where the sulfide has to diffuse through a biofilm to reach the photic zone – and such cases will be reviewed in Sections III and IV – the sulfide concentration in the photic zone may be <10% of the sulfide concentration in the wastewater contacting the biofilm. Sulfide availability has also been implicated in the regulation of acetate co-assimilation by one GSB strain, *Chlorobium phaeobacteroides* strain K1 (see below).

6. The Influence of the Presence of Volatile Fatty Acids on the Rate of Sulfide Oxidation

The presence in a cell's environment of organic compounds that the cell's major biosynthetic pathways can utilize with little if any modification represents a potential boon for the cell. It will take advantage of the organic compounds to the extent that its metabolism and the availability of light and other nutrients allow. Sugars and volatile fatty acids are to be found in many types of wastewater. It is to be apprehended that the cell's utilization of such sources of organic carbon could result in a lower cell-specific demand for CO_2, a lower cell-specific demand for reducing equivalents derived from sulfide, and, as a possible consequence, a lower cell-specific rate of sulfide oxidation, and a lower volumetric rate of sulfide oxidation. The potential for this type of effect would seem to be greatest with PNSB and least with GSB. Many PNSB strains can grow photoheterotrophically on a wide range of organic compounds (Imhoff, 2001a, b). Most GSB strains can by contrast assimilate very few

Table 2. Kinetic parameters for representative anoxygenic photosynthetic bacterial strains growing under sulfide limitation.

Study	Species (Family) (strain)	Comment	$\hat{\mu}$	$\hat{\mu}_{max}$ (h^{-1})	K_s (μM)	K_i (mM)
van Gemerden (1974)	*Chromatium vinosum* (PSB) (DSM 185) (now *Allochromatium vinosum*)	chemostat; pure culture; autotrophic growth	0.130	0.117	7	2.5
"	*Chromatium weissei* (PSB) (DSM 171)	"	0.050	0.040	10	0.7
Veldhuis and van Gemerden (1986)	*Thiocapsa roseopersicina* (PSB) (K2)	chemostat; pure cultures; concurrent growth on acetate and carbon dioxide	0.087	0.072	21	2.2
van Gemerden (1984)	*Chlorobium phaeobacteroides* (GSB) (K1)	"	0.096	0.093	0.8	2.5
"	*Ectothiorhodospira vacuolata* (PSB) (DSM 2111) (now *Ectothiorhodospira shaposhnikovii*)	chemostat; pure culture; autotrophic growth	0.138	0.129	3	2.8
van Gemerden and Beeftink (1981)	*Chlorobium limicola f. thiosulfatophilum* (GSB) (DSM 249) now *Chlorobaculum thiosulfatiphilum*	chemostat; pure culture; autotrophic growth	0.110	0.105	1.5	3
van Gemerden (1984)	*Rhodobacter capsulatus* (PNSB) (DSM 155)	chemostat; pure culture; autotrophic growth	0.145	0.132	2.3	1

organic compounds, e.g., acetate and pyruvate, and then only in the presence of CO_2 and an electron donor such as sulfide (Sadler and Stanier, 1960; Overmann, 2001).

Several reports have provided evidence that the cell-specific rate of sulfide oxidation by *Rhodobacter capsulatus*, *Thiocapsa roseopersicina* and *Chlorobium phaeobacteroides* is suppressed by the presence of acetate when the cell-specific growth rate, μ, is held constant (Wijbenga and van Gemerden, 1981; Hofman et al., 1985; Veldhuis and van Gemerden, 1986). Furthermore, when cultivated in the presence of acetate and sulfide, *Rhodobacter capsulatus* and *Thiocapsa roseopersicina* seem to prefer acetate to sulfide, and *Chlorobium phaeobacteroides* sulfide to acetate. If this behavior is indicative of the other strains in the respective groups, a PNSB/PSB-based phototrophic BSR system would seem to be susceptible to direct inhibition by organic compounds, particularly VFAs. By contrast, when *Chlorobium phaeobacteroides* was cultivated under sulfide-limited or light-limited continuous cultures, it was observed that the higher the sulfide concentration was, the lower the rate of acetate uptake was; and in batch experiments, practically no acetate was taken up if the sulfide concentration exceeded 0.8 mM.

The chief question regarding sulfide removal by the reactor is whether the volumetric rate of sulfide removal will be affected in the presence of organic compounds, i.e., whether the inhibition of sulfide oxidation, described above, will be offset by an increase in either the average cell concentration or μ (if μ is free to change). Hofman et al. (1985) indicated that the inclusion of acetate in the chemostat feed led to practically no change in the sulfide concentration in the chemostat effluent, a result implying that the volumetric rate of sulfide had changed negligibly. Presumably the cell concentration had increased to compensate for the lower cell-specific rate of sulfide oxidation, because the dilution rate, and hence μ, had not been changed. In biofilm systems, μ can vary independently of the dilution rate, but the average cell concentration may be restricted by the shear force exerted by the wastewater on the biofilm.

7. Conclusion

From a consideration of the above criteria and of what is known so far, it appears to us that if a BSR process is to be based on anoxygenic photosynthetic bacteria, then the GSB will in most cases be the most appropriate type to use. Almost without exception GSB oxidize sulfide to an extracellular intermediate that can be recovered as elemental sulfur separately from the cells. Their K_s values for sulfide oxidation are low relative to those of other anoxygenic photosynthetic bacteria, and they tolerate industrially relevant sulfide concentrations. Their ability to assimilate organic compounds is restricted to a few simple VFAs and is contingent on co-assimilating CO_2 and oxidizing a reduced inorganic compound, like sulfide. Hence, a suppression of the cell-specific sulfide removal rate is only to be anticipated in the presence of a few simple organic compounds. The GSB strongly absorb some wavelengths between 700 and 770 nm, and by virtue of this fact, the judicious control of the spectral composition of the light could be used as a means to select against purple bacteria, cyanobacteria and heliobacteria also. Although the matter is still to be comprehensively surveyed, GSB are believed to have the lowest quantum requirement during photoautotrophic growth, which strongly suggests that they have the lowest quantum requirement for sulfide oxidation. The quantum requirements for sulfide oxidation by heliobacteria and by sulfide-oxidizing *Chloroflexaceae* strains, which are also able to oxidize sulfide to an extracellular form of elemental sulfur, have not to our knowledge been reported.

A definitive declaration about the merit of seeking to base a BSR process on heliobacteria or *Chloroflexaceae* cannot be made until representative strains have been characterized with respect to the criteria discussed. The heliobacteria contain sulfide-oxidizing mesophiles and the *Chloroflexaceae* sulfide-oxidizing thermophiles (as do the GSB), which may lend themselves to a use for the treatment of warm to hot sulfidic wastewaters (Madigan and Brock, 1975; Bryantseva et al., 2000). The potential for basing BSR processes on heliobacteria, especially, may have to be reevaluated: they have an absorption spectrum distinct from that of all other known families of anoxygenic photosynthetic bacteria, which suggests that they could be selected for to the exclusion of other phototrophs by means of spectral control. (The feasibility of efficiently generating near monochromatic light will be considered further on.)

It should also be recalled that some cyanobacterial strains are reportedly capable of anoxygenic photosynthesis (with the oxidation of sulfide) in sulfidic environments, while others have been reported to maintain oxygenic photosynthesis in the presence of sulfide at a concentration of 1 mM (see Cohen and Gurevitz (1992) and references therein). Both types of cyanobacteria might in theory be used for a BSR system, but further characterization with respect to the points already elaborated is needed. In circumstances where the generation of oxygen is to be prevented or most carefully controlled, e.g., inside an anaerobic digester in which methane is also being evolved, the feasibility of selecting against oxygen-evolving strains and in favor of sulfide-oxidizing strains, or vice-versa if desired, would require careful study. Finally, if the wastewater were strongly alkaline or saline and contained little organic carbon, consideration should be given to trying to base the BSR process on *Ectothiorhodospiraceae* strains, as this family contains alkalophilic, halophilic and haloalkalophilic species capable of oxidizing sulfide. Globules of a zerovalent sulfur are formed outside the cell (Imhoff, 1999).

III.   Overview of Reactor Concepts

The purpose of this section is to draw the reader's attention to the major distinguishing features of the engineered systems that use anoxygenic phototosynthetic bacteria to remove sulfide from wastewater. The scene will thereby be set for a detailed review of some of these systems, which will be presented in the next section. Before proceeding toward this objective, we wish to mention, briefly, other engineered wastewater systems in which phototrophs play a leading role, but in which the phototrophs remove sulfide incidentally or with unpredictable effectiveness.

Full-scale wastewater treatment systems in which photosynthetic microorganisms play important roles have been described in several communications from Japan during the past three decades (Kobayashi, 1975; Kobayashi and Kobayashi, 1995). In those systems the anoxygenic photosynthetic bacteria serve to remove most of the *organic* pollutants not removed by an initial sedimentation step, which is quite a different situation from the BSR systems to be reviewed in this chapter. The wastewater in these Japanese wastewater treatment systems is not sterilized before treatment, and the reactors containing the anoxygenic photosynthetic bacteria may be open and aerated. It is hardly surprising, therefore, that PNSB appear to be the predominant phototrophs in those reactors (Hiraishi et al., 1995).

The presence of photosynthetic bacteria in the units of a conventional activated sludge wastewater treatment plant and in wastewater lagoons has been reported. The vast majority of the phototrophs found in the activated sludge were PNSB, and were believed to be deriving most of their energy by respiration (Siefert et al., 1978). In wastewater treatment lagoons receiving heavy organic loads, concentrations of sulfide in the water column can reach levels that lead to the proliferation of PSB and cause odors. Although PSB remove some sulfide, it is inherently difficult to ensure consistent and sufficient sulfide removal as the process is open to the atmosphere and dependent on sunlight (Houghton and Mara, 1992).

Concepts entailing the use of anoxygenic photosynthetic bacteria in engineered systems specifically for the removal of sulfide from wastewater have been proposed in patents and in a few well-documented scientific reports by various authors (see below). All reported studies have been carried out at the bench scale. The studies can be classified according to the types of phototrophic bacteria that have been used in these various studies (Table 3). Two striking features emerging from Table 3 are the prominent place played by GSB in nearly all engineered phototrophic BSR processes, and the minimal degree to which the application of other classes has been investigated. The phototrophic BSR processes described can also be usefully classified according to whether the organisms were freely suspended within the wastewater or attached to some solid surface (substratum), and whether there was a continuous flow of wastewater through the system. The research thus classified is summarized in Table 4. We have included in the Tables some references

to studies on the use of anoxygenic photosynthetic bacteria for desulfurizing a continuously flowing stream of gas.

In recent years research into the application of anoxygenic photosynthetic bacteria for removing sulfide from wastewater has been focused on attached-growth systems, i.e., biofilm systems, with light supplied either through the medium (Ferrera et al., 2004a, b) or through a transparent substratum (Kobayashi et al., 1983; Henshaw et al., 1998; Henshaw and Zhu, 2001; Syed and Henshaw, 2003, 2005; Hurse and Keller, 2004a, b, 2005). The performance of these attached-growth systems will now be described.

IV. Reactor Performance

In the preliminary remarks made at the beginning of Section II, two parameters were singled out as being of particular practical importance: the volumetric rate of sulfide removal in the reactor (R_V), and the amount of sulfide removed per unit of radiant energy supplied to the reactor (Ω). Their importance, it was explained, derives from their overwhelming influence on the whole-of-life cost of a phototrophic BSR system. It is only natural, therefore, to consider these parameters when comparing and contrasting the performance of reactor systems. In line with the other points made in Section II, when assessing the relative merits of different reactor systems one should also have regard to the sulfide oxidation products, and to whether the wastewater was of a type that could have allowed non-sulfide oxidizing phototrophs to develop, and thereby to divert some of the supplied radiant energy away from sulfide oxidation.

Both R_V and Ω can generally be computed from reported data by means of Eq. 3 and Eq. 5, respectively. Further useful information can be extracted from reported reactor performance figures if R_V is broken down into its component terms. R_V can be expressed as the product of the cell concentration and the cell-specific rate of sulfide removal (Eq. 4), where both quantities are to be understood as average values over the entire reactor volume. As these two quantities may be independent of each other, comparisons of the performance of different systems will be more informative if the cell concentration and

Table 3. Classification of researched phototrophic BSR systems on the basis of culture purity.

Mixed Culture[a]		Pure Culture[a]	
Researchers	Predominant phototrophic family represented	Researchers	Family
Kobayashi et al. (1983)	GSB and PNSB	Henshaw et al. (1998)	GSB
Khanna et al. (1996)	GSB	Henshaw and Zhu (2001)	GSB
Hurse and Keller (2004a, b, 2005)	GSB	Syed and Henshaw (2003, 2005)	GSB
Ferrera et al. (2004a, b)	PNSB, PSB and Chlorophyta	Ferrera et al. (2004b)	GSB
		Cork et al. (1983, 1985)	GSB
		Cork and Kenevan (1990)	GSB
		Maka and Cork (1990a, b)	GSB
		Kim and Chang (1991)	GSB
		Kim et al. (1992, 1996, 1997)	GSB
		Basu et al. (1996)	GSB
		Lee and Kim (1998)	GSB
		Sardesai et al. (2006)	heliobacteria

[a] Underlined studies pertain to wastewater treatment.

Table 4. Classification of researched phototrophic BSR systems on the basis of biomass distribution and flow.

Sulfidic stream	Reactor type	Biomass attached to surface	Biomass suspended in liquid
Wastewater	Continuous	Kobayashi et al. (1983)	Henshaw et al. (1998)
"	"	Henshaw and Zhu (2001)	Khanna et al. (1996)
"	"	Syed and Henshaw (2003, 2005)	
"	"	Hurse and Keller (2004a, b, 2005)	
"	"	Ferrera et al. (2004b)	
"	Batch		
Gas	Continuous		Henshaw et al. (1992)
"	"		Basu et al. (1996)
"	Fed batch	Kim and Chang (1991)	Kim and Chang (1991)
"	"		Cork et al. (1983,1985)
"	"		Cork and Kenevan (1990)
"	"		Maka and Cork (1990a, b)
"	"		Kim et al. (1992, 1996, 1997)
"	"		Lee and Kim (1998)

the cell-specific rate of sulfide removal of the respective systems are clearly distinguished. If R_v and the measured cell concentration are both known, Eq. 4 can be used to determine the cell-specific rate of sulfide removal.

For practical reasons an indirect rather than an actual measure of cell concentration is generally used as the rate standardizing parameter. Examples of these indirect measures include the mass of pigment or protein per unit volume of reactor. If the biomass exists largely in biofilms growing on illuminated surfaces, the area of illuminated surface per unit reactor volume could also be used as the rate standardizing parameter. (The use of surface area as a standardizing parameter for attached-growth wastewater treatment reactors is widespread.) In that case the standardized removal rate is the mass of sulfide removed per unit time per unit area of illuminated surface. We have found it convenient to express this quantity in units of g m^{-2} d^{-1}.

We have analyzed most of the recent reported studies of phototrophic BSR systems, and have derived the values of the various rates in accordance with the foregoing recommendations, if there was sufficient information in the reports for us to do so with confidence. Pertinent parameters of these BSR systems are summarized in Table 5. Several points in particular deserve emphasis.

Table 5. Performance characteristics of researched phototrophic BSR systems.

Researchers	Kobayashi et al.	Henshaw et al.	Henshaw and Zhu	Syed and Henshaw	Ferrera et al.	Hurse and Keller	Syed and Henshaw
Year	1983	1998	2001	2003	2004	2004a, b; 2005	2005
Biomass	Attached	Suspended	Attached	Attached	Attached	Attached	Attached
Reactor shape	Single tube	Chemostat	20 tubes in parallel	16 tubes in parallel	Packed tube and recirculation tank	Flat cylinder	16 tubes in parallel
Tube dimensions	3.2 mm i.d. 12.8 m long	Not applicable	3.0 mm i.d. 150 mm long	1.6–6.4 mm i.d. 150 mm long	15.6 mm i.d. 275 mm long	Not applicable	1.6 mm i.d. 150 mm long
Flow	Plug	Mixed	Plug	Plug	Recirculated	Mixed	Plug
Inoculum	Anaerobic digester effluent	*Chlorobium limicola* (*Chlorobium thiosulfatophilum*) (ATCC No. 17092)	*Chlorobium limicola* (DSM No. 257)	*Chlorobium limicola* (DSM No. 257)	(1) *Chlorobium limicola* UdG6008; (2) Sediment of sulfurous lake	Mixture of GSB enrichments from anaerobic digester and from lake water & sediment	*Chlorobium limicola* (DSM No. 257)
Phototrophic community during study	GSB and PNSB in overlapping zones	Not reported	Not reported	Not reported	(2) PNSB, PSB, Chlorophyta, Cyanobacteria	GSB	Not reported
Wastewater	Anaerobic digester effluent	Synthetic (after Madigan)	Synthetic (Pfennig's medium II)	Synthetic (Pfennig's medium II)	Synthetic (carbonate as C)	Pfennig's medium sometimes spiked with acetate or propionate	Synthetic (Pfennig's medium II)
Reactor volume	103 ml[a]	121, 13.71	21.2 ml	4.8–77 ml	1.81 (reservoir) 52.5 ml (tube)	295 ml	4.8 ml
Wastewater flow rate	17, 29 ml h^{-1}	70–310 ml h^{-1}	9–42 ml h^{-1}	34–85 ml h^{-1}	Varied but not reported	65–73 ml h^{-1}	(1) 17–24; (2) 10–20 ml h^{-1}
Hydraulic residence time	0.41–0.85 h	44–175 h	0.5–2.35 h	0.07–2.3 h	Varied but not reported	4.1–4.5 h	(1) 0.2–0.3; (2) 0.2–0.5 h
Sulfide concentration in influent	14–24 mg l^{-1}	89–627 mg l^{-1}	140–452 mg l^{-1}	75–311 mg l^{-1}	26–112 mg l^{-1} (800–3500 μM)	3–32 mg l^{-1}	(1) 68–70; (2) 63–71 mg l^{-1}
Sulfide concentration in reactor	0-inlet conc.	0–112 mg l^{-1}	0-inlet conc.	0-inlet conc.	1.6 mg l^{-1} (50 μM) (in stirred vessel)	2–17 mg l^{-1}	0-inlet conc.
Temperature	Not reported	30°C	27°C	27–29°C	Not reported	21–29°C	27–29°C
Incident irradiance (measurement band in nm)	350 ft-cd at water-bath surface	over window: 258 W m^{-2} (400–1000 nm); 115 W m^{-2} (700–975 nm)	25.4 W m^{-2} (400–1000 nm)	77–153 W m^{-2}	50–150 μE m^{-2} s^{-1}	0.2–2.1 W m^{-2}	16.5–33 W m^{-2} (400–1000 nm)
Light source	40 W tungsten bulb	2, 175 W incandescent IR-lamps	1, 250 W incandescent IR-lamp	1, 250 W incandescent IR-lamp	2, 60 W incandescent bulbs	Filtered light from slide projector	(1) LEDs; (2) 1, 250 W incandescent IR-lamp
Light bandwidth	Not reported	Not reported	Not reported	250–4000 nm	Not reported	710–790 nm	(1) 710–850; (2) 250–4000 nm

Volumetric sulfide loading	31–82 mg l^{-1} h^{-1b}	2.1–6.7 mg l^{-1} h^{-1}	107–328 mg l^{-1} h^{-1}	83–1451 mg l^{-1} h^{-1}	Not reported	0.6–7.6 mg l^{-1} h^{-1}	(1) 244–338; (2) 145–255 mg l^{-1} h^{-1}
Volumetric rate of sulfide removal	26–42 mg l^{-1} h^{-1}	2.0–5.2 mg l^{-1} h^{-1}	107–284 mg l^{-1} h^{-1}	83–1451 mg l^{-1} h^{-1}	131 mg l^{-1} h^{-1} based on tube volume	0.1–3.3 mg l^{-1} h^{-1}	(1) 244–338; (2) 145–255 mg l^{-1} h^{-1}
Biomass concentration	Not reported	2.8–13.5 mg l^{-1}	12.5–51.4 mg l^{-1}; 9.4–39 mg m^{-2}	363–1639 mg l^{-1}; 550–1065 mg m^{-2}	1660–4520 mg pigment m^{-2}	Not reported	(1) 1262–1306; (2) 1214–1295 mg l^{-1}. (1) 505–522; (2) 486–518 mg m^{-2}
Biomass specific sulfide removal rate	Not calculable	0.15–1.1 mg mg^{-1} BChl h^{-1}	3.0–9.2 mg mg^{-1} BChl h^{-1}	0.2–0.9 mg mg^{-1} BChl h^{-1}	0.34–0.36 µmol mg^{-1} protein h^{-1}; 0.072–0.31 mg mg^{-1} pigment h^{-1}	Not calculable	(1) 0.19–0.26 (2) 0.12–0.20 mg mg^{-1} BChl h^{-1}
Surface area of sulfide-removing zone^{c}	88–307 cm^{2}	Not applicable	283 cm^{2}	120–480 cm^{2}	135 cm^{2} (column wall); 325 cm^{2} (packing)	100 cm^{2}	121 cm^{2}
Areal rate of sulfide removal	0.49–0.82 g m^{-2} d^{-1}	Not applicable	1.9–5.1 g m^{-2} d^{-1}	3.2–14.1 g m^{-2} d^{-1}	0–12.3 g m^{-2} of column wall d^{-1}	0.1–2.4 g m^{-2} d^{-1}	(1) 2.3–3.2; (2) 1.4–2.4 g m^{-2} d^{-1}
% removal of sulfide	100	56–100	83–100	92–98	97.0–99.5	7–65	100
% conversion of removed sulfide to sulfur	8–12 (in effluent) 15–26 (in effluent or accumulated)	−9–123 (in effluent or accumulated)	75–95	75–95	Not reported	0 (in effluent)	Not reported
Radiant flux	Not reported	6.0 W (400–1000 nm band); 2.7 W (700–975 nm band)	0.23 W (400–1000 nm band)^{d}	0.3–2.4 W	Not calculable	0.002–0.021 W	0.063–0.127 W
Max. amount of sulfide removed per unit of supplied radiant energy (Ω)	Not calculable	0.29 g W^{-1} d^{-1e}	0.63 g W^{-1} d^{-1}	0.29 g W^{-1} d^{-1}	Not calculable	2.04 g W^{-1} d^{-1}	(1) 0.45; (2) 0.27 g W^{-1} d^{-1}

[a] This is the volume of the entire phototube. The volume associated with sulfide removal ranged between 7 and 25 ml.
[b] The loading is based on the assumption that the reactor is 1 m long. The HRT and removal rates have been based on the length at which the sulfide concentration reached 1 mg l^{-1}.
[c] Applies to the case of the biofilm reactors: unless otherwise indicated, it is the area over which the sulfide-removing biofilm was found.
[d] Not 0.15 W as stated in the article (P. Henshaw, personal communication).
[e] The irradiance in the 400–1000 nm band has been used because GSB can utilize some wavelengths in the visible spectrum (<700 nm) for photosynthesis.

A. Sulfide Concentrations that can be Treated in Current Phototrophic BSR Systems

Henshaw and coworkers have used substratum-irradiated biofilms of GSB in narrow tubular bioreactors ("phototubes") to reduce sulfide concentrations in synthetic wastewater from several hundred parts per million to zero with a hydraulic residence time of <10 min. Other groups have demonstrated that attached-growth reactors, containing axenic or mixed cultures, are also suitable for treating wastewater having sulfide concentrations in the order of 10 mg l^{-1} or less.

B. Conversion of Sulfide to Elemental Sulfur

In all the phototrophic BSR studies reviewed in Table 5 it has been demonstrated that elemental sulfur can be one of the products of oxidation. Henshaw and coworkers have reportedly been able to control the conditions within both suspended growth and attached growth reactors so that the sulfide is transformed practically quantitatively into an elemental sulfur that *leaves the reactor in the effluent* (Henshaw et al., 1998; Henshaw and Zhu, 2001; Syed and Henshaw, 2003, 2005). This technically and economically significant achievement has yet to be matched by others working with attached-growth systems. Data presented by Kobayashi et al. (1983) indicate that in their externally illuminated "phototube" reactor sulfide was mainly transformed into sulfate, and that <12% of the sulfide was accounted for as elemental sulfur *leaving the reactor*. The same data indicate that up to 18% of the sulfide-S entering the reactor was unaccounted for in the reactor effluent. Microscopic examination of the biofilm revealed the presence there of elemental sulfur, so if all the unaccounted for sulfur was actually elemental sulfur accumulated within the reactor, the actual conversion of sulfide to elemental sulfur could have been as much as 26%.

Presumably the explanation for the different degrees of sulfide conversion to free elemental sulfur reported by Henshaw and coworkers and by Kobayashi et al. is to be found in the differences between the two systems. Henshaw and coworkers worked with Tygon tubular reactors of various diameters, including, like Kobayashi et al. Tygon

tubes of 3.2 mm internal diameter – and in that case, liquid flow rates were also similar to those used by Kobayashi et al. Henshaw and coworkers used as reactor inocula pure cultures of *Chlorobium limicola* (ATCC No. 17092; DSM 257) and, as reactor feed, a synthetic medium practically devoid of organic-C, whereas the biofilm in Kobayashi et al.'s reactor was a mixed community and the wastewater was the effluent from an anaerobic digester. Henshaw and coworkers also tended to operate their biofilm reactors at considerably greater volumetric and areal rates of sulfide removal, and at considerably greater average sulfide concentrations, than Kobayashi et al. did. The light bulbs used by Henshaw and coworkers had a considerably higher wattage than those used by Kobayashi et al.; however, the characterizations of the light directed to the reactors are insufficient for us to comment on the equivalence or otherwise of the incident irradiance in the spectral band that GSB can utilize for photosynthesis. Thickness of the biofilm and the relative sizes of the photic and aphotic zones within the biofilm are probably relevant parameters, but these were neither reported nor deducible in most cases. In view of the uncertainty about the actual values of important operating conditions, we do not believe that a single convincing explanation for the differences between Kobayashi et al.'s system and those of Henshaw and coworkers can be offered.

Hurse and Keller, working with a reactor containing a planar, mixed biofilm community dominated by GSB and exposed to sulfide concentrations up to 11.5 mg l^{-1}, reported <1 mg l^{-1} if any elemental sulfur in the effluent, even though elemental sulfur was undoubtedly produced within the biofilm and the reactor was operated continuously for 8 months (Hurse and Keller, 2004b). Elemental sulfur was sometimes produced in such amounts as to turn the biofilm from green to yellow, and mass balance calculations for such periods confirm that more S was entering the reactor than leaving it, at least in the forms of sulfide, elemental sulfur, thiosulfate, sulfite and sulfate. The largest measured elemental sulfur content of a biofilm sample was 39% w/w (dry basis). Reasons for the failure to observe elemental sulfur in the reactor effluent, including the possible formation of polysulfides by reaction with sulfide, were proposed and discussed at length, but the issue still requires further investigation.

C. Cycling of Sulfur Species within the Biofilm

Kobayashi et al. (1983) have demonstrated that sulfide can be removed from anaerobic digester effluent by means of a substratum-irradiated biofilm in which GSB were predominant. The sulfide was removed from the wastewater by a green-colored biofilm extending some distance from the inlet of the phototube reactor. Further downstream, the green colour gradually gave way to a red, and from this red biofilm the investigators obtained various isolates identified as non-sulfide-oxidizing, purple non-sulfur bacteria.

In their report Kobayashi et al. associated the green photosynthetic bacteria with sulfide removal, and the red photosynthetic bacteria with the utilization of organic compounds. They did not, however, examine how exposure to organic compounds influenced the GSB; nor whether sulfidogenic microorganisms might have been involved in transformations of both S and organic compounds within the reactor, and might, thereby, have influenced the sulfide-removing performance of the reactor. Because the sulfide present in the anaerobic digester effluent fed to the phototube was generated in that digester, it stands to reason that sulfate- and sulfur-reducing organisms as well as volatile fatty acids were present in the anaerobic digester effluent. Because GSB and sulfidogenic chemotrophs are known to form stable associations in some natural biofilms, it seems quite plausible that the biofilm in the phototube would also have contained sulfate- and sulfur-reducing bacteria. By means of molecular methods Hurse et al. have confirmed that sulfidogens can be present in a sulfide-removing substratum-irradiated biofilm in which GSB are predominant (unpublished). In consequence, we hypothesize that sulfidogenesis and sulfide removal were occurring within the tubular photobioreactor studied by Kobayashi et al. and that the observed removal of sulfide from the wastewater by their photobioreactor was the result of sulfide removal's outpacing sulfide generation, i.e., the observed removal of sulfide was a net removal.

D. Effect of Exposure to VFAs

For the case of a GSB-dominated biofilm Hurse and Keller (2005) have reported that the addition of acetate or propionate to wastewater at dosages of the order of 20–40 mg l^{-1} causes the observed areal sulfide removal rate of the reactor to fall to 50% of the value observed in the absence of those VFAs but under the same incident irradiance, sulfide loading and wastewater flow rate. The inclusion of acetate or propionate in the wastewater also caused the net rate of sulfate formation to fall to practically zero, and led the biofilm to yellow noticeably, presumably due to elemental sulfur formation. This sulfur did not disengage from the biofilm. The extent to which the switch from sulfate to elemental sulfur production is due to an increase in the rate of sulfate reduction rather than to a suppression in the degree to which GSB oxidize sulfide has yet to be established.

E. Effect of Interruption of the Light

Preliminary investigations into the effects of stopping the irradiation of a GSB-dominated biofilm for a period of hours have been reported by Hurse and Keller (2005). Almost immediately after irradiation was discontinued, the biofilm became a net source of sulfide, and the rate of sulfide generation increased over the ensuing hours. The technological implications of these observations are discussed in Section VI. It may be thought that only the sulfidogenic chemotrophs could be responsible for sulfidogenesis in the dark, but the possible involvement of GSB has also been raised (Hurse and Keller, 2005). There are conflicting reports concerning the sulfidogenic potential of certain strains of GSB (Larsen, 1953; Gulanyan and Kurella, 1970; Paschinger et al., 1974; Brune and González, 1982; Trüper, 1984; Jeong and Kim, 1999; Pringault et al., 1999), and sulfidogenesis in the dark has been reported for other types of phototrophs (van Gemerden, 1968; Oren and Shilo, 1979; Glaeser and Overmann, 1999).

F. Sulfide Removal Rates

A wide range of areal and volumetric sulfide removal rates are implied by the observations reported by the workers referred to in Table 5. The maximum value of areal removal rate reported for a reactor so far is 14 g m^{-2} d^{-1}, which was achieved with a 1.6 mm i.d. phototube that reduced the sulfide concentration from 164 to 0 mg l^{-1} in about 7 min (Syed and Henshaw, 2003).

The corresponding value of Ω, the amount of sulfide removed per unit of supplied radiant energy, was $0.29\,\mathrm{g}\ \mathrm{W}^{-1}\ \mathrm{d}^{-1}$. In their research Henshaw and coworkers have concentrated on identifying the conditions that ensure a near quantitative conversion of sulfide to sulfur in a narrow diameter tubular reactor. In those studies, the hydrodynamic conditions, which may conceivably affect the structure of the biofilm and the removal of elemental sulfur from the biofilm, have not generally been held constant when sulfide concentration or sulfide loading have been varied. Furthermore, the concentration of sulfide in the wastewater, and hence the sulfide availability within the biofilm, necessarily vary with distance along the tube. Many processes are potentially occurring in these substratum irradiated biofilms. We believe that a clearer conceptual picture of the biofilms would emerge if the influences of incident irradiance and sulfide concentration were investigated with as little disturbance as possible to the hydrodynamic conditions inside the reactor.

Hurse and Keller (2004b) have investigated the effects of incident irradiance and sulfide concentration on the observed areal rate of sulfide removal by a planar, mixed biofilm community dominated by GSB, exposed to constant hydrodynamic conditions. The light used was filtered so as to comprise wavelengths between 725 and 775 nm. Their reactor was designed to enable investigation of the areal rate of sulfide removal under well defined conditions, rather than to maximize the volumetric rate of sulfide removal. The incident irradiance (E_i) ranged from 0.21 to $1.51\,\mathrm{W}\ \mathrm{m}^{-2}$ and the concentration of sulfide in the wastewater within the reactor (c_{bulk}) ranged from 1.6 to $11.5\,\mathrm{mg}\ \mathrm{l}^{-1}$. The observed areal rate of sulfide removal was positively correlated with both the sulfide concentration in the liquid bulk and the incident irradiance, i.e., there was a simultaneous, dual limitation. The maximum value of this areal net removal rate was $2.1\,\mathrm{g}\ \mathrm{m}^{-2}\ \mathrm{d}^{-1}$. The corresponding value of Ω was $1.4\,\mathrm{g}\ \mathrm{W}^{-1}\ \mathrm{d}^{-1}$.

G. The Effectiveness of Supplied Light for Sulfide Removal

In their investigation into the effects of c_{bulk} and E_i on sulfide removal by the planar GSB-dominated biofilm, Hurse and Keller (2004b) also found a pronounced positive correlation between Ω and c_{bulk}, a correlation that implied that light was more effectively used for sulfide removal at higher values of c_{bulk} than at lower values of c_{bulk}. On the other hand, they determined that at any given value of c_{bulk}, there exists a value of incident irradiance at which the radiant energy requirement for sulfide removal is a minimum (i.e., at which Ω is a maximum). This optimal incident irradiance was, furthermore, weakly dependent on the value of c_{bulk}. A similar trend is suggested in the results reported by Syed and Henshaw, (2005), although in those experiments incident irradiance was not varied independently of flow rate.

The maximum value of Ω reported so far for a phototrophic BSR system is $2.04\,\mathrm{g}\ \mathrm{W}^{-1}\ \mathrm{d}^{-1}$, achieved by Hurse and Keller (2004b) in the aforementioned investigation. As discussed in their report, a Ω value of $2.04\,\mathrm{g}\ \mathrm{m}^{-2}\ \mathrm{d}^{-1}$ implies, in those circumstances, that the quantum requirement for fixation of a molecule of CO_2 was less than 6 photons. This value is less than the requirement of 9–10 photons, determined by Larsen with suspended cultures of one strain of GSB, but is greater than the theoretical limit of between 3 1/3 and 4 1/2 postulated by Brune (see Section II.B.3). If Brune's calculations are correct, it would seem possible for Ω to reach a value between 2.5 and $3.4\,\mathrm{g}\ \mathrm{W}^{-1}\ \mathrm{d}^{-1}$ for light of the same spectral composition.

The making of comparisons between different phototrophic BSR systems on the basis of Ω values (or, for that matter, the quantum requirement for sulfide removal) is hampered by the generally inadequate descriptions, given in the reports, of the radiant energy used in the experiments. Not all wavelengths are photosynthetically active, and wavelengths active for some phototrophs are not active for others. Furthermore, it seems not to be generally appreciated that the irradiance measured also depends on the type of sensor employed; and that within the range of wavelengths detected by a sensor, the sensor's response may be wavelength dependent. It would be most helpful if all of the following were reported:

– Measurements of the **incident irradiance, in a defined spectral band, over the surface of the reactor.** The location and orientation of the sensor should be reported. If the spectral composition of the radiant energy is also reported, it does not matter whether the irradiance is reported as power per unit area or as quantum flux per unit area.

– The spectral composition of the radiant energy over a wavelength range sufficiently broad to include all photosynthetically active wavelengths and all wavelengths to which the irradiance meter responds.

– As many as possible of the following details about the irradiance sensor (make, model, field of view, calibration standard, spectral responsiveness).

It would be most helpful if the spectral composition of the light were represented graphically, as this would allow a stated value of incident irradiance in one defined spectral band to be converted into a value of incident irradiance in another spectral band.

V. Conceptual Model of the Substratum-Irradiated Biofilm

The following has been proposed by Hurse and Keller (2004b, 2005) as a conceptual model for the substratum-irradiated GSB-dominated biofilm that developed on a transparent substratum exposed to a sulfide-containing wastewater. In principle this model can be applied to substratum-irradiated sulfide-oxidizing-phototroph-dominated biofilms of other geometries. Because radiant energy is supplied through the transparent substratum and sulfide is supplied from the wastewater, opposing gradients of sulfide and light availabilities exist within the biofilm (Fig. 1).

The biofilm is regarded as comprising two distinct regions, known as the *photic* and *"dark"* zones. Only in the former is light availability sufficient to support growth of the phototrophs. Thus, the photic zone lies between the (irradiated) substratum and the "dark" zone. Substrates (including sulfide) diffuse from the wastewater across the "dark" zone to the photic zone, where the phototrophs oxidize the sulfide to sulfur and sulfate (Fig. 2). Sulfur- and sulfate-reduction mediated by chemotrophs occur in the photic and "dark" zones. There is a net removal of sulfide by the biofilm, however, and the sulfate not reduced by chemotrophs diffuses out of the biofilm into the wastewater.

The growth of the cells as well as the formation of elemental sulfur in the biofilm lead to a net outgrowth of the whole biofilm in the direction away from the irradiated substratum. This outgrowth is offset by a sloughing of biomass and elemental sulfur into the wastewater. Elemental sulfur is carried along with the outgrowing biofilm towards the wastewater. As the sulfur is displaced towards the wastewater, it encounters increasing sulfide concentrations, and passes through the dark zone. During its passage towards the wastewater, the sulfur may react chemically with sulfide to form polysulfides, and may be reduced to sulfide by the chemotrophic sulfur reducers in both zones, and by the phototrophs in the "dark" zone. In Hurse and Keller's investigation, very little if any of this elemental sulfur reached the liquid, and some of the S entering the reactor was unaccounted for under some conditions (described in Fig. 2 as "disappearing").

The presence of organic compounds in the wastewater may stimulate the activity of the sulfur- and sulfate-reducing bacteria, and these compounds may possibly be metabolized by the phototrophs as well (see Section II.B.6). As a consequence of their metabolism of the organic compounds, the phototrophs may alter their demand for sulfide, the extent to which they oxidize it, and possibly both. It is postulated that a cessation of irradiation causes the phototrophs to stop oxidizing sulfide, and to start producing VFAs and sulfide. The influences of organic compounds and darkness are also shown in Fig. 2.

VI. Implications for Process Scale-Up and Future Research Directions

Although there is no doubt that elemental sulfur can be the main product into which sulfide is transformed when removed, contrasting accounts have been published about the ease with which that elemental sulfur is actually disengaged from the biofilm. As planar biofilms have some advantages over biofilms growing inside narrow tubes (Hurse and Keller, 2004a), a close examination of the fate of the elemental sulfur formed in planar biofilms would be of interest, as too would further research into means of liberating the elemental sulfur, e.g., by increasing the shear force exerted by the wastewater on the biofilm, or by increasing the incident irradiance so as to make the biofilm grow out more quickly. The latter option may lead to the use of irradiances that are sub-optimal with respect to energy efficiency, but they would increase the areal and potentially the volumetric rates of sulfide removal, and allow the reactor vessel to be smaller and cheaper.

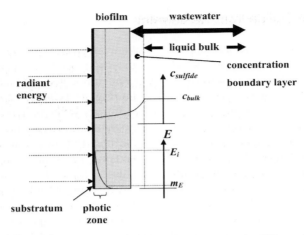

Fig. 1. Conceptual model of the biofilm. The representations of radiant energy and sulfide availabilities are approximate only. © 2004 Wiley Periodicals, Inc. Biotechnology & Bioengineering (2004) Vol. 87, No. 1, 14–23.

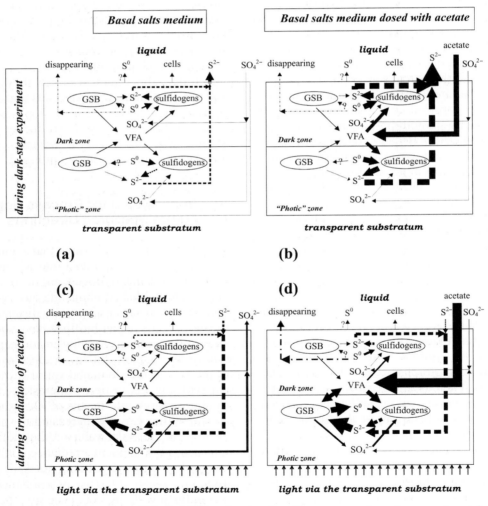

Fig. 2. Transformations of S occurring within the biofilm, showing the salient processes in various experimental situations. (top) operation in the dark with (a) VFA-free feed and (b) acetate-dosed feed; (bottom) operation in the light with (c) VFA-free feed and (d) acetate-dosed feed. Width of the arrow stem indicates the flux. © 2004 Wiley Periodicals, Inc. Biotechnology & Bioengineering (2005) Vol. 89, No. 2, 178–187.

The studies reported by Kobayashi et al. (1983) and by Hurse and Keller (2005) show that it is technically possible to use phototrophs to remove sulfide from wastewater undergoing or having undergone anaerobic treatment. The characteristic presence of VFAs in this type of wastewater is something that cannot be overlooked, for if the findings of Hurse and Keller's study into the effect of dosing wastewater with acetate and propionate are generally applicable, the presence of VFAs considerably suppresses observed rates of sulfide removal and the degree to which sulfide is oxidized. A broader investigation of the effects of VFAs on reactor performance is needed so as to assess the generality of these findings.

Tubular reactors of the diameter used by Henshaw and coworkers and by Kobayashi et al. are unlikely to be suitable for treating wastewaters containing solids. A report on the performance of tubular reactors of a larger diameter (say 2 cm) could be useful for judging their suitability for treating solids-containing wastewaters. Our understanding of the processes occurring within these tubular reactors would certainly be advanced if future experiments were designed so as to distinguish the individual effects of tube diameter, wastewater velocity, shear stress on the biofilm, incident irradiance, and sulfide concentration.

Phototubes should not be the only type of substratum-irradiated biofilm reactor researched. Plane transparent panels and light-diffusing optical fibers (LDOFs) have been proposed as alternative substrata to transparent tubes (Hurse and Keller, 2004a). A reactor filled with LDOFs has been used for the cultivation of phototrophs, including GSB for gas desulfurization (Matsunaga et al., 1991; Lee and Kim, 1998). Whereas the wetted perimeter of a biofilm growing inside a tube becomes smaller as the biofilm becomes thicker, the wetted perimeter of a biofilm growing on (the outside of) an LDOF becomes larger as the biofilm becomes thicker. It would be interesting to investigate whether the wetted surface of a biofilm growing on an LDOF will tend to develop splits that promote sloughing, and thereby facilitate release of the elemental sulfur formed in the biofilm.

The use of LDOFs as a substratum would also facilitate the use of compact, semi-conductor light sources, such as diode lasers or the increasingly powerful and economic LEDs. The attractions of such sources are multiple (Bertling et al., 2006): electrical energy is converted into photosynthetically active light with a higher efficiency than is possible with incandescent sources; there is less waste heat to dispose of; the sources are compact; etc. Syed and Henshaw (2005) have recently used LEDs as a light source for their phototubes, and Bertling et al. (2006) have demonstrated that *Rhodobacter capsulatus* can be cultivated with semi-conductor lasers emitting light strongly absorbed by the bacteria's light-harvesting BChl (in vivo). It is noteworthy that in both cases growth per unit energy supplied was up to three times greater with light from the semi-conductor than with light from an incandescent source. We urge researchers in phototrophic BSR to characterize completely the light used in their experimental systems (see Section IV.G). We also suggest that they form links with researchers in optics or electrical engineering laboratories, where spectral analyzers and power density meters are standard items, and where, most importantly, correct advice about radiometry and optics can be obtained.

It has been demonstrated experimentally with a mixed biofilm community dominated by GSB and containing some chemotrophic sulfidogens that the interruption of irradiation not only caused the biofilm to cease removing sulfide, but caused the biofilm to become a net source of sulfide. Such behavior has three clear implications for a full-scale embodiment of the technology. Firstly, if the sulfide-removal process were to be operated during night-time, an artificial light source would be necessary. Secondly, even if the process were not required to run during night-time, the sulfide formed overnight would have to be removed during the following day, and this would increase the long-term average radiant power requirement of the process. Finally, because an increased sulfide concentration could have an undesirable impact in the sewer downstream, it would be necessary to consider the risk posed by an interruption to the electricity supply. Hence, if this technology ever reaches the pilot stage, the sulfide-generating characteristics of the reactor in the dark should be thoroughly investigated.

References

Bardsley T (2002) Comparison of EPA approved odour measurement methods. SR1 Environment Protection Authority, Melbourne

Basu R, Clausen EC and Gaddy JL (1996) Biological conversion of hydrogen sulfide into elemental sulfur. Environ Prog 15: 234–238

Beeftink HH and van Gemerden H (1979) Actual and potential rates of substrate oxidation and product formation in continuous cultures of *Chromatium vinosum*. Arch Microbiol 121: 161–167

Bertling K, Hurse TJ, Kappler U and Rakic A (2006) Lasers – an effective artificial source of radiation for the cultivation of anoxygenic photosynthetic bacteria. Biotechnol Bioeng 94: 337–345

Biebl H and Pfennig N (1978) Growth yields of green sulfur bacteria in mixed cultures with sulfur and sulfate reducing bacteria. Arch Microbiol 117: 9–16

Brune DC (1989) Sulfur oxidation by phototrophic bacteria. Biochim Biophys Acta 975: 189–221

Brune DC and González I (1982) Measurements of photosynthetic sulfide oxidation by *Chlorobium* using a sulfide ion selective electrode. Plant Cell Physiol 23: 1323–28

Bryantseva IA, Gorlenko VM, Tourova TP, Kuznetsov BB, Lysenko AM, Bykova SA, Galchenko VF, Mityushina LL and Osipov GA (2000) *Heliobacterium sulfidophilum* sp. nov. and *Heliobacterium undosum* sp. nov.: sulfide-oxidizing Heliobacteria from thermal sulfidic springs. Microbiology 69: 325–334

Cohen Y and Gurevitz M (1992) The cyanobacteria – ecology, physiology, and molecular genetics. In: Balows A, Trüper HG, Dworkin M, Harder W and Schleifer K-H (eds) The Prokaryotes. Springer, New York

Cork DJ and Kenevan JR (1990) A model for the control of the Van Niel sulfur reaction in an anaerobic fed-batch reactor. Dev Ind Microbiol 31: 127–131

Cork DJ, Garunas R and Sajjad A (1983) *Chlorobium limicola* forma *thiosulfatophilum*: biocatalyst in the production of sulfur and organic carbon from a gas stream containing H_2S and CO_2. Appl Environ Microbiol 1983: 913–918

Cork D, Mathers J, Maka A and Srnak A (1985) Control of oxidative sulfur metabolism of *Chlorobium limicola* forma *thiosulfatophilum*. Appl Environ Microbiol 49: 269–272

Costigan MG (2003) Hydrogen sulfide: UK occupational exposure limits. Occup Environ Med 60: 308–312

Ferrera I, Sánchez O and Mas J (2004a) A new non-aerated illuminated packed-column reactor for the development of sulfide-oxidizing biofilms. Appl Microbiol Biotechnol 64: 659–664

Ferrera I, Massana R, Casamayor EO, Balagué V, Sánchez O, Pedrós-Alió C and Mas J (2004b) High-diversity biofilm for the oxidation of sulfide-containing effluents. Appl Microbiol Biotechnol 64: 726–734

French CS (1937) The quantum yield of hydrogen and carbon dioxide assimilation in purple bacteria. J General Physiol 20: 711–735

Glaeser J and Overmann J (1999) Selective enrichment and characterization of *Roseospirillum parvum*, gen. nov. and sp. nov., a new purple nonsulfur bacterium with unusual light absorption properties. Arch Microbiol 171: 405–416

Göbel F (1978) Quantum efficiencies of growth. In: Clayton RK and Sistrom WR (eds) The Photosynthetic Bacteria, pp 907–925. Plenum, New York

Gulanyan SA and Kurella GA (1970) Potentiometric investigation of the dynamics of the photooxidation of sulfide by photosynthesizing bacteria. Mikrobiologiya 39: 259–263

Guyoneaud R, Borrego CM, Martínez-Planells A, Buitenhuis ET and Garcia-Gil LJ (2001) Light responses in the green sulfur bacterium *Prosthecochloris aestuarii*: changes in prostecae length, ultrastructure, and antenna pigment composition. Arch Microbiol 176: 278–284

H_2S Technological Standing Committee (1989) The hydrogen sulphide control manual. Melbourne and Metropolitan Board of Works, Melbourne

Hanada S and Pierson BK (2002) The family Chloroflexaceae. In: Dworkin M (ed) The Prokaryotes: an Evolving Electronic Resource for the Microbiological Community, 3rd edn. Springer, New York (release 3.11. 22 November 2002)

Hansen TA and Van Gemerden H (1972) Sulfide utilization by purple nonsulfur bacteria. Arch Microbiol 86: 49–56

Henshaw PF and Zhu W (2001) Biological conversion of hydrogen sulphide to elemental sulphur in a fixed-film continuous flow photo-reactor. Water Res 35: 3605–3610

Henshaw PF, Bewtra JK, Biswas N and Franklin M (1992) Biological removal of hydrogen sulfide from refinery wastewater and conversion to elemental sulfur. Water Sci Technol 25: 65–267

Henshaw PF, Bewtra JK and Biswas N (1998) Hydrogen sulphide conversion to elemental sulphur in a suspended-growth continuous stirred tank reactor using *Chlorobium limicola*. Water Res 32: 1769–1778

Hiraishi A, Muramatsu K and Urata K (1995) Characterization of new denitrifying *Rhodobacter* strains isolated from photosynthetic sludge for wastewater treatment. J Ferment Bioeng 79: 39–44

Hofman PAG, Veldhuis MJW and van Gemerden H (1985) Ecological significance of acetate assimilation by *Chlorobium phaeobacteroides*. FEMS Microbiol Ecol 31: 271–278

Houghton SR and Mara DD (1992) The effects of sulfide generation in waste stabilization ponds on photosynthetic populations and effluent quality. Water Sci Technol 26: 1759–1768

Hurse TJ and Keller J (2004a) Reconsidering the use of photosynthetic bacteria for removal of sulfide from wastewater. Biotechnol Bioeng 85: 47–55

Hurse TJ and Keller J (2004b) Performance of a substratum-irradiated photosynthetic biofilm reactor for the removal of sulfide from wastewater. Biotechnol Bioeng 87: 14–23

Hurse TJ and Keller J (2005) Effects of acetate and propionate on the performance of a photosynthetic biofilm reactor for sulfide removal. Biotechnol Bioeng 89: 178–187

Imhoff JF (1999) The family Ectothiorhodospiraceae. In: Dworkin M (ed) The Prokaryotes: an Evolving Electronic Resource for the Microbiological Community, 3rd edn. Springer, New York (release December 7, 2000)

Imhoff JF (2001a) The phototrophic alpha-proteobacteria. In: Dworkin M (ed) The Prokaryotes: an Evolving Electronic Resource for the Microbiological Community, 3rd edn. Springer, New York (release 3.6. 22 June 2001)

Imhoff JF (2001b) The phototrophic beta-proteobacteria. In: Dworkin M (ed) The Prokaryotes: an Evolving Electronic Resource for the Microbiological Community, 3rd edn. Springer, New York (release 3.6. 22 June 2001)

Jeong G-S and Kim B-W (1999) The influence of light/dark cycle at low light frequency on the desulfurization by a photosynthetic microorganism. J Biosci Bioeng 87: 481–488

Keppen OI, Baulina OI, Lysenko AM and Kondrateva EN (1993) A new green bacterium belonging to the Chloroflexaceae family. Microbiology 62: 179–185

Khanna P, Rajkumar B and Jothikumar N (1996) Microbial recovery of sulfur from thiosulfate-bearing wastewater with phototrophic and sulfur-reducing bacteria. Curr Microbiol 32: 33–37

Kim BW and Chang HN (1991) Removal of hydrogen sulfide by Chlorobium thiosulfatophilum in immobilized-cell and sulfur-settling free-cell recycle reactors. Biotechnol Prog 1991: 495–500

Kim BW, Chang HN, Kim IK and Lee KS (1992) Growth kinetics of the photosynthetic bacterium Chlorobium thiosulfatophilum in a fed-batch reactor. Biotechnol Bioeng 40: 583–592

Kim BW, Chang KP and Chang HN (1997) Effect of light source on the microbiological desulfurization in a photobioreactor. Bioprocess Eng 17: 343–348

Kim YJ, Kim BW and Chang HN (1996) Desulfurization in a plate-type gas-lift photobioreactor using light emitting diodes. Kor J Chem Eng 13: 606–611

Kobayashi HA, Stenstrom M and Mah RA (1983) Use of photosynthetic bacteria for hydrogen sulfide removal from anaerobic waste treatment effluent. Water Res 17: 579–587

Kobayashi M (1975) Role of photosynthetic bacteria in foul water purification. Prog Wat Technol 17: 309–315

Kobayashi M and Kobayashi M (1995) Waste remediation and treatment using anoxygenic phototrophic bacteria. In: Blankenship RE, Madigan MT and Bauer CE (eds) Anoxygenic Photosynthetic Bacteria, pp 1269–1282, Vol 2 of Advances in Photosynthesis (Govindjee ed.). Kluwer Academic (now Springer), Dordrecht

Larsen H (1953) On the microbiology and biochemistry of the photosynthetic green sulfur bacteria. Bruns Bokhandel, Trondheim

Lee K-H and Kim B-W (1998) Enhanced microbial removal of H_2S using Chlorobium in an optical-fiber bioreactor. Biotechnol Lett 20: 525–529

Levenspiel O (1972) Chemical Reaction Engineering. Wiley, New York

Madigan MT (2001) The family Heliobacteriaceae. In: Dworkin M (ed) The Prokaryotes: an Evolving Electronic Resource for the Microbiological Community, 3rd edn. Springer, New York (release 3.5. 13 March 2001)

Madigan MT and Brock TD (1975) Photosynthetic sulfide oxidation by Chloroflexus aurantiacus, a filamentous, photosynthetic, gliding bacterium. J Bacteriol 122: 782–784

Maka A and Cork DJ (1990a) Introduction to the sulfur microorganisms and their applications in the environment and industry. Dev Ind Microbiol 31: 99–102

Maka A and Cork D (1990b) Quantum efficiency requirements for an anaerobic photobioreactor. J Ind Microbiol 5: 337–354

Matsunaga T, Takeyama H, Sudo H, Oyama N, Ariura S, Takano H, Hirano M, Burgess JG, Sode K and Nakamura N (1991) Glutamate production from CO_2 by marine cyanobacterium Synechococcus sp. using a novel biosolar reactor employing light-diffusing optical fibres. Appl Biochem Biotechnol 28/29: 157–167

Oren A and Shilo M (1979) Anaerobic heterotrophic dark metabolism in the cyanobacterium Oscillatoria limnetica: sulfur respiration and lactate fermentation. Arch Microbiol 122: 77–84

Overmann J (2000) The family Chlorobiaceae. In: Dworkin M (ed) The Prokaryotes: an Evolving Electronic Resource for the Microbiological Community, 3rd edn. Springer, New York. http://link.springer-ny.com/link/service/books/10125 (release 3.1. January 20, 2000)

Overmann J (2001) Green sulfur bacteria. In: Garrity GM (ed) Bergey's Manual of Systematic Bacteriology, pp 601–623. Springer, New York

Overmann J and Garcia-Pichel F (2000) The phototrophic way of life. In: Dworkin M (ed) The Prokaryotes: an Evolving Electronic Resource for the Microbiological Community, 3rd edn. Springer, New York. http://link.springer-ny.com/link/service/books/10125 (release 3.2. 25 July 2000)

Overmann J, Cypionka H and Pfennig N (1992) An extremely low-light-adapted phototrophic sulfur bacterium from the Black Sea. Limnol Oceanogr 37: 150–155

Paschinger H, Paschinger J and Gaffron H (1974) Photochemical disproportionation of sulfur into sulfide and sulfate by Chlorobium limicola forma thiosulfatophilum. Arch Microbiol 96:3 41–51

Pfennig N and Trüper HG (1992) The family Chromatiaceae. In: Balows A, Trüper HG, Dworkin M, Harder W and

Schleifer K-H (eds) The Prokaryotes, Vol III, pp 3200–3221. Springer, New York

Pringault O, Epping E, Guyoneaud R, Khalili A and Kühl M (1999) Dynamics of anoxygenic photosynthesis in an experimental green sulphur bacteria biofilm. Environ Microbiol 1: 295–305

Sadler WR and Stanier RY (1960) The function of acetate in photosynthesis by green bacteria. Proc Natl Acad Sci USA 46: 1328–34

Sardesai P, Seames W, Dronen L and Kozliak E (2006) Exploring the gas-phase anaerobic bioremoval of H_2S for coal gasification fuel cell feed streams. Fuel Processing Technol 87: 319–324

Siefert E, Irgens RL and Pfennig N (1978) Phototrophic purple and green bacteria in a sewage treatment plant. Appl Environ Microbiol 35: 38–44

Syed MA and Henshaw PF (2003) Effect of tube size on performance of a fixed-film tubular bioreactor for conversion of hydrogen sulfide to elemental sulfur. Water Res 37: 1932–1938

Syed MA and Henshaw PF (2005) Light emitting diodes and an infrared bulb as light sources of a fixed-film tubular photobioreactor for conversion of hydrogen sulfide to elemental sulfur. J Chem Technol Biotechnol 80: 119–123

Trüper HG (1984) Phototrophic bacteria and their sulfur metabolism. Stud Inorganic Chem 5: 367–382

Trüper HG and Pfennig N (1992) The family Chlorobiaceae. In: Balows A, Trüper HG, Dworkin M, Harder W and Schleifer K-H (eds) The Prokaryotes, Vol III, pp 3583–3592. Springer, New York

van Gemerden H (1968) On the ATP generation by *Chromatium* in darkness. Arch Microbiol 64: 118–124

van Gemerden H (1974) Coexistence of organisms competing for the same substrate: an example among the purple sulfur bacteria. Microbial Ecol 1: 104–119

van Gemerden H (1984) The sulfide affinity of phototrophic bacteria in relation to the location of elemental sulfur. Arch Microbiol 139: 289–294

van Gemerden H (1986) Production of elemental sulfur by green and purple sulfur bacteria. Arch Microbiol 146: 52–56

van Gemerden H and Beeftink HH (1981) Coexistence of *Chlorobium* and *Chromatium* in a sulfide-limited continuous culture. Arch Microbiol 129: 32–34

van Gemerden H and Mas J (1995) Ecology of phototrophic sulfur bacteria. In: Blankenship RE, Madigan MT and Bauer CE (eds) Anoxygenic Photosynthetic Bacteria, pp 49–85, Vol. 2 of Advances in Photosynthesis (Govindjee ed.). Kluwer Academic (now Springer), Dordrecht

Veldhuis MJW and van Gemerden H (1986) Competition between purple and brown phototrophic bacteria in stratified lakes: sulfide, acetate, and light as limiting factors. FEMS Microbiol Ecol 38: 31–38

Wassink EC, Katz E and Dorrestein R (1942) On photosynthesis and fluorescence of bacteriochlorophyll in Thiorhodaceae. Enzymologia 10: 285–354

Wijbenga D-J and van Gemerden H (1981) The influence of acetate on the oxidation of sulfide by *Rhodopseudomonas capsulata*. Arch Microbiol 129: 115–118

X-ray Absorption Spectroscopy as Tool for the Detection and Identification of Sulfur Compounds in Phototrophic Organisms

Alexander Prange*

The J. Bennett Johnston, Sr. Center for Advanced Microstructures and Devices (CAMD), Louisiana State University and Agricultural & Mechanical College, 6980 Jefferson Highway, Baton Rouge, LA 70806, USA
Mikrobiologie und Lebensmittelhygiene, Fachbereich Oecotrophologie, Hochschule Niederrhein – Niederrhein University of Applied Sciences, Rheydter Straße 277, D-41065 Mönchengladbach, Germany

Josef Hormes

The J. Bennett Johnston, Sr. Center for Advanced Microstructures and Devices (CAMD), Louisiana State University and Agricultural & Mechanical College, 6980 Jefferson Highway, Baton Rouge, LA 70806, USA
Physikalisches Institut, Rheinische Friedrich-Wilhelms-Universität Bonn, Nußallee 12, D-53115 Bonn, Germany

Hartwig Modrow

Physikalisches Institut, Rheinische Friedrich-Wilhelms-Universität Bonn, Nußallee 12, D-53115 Bonn, Germany

Dedicated to the memory of our colleague and friend Dr. Roland C. Tittsworth, a professor of research at CAMD, in appreciation of his enthusiastic work at *his* DCM beamline (*"What time is it? It's beam-time!"*), who died December 20th, 2005.

*Corresponding author, Phone: 49 2161 1865390, Fax: 49 2161 1865314, E-mail: A.Prange@gmx.de, alternative: Prange@lsu.edu

Rüdiger Hell et al. (eds.), Sulfur Metabolism in Phototrophic Organisms, 461–482.
© 2008 *Springer.*

Summary

X-ray absorption spectroscopy (XAS) is an *in situ* technique which combines the advantages of a local probe technique with the high penetration strength inherent to X-rays, such as no need for long range order and the ability to obtain information on selected sites of a given sample only. Therefore, this technique is applicable to a broad variety of scientific topics, including many applications to elucidate the chemical speciation of sulfur in phototrophic organisms. The first part of the chapter provides an elementary introduction to the physical background and method, whose application to a broad variety of problems is discussed in detail, followed by detailed examples and explanations on X-ray absorption near edge structure (XANES) measurements of sulfur compounds. In the second part, examples of successful applications of XANES analyses of bacterial sulfur globules of purple and green sulfur bacteria, the wheat gluten network, and sulfur in host–plant interactions are presented.

I. Introduction

X-ray absorption spectroscopy (XAS) has been used extensively in recent years to probe the chemical forms of various elements (especially sulfur) in different biological, agricultural and environmental systems by investigating X-ray absorption near edge structures (XANES) (e.g. Schulze and Bertsch 1995; Prange and Modrow 2002; Akabayov et al. 2005 and references therein) as well as for investigating topics related to structural biology using extended X-ray absorption fine structures (EXAFS) (e.g. Yachandra 1995; Penner-Hahn 2005 and references therein). Whereas ten years ago only a very few hits for the keywords "XANES, sulfur and biology" could be found in literature databases, today, some hundred references can be found. Thus, X-ray spectroscopy techniques have become more common in biological and applied sciences. Furthermore, in recent years different topics have been reported related to phototrophic organisms and also the investigation of special questions elucidating sulfur speciations in phototrophic bacteria and plants.

II. X-ray Absorption Spectroscopy – an Elementary Introduction

Whenever one plans to apply a spectroscopic technique to a biological system, there are three key difficulties to overcome: The first of these is

---

Abbreviations: EXAFS – Extended X-ray absorption fine structure; NEXAFS – Near edge X-ray absorption fine structure; UHV – Ultra high vacuum; XANES – X-ray absorption near edge structure; XAS – X-ray absorption spectroscopy; XPS – X-ray photoelectron spectroscopy; XRD – X-ray diffraction

understanding the physics behind the method, in order to understand what may (and, just as important, what may NOT) be derived from it. The aim of this section is to solve this difficulty with respect to X-ray absorption spectroscopy in a relatively "soft" way, i.e. using as little (mathematical) formalism as possible without sacrificing too much precision. More rigorous introductions to this topic can be found in the literature (e.g. Stoehr 1996; Behrens 1992a,b; Prange and Modrow 2002; Hormes and Modrow 2003; Modrow 2004). The second difficulty resides in the fact that biological samples tend to be complex. Therefore, in an ideal world one would like to apply a spectroscopic technique which allows for a selective gain of information on selected parts of the biomaterial under investigation. Last but not least, sample preparation is a critical issue more often than not, because usually it is difficult to make sure that the preparation does not affect the sample. Thus, in an ideal world, all investigations would be *in situ* (keeping the sample under conditions in which it really is found in nature) and involve as little "uncontrolled" sample manipulation (such as drying, freezing, crystallizing, concentrating, etc.) as possible while allowing at the same time for monitoring effects of controlled changes, for example, changing the pH value of a solution the biological molecule is in.

So why use X-ray absorption spectroscopy, which usually means that one has to write a proposal to obtain user time at and the need to travel for several days to one of the synchrotron radiation sources? The motivation to do so is that both the second and third problem can be met when using X-rays as an analytical tool. As a matter of fact, we all know from personal experience (airport security, medical X-ray imaging, …) that they feature high penetration strength and can distinguish between different elements, because at a given energy the probability for an X-ray photon to be absorbed by a different type of atom is different. "Absorbed" means that the energy contained in the photon is transferred completely to the atomic system, and the only thing that can happen in the atom with this energy input is an excitation process: One of the electrons is removed from its original state and transferred to an excited state, as schematically displayed in Fig. 1a. This process can occur whenever the photon energy is larger than the binding energy of the electron which is to be excited in the atomic system, and it happens dominantly to the electron whose binding energy is the "best match", i.e. the most tightly bound electron which can be excited. This implies that with increasing photon energy whenever a new excitation channel is opened, a notable increase in the probability that an absorption occurs, i.e. in the cross section, which is frequently called "edge jump". At the same time, the set of binding energies of the electrons in a given atom is as unique for an atom as a fingerprint for a human being (Mosley's law)! This is actually how the element selectivity comes into play: By varying the energy of the X-rays in the vicinity of the ionization threshold of the target-element on which information is to be gained, the change in the observed signal is dominated by the contribution of this target element. This is exactly what is done in an X-ray absorption spectroscopy experiment: The photoabsorption cross section is measured in small steps in the vicinity of an absorption edge.

To obtain an energy-dependent measurement of this entity, evidently an X-ray source with a continuous spectrum is needed. It should be noted that whereas conventional X-ray tubes do provide

this type of emission generated by bremsstrahlung, it is so weak that normally (e.g. XRD (X-ray diffraction) and/or XPS (X-ray photoelectron spectroscopy)) experiments involving this type of source use the characteristic emission lines (e.g. Cu Kα), which are discrete in energy space but have significantly higher intensity. Therefore, the standard X-ray absorption spectroscopy setup is found at synchrotron radiation sources that offer a continuous spectrum combined with high intensity. At the same time, each single data point must be correlated to one photon energy, thus the experimental setup must involve a "filter" which allows only the selected wavelength to enter the detection branch of the experiment. In the X-ray range, the method of choice to obtain this is Bragg-reflexion on a given set of crystal-planes with Miller-indices hkl and interplanar distance d_{hkl}, which connects the Bragg-angle ϑ_b, under which a reflection occurs, and the wavelength λ of the photon by the relation:

$$n\lambda = 2d_{hkl} \sin \vartheta_b$$

This implies that one obtains a discrete set of photon energies, corresponding to a base frequency and its integer multiples, which are called "harmonics". If one uses a single reflection in such a geometry, it would be necessary to rotate the entire experiment by 2φ whenever changing the Bragg angle by φ. In contrast to that, a double reflection setup allows to work with a fixed experimental setup, which is the reason why double crystal monochromators are the instruments of choice for monochromatization in an X-ray absorption experiment at synchrotron radiation facilities. A second advantage of working in a double reflection setup is that, both width and central position of the reflections of the harmonics vary, which allows for a removal of harmonics

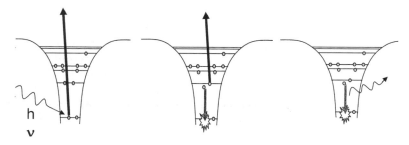

Fig. 1. Left to right: (a) illustration of the photoabsorption process, (b) illustration of the Auger process (non-radiative deexcitation), and (c) illustration of the fluorescence process (radiative deexcitation).

from the transmitted beam by slight detuning of the crystals. It should be noticed, however, that this implies a deterioration of the experimental resolution. More details on double crystal monochromators can be found, e.g., in Lemmonier et al. (1978).

The next question which arises is how to detect the absorption cross section μ. In the X-ray region, the most direct way to do so is to look for "missing photons", i.e. analyzing the monochromatic flux before (I_0) and after a sample (I), which are correlated by Beer's law

$$I = I_0 e^{-\mu d}$$

The device of choice for the measurement of I and I_0, respectively, is an ionization chamber in which electrodes at negative voltage collect the ions which are generated by the interaction between X-rays and a gas filling the chamber. It is crucial for this technique that the detected currents are not influenced by additional contributions. Such contributions can occur, e.g. if pinholes are present in the sample, or if harmonic contributions, whose mean free path in matter is significantly higher than the first order transmitted photon's and whose relative contribution is thus increasing with increasing sample thickness, are not sufficiently suppressed. As a consequence, the spectral features will appear significantly broadened, where the broadening is a function of the intensity of the feature. The more intense the absorption is at a given point, the greater the influence of the additional contribution to the current. This implies that there is a concentration-dependent limit that depends on the surrounding matrix, but lies at about 1%. Still, provided that sample are carefully prepared, then, even at low Z elements experiments can be performed in transmission mode, as discussed by comparison between measurements on in transmission mode and in the fluorescence mode described below.

An alternative approach to verify that an absorption process has occurred is the detection of secondary products of this process, which are directly proportional to the number of holes which has been generated (which in turn is dependent on the number of incoming photons – thus a measurement of I_0 using an ionization chamber remains a necessity). As displayed schematically in Fig. 1b and c, the absorption process leaves an unoccupied state in an inner shell of the atomic system, which is filled quickly by an electron from a higher shell. The energy gained by this process is used either for the ejection of another electron from the atom – the so-called Auger process – or the emission of a fluorescence photon. The branching ratio between these processes is a function of the atomic number Z of the absorbing atom, for low Z the radiationless decay is favored whereas for higher Z the radiative process gets dominant. For electron detection, He flow-chambers, channeltrons and/or multichannelplates are commonly used types of detectors. In principle, there are two possible flavors of electron yield measurements: Total yield, i.e. without using any electron energy filtering, and partial yield, e.g. just collecting electrons belonging to one given Auger transition. Whereas the first setup will provide much better count rates, the second offers the advantage of a much more defined sampling depth and homogeneity of the obtained in information. Further details on electron yield setups (as well as other setups) can be found, e.g., in Johnson (1983).

Two major difficulties are inherent to this detection mode: One has to make sure that the sample is not charged more and more during the measurement process, which would lead to a reduction of the probability for electron emission as a function of the measurement time and thus reduction of the detected current. This problem is mainly relevant for insulators and semiconducting material. Secondly, electrons have a low, energy dependent mean free path length in matter. This means that one often loses the possibility to perform experiments *in situ* and/or under defined conditions, because even transmission through a thin window is not possible for an electron beam. On the other hand, using this technique provides a good surface-sensitivity.

Unfortunately, fluorescence yield, i.e. detection of the fluorescence photons by a semiconductor detector such as a SiLi or a high purity Germanium detector or a Lytle chamber, can also be affected by a variation of the detection probability for the photons. This occurs, for example, in highly concentrated samples, where the absorption length for the fluorescence photon is lower or within an order of magnitude of the actual particle size; consequently self-absorption of the fluorescence photon in the sample occurs and structures are damped. Thus, fluorescence yield is the technique of choice for small concentrations of the target element within the sample. The lower concentration threshold is dependent on the actual beamline and the exact detection system which is used; measurements on target elements with concentrations in and even below the ppm-range are possible. It should be stressed that for homogenous samples of course measurements in different detection modes yield identical results, if one has carefully avoided the pitfalls connected to each of the individual detection modes.

Next, one has to understand what all the detailed structures visible in the real spectra mean. To facilitate this understanding, the result of such a measurement is displayed schematically in Fig. 2. In the insert at the top, the general behavior of the absorption cross section as discussed at the beginning of this chapter is sketched. However, investigating this behavior in more detail, the absorption edge, i.e. the onset of the increase in the absorption cross section, is found at lower energies than the ionization edge. This is not really surprising, as excitations into unoccupied bound states should be possible. Perhaps it is more surprising that even at energies higher than the ionization threshold (which is NOT directly observable in an X-ray absorption spectrum), one does not observe a monotonic decline of the absorption cross section; instead oscillatory behavior is observed. This observation can be explained in a simple model if one recalls that in fact the photoelectron propagating through matter can be treated as a spherical wave, whose wavenumber k is connected with the energy of the incoming photon E and the ionization energy E_0 by the relation

$$k = \sqrt{\frac{2m_e}{\hbar^2}(E - E_0)}$$

where m_e represents the electron mass and h the normalized Planck's constant.

This photoelectron wave, however, does not propagate through free space when the target atom is in a molecule or in a solid, but through an environment in which it is likely to undergo interaction – to be more precise, scattering processes. If now the outgoing wave and the backscattered wave interfere one observes an interference pattern (which is visualized easily in two dimensions by throwing a stone into a pond near the pillars of a bridge), as schematically displayed in Fig. 3a. On the other hand, varying the energy of the incoming photon varies the wavelength of the photoelectron, the interference pattern changes as

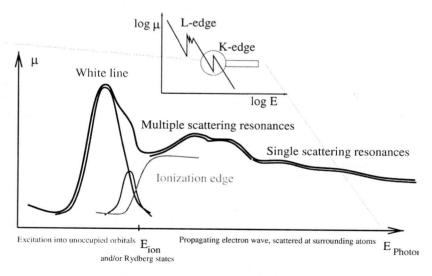

Fig. 2. Schematic sketch of an experimental X-ray absorption spectrum and the different physical processes which lead to the different elements of the observed fine structure. (Figure adopted from Prange & Modrow (2002) X-ray absorption spectroscopy and its application in biological, agricultural and environmental research. Re/Views Environ Sci Bio/Technol 1: 259–276, kindly permitted by Springer, Berlin.)

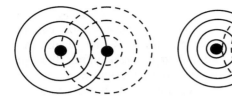

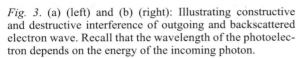

Fig. 3. (a) (left) and (b) (right): Illustrating constructive and destructive interference of outgoing and backscattered electron wave. Recall that the wavelength of the photoelectron depends on the energy of the incoming photon.

shown in Fig. 3b. Consequently, we may expect to see a series of constructive and destructive interference terms which contribute to the absorption cross section. Furthermore, the scattering probability will be a function of the energy of the photoelectron, consequently in the energy region just above the absorption edge multiple scattering will be possible, and a large number of strong interference terms will appear in the spectrum – and that, too, is just what is observed. So what we have just developed is some rough understanding of what makes the substructures appear in the spectrum: directly at the edge, one is dealing with transitions into unoccupied bound states, at higher energies a multiple scattering region with strong structures, finally a region in which the scattering probability gets lower and lower so that single scattering takes over. For historical reasons, the first two regions are known as X-ray absorption near edge structure (XANES) – also called near edge X-ray absorption fine structure (NEXAFS) – by people working in the vacuum-ultraviolet energy region, the latter is known as the extended X-ray absorption fine structure (EXAFS). More rigorous and homogenous approaches to the theoretical background of these effects are available in the literature (Rehr and Albers 2000; Modrow 2004).

What is now so "fine" about all this fine structure? Right at the absorption edge, excitations into unoccupied valence states are reflected by the spectrum. However, as chemistry modifies the valence state/the electronic structure of the elements, we should be able to learn something about the chemical environment of our absorber atom in the sample by analyzing that structure. On the other hand, the interference pattern is also characteristic for a given arrangement of the atoms surrounding the absorbing atom and allows for the extraction of information on its coordination geometry. So you do something that yields information similar to a combined XPS/XRD study – but due to the intrinsic penetration strength of X-rays you do so without any need for an UHV (ultra high vacuum) apparatus which is crucial for XPS or long range order without which XRD will not work. This means one can work *in situ*, phase-independent, and you do not have to worry about tedious creation of crystalline samples. By choosing the energy region of the scan, one can pick the element of interest, even if it is sitting somewhere in the bulk, buried under a pile of material (i.e. in cases where today's excellent microscopic techniques with atomic resolution such as tunneling or atomic force microscopy fail) – there are no additional, confusing contributions from other elements to the data. As discussed below in some more detail, by choosing between different methods of detection, even the probing depth/degree of surface sensitivity of the experiment can be chosen in many cases.

But how does one actually learn something from this signal? Possibly one of the reasons why XANES spectroscopy has developed to a rather popular experimental technique is that a vast amount of valuable information can be extracted based on what is called "fingerprinting" – the comparison of the spectrum of an unknown substance to known reference compounds. To illustrate this principle, let us first have a look at a complicated unknown substance: A piece of tire, which has been heated under air (as you do when hitting the brakes that hard that you slide over the road). The mechanical properties of a tire are mainly determined by number and length of the sulfur cross-links between its polymer backbones, and to check out what has happened to these sulfur atoms, one can run a XANES spectrum at the sulfur K-edge, which is shown in Fig. 4a. Next, one should think about what might be suitable reference compounds if a sulfur atom acting as a bridge between two polymer backbones is oxidized: One needs a sulfane/sulfide, a sulfoxide, a sulfone and a sulfate. Corresponding reference spectra are displayed in the lower part of Fig. 4a. Evidently, there is a clear trend for the absorption edge in these reference spectra to shift to higher energy with increasing formal oxidation state. This behavior is known as "chemical shift". The general principle behind this trend is easily

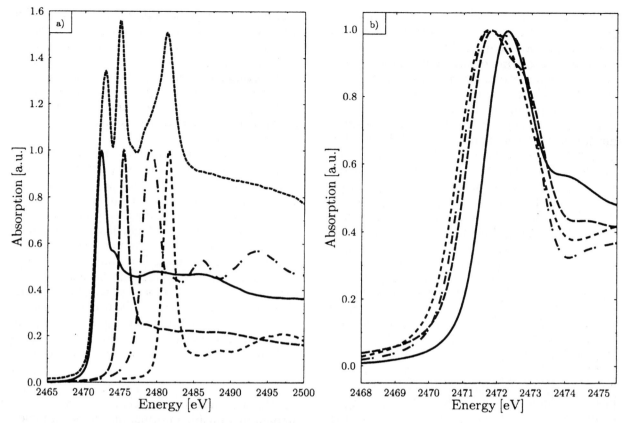

Fig. 4. (a) Sulfur K-edge XANES spectrum of a tire which has been subject to thermooxidative aging (top) and sulfide, sulfoxide, sulfone and sulfate-reference spectra (bottom, left to right). Note the increasing white line area with increasing formal oxidation state. (b) Sulfur K-XANES spectra of di-octenyl mono- (solid), di- (dash), tri- (dot) and tetra- (dash dot) sulfanes.

understood: If charge is withdrawn, the remaining electrons are bound more strongly to the core (as reflected by the fact that the second ionization energy is always higher than the first), resulting in an increase in excitation energy. Also, have a look at the relative intensity of the white line: With increasing formal oxidation number, its area increases. In fact, this area increase is what we should see according to general understanding of a chemical bond: The more electronegative the binding partner, the weaker the contribution of the sulfur atom to the occupied, binding orbitals. Consequently, there is a stronger contribution to the antibinding orbitals into which excitation can occur.

Both effects, chemical shift and the analysis of white line areas, can be utilized to obtain quantitative information on the formal valency of an unknown material: One obtains a set of refer-

ence samples with known valency, uses one or both criteria to create a "scale" and uses this scale to obtain a value for this property. Examples for this approach can be found, e.g., in Capeheart et al. (1995) and Pantelouris et al. (2004). It should be stressed, however, that these scales must be treated with care, because there are numerous exceptions to these rules, which are easily interpreted physically. For example, if one compares systems in which different types of hybridization are involved, one can run into problems.

The resolving power of this method with respect to the chemical environment of the absorbing atom goes much beyond the formal valency: In Fig. 4b, the spectra of reference compounds for the crosslinks are shown, whose chainlength varies. Evidently, it is easily possible to distinguish between the monosulfidic chain (monosulfane) and all the other chains, because in the former case only one type of bond is present (S-C σ^*, as calculations show), whereas in the other case also the S-S π^* bonds contribute. But even the varying chain length leads to clear and systematic differences in the corresponding absorption spectra. A more detailed discussion of the information which can be obtained, e.g., at the sulfur K-edge is given, e.g., by Vairavamurthy 1998; Chauvistré et al. 1997; Flemming et al. 2001; von Busch et al. 2003). From Fig. 4a it is also evident that the spectrum of the oxidized tire appears to be composed from the spectra of the oxidic reference spectra. This is due to the fact that XANES is a local probe technique and that XANES spectra from target atoms in different chemical environments are additive. It is even possible to quantify the respective contributions to the spectrum, e.g. by doing a least squares fit or principal component analysis. Detailed examples for the application of this approach on biological samples are discussed below, some other instructive studies on a broader class of materials are, e.g., Modrow et al. (2001) and Beauchemin et al. (2002).

Already at this point of our discussion, it should be stressed that this high sensitivity is not only a positive feature, but has a clear drawback, especially when one is operating on a mere fingerprinting basis, because a missing reference spectrum can lead to results which cannot be interpreted, even if from a biochemical point of view it may appear like a "near miss". That means, when opting for this kind of experiment, one should try to cover all reasonable reference compounds.

Still, this already extremely detailed information on the local electronic structure is not the only information which can be obtained from X-ray absorption spectroscopy; so far we have not yet made any use of the "interference pattern". To illustrate its importance, in Fig. 5 the (calculated) iron K-edge XANES spectra of α- (bcc, based centred cubic) and γ- (fcc, face centred cubic) iron are displayed. In both cases, iron is surrounded by iron atoms, but one is dealing with two different geometrical arrangements of iron atoms in cubic symmetry. Naturally, in both cases one is dealing with Fe(0)- so no notable shifts of the absorption edge are expected – which is confirmed. However, one observes changes in details in the electronic structure (which are induced by different hybridization of the atomic orbitals) as well as dramatic changes of the strong structures in the

multiple scattering region above the absorption edge, which for historical reasons are frequently called "shape resonances". For the analysis of this "geometry"-part of the near edge structure, fingerprinting can be applied, too, even though it is much less straightforward. In fact, Stoehr et al. (1984) have shown that in suitable systems even the bond length can be determined from XANES spectra with a ruler.

Even though a vast amount of qualitative evidence can often be derived using fingerprint techniques, there is some degree of dissatisfaction when doing so due to two unanswered questions: (1) How can one determine how suitable a reference material really is and whether a general trend can really be trusted? (2) How can one obtain suitable references, as there is a number of systems for which this is very difficult because the relevant chemical environment is stable only under very special conditions? Only if one can assign certain spectral features to a given type of bond or (to a change in the) atomic environment, it is possible to understand the observed changes in the

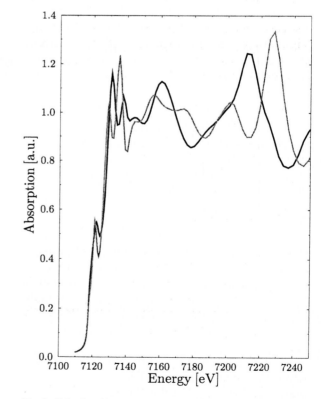

Fig. 5. Calculated iron K-edge XANES spectra of α- and γ-iron.

near-edge spectra in sufficient detail. To come to such a correlation, two general approaches can be followed, which rely on either molecular orbital (MO) or scattering theory, respectively. MO theory and XAS have been applied extremely successfully for hydrocarbons (Hormes and Modrow 2003 and references therein), but the inherent problem is that a precise calculation is rather costly for systems which contain a large number of electrons (in a Hartree-Fock method, it would scale with the fourth power of the number of electrons involved) and relativistic effects gain importance with increasing Z.

Also, MO methods show intrinsic weaknesses and a trend to basis-set induced artefacts when applied for the description of unoccupied states, and in most cases their application is only possible on a ground-state level, which makes a direct correlation between calculated states and the measured spectra difficult. To illustrate what we can learn from molecular orbital theory in spite of all of these drawbacks, let us compare the XANES spectra of two more sulfur-containing molecules, methyl-1,3-dithionate (MDT) and *cyclo*-hexenesulfide (CHS) as shown in Fig. 6. In the first case, two sulfur atom are integrated in non-neighboring positions together with four carbon atoms into a six-ring. In the *cyclo*-hexenesulfide molecule, one sulfur atom is connected to a ring consisting of six C atoms. Comparing the two spectra to each other, in both cases we observe a double structure, but in the case of the *cyclo*-hexenesulfide it is shifted significantly to lower energies (which is quite counterintuitive, as the environment should be more eager to draw electrons in this configuration, so any shift should go to the opposite direction). There are two evident questions: (a) Why do we see a doublet structure – after all, there should be one bond type, should it not? (b) Why is it shifted? When running density functional calculations using a B3LYP potential, first of all the local projected charge density at the sulfur atom(s) confirms that the energy shift is contraintuitive: In fact, in the MDT case there is a local positive charge of 0.19 whereas positive charge in the CHS case is 0.3. Thus, the chemical shift should occur to the other direction. However, the calculations reveal the origin of the second structure: apart from the σ-type S-C bond, the out-of plane S p-orbitals couple significantly into the delocalized ring-orbital, and in fact it

turns out that this orbital is energetically less favorable in the MDT case, but more favorable in the CHS case. This clarifies at once why we do not see the expected chemical shift: It can be observed between identical types of orbitals, but if their respective order is changed, the argument can break down. The calculated splitting between the respective orbitals within each molecule is about 2 eV and thus appears to be overestimated by the calculation, but the relative intermolecular shifts in the position of the respective upper and lower transitions are 0.6 and 0.8 eV respectively and thus close to the experimentally observed result. Thus, even though a direct reproduction of the near edge spectrum is not possible based on these ground-state calculations, a significantly improved understanding of the spectral features is achieved. Further examples can be found, e.g., in Flemming et al. (2001) and von Busch et al. (2003).

To illustrate the possibilities of interpretation using the scattering theoretical approach, let us analyze the spectra of the chromium(III)-

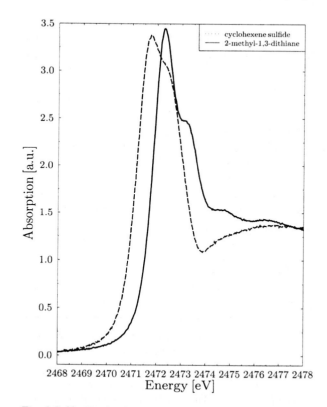

Fig. 6. Sulfur K-edge XANES spectra of methyl-1,3-,dithionate (solid) and *cyclo*-hexenesulfide (dashed).

compounds with varying electronegativity of their respective environment (the interested reader can read the entire story in Pantelouris et al. (2004)). On the fingerprint level, once again the shift towards higher energy with increasing donation of electrons to the more electronegative binding partner is easily verified. The same holds for the increase in white line intensity. All of these compounds feature an octahedral environment of the chromium atom. In fact, calculating the atom- and l-projected densities of states with the FEFF8 code (Ankudinov et al. 1998), as shown in Fig. 7, one can easily achieve a direct assignment of features in these plots to the splitting predicted from ligand field theory for an octahedral environment. Consequently, it is possible to assign spectral features to given bond-types using this type of approach. Based on this assignment for a set of suitable references it is then possible to interpret changes in also complex spectra for which fingerprinting is impossible.

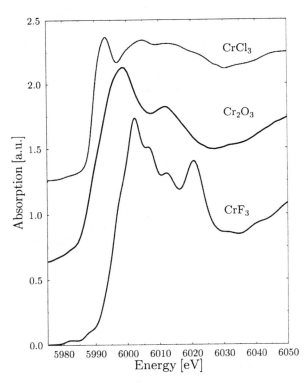

Fig. 7. Chromium K-edge XANES spectra of Cr(III) compounds with varying electronegativity of the binding partner. Note the increased relative intensity of the white line and its shift to higher energies with increasing electronegativity of the binding partner.

However, the analysis of these plots shows also another effect: With increasing bond strength, the splitting between the respective occupied and unoccupied orbitals increases contributing to an increasing chemical shift. But at the same time, even a simple Hartree-Fock based calculation shows that this variation of ligands leads to a change of several eV in the energy level of the 1s orbital. This is easily understood when bearing in mind that, e.g., the orbitals in an ionized system contract. Consequently, two different effects participate in the chemical shift. However, the ordering of molecular orbitals is strongly influenced by the coordination geometry, and consequently this has to be taken into consideration when trying to establish a connection between white line positions and formal valency.

You may ask now why anyone does EXAFS spectroscopy, bearing in mind the superb sensitivity of the near edge structure on the local geometry. In fact, whereas XANES can distinguish between a square and a rectangular arrangement of four atoms in identical distance from the absorber atom, EXAFS cannot. But the big drawback of working with the near edge structure for structural analysis is that presently there is no way for a direct *analytical* extraction of information on the geometric environment of the absorber atom. Partly, this is just due to this sensitivity: as many multiple scattering paths (only these contain angular information) contribute and electronic valence structure is important, the corresponding calculations are too demanding to allow for systematic optimization of the model.

In contrast to that, in the EXAFS region analytical extraction of structural parameters is possible, as shown by Sayers et al. in the seventies (Sayers et al. 1971; Koningsberger and Prins 1988 and references therein). At first sight, the formula which describes the oscillations in this region looks quite complex:

$$\chi(k) = \sum_j S_{0j}^2(k) \frac{N_j}{kr_j^2} F_j(k) e^{-2r_j/\lambda_j(k)} e^{-2\sigma_j^2 k^2}$$

$$\sin(2kr_j + 2\delta_c + \Phi_j(k))$$

This expression represents a summation over all different coordination shells j. Each of these coordination shells is connected with a set of specific variables, such as the number of atoms in the shell, the distance from the target atom to the scatterer r_j, a structural disorder parameter σ_j which is commonly called Debye-Waller Factor and the scattering phase and amplitude of the backscatterer Φ_j and F_j, from which the type of scattering atom can be deduced. Typically, distances can be determined to a precision of about 1%, coordination numbers to about 20% and the type of backscatterer to $Z + /-2$. As the following discussion of the application of X-ray absorption spectroscopy on sulfur in biological systems will focus on XANES spectra, we refer the more interested reader for more detailed discussion of the EXAFS formula and the details of the analysis and interpretation of EXAFS spectra the literature (e.g. Sayers et al. 1971; Koningsberger and Prins 1988).

As final issue, before turning to the discussion of the application of X-ray absorption spectroscopy for the detection and identification of sulfur compounds, one should stress once again that all of the information that can be derived from XAS is related to the *local* environment, i.e. the "information depth" of the technique is limited to a distance of about 5–8

Å from the target atom. On the one hand, this means that using this technique, e.g., distances between the atoms in a single molecule can be determined, implying that one can work with amorphous materials, liquids, etc. and no crystals are necessary.

III. X-ray absorption-near edge structure (XANES) Spectroscopy Measurements of Sulfur Compounds

In the following, the more general remarks from the previous section will be further clarified by discussing some specific examples of sulfur K-edge XANES spectroscopy. The first point to be discussed is the determination of valency for XANES spectra. Figure 8 shows the spectra of three sulfur containing compounds with the major valencies of sulfur, namely + II, + IV and + VI. As one can see from this figure there is a "huge" (more than 6 eV!) chemical shift between the maxima of the white lines for valency + II and valency + VI indicating already the pronounced sensitivity of sulfur K-XANES for the chemical environment. As was discussed before, this chemical shift can be used as a "ruler" that can

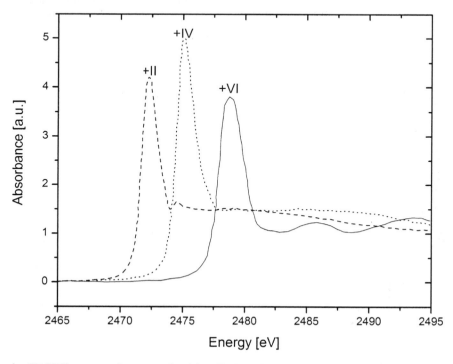

Fig. 8. Sulfur K-edge XANES spectra of compounds with sulfur in valency +II, +IV, and +VI.

be used to determine the valency of sulfur in an unknown compound just by using the position of the white line. The corresponding ruler – showing again a linear dependence between the formal valency and the energy position of the maximum of the white line – is shown in Fig. 9.

As was discussed before, XANES spectra do not just reflect the electronic structure of the atomic species of interest but also the geometric structure, for example the local symmetry. However, in some cases XANES spectra are also influenced by the medium range order arrangement around the atom of interest. The "standard form" of sulfur is the S_8 ring (*cyclo*-octasulfur), however, it is well known that sulfur can form also different ring structures with 6, 12 and 14 atoms (Steudel 2000) and sulfur is also found without a closed ring structure as "polymeric sulfur" with sulfur chains of different lengths. The sulfur K-edge XANES spectra of all these sulfur compounds are shown in Fig. 10. As the valency of the sulfur atoms and also the electronegativity of the neighbors is identical for all five compounds, the peak position of the white line is at the same energy for all compounds. However,

there are significant differences in the intensity of the white line where S_8 has the lowest and the polymeric sulfur the highest intensity. The intensity of the various ring structures is obviously not just a function of the number of atoms in the ring but it is determined by the details of the geometric arrangement. It should be noted that there are also differences in the fine structure of the white lines and in the region of the "shape resonances" reflect the geometric structure.

The extreme sensitivity of sulfur K-XANES against the chemical environment is on one hand a "blessing" as it offers the opportunity to derive detailed information from the spectra, on the other hand it is also a "curse" as it is necessary to have reference compounds with a chemical environment identical at least to the second coordination if one tries to do a quantitative analysis by fitting spectra.

The crucial importance of high quality reference compounds is illustrated by Fig. 11. This figure shows the sulfur K-edge XANES spectra of a group of disulfanes. All the spectra show – more or less clearly – two peaks from which the high energy one is assigned to S $1s \rightarrow \sigma^*$ (S-C)

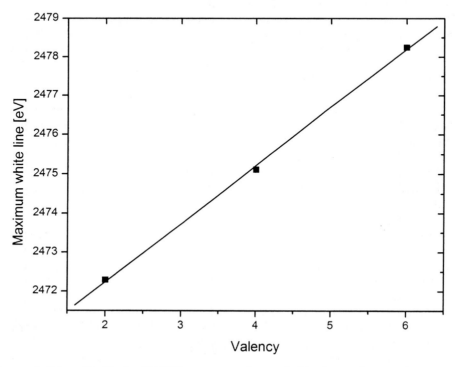

Fig. 9. The "valency ruler" for sulfur K-edge XANES spectra: maximum of white line vs. formal valency.

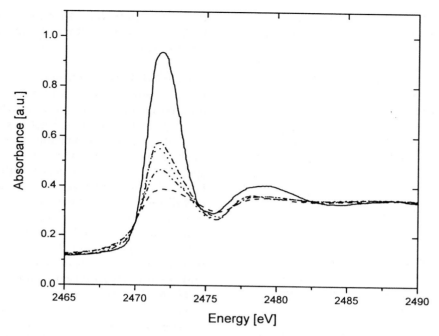

Fig. 10.   Sulfur K-edge XANES spectra of various sulfur rings and polymeric sulfur; from bottom to top: S_8 (dash), S_6 (dash dot dot), S_{14} (dot), S_{12} (dash dot), and polymeric sulfur (solid).

and the low energy one to S 1s → σ^* (S-S). As this fact in itself is again very positive as it allows to distinguish spectroscopically between C-S and S-S bonds which is very difficult – if possible at all – using other spectroscopic techniques – the spectra in Fig. 11 show also significant differences regarding (mainly) the energy difference between the two resonances. A detailed analysis of the spectra showed that the energy position of the σ^* (S-S) resonance is constant within an interval of less than 0.2 eV whereas the splitting of the two resonances is changing between 1.06 and 1.9 eV. So the differences in the spectra of Fig. 11 are caused by changes in the energy position of the σ^* (S-C) resonance. This is not really astonishing as the S-C bond is much more affected by a change of the substituents than the "central" S-S bond. Chauvistré et al. (1997) could actually show that this splitting is correlated with the S-C bond length or in other terms with the bond strength. Without further discussion, it is obvious from Fig. 11 that fitting the spectrum of an "unknown" compound with the goal of a quantitative analysis of sulfur chain lengths with the "wrong" reference compound will lead to com-

pletely false results. What was discussed here for the biologically more relevant case of disulfidic compounds is of course also true for compounds with longer sulfur chains that are required, for example, for the quantitative analysis of sulfur networks in rubber.

The molecules discussed by Chauvistré et al. had structural differences in the second coordination shell so that one could assume that a reference compound that models correctly at least that shell should be suitable for a quantitative analysis. However, there are several cases where even differences in the third coordination shell result in significantly different spectra. As changes are expected to be most obvious for the σ^* (S-C) such investigations have been carried out for a group of 4 symmetric monosulfanes, namely $(CH_2CH)_2S$ (vinyl monosulfane) $C_6H_5C_2)_2S$ (phenyl monosulfane), $((CH_3)_3SiC_2)_2S$ (silyl monosulfane) and $((CH3)3C3)S$ (t-butyl monosulfane). Figure 12 shows the sulfur K-edge XANES spectra of those compounds. At first sight, there are several striking differences between the spectra in Fig. 12 and also those in Fig. 11. There is a energy shift of about 0.7 eV between the maxima of the

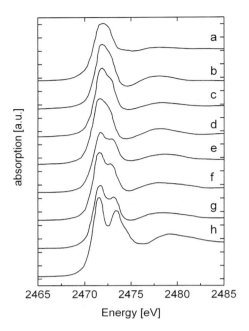

Fig. 11. Sulfur K-edge XANES spectra of various disulfanes; from top to bottom: di-benzyldisulfane (a), bis-(2,3-dimethyl-2-butenyl)disulfane (b), di-(2-butenyl)disulfane (c), di-tert-butyldisulfane (d), di-methyldisulfane (e), di-isopropyldisulfane (f), di-(butyl)sulfane (g), and di-phenyldisulfane (h).

IV. Application of XANES Spectroscopy to Investigate Sulfur in Phototrophic Organisms

In recent years, the application of X-ray absorption spectroscopy, XANES and EXAFS, to investigate different chemical elements related to different topics of phototrophic organisms has been reported, e.g. studies of heavy metal accumulation or coordination of metalloenzymes from plants. XANES spectroscopy has also been used for investigations of sulfur metabolism, sulfur in proteins and different sulfur compounds in different phototrophic organisms. For example, sulfur globules of phototrophic sulfur bacteria (Prange et al. 1999; 2002a,b,c; Pickering et al. 2001), elemental sulfur utilization by *Allochromatium vinosum* (Franz et al. 2007), sulfur in the wheat gluten network (Prange et al. 2001, 2003, 2005a), sulfur in onions (Yu et al. 2001; Lichtenberg, H., Prange, A., de Kok, L., Hormes, J. unpublished), sulfur speciation in coffee beans (Prange et al. 2005b; Lichtenberg et al. 2007), sulfur in host–plant interactions (Prange et al. 2005c), sulfur-containing species in horseradish and wasabi (Yu et al. 2001).

In the following sub-sections, three examples will show the usefulness of XANES spectroscopy to investigate sulfur speciation in complex biological systems: (1) purple and green sulfur bacteria, (2) gluten network and (3) host–plant interactions. Furthermore, the 'quantitative analysis' and spatially resolved XANES spectroscopy will be elucidated.

A. Sulfur in Sulfur Globules of Phototrophic Sulfur Bacteria

Reduced inorganic sulfur compounds such as sulfide, polysulfides, elemental sulfur, sulfite, polythionates and thiosulfate are oxidized and used as electron donors for energy generating systems by many phototrophic bacteria (e.g. Brune 1995; Dahl et al. 2002; Dahl and Prange 2006). In many cases, sulfur globules of sulfur in the "zero" oxidation state ("elemental sulfur") are formed as intermediate of reduced sulfur compound oxidation and are either stored inside (e.g. *Chromatiaceae*) or outside the (e.g. *Ectothiorhodospiraceae*, *Chlorobiaceae*) of the cells. Although the sulfur globules of purple sulfur bacteria and green sulfur bacteria have been known

white lines of the alcynilic sulfanes on the one and the vinylic sulfane on the other side – that can be explained by the differences in the chemical environment of the carbon atom located in the first coordination shell – and there are significant differences in spectra of the three alcynilic compound whereas for the sulfanes in Fig. 10 just small variations of shape and peak position on the white lines are visible. This extreme sensitivity of the alcynilic sulfanes can be understood as a direct consequence of the triple bond between the first and the second coordination shell carbon atom. Whereas in other cases the electronic effects caused by the modification of the end group are "split" between the other ligands and the sulfur atom, such a splitting is impossible in the alcinylic case. Though alcinylic molecules do not play an important role in biology, the examples discussed in Fig. 11 demonstrate clearly that extreme care is necessary when choosing reference samples for a fingerprint analysis of sulfur K-edge XANES spectra or the second and third coordination shell can also influence the spectra significantly.

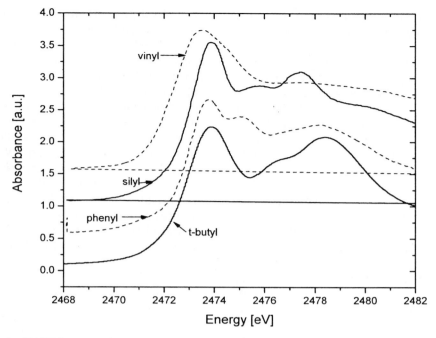

Fig. 12. Sulfur K-edge XANES spectra of (CH$_2$CH)$_2$S (= vinyl monosulfane), C$_6$H$_5$C$_2$)$_2$S (= phenyl monosulfane), ((CH$_3$)$_3$SiC$_2$)$_2$S (= silyl monosulfane), and ((CH3)3C3)S (= t-butyl monosulfane).

for more than a century (Winogradsky 1887; van Niel 1931; Trüper 1984, Dahl and Prange 2006) and although essential for a thorough understanding of the sulfur metabolism in these organisms, the exact chemical nature of the stored sulfur was not clarified until *in situ* XANES spectroscopy has been used to investigate the chemical speciation of the globules (Prange et al. 1999, 2002a, b, c). Prange et al. used sulfur K-edge XANES to probe the forms of sulfur *in situ* in different phototrophic bacteria as well as some chemotrophic sulfur bacteria. By fitting the XANES spectra with suitable reference compounds (see below), evidence for at least three different sulfur speciations in bacterial sulfur globules was obtained: *Cyclo*-octasulfur was found in *Beggiatoa alba* and *Thiomargarita namibiensis*, long-chain polythionates are present in the chemotrophic sulfur oxidizer *Acidithiobacillus ferrooxidans*, whereas the sulfur globules of the phototrophic sulfur bacteria consists of long sulfur chains, most probably terminated by organic residues (mono- or bis-organyl polysulfanes), irrespective of whether the sulfur globules are deposited inside (*Allochromatium vinosum, Thiocapsa roseopersicina, Marichromatium purpuratum*) or outside the cells (*Halorhodospira halophila,*

Halorhodospira abdelmalekii, Chlorobium vibrioforme) (Prange et al. 2002a). Furthermore, fitting the XANES spectra indicated that the organic residue seems to be a "glutathione group" (Prange et al. 2002a, b). From the results obtained by XANES spectroscopy and the quantitative analyses (see below), a clear correlation between sulfur speciation and occurrence of oxygen during growth of the bacteria can be deduced. While sulfur chains occur in the anaerobically grown phototrophic sulfur bacteria, *cyclo*-octasulfur and polythionates were found in aerobically grown organisms. Furthermore, sulfur of isolated sulfur globules from anaerobically grown *A. vinosum* cells was found as *cyclo*-octasulfur, indicating the influence of oxygen. The results indicate that the speciation of sulfur in the sulfur globules reflects the different ecological and physiological properties of different "types" of bacteria (Prange et al. 2002a) which is in good accordance with previous findings by, e.g., Hageage et al. (1970), Guerrero et al. (1984), Steudel and Albertsen (1999) and others. However, the investigation of sulfur globules of phototrophic bacteria by XANES spectroscopy by Prange et al. (1999, 2002a) and Pickering et al. (2001) yielded partly comparable experimental data but were interpreted in quite a different way by the two research groups. Pickering et al. (2001) concluded on the basis of theoretical considerations that the sulfur is "simply solid S_8". The discrepancies and discussion detail around this topic focuses on the measurement mode and physics (George et al. 2002; Prange et al. 2002c), however, Kleinjan et al. (2003) stated, that the model for the sulfur globules of *A. vinosum* by Prange et al. corresponds best with the available experimental data.

The oxidation of sulfide to "elemental sulfur globules" and the further oxidation of the stored sulfur to sulfate was investigated in more detail in the case of *A. vinosum* (Prange et al. 2002b). In batch culture, sulfide oxidation to sulfate by *A. vinosum* and also other phototrophic bacteria proceeds in a biphasic manner. Firstly, sulfur is accumulated in globules stored in the periplasm, as long as sulfide is present in the medium (Pattaragulwanit et al. 1998). Secondly, further oxidation of the stored sulfur starts and results in sulfate accumulation (Brune 1989). Figure 13 shows the XANES spectra and the corresponding fits of *A. vinosum* 1, 6 and 10 h after the cells (grown photoorganoheterotrophically) had been transferred to sulfide-containing medium; Table 1 shows the results of fitting the XANES spectra.

A set of six reference spectra was used (reduced and oxidized glutathione, di-methylsulfoxide, di-octenyltrisulfane (R-S_3-R, R = -$C_2H_2C_5H_{10}CH_3$), polymeric sulfur and zinc sulfate) for fitting the spectra using the MINUIT fitting routine (Prange et al. 2002b). A χ^2-criterium was used to find the linear combination of these spectra which reproduces the XANES spectra of *A. vinosum* with the highest probability. In general, non-statistical errors may occur from deviations in energy calibration, an incomplete set of reference spectra or from reference spectra which do not monitor exactly the local environment of sulfur; the errors of the percentage contributions can be estimated smaller than ±10% (absolute value) (details: Modrow 1999; Prange et al. 2002a). The XANES spectra of compounds which seem to be representative and yield the best fits were used for fitting the S K-edge XANES spectra to specify the forms of sulfur in a quantitative manner. However, it has to be kept in mind that the chosen spectra of reference compounds for the "quantitative analysis" play the "key role" and must be performed

very carefully with special regard to the atomic environment of interest.

By fitting the spectra, a quantitative description of the sulfur speciation during sulfide oxidation to sulfate in *A. vinosum* was obtained. In the sulfur globules of *A. vinosum* 1 hour after exposition to sulfide, polymeric sulfur is the dominant sulfur species (≈67%). Furthermore, the presence of the "initial, short chains" at the beginning of sulfide oxidation is indicated by the occurrence (≈32%) of the trisulfane. After one hour, sulfide oxidation and formation of the sulfur globules are still in progress (Prange et al. 1999). Thus, the presence of a considerable amount of short "initial chains" can be expected. After 6 h, the degradation of the sulfur globules is already in

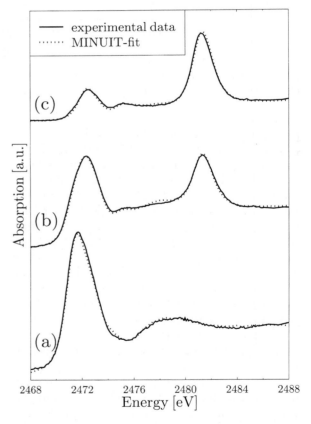

Fig. 13. Sulfur K-edge XANES spectra of *Allochromatium vinosum* (solid) recorded 1 hour (a), 6 hours (b) and 10 hours (c) after the cells had been transferred to sulfide-containing medium and corresponding MINUIT fits (dot). (Figure adopted from Prange et al. (2002) Investigation of S-H bonds in biologically important compounds by sulfur K-edge X-ray absorption spectroscopy. *Eur Phys J D* 20: 589–596, kindly permitted by Springer, Berlin.)

full progress, which is clearly indicated by the decreasing amount of sulfur chains (polymeric sulfur, ≈19%), the appearance (≈15%) of sulfate and the increasing amounts of reduced glutathione (≈37%) and oxidized glutathione (28%). The presence of oxidized glutathione (C-S-S-C) might either be related to small sulfur chains and to the complete shortening of chains in case glutathione is the terminating group of the sulfur chains (see above). Alternatively, a combination of both species (reduced and oxidized glutathione) might be use as an indicator for perthiol-groups. After 10 h, the final state of degradation, both polymeric sulfur and the initial chains have been removed completely and only the oxidized species and reduced glutathione contribute to the fit. Therefore, XANES yielded clear evidence that the sulfur chains stored inside the sulfur globules of *A. vinosum* are gradually shortened during oxidation of sulfur chains to sulfate.

B.   Sulfur in Wheat Gluten Proteins

The application of sulfur K-edge XANES spectroscopy for the investigation of sulfur in wheat gluten proteins is an example which demonstrates the usefulness of XANES to investigate a complex plant protein system. The gluten proteins mainly determine the baking quality of wheat and the rheological properties of doughs. They are classified by their molecular weight and amino acid sequences into sulfur-rich, sulfur-poor and high molecular weight (HMW) groups (Shewry and Miflin 1985). Especially cysteine containing components are important for the functionality of doughs forming the so-called "gluten network", a three-dimensional high molecular weight aggregate linked by intermolecular disulfide bonds (Köhler et al. 1991; Wieser et al. 1991). In freshly prepared flours about 95% of the cysteine residues of gluten proteins are present as disulfides and only 5% as free thiol groups (Shewry and Tatham 1997; Grosch and Wieser 1999). During dough mixing these thiol groups might catalyze the thiol/disulfide exchange reactions leading to glutenin. The resulting disulfide bonds are decisive for building up the gluten network, because their number determines the degree of polymerization, which directly influences the rheological parameters of doughs and the baking quality (Shewry and Tatham 1997). The total cysteine content of gluten protein components

can be determined by amino acid analysis and free SH- and SS-groups can be determined spectrophotometrically prior to and after reduction (Ellman 1959; Henschen 1986). However, the most important disadvantage of these methods is the limited information about the oxidation state of sulfur as only the SH- and SS-form can be determined and structural changes may have occurred.

Therefore, XANES spectroscopy at the sulfur K-edge was applied to characterize the speciation of sulfur in the gluten proteins on the basis of native samples to understand the gluten network in more detail. XANES spectroscopy showed the existence of disulfide bonds in oxidized (oxygen stream) glutenin subunits *in situ* for the first time and underlined their significance for the gluten network formation. Additionally, glutenin subunits, which were stored under ambient air and temperature conditions for two years, predominantly contained sulfur of higher oxidation states (sulfoxide, sulfonic acid state) and less sulfur in the disulfane (in the literature often named as disulfide) state (Prange et al. 2001). Furthermore, the sulfur speciation of glutenin subunits after reoxidation with potassium iodate and potassium bromate at different pH-values was investigated (Prange et al. 2003). Quantitative analyses of the XANES spectra showed that reoxidation with iodate and bromate leads to disulfane states but also to higher oxidation states (sulfoxide state, sulfonic acid state) which is in contrast to the traditional view. The results showed that the oxidation state of sulfur prior to oxidation (thiol, disulfane/disulfide) strongly influences sulfur speciation after oxidation. At low pH-values, the strongest oxidation occurred whereas the choice of the oxidizing reagent seems to be of minor importance (Prange et al. 2003). Another important effect on the structure and integrity of gluten network as well on the baking quality of wheat can be caused by different moulds. Field fungi of the genus *Fusarium* have hardly any influence on both the sulfur speciation of wheat gluten proteins and baking properties, whereas storage fungi of the genera *Aspergillus* and *Penicillium* have a direct influence (Prange et al. 2005a). An increased amount of sulfur in the sulfonic acid state was found in the respective XANES spectra, which is not available for thiol/disulfide exchange reactions in the gluten network, and thus leads to a considerably reduced baking volume (Prange et al. 2005a). For more details on the chosen reference compounds for the system "gluten network", quantitative analyses, including additional statistical information, the reader is referred to the cited articles by Prange and co-workers.

C. Sulfur in Plant–Fungus Interactions

Biotrophic fungal plant pathogens modify the metabolism of their host plants by inducing a metabolic sink in order to allocate a nutrient flow to the site of infection. Due to their biotrophic nature, the uptake of nutrients, especially amino acids and carbohydrates, by obligate pathogens can be studied only in the host–pathogen interaction. Typically only a small part of plant tissue is colonized by the pathogens and the differentiation of diseased and healthy tissue is crucial for investigations of the uptake and metabolism of nutrients by the pathogen. Prange et al. (2005c) firstly reported on the application of XANES spectroscopy in phytopathology used to elucidate, spatially resolved, the occurrence and distribution of sulfur speciation in biotrophic host–plant interactions. With spatially

resolved XANES spectroscopy, it is possible to investigate the metabolic activity of a plant pathogenic fungus colonizing the surface as well as vital tissue of leaves, i.e. the interactions in a host–pathogen system. Spatial resolution at the micrometer level allows the non-destructive assessment of heterogeneity in the distribution of various sulfur speciations induced during fungal pathogenesis in plants. In their "proof of principle" study, the characterization of chemical processes occurring at the sulfur K-edge within the interactions of the brown rust fungus, *Puccinia triticina*, colonizing wheat leaves is described. Using spatially resolved XANES spectroscopy as a fingerprint, differences in sulfur speciation at different areas of wheat leaves were observed and compared to isolated urediniosori. Three distinct leaf areas were investigated by XANES spectroscopy in fluorescence mode with a spatial resolution of about 100 μm × 100 μm of a wheat leaf infected with *P. triticina* (a. uninfected leaf areas, b. the center of urediniosori, and c. the non-visibly infected leaf tissue bordering the uredinal sorus). In Fig. 14, different sulfur K-edge XANES spectra of biological relevance

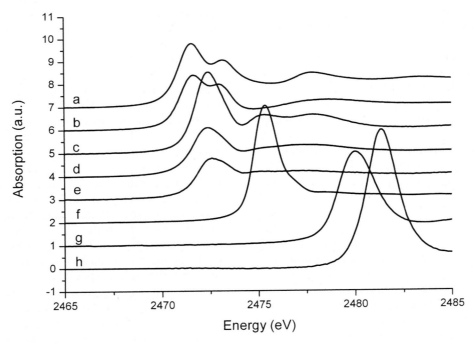

Fig. 14. Sulfur K-edge XANES spectra of reference compounds (top to bottom): cystine (a), glutathione (oxidized form) (b), glutathione (reduced form) (c), cysteine (d), methionine (e), dimethylsulfoxide (f), cysteic acid (g) and zinc sulfate (h); (a.u. = arbitrary units). (Figure adopted from Prange et al. (2005) Spatially resolved sulfur K-edge XANES spectroscopy for the *in situ*-characterization of the fungus-plant interaction *Puccinia triticina* and wheat leaves. *J Phytopathol* 153:627–632, kindly permitted by Blackwell Verlag, Berlin.)

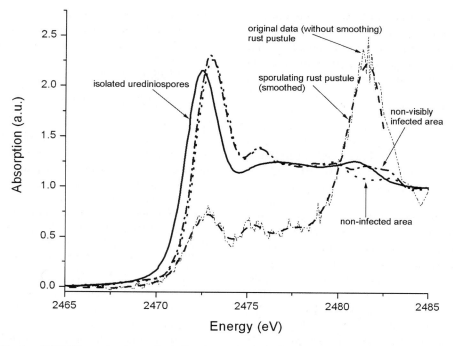

Fig. 15. Sulfur K-edge XANES spectra (smoothed, 7-point setting): isolated *Puccinia triticina* urediniospores (solid line), *P. triticina* sporulating rust pustule on the wheat leaf (dash line), non-visibly infected leaf tissue bordering the uredinial sorus (dash + dot line), non-infected area of wheat leaf (dot line). Original data without smoothing of *P. triticina* sporulating rust pustule on the wheat leaf (thin, dash + dot + dot line); (a.u. = arbitrary units). (Figure adopted from Prange et al. (2005) Spatially resolved sulfur K-edge XANES spectroscopy for the *in situ*-characterization of the fungus-plant interaction *Puccinia triticina* and wheat leaves. *J Phytopathol* 153:627–632, kindly permitted by Blackwell Verlag, Berlin.)

are displayed, which were used for comparison with the spectra obtained from the leaf (Fig. 14) (fingerprint analysis, cf. sections II, III).

As discussed earlier, the energy position of the main spectral features is a function of the formal valency of the neighboring atoms (cf. section III). The variation in the XANES spectra of the three wheat leaf regions of interest is given in Fig. 15; furthermore, the spectrum of isolated mature urediniospores is shown. In the center of sporulating rust pustules, a strong peak is observed in the spectrum at about 2481.4 eV. This structure is not present in either healthy leaf areas or symptomless colonized areas or in isolated urediniospores.

Comparing to the reference spectra (Fig. 14), it can be easily seen that the peak is correlated to an increased amount of sulfate (S^{+VI}) in the sample. The increase of sulfate in response to fungal development is confirmed when comparing the spectra for the region bordering the sporulating sorus. In this symptomless infected area, just a slight increase in the corresponding energy range

was observed in comparison to healthy leaf areas. As compared to healthy leaf areas. At this site, fungal structures were present in the mesophyll with haustoria formed also abundantly in the epidermal cells. As the spectra in this study show sulfur species resulting from the rust fungus and from the wheat leaf, an estimation on the probing depth must be kept in mind: A 2.4 keV photon beam in water is attenuated to 50% of its original intensity after passing through approximately 22.5 µm (cf. tools like on http://www.cxro.gov) which suggests that the probing depth in biological material (mostly water!) is of the order of tens of microns. Therefore, here the main contribution of sulfur species results from the epidermal cell layer and mesophyll cells up to a depth of approximately 50 µm. A clear assignment of the variation in the spectra of healthy and symptomless colonized leaf tissue to sub-surface fungal material is complicated by the fact that isolated spores on the intact leaf surface may be an additional source of sulfur-containing compounds causing modifications in the spectra. Apart from the sulfate, at all sites a resonance located at about 2472.9 eV contributes to the spectra, which can be assigned to the amino acids cysteine and methionine (C-S-H; C-S-C) of the proteins (cf. Fig. 14, e.g. Prange et al. 1999). However, the significant increase of sulfate in the center of sporulating rust pustules can only be explained as an effect of a specific fungus-host interaction. For more details on the presented example, the reader is referred to Prange et al. (2005c).

In conclusion, compared to non-infected leaf areas, minor changes in the spectra were observed for the non-visibly colonized tissue neighboring the rust sori. As the spectra for isolated urediniospores and the healthy leaf areas did not fit the spectra of the urediniosori, a significant impact of the biotrophic pathogen on sulfur metabolism of wheat has been shown. This example demonstrates that spatially resolved XANES spectroscopy will extend the range of qualitative and quantitative methods for *in situ* investigations of host–pathogen interactions and may help to increase knowledge about the metabolism of diseased plants.

Acknowledgements

The author's work has been funded by different funding agencies (Deutsche Forschungsgemeinschaft (DFG), Fonds der Chemischen Industrie (FCI), Bundesministerium für Bildung und Forschung (BMBF), National Science Foundation (NSF), Heinrich-Hertz-Stiftung Northrhine-Westphalia, which are gratefully acknowledged. Also, the work presented in the examples would have not been possible without the excellent cooperation with numerous collaborators from physics, chemistry and life sciences: R. Chauvistré, C. Dahl, H.-W. Dehne, E.-C. Oerke, U. Steiner, H.G. Trüper (University of Bonn); C.M. Bianchetti, R. Louis, G. Merchan, R.C. Tittsworth (†), A. Roy (Louisiana State University, Baton Rouge); P. Köhler (German Research Centre for Food Chemistry, Garching); R. Steudel (Technical University of Berlin); J. Hahn (University of Cologne).

References

Akabayov B, Doonan CJ, Pickering IJ, George GN and Sagi I (2005) Using softer X-ray absorption spectroscopy to probe biological systems. J Synchrotron Rad 12: 392–401

Ankudinov AL, Ravel B, Rehr JJ and Conradson SD (1998) Real-space multiple-scattering calculation and interpretation of X-ray absorption near-edge structure. Phys Rev B 58: 7565–7576

Beauchemin S, Hesterberg D and Beauchemin M (2002) Principal component analysis approach of modeling sulfur K-XANES spectra in humic acids. Soil Sci Soc Amer J 66: 83–91

Behrens P (1992a) X-ray absorption spectroscopy in chemistry. II. X-ray absorption near edge structure. Trends Anal Chem 11: 237–244

Behrens P (1992b) X-ray absorption spectroscopy in chemistry. I. Extended X-ray absorption fine structure. Trends Anal Chem 11: 218–222

Brune DC (1989) Sulfur oxidation by phototrophic bacteria. Biochim Biophys Acta 975: 189–221

Brune DC (1995) Sulfur compounds as photosynthetic electron donors. In: Blankenship RE, Madigan MT and Bauer CE (eds) Anoxygenic Photosynthetic Bacteria, pp 847–870. Kluwer Academic, Dordrecht

Capehart TW, Herbst JF, Mishra RK and Pinkerton FE (1995) X-ray absorption edge shifts in rare-earth-transition metal compounds. Phys Rev B 52: 7907–7914

Chauvistré R, Hormes J, Hartmann E, Etzenbach N, Hosch R and Hahn J (1997) Sulfur K-shell photoabsorption spectroscopy of the sulfanes R-S$_n$-R, n = 2–4. Chem Phys 223: 293–302.

Dahl C and Prange A (2006) Bacterial sulfur globules: Occurrence, structure and metabolism. In: Shively M (ed) Inclusions in Prokaryotes, chapter 2. Series "Microbiology Monographs", pp 21–51. Springer, Berlin Heidelberg,

Dahl C, Prange A and Steudel R (2002) Metabolism of natural polymeric sulfur compounds. In: Steinbüchel A and Matsumura S (eds) Biopolymers, Vol 6, Miscellaneous Biopolymers and Biodegradation of Synthetic Polymers, pp 35–62. Wiley-VCH, Weinheim

Ellman GL (1959) Tissue sulfhydryl groups. Arch Biochem Biophys 82: 70–77

Flemming B, Modrow H, Hallmeier K-H, Hormes J, Reinhold J and Szargan R (2001) Sulfur in different chemical surroundings – S K XANES spectra of sulfur-containing heterocycles and their quantum-chemically supported interpretation. Chem Phys 270: 405–413

Franz B, Lichtenberg H, Hormes J, Modrow H, Dahl C and Prange A (2007) Utilization of solid 'elemental' sulfur by the phototrophic purple sulfur bacterium *Allochromatium vinosum*: A sulfur K-edge X-ray absorption spectroscopy study. Microbiology-SGM 153: 1268–1274

George GN, Pickering IJ, Yu EY and Prince RC (2002) X-ray absorption spectroscopy of bacterial sulfur globules. Microbiology 148: 2267–2268

Grosch W and Wieser H (1999) Redox reactions in wheat dough as affected by ascorbic acid. J Cereal Sci 29: 1–16

Guerrero R, Mas J and Pedrós-Alió C (1984) Buoyant density changes due to intercellular content of sulfur in *Chromatium warmingii* and *Chromatium vinosum*. Arch Microbiol 137: 350–356

Hageage Jr GJ, Eanes ED and Gherna RL (1970) X-ray diffraction studies of the sulfur globules accumulated by *Chromatium* species. J Bacteriol 101: 464–469

Henschen A (1986) Analysis of cyst(e)ine residues, disulfide bridges and sulfhydryl groups in proteins. In: Wittmann-Liebold B, Salnikow J and Erdmann VA (eds) Advanced Methods in Protein Microsequence Analysis, pp 244–255. Springer, Berlin.

Hormes J and Modrow H (2003) X-ray absorption spectroscopy for the analysis of hydrocarbons and their chemistry In: Hsu CS (ed) Analytical Advances for Hydrocarbon Research, pp 421–454. Kluwer Academic/Plenum, Dordrecht.

Johnson RL (1983) Grating monochromators and optics for the VUV and soft X-ray region. In: Koch E (ed) Handbook on Synchrotron Radiation, Vol 1a, pp 173–260. North Holland, Amsterdam, New York

Kleinjan, W. E., de Keizer, A. and Janssen, A. J. H. (2003) Biologically produced sulfur In: Steudel R (ed) Elemental Sulfur and Sulfur-rich Compounds I, pp 167–187. Springer, Berlin

Köhler P, Belitz H-D and Wieser H (1991) Disulphide bonds in wheat gluten: Isolation of a cystine peptide from glutenin. Z Lebens Unter Forsch 192: 234–239

Koningsberger DC and Prins R (eds) (1988) X-ray Absorption: Principles, Applications, Techniques of EXAFS, SEXAFS and XANES. Wiley, New York

Lemonnier M, Collet O, Depautex C, Esteva JM and Raoux D (1978) High vacuum two crystal soft X-ray monochromator. Nucl Instr Meth A 152: 109–111

Lichtenberg H, Prange A, Modrow H and Hormes J (2007) Characterization of sulfur compounds in coffee beans by sulfur K-XANES spectroscopy. Amer Inst Phys Proc 882: 824–826

Modrow H (1999) Möglichkeiten und Grenzen der quantitativen Analyse von Röntgenabsorptionsspektren und ihre Anwendung auf Vulkanisation und thermo-oxidative Alterung von Kautschuken. PhD thesis, University of Bonn, Germany (in German)

Modrow H (2004) Tuning nanoparticle properties – the X-ray absorption spectroscopic point of view. Appl Spect Rev 39: 183–290

Modrow H, Hormes J, Visel F and Zimmer R (2001) Monitoring thermal oxidation of sulfur crosslinks in SBR eleastomers by quantitative analysis of sulfur K-edge XANES spectra. Rubber Chem Technol 74: 281–294

Pantelouris A, Modrow H, Pantelouris M, Hormes J and Reinen D (2004) The influence of coordination geometry and valency on the Cr K-edge XANES spectra of selected chromium compounds. Chem Phys 300: 13–22

Pattaragulwanit K, Brune DC, Trüper HG and Dahl C (1998) Molecular genetic evidence for extracytoplasmatic localization of sulfur globules in *Chromatium vinosum*. Arch Microbiol 169: 434–444

Penner-Hahn JE (2005) Characterization of "spectroscopically quiet" metals in biology. Coord Chem Rev 249: 161–177

Pickering IJ, George GN, Yu EY, Brune DC, Tuschak C, Overmann J, Beatty JT and Prince RC (2001) Analysis of sulfur biochemistry of sulfur bacteria using X-ray absorption spectroscopy. Biochemistry 40: 8138–8145

Prange A and Modrow H (2002) X-ray absorption spectroscopy and its application in biological, agricultural and environmental research. Rev Environ Sci Biotechnol 1: 259–276

Prange A, Arzberger I, Engemann C, Modrow H, Schumann O, Trüper HG, Steudel R, Dahl C and Hormes J (1999) *In situ* analysis of sulfur in the sulfur globules of phototrophic sulfur bacteria by X-ray absorption near edge spectroscopy. Biochim Biophys Acta 1428: 446–454

Prange A, Kühlsen N, Birzele B, Arzberger I, Hormes J, Antes S and Köhler P (2001) Sulfur in wheat gluten: *In situ* analysis by X-ray absorption near edge structure (XANES) spectroscopy. Eur Food Res Technol 212: 570–575

Prange A, Chauvistré R, Modrow H, Hormes J, Trüper HG and Dahl C (2002a) Quantitative speciation of sulfur in bacterial sulfur globules: X-ray absorption spectroscopy reveals at least three different species of sulfur. Microbiology 148: 267–276

Prange A, Dahl C, Trüper HG, Behnke M, Hahn J, Modrow H and Hormes J (2002b) Investigation of S-H bonds in biologically important compounds by sulfur K-edge X-ray absorption spectroscopy. Eur Phys J D 20: 589–596

Prange A, Dahl C, Trüper HG, Chauvistré R, Modrow H and Hormes J (2002c) X-ray absorption spectroscopy of bacterial sulfur globules: a detailed reply. Microbiology 148: 2268–2270

Prange A, Birzele B, Krämer J, Modrow H, Chauvistré R, Hormes J and Köhler P (2003) Characterization of sulfur speciation in low molecular weight subunits of glutenin after reoxidation with potassium iodate and potassium bromate at different pH values using X-ray absorption near-edge structure (XANES) spectroscopy. J Agric Food Chem 51: 7431–7438

Prange A, Modrow H, Hormes J, Krämer J and Köhler P (2005a) Influence of mycotoxin producing fungi (*Fusarium, Aspergillus, Penicillium*) on gluten proteins during suboptimal storage of wheat after harvest and competitive interactions between field and storage fungi. J Agric Food Chem 53: 6930–6938

Prange A, Hindorf H, Bianchetti CM, Tittsworth RC, Lichtenberg H, Modrow H and Hormes J (2005b) *In situ* analysis of sulfur in coffee by X-ray absorption near edge structure (XANES) spectroscopy: "A first impression of changes in sulfur speciation in green coffee and during roasting and brewing". In: Broussard LA and Scott JD (eds) CAMD 2004 Annual Report, pp 84–87. Center for Advanced Microstructures and Devices, Baton Rouge, LA, USA

Prange A, Oerke E-C, Steiner U, Bianchetti CM, Hormes J and Modrow H (2005c) Spatially resolved sulfur K-edge XANES spectroscopy for the *in situ*-characterization of the fungus-plant interaction *Puccinia triticina* and wheat leaves. J Phytopathol 153: 627–632

Rehr JJ and Albers RC (2000) Theoretical approaches to X-ray absorption fine structure. Rev Mod Phys 72: 621–654

Sayers DE, Stern EA and Lytle FW (1971) New technique for investigating noncrystalline structures: Fourier analysis of the extended X-ray absorption fine structure. Phys Rev Lett 27: 1204–1207

Schulze DG and Bertsch PM (1995) Synchrotron X-ray techniques in soil, plant, and environmental research. Adv Agronom 55: 1–66

Shewry PR and Miflin BJ (1985) Seed storage proteins of economically important cereals. In: Pomeranz Y (ed) Advances in Cereal Science and Technology VII, pp 1–83. American Association of Cereal Chemists: St. Paul, Minnesota

Shewry PR and Tatham AS (1997) Disulphide bonds in wheat gluten proteins. J Cereal Sci 25:207–227

Steudel R (2000) The chemical sulfur cycle. In: Lens P and Hulshof Pol L (eds) Environmental Technologies to Treat Sulfur Pollution, pp 1–32. IWA Publishing, London

Steudel R and Albertsen A (1999) The chemistry of aqueous sulfur sols – models for bacterial sulfur globules? In: Steinbüchel A (ed) Biochemical Principles and Mechanisms of Biosynthesis and Biodegradation of Polymers, pp 17–26. Wiley-VCH, Weinheim

Stoehr J, Sette F and Johnson AL (1984) Near-edge X-ray absorption fine-structure studies of chemisorbed hydrocarbons – Bond lengths with a ruler. Phys Rev Lett 53: 1684–1687

Stoehr J (1996) NEXAFS Spectroscopy. Springer Series in Surface Sciences Vol. 25, Springer, Berlin

Trüper HG (1984) Microorganisms and the sulfur cycle. In: Müller A and Krebs B (eds) Sulfur, its Significance for Chemistry, for the Geo-, Bio-, and Cosmosphere and Technology, pp 351–365. Elsevier SciencePublishers B.V., Amsterdam

Vairavamurthy A (1998) Using X-ray absorption to probe sulfur oxidation states in complex molecules. Spectrochim Acta A 54: 2009–2017

van Niel CB (1931) On the morphology and physiology of the purple and green sulfur bacteria. Arch Mikrobiol 3: 1–112

von Busch F, Hormes J, Modrow H and Nestmann MB (2003) Interaction of atomic core electrons with the molecular valence shell. In: Peyerimhoff SD (ed) Interactions in Molecules – Electronic and Steric Effects, pp 193–254. Wiley-VCH, Weinheim

Wieser H, Seilmeier W and Belitz H-D (1991) Klassifizierung der Proteinkomponenten des Weizenklebers. Getreide, Mehl und Brot 45: 35–38 (in German)

Winogradsky SN (1887) Über Schwefelbakterien. Bot Ztg 45:489–508 (in German)

Yachandra VK (1995) X-ray absorption spectroscopy and applications in structural biology. Meth Enzymol 246: 638–675

Yu EY, Pickering IJ, George GN and Prince RC (2001) *In situ* observation of the generation of isothiocyanates from sinigrin in horseradish and wasabi. Biochim Biophys Acta 1527: 156–160

Chapter 24

Imaging Thiol-Based Redox Processes in Live Cells

Andreas J. Meyer*

*Heidelberg Institute of Plant Sciences, University of Heidelberg, Im Neuenheimer Feld 360,
69120 Heidelberg, Germany*

Mark D. Fricker

Department of Plant Science, University of Oxford, South Parks Road, Oxford OX1 3RB, UK

Summary

Biotic and abiotic stresses often lead to transient oxidation of the internal milieu of cells. This oxidation is considered to be an essential signal for further downstream events within the cellular signaling network and therefore demands careful and, if possible, quantitative analysis. Fluorescent probes and confocal or multi-photon imaging offer almost unparalleled opportunities for visualization of an ever increasing range of compounds and even the dynamics of physiological processes within living cells and tissues with minimal perturbation. However, the utility of the methods relies on the development of probes and imaging protocols that can achieve sufficient specificity and sensitivity to give unambiguous physiological measurements. In situ estimates of glutathione have been undertaken on relatively few systems so far and in vivo measurements of redox state have only recently been made possible with

*Corresponding author, Fax: 0049 6221 545859, E-mail: ameyer@hip.uni-hd.de

Rüdiger Hell et al. (eds.), Sulfur Metabolism in Phototrophic Organisms, 483–501.
© 2008 *Springer.*

the advent of transgenic fluorescent reporters. Detection and quantification of reactive oxygen species (ROS) is also challenging, and a large number of different approaches have been developed to deduce information about components of this ROS-dependent signaling pathway. This review describes different imaging-based approaches for qualitative and quantitative measurement of glutathione and ROS in living plant cells. Particular attention is given to general strategies for probe design and application of quantitative confocal fluorescence imaging.

I. Introduction

The first simple life forms developed in a reducing environment. However, with the evolution of oxygenic photosynthesis the environment changed dramatically with the increasing levels of oxygen, potentially exposing cells to a severe stress. To maintain their predominantly reducing metabolism, cells had to develop additional systems to keep the internal medium reducing. One of the key metabolites used to achieve this was the tripeptide glutathione, which can be converted from its reduced form GSH to the oxidized glutathione disulfide GSSG. The GSH-GSSG redox couple is the most important redox buffer in virtually all eukaryotic cells, most Gram-negative bacteria, and a few Gram-positive bacteria (Sundquist and Fahey, 1989; Fahey and Sundquist, 1991; Newton et al., 1996). In some archebacteria, high concentrations of γ-glutamylcysteine appear to function instead of GSH and γ-EC is sufficient to buffer the cellular redox state (Newton and Javor, 1985; Sundquist and Fahey, 1989).

Abbreviations: BSO – L-buthionine-(S,R)-sulfoximine; CALI – chromophore-assisted light inactivation; CLSM – confocal laser scanning microscopy; DHE – dihydroethidium; DHR – 123 dihydrorhodamine 123; FLIP – fluorescence indicator protein; FRET – fluorescence resonance energy transfer; FWHM – full width at half maximum; GRX – glutaredoxin; GSB – glutathione *S*-bimane; GSH – reduced glutathione; GSSG – oxidized glutathione; GST – glutathione *S*-transferase; GSX – glutathione conjugate of xenobiotics 'X'; H$_2$DCFDA – dihydro-dichloro fluorescein diacetate; MCB – monochlorobimane; MPLSM – multiphoton laser scanning microscopy; NO – nitric oxide; PBP – periplasmic binding protein; PI – propidium iodide; PSF – point spread function; roGFP – reduction–oxidation sensitive GFP; ROS – reactive oxygen species; rxYFP – redox-sensitive YFP; SOSG – Singlet Oxygen Sensor Green

Severe stress of any description is usually accompanied by increasing concentrations of reactive oxygen species (ROS) (Apel and Hirt, 2004; Mittler et al., 2004). Thus, the redox buffer system not only helps to ameliorate the stress but may also act as a sensor and transmit these signals to the respective targets (May et al., 1998). These targets include both metabolic processes and transcriptional events. However, at this stage it has not yet been fully elucidated how the signals are transduced at the molecular level. Exploiting the redox system as part of a signal transduction network requires means to sense a temporary deviation from the resting level. By analogy with Ca^{2+} signal transduction (Hetherington and Brownlee, 2004) it can be envisaged that the amplitude, duration and location of such redox signatures might code for some detailed information about the particular stress.

To understand the different redox poise and sensitivity of different cells and to decode the redox signature it is of particular importance to be able to measure the redox signals in time and space. A number of different techniques have been developed to visualize different aspects of the cellular redox systems and deviations from resting levels in particular. These techniques range from histochemical localization of H$_2$O$_2$ as insoluble precipitates to the large number of fluorescent probes available to image different ROS. These studies have also highlighted the extent of stress-related formation of ROS, not least from wound reactions during simple preparation of the material for observation. Thus redox measurements need to be made on systems that are as intact as possible and operating in as close to their natural environment to minimize the perturbation associated with the measurement.

Live cell imaging currently offers the best possible solution when measurements are required at the level of single cells or defined tissues. Even so, quantitative live-cell imaging is technically challenging and requires careful attention to artifacts arising from preparation of the material, the perfusion systems used, and the potential for stress induced by the process of imaging itself (Fricker et al., 1999, 2006). For example, there are particular problems with microscopy of aerial tissues that are normally operating in contact with a gaseous phase, as this air gap significantly degrades image quality. Immersion in buffer or water significantly improves the imaging characteristics, but may substantially modify the local environment around the tissues. In non-photosynthetic tissues this may lead to reduced oxygen levels, conversely in photosynthetically active tissues, locally higher O_2 levels may result under microscope illumination.

II.   General Strategies of Probe Design to Image Components of Redox Pathways

Although useful measurements can be made using histochemical techniques, the vast majority of probes currently under development are fluorescent as these give high sensitivity and contrast and are compatible with confocal or multi-photon imaging systems that provide the best route to quantitative measurements in situ.

A limited number of cellular components are intrinsically fluorescent, which provides a straightforward route to measure their relative or absolute concentration, if suitable calibration standards are available. In a subset of these compounds, the fluorescence properties alter in response to the physiological environment, such as redox poise, and can be used to monitor dynamics of the corresponding parameter. For compounds that do not show useful auto-fluorescence, there are three main routes available to develop a fluorescent assay appropriate for fluorescence imaging and analysis. The preferred approach is to develop a probe that interacts specifically, but reversibly, with the target molecule, but does not react directly. This class of probes is typified by the ion reporters that reversibly bind to the ion of interest with a particular dissociation constant (k_d) and selectivity (Fricker and Meyer, 2001; Fricker et al., 2006). The second approach involves a probe that chemically interacts with the molecule of interest, but in a reversible reaction,

and is exemplified by the di-cysteine based redox indicators described later. In this case the target is also modified by the interaction with the probe. The third approach involves a probe that is itself a substrate for one of the enzymes in the pathway and exhibits a change in its fluorescent properties during the reaction. The target molecule is also irreversibly modified in this process.

A.   Intrinsically Fluorescent Probes

The redox state of the NADH pool has been measured from auto-fluorescence of reduced nucleotides using UV-CLSM or multi-photon laser scanning microscopy (MPLSM) (Masters and Chance, 1993; Piston et al., 1995; Patterson et al., 2000; Huang et al., 2002; Kasimova et al., 2006) and used to infer changes in glucose metabolism (Bennett et al., 1996). Using high-speed microscopy and microspectrophotometry it was shown that in living human neutrophils NAD(P)H waves travel through cells intracellularly at high speed ($\sim50\,\mu m\ s^{-1}$) and supply the membrane bound NADPH oxidase with pulses of substrate (Kindzelskii and Petty, 2002). Oscillatory NAD(P)H fluorescence was recently also observed in pollen tubes and shown to correlate with oscillatory tip growth (Cardenas et al., 2006). Changes in redox state can also be inferred from flavoprotein fluorescence (e.g. Rocheleau et al., 2004; Quesada et al., 2006), which also offers a more convenient set of excitation wavelengths than NAD(P)H. Given the steadily increasing interest in redox regulated processes and the awareness of the importance of even minor changes in different cellular redox compounds, NAD(P)H or flavoprotein redox imaging would thus provide a very powerful adjunct to many biochemical analyses. It should be noted, however, that the autofluorescence from reduced pyridine nucleotides is very weak. The major challenge therefore is to effectively separate these weak signals from the range of other UV-excitable compounds present in plants and to quantify them accurately.

B.   Reversible Probes

Probes that bind reversibly to their target can provide information on the steady-state concentration of the target, but not directly its rate of synthesis or consumption. Based on the precedents from the ion-imaging field, the most useful reversible probes, termed ratio probes, exhibit a shift in either their excitation or emission spectrum between the free and bound forms. The shift in spectrum can be readily measured as a ratio between images or average intensities measured at two wavelengths, typically corresponding to the peak wavelengths for the free and bound forms (Fricker et al., 2001). The ratio provides a convenient means to correct for changes in dye concentration, pathlength, photobleaching or dye leakage. Single wavelength probes only show changes in intensity on binding with no spectral shift. In these cases, it is more difficult to separate changes in fluorescence arising from the target versus the dye concentration, etc.

The number of chemical probes for cations is increasing as it is relatively easy to design specific binding sites to co-ordinate metal ion ligands. Progress has been less rapid for anions and essentially non-existent for other compounds as it is extremely difficult to design binding sites for small molecules that then also trigger a change in fluorescence. An alternative approach uses peptide-based binding domains to achieve specificity for different substrates and exploits any conformational change on binding to affect the amount of fluorescence resonance energy transfer (FRET) between spectral variants of GFP fused in-frame with the binding peptide(s). The first such transgenic ratio indicator designed was for calcium and was termed cameleon as it changed color and incorporated a calmodulin (CaM) binding domain as part of a linker between CFP and YFP (Miyawaki et al., 1997, 1999). The cameleon concept inspired the development of genetically-encoded FRET-based sensors for other metabolites. For example, Frommer and co-workers exploited the substrate-induced conformation change in bacterial periplasmic-binding proteins (PBPs) to construct fluorescent indicator protein (FLIP) sensors for maltose (Fehr et al., 2002), glucose (Fehr et al., 2003, 2004), ribose (Lager et al., 2003) and glutamate (Okumoto et al., 2005). Unlike most measurements of metabolites, these sensors report concentration directly from specific cellular compartments. As reaction rates and enzyme kinetics are concentration dependent, this should provide much better understanding of the control of metabolism in vivo. Sensitivity can be altered

by mutations in the binding site and sensors can be targeted to other compartments, such as the ER (M. Fehr and W. Frommer, personal communication). The ratio change for the first generation of these sensors was very small. However, recent improvements include replacing EYFP with Venus (Deuschle et al., 2005; Okumoto et al., 2005) and modification of the linker length and site of chromophore insertion to improve dipole-dipole coupling, giving a ~2-fold increase in response (Deuschle et al., 2005). Other PBP-based sensors for sugars, amino acids, sulfate and phosphate have already been synthesized using chemical coupling of fluorescent dyes to appropriate binding peptides (De Lorimier et al., 2002). Incorporation of the same binding peptides into the genetically-encoded FLIP sensors should yield equivalent transgenic probes and this approach has recently been shown to work for a phosphate binding PBP (Gu et al., 2006). These sensors can be expressed in plants (Fehr et al., 2004) but are subject to transgene silencing (Deuschle et al., 2006). Nevertheless, metabolite measurements can be made if the constructs are expressed in silencing mutant backgrounds (Deuschle et al., 2006). As yet, no probes have been developed for specific components of redox systems. However, the transgenic FRET approach is generic, and it is only a matter of time before suitable binding motifs are incorporated into the basic cassette to yield reversible sensors to report the concentration of individual components in different redox pathways.

C. Reversible Reactive Probes

Currently the most promising reactive, but reversible probes for dynamic redox imaging are redox-sensitive fluorescent proteins. Additional cysteine residues have been engineered into the protein barrel of both GFP and YFP, such that the two cysteines are located on two adjacent ß-strands (Østergaard et al., 2001; Dooley et al., 2004; Hanson et al., 2004). The two cysteine residues are sufficiently close to form disulfide bridges depending on the redox-environment of the protein. Formation or release of this disulfide bridge then leads to slight conformational changes in the protein barrel and alters the protonation state of the fluorophore. This change in protonation leads to a change in the absorption properties and hence in a change of fluorescence.

A range of different redox-sensitive GFP (roGFP) and YFP (rxYFP) versions have been generated, which all are slightly different in their redox and fluorescence properties (Østergaard et al., 2001, 2004; Hanson et al., 2004; Cannon and Remington, 2006).

In the case of rxYFP, the introduction of a pair of cysteine residues (N149C and S202C) confers reversible redox-dependent changes in fluorescence intensity (Østergaard et al., 2001, 2004). rxYFP appears to be specifically reporting the state of the GSH-GSSG redox couple in yeast cells, as it interacts with glutaredoxins (GRXs) which shuffle electrons between the cellular glutathione redox buffer and rxYFP (Østergaard et al., 2004). Thus, in this case rxYFP can be regarded as a bonafide GSH redox sensor.

Whilst rxYFP is a single wavelength redox sensor, the roGFPs (Dooley et al., 2004; Hanson et al., 2004) are particularly useful as they can be excited with two different wavelengths (405 and 488 nm) and the resulting fluorescence intensity of the two peaks changes relative to each other depending on the degree of disulfide bond formation between the two cysteines. This allows for ratiometric measurements, facilitating quantitative imaging of redox dynamics, which makes roGFPs potentially superior to single wavelength rxYFP (Björnberg et al., 2006). The structural similarities of rxYFP and roGFP suggested that roGFP might also be a specific glutathione redox sensor, provided a GRX is available for transferring electrons between glutathione and roGFP. At least in plants this is indeed likely because plants contain a large number of different GRXs, which exhibit a strong structural homology to the yeast GRX (Lemaire, 2004; Xing et al., 2006). In vitro studies with poplar GRX C4 confirmed this hypothesis and showed that only GRX together with GSH are capable of efficiently reducing the disulfide bond of roGFP while other reducing compounds like ascorbate or NADPH and thioredoxin were not able to reduce roGFP (Meyer et al., 2007). Thus it seems likely that roGFP also specifically reports on the state of the GSH-GSSG redox couple in vivo.

D. Irreversible Reactive Probes

In contrast to reversible interactions exemplified by the cameleons, biosensors and roGFPs,

it is possible to image some metabolites, such as glutathione, various ROS or nitric oxide (NO), following reaction in vivo to give a fluorescent product. The majority of the probes so far developed are fluorescent substrates or products for cleavage reactions catalyzed by esterases, lipases or proteases, rather than probes useful to track metabolic pathways. The major exception are probes for ROS, as it is possible to present reduced forms of the fluorescent probes that can be oxidized back to their fluorescent parent by different ROS with varying degrees of specificity.

For this type of probe, the fluorescent signal is a cumulative measure of the amount of target molecule that reacts, and is usually irreversible. Thus the signal increases over time at a rate dependent on the concentration of the target, the concentration of the probe and, in some cases, the presence of an enzyme catalyst. Interpretation of the kinetics, either from the gradient of the increase over time or as the amount present at a fixed point is therefore not straightforward. For example, in the un-catalyzed case, the reaction will be at least second order, depending on the stoichiometry of the reactants, and thus critically dependent on the local concentration of the probe. This is likely to be highly variable in intact tissues due to variation in access, tissue penetration rate, and the local environment. High concentrations of probe should give faster reactions and thus make the signal easier to detect, conversely, they will deplete the target more rapidly, perturbing any downstream processes. In the case of enzyme-catalyzed reactions, it is possible to use near saturating probe concentrations to make the kinetics pseudo-first order and thus provide a less ambiguous measure of the concentration of the target. However, high concentrations of the probe may themselves cause technical problems with self-absorbance and auto-fluorescence of the unreacted form. Even more worrying are reports that show decreases in fluorescence using these probes. As the reactions are essentially irreversible, this suggests these measurements are subject to extreme interference by other processes, such as photobleaching, sequestration in a low pH environment quenching the fluorescence, dye leakage to the medium or dye destruction by detoxification systems.

III. Quantitative Confocal Fluorescence Imaging

The requirement to reduce experimental manipulation of the tissue to a minimum places a premium on working with intact tissues. Currently the best approaches to image such material are confocal laser scanning microscopy (CLSM), multiphoton laser scanning microscopy (MPLSM), or confocal disk-scanning instruments. These systems can be used to collect optical sections, free from out-of-focus blur, from fluorescent probes distributed within living plant tissues with repeated sampling at sub-cellular resolution. The removal of out-of-focus blur allows discrimination of signals from sub-cellular domains, whilst contaminating signal from auto-fluorescence and/or stained tissues outside the focal plane is rejected. The increased contrast achieved in comparison to conventional wide-field fluorescence imaging, greatly improves visualization of cellular and sub-cellular morphology in thin optical sections.

The main advantages of confocal or multiphoton microscopy for quantitative fluorescence measurements arise from the well defined 3-D volume (voxel) that is sampled to form each 2-D picture element (pixel). The volume probed is always asymmetric, being at least 3–4 times longer in the axial (z) direction. The overall shape of the confocal probe in fluorescence is described by the point spread function (psf). Optical probe dimensions are usually given in terms of the full width at half the maximum height (FWHM) in a given direction along one of the orthogonal axes. Point-scanning confocal instruments can achieve a probe size of $0.2\,\mu m \times 0.2\,\mu m \times 0.6\,\mu m$, in x, y and z respectively, with a high NA (1.4) oil-immersion lens. However, typical probe dimensions for physiological measurements are likely to be larger, in the region of $0.4\,\mu m \times 0.4\,\mu m \times 1.2\,\mu m$, as longer-working distance, lower NA, water-immersion lenses are usually used, and the optical sectioning is often relaxed slightly by increasing the pinhole diameter to obtain sufficient signal within a useful sampling time.

The primary objective of quantitative physiological measurements is to maximize the signal to noise (S/N) ratio with minimal disruption to the cell physiology. In many instruments there is control over the degree of confocality, the area scanned, the scan speed and the number of frames averaged that provides the user with considerable flexibility in choice of sampling speed and the volume of specimen imaged. At one extreme, repeated sampling can be made of a single point in the 1 MHz range, whilst a 3-D volume of a thick tissue specimen, such as a root tip, can be sampled at high resolution in a few minutes. Thus different facets of a given biological question can be tackled on the same instrument.

To make absolute photometric measurements, the relationship between the image brightness and the concentration of the fluorophore has to be calibrated. One of the limitations of CLSM and MPLSM is the very low number of photons emanating from each voxel, making accurate measurements more difficult. Furthermore, quantitative measurements need to subtract a good estimate of the weak background signal from the specimen, auto-fluorescence in lens elements, sample chamber and perfusion medium. It is not usually possible to separate out and remove these different background noise components, thus noise from all sources contributes to an increase in the variance of the signal. To make an accurate estimate of the mean signal requires that both the background and the total intensity are recorded without clipping the intensity distributions (Fricker et al., 2001).

Although confocal and multiphoton imaging can work with intact tissues, they are still limited to measurements near the surface. One of the major optical consequences of imaging progressively deeper into tissues is blurring of the psf as the light rays travel through more and more refractive boundaries. Blurring of the probe results in a decrease in signal accepted by the confocal detector pinhole or reduces the probability of multi-photon excitation, leading to significant attenuation of fluorescence intensity with depth into the tissue. In theory, ratiometric measurements can compensate for such attenuation, provided it is equivalent at all wavelengths needed for the measurement. However, single wavelength measurements can often show several fold reduction in signal over a few tens of microns from the surface. Partial correction for tissue-dependent attenuation can be achieved by a pragmatic approach based on determination of the axial intensity profile of a permeabilised specimen filled with a fluorescent 'sea' (White et al., 1996; Fricker et al., 2000, 2001; Hartmann et al., 2003). The resultant response combines the effects of depth-dependent 'sea' response and the additional contribution of the permeabilised tissue.

With suitable calibration, the fluorescence intensity can be directly related to target concentration as the psf gives a defined sampling volume both in the tissue and in the calibration solutions. To relate in vivo measurements to other biochemical measurements, it is useful to be able to normalize the fluorescence intensity in a range of different units, such as µmole (mg protein)$^{-1}$. If particular tissue zones, cells or compartments are segmented from the image, results can be expressed on an appropriate database, such as pmole cell^{-1}. At the moment it is rather more complex to relate fluorescence levels to other parameters such as protein levels or DNA content. The basic approach involves collection of 3-D images of protein or DNA distribution from corresponding regions of the same or an equivalent specimen and measurement of the total amount of protein DNA^{-1} (Fricker et al., 2000).

IV. Application of Thiol-Based Redox Imaging in Plants

A. Measurement of Total Glutathione

Total cytoplasmic glutathione (GSH) can be determined in vivo following GST-catalyzed conjugation to monochlorobimane (MCB) to give a fluorescent glutathiones-bimane (GSB) adduct (Fig. 1). The strong preference of MCB for GSH over other low molecular weight thiols and protein thiols was initially shown for Arabidopsis suspension culture cells (Meyer et al., 2001). Recently, the specificity was further confirmed by the absence of labeling in homozygous *gsh1* knockout embryos, which cannot be labeled by the routine protocol used for labeling of cells (Cairns et al., 2006). The only exception where significant labeling of thiols other than GSH was observed in homozygous *gsh2* knockouts, which hyperaccumulate γ-EC to 5,000-fold normal levels in the absence of GSH (Pasternak et al., 2008). Imaging deep within tissues requires correction for depth-dependent attenuation (Fricker et al., 2000; Meyer and Fricker, 2000; Meyer et al., 2001), but has now been applied to GSH

measurements in several different types of tissues including roots (Fig. 2a and b) (Sánchez-Fernández et al., 1997; Fricker et al., 2000; Meyer and Fricker, 2000; Fricker and Meyer, 2001), suspension culture cells (Fig. 2c and d) (Meyer and Fricker, 2002), trichomes of Arabidopsis (Gutiérrez-Alcalá et al., 2000) and tobacco (Fig. 2g), and mesophyll, epidermal and guard cells of poplar leaves (Hartmann et al., 2003; see also Fig. 2e and f). Sequential imaging at different time points during incubation with MCB clearly revealed that conjugates with GSH are formed in the cytoplasm and subsequently sequestrated to the vacuole (Fig. 2). The time required for this sequestration might be different in different cell types depending on the activity of the vacuolar GSX conjugate pumps (see section IV.B). Figure 2b shows an example for Arabidopsis roots in which epidermal cells have already sequestered most of the conjugates whereas the underlying cortex cells show still mainly cytosolic fluorescence. For cells of the root cap it has also been shown that conjugate formed in the cytosol of the central columella cells can be transferred and sequestered in the vacuoles of the outermost cell layer (Fricker et al., 2000; A.J. Meyer and M.D. Fricker, unpublished results). Through competition for the glutathione moiety, labeling with MCB can also be used as an indirect assay for other factors that affect GSH levels such as heavy metals, herbicides, or even explosive compounds (Fig. 3) (Meyer and Fricker, 2002; Mezzari et al., 2005).

The MCB assay depletes the level of GSH and thus perturbs the system under study during the measurement. In some cases this can be used advantageously to follow the capacity of the system to respond to GSH depletion (Meyer and Fricker, 2002). If no internal pools for de novo synthesis of GSH are available direct monitoring of demand-driven GSH-biosynthesis can also be used to measure flux through synthesis pathway. For Arabidopsis suspension culture cells it was shown that de novo synthesis of GSH was dependent on external sulfate supply and thus the rate of de novo synthesis effectively reported flux through the entire sulfate assimilation pathway down to GSH (Meyer and Fricker, 2002). In theory, conjugation of GSH with MCB can also be used to give controlled titration of GSH levels to explore responses in plants where

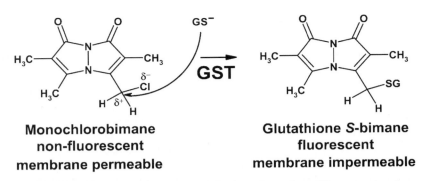

Fig. 1. Conjugation of GSH with monochlorobimane (MCB) leads to formation of fluorescent and membrane-impermeable glutathione S-bimane. Despite being reactive to all accessible thiols the conjugation to GSH is highly favored due to catalysis by glutathione S-transferases (GSTs).

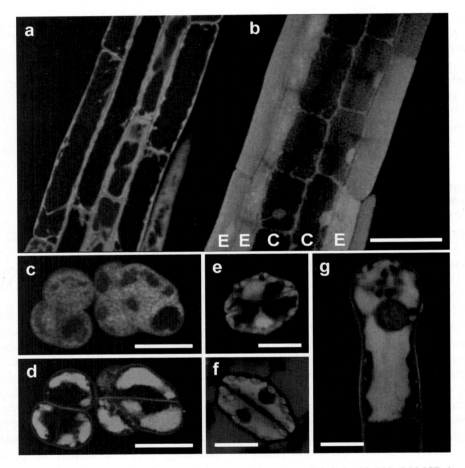

Fig. 2. In situ labeling of glutathione in plant cells. All cells and tissues were labeled with 100 μM MCB (green). (a) and (b) Optical sections through Arabidopsis roots show vacuolar sequestration of GSB. Initial labeling in the cytosol (a) is followed by vacuolar sequestration of GSB (b). The sequestration appears to occur faster in epidermal cell files (E) than in cortex cells (C). Bar = 50 μm. (c) and (d) Arabidopsis suspension culture cells stained with MCB and propidium iodide (PI) for cell walls and plasma membrane integrity (red). The MCB label appeared first in the cytosol after 5 min (c) and was completely sequestered to the vacuole after 60 min (d). Bars = 20 μm. (e) and (f) Tobacco guard cells stained with MCB. In addition to bimane autofluorescence from chloroplasts is detected (red). Initial labeling in the cytosol after 5 min (e) and complete vacuolar sequestration of GSB after 60 min (f). Bars = 20 μm. (g) Tobacco trichome labeled with MCB, 50 μM PI (red) and 10 μM Hoechst 33258 (blue) for 30 min. Bar = 10 μm (See Color Plates).

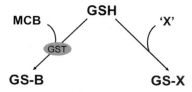

Fig. 3. Competition assay for indirect imaging of non-fluorescent glutathione conjugates. The amount of a GSH-reactive compound 'X' can be deduced from the reduction of the GSB signal compared to a control without 'X'.

overall redox protection capacity is reduced, although this approach has not yet been used to our knowledge.

B.   Analysis of the Glutathione-Dependent Xenobiotic Detoxification Pathway In Situ

The GSB is formed initially in the cytoplasm, but is subsequently transferred to the vacuole by GS-X conjugate pumps. This provides an opportunity to map the activity of the GSH-based detoxification pathway for a model substrate for each individual cell in a tissue. For example, the level of fluorescence initially increases in the cytoplasm of all cells in an Arabidopsis root tip observed in single optical sections followed over time (Fig. 2a and b). Most of the label is quickly transported into the vacuole, giving an increase in vacuolar fluorescence and eventually leading to a reduction in cytoplasmic fluorescence. In general, the smallest cells with the highest cytoplasm-to-vacuole ratios showed the greatest increase in vacuolar fluorescence. This dilution effect depending on volumes of subcellular compartments can also be observed in guard cells, which appear much brighter than surrounding epidermal cells.

To analyze the data, it is assumed that the conjugation and sequestration reactions can be described as a two step pathway (Fig. 4). The estimates of $[GSB]_{cyt}$ and $[GSB]_{vac}$ are based on the average fluorescence measured from manually-defined regions of interest (ROIs) in each compartment for each complete cell in the field of view (Fricker and Meyer, 2001). To express fluorescence levels in terms of GSB concentration requires subtraction of the average background signal, and calibration against the average fluorescence of GSB standards. The change in cytoplasmic GSB concentration over time will reflect the balance between the rate of conjugation and the rate of sequestration into the vacuole. In addition, as the cells under examination are still alive and elongating, cell expansion during the assay adds a volume-dependent decrease in the apparent concentration. The volumes of the cytoplasm and vacuole are therefore measured using stereological techniques (Meyer and Fricker, 2000) in parallel experiments. The change in vacuolar GSB concentration over time will reflect the vacuolar transport rate and the dilution (or concentration) arising from the difference in the relative volumes of cytoplasm and vacuole. The vacuolar concentration will also be affected by any increase in overall cell volume due to cell growth. With this approach, in vivo estimates for the kinetic parameters for the GST and the GSX pumps can be obtained for different cell types in the root tip (Fricker and Meyer, 2001).

Once inside the vacuole, glutathione conjugates are not just sequestered, but are also metabolized further to recover the amino acids used for the glutathione-tag. This degradation might be initiated from either end of the glutathione moiety followed by removal of the second terminal amino acid (Fig. 5). In both cases the initial two reactions lead to accumulation of cysteine conjugates. If MCB is used as a substrate for GSH conjugation, the resulting cysteine-bimane is still fluorescent. Because the fluorescence properties of the bimane-tag remain virtually unchanged during this degradation microscopic observation can help to define in which subcellular compartment degradation occurs, but can not provide evidence for either possible pathway. However, in combination with extraction and subsequent HPLC-analysis, accumulation of fluorescent intermediates might provide information about the order of the two initial reactions. In Arabidopsis, only Cys-B accumulation has been detected during a 48 h degradation period, indicating that the initial breakdown step is rate-limiting (Grzam et al., 2006). Recently, a γ-glutamyl transpeptidase was identified as the enzyme catalyzing this reaction (Grzam et al., 2007; Ohkama-Ohtsu et al., 2007). It is not clear at this stage, however, whether this also applies to other plants.

$$[GSH]_{cyt} + [MCB]_{cyt} \xrightarrow{\text{GST}} [GSB]_{cyt} \xrightarrow{\text{GSX pump}} [GSB]_{vac}$$

Fig. 4. Two-step pathway for glutathione-dependent conjugation and vacuolar sequestration of xenobiotics. Monochlorobimane (MCB) can be used as a visible model xenobiotics that can be followed microscopically during these reactions.

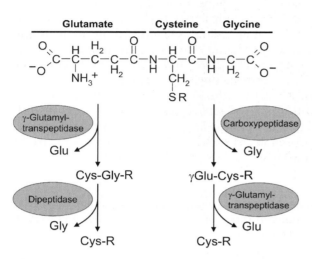

Fig. 5. Possible metabolic steps for the vacuolar degradation of glutathione conjugates in plants. Degradation is initiated at the termini of the glutathione moiety and leads to accumulation of cysteine conjugates (Cys-R), but so far none of the involved enzymes have been characterized in detail.

C.   Measurement of the Glutathione Redox Potential In Vivo

Whilst knowledge of the total GSH pool or the GSH detoxification pathway gained from the bimane imaging approach are useful, there is perhaps more interest in the redox poise of the glutathione pool. From the Nernst-equation it is apparent that the redox potential of the gluta-thione pool is not dependent on the [GSH]/[GSSG] ratio, which is the form normally reported in the literature, but rather the ratio $[GSH]^2/[GSSG]$. The redox potential is thus dependent on both the total concentration of glutathione and the degree of oxidation of the glutathione pool (Meyer and Hell, 2005). Until recently it was only possible to determine GSH and GSSG simultaneously by destructive sampling and chemical analysis. However, the redox sensitive fluorescent proteins, rxYFP and roGFP, can be expressed in plants and appear to function properly (Jiang et al., 2006; Meyer et al., 2007), facilitating in vivo measurements of the

GSH-GSSG redox state in vivo for different cell types and even different cell compartments.

Under non-stressed conditions, the probes are almost completely reduced (Fig. 6a). Based on midpoint redox potentials estimated in vivo around −280 to −290 mV (Dooley et al., 2004; Hanson et al., 2004), it can be deduced that the steady state redox potential in the cytosol of plant cells is around −320 mV (Meyer et al., 2007). Substantial shifts in redox potential can be achieved by H_2O_2 and DTT and the probes can be used to monitor redox differences in different cells along the root axis (Jiang et al., 2006). Further in vivo experiments in which the cellular glutathione level was altered through either pharmacological or genetic approaches confirmed the direct link between the poise of the glutathione redox buffer and roGFP fluorescence (Meyer et al., 2007). For example, cellular glutathione can be effectively depleted with L-buthionine-(S,R)-sulfoximine (BSO) as a specific inhibitor of glutamate-cysteine ligase (GSH1), the first enzyme of glutathione biosynthesis. Germination of Arabidopsis seeds on medium containing 1 mM BSO thus leads to a seedling remarkably similar in phenotype as the GSH-deficient *rml1* mutant (Vernoux et al., 2000; Cairns et al., 2006). Germination of roGFP-expressing Arabidopsis seed on BSO leads to almost complete oxidation of the probe compared to the reduced roGFP in control seedlings (Fig. 6a and b).

Like all other genetically encoded probes, roGFPs can be targeted to different subcellular compartments, such as the ER, mitochondria or chloroplasts (M. Schwarzländer et al., submitted). Thus, it is now feasible to study the subcellular compartmentation of glutathione in vivo and to address questions regarding intracellular and intercellular transport and turnover of the glutathione pool in living cells.

The link between the glutathione redox buffer and roGFP with GRXs as mediators also suggest that changes in glutathione redox potential would trigger modification of native target proteins either through glutathionylation or formation of disulfide bridges. One important goal for future research thus is the identification of such GRXs targets. A strategy for the identification of proteins targeted by thioredoxin combines two-dimensional gel-based separation methods for proteins with fluorescent tagging of accessible

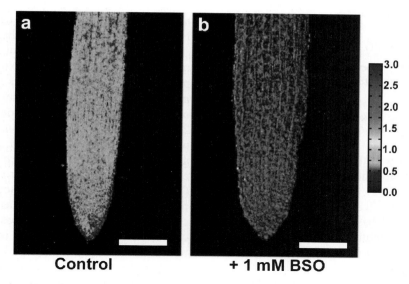

Fig. 6. Inhibition of glutathione biosynthesis causes oxidation of roGFP2. Arabidopsis seeds transformed with roGFP2 were germinated on Agar plates with or without 1 mM BSO for 7 days and imaged for the redox status of roGFP2 and GSH levels. All images are maximum projections from stacks of serial optical sections. The ratio images show analysis of control roots (a) and roots grown on 1 mM BSO (b). The color scale shows the pseudocolor coding for reduced (blue) and oxidized roGFP2 (red). Bars = 100 μm (*See Color Plates*).

protein thiol-groups with monobromobimane, a more reactive homologue of MCB (Yano et al., 2001; Ströher and Dietz, 2006). Similar gel-based approaches using other reactive probes have also been applied successfully to monitor protein thiol modifications in vivo (Leichert and Jakob, 2004, 2006) and it can be envisaged that such an approach might help identifying specific GRX targets.

V.   Probes for Components that Interact with Thiols

A.   Measurement of Reactive Oxygen Species (ROS) and Nitric Oxide (NO)

Production of several ROS is of interest both in terms of monitoring their destructive effects during photo-damage or pathogen attack and also because of increasing evidence that ROS might act as signals in their own right (Buchanan and Balmer, 2005; Foyer and Noctor, 2005). A common form of ROS in plant cells is singlet oxygen ($^1O_2^*$) which is formed when activation energy is directly transferred from the triplet-state of chlorophyll to molecular oxygen. Formation of other ROS results from sequential transfer of electrons to molecular oxygen (Fig. 7). Formation of superoxide ($\bullet O_2^-$) is a common process and plants have evolved efficient systems for detoxification of $\bullet O_2^-$. Superoxide dismutases, as a rapid defense against ROS, are capable of generating molecular oxygen and the far less reactive hydrogen peroxide (H_2O_2) from two $\bullet O_2^-$ molecules (Grene, 2002). Hydrogen peroxide can then be further detoxified through different peroxidases or the glutathione-ascorbate-cycle (Noctor and Foyer, 1998). Under stress conditions, however, the capacity for detoxification of ROS might not be sufficient, resulting in formation of a multitude of different ROS within the same cell or subcellular compartment. This mixture of ROS in most cases makes it difficult to unequivocally detect a particular form of ROS.

A number of chromogenic probes for ROS have been widely used for a broad range of different applications including toxicity assays and formation of ROS after pathogen attack. Nitroblue tetrazolium salt (NBT) reacts with $\bullet O_2^-$, whilst 3,3′-diaminobenzidine (DAB) gives

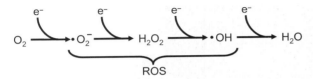

Fig. 7. Sequential transfer of electrons to molecular oxygen and formation of reactive oxygen species (ROS).

brown precipitates with H_2O_2 in the presence of ascorbate peroxidase (Thordal-Christensen et al., 1997; Fryer et al., 2002).

Singlet oxygen specifically quenches the fluorescence from the dansyl-based fluorophore DanePy (5-Dimethylamino-naphthalene-1- sulfonic acid (2-diethylamino-ethyl)-(2,2,5,5-tetramethyl-2, 5-dihydro-1H-pyrrol-3-ylmethyl)-amide), whilst another dansyl-based compound, HO-1889NH, is quenched by both $^1O_2^*$ and $\cdot O_2^-$ (Hideg et al., 2001, 2002, 2006). These compounds were designed by chemically adding a spin trap to the fluorophore dansyl-chloride, which results in partial quenching of fluorescence in the presence of ROS. With these compounds ROS production can be measured as the relative decrease in sensor fluorescence after leaf infiltration and corrected for blue-green autofluorescence of leaves or another dye. Dansyl-chloride on its own can be used as a control as it has the same fluorophore, but lacks the spin trap (Hideg et al., 2001). These dyes need to be vacuum infiltrated into detached leaves or taken up via the transpiration stream, which might impose problems with even distribution of the dye within the tissue. In addition, dansyl-based fluorophores accumulate preferentially in chloroplasts in intact leaves (Hideg et al., 2001, 2002; Fryer et al., 2002). A better alternative to DanePy is Singlet Oxygen Sensor Green (SOSG) which is able to diffuse through cuticles and membranes and thus can be painted directly on leaves. SOSG shows a high preference for $^1O_2^*$ and has been used to show that $^1O_2^*$ is produced by wounding even in the dark (Flors et al., 2006).

Fluorescein derivatives and dihydrodichlorofluorescein diacetate (H_2DCFDA) in particular are currently the most widely used fluorescent probes for ROS. The dihydro-derivatives of fluorescein are non-fluorescent until oxidized to the brightly fluorescent dichlorofluorescein (DCF). For a long time the oxidation of H_2DCFDA was thought to be relatively specific for H_2O_2, but it is now known that peroxidases are capable of inducing H_2DCFDA oxidation even in the absence of H_2O_2 (Rota et al., 1999; Tarpey et al., 2004). Furthermore, the H_2O_2-dependent oxidation needs the presence of ferrous iron and is greatly enhanced in the presence of heme-containing compounds or other redox-active metal ions (Tarpey et al., 2004). Although this oxidation in living cells is preferentially mediated by H_2O_2, these probes are also reactive towards other ROS, with highest sensitivity towards the hydroxyl radical ($\cdot$OH) (Table 1). The oxidized dye accumulates in different organelles like chloroplasts, mitochondria, cytoplasm and vacuole, imposing problems in quantifying the response and inferring information about ROS production in distinct subcellular compartments. Possible problems related to dye leaking out of the labeled cell can be partially avoided by using the derivative carboxy-H_2-DCFDA, which is suggested to be better retained within the cell.

Dihydrorhodamine 123 (DHR 123) is a structural analog of H_2DCF and is also responsive to ROS through oxidation to the fluorescent rhodamine 123. DHR 123 is membrane permeable and accumulates in the mitochondria. Oxidation by H_2O_2 requires the presence of heme-containing peroxidases or other heme-compounds like cytochrome c (Royall and Ischiropoulos, 1993).

Dihydrocalcein is another probe that becomes fluorescent after oxidation and is recommended to be superior to the other ROS probes because the oxidation product calcein is not thought to leak out of cells. Oxidation appears to take place in the mitochondria where calcein accumulates. However, it was also shown that dihydrocalcein inhibits the activity of complex I of the mitochondrial electron transport chain (Keller et al., 2004). Independent of whether electrons for reduction of dihydrocalcein are derived from complex I or whether the probe is only inhibiting normal function of the electron transport chain this interference puts major doubts on the usefulness of this probe for reliable ROS imaging.

In contrast to the initially colorless probes discussed before, dihydroethidium (DHE) is already fluorescent in its reduced form. Dihydroethidium has been reported to be sensitive to $\cdot O_2^-$, but there is some controversy about whether the oxidation product is ethidium or a distinctly different form (Zhao et al., 2003). In any case the oxida-

Table 1. Reactivity of selected fluorescent probes for live cell imaging towards different reactive oxygen species.

	$^1O_2^*$	$\cdot O_2^-$	H_2O_2	$\cdot OH$
DanePy	X			
HO-1889NH	X	X		
SOSG	X			
MitoSOX		X		
H$_2$DCFDA	X	X	X	XX
DHR 123			X	
Dihydrocalcein			X	
Dihydroethidium		X		
RedoxSensor Red CC-1		X	X	

DanePy: 5-Dimethylamino-naphthalene-1-sulfonic acid (2-diethylamino-ethyl)-(2,2,5,5-tetramethyl-2,5-dihydro-1*H*-pyrrol-3-ylmethyl)-amide;   HO-1889NH: 5-Dimethylamino-naphthalene-1-sulfonic acid (2,2,5,5-tetramethyl-2,5-dihydro-1*H*-pyrrol-3-ylmethyl)-amide; SOSG: Singlet Oxygen Sensor Green; H$_2$DCFDA: dihydro-dichloro fluorescein diacetate; DHR 123: Dihydrorhodamine 123.

tion product binds to DNA resulting in significant increase in fluorescence. In plants DHE has been used to observe $\cdot O_{2-}$ production under conditions of heavy metal stress (Yamamoto et al., 2002; Rodriguez-Serrano et al., 2006), but can be difficult to introduce into different cell types (M.D. Fricker, unpublished).

RedoxSensor Red CC1 is unique in that the subcelluar localization of fluorescence appears to be dependent of the redox environment in the cytosol. In animal cells the non-fluorescent probe is either oxidized in the cytosol followed by accumulation of the fluorescent product in the mitochondria or alternatively the non-fluorescent probe is sequestered to the lysosome where it is subsequently oxidized (Haugland, 2005). For plant cells a differential localization of the probe between mitochondria and the vacuole has been observed (S. Jeridi and A.J. Meyer, unpublished).

Considering the problem highlighted in relation to compartmentation of the dyes, lack of specificity to a particular ROS and possible interference with biological compounds leading to artifactual fluorescence, all these probes can only be used as qualitative indicators of oxidative stress rather than quantitative sensors for the rate of H_2O_2 formation. In addition, the accumulation of H_2O_2 during oxidative stress is thought to be transient, which requires detailed analysis of the time-course of fluorescence increase to quantify the kinetics.

Another important reactive signaling molecule that interacts with thiol groups is nitric oxide (NO) (Lamattina et al., 2003; del Rio et al., 2004; Crawford and Guo, 2005). The *S*-nitrosation reaction and its reverse reaction represent the most convenient general way to store, to transport and finally to release NO and at the same time modify potential target proteins. A number of fluorescent probes have been developed that allow visualization of NO in living cells. Cellular imaging has mainly used the increase in fluorescence following reaction of NO oxidation products with vicinal amine groups of diaminofluorescein (DAF-2) or the less pH sensitive fluoromethyl derivative (DAF-FM) to give a fluorescent triazole product (Kojima et al., 1998, 1999). These dyes are loaded into cells as membrane permeant diacetate derivatives (DA), which are then deacetylated to release the free dye. Reaction with auto-oxidation products of NO, such as N_2O_3, gives a 160-fold increase in fluorescence from a low background. Both dyes have been used in plants as qualitative indicators of NO production (Foissner et al., 2000).

Quantitative analysis is more difficult. The relationship between fluorescence and the level of NO is not straightforward. NO requires oxidation to N_2O_3 to react, which may be slow and depends on the level of O_2 and other NO scavenging systems. Substantial intracellular accumulation of the free dye to mM concentrations increases the probability of reaction, but can also give sufficient fluorescence from the unreacted dye to mask low levels of triazole formation (Rodriguez et al., 2005). The measurement itself may trigger a marked fluorescent increase from photoactivation of the dye (Broillet et al., 2001; Balcerczyk et al., 2005) or stimulation of NO release from artificial NO donors. Activation by oxidizing agents like HRP/H_2O_2 may promote direct reaction with NO rather than N_2O_3 (Jourd'heuil, 2002). The effect of reducing agents is also complex. Ascorbic acid (AA) and dehydroascorbate (DHA) compete with the dye for N_2O_3, reducing the fluorescence yield (Rodriguez et al., 2005), but may themselves react with the dye to give fluorescent products that are spectrally similar, giving increases in fluorescence that are independent of NO levels (Zhang et al., 2002). One useful control is to use mono-amine derivatives such as 4-aminofluorescein (4-AF) that do not react with NO. Substantial increases in 4-AF fluorescence in some tissues have been used as a criterion to reject the use of DAF probes to measure NO (Beligni et al., 2002). In this case, the rhodamine-based di-amine NO probes were more appropriate.

To ensure specificity for NO, the experimental protocol typically includes abolition of the signal in the presence of NO scavengers, such as cPTIO, inhibition of NO synthesis by non-metabolized analogues, and artificial elevation of NO by NO donors. In the latter case there is some evidence that the rate of detectable NO production is itself dependent on Ca^{2+} (Broillet et al., 2001).

Unlike reversible binding exploited in the development of ion indicators, for example, the absolute fluorescent signal integrates the total level of triazole formed during the time course. The instantaneous gradient should give an indirect measure of the NO concentration, although the calibration is essentially unknown because of the involvement of additional reaction intermediates. To measure rates requires more detailed timecourse measurements, which have not usually been attempted in plants. In some reports that do include time-courses, the level of triazole fluorescence apparently declines. As triazole formation is essentially irreversible, a decrease in fluorescence does not mean the level of NO production has changed, but that there are other experimental artifacts affecting the signal, such as redistribution of the dye outside the field of view, photobleaching,

sequestration in a low pH environment quenching the fluorescence, leakage to the medium or dye destruction by detoxification systems.

If the dyes are working properly and react with a significant proportion of the NO produced, they should also deplete NO levels and interfere with the downstream signaling pathway. Whether this occurs in vivo has not been reported.

B. Ratioable Transgenic Probes for H_2O_2

Another genetically encoded probe with great potential for future research is the recently developed H_2O_2-sensitive HyPer (Belousov et al., 2006). This probe consists of a circularly permutated YFP inserted into the regulatory domain of the *Escherichia coli* protein OxyR (OxyR-RD). OxyR is a redox-responsive transcription factor that undergoes significant conformational changes after H_2O_2 binding which enables DNA binding (Choi et al., 2001). In the fusion with YFP the conformational change after H_2O_2 binding is transmitted to YFP and changes the excitation properties of the fluorophore. Therefore, HyPer is also a ratiometric probe, with excitation peaks at 420 nm and 500 nm and emission at 516 nm. In the free form the 420 nm excitation peak is predominant, whereas upon binding of H_2O_2 this peak decreases and the 500 nm excitation peak increases proportionally. HyPer is sensitive to submicromolar changes in H_2O_2 and thus is expected to allow visualization of physiological H_2O_2 concentrations. The fully reversible probe should also enable dynamic H_2O_2 measurements in cells exposed to biotic or abiotic stress.

C. Experimental Induction of ROS

In contrast to the imaging of ROS generated during environmental stress it might also be of interest to deliberately generate ROS to validate and possibly calibrate the fluorescence measured with ROS-specific dyes or to generate ROS to initiate localized transduction cascades or destroy cellular components in a targeted way. Photosensitizers, such as Rose Bengal, are chromophores that generate ROS upon light irradiation. More precise inactivation of selected proteins can be achieved by fusing the targeted protein with a photosensitizer in a technique termed chromophore-assisted light inactivation (CALI;

Tour et al., 2003). Initially, this was achieved through addition of a small tetra-cysteine peptide motif (Cys-Cys-Pro-Gly-Cys-Cys) that binds fluorescent bis-arsenical ligands with high affinity ($\sim10^{-11}$ M) to give a fluorescent complex (Griffin et al., 1998; Adams et al., 2002). Three fluorescent ligands, with different spectra based on xanthene (CHoXAsH), fluorescein (FlAsH) or resorufin (ReAsH), are now commercially available (Invitrogen, Lumio series). The ligands are applied as membrane-permeant complexes with 1,2-ethanedithiol (EDT). Once in the cell, the tetracysteine hairpin displaces the EDT to generate a minimally disrupted fluorescent fusion protein. Under high levels of illumination, these fluorophores produce singlet oxygen that is capable of inactivating proteins in the immediate vicinity. More recently, the protein KillerRed has been developed in which the entire photosensitizer is genetically encoded and can be transcriptionally fused to the target (Bulina et al., 2006). This protein has been developed on the basis of the Anthomedusa chromoprotein anm2CP (Shagin et al., 2004) and generates ROS upon irradiation with green light. Due to the high reactivity of ROS the localized ROS production then will lead to selective damage of adjacent molecules.

VI.   Conclusion and Future Perspectives

Generation of ROS is a normal feature of metabolism that occurs at tightly controlled rates under non-stress conditions, but can be dramatically increased under conditions of oxidative stress. This deviation from steady state directly of indirectly affects the oxidation of different cellular components and cysteine residues in particular. Glutathione is a key metabolite for redox control, and it is likely that the glutathione redox potential is directly exploited to alter the oxidation state and thus the activity of effector proteins, with GRXs as important mediators to ensure tight coupling and specificity. Thus, the formation of ROS, the glutathione redox buffer, GRXs and target proteins are all components of the same signaling pathway. To understand the dynamics of this pathway, particularly as many of the signals are themselves short-lived and highly localized, it is essential to quantitatively analyze these components in living cells with good spatial and temporal resolution.

Formation of ROS in non-stressed cells is mainly restricted to organelles like mitochondria and chloroplasts. Micro-compartmentation is also likely to occur during many forms of stress and biotic interactions in particular. Analysis of micro-compartmentation of oxidative stress and signaling as well as dissection of redox signals at high spatial and temporal resolution can only be achieved through imaging approaches applied to living cells. Another challenge will be to understand more about the way plants regulate the amount of ROS at the cellular and subcellular level. In this context reversible genetically encoded probes that can be targeted to different organelles currently have the greatest potential for providing new insights into the cellular redox network. In combination with thiol redox proteomics the imaging approaches will provide detailed insights into the complex redox network.

As already noted the dynamic measurement of ROS is still in its infancy. Nevertheless, with the development of new GFP-based probes it now seems feasible to get new insights into the dynamics of redox metabolism and to dissect ROS-dependent signaling pathways. Thus, it is expected that broad application of these novel imaging technologies will lead to a dramatic increase of our knowledge about redox signaling.

Increasing evidence points at significant crosstalk between redox-dependent pathways and other signaling pathways. This has already been shown for ROS and Ca^{2+} (Gomez et al., 2004), abscisic acid (Pei et al., 2000), ethylene (Moeder et al., 2002), salicylic acid (Mateo et al., 2006), and is also likely to occur for pathways dependent on other signaling molecules. Integration of all these signals into the entire cellular signaling network further highlights the necessity for a dynamic and quantitative analysis of signals in living cells. Better understanding of this network is expected to aid the improvement of plants towards environmental stress.

Acknowledgements

This work was supported in parts by grants the Deutsche Forschungsgmeinschaft (grant ME1567/3-2) and the University of Heidelberg to AJM, and the Biotechnology and Biological Sciences Research Council and Rhône-Poulenc Agro to MDF.

References

Adams SR, Campbell RE, Gross LA, Martin BC, K. WG, Yao Y, Llopis J and Tsien RY (2002) New biarsenical ligands and tetracysteine motifs for protein labeling *in vitro* and *in vivo*: Synthesis and biological applications. J Am Chem Soc 124: 6063–6076

Apel K and Hirt H (2004) Reactive oxygen species: Metabolism, oxidative stress, and signal transduction. Annu Rev Plant Biol 55: 373–399

Balcerczyk A, Soszynski M and Bartosz G (2005) On the specificity of 4-amino-5-methylamino-2′,7′-difluorofluorescein as a probe for nitric oxide. Free Radic Biol Med 39: 327–335

Beligni MV, Fath A, Bethke PC, Lamattina L and Jones RL (2002) Nitric oxide acts as an antioxidant and delays programmed cell death in barley aleurone layers. Plant Physiol 129: 1642–1650

Belousov V, Fradkov A, Lukyanov K, Staroverov D, Shakhbazov K, Terskikh A and Lukyanov S (2006) Genetically encoded fluorescent indicator for intracellular hydrogen peroxide. Nat Methods 3: 281–286

Bennett BD, Jetton TL, Ying G, Magnuson MA and Piston DW (1996) Quantitative subcellular imaging of glucose metabolism within intact pancreatic islets. J Biol Chem 271: 3647–3651

Björnberg O, Østergaard H and Winther J (2006) Measuring intracellular redox conditions using GFP-based sensors. Antioxid Redox Signal 8: 354–361

Broillet M, Randin O and Chatton J (2001) Photoactivation and calcium sensitivity of the fluorescent NO indicator 4,5-diaminofluorescein (DAF-2): Implications for cellular NO imaging. FEBS Lett 491: 227–232

Buchanan BB and Balmer Y (2005) Redox regulation: A broadening horizon. Annu Rev Plant Biol 56: 187–220

Bulina M, Chudakov D, Britanova O, Yanushevich Y, Staroverov D, Chepurnykh T, Merzlyak E, Shkrob M, Lukyanov S and Lukyanov K (2006) A genetically encoded photosensitizer. Nat Biotechnol 24: 95–99

Cairns NG, Pasternak M, Wachter A, Cobbett CS and Meyer AJ (2006) Maturation of Arabidopsis seeds is dependent on glutathione biosynthesis within the embryo. Plant Physiol 141: 446–455

Cannon MB and Remington SJ (2006) Re-engineering redox-sensitive green fluorescent protein for improved response rate. Protein Sci 15: 45–57

Cardenas L, McKenna ST, Kunkel JG and Hepler PK (2006) NAD(P)H Oscillates in pollen tubes and is correlated with tip growth. Plant Physiol 142: 1460–1468

Choi H, Kim S, Mukhopadhyay P, Cho S, Woo J, Storz G and Ryu S (2001) Structural basis of the redox switch in the OxyR transcription factor. Cell 105: 103–113

Crawford N and Guo F (2005) New insights into nitric oxide metabolism and regulatory functions. Trends Plant Sci 10: 195–200

De Lorimier RM, Smith JJ, Dwyer MA, Looger LL, Sali KM, Paavola CD, Rizk SS, Sadigov S, Conrad DW, Loew L and Hellinga HW (2002) Construction of a fluorescent biosensor family. Protein Sci 11: 2655–2675

del Rio L, Corpas F and Barroso J (2004) Nitric oxide and nitric oxide synthase activity in plants. Phytochemistry 65: 783–792

Deuschle K, Okumoto S, Fehr M, Looger LL, Kozhukh L and Frommer WB (2005) Construction and optimization of a family of genetically encoded metabolite sensors by semirational protein engineering. Protein Sci 14: 2304–2314

Deuschle K, Chaudhuri B, Okumoto S, Lager I, Lalonde S and Frommer WB (2006) Rapid metabolism of glucose detected with FRET glucose nanosensors in epidermal cells and intact roots of Arabidopsis RNA-silencing mutants. Plant Cell 18: 2314–2325

Dooley CT, Dore TM, Hanson GT, Jackson WC, Remington SJ and Tsien RY (2004) Imaging dynamic redox changes in mammalian cells with green fluorescent protein indicators. J Biol Chem 279: 22284–22293

Fahey RC and Sundquist AR (1991) Evolution of glutathione metabolism. Adv Enzymol RAMB 64: 1–53

Fehr M, Frommer WB and Lalonde S (2002) Visualization of maltose uptake in living yeast cells by fluorescent nanosensors. Proc Natl Acad Sci U S A 99: 9846–9851

Fehr M, Lalonde S, Lager I, Wolff MW and Frommer WB (2003) *In vivo* imaging of the dynamics of glucose uptake in the cytosol of COS-7 cells by fluorescent nanosensors. J Biol Chem 278: 19127–19133

Fehr M, Lalonde S, Ehrhardt D and Frommer W (2004) Live imaging of glucose homeostasis in nuclei of COS-7 cells. J Fluoresc 14: 603–609

Flors C, Fryer MJ, Waring J, Reeder B, Bechtold U, Mullineaux PM, Nonell S, Wilson MT and Baker NR (2006) Imaging the production of singlet oxygen *in vivo* using a new fluorescent sensor, singlet oxygen sensor green. J Exp Bot 57: 1725–1734

Foissner I, Wendehenne D, Langebartels C and Durner J (2000) *In vivo* imaging of an elicitor-induced nitric oxide burst in tobacco. Plant J 23: 817–824

Foyer CH and Noctor G (2005) Redox homeostasis and antioxidant signaling: A metabolic interface between stress perception and physiological responses. Plant Cell 17: 1866–1875

Fricker MD and Meyer AJ (2001) Confocal imaging of metabolism *in vivo*: Pitfalls and possibilities. J Exp Bot 52: 631–640

Fricker MD, Plieth C, Knight H, Blancaflor E, Knight MR, White NS and Gilroy S (1999) Fluorescence and luminescence techniques to probe ion activities in living plant cells. In: Mason WT (ed) Fluorescent and luminescent probes, pp 569–596. Academic, London

Fricker MD, May M, Meyer AJ, Sheard N and White NS (2000) Measurement of glutathione levels in intact roots of Arabidopsis. J Microsc (Oxf) 198: 162–173

Fricker MD, Parsons A, Tlalka M, Blancaflor E, Gilroy S, Meyer AJ and Plieth C (2001) Fluorescent probes for living plant cells. In: Hawes C and Satiat-Jeunemaitre B (eds) Plant Cell Biology: A Practical Approach, Vol 250, pp 35–84. Oxford University Press, Oxford

Fricker M, Runions J and Moore I (2006) Quantitative fluorescence microscopy: From art to science. Annu Rev Plant Biol 57: 79–107

Fryer MJ, Oxborough K, Mullineaux PM and Baker NR (2002) Imaging of photo-oxidative stress responses in leaves. J Exp Bot 53: 1249–1254

Gomez LD, Noctor G, Knight MR and Foyer CH (2004) Regulation of calcium signalling and gene expression by glutathione. J Exp Bot 55: 1851–1859

Grene R (2002) Oxidative stress and acclimation mechanisms in plants. In: CR Somerville, EM Meyerowitz (eds) The Arabidopsis Book. American Society of Plant Biologists, Rockville, MD, doi: 10.1199/tab.0036.1, www.aspb.org/publications/arabidopsis/

Griffin BA, Adams SR and Tsien RY (1998) Specific covalent labeling of recombinant protein molecules inside live cells. Science 281: 269–272

Grzam A, Tennstedt P, Clemens S, Hell R and Meyer AJ (2006) Vacuolar sequestration of glutathione S-conjugates outcompetes a possible degradation of the glutathione moiety by phytochelatin synthase. FEBS Lett 580: 6384–6390

Grzam A, Martin M, Hell R and Meyer A (2007) Gamma-glutamyl transpeptidase GGT4 initiates vacuolar degradation of glutathione S-conjugates in Arabidopsis. FEBS Lett 581: 3131–3138

Gu H, Lalonde S, Okumoto S, Looger LL, Scharff-Poulsen AM, Grossman AR, Kossmann J, Jakobsen I and Frommer WB (2006) A novel analytical method for in vivo phosphate tracking. FEBS Letters 580: 5885–5893

Gutiérrez-Alcalá G, Gotor C, Meyer AJ, Fricker M, Vega JM and Romero LC (2000) Glutathione biosynthesis in Arabidopsis trichome cells. Proc Natl Acad Sci U S A 97: 11108–11113

Hanson GT, Aggeler R, Oglesbee D, Cannon M, Capaldi RA, Tsien RY and Remington SJ (2004) Investigating mitochondrial redox potential with redox-sensitive green fluorescent protein indicators. J Biol Chem 279: 13044–13053

Hartmann TN, Fricker MD, Rennenberg H and Meyer AJ (2003) Cell-specific measurement of cytosolic glutathione in poplar leaves. Plant Cell Environ 26: 965–975

Haugland RP (2005) The Handbook. A guide to fluorescent probes and labeling technologies. Invitrogen Corp., Eugene

Hetherington A and Brownlee C (2004) The generation of Ca^{2+} signals in plants. Annu Rev Plant Biol 55: 401–427

Hideg E, Ogawa Ki, Kalai T and Hideg K (2001) Singlet oxygen imaging in Arabidopsis thaliana leaves under photoinhibition by excess photosynthetically active radiation. Physiol Plant 112: 10–14

Hideg E, Barta C, Kalai T, Vass I, Hideg K and Asada K (2002) Detection of singlet oxygen and superoxide with fluorescent sensors in leaves under stress by photoinhibition or UV radiation. Plant Cell Physiol 43: 1154–1164

Hideg E, Kalai T, Kos P, Asada K and Hideg K (2006) Singlet oxygen in plants – its significance and possible detection with double (fluorescent and spin) indicator reagents. Photochem Photobiol 82: 1211–1218

Huang S, Heikal AA and Webb WW (2002) Two-photon fluorescence spectroscopy and microscopy of NAD(P)H and flavoprotein. Biophys J 82: 2811–2825

Jiang K, Schwarzer C, Lally E, Zhang S, Ruzin S, Machen T, Remington SJ and Feldman L (2006) Expression and characterization of a redox-sensing green fluorescent protein (reduction–oxidation-sensitive green fluorescent protein) in Arabidopsis. Plant Physiol 141: 397–403

Jourd'heuil D (2002) Increased nitric oxide-dependent nitrosylation of 4,5-diaminofluorescein by oxidants: Implications for the measurement of intracellular nitric oxide. Free Radic Biol Med 33: 676–684

Kasimova MR, Grigiene J, Krab K, Hagedorn PH, Flyvbjerg H, Andersen PE and Moller IM (2006) The Free NADH Concentration is kept constant in plant mitochondria under different metabolic conditions. Plant Cell 18: 688–698

Keller A, Mohamed A, Drose S, Brandt U, Fleming I and Brandes R (2004) Analysis of dichlorodihydrofluorescein and dihydrocalcein as probes for the detection of intracellular reactive oxygen species. Free Radic Res 38: 1257–1267

Kindzelskii AL and Petty HR (2002) Apparent role of traveling metabolic waves in oxidant release by living neutrophils. Proc Natl Acad Sci U S A 99: 9207–9212

Kojima H, Nakatsubo N, Kikuchi K, Kawahara S, Kirino Y, Nagoshi H, Hirata Y and Nagano T (1998) Detection and imaging of nitric oxide with novel fluorescent indicators: Diaminofluoresceins. Anal Chem 70: 2446–2453

Kojima H, Urano Y, Kikuchi K, Higuchi T, Hirata Y and Nagano T (1999) Fluorescent indicators for imaging nitric oxide production. Angew Chem Int Ed Engl 38: 3209–3212

Lager I, Fehr M, Frommer W and Lalonde S (2003) Development of a fluorescent nanosensor for ribose. FEBS Lett 553: 85–89

Lamattina L, Garcia-Mata C, Graziano M and Pagnussat G (2003) Nitric oxide: The versatility of an extensive signal molecule. Annu Rev Plant Biol 54: 109–136

Leichert L and Jakob U (2004) Protein thiol modifications visualized in vivo. PLoS Biol 2: e333

Leichert L and Jakob U (2006) Global methods to monitor the thiol-disulfide state of proteins in vivo. Antioxid Redox Signal 8: 763–772

Lemaire SD (2004) The glutaredoxin family in oxygenic photosynthetic organisms. Photosynth Res 79: 305–318

Masters BR and Chance B. 1993. Redox confocal imaging: Intrinsic fluorescent probes of cellular metabolism. In:

Mason WT (ed) Fluorescent and luminescent probes for biological activity, pp 44–57. Academic, London

Mateo A, Funck D, Muhlenbock P, Kular B, Mullineaux PM and Karpinski S (2006) Controlled levels of salicylic acid are required for optimal photosynthesis and redox homeostasis. J Exp Bot 57: 1795–1807

May M, Vernoux T, Leaver C, Van Montagu M and Inze D (1998) Glutathione homeostasis in plants: Implications for environmental sensing and plant development. J Exp Bot 49: 649–667

Meyer AJ and Fricker MD (2000) Direct measurement of glutathione in epidermal cells of intact Arabidopsis roots by two-photon laser scanning microscopy. J Microsc (Oxf) 198: 174–181

Meyer AJ and Fricker MD (2002) Control of demand-driven biosynthesis of glutathione in green Arabidopsis suspension culture cells. Plant Physiol 130: 1927–1937

Meyer AJ and Hell R (2005) Glutathione homeostasis and redox-regulation by sulfhydryl groups. Photosynth Res 86: 435–457

Meyer AJ, May MJ and Fricker M (2001) Quantitative *in vivo* measurement of glutathione in Arabidopsis cells. Plant J 27: 67–78

Meyer AJ, Brach T, Marty L, Kreys S, Rouhier N, Jacquol J-P and Hell R (2007) Redox-sensitive GFP in *Arabidopsis thaliana* is a quantitative biosensor for the redox potential of the cellular glutathione redox buffer. Plant J 52: 973–986

Mezzari MP, Walters K, Jelinkova M, Shih M-C, Just CL and Schnoor JL (2005) Gene expression and microscopic analysis of Arabidopsis exposed to chloroacetanilide herbicides and explosive compounds. A phytoremediation approach. Plant Physiol 138: 858–869

Mittler R, Vanderauwera S, Gollery M and Van Breusegem F (2004) Reactive oxygen gene network of plants. Trends Plant Sci 9: 490–498

Miyawaki A, Llopis J, Heim R, McCaffery JM, Adams JA, Ikura M and Tsien RY (1997) Fluorescent indicators for Ca^{2+} based on green fluorescent proteins and calmodulin. Nature 388: 882–887

Miyawaki A, Griesbeck O, Heim R and Tsien RY (1999) Dynamic and quantitative Ca^{2+} measurements using improved cameleons. Proc Natl Acad Sci U S A 96: 2135–2140

Moeder W, Barry CS, Tauriainen AA, Betz C, Tuomainen J, Utriainen M, Grierson D, Sandermann H, Langebartels C and Kangasjarvi J (2002) Ethylene synthesis regulated by biphasic induction of 1-aminocyclopropane-1-carboxylic acid synthase and 1-aminocyclopropane-1-carboxylic acid oxidase genes is required for hydrogen peroxide accumulation and cell death in ozone-exposed tomato. Plant Physiol 130: 1918–1926

Newton GL and Javor B (1985) Gamma-glutamylcysteine and thiosulfate are the major low-molecular-weight thiols in halobacteria. J Bacteriol 161: 438–441

Newton G, Arnold K, Price M, Sherrill C, Delcardayre S, Aharonowitz Y, Cohen G, Davies J, Fahey R and Davis C (1996) Distribution of thiols in microorganisms: Mycothiol is a major thiol in most actinomycetes. J Bacteriol 178: 1990–1995

Noctor G and Foyer CH (1998) Ascorbate and glutathione: Keeping active oxygen under control. Annu Rev Plant Physiol Plant Mol Biol 49: 249–279

Ohkama-Ohtsu N, Zhao P, Xiang C and Oliver D (2007) Glutathione conjugates in the vacuole are degraded by gamma-glutamyl transpeptidase GGT3 in Arabidopsis. Plant J 49: 878–888

Okumoto S, Looger LL, Micheva KD, Reimer RJ, Smith SJ and Frommer WB (2005) Detection of glutamate release from neurons by genetically encoded surface-displayed FRET nanosensors. Proc Natl Acad Sci U S A 102: 8740–8745

Østergaard H, Henriksen A, Hansen FG and Winther JR (2001) Shedding light on disulfide bond formation: Engineering a redox switch in green fluorescent protein. EMBO J 20: 5853–5862

Østergaard H, Tachibana C and Winther JR (2004) Monitoring disulfide bond formation in the eukaryotic cytosol. J Cell Biol 166: 337–345

Pasternak M, Lim B, Wirtz M, Hell R, Cobbett CS and Meyer AJ (2008) Restricting glutathione biosynthesis to the cytosol is sufficient for normal plant development. Plant J, in press

Patterson GH, Knobel SM, Arkhammar P, Thastrup O and Piston DW (2000) Separation of the glucose-stimulated cytoplasmic and mitochondrial NAD(P)H responses in pancreatic islet beta cells. Proc Natl Acad Sci U S A 97: 5203–5207

Pei ZM, Murata Y, Benning G, Thomine S, Klusener B, Allen GJ, Grill E and Schroeder JI (2000) Calcium channels activated by hydrogen peroxide mediate abscisic acid signalling in guard cells. Nature 406: 731–734

Piston D, Masters B and Webb W (1995) Three-dimensionally resolved NAD(P)H cellular metabolic redox imaging of the *in situ* cornea with two-photon excitation laser scanning microscopy. J Microsc (Oxf) 178: 20–27

Quesada I, Todorova MG and Soria B (2006) Different metabolic responses in alpha-, beta-, and delta-cells of the islet of Langerhans monitored by redox confocal microscopy. Biophys J 90: 2641–2650

Rocheleau JV, Head WS and Piston DW (2004) Quantitative NAD(P)H/flavoprotein autofluorescence imaging reveals metabolic mechanisms of pancreatic islet pyruvate response. J Biol Chem 279: 31780–31787

Rodriguez J, Specian V, Maloney R, Jourd'heuil D and Feelisch M (2005) Performance of diamino fluorophores for the localization of sources and targets of nitric oxide. Free Radic Biol Med 38: 356–368

Rodriguez-Serrano M, Romero-Puertas M, Zabalza A, Corpas F, Gomez M, Del Rio L and Sandalio L (2006) Cadmium effect on oxidative metabolism of pea (*Pisum sativum* L.)

roots. Imaging of reactive oxygen species and nitric oxide accumulation *in vivo*. Plant Cell Environ 29: 1532–1544

Rota C, Fann YC and Mason RP (1999) Phenoxyl free radical formation during the oxidation of the fluorescent dye 2′,7′-dichlorofluorescein by horseradish peroxidase. Possible consequences for oxidative stress measurements. J Biol Chem 274: 28161–28168

Royall J and Ischiropoulos H (1993) Evaluation of 2′,7′-dichlorofluorescin and dihydrorhodamine 123 as fluorescent probes for intracellular H_2O_2 in cultured endothelial cells. Arch Biochem Biophys 302: 348–355

Sánchez-Fernández R, Fricker M, Corben LB, White NS, Sheard N, Leaver CJ, Van Montagu M, Inze D and May MJ (1997) Cell proliferation and hair tip growth in the Arabidopsis root are under mechanistically different forms of redox control. Proc Natl Acad Sci U S A 94: 2745–2750

Shagin DA, Barsova EV, Yanushevich YG, Fradkov AF, Lukyanov KA, Labas YA, Semenova TN, Ugalde JA, Meyers A, Nunez JM, Widder EA, Lukyanov SA and Matz MV (2004) GFP-like proteins as ubiquitous metazoan superfamily: Evolution of functional features and structural complexity. Mol Biol Evol 21: 841–850

Ströher E and Dietz K-J (2006) Concepts and approaches towards understanding the cellular redox proteome. Plant Biol 8: 407–418

Sundquist A and Fahey R (1989) The function of γ-glutamylcysteine and bis-γ-glutamylcystine reductase in *Halobacterium halobium*. J Biol Chem 264: 719–725

Tarpey MM, Wink DA and Grisham MB (2004) Methods for detection of reactive metabolites of oxygen and nitrogen: *In vitro* and *in vivo* considerations. Am J Physiol-Reg I 286: R431–444

Thordal-Christensen H, Zhang Z, Wei Y and Collinge DB (1997) Subcellular localization of H_2O_2 in plants. H_2O_2 accumulation in papillae and hypersensitive response during the barley-powdery mildew interaction. Plant J 11: 1187–1194

Tour O, Meijer R, Zacharias D, Adams S and Tsien R (2003) Genetically targeted chromophore-assisted light inactivation. Nat Biotechnol 21: 1505–1508

Vernoux T, Wilson RC, Seeley KA, Reichheld JP, Muroy S, Brown S, Maughan SC, Cobbett CS, Van Montagu M, Inze D, May MJ and Sung ZR (2000) The *root meristemless1/cadmium sensitive2* gene defines a glutathione-dependent pathway involved in initiation and maintenance of cell division during postembryonic root development. Plant Cell 12: 97–110

White NS, Errington RJ, Fricker MD and Wood JL (1996) Aberration control in quantitative imaging of botanical specimens by multidimensional fluorescence microscopy. J Microsc (Oxf) 181: 99–116

Xing S, Lauri A and Zachgo S (2006) Redox regulation and flower development: A novel function for glutaredoxins. Plant Biol 8: 547–555

Yamamoto Y, Kobayashi Y, Devi SR, Rikiishi S and Matsumoto H (2002) Aluminum toxicity is associated with mitochondrial dysfunction and the production of reactive oxygen species in plant cells. Plant Physiol 128: 63–72

Yano H, Wong JH, Lee YM, Cho M-J and Buchanan BB (2001) A strategy for the identification of proteins targeted by thioredoxin. Proc Natl Acad Sci U S A 98: 4794–4799

Zhang X, Kim W-S, Hatcher N, Potgieter K, Moroz LL, Gillette R and Sweedler JV (2002) Interfering with nitric oxide measurements. 4,5-diaminofluorescein reacts with dehydroascorbic acid and ascorbic acid. J Biol Chem 277: 48472–48478

Zhao H, Kalivendi S, Zhang H, Joseph J, Nithipatikom K, Vasquez-Vivar J and Kalyanaraman B (2003) Superoxide reacts with hydroethidine but forms a fluorescent product that is distinctly different from ethidium: Potential implications in intracellular fluorescence detection of superoxide. Free Radic Biol Med 34: 1359–1368

Index

Advances in Photosynthesis and Respiration

Series editor: Govindjee, University of Illinois, Urbana, Illinois, U.S.A.

Advances in Photosynthesis and Respiration

For further information about the series and how to order please visit our Website
http://www.springer.com